FOUNDATIONS OF ASTRONOMY

EIGHTH EDITION

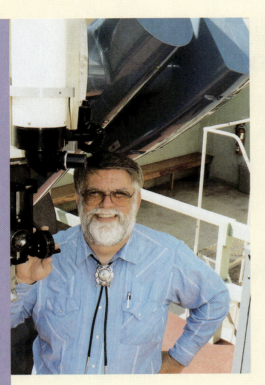

Mike Seeds is Professor of Astronomy at Franklin and Marshall College, where he has taught astronomy since 1970. His research interests have focused on peculiar variable stars and the automation of astronomical telescopes. He is continuing his research by serving as Principal Astronomer in charge of the Phoenix 10, the first fully robotic telescope, located in southern Arizona. In 1989, he received the Christian R. and Mary F. Lindback Award for Distinguished Teaching. In addition to teaching, writing, and research, Mike has published educational tools for use in computer-smart classrooms. His interest in the history of astronomy led him to offer upper-level courses "Archaeoastronomy" and "Changing Concepts of the Universe," which he continues to develop. He has also published educational software for preliterate toddlers. Mike was Senior Consultant in the creation of the 26-episode telecourse *Universe: The Infinite Frontier.* He is the author of *Astronomy: The Solar System and Beyond,* Fourth Edition (2005), and *Horizons: Exploring the Universe,* Eighth Edition (2004) published by Brooks/Cole.

ABOUT THE COVER

An infrared image of our Milky Way Galaxy stretches diagonally in the distance. Although this image is a panorama recorded from Earth's position inside our galaxy, our home galaxy must look much like this from a distance. Billions of stars produce a fog of starlight, and the dust clouds, usually seen as black obscuring masses, here glow red with their meager warmth. As we explore the universe with Earth-based telescopes and with spacecraft, we explore the cycle of star formation from the birth of the universe to the origin of Earth to the genealogy of the atoms of which we are made.

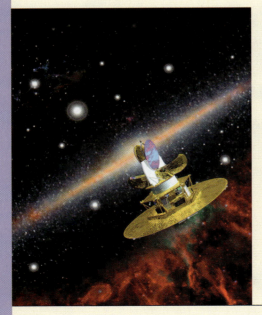

FOUNDATIONS
OF ASTRONOMY

EIGHTH EDITION

Michael A. Seeds
Joseph R. Grundy Observatory, Franklin and Marshall College

THOMSON

BROOKS/COLE

Australia • Canada • Mexico • Singapore • Spain
United Kingdom • United States

For Emery and Helen Seeds

THOMSON

BROOKS/COLE

Astronomy Editor: *Keith Dodson*
Assistant Editor: *Carol Benedict*
Editorial Assistant: *Melissa Newt*
Publisher/Executive Editor: *David Harris*
Development Editor: *Marie Carigma-Sambilay*
Technology Project Manager: *Sam Subity*
Marketing Managers: *Erik Evans and Kelley McAllister*
Marketing Assistant: *Leyla Jowza*
Advertising Project Manager: *Nathaniel Bergson-Michelson*
Project Manager, Editorial Production: *Lisa Weber*
Executive Art Director: *Vernon Boes*
Print Buyer: *Karen Hunt*

Permissions Editor: *Kiely Sexton*
Production Service: *Heckman & Pinette*
Text Designer: *Jeanne Calabrese*
Copy Editor: *Margaret Pinette*
Illustrator: *Precision Graphics*
Cover Designer: *Denise Davidson*
Cover Images: *Composite images courtesy of NASA,
    Hubble Space Telescope, and the Wilkinson Microwave
    Anisotropy Probe (WMAP); digital illustration, image
    design, and composite, Bob Western*
Compositor: *Thompson Type*
Text and Cover Printer: *Transcontinental Printing/Interglobe*

Printed in Canada
1 2 3 4 5 6 7 08 07 06 05 04

For more information about our products, contact us at:
Thomson Learning Academic Resource Center
1-800-423-0563
For permission to use material from this text, contact us by:
Phone: 1-800-730-2214
Fax: 1-800-730-2215
Web: http://www.thomsonrights.com

Library of Congress Control Number: 2003114972

Student Edition with InfoTrac College Edition: ISBN 0-534-42120-2
Instructor's Edition: ISBN 0-534-42121-0
International Student Edition: ISBN 0-534-40975-X (Not for sale in
the United States)

Brooks/Cole–Thomson Learning
10 Davis Drive
Belmont, CA 94002
USA

Asia
Thomson Learning
5 Shenton Way #01-01
UIC Building
Singapore 068808

Australia/New Zealand
Thomson Learning
102 Dodds Street
Southbank, Victoria 3006
Australia

Canada
Nelson
1120 Birchmount Road
Toronto, Ontario M1K 5G4
Canada

Europe/Middle East/Africa
Thomson Learning
High Holborn House
50/51 Bedford Row
London WC1R 4LR
United Kingdom

Latin America
Thomson Learning
Seneca, 53
Colonia Polanco
11560 Mexico D.F.
Mexico

Spain/Portugal
Paraninfo
Calle/Magallanes, 25
28015 Madrid, Spain

BRIEF CONTENTS

CONTENTS

A NOTE TO THE STUDENT

FROM MIKE SEEDS

Astronomy is about us. Although an astronomy course covers planets and stars and galaxies, it is really about you and me. Astronomy helps us answer the ultimate question of human existence: What are we? Great minds like Plato, Beethoven, and Hemingway have tried to give their personal answers, but we must each find our own answer. Astronomy helps us understand the meaning of our own existence.

Of course, part of our personal answer must include philosophical and cultural issues. We define ourselves by what we create, what we worship, what we admire, and what we expect of each other. But a major part of our answer is embedded in the physical universe. We cannot hope to find a complete answer until we understand how the birth of the universe, the origin of galaxies, the evolution of stars, and the formation of planets created and brought together the atoms of which we are made. A basic understanding of astronomy is fundamental to knowing what we are. Astronomy is exciting because it is about us.

Part of the excitement of astronomy is the discovery of new things. Astronomers expect to be astonished. That excitement is reflected in this book in the form of new images, new discoveries, and new understandings that take us, in an introductory course, to the frontier of human knowledge. Here we explore the new evidence of an ancient ocean on Mars and a modern ocean on Jupiter's moon Europa. We will study the newest images of dying stars, and we will struggle to understand new evidence that the expansion of the universe is speeding up. The Hubble Space Telescope, spacecraft orbiting Mars and Jupiter, the Chandra X-Ray Observatory, and giant Earth-based telescopes on remote mountaintops provide us with a daily dose of excitement that goes far beyond sensationalism. These new discoveries in astronomy are exciting because they are about us. They tell us more and more about what we are.

I hope you will share in the excitement and satisfaction that scientists feel as they puzzle out nature's secrets. That is why I have created 27 special art spreads located throughout this book. Each presents a carefully selected subject visually, not only to make it exciting but to challenge you to synthesize your own undersanding. Rather than memorize facts and predigested descriptions of nature, you can create your own understanding through the guided-discovery design of these special features. The understanding we create ourselves is the most exciting and the most durable learning that we will ever experience.

In addition, I have used the principles of guided discovery to create over 22 new figures that will guide you to discover for yourself some of nature's most interesting secrets.

As a teacher, my quest is simple. I want you to understand your place in the universe—not just your location in space, but your location in the unfolding history of the physical universe. Not only do I want you to know where you are and what you are in the universe, but I want you to understand how we know. By the end of this book, I want you to know that the universe is very big, but that it is described by a small set of rules and that we have found a way to figure out the rules—a method called science.

The most important concept in introductory astronomy is the process of science, the process by which scientists ask questions of nature and gradually puzzle out the beautiful secrets of the physical world. You should not memorize facts or believe principles because some astronomer says it is so. Of course not! To be a participant in the adventure and not a mere observer, you need to understand how evidence and hypotheses interact to create new understanding. Those who understand how science works are part of the adventure.

Another reason you need to understand how science works is that you live in a technological world that is changing rapidly through the practical application of new scientific discoveries. To control your life in such a world, to guide and care for others, you need to understand the power and the limitation of scientific study.

Science is based on the interplay of evidence and hypothesis, and that interplay is the principal organizing theme for this book at every level—from the individual sentences to the order of the chapters. We will deal with each topic by surveying basic observational facts, synthesizing hypotheses, and then discussing further observations as evidence that supports or contradicts those hypotheses.

When we compare evidence and hypotheses, we make science logical and easy to understand, and that approach is reflected at every level in this book. At the highest level, we discuss stars and galaxies first, and then you can understand the logical place of the solar system in the evolving universe. At a lower level, Chapter 19, The Origin of the Solar System, will help you organize the evidence that appears in Chapters 20 through 25. At a still lower level, individual chapters, such as Chapter 14, Neutron Stars and Black Holes, are organized to explain what we know and compare that with theories. Look at the section on Jupiter's rings and notice how it, like most of the sections in this book, discusses observations compared with theories. Even at the level of single paragraphs, this book avoids a dry recitation of facts and instead compares observations with hypotheses.

Part of the logic of science is understanding how we know what we know. Rather than asking you to remember that Jupiter's rings are made of dust, this book explains how we know that fact. We can't use facts if we can't depend on them, and we can't depend on facts if we don't know how they were understood. Notice how this book helps you understand how we know.

I want you to understand that science is nothing more than a logical attempt to understand nature.

Scientists think in logical arguments, and this book presents the usual facts in carefully organized logical arguments, not because the facts are easier to remember that way, but because the real goal is understanding nature, not remembering facts.

To help you grasp the logic of astronomy, I have created a number of features within the book.

- **Special art spreads,** mentioned earlier, provide an opportunity for you to create your own understanding and share in the satisfaction that scientists feel as they uncover the secrets of natural processes.

- **Guided discovery figures** illustrate important ideas visually and guide students to understand relationships and contrasts interactively.

- **Windows on Science** provide a parallel commentary on how all of science works. For example, the Windows point out where we are using statistical evidence, where we are reasoning by analogy, and where we are building a scientific model rather than a scientific hypothesis.

- **Guideposts** on the opening page of each chapter help you see the organization of the book. The Guidepost connects the chapter with preceding and following chapters to provide an overall organizational guide.

- **Review/Critical Inquiry** at the end of each text section is a carefully designed question to help you review and synthesize concepts and understand the scientific procedures used in the section. A short answer follows to show how scientists construct logical arguments (from observations, evidence, theories, and natural laws) that lead to a conclusion. A further question then gives you a chance to construct your own argument on a related issue.

- **End-of-Chapter Review Questions** are designed to help you review and test your understanding of the material.

- **End-of-Chapter Discussion Questions** go beyond the text and invite you to think critically and creatively about scientific questions. You can think about these questions yourself or discuss them in class.

- **Critical Inquiries for the Web** conclude each chapter by challenging you to use the World Wide Web to explore further and to think creatively and analytically about astronomy.

- **Exploring TheSky** are experiments you can perform with the software *TheSky,* experiments that will allow you to see the sky and celestial bodies in new ways on your own computer screen.

As additional aids, be aware that this book also includes, free of charge, the following electronic enhancements:

- **Virtual Astronomy Laboratories.** This set of 20 online labs is free with a passcode included with every new copy of this textbook. The labs cover topics from helioseismology to dark matter and allow you to submit your results electronically to your instructor or print them out to hand in. The first page of each chapter in this textbook notes which labs correlate to that chapter.

- **AceAstronomy.** Take charge of your learning with the first assessment-centered student learning tool for astronomy. Access AceAstronomy free via the Web with the passcode card bound into this book and begin to maximize your study time with a host of interactive tutorials and quizzes that help you focus on what you need to learn to master astronomy.

- **TheSky Student Edition CD-ROM.** With this CD-ROM, a personal computer becomes a powerful personal planetarium. Loaded with data on 118,000 stars and 13,000 deep-sky objects with images, it allows you to view the universe at any point in time from 4000 years ago to 8000 years in the future, to see the sky in motion, to view constellations, to print star charts, and much more.

- **InfoTrac® College Edition.** You receive free unlimited access for one academic term to this online database of complete articles and images from more than 700 popular and scholarly periodicals of the past four years, updated daily. An online student guide gives you tips for using InfoTrac and correlates each chapter in this book to InfoTrac articles. (Check with your instructor for availability outside North America.)

- **The Brooks/Cole Astronomy Resource Center (www.brookscole.com/astronomy).** Thousands of students and instructors visit this free Web site every week. The Astronomy Resource Center offers—among other things—online study aids, virtual field trips, and links to other sites.

Do not be humble. Astronomy tells us that the universe is vast and powerful, but it also tells us that we are astonishing creatures. We humans are the parts of the universe that think. The human brain is the most complex piece of matter known, so as you explore the universe, remember that it is your human brain that is capable of understanding the depth and beauty of the cosmos.

To appreciate your role in this beautiful universe, you must learn more than just the facts of astronomy. You must understand what we are and how we know. Every page of this book reflects that ideal.

Mike Seeds
mike.seeds@fandm.com

ACKNOWLEDGMENTS

I started writing astronomy textbooks in 1973, and over the years I have had the guidance of a great many people who care about astronomy and teaching. I would like to thank all of the students and teachers who have responded so enthusiastically to *Foundations of Astronomy*. Their comments and suggestions have been very helpful in shaping this book. I would especially like to thank the reviewers whose careful analysis and thoughtful suggestions have been invaluable in completing this new edition: Dr. Martha S. Gilmore, Wesleyan University; Dr. Charles Lindsey, Solar Physics Research Group; Dr. Rebecca Eby Williams, Malin Space Science Systems; Dr. David Wilner, Harvard University.

The following institutions have provided the many photos and diagrams in this book: 2dF Galaxy Redshift Survey Team, 2Mass Sky Survey, Anglo-Australian Telescope Board, Arizona State Computer Science and Geography Departments, Associated Universities for Research in Astronomy, AT&T Archives, BOOMERANG Collaboration, California Association for Research in Astronomy, Caltech, Celestron International, Chandra X-Ray Observatory Center, European Southern Observatory, Gemini Observatory, Goddard Space Flight Center, High Altitude Observatory, Hubble Heritage Team, Institute for Astronomy, IPAC, Johns Hopkins Applied Physics Laboratory, JPL, Keck Observatories, The Large Binocular Telescope Project, Malin Space Science Systems, Max-Planck Institute for Extraterrestrial Physics, National Aeronautics and Space Administration, National Geophysical Data Center, National Museum of Natural History, National Oceanic and Atmospheric Administration, National Optical Astronomy Observatories, National Radio Astronomy Observatory, National Science Foundation, New England Meteoritical, The Observatories of the Carnegie Institution of Washington, Royal Swedish Academy of Sciences, Sacramento Peak Observatory, SETI Institute, Sloan Digital Sky Survey, Solar and Heliospheric Observatory, Space Telescope Science Institute, Steward Observatory, Sudbury Neutrino Observatory, Transition Region and Coronal Explorer, United States Geological Survey, University of Hawaii, and Yerkes Observatory.

I would especially like to thank the following individuals who went out of their way to help me locate materials for *Foundations of Astronomy:* H. A. Abt, David P. Anderson, Andy Aperala, Anthony Aveni, Dana Backman, Mike Bailey, Sam Barden, Allen Beechel, Charles Blue, David Bradstreet, Jack Burns, Linda Bustos, Bruce Campbell, Duncan Chesley, Chip Clark, Jan Curtis, David Hathaway, Cheryl and Don Davis, Frank Drake, Reginald Dufour, John A. Eddy, Michelangelo Fazio, Robert Fosbury, Sidney Fox, Henry Freudenreich, Bruce Fryxell, Dan Gezari, Daniel Good, Bradford Greeley, Randall Grubbs, John Grula, Calvin J. Hamilton, Kevin Hand, F. R. Harnden, William Hartmann, John Harvey, Phil Harvey, John Hayes, Floyd Herbert, Jeff Hester, John Hill, John P. Hughes, Shawn Hughes, Vincent Icke, George Jacoby, Paul Kalas, Victoria Kaspi, William Keel, Russell Kempton, Dean Ketelsen, Stephen M. Larson, Tod Lauer, Deborah Levine, Elsa Luna, Kwok-Yung Lo, John W. Mackenty, David Malin, Jack B. Marling, Ursula B. Marvin, Alfred S. McEwen, Stanley Miller, Reinhard Mundt, Harold Nations, David Parker, Richard Perley, Carolyn C. Porco, Ron Reese, Vera Rubin, Laurence Rudnick, Rudolph E. Schild, Maarten Schmidt, J. William Schopf, François Schweizer, Michael Seldner, Nigel Sharp, Seth Shostak, L. D. Simon, Damon Simonelli, Mike Skrutskie, T. Scott Smith, Steve Snowden, Hyron Spinrad, Alan Toomre, Joachim Trümper, Tony Tyson, Joseph Ververka, Mike Werner, Michael J. West, Gart Westerhout, Jimmy Westlake, Richard E. White, Simon White, Reed Wicander, and Robin Witmore.

Certain unique diagrams in Chapters 11, 12, 13, and 14 are based on figures I designed for my article "Stellar Evolution," which appeared in *Astronomy,* February 1979.

I am happy to acknowledge the use of images and data from a number of important programs. In preparing materials for this book I used NASA's Sky View facility located at NASA Goddard Space Flight Center. I have used atlas images and mosaics obtained as part of the Two Micron All Sky Survey (2MASS), a joint project of the University of Massachusetts and the Infrared Processing and Analysis Center/California Institute of Technology, funded by the National Aeronautics and Space Administration and the National Science Foundation. A number of solar images are used by the courtesy of the SOHO consortium, a project of international cooperation between ESA and NASA.

It is always a pleasure to work with the Thomson Learning team at Brooks/Cole and Wadsworth. Special thanks go to all of the people who have contributed to this project including Kelley McAllister, Leyla Jowza, David Harris, Carol Benedict, Melissa Newt, Nathaniel Bergson-Michelson, Sam Subity, Lisa Torri, and Lisa Weber. I have enjoyed working on production with Margaret Pinette and Bill Heckman of Heckman & Pinette, and I appreciate their understanding and goodwill. I would especially like to thank my editor Keith Dodson for his help and guidance on this project.

Most of all, I would like to thank my wife, Janet, and my daughter, Kate, for putting up with "the books." They know all too well that textbooks are made of time.

Mike Seeds

NEW MEDIA

We encourage you to use the media resources that accompany this textbook. It is far easier to understand astronomy if you see it in action and these new materials will enable you to become a part of that action. The media listed below can be found at **http://val.brookscole.com** (Virtual Astronomy Labs) and **http://astronomy.brookscole.com/seeds8e** (AceAstronomy).

ANIMATIONS

Chapter 2
Daily Motion in the Sky
Constellations from Different
  Latitudes
Constellations in Different
  Seasons
The Celestial Sphere
Seasons
Path of the Sun

Chapter 3
Small-Angle Formula
Lunar Phases

Chapter 4
Epicycles
Eratosthenes

Chapter 5
Gravity and Orbits
Orbital Motion
Center of Mass
Satellites

Chapter 6
Different Telescopes
Resolution and Telescopes

Chapter 7
Kirchhoff's Laws

Chapter 9
Apparent Brightness

Chapter 10
Scattering

Chapter 11
Conduction, Convection,
  and Radiation

Chapter 12
Cutaway of the Sun
The H–R Diagram
Cluster Turnoff
Future of the Sun

Chapter 13
Cutaway of High/Low Mass
  Stars

Chapter 14
Schwarzschild Radius
Neutron Star

Chapter 15
Explorable Milky Way

Chapter 16
Galaxy Types
Hubble

Chapter 17
Jet Deflection

Chapter 18
Hubble
The Raisin Bread Model
Cosmic Microwave Background

Chapter 19
Radioactive Decay
Circumstellar Disk

Chapter 20
Planetary Atmospheres
Seismic Waves

VIRTUAL ASTRONOMY LABS

THE SCALE OF THE COSMOS

The longest journey begins with a single step.

Confucius

© Anglo-Australian Telescope Board

GUIDEPOST

How can we study something so big it includes everything, even us? The cosmos, or the universe as it is more commonly called, is our subject in astronomy. Perhaps the best way to begin our study is to grab a quick impression as we zoom from things our own size up to the largest things in the universe.

That cosmic zoom, the subject of this chapter, gives us our first glimpse of the objects we will study in the rest of this book. In the next chapter, we will return to Earth to think about the appearance of the sky, and subsequent chapters will discuss stars, galaxies, and worlds that fill our universe.

Our quick survey of the universe in this chapter is more than just a listing of objects. It will illustrate the relations between objects—which are big, which are small, and which are contained inside others. In other words, this chapter will give us perspective for all of our exploration to follow. It is easy to learn a few facts, but it is the relationships between facts that are interesting. The relationships illustrated in this chapter will give us a perspective on our place in the cosmos.

While we study the cosmos, we will observe the process by which we learn. That process, science, gives us a powerful way to understand not only the universe but also ourselves.

VIRTUAL LABORATORY

UNITS, MEASURE, AND UNIT CONVERSION

We are going on a voyage out to the end of the universe. Marco Polo journeyed east, Columbus west, but we will travel outward away from our home on Earth, out past the moon, sun, and other planets, past the stars we see in the sky, and past billions more that we cannot see without the aid of the largest telescopes. We will journey through great whirlpools of stars to the most distant galaxies visible from Earth—and then we will continue on, carried only by experience and imagination—looking for the structure of the universe itself.

Besides journeying through space, we will also travel in time. We will explore the past to see the sun and planets form and search for the formation of the first stars and the origin of the universe. We will also explore forward in time to watch the sun die and Earth wither. Our imagination will become a scientific time machine searching for the ultimate end of the universe.

Although we will discover a beginning for our universe, a time when it came into existence, we will not find an edge in space. However far we voyage, we will not find a limit, and we will see evidence that our universe may be infinite and extends in all directions without limit. Such vastness dwarfs our human dimensions but not our human intelligence or imagination.

Astronomy—this imagined voyage—is more than the study of stars and planets. It is the study of the universe in which we exist. Our personal lives are confined to a small part of a small planet circling a small sun drifting through the universe, but astronomy can take us out of ourselves and thus help us understand what we are.

We live small lives on our planet, but our study of astronomy introduces us to sizes, distances, and times far beyond our usual experience. Our task in this chapter is to grasp the meaning of these unfamiliar sizes, distances, and times. The solution lies in a single word, *scale.* We will compare objects of different sizes to grasp the scale of the universe.

We begin with a human figure whose size we recognize. ▌ Figure 1-1 shows a region about 52 feet across occupied by a human being, a sidewalk, and a few trees—all objects whose size we can understand. Each successive picture in this chapter will show a region of the universe that is 100 times wider than the preceding picture. That is, each step will widen our view by a factor of 100.

In ▌ Figure 1-2 we widen our field of view by a factor of 100 and see an area 1 mile in diameter. The arrow points to the scene shown in the preceding photograph. People, trees, and sidewalks have vanished, but now we can see a college campus and the surrounding streets and houses. The dimensions of houses and streets are familiar to us. We have been in houses, crossed streets, and walked or run a mile. This is the familiar world around us, and we can relate such objects to the scale of our bodies.

FIGURE 1-2
(USGS)

FIGURE 1-1
(Michael A. Seeds)

Ace⊚Astronomy™ The AceAstronomy icon throughout the text indicates an opportunity for you to test yourself on key concepts and to explore animations and interactions on the AceAstronomy Web site at **http://astronomy.brookscole .com/seeds8e**

Infrared-wavelength image

FIGURE 1-3
(NASA infrared photograph)

sensitive only to a narrow range of colors. As we explore the universe, we must learn to use a wide range of "colors," which astronomers refer to as wavelengths. We will discuss wavelengths in detail in a later chapter, but for the moment we should note that Figure 1-3 was made at infrared wavelengths. Photographs in this book are carefully labeled to show the approximate wavelength at which the images were made. Astronomical images range from X-ray to radio wavelengths and reveal sights invisible to our eyes.

At the next step in our journey, we see our entire planet (▌Figure 1-4), which is 12,756 km in diameter. Earth rotates on its axis once a day, exposing half of

Visual-wavelength image

FIGURE 1-4
(NASA)

We have begun our adventure using feet and miles, but we should use the metric system of units. Not only is it used by all scientists around the world, but it makes calculations much easier. If you are not already familiar with the metric system, or need a review, study Appendix A before reading on.

The photo in Figure 1-2 is 1 mile in diameter. A mile equals 1.609 kilometers, so we can see in the photo that a kilometer is a bit over two thirds of a mile—a short walk across a neighborhood.

Our view in ▌Figure 1-3 spans 160 kilometers. Green foliage of different kinds appears as shades of red in this infrared photo. The college campus is now invisible. The gray blotches are small cities, and the suburbs of Philadelphia are visible at the lower right. At this scale, we see the natural features of Earth's surface. The Allegheny Mountains of southern Pennsylvania cross the image in the upper left, and the Susquehanna River flows southeast into Chesapeake Bay. What appear as white bumps are a few puffs of clouds.

These features remind us that we live on the surface of a changing planet. Forces in Earth's crust pushed the mountain ranges up into parallel folds, like a rug wrinkled on a polished floor. The clouds remind us that Earth's atmosphere is rich in water, which falls as rain and erodes the mountains, washing material down the rivers and into the sea. The mountains and valleys that we know are only temporary features; they are constantly changing. As we explore the universe, we will see that it, like Earth's surface, is evolving. Change is the norm in the universe.

Look again at Figure 1-3 and note the red color. This photo is an infrared photograph in which healthy green leaves and crops show up as red. Our eyes are

its surface to daylight at any particular moment. The photo shows most of the daylight side of the planet. The blurriness at the extreme right is the sunset line. The rotation of Earth carries us eastward, and as we cross the sunset line into darkness, we say the sun has set. Thus, the rotation of our planet causes the cycle of day and night.

We know that Earth's interior is made mostly of iron and nickel and that its crust is mostly silicate rocks. Only a thin layer of water makes up the oceans, and the atmosphere is only a few hundred kilometers deep. On the scale of this photograph, the depth of the atmosphere on which our lives depend is less than the thickness of a piece of thread.

Again we enlarge our field of view by a factor of 100, and we see a region 1,600,000 km wide (▌Figure 1-5). Earth is the small white dot in the center, and the moon, whose diameter is only one-fourth that of Earth, is an even smaller dot along its orbit 380,000 km from Earth.

These numbers are so large that it is inconvenient to write them out. Astronomy is the science of big numbers, and we will use numbers much larger than these to discuss the universe. Rather than writing out these numbers as in the previous paragraph, it is convenient to write them in **scientific notation.** This is nothing more than a simple way to write numbers without writing a great many zeros. In scientific notation, we write 380,000 as 3.8×10^5. If you are not familiar with scientific notation, read the section on Powers of 10 Notation in the Appendix. The universe is too big to discuss without using scientific notation.

When we once again enlarge our field of view by a factor of 100 (▌Figure 1-6), Earth, the moon, and the moon's orbit all lie in the small red box at lower left. But now we see the sun and two other planets that are part of our solar system. Our **solar system** consists of the sun, its family of planets, and some smaller bodies such as moons and comets.

Like the Earth, Venus and Mercury are **planets,** small, nonluminous bodies that shine by reflected light. Venus is about the size of Earth, and Mercury is a bit larger than our moon. On this diagram, they are both too small to be seen as anything but tiny dots. The sun is a **star,** a self-luminous ball of hot gas that generates

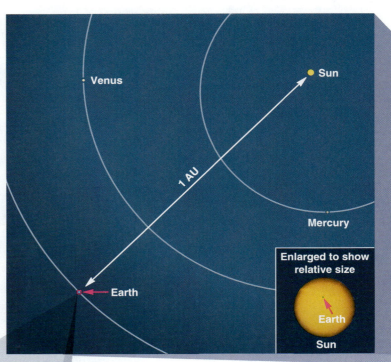

FIGURE 1-6
(NOAO)

FIGURE 1-5
(NASA)

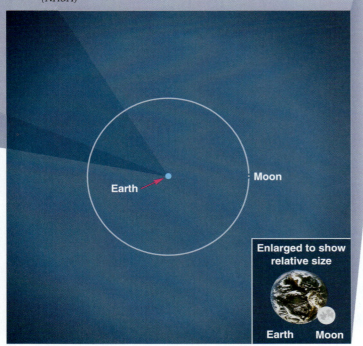

its own energy. The sun is 109 times larger in diameter than Earth (inset), but it too is nothing more than a dot in this diagram.

This diagram has a diameter of 1.6×10^8 km. One way astronomers deal with large numbers is to define new units. The average distance from Earth to the sun is a unit of distance called the **astronomical unit (AU),** a distance of 1.5×10^{11} m. Using this unit we can say that the average distance from Venus to the sun is about 0.7 AU. The average distance from Mercury to the sun is about 0.39 AU.

The orbits of the planets are not perfect circles, and this is particularly apparent for Mercury. Its orbit carries it as close to the sun as 0.307 AU and as far away as 0.467 AU. Earth's orbit is more circular, and its distance from the sun varies by only a few percent.

Our first field of view was only 52 feet (about 16 m) in width. After only six steps of enlarging by a factor of 100, we now see the entire solar system (▌Figure 1-7). Our field of view is 1 trillion (10^{12}) times wider than in our first view. The details of the preceding figure are now lost in the red square at the center of this diagram. We see only the brighter, more widely separated objects as we enlarge our view. The sun, Mercury, Venus, and Earth lie so close together that we cannot separate them at this scale. Mars, the next outward planet, lies only 1.5 AU from the sun. In contrast, Jupiter, Saturn, Uranus, Neptune, and Pluto are so far from the sun that they are easy to place in this diagram. These are cold worlds far from the sun's warmth. Light from the sun reaches Earth in only 8 minutes, but it takes over 4 hours to reach Neptune. Notice that Pluto's orbit is so elliptical that Pluto can come closer to the sun than Neptune does, as Pluto did between 1979 and 1999.

When we again enlarge our field of view by a factor of 100, our solar system vanishes (■ Figure 1-8). The sun is only a point of light, and all the planets and their orbits are now crowded into the small, red square at the center. The planets are too small and reflect too little light to be visible so near the brilliance of the sun.

Nor are any stars visible except for the sun. The sun is a fairly typical star, and it seems to be located in a fairly average neighborhood in the universe. Although

nearest stars. These stars are so distant that it is not reasonable to give their distances in astronomical units. We must define a new unit of distance, the light-year. One **light-year (ly)** is the distance that light travels in one year, roughly 10^{13} km or 63,000 AU. The diameter of our field of view is, in our new unit, 17 ly. The nearest star to the sun, Proxima Centauri, is 4.2 ly from Earth. In other words, light from Proxima Centauri takes 4.2 years to reach Earth.

Although these stars are roughly the same size as the sun, they are so far away that we cannot see them as anything but points of light. Even with the largest telescopes on Earth, we still see only points of light when we look at stars. Many of these stars probably have planets, but the planets reflect too little light and are too close to their bright stars to be visible from Earth. Astronomers have found over 100 such planets by indirect means, but planets would not be visible in Figure 1-9.

In our diagram, the sizes of the dots represent not the size of the stars but their brightness. This is the custom in astronomical diagrams, and it is also how star images are recorded on photographs. Bright stars make larger spots on a photograph than faint stars. Thus, the size of a star image in a photograph tells us not how big the star is but only how bright it looks.

In ■ Figure 1-10, we expand our field of view by another factor of 100, and the sun and its neighboring stars vanish into the background of

FIGURE 1-7

FIGURE 1-8

there are many billions of stars like the sun, none is close enough to be visible in this diagram, which shows an area only 11,000 AU in diameter. The stars are typically separated by distances about 10 times larger than the distance represented in this diagram. We will see stars in our next field of view, but, except for the sun at the center, this diagram is empty.

It is difficult to grasp the isolation of the stars. If the sun were represented by a golf ball in New York City, the nearest star would be another golf ball in Chicago. Except for the widely scattered stars and a few atoms of gas drifting between the stars, the universe is nearly empty.

In ■ Figure 1-9, our field of view has expanded to a diameter a bit over 1 million AU. The sun is at the center, and we see a few of the

FIGURE 1-9

photo, but if we could see it, we would find it in the outer parts of the disk of the galaxy.

Our galaxy, like many others, has graceful **spiral arms** winding outward through the disk; we will discover that stars are born in great clouds of gas and dust as they pass through the spiral arms.

Ours is a fairly large galaxy, roughly 75,000 ly in diameter. Only a century ago astronomers thought it was the entire universe—an island universe of stars in an otherwise empty vastness. Now we know that our galaxy is not unique. Indeed ours is only one of many billions of galaxies scattered throughout the universe.

As we expand our field of view by another factor of 100, our galaxy appears as a tiny lumi-

thousands of stars. The field of view is now 1700 ly in diameter. Of course, no one has ever journeyed thousands of light-years from Earth to look back and photograph the solar neighborhood, so we use a representative photo of the sky. The sun is a relatively faint star that would not be easily located in a photo at this scale. Looking at this photograph, we notice a loose cluster of stars in the lower left quarter of the image. We will discover that many stars are born in clusters and that both old and young clusters exist in the sky. Star clusters are forming right now.

What we do not see is critically important. We do not see the thin gas that fills the spaces between the stars. Although those clouds of gas are thinner than the best vacuum on Earth, it is those clouds that give birth to new stars. Our sun formed from such a cloud about 5 billion years ago. We will see more star formation in our next view.

If we expand our field of view by a factor of 100, we see our galaxy (❙ Figure 1-11). A **galaxy** is a great cloud of stars, gas, and dust bound together by the combined gravity of all the matter. Galaxies range from 1500 to over 300,000 ly in diameter and can contain over 100 billion stars. In the night sky, we see our galaxy as a great, cloudy wheel of stars ringing the sky as the **Milky Way,** and we refer to our galaxy as the **Milky Way Galaxy.** Of course, no one can journey far enough into space to look back and photograph our galaxy, so the photo in Figure 1-11 shows a galaxy similar to our own. Our sun would be invisible in such a

Visual-wavelength image

FIGURE 1-10

This box ■ represents the relative size of the previous frame. (NOAO)

nous speck surrounded by other specks (❙ Figure 1-12). This diagram includes a region 17 million ly in diameter, and each of the dots represents a galaxy. Notice that our galaxy is part of a cluster of a few dozen galaxies. We will find that galaxies are commonly grouped together in clusters. Some of these galaxies have beautiful spiral patterns like our own galaxy, but others do not. Some are strangely distorted. One of the mysteries we will try to solve is what produces these differences among the galaxies.

If we again expand our field of view, we see that the clusters of galaxies are connected in a vast network (❙ Figure 1-13). Clusters are grouped into superclus-

Our problem in studying astronomy is to keep a proper sense of scale. Remember that each of the billions of galaxies contains billions of stars. Most of those stars probably have families of planets like our solar system, and on some of those billions of planets liquid-water oceans and protective atmospheres may have sheltered the spark of life. It is possible that some other planets are inhabited by intelligent creatures who share our curiosity and our wonder at the scale of the cosmos.

FIGURE 1-13

This box ▪ represents the relative size of the previous frame. *(Detail from galaxy map from M. Seldner, B. L. Siebers, E. J. Groth, and P. J. E. Peebles,* Astronomical Journal 82 *[1977])*

ters—clusters of clusters—and the superclusters are linked to form long filaments and walls outlining voids that seem nearly empty of galaxies. These appear to be the largest structures in the universe. Were we to expand our field of view another time, we would probably see a uniform fog of filaments and voids. When we puzzle over the origin of these structures, we are at the frontier of human knowledge.

Milky Way Galaxy ⟶

FIGURE 1-12

SUMMARY

Our goal in this chapter is to preview the scale of astronomical objects. To do so, we journey outward from a familiar campus scene by expanding our field of view by factors of 100. Only 12 such steps take us to the largest structures in the universe.

The numbers in astronomy are so large it is not convenient to express them in the usual way. Instead, we use the metric system to simplify our calculations and scientific notation to simplify the writing of big numbers. The metric system and scientific notation are discussed in Appendix A.

We live on the rotating planet Earth, which orbits a rather typical star we call the sun. A unit of distance convenient for discussing our solar system, the astronomical unit (AU), equals the average distance from Earth to the sun. Of the other planets in our solar system,

Visual-wavelength image

FIGURE 1-11

(© Anglo-Australian Telescope Board)

Mercury is closest to the sun, only 0.39 AU away, and Neptune is about 30 AU.

The sun, like most stars, is very far from its neighboring stars, and this leads us to use another unit of distance, the light-year. A light-year is the distance light travels in one year. The star nearest to the sun is Proxima Centauri, at a distance of 4.2 ly.

As we enlarge our field of view, we discover that the sun is only one of about 100 billion stars in our galaxy and that our galaxy is only one of many billions of galaxies in the universe. Galaxies appear to be grouped together in clusters, superclusters, and filaments, the largest structures known.

As we explore, we note that the universe is evolving. Earth's surface is evolving, and so are stars. Stars form from the gas in space, grow old, and eventually die.

Among the billions of stars in each of the billions of galaxies, many probably have planets, but detecting planets orbiting other stars is very difficult, and only a few more than a hundred have been found. We suppose that among the many planets in the universe, there must be some that are like Earth. We wonder if a few are inhabited by intelligent beings like ourselves.

NEW TERMS

scientific notation	light-year (ly)
solar system	galaxy
planet	Milky Way
star	Milky Way Galaxy
astronomical unit (AU)	spiral arm

REVIEW QUESTIONS

Ace⊘Astronomy™ Assess your understanding of this chapter's topics with additional quizzing and animations at **http:// astronomy.brookscole.com/seeds8e**

1. What is the largest dimension you have personal knowledge of? Have you run a mile? Hiked 10 miles? Run a marathon?

2. In Figure 1-4, the division between daylight and darkness is at the right on the globe of Earth. How do we know this is the sunset line and not the sunrise line?

3. What is the difference between our solar system, our galaxy, and the universe?

4. Look at Figure 1-6. How can you tell that Mercury follows an elliptical orbit? Can you detect the elliptical shape of any other orbits in this figure or the next?

5. Which is the outermost planet in our solar system? Why does that change?

6. Why are light-years more convenient than miles, kilometers, or astronomical units for measuring certain distances?

7. Why is it difficult to detect planets orbiting other stars?

8. What does the size of the star image in a photograph tell us?

9. What is the difference between the Milky Way and the Milky Way Galaxy?

10. What are the largest known structures in the universe?

PROBLEMS

1. The diameter of Earth is 7928 miles. What is its diameter in inches? In yards?

2. If a mile equals 1.609 km and the moon is 2160 miles in diameter, what is its diameter in kilometers?

3. One astronomical unit is about 1.5×10^8 km. Explain why this is the same as 150×10^6 km.

4. Venus orbits 0.7 AU from the sun. What is that distance in kilometers?

5. Light from the sun takes 8 minutes to reach Earth. How long does it take to reach Mars?

6. The sun is almost 400 times farther from Earth than is the moon. How long does light from the moon take to reach Earth?

7. If the speed of light is 3×10^5 km/s, how many kilometers are in a light-year? How many meters?

8. How long does it take light to cross the diameter of our Milky Way Galaxy?

9. The nearest galaxy to our own is about 2 million light-years away. How many meters is that?

10. How many galaxies like our own would it take laid edge to edge to reach the nearest galaxy? (*Hint:* See Problem 9.)

CRITICAL INQUIRIES FOR THE WEB

1. Locate photographs of Earth taken from space. What do cities look like? Can you see highways? Is the presence of our civilization detectable from space?

2. Locate photographs of nearby galaxies and compare them with photos of very distant galaxies. What kind of detail is invisible for distant galaxies?

3. One of the biggest clusters of galaxies is the Virgo cluster. Find out how many and what kind of galaxies are in the cluster. Is it nearby or far away?

EXPLORING *THESKY*

1. Locate and center one example of each of three different types of objects:

   a. A planet, such as Saturn. Find its rising and setting time. Such objects have distances measured in astronomical units (AU).

   How to proceed: Decide on the object you want to locate. Then find and center the object by pressing the **Find** button on the **Object Toolbar.** The second method is to press the **F** key. The third is to click **Edit,** then

Find. Once you have the **Object Information** window, press the **center** button.

b. A star. All stars in *TheSky* belong to our Milky Way Galaxy. Give the star's name, its magnitude, and its distance in light-years.

How to proceed: Click on any star, which brings up an **Object Information** window containing a variety of information about the star.

c. A galaxy; give its name and/or its designation.

How to proceed: To show galaxies, click on the **Galaxies** button in the **Object Toolbar,** then click on any galaxy. Distances to galaxies are on the order of millions and billions of light years.

2. Look at the solar system from beyond Pluto by clicking on **View** and then on **3D Solar System Mode.** Tip the solar system edge-on and then face-on. Zoom in to see the inner planets. Under **Tools,** set the **Time Skip Increment** to 1 day and then go **forward in time** to watch the planets move.

3. Identify some of the brightest constellations located along the Milky Way. (*Hint:* See **View, Reference Lines.**)

 Visit the Seeds *Foundations of Astronomy* companion Web site for critical thinking exercises, articles, and additional readings from InfoTrac College Edition, Brooks/Cole's online student library.

The Southern Cross I saw every night abeam. The sun every morning came up astern; every evening it went down ahead. I wished for no other compass to guide me, for these were true.

Captain Joshua Slocum, Sailing Alone Around the World

Chapter 2

William Hartmann

GUIDEPOST

Astronomy is about us. As we learn about astronomy, we learn about ourselves. We search for an answer to the question "What are we?" The quick answer is that we are thinking creatures living on a planet that circles a star we call the sun. In this chapter, we begin trying to understand that answer. What does it mean to live on a planet?

The preceding chapter gave us a quick overview of the universe, and chapters later in the book will discuss the details. This chapter and the next help us understand what the universe looks like seen from the surface of our spinning planet.

But appearances are deceiving. We will see in Chapter 4 how difficult it has been for humanity to understand what we see in the sky every day. In fact, we will discover that modern science was born when people tried to understand the appearance of the sky.

The night sky is the rest of the universe as seen from our planet. When we look up at the stars, we look out through a layer of air only a few hundred kilometers deep. Beyond that, space is nearly empty, with the stars scattered light-years apart. We begin our search for the natural laws that govern the universe by trying to understand what that universe looks like.

What we see in the sky is influenced dramatically by a simple fact. We live on a planet. The stars are scattered into the void all around us, most very distant and some closer. Our planet rotates on its axis once a day, so from our viewpoint the sky appears to rotate around us each day. Not only does the sun rise in the east and set in the west, but so also do the stars. That apparent daily motion of the sky is caused by the rotation of our planet. As you read this chapter and learn about the sky, keep in mind that you are also learning what it means to live on a planet.

Ace Astronomy™ Go to AceAstronomy and click Active Figures to see "Rotation of the Sky." Notice the angles at which stars rise in the east and set in the west.

2-1 THE STARS

On a dark night far from city lights, we can see a few thousand stars in the sky. Like ancient astronomers, we will try to organize what we see by naming groups of stars and individual stars and by specifying the brightness of individual stars.

CONSTELLATIONS

All around the world, ancient cultures celebrated heroes, gods, and mythical beasts by naming groups of stars—**constellations** (▌ Figure 2-1). We should not be surprised that the star patterns do not look like the creatures they represent any more than Columbus, Ohio, looks like Christopher Columbus. The constellations "celebrate" the most important mythical figures in each culture. The constellations we know in Western culture originated in Mesopotamia over 5000 years ago, with other constellations added by Babylonian, Egyptian, and Greek astronomers during the classical age. Of these ancient constellations, 48 are still in use.

Ace Astronomy™ The AceAstronomy icon throughout the text indicates an opportunity for you to test yourself on key concepts and to explore animations and interactions on the AceAstronomy Web site at **http://astronomy.brookscole .com/seeds8e**

FIGURE 2-1

The constellations are an ancient heritage handed down for thousands of years as celebrations of great heroes and mythical creatures. Here Sagittarius and Scorpius hang above the southern horizon.

Different cultures grouped stars and named constellations differently. The constellation we know as Orion was known as Al Jabbar, the giant, to the ancient Syrians, as the White Tiger to the Chinese, and as Prajapati in the form of a stag in ancient India. The Pawnee Indians knew the constellation Scorpius as two groupings. The long tail of the scorpion was the Snake, and the two bright stars at the tip of the scorpion's tail were the Two Swimming Ducks.

Many ancient cultures around the world, including the Greeks, northern Asians, and Native Americans, associated the stars of the Big Dipper with a bear. The concept of the celestial bear may have crossed the land bridge into North America with the first Americans over 10,000 years ago. Some of the constellations we know today may be among the oldest surviving traces of human culture.

To the ancients, a constellation was a loose grouping of stars. Many of the fainter stars were not included in any constellation, and regions of the southern sky not visible to the ancient astronomers of northern latitudes were not identified with constellations. Constellation boundaries, when they were defined at all, were only approximate (▌ Figure 2-2a), so a star like Alpheratz could be thought of as part of Pegasus or part of Andromeda. In recent centuries astronomers have added 40 modern constellations to fill gaps, and in 1928 the International Astronomical Union established 88 official constellations with clearly defined boundaries (Figure 2-2b). Thus a constellation now represents not a group of stars but an area of the sky, and any star within the region belongs to one and only one constellation.

a

b

Andromeda

Alpheratz

Pegasus

Great square
of Pegasus

FIGURE 2-2

(a) In antiquity, constellation boundaries were poorly defined, as shown on this map by the curving dotted lines that separate Pegasus from Andromeda. *(From Duncan Bradford,* Wonders of the Heavens, *Boston: John B. Russell, 1837)* (b) Modern constellation boundaries are precisely defined by international agreement.

In addition to the 88 official constellations, the sky contains a number of less formally defined groupings called **asterisms.** The Big Dipper, for example, is a well-known asterism that is part of the constellation Ursa Major (the Great Bear). Another asterism is the Great Square of Pegasus (Figure 2-2b), which includes three stars from Pegasus and one from Andromeda. The star charts at the end of this book will introduce you to the brighter constellations and asterisms.

Although we name constellations and asterisms, most are made up of stars that are not physically associated with one another. Some stars may be many times farther away than others and moving through space in different directions. The only thing they have in common is that they lie in approximately the same direction from Earth (❙ Figure 2-3).

THE NAMES OF THE STARS

The brightest stars were named thousands of years ago, and these names, along with the names of the constellations, found their way into the first catalogs of stars. We take our constellation names from Greek versions translated into Latin—the language of science from the fall of Rome to the 19th century—but most star names come from ancient Arabic. Names such as Sirius (the Scorched One), Capella (the Little She Goat), and Aldebaran (the Follower of the Pleiades) are beautiful additions to the mythology of the sky.

Giving the stars individual names is not very helpful, because we see thousands of stars, and these names do not help us locate the star in the sky. In which constellation is Antares, for example? In 1603, Bavarian lawyer Johann Bayer published an atlas of the sky called *Uranometria* in which he assigned lowercase Greek letters to the brighter stars of each constellation in approximate order of brightness. Astronomers have used those Greek letters ever since. (See the Appendix table with the Greek alphabet.) Thus, the brightest star is usually designated α (alpha), the second-brightest β (beta), and so on (❙ Figure 2-4). To identify a star in this way, we give the Greek letter followed by the possessive form of the constellation name, such as α Scorpii (for Antares). Now we know that Antares is in the constellation Scorpius and that it is probably the brightest star in that constellation.

This method of identifying a star's brightness, however, is only approximate. To be precise, we must have an accurate way of referring to the brightness of stars, and for that we must consult one of the first great astronomers.

THE BRIGHTNESS OF STARS

Astronomers measure the brightness of stars using the **magnitude scale,** a system that first appeared in the writings of the ancient astronomer Claudius Ptolemy about 140 AD. The system may have originated earlier than Ptolemy, and most astronomers attribute it to the Greek astronomer Hipparchus (160–127 BC) (❙ Figure 2-5).

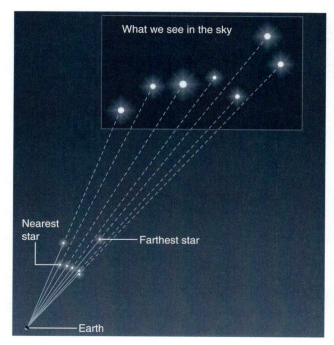

FIGURE 2-3

The stars we see in the Big Dipper—the brighter stars of the constellation Ursa Major, the Great Bear—are not at the same distance from Earth. We see the stars in a group in the sky because they lie in the same general direction as seen from Earth. The size of the star dots in the star chart represents the apparent brightness of the stars.

FIGURE 2-5

Hipparchus (2nd century BC) was the first great observational astronomer. He is believed to have compiled the first star catalog and may have invented the magnitude scale. He is honored here on a Greek stamp, which also shows one of his observing instruments.

The ancient astronomers divided the stars into six classes. The brightest were called first-class stars and those that were fainter, second-class. The scale continued downward to sixth-class stars, the faintest visible to the human eye. Thus, the larger the magnitude number, the fainter the star. This makes sense if we think of the bright stars as first-class stars and the faintest stars visible as sixth-class stars.

Hipparchus is believed to have compiled the first star catalog, and he may have used the magnitude system in that catalog. Almost 300 years later Ptolemy used the magnitude system in his own catalog, and successive generations of astronomers have continued to use the system. Later astronomers had to revise the ancient magnitude system to include very faint and very bright stars. Telescopes reveal many stars fainter than those the eye can detect, so the magnitude scale was extended to include these faint stars. The Hubble Space Telescope, currently the most sensitive astronomical telescope, can detect stars

The brighter stars in a constellation are usually given Greek letters in order of decreasing brightness.

Orion

α Orionis is also known as Betelgeuse.

β Orionis is also known as Rigel.

In Orion β is brighter than α, and κ is brighter than η. Fainter stars do not have Greek letters or names, but if they are located inside the constellation boundaries, they are part of the constellation.

FIGURE 2-4

Stars in a constellation can be identified by Greek letters and by names derived from Arabic. *(William Hartmann)*

FIGURE 2-6

The scale of apparent visual magnitudes extends into negative numbers to represent the brightest objects and to positive numbers larger than 6 to represent objects fainter than the human eye can see.

■ Figure 2-6

TABLE 2-1
Magnitude and Intensity

Magnitude Difference	Intensity Ratio
0	1
1	2.5
2	6.3
3	16
4	40
5	100
6	250
7	630
8	1600
9	4000
10	10,000
⋮	⋮
15	1,000,000
20	100,000,000
25	10,000,000,000
⋮	⋮

fainter than +28 magnitude. In contrast, the brightest stars are actually brighter than the first brightness class in the ancient magnitude system. Thus, modern astronomers extended the magnitude system into negative numbers to account for brighter objects. For instance, Vega (alpha Lyrae) is almost zero magnitude at 0.04, and Sirius, the brightest star in the sky, has a magnitude of −1.42. We can even place the sun on this scale at −26.5 and the moon at −12.5 (■ Figure 2-6).

These numbers are known as **apparent visual magnitudes (m_v)**, and they describe how the stars look to human eyes observing from Earth. Although some stars emit large amounts of infrared or ultraviolet light, we can't see it, and it is not included in the apparent visual magnitude. The subscript "v" reminds us that we are only including light we can see. Another problem is the distance to the stars. Very distant stars look fainter, and nearby stars look brighter. Apparent visual magnitude ignores the effect of distance and tells us only how bright the star looks as seen from Earth.

MAGNITUDE AND INTENSITY

Nearly every star catalog from today back to the time of the ancients uses the magnitude scale of stellar brightness. However, brightness is subjective, depending on such things as the physiology of the eye and the psychology of perception. To be accurate in measurements and calculations astronomers use the more precise term *intensity*—a measure of the light energy from a star that hits 1 square meter in 1 second. Thus astronomers need a way to convert between magnitudes and intensities.

The human eye senses the brightness of objects by comparing the ratios of their intensities. If two stars have intensities I_A and I_B, we can compare them by writing the ratio of their intensities, I_A/I_B. Hipparchus's magnitude system is based on a constant intensity ratio of about 2.5 for each magnitude. That is, if two stars differ by 1 magnitude, then they have an intensity ratio of about 2.5. If they differ by 2 magnitudes, their ratio is 2.5 × 2.5, and so on.

When 19th-century astronomers began measuring starlight, they realized they needed to define a mathematical magnitude system that was precise, but for convenience they wanted a system that agreed at least roughly with that of Hipparchus. The stars that ancient astronomers classified as first and sixth differ by 5 magnitudes and have an intensity ratio of almost exactly 100, so the modern system of magnitudes specifies that a magnitude difference of 5 magnitudes corresponds to an intensity ratio of 100. That means that 1 magnitude corresponds to an intensity ratio of precisely 2.512, the fifth root of 100. That is, $100 = (2.512)^5$.

Now we can compare the light we receive from two stars. The light from a first-magnitude star is 2.512 times more intense than that from a second-magnitude star. The light from a third-magnitude star is $(2.512)^2$ times more intense than the light from a fifth-magnitude star. In general, the intensity ratio equals 2.512 raised to the power of the magnitude difference. That is:

$$I_A/I_B = (2.512)^{(m_B - m_A)}$$

For instance, if two stars differ by 6.32 magnitudes, we can calculate the ratio of their intensities as $2.512^{6.32}$. A pocket calculator tells us that the ratio is 337.

This magnitude system has some advantages. It compresses a tremendous range of intensity into a small range of magnitudes (■ Table 2-1) and makes it possible for modern astronomers to compare their measurements with all the recorded measurements of the past, right back to the first star catalogs over 2000 years ago.

In this section, we have found a way to identify stars by name and by brightness. Next we must look at the sky as a whole and notice its motion.

2-2 THE SKY AND ITS MOTION

The sky above us seems to be a great blue dome in the daytime and a sparkling ceiling at night. Learning to look at the sky requires that we begin thousands of years ago.

THE CELESTIAL SPHERE

Ancient astronomers believed the sky was a great sphere surrounding Earth with the stars stuck on the inside like thumbtacks in the ceiling. Modern astronomers know that the stars are scattered through space at different distances, but it is still convenient to think of the sky as a great celestial sphere.

As you study ▌"The Sky Around Us" on pages 16 and 17, notice two important points. First, the sky appears to rotate westward around Earth each day, but that is a consequence of the eastward rotation of Earth. Second, what we can see of the sky depends on where we are on Earth. If we lived in Australia, we would see many constellations and asterisms invisible from North America, but we would never see the Big Dipper.

The celestial sphere is an example of a scientific model, a common feature of scientific thought (Window on Science 2-1). Notice that a scientific model does not have to be true to be useful. We will discuss many scientific models in the chapters that follow.

In addition to the daily motion of the sky, Earth's rotation adds a second motion to the sky that can be detected only over centuries.

PRECESSION

Over 2000 years ago, Hipparchus compared a few of his star positions with those made nearly two centuries before and realized that the celestial poles and equator were slowly moving across the sky. Later astronomers understood that this motion is caused by Earth's top-like motion.

WINDOW ON SCIENCE 2-1

Frameworks for Thinking About Nature: Scientific Models

In everyday language, we use the word *model* in various ways—fashion model, model airplane, model student—but scientists use the word in a very specific way. A **scientific model** is a carefully devised mental conception of how something works, a framework that helps scientists think about some aspect of nature. For example, astronomers use the celestial sphere as a way to think about the rotations of the sky, sun, moon, and stars.

Although we will think of a scientific model as a mental conception, it can take many forms. Some models are quite abstract—the psychologist's model of how the human mind processes visual information into images, for instance. But other models are so specific that they can be expressed as a set of mathematical equations. For example, an astronomer might use a set of equations to describe in detail how gas falls into a black hole. We could refer to such a calculation as a model. Of course, we could use metal and plastic to build a celestial globe, but the thing we build wouldn't really be the model any more than the equations are a model. A scientific model is a mental conception, an idea in our heads, that helps us think about nature.

On the other hand, a model is not meant to be a statement of truth. The celestial sphere is not real; we know the stars are scattered through space at various distances, but we can imagine a celestial sphere and use it to help us think about the sky. A scientific model does not have to be true to be useful.

Chemists, for example, think about the atoms in molecules by visualizing them as balls joined together by bonds. This model of a molecule is not fully correct, but it is a helpful way to think about molecules; it gives chemists a framework within which to organize their ideas.

Because scientific models are not meant to be totally correct, we must always remember the assumptions on which they are based. If we begin to think a model is true, it can mislead us instead of helping us. The celestial sphere, for instance, can help us think about the sky, but we must remember that it is only a mental crutch. The universe is much larger and much more interesting than this ancient scientific model of the heavens.

The Sky Around Us

The sky above appears to be part of a sphere, the **celestial sphere,** with Earth at its center. From any location on Earth we see only half of the celestial sphere, the half above our **horizon.** The **zenith** marks the top of the sky above our head, and the **nadir** marks the bottom of the sky directly under our feet. What we see in the sky depends on where we are on Earth. The drawing at right shows the view for an observer in North America. An observer in South America would have a dramatically different horizon, zenith, and nadir.

The eastward rotation of Earth causes the sun, moon, and stars to move westward in our sky as if the celestial sphere were rotating westward around Earth. The apparent pivot points are the **north celestial pole** and the **south celestial pole** located directly above Earth's north and south poles. Halfway between the celestial poles lies the **celestial equator.** Earth's rotation defines the directions we use every day. The **north point** and **south point** are the points on the horizon closest to the celestial poles. The **east point** and the **west point** lie halfway between the north and south points. The celestial equator always touches the horizon at the east and west points.

AURA/NOAO/NSF

This time exposure of about 30 minutes shows stars as streaks, called star trails, rising behind an observatory dome. The camera was facing northeast to take this photo. The motion we see in the sky depends on which direction we look, as shown at right. Looking north, we see the star Polaris, the North Star, located near the north celestial pole. As the sky appears to rotate westward, Polaris hardly moves, but other stars circle the celestial pole. Looking south from a location in North America, we can see stars circling the south celestial pole, which is invisible below our southern horizon.

Looking north

Looking east

Looking south

Astronomers measure distance across the sky as angles.

Angular distance

Astronomers might say, "The star was only 2 degrees from the moon." Of course, the stars are much farther away than the moon, but when we think of the celestial sphere, we can measure distances *on the sky* as **angular distances** in degrees, minutes of arc, and seconds of arc. A **minute of arc** is 1/60th of a degree, and a **second of arc** is 1/60th of a minute of arc. Then the **angular diameter** of an object is the angular distance from one edge to the other. The sun and moon are each about half a degree in diameter, and the bowl of the Big Dipper is about 10° wide.

Ace ✪ Astronomy™
Go to AceAstronomy and click Active Figures to see "Constellations from Different Latitudes" to see the sky from different latitudes.

Circumpolar constellations are those that never rise or set. From mid-northern latitudes, as shown below, we see a number of familiar constellations circling Polaris and never dipping below the horizon. As the sky rotates, the pointer stars at the front of the Big Dipper always point toward Polaris. Circumpolar constellations near the south celestial pole never rise as seen from mid-northern latitudes. From a high latitude such as Norway, we would have more circumpolar constellations, and from Quito, Ecuador, located on Earth's equator, we would have no circumpolar constellations at all.

A few circumpolar constellations

Cepheus

Cassiopeia

Perseus

Rotation of sky

Rotation of sky

Polaris

Ursa Minor

Ursa Major

North celestial pole
Zenith
Latitude 90°

Zenith
North celestial pole
S — — N
Latitude 60° E

Zenith
North celestial pole
W
L
S — — N
Latitude 30° E

Zenith
North celestial pole
W
S — — N
Latitude 0° E

South celestial pole
Zenith
W
S — — N
Latitude −30° E

What we see in the sky depends on our latitude as shown above. Imagine that we begin a journey in the ice and snow at Earth's North Pole with the north celestial pole directly overhead. As we walk southward the celestial pole moves toward the horizon and we can see further into the southern sky. The angular distance from the horizon to the north celestial pole always equals our latitude (L)—the basis for celestial navigation. As we cross Earth's equator, the celestial equator would pass through our zenith, and the north celestial pole would sink below our northern horizon

If you have ever played with a gyroscope or top, you have seen how the spinning mass resists any change in the direction of its axis of rotation. The more massive the top and the more rapidly it spins, the more difficult it is to change the direction of its axis of rotation. But you probably recall that the axis of even the most rapidly spinning top sweeps around in a conical motion. The weight of the top tends to make it tip over, and this combines with its rapid rotation to make its axis sweep around in a conical motion called **precession** (❙ Figure 2-7a).

Earth spins like a giant top, but it does not spin upright in its orbit; it is inclined 23.5° from vertical. Earth's large mass and rapid rotation keep its axis of rotation pointed toward a spot near the star Polaris, and the axis would not wander if Earth were a perfect sphere. However, Earth, because of its rotation, has a slight bulge around its middle, and the gravity of the sun and moon pull on this bulge, tending to twist Earth upright in its orbit. The combination of these forces and Earth's rotation causes Earth's axis to precess in a conical motion, taking about 26,000 years for one cycle (Figure 2-7b).

Because the celestial poles and equator are defined by Earth's rotational axis, precession moves these reference marks. We notice no change at all from night to night or year to year, but precise measurements reveal the precessional motion of the celestial poles and equator.

Over centuries, precession has dramatic effects. Egyptian records show that 4800 years ago the north celestial pole was near the star Thuban (α Draconis). The pole is now approaching Polaris and will be closest to it about 2100. In about 12,000 years, the pole will have moved to within 5° of Vega (α Lyrae). Figure 2-7c shows the path followed by the north celestial pole. We will discover in later chapters that precession is common among rotating celestial bodies.

As you study astronomy, notice the special terms used to describe such things as precession and the celestial sphere. We need to know those terms, but science is about understanding nature and not about naming its parts (Window on Science 2-2). Science is more than just vocabulary.

a

b

c

REVIEW CRITICAL INQUIRY

Does everyone see the same circumpolar constellations? Here we can use the celestial sphere as a convenient model of the sky. A circumpolar constellation is one that does not set or rise. Which constellations are circumpolar depends on our latitude. If we live on Earth's equator, we see all the constellations rising and setting, and there are no circumpolar constellations at all. If we live at Earth's North Pole, all the constellations north of the celestial equator never set, and all the constellations south of the celestial equator never rise. In that case, every constellation is circumpolar. At intermediate latitudes, the circumpolar regions are caps whose angular radius equals the latitude of the observer. If we live in Iceland, the caps are very large, and if we live in Egypt, near the equator, the caps are much smaller.

Locate Ursa Major and Orion on the star charts at the end of this book. For people in Canada, Ursa Major is circumpolar, but people in Mexico see most of this constellation slip below the horizon. From much of the United States, some of the stars of Ursa Major set, and some do not. In contrast, Orion rises and sets as seen from nearly everywhere on Earth. Explorers at Earth's poles, however, never see Orion rise or set. How would you improve the definition of a circumpolar constellation to clarify the status of Ursa Major? Would your definition help in the case of Orion?

FIGURE 2-7

Precession. (a) A spinning top precesses in a conical motion around the perpendicular to the floor because its weight tends to make it fall over. (b) Earth precesses around the perpendicular to its orbit because the gravity of the sun and moon tend to twist it upright. (c) Precession causes the north celestial pole to drift among the stars, completing a circle in 26,000 years.

The motions of the celestial sphere may seem to have little to do with our lives, but at the end of this chapter we will see how precession may be one of the causes of the ice ages. Before we can think about ice ages, however, we must consider Earth's orbital motion around the sun and the resulting apparent motion of the sun.

2-3 THE CYCLES OF THE SUN

Perhaps the most obvious cycles in the sky are those of the sun, but those apparent motions are produced by Earth's rotation and revolution. **Rotation** is the turning of a body about an axis through its center. Thus we say that Earth rotates on its axis once a day. **Revolution** is the motion of a body around a point located outside the body, and we say that Earth revolves around the sun once a year.

The cycle of day and night is caused by the rotation of Earth on its axis. In ▌Figure 2-8 we see that four people in different places on Earth have different times of day, and that Earth's rotation carries us across the daylight side and then across the night side of Earth, producing the cycle of day and night.

Earth's annual revolution around the sun produces a cycle in the sky that we experience as the seasons. To understand that motion, we must imagine that we can make the sun fainter.

THE ANNUAL MOTION OF THE SUN

Even in the daytime, the sky is filled with stars, but the glare of sunlight fills our atmosphere with scattered light, and we can only see the brilliant sun. If the sun were fainter and we could see the stars, we would no-

tice that day by day the sun was moving slowly eastward against the background of stars. This motion is caused by Earth's motion as it revolves around the sun. In ▌Figure 2-9, we can see how the sun would appear in front of the stars of the constellation Sagittarius on January 1 and how Earth's motion along its orbit would cause the sun to appear to move eastward against the constellations. By March 1 the sun appears to be in Aquarius.

Of course, it is not quite correct to say that the sun is "in Aquarius." The sun is only 93 million miles away, and the stars of Aquarius are at least a million times farther away. But in March of each year, we would see the sun against the background of the stars in Aquarius, and thus we could say, "The sun is in Aquarius."

If we continue watching the sun against the background of stars throughout the year, we could plot its path on a star chart. After one full year, we would see the sun begin to retrace this line as it continued its

FIGURE 2-8

Looking down on Earth from above the North Pole shows how the time of day or night depends on our location on Earth.

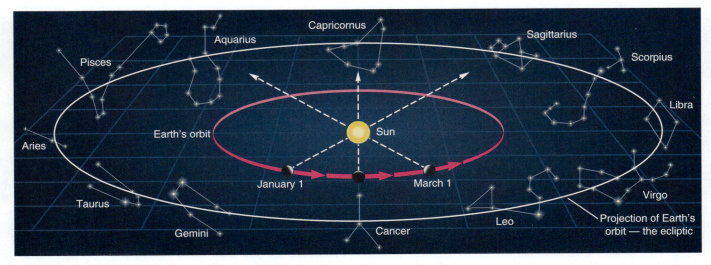

ACTIVE FIGURE 2-9

Earth's motion around the sun makes the sun appear to move against the background of the stars. Earth's circular orbit is thus projected on the sky as the circular path of the sun, the ecliptic.

Ace ✺ Astronomy™ Go to AceAstronomy and click Active Figures to see "Constellations in Different Seasons." Notice how the visible constellations change as the sun moves eastward along the ecliptic.

annual cycle of motion around the sky. This line, the apparent path of the sun in its yearly motion around the sky, is called the **ecliptic.** Another way to define the ecliptic is to say it is the projection of Earth's orbit on the sky. If the sky were a great screen illuminated by the sun at the center, then the shadow cast by Earth's orbit would be the ecliptic. Yet a third way to define the ecliptic is to refer to it as the plane of Earth's orbit. These three definitions of the ecliptic are equivalent, and it is worth considering them all because the ecliptic is one of the most important reference lines on the sky.

Earth circles the sun in 365.25 days, and consequently the sun appears to circle the sky in the same period. That means the sun, traveling 360° around the ecliptic in 365.25 days, travels about 1° eastward each day, about twice its angular diameter. We don't notice this motion because we can't see the stars in the daytime, but the motion of the sun has an important consequence that we do notice—the seasons.

THE SEASONS

The seasons are caused by a simple fact: Earth's axis of rotation is tipped 23.5° from the perpendicular to its orbit (▌ Figure 2-10). As you study "▌ The Cycle of the Seasons" on pages 22 and 23, notice two important principles. First, the seasons are not caused by any variation in the distance from Earth to the sun. Earth's orbit is nearly circular, so it is always about the same distance from the sun. Second, the seasons are caused by the changes in solar energy that Earth's northern and southern hemispheres receive at different times of the year. Because of circulation patterns in Earth's atmosphere, the northern and southern hemispheres are mostly isolated from each other and exchange little heat. When one hemisphere receives more solar energy than the other, it grows rapidly warmer.

In ancient times, the cycle of the seasons and the solstices and equinoxes were celebrated with rituals and festivals. Shakespeare's play *A Midsummer Night's Dream* describes the enchantment of the summer solstice night. (In Shakespeare's time, the equinoxes and solstices were taken to mark the midpoint of the seasons.) Many North American Indians marked the summer solstice with ceremonies and dances. Early church officials placed Christmas day in late December to coincide with an earlier pagan celebration of the winter solstice.

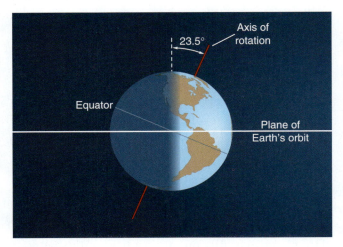

FIGURE 2-10

Earth's axis is inclined 23.5° from the perpendicular to the plane of its orbit. As it moves along its orbit, its axis of rotation remains fixed, pointing at the North Star.

If Earth had a significantly elliptical orbit, how would its seasons be different?

Suppose Earth had an elliptical orbit so that perihelion occurred in July and aphelion in January. At perihelion, Earth would be closer to the sun, and the entire surface of Earth would be a bit warmer. If that happened in July, it would be summer in the northern hemisphere and winter in the southern hemisphere, and both would be warmer than they now are. It could be a dreadfully hot summer in Canada, and it might not snow at all in southern Argentina. Six months later, aphelion would occur in January, which means Earth would be farther from the sun. Winter in northern latitudes would be frigid, and summers in Argentina would be cool.

Of course, this doesn't happen. Earth's orbit is nearly circular, and the seasons are caused not by a variation in the distance of Earth from the sun but by the inclination of Earth in its orbit.

Nevertheless, Earth's orbit is slightly elliptical. Earth passes perihelion about January 4 and aphelion about July 4. Although Earth's oceans tend to store heat and reduce the importance of this effect, this very slight variation in distance does affect the seasons. Does it make your winters warmer or cooler?

The sun is not the only body that moves along the ecliptic. The moon and planets maintain constant traffic along the sun's highway.

2-4 THE MOTION OF THE PLANETS

The ecliptic is important to us because it is the path of the sun around the sky, and that motion gives rise to the seasons. In a later chapter, we will consider the moon's cycle around the ecliptic, but here it is worthwhile to mention the motion of the planets. Not only are most of the planets visible to the naked eye, but their motion is the basis for one of the longest-living superstitions in human history—that our fate is written in the stars.

THE MOVING PLANETS

Most of the planets of our solar system are visible to the unaided eye, though they produce no light of their own. We see them by reflected sunlight. Mercury, Venus, Mars, Jupiter, and Saturn are all visible to the naked eye, but Uranus is usually too faint to be seen (magnitude 5.6 at its brightest), and Neptune is never bright enough. Pluto is even fainter, and we need a large telescope to find it. Although Uranus, Neptune, and Pluto are usually too faint to see, their motions are the same as those of the other planets.

All the planets of our solar system move in nearly circular orbits around the sun. If we were looking down on the solar system from the north celestial pole, we would see the planets moving in the same counterclockwise direction around their orbits (Chapter 1). The farther they are from the sun, the more slowly the planets move.

When we look for planets in the sky, we always find them near the ecliptic, because their orbits lie in nearly the same plane as the orbit of Earth. As they orbit the sun, they appear to move generally eastward along the ecliptic.* In fact, the word *planet* comes from a Greek word meaning "wanderer." Mars moves completely around the ecliptic in slightly less than two years, but Saturn, being farther from the sun, takes nearly 30 years.

As seen from Earth, Venus and Mercury can never move far from the sun because their orbits are inside Earth's orbit. They sometimes appear near the western horizon just after sunset or near the eastern horizon just before sunrise. Venus is easier to locate because its larger orbit carries it higher above the horizon than Mercury (❙ Figure 2-11). Mercury's orbit is so small that it can never get farther than about 28° from the sun. Consequently, it is usually hard to see against the sun's glare and is often hidden in the clouds and haze near the horizon. At certain times when it is farthest from the sun, however, Mercury shines brightly and can be located near the horizon in the evening or morning sky. (See the Appendix for the best times to observe Venus and Mercury.)

By tradition, any planet visible in the evening sky is called an **evening star,** although planets are not stars. Any planet visible in the sky shortly before sunrise is called a **morning star.** Perhaps the most beautiful is Venus, which can become as bright as minus fourth magnitude. As Venus moves around its orbit, it can dominate the western sky each evening for about half the year, but eventually its orbit carries it back toward the sun, and it is lost in the haze near the horizon. In a few weeks it reappears in the dawn sky as a brilliant morning star.

ASTROLOGY

Seen from Earth, the planets move gradually eastward along the ecliptic, but they don't follow the ecliptic exactly. Also, each travels at its own pace and seems to speed up and slow down at various times. To the ancients, this complex motion reflected the moods of the sky gods, and astrology was born.

Ancient astrologers defined a **zodiac,** a band 18° wide centered on the ecliptic, as the highway the planets follow. They divided this band into 12 segments named for the constellations along the ecliptic—the

---

*We will discuss occasional exceptions to this eastward motion in Chapter 4.

The Cycle of the Seasons

We can use the celestial sphere to help us think about the seasons. The celestial equator is the projection of Earth's equator on the sky, and the ecliptic is the projection of Earth's orbit on the sky. Because Earth is tipped in its orbit, the ecliptic and equator are inclined to each other by 23.5°. As the sun moves eastward around the sky, it spends half the year in the southern half of the sky and half of the year in the northern half. That causes the seasons.

The sun crosses the celestial equator going northward at the point called the **vernal equinox**. The sun is at its farthest north at the point called the **summer solstice**. It crosses the celestial equator going southward at the **autumnal equinox**, and reaches its most southern point at the **winter solstice**.

The seasons are defined by the dates when the sun crosses these four points, as shown in the table at the right. *Equinox* comes from the word for "equal"; on the day of an equinox, we have equal amounts of daylight and darkness. *Solstice* comes from the words meaning "sun" and "stationary." *Vernal* comes from the word for "green." The "green" equinox marks the beginning of spring.

Event	Date*	Season
Vernal equinox	March 20	Spring begins
Summer solstice	June 22	Summer begins
Autumnal equinox	September 22	Autumn begins
Winter solstice	December 22	Winter begins

* Give or take a day due to leap year and other factors.

Ace Astronomy™
Go to AceAstronomy and click Active Figures to see "Seasons" and watch Earth orbiting the sun.

On the day of the summer solstice in late June, Earth's northern hemisphere is inclined toward the sun, and sunlight shines almost straight down at northern latitudes. At southern latitudes, sunlight strikes the ground at an angle and spreads out. North America has warm weather, and South America has cool weather.

Earth's axis of rotation points toward Polaris, and, like a top, the spinning Earth holds its axis fixed as it orbits the sun. On one side of the sun, Earth's northern hemisphere leans toward the sun; on the other side of its orbit, it leans away. However, the direction of the axis of rotation does not change.

Earth at summer solstice

Summer solstice light

Winter solstice light

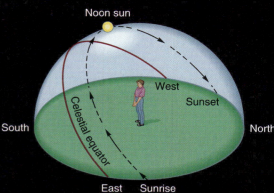

At summer solstice

Light striking the ground at a steep angle spreads out less than light striking the ground at a shallow angle. Light from the summer-solstice sun strikes northern latitudes from nearly overhead and is concentrated. Light from the winter-solstice sun strikes northern latitudes at much steeper angle and spreads out. The same amount of energy is spread over a larger area, so the ground receives less energy from the winter sun.

The two causes of the seasons are shown at right for someone in the northern hemisphere. First, the noon summer sun is higher in the sky and the winter sun is lower, as shown by our longer winter shadows. Thus winter sunlight is more spread out. Second, the summer sun rises in the northeast and sets in the northwest, spending more than 12 hours in the sky. The winter sun rises in the southeast and sets in the southwest, spending less than 12 hours in the sky. Both of these effects mean that northern latitudes receive more energy from the summer sun, and summer days are warmer than winter days.

At winter solstice

Ace ◯ Astronomy™

Go to AceAstronomy and click Active Figures to see "Path of the Sun" and see this figure from the inside.

Sunlight spread out on northern latitudes →

23.5°

To Polaris

40° N latitude

← To sun

Equator

Sunlight nearly direct on southern latitudes →

40° S latitude

On the day of the winter solstice in late December, Earth's northern hemisphere is inclined away from the sun, and sunlight strikes the ground at an angle and spreads out. At southern latitudes, sunlight shines almost straight down and does not spread out. North America has cool weather and South America has warm weather.

Earth's orbit is only very slightly elliptical. About January 4th, Earth is at **perihelion**, its closest point to the sun, when it is only 1.7 percent closer than average. About July 4th, Earth is at **aphelion**, its most distant point from the sun, when it is only 1.7 percent farther than average. This small variation does not significantly affect the seasons.

Earth at winter solstice

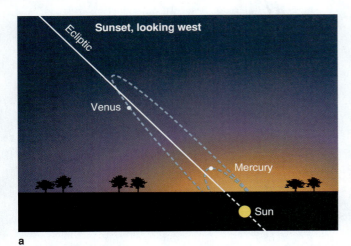

a

b

FIGURE 2-11
Mercury and Venus are sometimes visible in the western sky just after sunset (a) or in the eastern sky just before sunrise (b).

or between November 30 and December 17, the sun was passing through a corner of the nonzodiacal constellation Ophiuchus, and you have no official zodiacal sign.* Furthermore, astronomers like to point out, there is no mechanism by which the planets could influence us. The gravitational influence of a doctor who is delivering a baby is many times more powerful than the gravitational influence of the planets.

The arguments in the preceding paragraph actually miss the point. Astrology is not related to the physical world at all. It does not matter what constellation the sun occupies, because astrology divides the zodiac into equal mathematical sections, sometimes called houses, and it does not matter to the believer in astrology that the constellations don't match. Furthermore, the physical mechanism is beside the point for the true believer. Astrology is not so much an astronomical superstition as it is a mathematical superstition.

Astrology makes sense only when we think of the world as the ancients did. They believed in multiple sky gods whose moods altered events on Earth. They believed in mystical influences between natural events and human events. The ancients did not understand natural forces such as gravity as the mechanisms that cause things to happen, and thus they did not believe in cause and effect as we do. If a house burned, they might conclude that it caught fire because it was cursed and not because someone was careless with an oil lamp.

Modern science left astrology behind centuries ago, but it survives as a fascinating part of human history—an early attempt to understand the meaning of the sky.

*The author of this book was born on December 14 and thus has no astrological sign. An astronomer friend claims that the author must therefore have no personality.

signs of the zodiac. A **horoscope** shows the location of the sun, moon, and planets among the zodiacal signs with respect to the horizon at the moment of a person's birth as seen from that longitude and latitude. Even if astrology worked, the generalized horoscopes published in newspapers and tabloids can't have been calculated accurately for individual readers.

Astrology buffs argue that a person's personality, life history, and fate are revealed in his or her horoscope, but the evidence contradicts this belief. Astrology has been tested many times over the centuries, and it just doesn't work. Believers, however, don't give up on it. Thus, astrology is a superstition that depends on blind belief and not a science that depends on evidence (Window on Science 2-3).

One reason astronomers find astrology irritating is that it has no link to the physical world. For example, precession has moved the constellations so that they no longer match the zodiacal signs. Whatever sign you were "born under," the sun was probably in the previous zodiacal constellation. In fact, if you were born on

REVIEW CRITICAL INQUIRY

Planets like Mars can sometimes be seen rising in the east as the sun sets. Why is the same not true for Mercury or Venus?

Mars has an orbit outside the orbit of Earth, and that means it can reach a location opposite the sun in the sky. As the sun sets in the west, Mars rises in the east. But Mercury and Venus follow orbits that are smaller than the orbit of Earth, so they can never reach a point opposite the sun in the sky. As they follow their orbits around the sun, we on Earth see them gradually swing out on the east side of the sun, reach a maximum distance from the sun, swing back toward the sun, move out to the west side of the sun, reach a maximum distance west, and then move back toward the sun again. Because they never get very far from the sun in the sky, they can never be seen rising in the east as the sun sets.

The planetary motions we see in the sky are produced by the orbital motions of the planets around the sun. What would we see if the planets followed orbits that were all in exactly the same plane as the orbit of Earth?

Modern astronomers have a low tolerance for astrological superstition, yet the motions of the heavenly bodies do affect our lives. The motion of the sun produces the seasons, and, as we will see in Chapter 3, the moon governs the tides. In addition, small changes in Earth's orbit may partially control global climate.

2-5 ASTRONOMICAL INFLUENCES ON EARTH'S CLIMATE

Weather is what happens today; climate is the average of what happens over decades and centuries. We know that Earth has gone through past episodes, called ice ages, when the worldwide climate was cooler and dryer and thick layers of ice covered northern latitudes. The earliest known ice age occurred about 570 million years ago and the next about 280 million years ago. The most recent ice age began only about 3 million years ago and is still going on. We are living in one of the periodic episodes when the glaciers melt and Earth grows slightly warmer. The current warm period began about 20,000 years ago.

Ice ages seem to occur with a period of roughly 250 million years, and cycles of glaciation within ice ages

occur with a period of about 40,000 years. These cyclic changes have an astronomical origin.

THE HYPOTHESIS

Sometimes a theory or hypothesis is proposed long before scientists can find the critical evidence to test it. That situation happened in 1920 when Yugoslavian meteorologist Milutin Milankovitch proposed what became known as the **Milankovitch hypothesis**—that changes in the shape of Earth's orbit, in precession, and in inclination affect Earth's climate and trigger ice ages. We will examine each of these three motions in turn.

First, astronomers know that the elliptical shape of Earth's orbit varies slightly over a period of about 100,000 years. At present, Earth's orbit carries it 1.7 percent closer than average to the sun during northern-hemisphere winters and 1.7 percent farther away in northern-hemisphere summers. This makes the northern winters warmer, and that is critical—most of the land mass where ice can accumulate is in the northern hemisphere. If Earth's orbit became more elliptical, Milankovitch suggested, northern winters might be warm enough to prevent the accumulation of snow and ice from forming glaciers, and Earth's climate would warm.

A second factor is also at work. Precession causes Earth's axis to sweep around a cone with a period of about 26,000 years, and that changes the location of the

The Foundation of Science: Evidence

Science is based on evidence. Every theory and conclusion must be supported by evidence obtained from experiments or from observation. If a theory is supported by many pieces of evidence but is clearly contradicted by a single experiment or observation, scientists quickly abandon it. For a theory to be true, there can be no contradictory evidence.

Of course, scientists argue about the significance of particular evidence and often disagree on the interpretation of evidence. Some observations may seem significant at first glance, but upon closer examination we may find that the procedure was flawed and so the piece of evidence is not important. Or we might conclude that the observational fact is correct but is being misinterpreted. The observation may not mean what it seems. Some of the most famous disagreements in science, such as those surrounding Galileo and Darwin, have arisen over the interpretation of well-established factual evidence.

Furthermore, scientists are not allowed to be selective in considering evidence. A lawyer in court can call a certain witness and intentionally fail to ask a critical question that would reveal evidence harmful to the lawyer's case. But a scientist may not ignore any known evidence. The difference in their methods is revealing. The lawyer is attempting to prove only one side of the case and rightly may ignore contradictory evidence. The scientist, however, is searching for the truth and so must test any theory against all available evidence. In a sense, the scientist, in dealing with evidence, must act as both the prosecution and the defense.

As you read about any science, look for the evidence in the form of measurements or observations. Every theory or conclusion should have supporting evidence. If you can find and understand the evidence, the science will make sense. All scientists, from astronomers to zoologists, demand evidence. You should, too.

seasons around Earth's orbit. Northern winters now occur when Earth is 1.7 percent closer to the sun, but in 13,000 years northern winters will occur on the other side of Earth's orbit where Earth is farther from the sun. Northern winters will be colder, and glaciers may grow.

The third factor is the inclination of Earth's equator to its orbit. Currently at 23.5°, this angle varies from 22° to 24° with a period of roughly 41,000 years. When the inclination is greater, seasons are more severe.

In 1920, Milankovitch proposed that these three factors cycled against each other to produce complex periodic variations in Earth's climate and the advance and retreat of glaciers (❚ Figure 2-12a). But no evidence was available to test the theory in 1920, and scientists treated it with skepticism. Many thought it was laughable.

THE EVIDENCE

By the middle 1970s, Earth scientists could collect the data that Milankovitch needed. Oceanographers could drill deep into the seafloor and collect samples, and geologists could determine the age of the samples from the natural radioactive atoms they contained. From all this, scientists constructed a history of ocean temperatures that convincingly matched the predictions of the Milankovitch hypothesis (Figure 2-12b).

The evidence seemed very strong, and by the 1980s, the Milankovitch hypothesis was widely discussed as the leading hypothesis. But science follows a mostly unstated set of rules that holds that a hypothesis must be tested over and over against all available evidence (Window on Science 2-4). In 1988, scientists discovered contradictory evidence.

While scuba diving in a water-filled crack in Nevada called Devil's Hole, scientists drilled out samples of calcite, a mineral that contains oxygen atoms. For 500,000 years, layers of calcite have built up in Devil's Hole, recording in their oxygen atoms the temperature of the atmosphere when rain fell there. Finding the ages of the mineral samples was difficult, but the results seemed to show that the previous ice age ended thousands of years too early to have been caused by Earth's motions.

These contradictory findings are irritating because we naturally prefer certainty, but such circumstances are common in science. The disagreement between ocean floor samples and Devil's Hole samples triggered a scramble to understand the problem. Were the ages of one or the other set of samples wrong? Were the ancient temperatures wrong? Or were scientists misunderstanding the significance of the evidence?

In 1997, a new study of the ages of the samples confirmed that those from the ocean floor are correctly dated. This seems to give scientists renewed confidence in the Milankovitch hypothesis. But the same study found that the ages of the Devil's Hole samples are also correct. Evidently the temperatures at Devil's Hole tell us about local climate changes in the region that became the southwestern United States. The ocean floor samples tell us about global climate changes, and they fit well with the Milankovitch hypothesis. In this way, the Milankovitch hypothesis, though widely accepted today, is still being tested as we try to understand the world we live on.

Earth temperatures predicted
from the Milankovitch effect

a 25,000 years ago   10,000 years ago

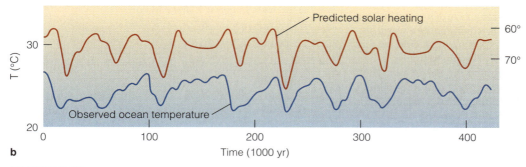

b

FIGURE 2-12

(a) Mathematical models of the Milankovich effect can be used to predict temperatures on Earth over time. In these Earth globes, cool temperatures are represented by violet and blue and warm temperatures by yellow and red. *(Courtesy Arizona State University, Computer Science and Geography Departments)* (b) Over the last 400,000 years, changes in ocean temperatures measured from fossils found in sediment layers from the seabed match calculated changes in solar heating. *(Adapted from Cesare Emiliani)*

REVIEW CRITICAL INQUIRY

How do precession and the shape of Earth's orbit interact to affect Earth's climate?

One technique that is useful in the critical analysis of an idea is exaggeration. If we exaggerate the variation in the shape of Earth's orbit, we can see dramatically the influence of precession. At present, Earth reaches perihelion during winter in the northern hemisphere and aphelion during summer. The variation in distance is only about 1.7 percent, and that difference doesn't cause much change in the severity of the seasons. But if Earth's orbit were much more elliptical, then winter in the northern hemisphere would be much warmer, and summer would be much cooler.

Now we can see the importance of precession. As Earth's axis precesses, it points gradually in different directions, and the seasons occur at different places in Earth's orbit. In 13,000 years, northern winter will occur at aphelion, and if Earth's orbit were highly elliptical, northern winter would be terrible. Similarly, summer would occur at perihelion, and the heat would be awful. Such extremes might deposit large amounts of ice in the winter but then melt it away in the hot summer, thus preventing the accumulation of glaciers.

Continue this analysis by exaggeration. What effect would precession have if Earth's orbit were more circular?

This chapter is about the sky, but it has told us a great deal about Earth. We have discovered that Earth's orbital motion produces the apparent motion of the sun along the ecliptic, causes the seasons, and may even cause the cycle of the ice ages. As is often the case, astronomy helps us understand what we are and where we are. In this case, our study of the sky has helped us understand what it means to live on a planet. However, we have omitted one of the most dramatic objects in the sky—the moon. We will repair that omission in the next chapter.

SUMMARY

Astronomers divide the sky into 88 areas called constellations. Although the ancient constellations originated in Greek mythology, the names are Latin. Even the modern constellations, added to fill in the areas between the ancient figures, have Latin names. The names of stars usually come from ancient Arabic, though modern astronomers often refer to a star by constellation and Greek letters assigned according to brightness within each constellation.

The magnitude system is the astronomers' brightness scale. First-magnitude stars are brighter than second-magnitude stars, which are brighter than third-magnitude stars, and so on. The magnitude we see when we look at a star in the sky is its apparent visual magnitude.

The celestial sphere is a model of the sky, carrying the celestial objects around Earth. Because Earth rotates eastward, the celestial sphere appears to rotate westward on its axis. The northern and southern celestial poles are the pivots on which the sky appears to rotate. The celestial equator, an imaginary line around the sky above Earth's equator, divides the sky in half.

Because Earth orbits the sun, the sun appears to move eastward around the sky following the ecliptic. Because the ecliptic is tipped 23.5° to the celestial equator, the sun spends half the year in the northern celestial hemisphere and half the year in the southern celestial hemisphere, producing the seasons. The seasons are reversed south of Earth's equator. That is, while the northern hemisphere is experiencing a warm season, the southern hemisphere is experiencing a cold season.

Of the nine planets in our solar system, Mercury, Venus, Mars, Jupiter, and Saturn are visible to the naked eye. Their orbital motion around the sun carries them along the zodiac, a band 18° wide centered on the ecliptic. The positions of the sun, the moon, and these five planets form the basis of astrology, an ancient superstition that originated in Babylonia about 1000 BC.

Earth's motion may change in ways that can affect the climate. Changes in orbital shape, in precession, and in axial tilt can alter the planet's heat balance and may be responsible for the ice ages and glacial periods.

NEW TERMS

constellation

asterism

magnitude scale

apparent visual magnitude (m_v)

celestial sphere

horizon

zenith

nadir

north celestial pole

south celestial pole

celestial equator

north point

south point

east point

west point

angular distance

minute of arc

second of arc

angular diameter

circumpolar constellation

scientific model

precession

rotation

revolution

ecliptic

vernal equinox

summer solstice

autumnal equinox

winter solstice

perihelion

aphelion

evening star

morning star

zodiac

horoscope

Milankovitch hypothesis

REVIEW QUESTIONS

Ace◉Astronomy™ Assess your understanding of this chapter's topics with additional quizzing and animations at **http://astronomy.brookscole.com/seeds8e**

1. Why are most modern constellations composed of faint stars or located in the southern sky?

2. What does a star's Greek-letter designation tell us that its ancient Arabic name does not?

3. From your knowledge of star names and constellations, which of the following stars in each group is the brighter and which is the fainter? Explain your answers.

   a. α Ursae Majoris; θ Ursae Majoris

   b. λ Scorpii; β Pegasus

   c. β Telescopium; β Orionis

4. Give two reasons why the magnitude scale might be confusing.

5. Why do modern astronomers continue to use the celestial sphere when they know that stars are not all at the same distance?

6. How do we define the celestial poles and the celestial equator?

7. From what locations on Earth is the north celestial pole not visible? the south celestial pole? the celestial equator?

8. If Earth did not turn on its axis, could we still define an ecliptic? Why or why not?

9. Give two reasons why winter days are colder than summer days.

10. How do the seasons in Earth's southern hemisphere differ from those in the northern hemisphere?

11. Why should the eccentricity of Earth's orbit make winter in Earth's northern hemisphere different from winter in the southern hemisphere?

DISCUSSION QUESTIONS

1. Have you thought of the sky as a ceiling? as a dome overhead? as a sphere around Earth? as a limitless void?

2. How would the seasons be different if Earth were inclined 90° instead of 23.5°? 0° instead of 23.5°?

PROBLEMS

1. If one star is 6.3 times brighter than another star, how many magnitudes brighter is it?

2. If one star is 40 times brighter than another star, how many magnitudes brighter is it?

3. If two stars differ by 7 magnitudes, what is their intensity ratio?

4. If two stars differ by 8.6 magnitudes, what is their intensity ratio?

5. If star A is third magnitude and star B is fifth magnitude, which is brighter and by what factor?

6. If star A is magnitude 4 and star B is magnitude 9.6, which is brighter and by what factor?

7. By what factor is the sun brighter than the full moon? (*Hint:* See Figure 2-6.)

8. What is the angular distance from the north celestial pole to the summer solstice? to the winter solstice?

9. As seen from your latitude, what is the angle between the north celestial pole and the northern horizon? between the southern horizon and the noon sun at the summer solstice?

CRITICAL INQUIRIES FOR THE WEB

1. Nearly all cultures have populated the sky with gods, heroes, animals, and objects. What can you learn on the Web about non-Western constellations?

2. Who was Orion? How is he related to the scorpion in the sky?

3. What holidays, rituals, special foods, and beliefs are associated with the winter solstice?

4. What can you find out about Milutin Milankovitch? What is the latest news about the Milankovitch hypothesis?

EXPLORING *THESKY*

1. As discussed in this chapter, Earth's rotation about its own axis gives us the impression that the whole sky rotates around the north celestial pole in a period of one day. This apparent motion of the celestial sphere is difficult to notice because it happens so slowly. However, *TheSky* makes it possible to simulate this motion at a pace that is easy to observe by using a feature called **Time Skip.** Observe and describe the apparent motion of the sky as you see it looking north, east, south, and west. How to proceed: Set the **Time Skip Increment** (a dropdown menu on the **Time Skip Toolbar**) to 1 minute, and click on the **Go Forward** button to begin the simulation. View the sky from the four cardinal directions, due north, south, east, and west. (You'll find **Time Skip** under the **Tools** menu as well.)

2. In which constellation was the sun located on the date of your birth? (*Hint:* Click **Data** and then **Site Information** to set the location, date, and time of your birth. Turn on constellation figures, constellation boundaries, and labels. Then find the sun.)

3. Set your location to Earth's North Pole and the date to the summer solstice. Turn on the ecliptic and then step forward in time through a day to see what happens to the sun. Repeat for the autumnal equinox and the winter solstice.

4. Repeat Problem 3 above for a location on Earth's equator.

 Visit the Seeds *Foundations of Astronomy* companion Web site for critical thinking exercises, articles, and additional readings from InfoTrac College Edition, Brooks/Cole's online student library.

Even a man who is pure in heart
And says his prayers by night
May become a wolf when the wolfbane blooms
And the moon shines full and bright.

Proverb from old Wolfman movies

Chapter 3

Daniel Good

GUIDEPOST

In the preceding chapter, we saw how the sun dominates our sky and determines the seasons. The moon is not as bright as the sun, but the moon passes through dramatic phases and occasionally participates in eclipses. The sun dominates the daytime sky, but the moon rules the night.

As we try to understand the appearance and motions of the moon in the sky, we discover that what we see is a product of light and shadow. To understand the appearance of the universe, we must understand light. Later chapters will show that much of astronomy hinges on the behavior of light.

In the next chapter, we will see how Renaissance astronomers found a new way to describe the appearance of the sky and the motions of the sun, moon, and planets.

VIRTUAL LABORATORY

TIDES AND TIDAL FORCES

"Don't stare at the moon—you'll go crazy," more than one child has been warned.* The word *lunatic* comes from a time when even doctors thought that the insane were "moonstruck." A "mooncalf" is someone who has been crazy since birth, and the word is probably related to the belief that moonlight can harm unborn children. Everyone "knows" that people act less rationally when the moon is bright, but everyone is wrong. Careful studies of hospital and police records show that there is no real correlation between the moon and erratic behavior. The moon is so bright and its cycles through the sky are so dramatic we *expect* it to influence us, and many people are disappointed to learn the moon does not influence us. Some simply refuse to believe the evidence.

The cycles of the moon are indeed dramatic. Many different cultures around the world explain eclipses of

*When I was very small, my grandmother told me if I gazed at the moon, I might go crazy. But it was too beautiful, and I ignored her warning. I secretly watched the moon from my window, became fascinated by the sky, and became an astronomer.

the sun as invisible monsters devouring the sun (■ Figure 3-1). A Chinese story tells of two astronomers, Hsi and Ho, who were too drunk to predict the solar eclipse of October 22, 2137 BC, or perhaps failed to conduct the proper ceremonies to scare away the dragon snacking on the sun. When the emperor recovered from the terror of the eclipse, he had the two astronomers beheaded. The cycles of the moon provide a wealth of beauty in the sky and an important part of the cultural beliefs of Earth's peoples.

This chapter discusses the lunar cycles of phases, tides, and eclipses. Studying these events will help us understand what we see in the night sky. They will also

Ace⟳Astronomy™ The AceAstronomy icon throughout the text indicates an opportunity for you to test yourself on key concepts and to explore animations and interactions on the AceAstronomy Web site at **http://astronomy.brookscole .com/seeds8e**

a

b

c

FIGURE 3-1

(a) A 12th-century Mayan symbol believed to represent a solar eclipse. The black-and-white sun symbol hangs from a rectangular sky symbol, and a voracious serpent approaches from below. (b) Chinese representation of a solar eclipse as a dragon flying in front of the sun. *(From the collection of Yerkes Observatory)* (c) A wall carving from the ruins of a temple at Vijayanagara in southern India. It symbolizes a solar eclipse as two snakes approaching the disk of the sun. *(T. Scott Smith)*

Gravity: The Universal Force

Isn't it weird that Isaac Newton is said to have "discovered" gravity in the late 17th century—as if people didn't have gravity before that, as if people in the 16th century floated around holding onto tree branches? Newton's accomplishment was that he realized that gravity was universal. The same force that makes an apple fall holds the moon in its orbit, guides the planets around the sun, and shapes the motion of the universe. We will discuss Newton's accomplishment in its proper historical place in Chapter 5, but here we need to recognize gravity for what it is, the master of the universe.

Newton realized that gravity was a property of matter, and consequently every object made of matter has to exert a gravitational attraction on every other object. Earth attracts the moon and the moon attracts Earth, but they also attract you and your book. Further, you and your book exert a gravitational force on each other and on every other object in the universe.

The size of an object's gravitational effect depends on the amount of matter it contains. A small amount of matter has a small gravitational effect, which is why you don't have satellites orbiting around you. Most objects in astronomy contain more matter, and their gravitational effects are larger. Moons, planets, stars, and galaxies each exert their gravitational attraction in proportion to the amount of matter they contain.

Of course, objects that are far away have less of an effect than objects that are nearby. That is another important point that Newton recognized. Earth's gravitational attraction presses us to our chairs, and the gravitational attraction of a distant galaxy containing much more matter than Earth is undetectably small because the galaxy is so far away.

The universe is filled with these gravitational influences, and every object in the universe moves under their guidance. The moon orbits Earth, and Earth orbits the sun. The sun orbits our galaxy, which moves under the influence of the gravitation of all the other galaxies in the universe. The universe is a swirling waltz of matter dancing to the music of gravity.

---

introduce us to some of the most basic concepts of astronomy. We will discover how gravity produces tides, how light and shadow move through space, how the size and distance of an object affect what we see, and how cyclic events can be analyzed by searching for their patterns. Most of all, this chapter forces us to begin thinking of our home as a world in space.

3-1 THE CHANGEABLE MOON

Starting this evening, look for the moon in the sky. If it is a cloudy night or if the moon is in the wrong part of its orbit, you may not see it, but keep trying on successive evenings, and within a week or two you will see the moon. Then watch for the moon on following evenings, and you will see it following its orbit around Earth and cycling through its phases as it has done for billions of years (Window on Science 3-1).

THE MOTION OF THE MOON

When we watch the moon night after night, we notice two things about its motion. First, we see it moving eastward against the background of stars; second, we notice that the markings on its face don't change. These two observations help us understand the motion of the moon and the origin of the moon's phases.

The moon moves rapidly among the constellations. If you watch the moon for just an hour, you can see it move eastward against the background of stars by slightly more than its angular diameter. In the previous chapter, we discovered that the moon is about 0.5° in angular diameter, so it moves eastward a bit more than 0.5° per hour. In 24 hours, it moves 13°. Each night when we look at the moon, we see it about 13° eastward of its location the night before. This eastward movement is the result of the motion of the moon along its orbit around Earth.

THE CYCLE OF PHASES

The changing shape of the moon as it orbits Earth is one of the most easily observed phenomena in astronomy. We have all noticed the full moon rising dramatically or a thin crescent moon hanging in the evening sky.

Study "▌ The Phases of the Moon" on pages 34 and 35 and notice three important points. First, the moon always keeps the same side facing Earth, and we never see the far side of the moon. "The Man in the Moon" is produced by the familiar features on the moon's near side. Second, the changing shape of the moon as it passes through its cycle of phases is produced by sunlight illuminating different parts of the side of the moon we can see. The third thing to notice is the difference between the orbital period of the moon around Earth and the length of the lunar phase cycle. That difference is a good illustration of how our view from Earth is produced by the combined motions of Earth and other heavenly bodies such as the sun and moon.

You can make a moon-phase dial from the middle diagram on page 34 by covering the lower half of the moon's orbit with a sheet of paper, aligning the edge of the paper to pass through the word "Full" at the left and the word "New" at the right. Push a pin through the edge of the paper at Earth's North Pole to make a pivot, and under the word "Full" write on the paper

"Eastern Horizon." Under the word "New" write "Western Horizon." The paper now represents the horizon we see facing south.

You can set your moon-phase dial for a given time by rotating the diagram behind the horizon paper. Set the dial to sunset by turning the diagram until the human figure labeled "sunset" is standing at the top of the Earth globe; the dial shows, for example, that the full moon at sunset would be at the eastern horizon.

REVIEW **CRITICAL INQUIRY**

Why do we sometimes see the moon in the daytime?

The full moon rises at sunset and sets at sunrise, so it is visible in the night sky but never in the daytime sky. But at other phases, it is possible to see the moon in the daytime. For example, when the moon is a waxing gibbous moon, it rises a few hours before sunset. If you look in the right spot in the sky, you can see it in the late afternoon in the southeast sky. It looks pale and washed out because of sunlight in our atmosphere, but it is quite visible once you notice it. You can also locate the waning gibbous moon in the morning sky. It sets a few hours after sunrise, so you would look for it in the southwestern sky in the morning.

If you look in the right place, you might even see the first- or third-quarter moon in the daytime sky, but you have probably never seen a crescent moon in the daytime. Why not? Where would you have to look?

The phases of the moon don't affect us directly—we don't act crazier than usual at full moon. But the moon does have an important influence on Earth—the periodic advance and retreat of the ocean tides.

3-2 THE TIDES

Anyone who lives near the sea is familiar with a dramatic lunar effect—the ebb and flow of the tides. These periodic changes in the ocean are caused by the moon's gravity.

THE CAUSE OF THE TIDES

We feel Earth's gravity drawing us downward with a force we refer to as our weight, but the moon also exerts a gravitational force on us. Because the moon is less massive and more distant, its gravitational force acting on an object on Earth's surface is only about 0.0003 percent of Earth's gravitational force. That is a tiny force, but it is enough to cause tides in Earth's oceans.

Tides are produced by a *difference* between the gravitational force acting on different parts of an object.

FIGURE 3-2

(a) Because one side of Earth is closer to the moon than the other side, the moon's gravity does not attract all parts of Earth with the same force. (b) Relative to the center of Earth, small differences in the forces cause tidal bulges (much exaggerated here) on the sides of Earth facing toward and away from the moon.

The side of Earth facing the moon is about 6400 km (4000 miles) closer to the moon than is Earth's center, and the moon's gravity pulls more strongly on the oceans on the near side than on Earth's center. The difference is small, only about 3 percent of the moon's total gravitational force on Earth, but it is enough to make the ocean waters flow into a bulge on the side of Earth facing the moon.

A bulge also forms on the side of Earth facing away from the moon. Earth's far side is about 6400 km farther from the moon than is Earth's center, and the moon's gravity pulls on it less strongly than it does on Earth's center. Thus, relative to Earth's center, a small force makes the ocean waters on Earth's far side flow away from the moon (❙ Figure 3-2).

We can see dramatic evidence of tidal forces if we watch the ocean shore for a few hours. Although Earth rotates on its axis, the tidal bulges remain fixed along the Earth–moon line. As the turning Earth carries us into a tidal bulge, the ocean water deepens, and the tide crawls up the beach. To be precise, the tide does not "come in." Rather, we are "carried into" the tidal bulge. Because there are two bulges on opposite sides of Earth, the tides rise and fall twice a day, and the times of high and low tide depend on the phase of the moon.

In reality, the tide cycle at any given location can be quite complex, depending on the latitude of the site, the shape of the shore, the north–south location of the moon, and so on. Tides in the Bay of Fundy occur twice a day and can exceed 12 m. The northern coast of the Gulf of Mexico has only one tidal cycle of roughly 30 cm each day.

The Phases of the Moon

We see the moon go through a month-long cycle of phases. This cycle is caused by the play of light and shadow on the moon as it follows its orbit around Earth.

As the moon orbits Earth, it rotates to keep the same side facing Earth. Consequently we always see the same features on the moon, and we never see the far side of the moon. A mountain on the moon that points at Earth will always point at Earth as the moon revolves and rotates.

(Not to scale)

Sunlight always illuminates half of the moon. Because we see different amounts of this sunlit side, we see the moon cycle through phases. At the phase called "new moon," sunlight illuminates the far side of the moon, and the side we see is in darkness. At new moon we see no moon at all. At full moon, the side we see is fully lit, and the far side is in darkness. How much we see depends on where the moon is in its orbit.

Notice that there is no such thing as the "dark side of the moon." All parts of the moon experience day and night in a month-long cycle.

In the diagram at the left, we see that the new moon is close to the sun in the sky, and the full moon is opposite the sun. The time of day depends on the observer's location on Earth. (Compare with Figure 2-8.)

First quarter

Waxing gibbous

Waxing crescent

Sunset

North Pole

Midnight

Noon

Earth's rotation

Sunrise

New

SUNLIGHT

Full

Waning gibbous

Third quarter

Waning crescent

Ace Astronomy™

Go to AceAstronomy and click Active Figures to see "Lunar Phases" and watch the moon orbit Earth.

Gibbous comes from the Latin word for humpbacked.

The first quarter moon is one week through its 4-week cycle.

The first two weeks of the cycle of the moon is shown below by its position at sunset on 14 successive evenings. As the moon grows fatter from new to full, it is said to wax.

The full moon is two weeks through its 4-week cycle.

Waxing gibbous

Waxing crescent

THE SKY AT SUNSET

Full moon rises at sunset

New moon is invisible near the sun

East

South

West

New moon

Ecliptic

Sun

New moon

Sagittarius

Scorpius

The sun and moon are near each other at new moon.

One sidereal period after new moon

Ecliptic

Sun

Sagittarius

Moon

Scorpius

One sidereal period after new moon, the moon has returned to the same place among the stars, but the sun has moved on along the ecliptic.

One synodic period after new moon

Ecliptic

Sun

New moon

Sagittarius

Scorpius

One synodic period after new moon, the moon has caught up with the sun and is again at new moon.

The moon orbits eastward around Earth in 27.32 days, its **sidereal period.** This is how long the moon takes to circle the sky once and return to the same position among the stars.

A complete cycle of lunar phases takes 29.53 days, the moon's **synodic period.** (Synodic comes from the Greek words for "together" and "path.")

To see why the synodic period is longer than the sidereal period, study the star charts at the left.

Although we think of the lunar cycle as being about 4 weeks long, it is actually 1.53 days longer than 4 weeks. Our calendar divides the year into 30-day periods called months (literally "moonths") in recognition of the 29.53 day synodic cycle of the moon.

You can use the diagram on the opposite page to determine when the moon rises and sets at different phases.

Earth   Moon

This diagram is to scale. The moon's orbit is about 30 Earth diameters in radius. Because the orbit is slightly elliptical, the moon's angular diameter in the sky can vary by plus or minus 6 percent. Because the orbit is tipped a bit over 5 degrees, the moon does not follow the ecliptic exactly.

TIMES OF MOONRISE AND MOONSET

Phase	Moonrise	Moonset
New	Dawn	Sunset
First quarter	Noon	Midnight
Full	Sunset	Dawn
Third quarter	Midnight	Noon

The third quarter moon is 3 weeks through its 4-week cycle.

The last two weeks of the cycle of the moon is shown below by its position at sunrise on 14 successive mornings. As the moon shrinks from full to new, it is said to wane.

Waning crescent

Waning gibbous

New moon is invisible near the sun

Full moon sets at sunrise

THE SKY AT SUNRISE

East   South   West

a

b

FIGURE 3-3
Looking down on Earth from above the North Pole, we can draw the shape of the tidal bulges (exaggerated here for clarity). (a) When the moon and sun pull along the same line, their tidal forces combine, and the tidal bulges are larger. (b) When the moon and sun pull at right angles, their tidal bulges do not combine, and the tidal bulges are smaller.

The sun, too, produces tides on Earth. The sun is roughly 27 million times more massive than the moon, but it lies almost 400 times farther from Earth. Consequently, tides on Earth caused by the sun are less than half those caused by the moon. At new moon and full moon, the moon and sun produce tidal bulges that join together to cause extreme tidal changes (❙ Figure 3-3a); high tide is very high, and low tide is very low. Known as **spring tides,** these occur at every new and full moon. **Neap tides** occur at first- and third-quarter moon, when the moon and sun pull at right angles to each other. Then the smaller tide caused by the sun slightly reduces the tide caused by the moon, and the resulting high and low tides are less extreme (Figure 3-3b).

Though the ocean has been used as an example, you should note that tides would occur even if Earth had no oceans. The difference in the gravitational forces acting on different parts of Earth causes slight bulges in the rocky shape of Earth itself. We do not feel the mountains and plains rising and falling by a few centimeters; the changes in the fluid oceans are much more obvious.

TIDAL EFFECTS

Tidal forces can have surprising effects on both rotation and orbital motion. The friction of Earth's ocean

FIGURE 3-4
Earth's rotation drags the tidal bulges ahead of the Earth–moon line (exaggerated here). The gravitational attraction of these masses of water does not pull directly toward Earth's center but rather pulls the moon slightly forward in its orbit (green arrow), forcing its orbit to grow larger.

waters against the seabeds slows Earth's rotation, and our days are getting longer by 0.0015 second per century. Some marine animals deposit layers in their shells in phase with the tides and the day–night cycle, and fossils of these shells confirm that only 900 million years ago Earth's day was 18 hours long.

In addition, Earth's gravitational field exerts tidal forces on the moon; and, although there are no oceans on the moon, tides do flex its rocky bulk. The resulting friction in the rock has slowed the moon's rotation. It once rotated much faster, but the tides caused by Earth have slowed the moon until it now rotates to keep one side permanently facing Earth. It is thus tidally locked to Earth. We will discover that most of the moons in the solar system are tidally locked to their planets.

Tidal forces can also affect orbital motion. For example, friction with the rotating Earth drags the tidal bulges eastward out of a direct Earth–moon line (❙ Figure 3-4). These tidal bulges contain a large amount of mass, and the gravitation of the bulge near the moon pulls the moon slightly forward in its orbit. The bulge on Earth's far side pulls the moon slightly backward in its orbit, but because it is farther from the moon, that bulge is less effective. The net effect is to drag the moon forward in its orbit. As a result, the moon's orbit is growing larger, and the moon is receding from Earth at about 3.8 cm per year, an effect that astronomers can measure by bouncing laser beams off reflectors left on the lunar surface by the Apollo astronauts. Thus, we have clear evidence that tides are changing the moon's orbit.

Tides and tidal forces are important in many areas of astronomy. In later chapters, we will see how tidal forces can pull gas away from stars, rip galaxies apart, and melt the interiors of moons orbiting near massive planets. Now, however, we must consider yet another kind of lunar phenomenon—eclipses.

3-3 LUNAR ECLIPSES

A lunar eclipse occurs at full moon when the moon moves through Earth's shadow. Because the moon shines only by reflected sunlight, it gradually darkens as it enters the shadow.

EARTH'S SHADOW

Earth's shadow consists of two parts. The **umbra** is the region of total shadow. If we were floating in space in the umbra of Earth's shadow, we would see no portion of the sun. However, if we moved into the **penumbra,** we would be in partial shadow and would see part of the sun peeking around Earth's edge. In the penumbra, the sunlight is dimmed but not extinguished.

We can construct a model of this by pressing a map tack into the eraser of a pencil and holding the tack between a lightbulb a few feet away and a white cardboard screen (❚ Figure 3-5). The lightbulb represents the sun, and the map tack represents Earth. When we hold the screen close to the tack, we see that the umbra is nearly as large as the tack and that the penumbra is only slightly larger. However, as we move the screen away from the tack, the umbra shrinks and the penumbra expands. Beyond a certain point, the shadow has no dark core at all, indicating that the screen is beyond the end of the umbra.

The umbra of Earth's shadow is about 1.4 million km (860,000 miles) long and points directly away from the sun. A giant screen placed in the shadow at the average distance of the moon would reveal a dark umbra about 9000 km (5700 miles) in diameter, and the faint outer edges of the penumbra would mark a circle about 16,000 km (10,000 miles) in diameter. For comparison, the moon's diameter is only 3476 km (2160 miles). Thus, when the moon's orbit carries it through the umbra, it has plenty of room to become completely immersed in shadow.

TOTAL LUNAR ECLIPSES

A lunar eclipse occurs when the moon passes through Earth's shadow and grows dark. If the moon passes through the umbra and no part of the moon remains outside the umbra in the partial sunlight of the penumbra, we say the eclipse is a **total lunar eclipse.**

❚ Figure 3-6 illustrates the stages of a total lunar eclipse as a three-dimensional diagram (Window on Science 3-2) showing Earth, its shadows, and the path of the moon. As the moon begins to enter the penumbra it is only slightly dimmed, and a casual observer may not notice anything odd. After an hour, the moon

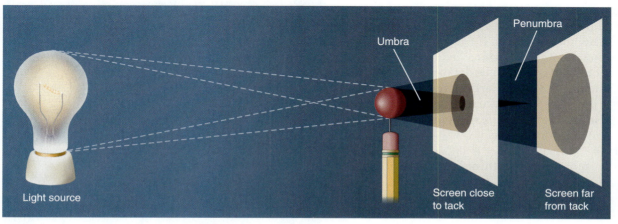

FIGURE 3-5
The shadows cast by a map tack resemble those of Earth and the moon. The umbra is the region of total shadow; the penumbra is the region of partial shadow.

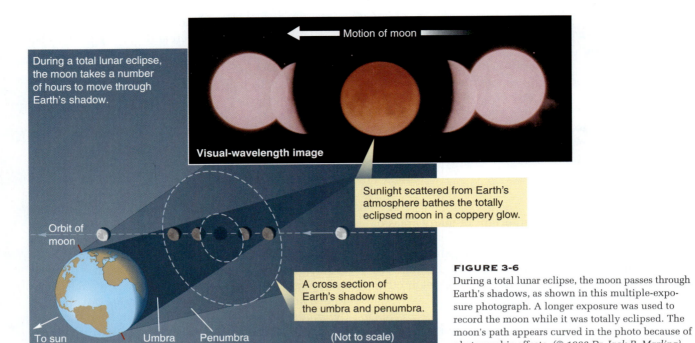

During a total lunar eclipse, the moon takes a number of hours to move through Earth's shadow.

Motion of moon

Visual-wavelength image

Sunlight scattered from Earth's atmosphere bathes the totally eclipsed moon in a coppery glow.

Orbit of moon

A cross section of Earth's shadow shows the umbra and penumbra.

To sun   Umbra   Penumbra   (Not to scale)

FIGURE 3-6

During a total lunar eclipse, the moon passes through Earth's shadows, as shown in this multiple-exposure photograph. A longer exposure was used to record the moon while it was totally eclipsed. The moon's path appears curved in the photo because of photographic effects. (© 1982 Dr. Jack B. Marling)

is deeper in the penumbra and dimmer; and, once it begins to enter the umbra, we see a dark bite on the edge of the lunar disk. The moon travels its own diameter in an hour, so it takes about an hour to enter the umbra completely and become totally eclipsed.

Even when the moon is totally eclipsed, it does not disappear completely. Sunlight, bent by Earth's atmosphere, leaks into the umbra and bathes the moon in a faint glow. Because blue light is scattered by Earth's atmosphere more easily than red light, it is red light that penetrates to illuminate the moon in a coppery glow. If we were on the moon during totality and looked back at Earth, we would not see any part of the sun because it would be entirely hidden behind Earth. However, we would see Earth's atmosphere illuminated from behind by the sun in a spectacular sunset completely ringing Earth. It is the red glow from this sunset that gives the totally eclipsed moon its reddish color.

How dim the totally eclipsed moon becomes depends on a number of things. If Earth's atmosphere is especially cloudy in those regions that must bend light into the umbra, the moon will be darker than usual. An unusual amount of dust in Earth's atmosphere (from volcanic eruptions, for instance) also causes a dark eclipse. Also, total lunar eclipses tend to be darkest when the moon's orbit carries it through the center of the umbra (▌Figure 3.7).

As the moon moves through Earth's umbral shadow, we can see that the shadow is circular. From this the Greek philosopher Aristotle (384–322 BC) concluded that Earth had to be a sphere, because only a sphere could cast a shadow that was always circular.

Depending on the geometry of the eclipse, the moon can take as long as 1 hour 40 minutes to cross the umbra

and another hour to emerge into the penumbra. Still another hour passes as it emerges into full sunlight. A total eclipse of the moon, including the penumbral stage, can take almost six hours from start to finish.

PARTIAL AND PENUMBRAL LUNAR ECLIPSES

Not all eclipses of the moon are total. Because the moon's orbit is inclined a bit over 5° to the plane of Earth's or-

FIGURE 3-7

During a total lunar eclipse, the moon turns coppery red. In this photo, the moon is darkest toward the lower right, the direction toward the center of the umbra. The edge of the moon at upper left is brighter because it is near the edge of the umbra. (Celestron International)

3-D Relationships

Much of science is explained in diagrams; one particular type of diagram can be confusing. When artists draw three-dimensional diagrams on flat sheets of paper, they use lots of artistic clues that have been discovered over the centuries. Perspective, shading, color, and shadows help us see the three-dimensional figure if the drawing is familiar—a house, for example. But when a drawing shows something that is unfamiliar, as is often the case in sci-ence, the artist's clues don't work as well. When we look at drawings of a molecule, a nerve cell, layers of rock under the ocean, or Earth and its shadows, we must pay special attention to the three-dimensional nature of the figure.

When you see a three-dimensional diagram in any science book, it helps to decide what the point of view is. For example, if the diagram were a photograph, where was the camera? In the drawing in Figure 3-6, the camera would have to have been in space looking back at Earth and the moon. Of course, astronauts have never been that far from Earth, but we can make such a voyage in our imagination, and that helps us understand the geometry of eclipses. Other three-dimensional diagrams have other points of view, so always be sure to first determine the point of view when you look at any three-dimensional diagram.

---

bit, the moon might not pass through the center of the umbra.

If the moon's orbit carries the full moon too far north or south of the umbra, the moon may only partially enter the umbra. The resulting **partial lunar eclipse** is usually not as dramatic as a total lunar eclipse. Because part of the moon remains outside the umbra, it receives some sunlight and looks bright in contrast with the dark part of the moon inside the umbra. Unless the moon almost completely enters the umbra, the glare from the illuminated part of the moon drowns out the fainter red glow inside the umbra. Partial lunar eclipses are interesting because part of the full moon is darkened, but they are not as beautiful as a total lunar eclipse.

If the orbit of the moon carries the moon far enough north or south of the umbra, the moon may only pass through the penumbra and never reach the umbra. Such **penumbral eclipses** are not dramatic at all. In the partial shadow of the penumbra, the moon is only partially dimmed. Most people glancing at a penumbral eclipse would not notice any difference from a full moon.

Total, partial, or penumbral, lunar eclipses are interesting events in the night sky and are not difficult to observe. When the moon passes through Earth's shadow, the eclipse is visible from anywhere on Earth's dark side. One or two lunar eclipses occur in most years. Consult ▌ Table 3-1 to find the next lunar eclipse visible in your part of the world.

TABLE 3-1
Total and Partial Eclipses of the Moon, 2004 to 2013

Year	Date	Time* of Mideclipse (GMT)	Length of Totality (Hours:Min)	Length of Eclipse† (Hours:Min)
2004	May 4	20:32	1:16	3:22
2004	Oct. 28	3:05	1:20	3:38
2005	Oct. 17	12:04	Partial	0:56
2006	Sept. 7	18:52	Partial	1:30
2007	Mar. 3	23:22	1:14	3:40
2007	Aug. 28	10:38	1:30	3:32
2008	Feb. 21	3:27	0:50	3:24
2008	Aug. 16	21:11	Partial	3:08
2009	Dec. 31	19:24	Partial	1:00
2010	June 26	11:40	Partial	2:42
2010	Dec. 21	8:18	1:12	3:28
2011	June 15	20:13	1:40	3:38
2011	Dec. 10	14:33	0:50	3:32
2012	June 4	11:03	Partial	2:08
2013	April 25	20:10	Partial	0:28

*Times are Greenwich Mean Time. Subtract 5 hours for Eastern Standard Time, 6 hours for Central Standard Time, 7 hours for Mountain Standard Time, and 8 hours for Pacific Standard Time. From your time zone, lunar eclipses that occur between sunset and sunrise will be visible, and those at midnight will be best placed.

†Does not include penumbral phase.

REVIEW CRITICAL INQUIRY

Why doesn't Earth's shadow on the moon look red during a partial lunar eclipse?

During a partial lunar eclipse, part of the moon protrudes from Earth's umbral shadow into sunlight. This part of the moon is very bright compared to the fainter red light inside Earth's shadow, and the glare of the reflected sunlight makes it difficult to see the red glow. If a partial eclipse is almost total, so that only a small sliver of moon extends out of the shadow into sunlight, we can sometimes detect the red glow in the shadow.

Of course, this red glow does not happen for every planet–moon combination in the universe. Suppose a planet had a moon but no atmosphere. Would the moon glow red during a total eclipse? Why not?

Lunar eclipses are slow and stately. For drama and excitement, there is nothing like a solar eclipse.

3-4 SOLAR ECLIPSES

A solar eclipse occurs when the moon moves between Earth and the sun. If the moon covers the disk of the sun completely, the eclipse is a **total solar eclipse.** If the moon covers only part of the sun, the eclipse is a **partial solar eclipse.** During a particular solar eclipse, people in one place on Earth may see a total eclipse, while people only a few hundred kilometers away see a partial eclipse.

These spectacular sights are possible because we on Earth are very lucky. Our moon has the same angular diameter as our sun, so it can cover the sun almost exactly. That lucky coincidence allows us to see total solar eclipses.

THE ANGULAR DIAMETER OF THE SUN AND MOON

We discussed the angular diameter of an object in Chapter 2; now we need to think carefully about how the size and distance of an object like the moon determine its angular diameter. This is the key to understanding solar eclipses.

Linear diameter is simply the distance between an object's opposite sides. We use linear diameter when we order a 16-inch pizza—the pizza is 16 inches in diameter. The linear diameter of the moon is 3476 km. The angular diameter of an object is the angle formed by lines extending toward us from opposite sides of the object and meeting at our eye (∎ Figure 3-8). Clearly, the farther away an object is, the smaller its angular diameter.

The **small-angle formula** gives us a way to figure out the angular diameter of any object, whether it is a pizza, the moon, or a galaxy. In the small-angle formula, we always express angular diameter in seconds of arc,* and we always use the same units for distance and linear diameter:

$$\frac{\text{angular diameter}}{206{,}265''} = \frac{\text{linear diameter}}{\text{distance}}$$

Of course, we can use this formula to find any one of these three quantities if we know the other two; here we are interested in finding the angular diameter of the moon.

The moon has a linear diameter of 3476 km and a distance from Earth of about 384,000 km. What is its angular diameter? The moon's linear diameter and distance are both given in the same units, so we can put them directly into the small-angle formula:

$$\frac{\text{angular diameter}}{206{,}265''} = \frac{3476 \text{ km}}{384{,}000 \text{ km}}$$

*The number 206,265″ is the number of seconds of arc in a radian. When we divide by 206,265″, we convert the angle from seconds of arc to radians.

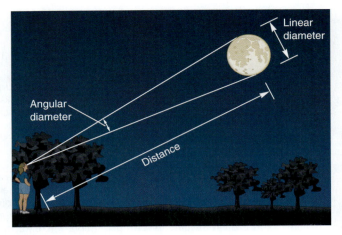

ACTIVE FIGURE 3-8
Angular diameter is the angle formed by lines extending from our eye to opposite sides of the object—in this figure, the moon. Linear diameter and distance are typically measured in kilometers or meters.

Ace◯Astronomy™ Go to AceAstronomy and click Active Figures to see "Small-Angle Formula." You can take control of this diagram.

To solve for angular diameter, we multiply both sides by 206,265 and find that the angular diameter is 1870 seconds of arc. If we divide by 60, we get 31 minutes of arc or, dividing by 60 again, about 0.5°. The moon's orbit is slightly elliptical, so it can sometimes look a bit larger or smaller, but its angular diameter is always close to 0.5°.

We can repeat this calculation for the angular diameter of the sun. The sun is 1.39×10^6 km in linear diameter and 1.50×10^8 km from Earth. Using the small-angle formula, we discover that the sun has an angular diameter of 1900 seconds of arc, which is 32 minutes of arc or about 0.5°. Earth's orbit is slightly elliptical, and thus the sun can sometimes look slightly larger or smaller, but it, like the moon, is always close to 0.5° in angular diameter.

By fantastic good luck, we live on a planet with a moon that is almost exactly the same angular diameter as our sun. When the moon passes in front of the sun, it is almost exactly the right size to block the brilliant surface of the sun. Then we see the most exciting sight in astronomy—a total solar eclipse. There are few other worlds where this can happen, because the angular diameters of the sun and a moon rarely match so closely. To see this beautiful sight, all we have to do is arrange to be in the moon's shadow when the moon crosses in front of the sun.

THE MOON'S SHADOW

Like Earth's shadow, the moon's shadow consists of a central umbra of total shadow and a penumbra of par-

TABLE 3-2
Total and Annular Eclipses of the Sun, 2004** to 2014

Date	Total/Annular (T/A)	Time of Mideclipse* (GMT)	Maximum Length of Total or Annular Phase (Min:Sec)	Area of Visibility
2005 Apr. 8	AT	21^h	0:42	Pacific, N. of S. America
2005 Oct. 3	A	11^h	4:32	Atlantic, Spain, Africa
2006 Mar. 29	T	10^h	4:07	Atlantic, Africa, Turkey
2006 Sept. 22	A	12^h	7:09	N.E. of S. America, Atlantic
2008 Feb. 7	A	4^h	2:14	S. Pacific, Antarctica
2008 Aug. 1	T	10^h	2:28	Canada, Arctic, Siberia
2009 Jan. 26	A	8^h	7:56	S. Atlantic, Indian Ocean
2009 July 22	T	3^h	6:40	Asia, Pacific
2010 Jan. 15	A	7^h	11:10	Africa, Indian Ocean
2010 July 11	T	20^h	5:20	Pacific, S. America
2012 May 20	A	23^h	5:46	Japan, N. Pacific, W. US
2012 Nov. 13	T	22^h	4:02	N. Australia, S. Pacific
2013 May 10	A	0^h	6:04	Australia, Pacific
2013 Nov. 3	AT	13^h	1:40	Atlantic, Africa

The next major total solar eclipse visible from the United States will occur on August 21, 2017.

*Times are Greenwich Mean Time. Subtract 5 hours for Eastern Standard Time, 6 hours for Central Standard Time, 7 hours for Mountain Standard Time, and 8 hours for Pacific Standard Time.

^{h}hours.

**There are no total or annular eclipses of the sun during 2004 or 2014.

tial shadow. What we see when the moon crosses in front of the sun depends on where we are in the moon's shadow. The moon's umbral shadow produces a spot of darkness roughly 269 km (167 miles) in diameter on Earth's surface (Figure 3-9). (The exact size of the umbral shadow depends on the location of the moon in its elliptical orbit and the angle at which the shadow strikes Earth.) If we are in this spot of total shadow, we see a total solar eclipse. If we are just outside the umbral shadow but in the penumbra, we see part of the sun peeking around the moon, and the eclipse is partial. Of course, if we are outside the penumbra, we see no eclipse at all. Because of the orbital motion of the moon, its shadow sweeps across Earth at speeds of at least 1700 km/h (1060 mph). To be sure of seeing a total solar eclipse, we must select an appropriate eclipse (Table 3-2), plan far in advance, and place ourselves in the **path of totality,** the path swept out by the umbral spot.

TOTAL SOLAR ECLIPSES

A total solar eclipse begins when we first see the edge of the moon encroaching on the sun. This is the moment when the edge of the penumbra sweeps over our location.

During the partial phase, part of the sun remains visible, and it is hazardous to look at the eclipse without protection. Dense filters and exposed film do not necessarily provide protection, because some do not block the invisible heat radiation (infrared) that can burn the retina of our eyes. This has led officials to warn the public not to look at solar eclipses and has even frightened some people into locking themselves and their children into windowless rooms during eclipses. In fact, the sun is a bit less dangerous than usual during an eclipse because part of the bright surface is covered by the moon. But an eclipse is dangerous

FIGURE 3-9
Observers in the path of totality see a total solar eclipse when the umbral shadow sweeps over them. Those in the penumbra see a partial eclipse.

Sunlight

Pinhole

Image of partially
eclipsed sun

FIGURE 3-10

A safe way to view the partial phases of a solar eclipse. Use a pinhole in a card to project an image of the sun on a second card. The greater the distance between the cards, the larger (and fainter) the image will be.

in that it can tempt us to look at the sun directly and burn our eyes.

The safest and simplest way to observe the partial phases of a solar eclipse is to use pinhole projection. Poke a small pinhole in a sheet of cardboard. Hold the sheet with the hole in the sunlight, and allow light to pass through the hole to a second sheet of cardboard (▌ Figure 3-10). On a day when there is no eclipse, the result is a small, round spot of light that is an image of the sun. During the partial phases of a solar eclipse, the image shows the dark silhouette of the moon obscuring part of the sun. These pinhole images of the partially eclipsed sun can also be seen in the shadows of trees as the sunlight peeks through the tiny openings between the leaves and branches. This can produce an eerie effect just before totality as the remaining sliver of sun produces thin crescents of light on the ground under trees.

Throughout the partial phases of a solar eclipse, the moon gradually covers the bright disk of the sun (▌ Figure 3-11). Totality begins as the last sliver of the sun's bright surface disappears behind the moon. This is the moment when the edge of the umbra sweeps over our location. So long as any of the sun is visible, the countryside is bright; but, as the last of the sun disappears, dark falls in a few seconds. Automatic streetlights come on, car drivers switch on their headlights, and birds go to roost. The darkness of totality depends on a number of factors, including the weather at the observing site, but it is usually dark enough to make it difficult to read the settings on cameras.

The totally eclipsed sun is a spectacular sight. With the moon covering the bright disk of the sun, called the

A Total Solar Eclipse

The moon moving from the right just begins to cross in front of the sun.

The disk of the moon gradually covers the disk of the sun.

Sunlight begins to dim as more of the sun's disk is covered.

During totality, pink prominences are often visible.

A longer-exposure photograph during totality shows the fainter corona.

FIGURE 3-11

This sequence of photos shows the first half of a total solar eclipse. *(Daniel Good)*

photosphere, * we can see the sun's faint outer atmosphere, the **corona,** glowing with a pale, white light so faint we can safely look at it directly. This corona is

*The photosphere, corona, chromosphere, and prominences will be discussed in detail in Chapter 8. Here the terms are used as the names of features we see during a total solar eclipse.

a

b

FIGURE 3-12

(a) During a total solar eclipse, the moon covers the photosphere, and the white corona and pink prominences are visible. Note the streamers in the corona caused by the sun's magnetic field. *(Daniel Good)*
(b) The diamond ring effect can sometimes occur momentarily at the beginning or end of totality if a small segment of the photosphere peeks out through a valley at the edge of the lunar disk. *(National Optical Astronomy Observatory)*

made of low-density, hot gas, which is given a wispy appearance by the solar magnetic field as shown in the last frame of Figure 3-11. Also visible just above the photosphere is a thin layer of bright gas called the **chromosphere.** The chromosphere is often marked by eruptions on the solar surface called **prominences (**❚ Figure 3-12a), which glow with a clear, pink color due to the high temperature of the gases involved. The small-angle formula tells us that a large prominence is about 3.5 times the diameter of Earth.

Totality cannot last longer than 7.5 minutes under any circumstances, and the average is only 2 to 3 minutes. Totality ends when the sun's bright surface reappears at the trailing edge of the moon. This corresponds to the moment when the trailing edge of the moon's umbra sweeps over the observer.

Just as totality begins or ends, a small part of the photosphere can peek out from behind the moon through a valley at the edge of the lunar disk. Although it is intensely bright, such a small part of the photosphere does not completely drown out the fainter corona, which forms a silvery ring of light with the brilliant spot of photosphere gleaming like a diamond (Figure 3-12b). This **diamond ring effect** is one of the most spectacular of astronomical sights, but it is not visible during every solar eclipse. Its occurrence depends on the exact orientation and motion of the moon.

Once totality is over, daylight returns quickly, and the corona and chromosphere vanish. Astronomers travel great distances to place their instruments in the path of totality to study the faint outer corona and make other measurements possible only during the few minutes of a total solar eclipse.

Sometimes when the moon crosses in front of the sun, it is too small to fully cover the sun, and we see an **annular eclipse,** a solar eclipse in which a ring (or annulus) of the photosphere is visible around the disk of the moon (❚ Figure 3-13). With a portion of the photosphere visible, the eclipse never becomes total; it never quite gets dark; and we can't see the prominences, chromosphere, and corona. Annular eclipses occur because the moon follows a slightly elliptical orbit around Earth, and thus its angular diameter can vary (❚ Figure 3-14). When it is at **perigee,** its point of closest approach to Earth, it looks significantly larger than when it is at **apogee,** the most distant point in its orbit. Furthermore, Earth's orbit is slightly elliptical, so the Earth–sun distance varies slightly, and thus the diameter of the solar disk varies slightly. If the moon is in the farther part of its orbit during totality, its angular diameter will be less than the angular diameter of the sun, and thus we see the annular eclipse. Such an annular eclipse of the sun swept across the United States on May 10, 1994.

a

b

FIGURE 3-13

(a) The annular eclipse of 1994. A bright ring of photosphere remains visible around the moon, and the corona and prominences are not visible. *(Daniel Good)* (b) An annular eclipse occurs when the moon is in the farther part of its orbit and its umbral shadow does not reach Earth. From Earth, we see an annular eclipse because the moon's angular diameter is smaller than the angular diameter of the sun.

REVIEW CRITICAL INQUIRY

If people on Earth were seeing a total solar eclipse, what would astronauts on the moon see when they looked at Earth?

Astronauts on the moon could see Earth only if they were on the side that faces Earth. Because solar eclipses always happen at new moon, the near side of the moon would be in darkness, and the far side of the moon would be in full sunlight. The astronauts would be standing in darkness, and they would be looking at the fully illuminated side of Earth. They would see a "full Earth." The moon's shadow would be crossing Earth; and, if the astronauts looked closely, they might be able to see the spot of darkness where the moon's umbral shadow touched Earth. It would take hours for the shadow to cross Earth.

Standing on the moon and watching the moon's umbral shadow sweep across Earth would be a cold, tedious assignment. Perhaps it would be more interesting for astronauts on the moon to watch Earth while people on Earth were seeing a total lunar eclipse. What would the astronauts see then?

Eclipses of the sun and moon are often dramatic and mysterious. Ancient astronomers studied them and found ways to predict the coming of an eclipse. In the section that follows, we will see how eclipses can be predicted from a basic understanding of the motion of the sun and the moon.

3-5 PREDICTING ECLIPSES

To make exact eclipse predictions, you would need to calculate the precise motions of the sun and moon, and that requires a computer and proper software. Such software is available for desktop computers, but it isn't necessary if you are satisfied with making less exact predictions. In fact, many primitive peoples, such as the builders of Stonehenge and the ancient Maya, are believed to have made eclipse predictions.

We examine eclipse prediction for three reasons. First, it is an important part of the history of science. Second, it illustrates how apparently complex phenomena can be analyzed in terms of cycles. Third, eclipse prediction exercises our mental muscles and

FIGURE 3-14

The angular diameter of the moon (left) varies by a total of almost 12 percent, because its orbit is elliptical, and its distance from Earth varies from perigee (closest) to apogee (farthest). The angular diameter of the sun (right) varies by a total of only 3.4 percent, because Earth's orbit is more nearly circular, and its distance from the sun varies only slightly from perihelion (closest) to aphelion (farthest).

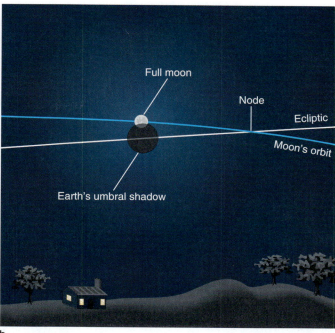

FIGURE 3-15

Eclipses can occur only near the nodes of the moon's orbit. (a) A solar eclipse occurs when the moon meets the sun near a node. (b) A lunar eclipse occurs when the sun and moon are near opposite nodes. Partial eclipses are shown here for clarity.

forces us to see Earth, the moon, and the sun as objects moving through space.

CONDITIONS FOR AN ECLIPSE

We can predict eclipses by understanding the conditions that make them possible. As we begin to think about these conditions, we must be sure we understand our point of view. (See Window on Science 3-2.) Later we will change our point of view, but to begin we will imagine that we can look up into the sky from our home on Earth and see the sun moving along the ecliptic and the moon moving along its orbit.

The orbit of the moon is tipped 5°8′43″ to the plane of Earth's orbit, so we see the moon follow a path tipped by that angle to the ecliptic. Each month, the moon crosses the ecliptic at two points called **nodes.** At one node it crosses going southward, and two weeks later it crosses at the other node going northward.

Eclipses can only occur when the sun is near one of the nodes of the moon's orbit. A solar eclipse is caused by the moon passing in front of the sun. Most new moons pass too far north or too far south of the sun to cause an eclipse. Only when the sun is near a node in the moon's orbit can the moon cross in front of the sun, as shown in ❙ Figure 3-15a. A lunar eclipse doesn't happen at every full moon because most full moons pass too far north or too far south of the ecliptic and miss Earth's shadow. The moon can enter Earth's shadow

only when the shadow is near a node in the moon's orbit, and that means the sun must be near the other node. This is shown in Figure 3-15b.

Thus, there are two conditions for an eclipse: The sun must be crossing a node, and the moon must be crossing either the same node (solar eclipse) or the other node (lunar eclipse). Clearly, solar eclipses can occur only when the moon is new, and lunar eclipses can occur only when the moon is full.

An **eclipse season** is the period during which the sun is close enough to a node for an eclipse to occur. For solar eclipses, an eclipse season is about 32 days. Any new moon during this period will produce a solar eclipse. For lunar eclipses, the eclipse season is a bit shorter, about 22 days. Any full moon in this period will be eclipsed.

This makes eclipse prediction easy. We simply keep track of where the moon crosses the ecliptic, and when the sun is near one of these nodes we predict that the nearest new moon will cause a solar eclipse and the nearest full moon will cause a lunar eclipse. This system works fairly well, and ancient astronomers such as the Maya may have used such a system. But we can do better if we change our point of view.

THE VIEW FROM SPACE

Let us change our point of view and imagine that we are looking at the orbits of Earth and the moon from a

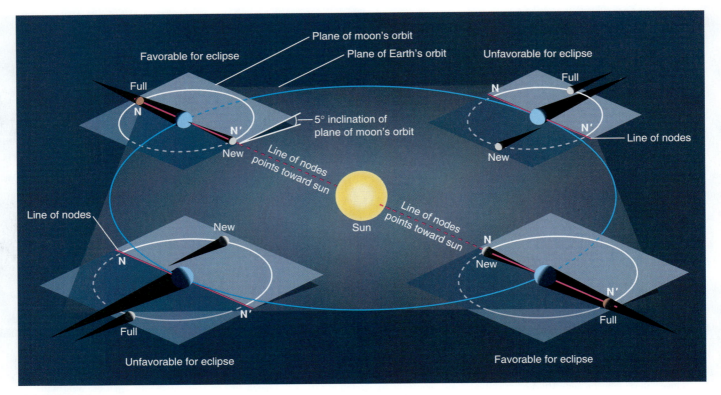

FIGURE 3-16
The moon's orbit is tipped about 5° to Earth's orbit. The nodes N and N' are the points where the moon passes through the plane of Earth's orbit. If the line of nodes does not point at the sun, the shadows miss, and there are no eclipses at new moon and full moon. At those parts of Earth's orbit where the line of nodes points toward the sun, eclipses are possible at new moon and full moon.

point far away in space. We see the moon's orbit as a smaller disk tipped at an angle to the larger disk of Earth's orbit. As Earth orbits the sun, the moon's orbit remains fixed in direction. The nodes of the moon's orbit are the points where it passes through the plane of Earth's orbit; an eclipse season occurs each time the line connecting these nodes, the **line of nodes,** points toward the sun (▌ Figure 3-16).

The shadows of Earth and moon, seen from space, are very long and thin (▌ Figure 3-17). Only at the time of an eclipse season, when the line of nodes points toward the sun, do the shadows produce eclipses.

From our point of view in space, we would see the orbit of the moon precess like a hubcap spinning

on the ground. This precession is caused mostly by the gravitational influence of the sun, and it makes the line of nodes rotate once every 18.6 years. People back on Earth see the nodes slipping westward along the ecliptic 19.4° per year, and the sun takes only 346.62 days (an **eclipse year**) to return to a node. This means that, according to our calendar, the eclipse seasons begin about 19 days earlier every year (▌ Figure 3-18).

The cyclic pattern of eclipses shown in Figure 3-18 makes eclipse prediction simple. We know that the eclipse seasons occur 19 days earlier each year because the moon's orbit precesses. Any new moon during an eclipse season will cross in front of the sun and cause

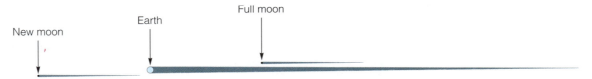

FIGURE 3-17
Umbral shadows of Earth and the moon. Notice how easy it is for the shadows to miss their mark at full moon and at new moon and fail to produce eclipses. That is, it is easy for the moon to reach full phase and not enter Earth's shadow. It is also easy for the moon to reach new phase and not cast its shadow on Earth. (The diameters of Earth and the moon are exaggerated by a factor of 2 for clarity.)

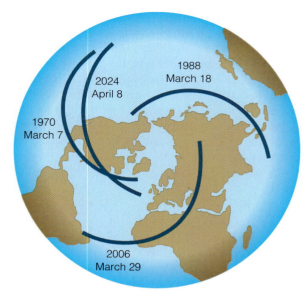

Total and annular solar eclipses   • Lunar eclipses

FIGURE 3-18

A calendar of eclipse seasons. Each year the eclipse seasons begin about 19 days earlier. Any new moon or full moon that occurs during an eclipse season results in an eclipse. Not all eclipses are shown here.

FIGURE 3-19

The saros cycle at work. The total solar eclipse of March 7, 1970, recurred after 18 years $11\frac{1}{3}$ days over the Pacific Ocean. After another interval of 18 years $11\frac{1}{3}$ days, the same eclipse will be visible from Asia and Africa. After a similar interval, the eclipse will again be visible from the United States.

a solar eclipse, and any full moon will enter Earth's shadow and cause a lunar eclipse.

THE SAROS CYCLE

Ancient astronomers could predict eclipses in a crude way using the eclipse seasons, but they could have been much more accurate if they recognized that eclipses occur following certain patterns. The most important of these is the **saros cycle** (sometimes referred to simply as the saros). After one saros cycle of 18 years $11\frac{1}{3}$ days, the pattern of eclipses repeats. In fact, *saros* comes from a Greek word that means "repetition."

The eclipses repeat because, after one saros cycle, the moon and the nodes of its orbit return to the same place with respect to the sun. One saros contains 6585.321 days, which is equal to 223 lunar months.

Therefore, after one saros cycle the moon is back to the same phase it had when the cycle began. But one saros is nearly equal to 19 eclipse years. After one saros cycle, the sun has returned to the same place it occupied with respect to the nodes of the moon's orbit when the cycle began. If an eclipse occurs on a given day, then 18 years $11\frac{1}{3}$ days later the sun, the moon, and the nodes of the moon's orbit return to nearly the same relationship, and the eclipse occurs all over again.

Although the eclipse repeats almost exactly, it is not visible from the same place on Earth. The saros cycle is one-third of a day longer than 18 years 11 days. When the eclipse recurs, Earth will have rotated one-third of a turn farther east, and the eclipse will occur eight hours of longitude west of its earlier location (❙ Figure 3-19). Thus, after three saros cycles—a period of 54 years 1 month—the same eclipse occurs in the same part of Earth.

One of the most famous predictors of eclipses was Thales of Miletus (about 640–546 BC), who supposedly learned of the saros cycle from the Chaldeans, who had discovered it. No one knows which eclipse Thales predicted, but some scholars suspect the eclipse of May 28, 585 BC. In any case, the eclipse occurred at the height of a battle between the Lydians and the Medes, and the mysterious darkness in midafternoon so startled the two factions that they concluded a truce.

In fact, many historians doubt that Thales actually predicted the eclipse. It would have been very difficult to gather enough information about past eclipses, because total solar eclipses are rare. If you stay in one city, you will see a total solar eclipse about once in 360 years. Also, 585 BC is very early for the Greeks to have known of the saros cycle. The important point is not that Thales did it, but that he could have done it. If he had had records of past eclipses of the sun visible from the area, he could have discovered that they tended to recur with a period of 54 years 1 month (three saros cycles). Indeed, he could have predicted the eclipse without ever understanding what the sun and moon were or how they moved.

REVIEW CRITICAL INQUIRY

Why can't two successive full moons be totally eclipsed?
A total lunar eclipse occurs when the moon passes through Earth's shadow, and that can happen only when the sun is near one node and the moon crosses the other node. Earth's shadow always points toward the ecliptic exactly opposite the sun, so most of the time the moon will pass north or south of Earth's shadow, and there will be no eclipse. Only when the shadow is near one of the nodes of the moon's orbit can the moon enter the shadow and be eclipsed.

An eclipse season for a total lunar eclipse is only 22 days long. If the moon crosses a node more than 11 days before or after the sun crosses the other node, there will be no eclipse. The moon takes 29.5 days to go from one full moon to the next, so if one full moon is totally eclipsed, the next full moon 29.5 days later will occur too late for an eclipse.

Use your knowledge of the cycles of the sun and moon to explain why the sun can be eclipsed by two successive new moons.

Predicting eclipses isn't very hard, and many ancient peoples were probably familiar enough with the cycles of the sun and moon to know when eclipses were likely. We might call those first predictions early science, but science involves understanding nature, not just predicting events. In the next chapter, we will see how modern science was born out of astronomers' attempts to understand the cycles they saw in the sky.

SUMMARY

Because we see the moon by reflected sunlight, its shape appears to change as it orbits Earth. The lunar phases wax from new moon to first quarter to full moon and wane from full moon to third quarter to new moon. A complete cycle of lunar phases takes 29.53 days.

The moon's gravitational field exerts tidal forces on Earth that pull the ocean waters up into two bulges, one on the side of Earth facing the moon and the other on the side away from the moon. As the rotating Earth carries the continents through these bulges of deeper water, the tides rise and fall. Friction

with the seabeds slows Earth's rotation, and the gravitational force the bulges exert on the moon forces its orbit to grow larger.

A lunar eclipse occurs when the moon enters Earth's shadow. If the moon becomes completely immersed in the umbra, the central shadow, the eclipse is termed total. The moon glows coppery red due to sunlight bent as it passes through Earth's atmosphere. If the moon enters the umbral shadow only partly, the eclipse is termed partial, and the reddish glow is not visible. A penumbral eclipse occurs when the moon passes through the penumbra, the region of partial shadow. Penumbral eclipses are not very noticeable.

A solar eclipse occurs when Earth passes through the moon's shadow. To see a total solar eclipse, observers must place themselves in the path of totality, the path swept by the umbra of the moon. As the umbra sweeps over the observers, they see the bright surface of the sun, the photosphere, blotted out by the moon. Then the fainter chromosphere and corona, higher layers of the sun's atmosphere, become visible. Eruptions on the solar surface, called prominences, may be visible peeking around the edge of the moon.

The corona, chromosphere, and prominences are not visible to observers outside the path of totality. Observers in the path of the penumbra of the moon will see a partial solar eclipse, but the photosphere will never be completely hidden.

If an eclipse of the sun occurs when the moon is not in the nearer part of its orbit, the moon's umbra does not reach Earth's surface. In these cases, the eclipse is not total. The moon's angular diameter is too small to cover the sun completely. Even at maximum eclipse, therefore, a bright ring, or annulus, of the photosphere is visible around the edge of the moon. This annular eclipse is not as dramatic as a total eclipse.

Ancient astronomers could predict eclipses because they occur not randomly but in a pattern. Two conditions must be met if an eclipse is to occur. First, the moon must be on or near the ecliptic. The two points where the moon crosses the ecliptic are called the nodes of the moon's orbit. The second condition is that the sun must be at or near one of the nodes. This means that eclipses can occur only at new moon or full moon during two eclipse seasons that are about a month long and that occur almost 6 months apart.

Because the moon's orbit precesses, the nodes slip westward along the ecliptic, and the eclipse seasons begin 19 days earlier each year. The moon, sun, and nodes return to the same relative positions every 18 years $11\frac{1}{3}$ days in what is called the saros cycle. After the passage of a saros cycle, the pattern of eclipses begins to repeat. This means that ancient astronomers could predict eclipses just by examining the dates of previous eclipses. The Chaldeans discovered the saros cycle, and the ancient Greeks used it. Many other primitive cultures also used this method.

NEW TERMS

sidereal period	penumbra
synodic period	total eclipse (lunar or solar)
spring tides	partial eclipse (lunar or solar)
neap tides	
umbra	penumbral eclipse

small-angle formula

path of totality

photosphere

corona

chromosphere

prominence

diamond ring effect

annular eclipse

perigee

apogee

node

eclipse season

line of nodes

eclipse year

saros cycle

REVIEW QUESTIONS

Ace Astronomy™ Assess your understanding of this chapter's topics with additional quizzing and animations at **http:// astronomy.brookscole.com/seeds8e**

1. Which lunar phases would be visible in the sky at dawn? at midnight?

2. If you looked back at Earth from the moon, what phase would you see when the moon was full? new? a first-quarter moon? a waxing crescent?

3. Give examples to show how tides can alter the rotation and revolution of celestial bodies. (*Hint:* Recall from Chapter 2 the difference between rotation and revolution.)

4. Could a solar-powered spacecraft generate any electricity while passing through Earth's umbral shadow? the penumbral shadow?

5. Draw the umbral and penumbral shadows onto the diagram in the middle of page 34. Explain why lunar eclipses can occur only at full moon and solar eclipses can occur only at new moon.

6. How did lunar eclipses lead Aristotle to conclude that Earth was a sphere?

7. Why isn't the corona visible during partial or annular solar eclipses?

8. Why can't the moon be eclipsed when it is halfway between the nodes of its orbit?

9. Why aren't solar eclipses separated by one saros cycle visible from the same location on Earth?

10. How could Thales of Miletus have predicted the date of a solar eclipse without observing the location of the moon in the sky?

DISCUSSION QUESTIONS

1. If the moon were closer to Earth such that it had an orbital period of 24 hours, what would the tides be like?

2. How would eclipses be different if the moon's orbit were not tipped with respect to the plane of Earth's orbit?

3. Are there other planets in our solar system from whose surface we could see a lunar eclipse? a total solar eclipse?

4. Can you detect the Saros cycle in Figure 3-18?

PROBLEMS

1. Identify the phases of the moon if on March 21 the moon is located at the point on the ecliptic called (a) the vernal equinox, (b) the autumnal equinox, (c) the summer solstice, (d) the winter solstice.

2. Identify the phases of the moon if at sunset the moon is (a) near the eastern horizon, (b) high in the southern sky, (c) in the southeastern sky, (d) in the southwestern sky.

3. About how many days must elapse between first-quarter moon and third-quarter moon?

4. How many hours would elapse between successive high tides if tides at a given location were caused only by the sun's gravity? only by the moon's gravity? Why is there a difference?

5. How many times larger than the moon is the diameter of Earth's umbral shadow at the moon's distance? (*Hint:* See Figure 3-6.)

6. Use the small-angle formula to calculate the angular diameter of Earth as seen from the moon.

7. During solar eclipses, large solar prominences are often seen extending 5 minutes of arc from the edge of the sun's disk. How far is this in kilometers? in Earth diameters?

8. If a solar eclipse occurs on October 3: (a) Why can't there be a lunar eclipse on October 13? (b) Why can't there be a solar eclipse on December 28?

9. A total eclipse of the sun was visible from Canada on July 10, 1972. When did this eclipse occur next? From what part of Earth was it total?

10. When will the eclipse described in Problem 9 next be total as seen from Canada?

11. When will the eclipse seasons occur during the current year? What eclipse(s) will occur?

CRITICAL INQUIRIES FOR THE WEB

1. Search the Web for myths about the moon. What common themes do different cultures ascribe to the moon and its cycle of phases?

2. A total solar eclipse swept across Europe and into Asia in the summer of 1999. Search for Web pages showing photos and observations of the eclipse. Can you find Web pages that show cultural responses to the eclipse such as celebrations or religious ceremonies?

3. Most people see more total lunar eclipses in a lifetime than total solar eclipses. Why is this so? Compare the regions of visibility for a number of past and upcoming eclipses and determine which future events will be visible from your area.

4. How do the tides on Earth vary with the phases of the moon? Use the Internet to explore the range of tides for a particular location through the next few weeks. Also, look up moon phase information for that same interval. How do the tidal amplitudes correlate with moon phase during that period?

EXPLORING *THESKY*

1. How many days must elapse between new moons? Use the **Moon Phase Calendar** under the **Tools** menu to locate new moons.

2. Where would you have to go to see a total solar eclipse in the next year or two? Use the **Eclipse Finder** under the **Tools** menu to select a total solar eclipse. Check **Show Path of Totality** to see a map of Earth.

3. View the total solar eclipse you located in activity 2. In the **Eclipse Finder,** click **View** to see the sun and moon from your present location. Then use **Site Information** under the **Data** menu to change your location to a spot in the path of totality. Set the **Time Step** to 5 minutes and click the **arrows** to move the moon across the sun.

4. Locate and observe an annular eclipse.

5. Locate and observe a total lunar eclipse.

 Visit the Seeds *Foundations of Astronomy* companion Web site for critical thinking exercises, articles, and additional readings from InfoTrac College Edition, Brooks/Cole's online student library.

THE ORIGIN OF MODERN ASTRONOMY

*How you would burst out laughing, my dear Kepler, if you would hear
what the greatest philosopher of the Gymnasium told the Grand Duke about me . . .*

From a letter by Galileo Galilei

Chapter 4

GUIDEPOST

The sun, moon, and planets sweep out a beautiful and complex dance across the heavens. Previous chapters have described that dance; this chapter describes how astronomers learned to understand what they saw in the sky and how that changed humanity's understanding of what we are.

In learning to interpret what they saw, Renaissance astronomers invented a new way of knowing about

nature, a way of knowing that we recognize today as modern science.

This chapter tells the story of heavenly motion from a cultural perspective. In the next chapter, we will give meaning to the motions in the sky by adding the ingredient Renaissance astronomers were missing—gravity.

The history of astronomy is more like a soap opera than a situation comedy. In a simple half-hour sitcom, only a few characters carry the story, but the plot of a soap opera is so complex, so filled with characters and subplots only tenuously related to one another, that the action jumps from character to character, from subplot to subplot, with almost no connection. The history of astronomy is similarly complex, filled with brilliant people who lived at different times and worked in different parts of the world.

Two subplots twine through our story. One is the human quest to understand the place of the Earth. That plot will lead us from Aristotle to Copernicus and finally to the trial of Galileo before the Inquisition. But the second subplot, the puzzle of planetary motion, is equally important to our story. That plot will involve an astronomer with a false nose and another whose mother was tried for witchcraft. We will see that the true place of the Earth could be understood only when the puzzle of planetary motion was solved.

Our story goes far beyond the birth of astronomy. The quest for the place of the Earth and the puzzle of planetary motion became, in the 16th and 17th centuries, the center of a storm of controversy over the best way to understand our world and ourselves. That conflict is now known as the scientific revolution. Thus, our story describes not only the birth of astronomy but also the birth of modern science.

Of course, astronomy had its beginnings long before the 16th century. It began when the first semihuman creature looked up at the moon and stars and wondered what they were. Since then, hundreds of generations of talented astronomers have lived and worked on this planet and left almost no record of their accomplishments. Only through archaeology can we get a hint of their insights and triumphs. Other cultures, such as ancient Greece, have left written records from which we can reconstruct the sophisticated astronomy of the ancient world.

4-1 THE ROOTS OF ASTRONOMY

Astronomy has its origin in that most noble of all human traits, curiosity. Just as modern children ask their parents what the stars are and why the moon changes, so did ancient humans ask themselves these questions. The answers, often couched in mythical or religious terms, reveal a great reverence for the order of the heavens.

Ace✪Astronomy™ The AceAstronomy icon throughout the text indicates an opportunity for you to test yourself on key concepts and to explore animations and interactions on the AceAstronomy Web site at http://astronomy.brookscole.com/seeds8e

ARCHAEOASTRONOMY

The study of the astronomy of ancient peoples, **archaeoastronomy,** came to public attention in 1965 when Gerald Hawkins published a book called *Stonehenge Decoded.* He reported that Stonehenge, the prehistoric ring of stones, was a sophisticated astronomical observatory.

Stonehenge, standing on Salisbury Plain in southern England, was built in stages from about 3000 BC to about 1800 BC, a period extending from the late Stone Age into the Bronze Age. Though the public is most familiar with the massive stones of Stonehenge (❙ Figure 4-1), those were added late in its history. In its first stages, Stonehenge consisted of a circular ditch slightly larger in diameter than the length of a football field, with a concentric bank just inside the ditch and a long avenue leading away toward the northeast. A massive stone, the Heelstone, stood then, as it does now, outside the ditch in the opening of the avenue.

As early as AD 1740, the English scholar W. Stukely suggested that the avenue pointed toward the rising sun at the summer solstice, but few accepted the idea. More recently, astronomers have recognized significant astronomical alignments at Stonehenge. Seen from the center of the monument, the summer-solstice sun rises behind the Heelstone. Other sight lines point toward the most northerly and most southerly risings of the moon.

The significance of these alignments has been debated. Hawkins claimed that the Stone Age people who built Stonehenge were using it as a device to predict lunar eclipses. Others have had more conservative notions, but the truth may never be known. The builders of Stonehenge had no written language and left no records of their intentions. Nevertheless, the presence of solar and lunar alignments at Stonehenge and at many other Stone Age monuments dotting England and continental Europe shows that so-called primitive peoples were paying detailed attention to the sky. The roots of astronomy lie not in sophisticated science and logic but in human curiosity and wonder.

The early inhabitants of North America were also interested in astronomy. The Big Horn Medicine Wheel,* located on a 3000-m shoulder of the Big Horn Mountains of Wyoming, was used by the Plains Indians about AD 1500–1750. This arrangement of rocks marks a 28-spoke wheel about 27 m in diameter. Piles of rocks, called cairns, mark the center and six locations on the circumference. Astronomer John Eddy has discovered that these cairns mark sight lines toward a number of important points on the eastern horizon and could have been used as a calendar to help schedule hunting, planting, harvesting, and celebrations (❙ Figure 4-2).

Many other American Indian sites have astronomical alignments. More than three dozen medicine wheels are known, although most do not have the sophistication of the Big Horn Medicine Wheel. The Moose Mountain Wheel in Saskatchewan, Canada, for in-

*Medicine is used here to mean magical power.

a

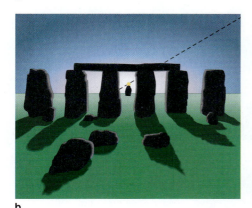

b

FIGURE 4-1

(a) A stamp issued to mark the return of Comet Halley shows the central horseshoe of upright stones at Stonehenge. (b) The best-known astronomical alignment at Stonehenge is the summer solstice sun rising over the Heelstone. (c) Although a number of astronomical alignments have been found at Stonehenge, experts debate their significance.

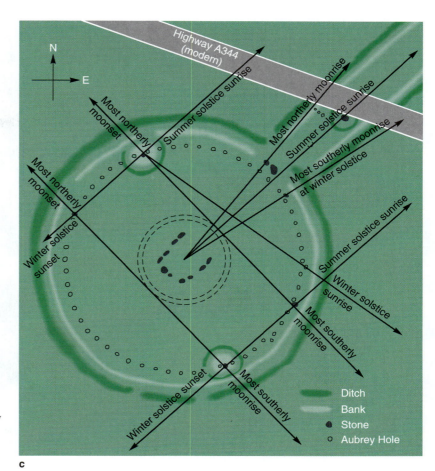

c

stance, seems to have been in use as early as AD 100. Alignments at some Mound Builder sites of the midwest show that these peoples were familiar with the sky. In the southwest, the Anasazi Indian ruins in Chaco Canyon, New Mexico, among others, have alignments that point toward the rising and setting of the sun at the summer and winter solstices.

Primitive astronomy also flourished in Central and South America. The Mayan and Aztec empires built many temples aligned with the solstice rising of the sun and with the extreme points where Venus rose and set. The Caracol temple in the Yucatán is a good example (▌Figure 4-3). It is a circular tower containing complicated passageways and a spiral staircase (thus the name *Caracol*—"the snail's shell"). The tubelike windows at the top of the tower point toward the equinox sunset point and the most northerly and most southerly setting points of Venus. Unfortunately, only about one-third of the tower top survives, so the directions of any other windows are forever lost.

Archaeoastronomers are uncovering the remains of ancient astronomical observatories around the world. Some temples in the jungles of southeast Asia, for instance, have astronomical alignments.

Other scholars are looking not at temples but at small artifacts from thousands of years ago. Scratches on certain bone and stone implements seem to follow a pattern and may be an attempt to keep a record of the phases of the moon (▌Figure 4-4). Some scientists contend that humanity's first attempts at writing were stimulated by a desire to record and predict lunar phases.

Archaeoastronomy is uncovering the earliest roots of astronomy and simultaneously revealing some of the first human efforts at systematic inquiry. The most important lesson of archaeoastronomy is that humans don't have to be technologically sophisticated to admire and study the universe.

One thing about archaeoastronomy is especially sad. Although we are learning how ancient people observed the sky, we may never know what they thought about their universe. Many had no written language. In other cases, the written record has been lost. Dozens, perhaps hundreds, of beautiful Mayan manuscripts, for instance, were burned by Spanish missionaries who believed that the books were the work of Satan. Only four of these books have survived, and all four contain astronomical references. One contains sophisticated tables that allowed the Maya to predict the motion of

FIGURE 4-2

The Big Horn Medicine Wheel in Wyoming was built by Native Americans before the westward spread of European settlers. Its rock cairns appear to be aligned with the rising and setting of the summer solstice sun (red) and with the rising of three bright stars (blue). Such connections with the sky are found in many Native American structures and ceremonies. *(John A. Eddy)*

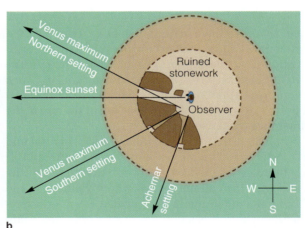

FIGURE 4-3

The Caracol at the Mayan city of Chichén Itzá (a) is a 1000-year-old observatory. The remaining windows in the partially ruined tower contain sight lines that point toward the equinox sunset, the setting places of Venus at its most northerly and most southerly positions, and the setting place of the star Achernar (b). Various evidence suggests that the Caracol was directly associated with the worship of the Venus god, Quetzalcoatl. *(Anthony Aveni)*

FIGURE 4-4
A fragment of a 27,000-year-old mammoth tusk found at Gontzi in Ukraine contains scribe marks on its edge, simplified in this drawing. These markings have been interpreted as a record of four cycles of lunar phases. Although controversial, such finds suggest that some of the first human attempts at recording events in written form were stimulated by astronomical phenomena

Venus and eclipses of the moon. We will never know what was burnt.

The fate of the Mayan books illustrates one reason why histories of astronomy usually begin with the Greeks. Some of their writing has survived, and we can discover what they thought about the shape and motion of the heavens.

THE ASTRONOMY OF GREECE

Greek astronomy was based on the astronomy of Babylonia and Egypt, but these astronomies were heavily influenced by religion and astrology. Those ancient astronomers studied the motions of the heavens as a way of worship and divination. The Greek astronomers studied astronomy in an entirely new way—they tried to understand the universe.

This new attitude toward the heavens, a truly scientific attitude, was made possible by two early Greek philosophers. Thales of Miletus (c. 624–547 BC) lived and worked in what is now Turkey. He taught that the universe is rational and that the human mind can understand why the universe works the way it does. This view contrasts sharply with those of earlier cultures, which believed that the ultimate causes of things are mysteries beyond human understanding. To Thales and his followers, the mysteries of the universe are mysteries because they are unknown, not because they are unknowable.

The second philosopher who made the new scientific attitude possible was Pythagoras (c. 570–500 BC). He and his students noticed that many things in nature seem to be governed by geometrical or mathematical relations. Musical notes, for example, are related in a regular way to the lengths of plucked strings. This led Pythagoras to propose that all nature was underlain by musical principles, by which he meant mathematics. One result of this philosophy was the later belief that the

harmony of the celestial movements produced actual music, the music of the spheres. But at a deeper level, the teachings of Pythagoras made Greek astronomers look at the universe in a new way. Thales said that the universe could be understood, and Pythagoras said that the underlying rules were mathematical.

In trying to understand the universe, Greek astronomers did something that Babylonian astronomers had never done—they tried to construct descriptions based on geometrical forms. Anaximander (c. 611–546 BC) described a universe made up of wheels filled with fire: The sun and moon are holes in the wheels through which we see the flames. Philolaus (5th century BC) argued that Earth moves in a circular path around a central fire (not the sun), which is always hidden behind a counterearth located between the fire and Earth. This was the first theory to suppose that Earth is in motion.

Plato (428–347 BC) was not an astronomer, but his teachings influenced astronomy for 2000 years. Plato argued that the reality we see is only a distorted shadow of a perfect, ideal form. If our observations are distorted, then observation can be misleading, and the best path to truth is through pure thought on the ideal forms that underlie nature.

Plato also argued that the most perfect form was the circle and that therefore all motions in the heavens should be made up of combinations of circular motion. Because the most perfect motion is uniform motion, later astronomers tried to describe the motions of the heavens using the principle of uniform circular motion.

Pythagoras had taught that Earth is a sphere and that the other heavenly bodies were divine, perfect spheres moving in perfect circles. Eudoxus of Cnidus (409–356 BC), a student of Plato, combined a system of 27 nested spheres rotating at different rates about different axes with the concept of uniform circular motion to produce a mathematical description of the motions of the universe (❙ Figure 4-5).

At the time of the Greek philosophers, it was common to refer to systems such as that of Eudoxus as descriptions of the world, where the word *world* included not only Earth but all of the heavenly spheres. The reality of these spheres was open to debate. Some thought of the spheres as nothing more than mathematical ideas that described motion in the world model, while others began to think of the spheres as real objects made of perfect celestial material. Aristotle, for example, seems to have thought of the spheres as real.

Aristotle (384–322 BC) taught and wrote on philosophy, history, politics, ethics, poetry, drama, and so on (❙ Figure 4-6). Because of his sensitivity and insight, he became the great authority of antiquity, and astronomers for almost 2000 years cited him as their authority in adopting the Greek model of the universe.

Much of what Aristotle wrote about scientific subjects was wrong, but he was not a scientist. He was trying to understand the universe by creating a system of

FIGURE 4-5
The spheres of Eudoxus explain the motions in the heavens by means of nested spheres rotating about various axes at different rates. Earth is located at the center.

formal philosophy. Modern science depends on evidence and hypothesis, but scientific methods had not been invented at the time of Aristotle. Rather, he attempted to use the most basic observations combined with first principles (ideas that he believed were obviously true) to understand the world. The perfection of the heavens was, for Aristotle, a first principle.

Aristotle believed that the universe was divided into two parts—Earth, corrupt and changeable; and the heavens, perfect and immutable. Like most of his predecessors, he believed that Earth was the center of the universe, so his model is called a geocentric (Earth-centered) universe. The heavens surrounded Earth, and he added more crystalline spheres to bring the total to 55. The lowest sphere, that of the moon, marked the boundary between the imperfect region of Earth and the perfection of the celestial realm above the moon.

Because he believed Earth to be immobile, he had to make these spheres whirl westward around Earth each day and move with respect to one another to produce the motions of the sun, moon, and planets. Because his model was geocentric, he taught that Earth could be the only center of motion. All of his whirling

spheres had to be centered on Earth. Like most other Greek philosophers, Aristotle viewed the universe as a perfect heavenly machine that was not many times larger than Earth itself.

About a century after Aristotle, the Alexandrian philosopher Aristarchus proposed a theory that Earth rotated on its axis and revolved around the sun. This theory is, of course, correct, but most of the writings of Aristarchus were lost, and his theory was not well known. Later astronomers rejected any suggestion that Earth could move, because it conflicted with the teachings of Aristotle.

Aristotle had taught that Earth had to be a sphere because it always casts a round shadow during lunar eclipses, but he could only estimate its size. About 200 BC, Eratosthenes, working in the great library in Alexandria, found a way to calculate Earth's radius. He learned from travelers that the city of Syene (Aswan) in southern Egypt contained a well into which sunlight shone vertically on the day of the summer solstice. Thus, the sun was at the zenith at Syene, but on that same day in Alexandria, he noted that the sun was one-fiftieth of the circumference of the sky (about 7°) south of the zenith. Because sunlight comes from such a great distance, its rays arrive at Earth traveling almost parallel. Thus, Eratosthenes could conclude from simple geometry (❙ Figure 4-7) that the distance from Alexandria to Syene was one-fiftieth Earth's circumference.

To find Earth's circumference, Eratosthenes had to know the distance from Alexandria to Syene. Travelers told him it took 50 days to cover the distance, and he knew that a camel can travel about 100 stadia per day. Thus, the total distance was about 5000 stadia. If 5000 stadia is one-fiftieth Earth's circumference, then Earth must be 250,000 stadia around, and, dividing by 2π, Eratosthenes found Earth's radius to be 40,000 stadia.

We don't know how accurate Eratosthenes was. The stadium (singular of *stadia*) had different lengths in ancient times. If we assume 6 stadia to the kilometer, then Eratosthenes's result was too big by only 4 percent. If he used the Olympic stadium, his result was 14 percent too big. In any case, this was a much better measurement of Earth's radius than Aristotle's estimate, which was only about 40 percent of the true radius.

The greatest of the ancient observers was Hipparchus (see Figure 2-5), who lived during the 2nd century BC, about two centuries after Aristotle. He is usually credited with the invention of trigonometry, he compiled the first star catalog, and he discovered precession (Chapter 2). Instead of describing the motion of the sun and moon using nested spheres, as most Greek philosophers did, Hipparchus proposed that the sun and moon traveled around circles with Earth near, but not at, their centers. These off-center circles are now known as **eccentrics.** Hipparchus recognized that he could produce the same motion by having the sun, for instance, travel around a small circle that followed a larger circle around Earth. The compounded circular

FIGURE 4-6
Aristotle, honored on this Greek stamp, wrote on such a wide variety of subjects and with such deep insight that he became the great authority on all matters of learning. His opinions on the nature of Earth and the sky were widely accepted for almost two millennia.

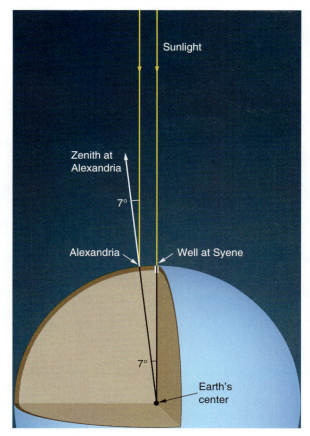

ACTIVE FIGURE 4-7
On the day of the summer solstice, sunlight fell to the bottom of a well at Syene, but the sun was about one-fiftieth of a circle (7°) south of the zenith at Alexandria. From this Eratosthenes was able to calculate Earth's radius.

Ace✺Astronomy™ Go to AceAstronomy and click Active Figures to see "Eratosthenes' Experiment."

motion that he devised became the key element in the masterpiece of the last great astronomer of classical times, Ptolemy.

THE PTOLEMAIC UNIVERSE

Claudius Ptolemaeus was one of the great astronomer-mathematicians of antiquity. His nationality and birth date are unknown, but he lived and worked in the Greek settlement at Alexandria about AD 140. He ensured the survival of Aristotle's universe by fitting to it a sophisticated mathematical model.

Study "▌ The Ancient Universe" on pages 58 and 59 and notice three important ideas. First, ancient philosophers and astronomers accepted as a first principle that Earth was located at the center of the universe. Although a few writers mentioned the possibility that Earth might move, they did so mostly to point out the obvious objections.

Second, notice that the ancients accepted without question that the heavens were perfect and that Earth was imperfect. This was a first principle of ancient as-

tronomy, and it led to the conclusion that the motions of the heavens must be made up of combinations of circles rotating uniformly.

The third thing to notice is how Ptolemy weakened the first principles of Plato and Aristotle by moving Earth slightly off-center and adjusting the speed of the planets as they circled Earth. Ptolemy lived roughly five centuries after Aristotle, and although Ptolemy believed in the Aristotelian universe, he was interested in a different problem. He was a brilliant mathematician, and he was mainly interested in creating a mathematical description of the motion of the planets. For him, first principles took second place to mathematical precision.

Aristotle's universe, as embodied in the mathematics of Ptolemy, dominated ancient astronomy, but it was wrong. The planets don't follow circles at uniform speeds. At first the Ptolemaic system predicted the positions of the planets well; but, as centuries passed, errors accumulated. A watch that gains only one second a year will keep time well for many years, but the error gradually accumulates. After a century the watch will be 100 seconds fast. So, too, did the errors in the Ptolemaic system gradually accumulate as the centuries passed. Islamic and later European astronomers tried to update the system, computing new constants and adjusting epicycles. In the middle of the 13th century, a team of astronomers supported by King Alfonso X of Castile studied the *Almagest* for 10 years. Although they did not revise the theory very much, they simplified the calculation of the positions of the planets using the Ptolemaic system and published the result as the *Alfonsine Tables,* the last great attempt to make the Ptolemaic system of practical use.

REVIEW CRITICAL INQUIRY

How did the astronomy of Hipparchus and Ptolemy violate the principles of the early Greek philosophers Plato and Aristotle?

Hipparchus and Ptolemy lived very late in the history of classical astronomy, and they concentrated more on the mathematical problems and less on philosophical principles. They replaced the perfect spheres of Plato with nested circles in the form of epicycles and deferents. Earth was moved slightly away from the center of the deferent, so their models of the universe were not exactly geocentric, and the epicycles moved uniformly only as seen from the equant. The celestial motions were no longer precisely uniform, and the principles of geocentrism and uniform circular motion were weakened.

The work of Hipparchus and Ptolemy led eventually to a new understanding of the heavens, but first astronomers had to abandon the principle of uniform circular motion. The great philosopher Plato had argued for uniform circular motion from what seemed the most basic of first principles. What were those first principles?

The Ancient Universe

For 2000 years, the minds of astronomers were shackled by a pair of ideas. The Greek philosopher Plato argued that the heavens were perfect. Because the only perfect geometrical shape is a circle and the only perfect motion is uniform motion, Plato concluded that all motion in the heavens must be made up of combinations of circles turning at uniform rates. This was called *uniform circular motion.*

Plato's student Aristotle argued that Earth was imperfect and lay at the center of the universe. That is, he argued for a *geocentric universe.* He devised a model universe with 55 spheres turning at different rates and at different angles to carry the seven known planets (the moon, Mercury, Venus, the sun, Mars, Jupiter, and Saturn) across the sky.

Aristotle was known as the greatest philosopher in the ancient world, and his authority, lasting for 2000 years, chained the minds of astronomers and forced them to expect the universe to be geocentric and to move in uniform circular motion.

From *Cosmographica* by Peter Apian (1539).

Seen by left eye

Seen by right eye

Ancient astronomers believed that Earth did not move because they saw no **parallax,** the apparent motion of an object because of the motion of the observer. To demonstrate parallax, close one eye and cover a distant object with your thumb held at arm's length. Switch eyes, and your thumb appears to shift position. If Earth moves, ancient astronomers reasoned, we should see the sky from different locations at different times of the year, and we should see parallax distorting the shapes of the constellations. They saw no parallax, so they concluded Earth could not move. Actually, the parallax of the stars is too small to see with the unaided eye.

Planetary motion was a big problem for ancient astronomers. In fact, the word *planet* comes from the Greek word for "wanderer," referring to the eastward motion of the planets against the background of the fixed stars. The planets did not, however, move at a constant rate, and they could occasionally stop and move westward for a few months before resuming their eastward motion. This backward motion is called **retrograde motion.**

East

Ecliptic

Oct. 3, 2005

Dec. 12, 2005

Taurus

West

Position of Mars at 5-day intervals

Orion

Every 2.14 years, Mars passes through a retrograde loop. Shown here is its path through Orion and Taurus during the fall of 2005.

Simple uniform circular motion centered on Earth could not explain retrograde motion, so ancient astronomers combined uniformly rotating circles much like gears in a machine to try to reproduce the motion of the planets.

Planet

Epicycle

Earth + Equant

Retrograde motion occurs here

Deferent

Uniformly rotating circles were key elements of ancient astronomy. Claudius Ptolemy created a mathematical model of the Aristotelian universe in which the planet followed a small circle called the **epicycle** which slid around a larger circle called the **deferent.** By adjusting the size and rate of rotation of the circles, he could approximate the retrograde motion of a planet.

To adjust the speed of the planet, Ptolemy supposed that Earth was slightly off center and that the center of the epicycle moved such that it appeared to move at a constant rate as seen from the point called the **equant.**

To further adjust his model, Ptolemy added small epicycles riding on top of larger epicycles (not shown here), producing a highly complex model.

Ace ☯ Astronomy™

Go to AceAstronomy and click Active Figures to see "Epicycles" and watch how the epicycle moved around the deferent.

Ptolemy's great book *Mathematical Syntaxis* (c. AD 140) contained the details of his model. Islamic astronomers preserved and studied the book through the Middle Ages, and they called it *Al Magisti* (The Greatest). When the book was found and translated from Arabic to Latin in the 12th century, it became known as *Almagest*.

Mars

Jupiter

Sphere of fixed stars

Sun

Saturn

Mercury

Venus

Earth

Moon

The Ptolemaic model of the universe was geocentric and based on uniform circular motion. Note that Mercury and Venus were treated differently from the rest of the planets. The centers of the epicycles of Mercury and Venus had to remain on the Earth–Sun line as the sun circled Earth through the year.

Equants and smaller epicycles are not shown here. Some versions contained nearly 100 epicycles as generations of astronomers tried to fine-tune the model to better reproduce the motion of the planets.

Notice that this modern illustration shows rings around Saturn and sunlight illuminating the globes of the planets, features that could not be known before the invention of the telescope.

Ptolemy was the last of the great classical astronomers, and his work dominated astronomical thought for almost 1500 years. The collapse of the Ptolemaic model and the rise of a new model of the universe make an exciting story, but it is not just the story of an astronomical idea. It includes the invention of science as a new way of knowing and understanding what we are and where we are.

4-2 THE COPERNICAN REVOLUTION

Nicolaus Copernicus (❙ Figure 4-8) triggered an Earth-shaking revision in human thought by proposing that the universe is not centered on Earth, but rather is centered on the sun. Such a model is called a **heliocentric universe.** That idea eventually brought Galileo before the Inquisition and changed forever how we think of our world and ourselves. This Copernican Revolution was much more than an upheaval in astronomy; it marks the birth of modern science.

COPERNICUS THE REVOLUTIONARY

When Copernicus proposed that the universe was heliocentric, he risked controversy. According to Aristotle, the most perfect region was the starry sphere, and the most imperfect was Earth's center. Thus, the classical geocentric universe matched the commonly held Christian geometry of heaven and hell, and anyone who criticized the geometry of the Aristotelian universe challenged Christian belief and thus risked a charge of heresy.

Throughout his life, Copernicus was associated with the church. His uncle, by whom he was raised and educated, was an important bishop in Poland; and, after studying canon law and medicine in some of the finest universities in Europe, Copernicus became a canon of the church at the unusually young age of 24. He was secretary and personal physician to his powerful uncle for 15 years. When his uncle died, Copernicus moved to quarters adjoining the cathedral in Frauenburg. No doubt his long association with the church added to his reluctance to publish controversial ideas.

He first wrote about a heliocentric universe sometime before 1514 in a short pamphlet that he distributed in handwritten form and in some cases anonymously. Over the years, Copernicus worked on his book *De Revolutionibus Orbium Coelestium* (❙ Figure 4-9a), but he hesitated to publish it even though other astronomers knew of his theories.

This was a time of rebellion—Martin Luther was speaking harshly about fundamental church teachings, and others, scholars and scoundrels, questioned the authority of the church. Even astronomy could stir argument; moving Earth from its central place was a controversial and perhaps heretical idea. Copernicus was concerned about how his ideas would be received, but in 1540 he allowed the visiting astronomer Joachim Rheticus (1514–1576) to publish an account of the Copernican universe in Rheticus's book *Prima Narratio* (*First Narrative*). In 1542, Copernicus sent the manuscript for *De Revolutionibus* off to be printed. He died in the spring of 1543 before the printing was completed.

The most important idea in the book was the placement of the sun at the center of the universe. That single innovation had an astonishing consequence—the retrograde motion of the planets was immediately explained in a straightforward way without the large epicycles that Ptolemy used. In the Copernican system, Earth moves faster along its orbit than the planets that lie further from the sun. Consequently, Earth periodically overtakes and passes these planets, and they appear to slow and fall behind (❙ Figure 4-10). Because the planetary orbits do not lie in precisely the same plane, a planet does not resume its eastward motion in precisely the same path it followed earlier. Consequently, it describes a loop whose shape depends on the angle between the orbital planes.

Copernicus could explain retrograde motion without epicycles, and that was impressive. But he could not do away with epicycles completely. Copernicus, a classical astronomer, had tremendous respect for the old concept of uniform circular motion. In fact, he objected strongly to Ptolemy's use of the equant. It seemed arbitrary to Copernicus, a direct violation of the elegance of Aristotle's philosophy of the heavens. Copernicus called equants "monstrous" in that they violated both geocentrism and uniform circular motion. In his model, Copernicus returned to a strong belief in uniform circular motion. Although he did not need epicycles to explain retrograde motion, Copernicus discovered that the sun, moon, and planets suffered small variations in their motions—variations that he could not explain

FIGURE 4-8
Copernicus proposed that the sun and not Earth was the center of the universe. Notice the heliocentric model on this stamp issued in 1973 to commemorate the 500th anniversary of his birth.

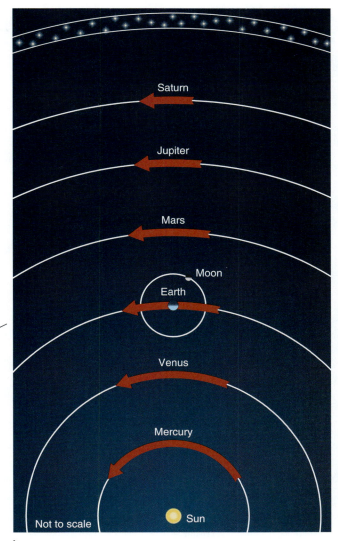

FIGURE 4-9

(a) The Copernican universe as reproduced in *De Revolutionibus.* Earth and all the known planets revolve in separate circular orbits about the sun (Sol) at the center. The outermost sphere carries the immobile stars of the celestial sphere. Notice the orbit of the moon around Earth. *(Yerkes Observatory)* (b) The model was elegant not only in its arrangement of the planets but in their motions. Orbital velocities (red arrows) decreased from Mercury, the fastest, to Saturn, the slowest. Compare the elegance of this model with the complexity of the Ptolemaic model on page 59.

with the concept of uniform circular motion centered on the sun. Today, we recognize those variations as typical of objects following elliptical orbits, but Copernicus held firmly to uniform circular motion, so he had to use small epicycles to produce these minor variations in the motions of the sun, moon, and planets.

Because Copernicus imposed uniform circular motion on his model, it could not accurately predict the motions of the planets. The *Prutenic Tables* (1551) were based on the Copernican model, and they were not significantly more accurate than the *Alfonsine Tables* (1251), which were based on Ptolemy's model. Both could be in error by as much as 2°, four times the angular diameter of the full moon.

The Copernican *model* was inaccurate, but the Copernican *hypothesis* that the universe is heliocentric was correct. There are probably a number of reasons why the hypothesis gradually won acceptance in spite of the inaccuracy of the epicycles and deferents. The most important factor may be the elegance of the idea. Placing the sun at the center of the universe produced a symmetry among the motions of the planets that was pleasing to the eye and to the intellect (Figure 4-9b). All of the planets moved in the same way at speeds that were simply related to their distance from the sun. In the Ptolemaic model, Venus and Mercury were treated differently from the other planets. In the Copernican model, all planets were treated the same. Thus, the model may have won support not for its accuracy, but for its elegance.

The most astonishing consequence of the Copernican hypothesis was not what it said about the sun, but

East   West

Apparent path of Mars
as seen from Earth

Mars

e   d
f   c
         b
g   Sun   a   Earth

FIGURE 4-10

The Copernican explanation of retrograde motion. As Earth overtakes
Mars (a–c), Mars appears to slow its eastward motion. As Earth
passes Mars (d), Mars appears to move westward. As Earth draws
ahead of Mars (e–g), Mars resumes its eastward motion against the
background stars. Compare with the illustration of retrograde motion
on page 58. The positions of Earth and Mars are shown at equal
intervals of one month.

what it said about Earth. By placing the sun at the center, Copernicus made Earth move along an orbit just as
the other planets did. By making Earth a planet, Copernicus revolutionized humanity's view of our place in
the universe and triggered a controversy that would
eventually bring Galileo before the Inquisition.

Although astronomers throughout Europe read and
admired *De Revolutionibus,* they did not usually accept the Copernican hypothesis. The mathematics was
elegant, and the astronomical observations and calculations were of tremendous value. Yet few astronomers
believed, at first, that the sun actually was the center of
the planetary system and that Earth moved. How the
Copernican hypothesis became gradually recognized
as correct has been named the Copernican Revolution
because it involved not just the adoption of a new idea,

but a total revolution in the way astronomers thought
about the place of the Earth (Window on Science 4-1).

GALILEO THE DEFENDER

Most people know two facts about Galileo, and both
facts are wrong. We should begin our discussion by getting those facts right: Galileo did not invent the telescope, and he was not condemned by the Inquisition
for believing that Earth moved around the sun. Then
why is Galileo so important that in 1979, almost 400
years after his trial, the Vatican reopened his case? As
we discuss Galileo, we will discover that his trial concerned not just the place of the Earth but a new way of
finding truth, a method known today as science.

Galileo Galilei (❚ Figure 4-11) was born in Pisa (in
what is now Italy) in 1564, and he studied medicine at
the university there. His true love, however, was mathematics, and although he had to leave school early because of financial difficulties, he returned only four
years later as a professor of mathematics. Three years
later, he became professor of mathematics at the university at Padua. He remained there for 18 years.

During this time, Galileo seems to have adopted
the Copernican model, although he admitted in a 1597
letter to Kepler, a German astronomer, that he did not
support Copernicanism publicly because of the criti-
cism such a declaration would bring. It was the telescope that drove Galileo to publicly defend the heliocentric model.

Galileo did not invent the telescope. It was apparently invented around 1608 by lens makers in Holland.
Galileo, hearing descriptions in the fall of 1609, was
able to build telescopes in his workshop. Galileo was
not the first person to look at the sky through a telescope, but he was the first person to observe the sky
systematically and apply his observations to the theoretical problem of the day—the place of the Earth.

What Galileo saw through his telescopes was so
amazing he rushed a small book into print. *Sidereus
Nuncius* (*The Sidereal Messenger*) reported three major
discoveries. First, the moon was not perfect. It had
mountains and valleys on its surface, and Galileo used
the shadows to calculate the height of the mountains.
The Ptolemaic model held that the moon was perfect,
but Galileo showed that it was not only imperfect, it
was a world like Earth.

The second discovery reported in the book was
that the Milky Way was made up of a myriad of stars
too faint to see with the unaided eye. While intriguing,
this discovery could not match the third discovery.
Galileo's telescope revealed four new "planets" circling
Jupiter, planets known today as the Galilean moons of
Jupiter.

The moons of Jupiter supported the Copernican
model over the Ptolemaic model. Critics of Copernicus

Creating New Ways to See Nature: Scientific Revolutions

The Copernican Revolution is often cited as the perfect example of a scientific revolution. Over a few decades, astronomers abandoned a way of thinking about the universe that was almost 2000 years old and adopted a new set of ideas and assumptions. The American philosopher of science Thomas Kuhn has referred to a commonly accepted set of scientific ideas and assumptions as a scientific **paradigm.** Thus, the pre-Copernican astronomers had a geocentric paradigm that included uniform circular motion and the perfection of the heavens. That paradigm survived for many centuries until a new generation of astronomers was able to overthrow the old paradigm

and establish a new paradigm that included heliocentrism and Earth's motion.

A scientific paradigm is powerful because it shapes our perceptions. It determines what we judge to be important questions and what we judge to be significant evidence. Thus, it is often difficult for us to recognize how our paradigms limit what we can understand. For example, the geocentric paradigm contained problems that seem obvious to us, but because astronomers before Copernicus lived and worked inside that paradigm, they had difficulty seeing the problems. Overthrowing an outdated paradigm is not easy, because we must learn to see nature in an en-

tirely new way. Galileo, Kepler, and Newton saw nature from a new paradigm that would have been almost incomprehensible to astronomers of earlier centuries.

We can find examples of scientific revolutions in many fields, including biology, geology, genetics, and psychology. These scientific revolutions have been difficult and controversial because they have involved the overthrow of accepted paradigms. But that is why scientific revolutions are exciting. They give us not just a new idea or a new theory, but an entirely new insight into how nature works—a new way of seeing the world.

---

had said Earth could not move because the moon would be left behind. But Jupiter moved yet kept its satellites, so Galileo's discovery proved that Earth, too, could move and keep its moon. Also, the Ptolemaic model included the Aristotelian belief that all heavenly motion was centered on Earth at the center of the universe.

Galileo showed that Jupiter's moons revolve around Jupiter, so there could be other centers of motion.

Sometime after *Sidereus Nuncius* was published, Galileo made another discovery that made Jupiter's moons even stronger evidence supporting the Copernican universe. Eventually, he was able to measure the

FIGURE 4-11
Galileo, remembered as the great defender of Copernicanism, also made important discoveries in the physics of motion. He is honored here on an Italian 2000-lira note. The reverse side shows one of his telescopes at lower right and a modern observatory above it.

FIGURE 4-12

(a) If Venus moved in an epicycle centered on the Earth–sun line (see page 59), it would always appear as a crescent. (b) Galileo's telescope showed that Venus goes through a full set of phases, proving that it must orbit the sun.

orbital periods of the four moons, and he found that the innermost moon had the shortest period and that the moons further from Jupiter had proportionally longer periods. In this way, Jupiter's moons made up a harmonious system ruled by Jupiter, just as the planets in the Copernican universe were a harmonious system ruled by the sun (Figure 4-9b). The similarity isn't proof, but Galileo saw it as an argument that the solar system was sun centered and not Earth centered.

Soon after *Sidereus Nuncius* was published, Galileo made two additional discoveries. When he observed the sun, he discovered sunspots, raising the suspicion that the sun was less than perfect. Further, by noting the movement of the spots, he concluded that the sun was a sphere and that it rotated on its axis. When he observed Venus, Galileo saw that it was going through phases like those of the moon. In the Ptolemaic model, Venus moves around an epicycle centered on a line between Earth and the sun. Thus, it would always be seen as a crescent (Figure 4-12a). But Galileo saw Venus go through a complete set of phases, proving that it did indeed revolve around the sun (Figure 4-12b).

Sidereus Nuncius was very popular and made Galileo famous. He became chief mathematician and philosopher to the Grand Duke of Tuscany in Florence. In 1611, Galileo visited Rome and was treated with great respect. He had long, friendly discussions with the powerful Cardinal Barberini, but he also made enemies. Personally, Galileo was outspoken, forceful, and sometimes tactless. He enjoyed debate, but most of all he enjoyed being right. Thus, in lectures, debates, and letters he offended important people who questioned his telescopic discoveries.

By 1616, Galileo was the center of a storm of controversy. Some critics said he was wrong, and others said he was lying. Some refused to look through a tele-

scope lest it mislead them, and others looked and claimed to see nothing (hardly surprising given the awkwardness of those first telescopes) (Figure 4-13). Pope Paul V decided to end the disruption, so when Galileo visited Rome in 1616 Cardinal Bellarmine interviewed him privately and ordered him to cease debate. There is some controversy today about the nature of Galileo's instructions, but he did not pursue astronomy for some years after the interview. Books relevant to Copernicanism, including *De Revolutionibus,* were banned.

In 1621 Pope Paul V died, and his successor, Pope Gregory XV, died in 1623. The next pope was Galileo's friend Cardinal Barberini, who took the name Urban VIII. Galileo rushed to Rome hoping to have the prohibition of 1616 lifted; and, although the new pope did not revoke the orders, he did encourage Galileo. Thus, Galileo began to write his great defense of the Copernican model, finally completing it on December 24, 1629. After some delay, the book was approved by both the local censor in Florence and the head censor of the Vatican in Rome. It was printed in February 1632.

Called *Dialogo Dei Due Massimi Sistemi (Dialogue Concerning the Two Chief World Systems),* it confronts the ancient astronomy of Aristotle and Ptolemy with the Copernican model and with telescopic observations as evidence. Galileo wrote the book as a debate among three friends. Salviati, a swift-tongued defender of Copernicus, dominates the book; Sagredo is intelligent but largely uninformed. Simplicio is the dismal defender of Ptolemy. In fact, he does not seem very bright.

The publication of *Dialogo* created a storm of controversy, and it was sold out by August 1632, when the Inquisition ordered sales stopped. The book was a clear defense of Copernicus, and, either intentionally or unintentionally, Galileo exposed the pope's authority to

FIGURE 4-13
Galileo's telescope made him famous, and he demonstrated his telescope and discussed his observations with powerful people. Some thought the telescope was the work of the devil and would deceive anyone who looked. In any case, Galileo's discoveries produced intense, and in some cases angry, debate. *(Yerkes Observatory)*

ridicule. Urban VIII was fond of arguing that, as God was omnipotent, He could construct the universe in any form while making it appear to us to have a different form, and thus we could not deduce its true nature by mere observation. Galileo placed the pope's argument in the mouth of Simplicio, and Galileo's enemies showed the passage to the pope as an example of Galileo's disrespect. The pope thereupon ordered Galileo to face the Inquisition.

Galileo was interrogated by the Inquisition four times and was threatened with torture. He must have thought often of Giordano Bruno, tried, condemned, and burned at the stake in Rome in 1600. One of Bruno's offenses had been Copernicanism. But the trial did not center on Galileo's belief in Copernicanism. *Dialogo* had been approved by two censors. Rather, the trial centered on the instructions given Galileo in 1616. From his file in the Vatican, his accusers produced a record of the meeting between Galileo and Cardinal Bellarmine that included the statement that Galileo was "not to hold, teach, or defend in any way" the principles of Copernicus. Some historians believe that this document, which was signed neither by Galileo nor by Bellarmine nor by a legal secretary, was a forgery. Others suspect it may be a draft that was never used. In any case, it is possible that Galileo's true instructions were much less restrictive. But Bellarmine was dead and could not testify at Galileo's trial.

The Inquisition condemned him not for heresy but for disobeying the orders given him in 1616. On June 22, 1633, at the age of 70, kneeling before the Inquisition, Galileo read a recantation admitting his errors. Tradition has it that as he rose he whispered *"E pur si muove"* ("Still it moves"), referring to Earth. Although he was sentenced to life imprisonment, he was actually confined at his villa for the next 10 years, perhaps through the intervention of the pope. He died there on January 8, 1642, 99 years after the death of Copernicus.

Galileo was not condemned for heresy, nor was the Inquisition interested when he tried to defend Copernicanism. He was tried and condemned on a charge we might call a technicality. Then why is his trial so important that historians have studied it for almost four centuries? Why have some of the world's greatest authors, including Bertolt Brecht, written about Galileo's trial? Why in 1979 did Pope John Paul II create a commission to reexamine the case against Galileo?

To understand the trial, we must recognize that it was the result of a conflict between two ways of understanding our universe. Since the Middle Ages, scholars had taught that the only path to true understanding was through religious faith. St. Augustine (AD 354–430) wrote *"Credo ut intelligame,"* which can be translated as "Believe in order to understand." But Galileo and other scientists of the Renaissance used their own observations to try to understand the universe; and, when their observations contradicted Scripture, they assumed their observations of reality were correct (█ Figure 4-14). Galileo paraphrased Cardinal Baronius in saying, "The Bible tells us how to go to heaven, not how the heavens go."

Various passages of Scripture seemed to contradict observation. For example, Joshua is said to have commanded the sun to stand still, not Earth to stop rotating (Joshua 10:12–13). In response to such passages, Galileo argued that we should "read the book

FIGURE 4-14

Although he did not invent it, Galileo will always be associated with the telescope because it was the source of the observational evidence he used to try to understand the universe. By depending on evidence instead of first principles, Galileo led the way to the invention of modern science as a way to know about the natural world.

of nature"—that is, we should observe the universe with our own eyes.

The trial of Galileo was not about the place of the Earth. It was not about Copernicanism. It wasn't really about the instructions Galileo received in 1616. It was about the birth of modern science as a rational way to understand our universe. The commission appointed by John Paul II in 1979, reporting its conclusions in October 1992, said of Galileo's inquisitors, "This subjective error of judgment, so clear to us today, led them to a disciplinary measure from which Galileo 'had much to suffer.'" Galileo was not found innocent in 1992 so much as the Inquisition was forgiven for having charged him in the first place.

The gradual change from reliance on personal faith to scientific observation came to a climax with the trial of Galileo. Since that time, scientists have increasingly reserved Scripture and faith for ethical guidance and personal comfort and have depended on systematic observation to describe the physical world. The final verdict of 1992 was an attempt to bring some balance to this conflict. In his remarks on the decision, Pope John

Paul II said, "A tragic mutual incomprehension has been interpreted as the reflection of a fundamental opposition between science and faith. . . . this sad misunderstanding now belongs in the past." Galileo's trial is over, but it continues to echo through history as part of our struggle to understand the place of the Earth in our universe.

REVIEW CRITICAL INQUIRY

How were Galileo's observations of the moons of Jupiter evidence against the Ptolemaic model?

Galileo presented his arguments in the form of evidence and conclusions, and the moons of Jupiter were key evidence. Ptolemaic astronomers argued that Earth could not move or it would lose its moon, but even in the Ptolemaic universe Jupiter moved and kept its moons. Evidently, Earth could move and not leave its moon behind. Furthermore, moons circling Jupiter did not fit the classical belief that all motion was centered on Earth. Obviously there could be other centers of motion. Finally, the orbital periods of the moons were related to their distance from Jupiter, just as the orbital periods of the planets were, in the Copernican system, related to their distance from the sun. This similarity suggested that the sun rules its harmonious family of planets just as Jupiter rules its harmonious family of moons.

Of all of Galileo's telescopic observations, the moons of Jupiter caused the most debate. But the craters on the moon and the phases of Venus were also critical evidence. How did they argue against the Ptolemaic model?

Galileo faced the Inquisition because of a conflict between two ways of knowing and understanding the world. The Church taught faith and understanding through revelation, but the scientists of the age were inventing a new way to understand nature that relied on evidence. In astronomy, evidence means observations, so our story turns now from the philosophical problem of the place of the Earth to the observational problem of the motion of the planets.

4-3 THE PUZZLE OF PLANETARY MOTION

While Galileo was demonstrating his telescope in Florence and Rome, two other astronomers were working in northern Europe, beyond the sway of the Inquisition. Although they, too, struggled to understand the place of the Earth, they approached the problem differently. They used the most accurate observations of the positions of the planets to try to discover the rules that govern planetary motion. When that puzzle was finally solved, astronomers understood why the Ptolemaic model of the universe does not work well and how to

FIGURE 4-15
Tycho Brahe (1546–1601), a Danish nobleman, established an observatory at Hveen and measured planetary positions with high accuracy. His artificial nose is suggested in this engraving.

make the Copernican universe a precise predictor of planetary motion.

TYCHO THE OBSERVER

The great observational astronomer of our story is a Danish nobleman, Tycho Brahe, born December 14, 1546, only three years after the publication of *De Revolutionibus* (Figure 4-15). As a son of a noble family, he was well known for his vanity and lordly manners.

Tycho's college days were eventful. He was officially studying law with the expectation that he would enter Danish politics, but he made it clear to his family that his real interest was astronomy and mathematics. It was also during his college days that he fought a duel and received a wound that disfigured his nose. For the rest of his life, he wore false noses made of gold and silver and stuck on with wax. The disfigurement probably did little to improve his disposition.

Tycho's first astronomical observations were made while he was a student. In 1563, Jupiter and Saturn passed very near each other in the sky, nearly merging into a single point on the night of August 24. Tycho found that the *Alfonsine Tables* were a full month in error and that the *Prutenic Tables* were in error by a number of days. These discrepancies dismayed Tycho and sparked his interest in the motions of the planets.

In 1572, a brilliant "new star" (now called Tycho's supernova) appeared in the sky. Such changes in the stars puzzled classically trained astronomers. Aristotle had argued that the starry sphere was perfect and unchanging, and therefore such new stars had to lie closer to Earth than the moon. Layers below the moon were thought to be less perfect and thus more changeable.

Tycho carefully measured the position of the star time after time through the night and found that it displayed no parallax. To understand the significance of this observation, we must note that Tycho believed the heavens rotated westward around Earth. The new star, according to classical astronomy, represented a change

in the heavens and therefore had to lie below the sphere of the moon. In that case, Tycho reasoned, the new star would appear slightly too far east as it rose and slightly too far west as it set (Figure 4-16). That is parallax. Tycho saw no parallax in the position of the new star, so he concluded that it must lie above the sphere of the moon and was probably on the starry sphere itself. This contradicted Aristotle's belief that the starry sphere was perfect and unchanging.

No one before Tycho could have made this discovery because no one had ever measured the positions of stars so accurately. Tycho had great confidence in the precision of his measurements, so when he failed to detect parallax for the new star, he knew it was important evidence against the Ptolemaic theory. He announced his discovery in a small book, *De Stella Nova* (*The New Star*), published in 1573.

The book attracted the attention of astronomers throughout Europe, and soon Tycho was summoned to the court of the Danish King Frederik II and offered funds to build an observatory on the island of Hveen just off the Danish coast. Tycho also received a steady source of income as landlord of a coastal district from which he collected rents. (He was not a popular landlord.) On Hveen, Tycho constructed a luxurious home with six towers specially equipped for astronomy and populated it with servants, assistants, and a dwarf to act as jester. Soon Hveen was an international center of astronomical study.

TYCHO BRAHE'S LEGACY

Tycho Brahe made no direct contribution to astronomical theory. Because he could measure no parallax for the stars, he concluded that Earth had to be stationary, thus rejecting the Copernican hypothesis. However, he

FIGURE 4-16
If an object lay lower than the celestial sphere, reasoned Tycho Brahe, then it should be seen east of its average position as it was rising and west of its average position when it was setting. Because he did not detect this daily parallax, he concluded the new star of 1572 had to lie on the celestial sphere.

a

b

FIGURE 4-17

(a) Tycho Brahe's model of the universe held that Earth was fixed at the center of the starry sphere. The moon and sun circled Earth while the planets circled the sun. (b) Much of Tycho's success was due to his skill in designing large, accurate instruments. In this engraving of his mural quadrant, the figure of Tycho, his dog, and the background scene are painted on the wall within the arc of the quadrant. The observer (Tycho himself) at the extreme right peers through a sight out the loophole in the wall at the upper left to measure an object's angular distance above the horizon. *(Granger Collection, New York)*

also rejected the Ptolemaic model because of its inaccurate predictions. Instead, he devised a complex model in which Earth was the immobile center of the universe around which the sun and moon moved. The other planets circled the sun. This model thus incorporated part of the Copernican model, but Earth—rather than the sun—was stationary. In this way, Tycho preserved the central, immobile Earth described in scripture (▌ Figure 4-17a). Although Tycho's model, the Tychonic Universe, was very popular at first, the Copernican model replaced it within a century.

The true value of Tycho's work was observational. Because he was able to devise new and better instruments, he was able to make highly accurate observations of the positions of the stars, sun, moon, and planets. Tycho had no telescopes—they were not invented until the next century—so his observations were made by the naked eye peering along sights. All of his instruments were designed to measure angles in the sky. For example, his quadrant could measure angles up to 90° above the horizon (Figure 4-17b). By designing and building large instruments with great care, he was able to measure angles to high precision. He measured the positions of 777 stars to better than 4 minutes of arc and regularly measured the positions of the sun, moon, and planets for the 20 years he stayed on Hveen.

Unhappily for Tycho, King Frederik II died in 1588, and his young son took the throne. Suddenly Tycho's temper, vanity, and noble presumptions threw him out of favor. In 1596, taking most of his instruments and books of observations, he went to Prague, the capital of Bohemia, and became imperial mathematician to the Holy Roman Emperor Rudolph II. His goal was to revise the *Alfonsine Tables* and publish the revision as a monument to his new patron. It would be called the *Rudolphine Tables.*

Tycho did not intend to base the *Rudolphine Tables* on the Ptolemaic system but rather on his own Tychonic system, proving once and for all the validity of his hypothesis. To assist him, he hired a few mathematicians and astronomers, including one Johannes Kepler. Then in November 1601, Tycho collapsed at a nobleman's home. Before he died, 11 days later, he asked Rudolph II to make Kepler imperial mathematician. Thus the newcomer, Kepler, became Tycho's replacement (though at one-sixth Tycho's salary).

KEPLER THE ANALYST

No one could have been more different from Tycho Brahe than Johannes Kepler (▌ Figure 4-18). He was born December 27, 1571, to a poor family in a region

FIGURE 4-18

Johannes Kepler (1571–1630) derived three laws of planetary motion from Tycho Brahe's observations of the positions of the planets. This Romanian stamp commemorates the 400th anniversary of Kepler's birth. Ironically, it contains an error—the horns of the crescent moon should be inclined upward from the horizon.

FIGURE 4-19

In this diagram, based on one drawn by Kepler, the sizes of the celestial spheres carrying the outer three planets, Saturn, Jupiter, and Mars, are determined by spacers (blue) consisting of a cube and a tetrahedron. Inside the sphere of Mars, the remaining regular solids (not shown here) separate the spheres of the Earth, Venus, and Mercury. The sun lay at the very center of this Copernican universe based on geometrical spacers.

now included in southwestern Germany. His father was unreliable and shiftless, principally employed as a mercenary soldier fighting for whoever paid enough. He finally failed to return from a military expedition, either because he was killed or because he found circumstances more to his liking. Kepler's mother was apparently an unpleasant and unpopular woman. She was accused of witchcraft in her later years, and Kepler had to defend her in a trial that dragged on for three years. She was finally acquitted but died the following year.

Kepler was the oldest of six children, and his childhood was no doubt unhappy. The family was not only poor, but suffered from an absentee father whom Kepler described as "vicious, inflexible, quarrelsome, and doomed to a bad end." In addition, Kepler was never healthy, even as a child, so it is surprising that he did well in school, eventually winning promotion to a Latin school and finally a scholarship to the university at Tübingen, where he studied to become a Lutheran pastor.

During his last year of study, Kepler accepted a job in Graz teaching mathematics and astronomy. Evidently, he was not a good teacher. He had few students his first year and none at all his second. His superiors put him to work teaching a few introductory courses and preparing an annual almanac that contained astronomical, astrological, and weather predictions. Through good luck, in 1595 some of his weather predictions were fulfilled, and he gained a reputation as an astrologer and seer; even in later life he earned money from his almanacs.

While still a college student, Kepler had become a believer in the Copernican theory, and at Graz he used his extensive spare time to study astronomy. By 1596, the same year Tycho left Hveen, Kepler was ready to solve the mystery of the universe. That year he pub-

lished a book called *The Forerunner of Dissertations on the Universe, Containing the Mystery of the Universe.* Like nearly all scientific works, the book was in Latin, and it is now known as *Mysterium Cosmographicum.*

The book begins with a long appreciation of Copernicanism and then goes on to speculate on the spacing of the planetary orbits. Kepler, as a Copernican, knew of six planets circling the sun—Mercury, Venus, Earth, Mars, Jupiter, and Saturn. According to his model, the spheres containing the six planets were separated from one another, and their relative sizes were fixed by the five regular solids*—the cube, tetrahedron, dodecahedron, icosahedron, and octahedron (❙ Figure 4-19). Kepler gave astrological, numerological, and musical arguments for his theory and so followed the tradition of Pythagoras in believing that the order of the universe is underlain by musical (meaning mathematical) principles.

In the second half of the book, Kepler tried to fit the five solids to the planetary orbits and so demonstrated that he was a talented mathematician well versed in astronomy. He sent copies to Tycho and to Galileo, and both recognized his talents.

Life was unsettled for Kepler because of the persecution of Protestants in the region, so when Tycho invited him to Prague in 1600, he went readily, anxious to work with the famous astronomer. Tycho's sudden death in 1601 left Kepler in a position to use the obser-

---

*A regular solid is a three-dimensional body each of whose faces is the same. For example, a cube is a regular solid each of whose faces is a square.

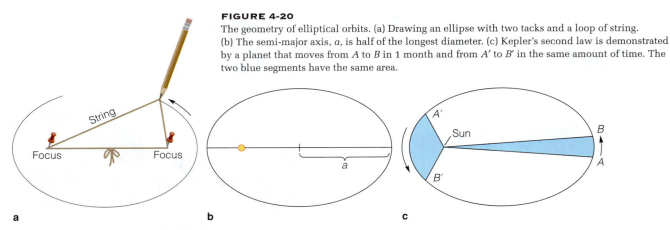

FIGURE 4-20

The geometry of elliptical orbits. (a) Drawing an ellipse with two tacks and a loop of string. (b) The semi-major axis, *a,* is half of the longest diameter. (c) Kepler's second law is demonstrated by a planet that moves from *A* to *B* in 1 month and from *A'* to *B'* in the same amount of time. The two blue segments have the same area.

vations from Hveen to analyze the motions of the planets and complete the *Rudolphine Tables.* Tycho's family, recognizing that Kepler was a Copernican and guessing that he would not follow the Tychonic system in completing the *Rudolphine Tables,* sued to recover the instruments and books of observations. The legal wrangle went on for years. The family did recover the instruments Tycho had brought to Prague, but Kepler had the books, and he kept them.

Whether Kepler had any legal right to Tycho's records is debatable, but he put them to good use. He began by studying the motion of Mars, trying to deduce from the observations how the planet moves. By 1606, he had solved the puzzle of planetary motion. The orbit of Mars is an ellipse and not a circle. Thus he abandoned the 2000-year-old belief in the circular motion of the planets. But the mystery was even more complex. The planets do not move at a uniform speed along their elliptical orbits. Kepler recognized that they move faster when close to the sun and slower when farther away. Kepler therefore abandoned both circular motion and uniform motion.

Kepler published his results in 1609 in a book called *Astronomia Nova* (*New Astronomy*). Like Copernicus's book, *Astronomia Nova* did not become an instant success. It is written in Latin for other scientists and is highly mathematical. In some ways, the book is surprisingly advanced. For instance, Kepler discusses the force that holds the planets in their orbits, a question that Isaac Newton considered later in his recognition of gravity as a natural force.

Despite the abdication of Rudolph II in 1611, Kepler continued his astronomical studies. He wrote about a supernova that had appeared in 1604 (now known as Kepler's supernova) and about comets, and he wrote a textbook about Copernican astronomy. In 1619, he published *Harmonice Mundi* (*The Harmony of the World*), in which he returned to the cosmic mysteries of *Mysterium Cosmographicum.* The only thing of note in *Harmonice Mundi* is his discovery that the radii of the planetary orbits are related to the planets' orbital periods. That and his two previous discoveries

are now recognized as Kepler's three laws of planetary motion.

KEPLER'S THREE LAWS OF PLANETARY MOTION

Although Kepler dabbled in the philosophical arguments of the day, he was a mathematician, and his triumph was the solution of the problem of the motion of the planets. The key to his solution was the ellipse.

An **ellipse** is a figure drawn around two points called the *foci* in such a way that the distance from one focus to any point on the ellipse and back to the other focus equals a constant. This makes it easy to draw ellipses with two thumbtacks and a loop of string. Press the thumbtacks into a board, loop the string about the tacks, and place a pencil in the loop. If you keep the string taut as you move the pencil, it traces out an ellipse (■ Figure 4-20a).

The geometry of an ellipse is described by two simple numbers. The **semimajor axis, *a,*** is half of the longest diameter. The **eccentricity** of an ellipse, *e,* is the distance from either focus to the center of the ellipse divided by the semimajor axis. To draw a circle with the string and tacks as shown in Figure 4-20a, we would move the two thumbtacks together, which shows that a circle is really just an ellipse with eccentricity equal to zero. As we move the thumbtacks farther apart, the ellipse becomes flatter, and the eccentricity moves closer to 1.

Kepler used ellipses to describe the motion of the planets in three fundamental rules that have been tested and confirmed so many times that astronomers now refer to them as natural laws (Window on Science 4-2). They are commonly called Kepler's laws of planetary motion (■ Table 4-1).

Kepler's first law states that the orbits of the planets around the sun are ellipses with the sun at one focus. Thanks to the precision of Tycho's observations and the sophistication of Kepler's mathematics, Kepler was able to recognize the elliptical shape of the orbits, even though they are nearly circular. Of the planets known to

Hypothesis, Theory, and Law: Levels of Confidence

Even scientists misuse the words *hypothesis, theory,* and *law.* We must try to distinguish these terms from one another because they are key elements in science.

A **hypothesis** is a single assertion or conjecture that must be tested. It could be true or false. "All Texans love chili" is a hypothesis. To know whether it is true or false, we need to test it against reality by making observations or performing experiments. Copernicus asserted that the universe was heliocentric; his assertion was a hypothesis subject to testing.

In Chapter 2, we saw that a model (Window on Science 2-1) is a description of some natural phenomenon; it can't be right or wrong. A model is not a conjecture of truth but merely a convenient way to think about a natural phenome-non. Consequently, a model such as the celestial sphere cannot be a hypothesis. Copernicus used his hypothesis to build a model, but they are not the same thing.

A **theory** is a system of rules and principles that can be applied to a wide variety of circumstances. A theory may have begun as one or more hypotheses, but it has been tested, expanded, and generalized. Many textbooks refer to the "Copernican theory," but some historians argue that it was not complete and had not been tested enough to be a theory. It is probably better to call it the Copernican hypothesis.

A **natural law** is a theory that has been refined, tested, and confirmed so often scientists have great confidence in it. Natural laws are the most fundamen-tal principles of scientific knowledge. Kepler's laws are good examples.

Confidence is the key to understanding these terms. Scientists have more confidence in a theory than in a hypothesis and great confidence in a natural law. Nevertheless, scientists are not always consistent about these words. For example, Einstein's theory of relativity is much more accurate than Newton's laws of gravity and motion, but tradition dictates that we do not refer to Einstein's laws and Newton's theories. Darwin's theory of evolution has been tested many times, and scientists have great confidence in it, but no one refers to Darwin's law. These distinctions are subtle and sometimes depend more on custom than on levels of confidence.

Kepler, Mercury has the most elliptical orbit, but even it deviates only slightly from a circle (❚ Figure 4-21).

Kepler's second law states that a line from the planet to the sun sweeps over equal areas in equal intervals of time. This means that when the planet is closer to the sun and the line connecting it to the sun is shorter, the planet moves more rapidly to sweep over the same area that is swept over when the planet is farther from the sun. Thus the planet in Figure 4-20c would move from point A' to point B' in one month, sweeping over the area shown. But when the planet is farther from the sun, one month's motion would be shorter, from A to B.

The time that a planet takes to travel around the sun once is its orbital period, P, and its average distance from the sun equals the semimajor axis of its orbit, a. Kepler's third law tells us that these two quantities are related; the orbital period squared is proportional to the semimajor axis cubed. If we measure P in years and a in astronomical units, we can summarize the third law as:

$$P_y^2 = a_{AU}^3$$

The subscripts remind us that we must measure the period in years and the semimajor axis in astronomical units (AU).

We can use Kepler's third law to make simple calculations. For example, Jupiter's average distance from the sun is 5.20 AU. What is its orbital period? If a equals 5.20, then a^3 equals 140.6. The orbital period

---

TABLE 4-1
Kepler's Laws of Planetary Motion

I. The orbits of the planets are ellipses with the sun at one focus.

II. A line from a planet to the sun sweeps over equal areas in equal intervals of time.

III. A planet's orbital period squared is proportional to its average distance from the sun cubed:

$$P_y^2 = a_{AU}^3$$

FIGURE 4-21

The orbits of the planets are nearly circular. In this scale drawing of the orbit of Mercury, the horizontal diameter and the vertical diameter of the orbit differ by less than 3 percent.

must be the square root of 140.6, which equals about 11.8 years.

It is important to notice that Kepler's three laws are empirical. That is, they describe a phenomenon without explaining why it occurs. Kepler derived them from Tycho's extensive observations, not from any fundamental assumption or theory. In fact, Kepler never knew what held the planets in their orbits or why they continued to move around the sun. His books are a fascinating blend of careful observation, mathematical analysis, and mystical theory.

THE *RUDOLPHINE TABLES*

In spite of Kepler's recurrent involvement with astrology and numerology, he continued to work on the *Rudolphine Tables*. At last, in 1627, they were ready, and he financed their printing himself, dedicating them to the memory of Tycho Brahe. In fact, Tycho's name appears in larger type on the title page than Kepler's own. This is especially surprising when we recall that the tables were based on the heliocentric model of Copernicus and the elliptical orbits of Kepler and not on the Tychonic system. The reason for Kepler's evident deference was Tycho's family, still powerful and still intent on protecting the memory of Tycho. They even demanded a share of the profits and the right to censor the book before publication, though they changed nothing but a few words on the title page and added an elaborate dedication to the emperor.

The *Rudolphine Tables* were Kepler's masterpiece. They could predict the positions of the planets 10 to 100 times more accurately than previous tables. Kepler's tables were the precise model of planetary motion that Copernicus had sought but failed to find. The accuracy of the *Rudolphine Tables* was strong evidence that both Kepler's model for planetary motion and the Copernican hypothesis for the place of Earth were correct. Copernicus would have been pleased.

Kepler solved the problem of planetary motion, and his *Rudolphine Tables* demonstrated his solution. Although he did not understand why the planets moved or why they followed ellipses, insights that had to wait half a century for Isaac Newton, Kepler's three rules worked. In science the only test of a theory is, "Does it describe reality?" Kepler's laws have been used for almost four centuries as a true description of orbital motion.

REVIEW CRITICAL INQUIRY

What were the main differences among the *Alfonsine Tables*, the *Prutenic Tables*, and the *Rudolphine Tables*?

All three of these tables predicted the motions of the heavenly bodies, but only the *Rudolphine Tables* proved accurate. The *Alfonsine Tables*, produced in Toledo around AD 1250, were based on the Ptolemaic model, so they were geocentric and used uniform circular motion; consequently, they were not very accurate. The *Prutenic Tables*, published in 1551, were based on the Copernican model and so were heliocentric. But because the *Prutenic Tables* included the classical principle of uniform circular motion, they were no more accurate than the *Alfonsine Tables*. Kepler's *Rudolphine Tables* were Copernican in that they were heliocentric, but they included Kepler's three laws of planetary motion and thus could predict the positions of the planets 10 to 100 times more accurately than previous tables.

One of the main reasons for the success of the Copernican hypothesis was not that it was accurate but that it was elegant. Compare the motions of the planets and the explanation of retrograde motion in the Copernican model with those in the Ptolemaic and Tychonic models. How was the Copernican model more elegant?

Kepler died November 15, 1630. During his life he had been an astrologer, a mystic, a numerologist, and a seer, but he became one of the world's great astronomers. His work not only overthrew the concept of uniform circular motion that had shackled the minds of astronomers for over 2000 years but also opened the way to the empirical study of planetary motion, a study that lies at the heart of modern astronomy.

4-4 MODERN ASTRONOMY

We date the origin of modern astronomy from the 99 years between the deaths of Copernicus and Galileo (1543 to 1642) because it was an age of transition. That period marked the transition between the Ptolemaic model of the universe and the Copernican with the attendant controversy over the place of the Earth. But that same period also marked a transition in the nature of astronomy in particular and science in general, a transition illustrated in the resolution of the puzzle of planetary motion. The puzzle was solved by precise observation and careful computation, techniques that are the foundation of modern science.

The discoveries of Kepler and Galileo found acceptance in the 1600s because the world was in transition. Astronomy was not the only thing changing during this period. The Renaissance is commonly taken to be the period between 1300 and 1600, and thus these 99 years of astronomical history lie at the culmination of the reawakening of learning in all fields (▌Figure 4-22). The world was open to new ideas and new observations. Martin Luther remade religion, and other philosophers and scholars re-formed their areas of human knowledge. Had Copernicus not published his hypothesis, someone else would have suggested that the uni-

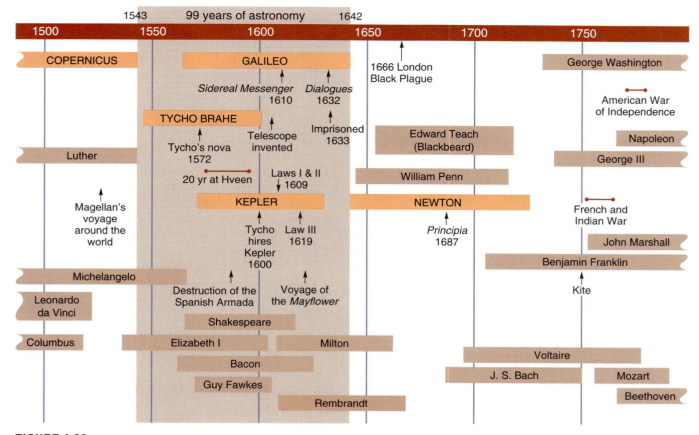

FIGURE 4-22

The 99 years between the death of Copernicus in 1543 and the birth of Newton in 1642 marked the transition from the ancient astronomy of Ptolemy and Aristotle to the revolutionary theory of Copernicus. This period saw the birth of modern scientific astronomy.

verse is heliocentric. History was ready to shed the Ptolemaic system.

In addition, this period marks the beginning of the modern scientific method. Beginning with Copernicus, Tycho, Kepler, and Galileo, scientists depended more and more on evidence, observation, and measurement. This, too, is coupled to the Renaissance and its advances in metalworking and lens making. Before our story began, no astronomer had looked through a telescope, because one could not be made. By 1642, not only telescopes but also other sensitive measuring instruments had transformed science into something new and precise. Also, the growing number of scientific societies increased the exchange of observations and hypotheses among scientists and stimulated more and better work. However, the most important advance was the application of mathematics to scientific questions. Kepler's work demonstrated the power of mathematical analysis, and as the quality of these numerical tools improved, the progress of science accelerated. Our story, therefore, is the story of the birth not just of modern astronomy but of modern science as well.

We will continue our story in the next chapter with the life of Isaac Newton. We will see how his accomplishments had their origins in the work of Galileo and Kepler.

SUMMARY

Archaeoastronomy, the study of the astronomy of ancient cultures, is revealing that primitive astronomers could make sophisticated astronomical observations. The Stone Age people of Britain who built Stonehenge aligned its long avenue and Heelstone to point toward the rising sun at the summer solstice. Other stones mark sight lines toward the rising moon. It seems clear that Stonehenge was a calendrical device, and some believe that its builders used it to predict eclipses.

Other archaeoastronomy sites span the world. A great many stone circles dot England and northern Europe, and related patterns, called medicine wheels, show that North American Indians also practiced astronomy. The Aztecs, Maya, and Incas of Central and South America also studied astronomy.

Traditional histories of astronomy usually begin with the Greeks. The Greek philosopher Aristotle held that Earth is fixed at the center of the universe, and Ptolemy based a mathematical model of the moving planets on this geocentric universe. To preserve the concept of uniform circular motion, he used epicycles, deferents, and equants for each of the planets, the sun, and the moon.

Near the end of his life, in 1543, Nicolaus Copernicus published his hypothesis that the sun is the center of the universe. Because the teachings of Aristotle had been adopted as official church doctrine, the heliocentric universe of Copernicus was considered radical.

Galileo Galilei was the first astronomer to use the newly invented telescope for a systematic study of the heavens, and he found that the appearance of the sun, moon, and planets supported the heliocentric model. After he described his findings in a book, the church, in 1616, ordered him to stop teaching Copernicanism. After he published another book on the subject in 1632, he was tried by the Inquisition and ordered to recant. He was imprisoned in his home until his death in 1642.

In northern Europe, beyond the power of the Church, Tycho Brahe rejected the Ptolemaic and Copernican systems. In Tycho's own system, Earth is fixed at the center of the universe, and the sun and moon revolve around Earth. The planets orbit the moving sun. Although his hypothesis is wrong, he was a brilliant observer and accumulated 20 years of precise measurements of the positions of the sun, moon, and planets.

At Tycho's death, one of his assistants, Johannes Kepler, became his successor. Analyzing Tycho's observations, Kepler finally discovered how the planets move. His three laws of planetary motion overthrew uniform circular motion. The first law says the orbits of the planets are elliptical rather than circular, and the second law says the planets do not move at a uniform rate but move faster when near the sun and slower when more distant. The third law relates orbital period to orbital size. These studies of the place of the Earth and the nature of planetary motion led to the birth of modern astronomy.

NEW TERMS

archaeoastronomy

eccentric

uniform circular motion

geocentric universe

parallax

retrograde motion

epicycle

deferent

equant

heliocentric universe

paradigm

ellipse

semimajor axis, *a*

eccentricity, *e*

hypothesis

theory

natural law

REVIEW QUESTIONS

Ace❖Astronomy™ Assess your understanding of this chapter's topics with additional quizzing and animations at **http:// astronomy.brookscole.com/seeds8e**

1. What evidence do we have that early human cultures observed astronomical phenomena?

2. Why did Plato propose that all heavenly motion was uniform and circular?

3. How do the epicycles of Mercury and Venus differ from those of Mars, Jupiter, and Saturn?

4. Why did Copernicus have to keep small epicycles in his system?

5. Explain how each of Galileo's telescopic discoveries contradicted the Ptolemaic theory.

6. Galileo was condemned, but Kepler, also a Copernican, was not. Why not?

7. When Tycho observed the new star of 1572, he could detect no parallax. Why did that result undermine belief in the Ptolemaic system?

8. Does Tycho's model of the universe explain the phases of Venus that Galileo observed? Why or why not?

9. How do the first two of Kepler's three laws overthrow one of the basic beliefs of classical astronomy?

10. What is the difference between a hypothesis, a theory, and a law?

11. How did the *Alfonsine Tables,* the *Prutenic Tables,* and the *Rudolphine Tables* differ?

DISCUSSION QUESTIONS

1. Historian of science Thomas Kuhn has said that *De Revolutionibus* was a revolution-making book but not a revolutionary book. How was it an old-fashioned, classical book?

2. Why might Tycho Brahe have hesitated to hire Kepler? Why do you suppose he appointed Kepler his scientific heir?

3. How does the modern controversy over creationism and evolution reflect two ways of knowing about the physical world?

PROBLEMS

1. Draw and label a diagram of the eastern horizon from northeast to southeast and label the rising point of the sun at the solstices and equinoxes. (See page 23 and Figure 4-1).

2. If you lived on Mars, which planets would describe retrograde motion? Which would never be visible as crescent phases?

3. Galileo's telescope showed him that Venus has a large angular diameter (61 seconds of arc) when it is a crescent and a small angular diameter (10 seconds of arc) when it is nearly full. Use the small-angle formula to find the ratio of its maximum distance to its minimum distance. Is this ratio compatible with the Ptolemaic universe shown in on page 59?

4. Galileo's telescopes were not of high quality by modern standards. He was able to see the moons of Jupiter, but he never reported seeing features on Mars. Use the small-

angle formula to find the angular diameter of Mars when it is closest to Earth. How does that compare with the maximum angular diameter of Jupiter?

5. If a planet has an average distance from the sun of 4 AU, what is its orbital period?

6. If a space probe is sent into an orbit around the sun that brings it as close as 0.5 AU and as far away as 5.5 AU, what will be its orbital period?

7. Pluto orbits the sun with a period of 247.7 years. What is its average distance from the sun?

CRITICAL INQUIRIES FOR THE WEB

1. The trial of Galileo is an important event in the history of science. We now know, and the Roman Catholic Church now recognizes, that Galileo's view was correct, but what were the arguments on both sides of the issue as it was unfolding? Research the Internet for documents that chronicle the trial, Galileo's observations and publications, and the position of the Church. Use this information to outline the case for and against Galileo in the context of the times in which the trial occurred.

2. Take a "virtual tour" of Stonehenge by browsing the Web for information related to this megalithic site and the possibility that it was used for astronomical purposes.

(Be careful to use legitimate scientific and historical Web sites in your survey.) Summarize the various astronomical alignments evident in the layout of the site.

EXPLORING *THESKY*

1. Go to Stonehenge in southern England and watch the sun rise on the morning of the summer solstice. Where does it rise on the morning of the winter solstice?

2. Observe Mars going through its retrograde motion. (*Hint:* Use **Reference Lines** under the **View** menu to turn on the ecliptic. Be sure you are in **Free Rotation** under the **Orientation** menu. Locate Mars and use the time skip arrows to watch it move.)

3. Compare the size of the retrograde loops made by Mars, Jupiter, and Saturn.

4. Can you recognize the effects of Kepler's second law in the orbital motion of any of the planets? (*Hint:* Use **3D Solar System Mode** under the **View** menu.)

5. Can you recognize the effects of Kepler's third law in the orbital motion of the planets?

 Visit the Seeds *Foundations of Astronomy* companion Web site for critical thinking exercises, articles, and additional readings from InfoTrac College Edition, Brooks/Cole's online student library.

NEWTON, EINSTEIN, AND GRAVITY

Nature and Nature's laws lay hid in night:
God said, "Let Newton be!" and all was light.

Alexander Pope

Chapter 5

GUIDEPOST

Astronomers are gravity experts. All of the heavenly motions described in the preceding chapters are dominated by gravitation. Isaac Newton gets the credit for discovering gravity, but even Newton couldn't explain what gravity was. Einstein proposed that gravity is a curvature of space, but that only pushes the mystery further away. "What is curvature?" we might ask.

This chapter shows how scientists build theories to explain and unify observations. Theories can give us entirely new ways to understand nature, but no theory is an end in itself. Astronomers continue to study Einstein's theory, and they wonder if there is an even better way to understand the motions of the heavens.

The principles we discuss in this chapter will be companions through the remaining chapters. Gravity is universal.

Isaac Newton was born in Woolsthorpe, England, on December 25, 1642, and on January 4, 1643. This was not a biological anomaly but a calendrical quirk. Most of Europe, following the lead of the Catholic countries, had adopted the Gregorian calendar, but Protestant England continued to use the Julian calendar. Thus, December 25 in England was January 4 in Europe. If we take the English date, then Newton was born in the same year that Galileo Galilei died.

Newton went on to become one of the greatest scientists who ever lived (∎ Figure 5-1), but even he admitted the debt he owed to those who had studied nature before him. He said, "If I have seen farther than other men, it is because I stood on the shoulders of giants."

One of those giants was Galileo (∎ Figure 5-2). Although we remember Galileo as the defender of Copernicanism, he was also a talented scientist who studied the motions of falling bodies.

Newton based his own work on the discoveries of Galileo and others. During his life he studied optics, invented calculus, developed three laws of motion, and discovered the principle of mutual gravitation. Of these, the last two are the most important to us in this chapter because they make it possible to understand the orbital motion of the moon and planets.

Newton's laws of motion and gravity were astonishingly successful in describing the heavens. A friend of Newton, Edmond Halley, used Newton's laws to calculate the orbits of comets and discovered that certain comets that had been seen throughout recorded history were actually a single object returning every 77 years, now known as Comet Halley.

For over two centuries following the publication of Newton's works, astronomers used his laws to describe the universe. Then, early in the 20th century, Albert Einstein proposed a new way to describe gravity. The new theory did not replace Newton's laws but rather showed that they were only approximately correct and could be seriously in error under special circumstances. We will see how Einstein's theories further extend our understanding of the nature of gravity. Just as Newton had stood on the shoulders of Galileo, Einstein stood on the shoulders of Newton.

5-1 GALILEO AND NEWTON

Johannes Kepler discovered three laws of planetary motion, but he never understood why the planets move along their orbits. At one place in his writings, he wonders if they are pulled along by magnetic forces

FIGURE 5-1
Isaac Newton (1642–1727) was the founder of modern physics. He made important discoveries in optics, developed three laws of motion, invented differential calculus, and discovered the law of mutual gravitation. These stamps were issued in 1987 to mark the 300th anniversary of the publication of his book *Principia*.

emanating from the sun. At another place, he considers and dismisses the idea that the planets are pushed along their orbits by angels.

Newton refined Kepler's model of planetary motion but did not perfect it. In science, a model is an intellectual conception of how nature works (Window on Science 2-1). No model is perfect. Kepler's model was better than Aristotle's, but Newton improved Kepler's model

FIGURE 5-2
Although Galileo is often associated with the telescope, as on this Italian stamp, he also made systematic studies of the motion of falling bodies and made discoveries that led to the law of inertia.

Ace⊙Astronomy™ The AceAstronomy icon throughout the text indicates an opportunity for you to test yourself on key concepts and to explore animations and interactions on the AceAstronomy Web site at **http://astronomy.brookscole .com/seeds8e**

by expanding it into a general theory of motion and gravity. In fact, most scientists now refer to Newton's *law* of gravity. Whether we call it a model, a theory, or a law, we must recognize that it was not perfect. Newton never understood what gravity was. It was as mysterious as an angel pushing the moon inward toward Earth instead of forward along the moon's orbit.

To understand science, we must understand the nature of scientific descriptions. The scientist studies nature by either creating new theories or refining old theories. Yet a theory can never be perfect, because it can never represent the universe in all its intricacies. Instead, a theory must be a limited approximation of a single phenomenon, such as orbital motion. It is fitting that Newton's discoveries all began with Kepler's fellow Copernican, Galileo.

GALILEO AND MOTION

Even before Galileo built his first telescope, he had begun studying the motion of freely moving bodies. After the Inquisition condemned and imprisoned him in 1633, he continued his study of motion. He seems to have realized that he would have to understand motion before he could truly understand the Copernican system. That he was eventually able to formulate principles that later led Newton to the laws of motion and the theory of gravity is a tribute to Galileo's ability to set aside authority and think for himself.

The authority of the age was Aristotle, whose ideas on motion were hopelessly confused. Aristotle said that the world is made up of four classical elements: earth, water, air, and fire. According to this idea, all earthly things—wood, rock, flesh, bone, metals, and so on—are made up of different mixtures of the four classical elements, with earth and water dominating. The motions of bodies are determined by the natural tendencies of the four elements to move toward their proper place in the cosmos. All objects dominated by earth and water, for example, move toward the center of the cosmos, which, in Aristotle's geocentric universe, is Earth's center. Thus, objects fall downward because they are moving toward their proper place.* On the other hand, fire and air move upward toward the heavens, their proper place in the cosmos.

Aristotle called these motions **natural motions** to distinguish them from **violent motions** produced, for instance, when we push on an object and make it move other than toward its proper place. According to Aristotle, such motions stop as soon as the force is removed. To explain how an arrow could continue to move upward even after it had left the bowstring, he said currents in the air around the arrow carried it forward even though the bowstring was no longer pushing it.

*This is one reason why Aristotle had to have a geocentric universe. If Earth's center had not also been the center of the cosmos, his explanation of gravity would not have worked.

These ideas about the proper places of objects, natural and violent motion, and the necessity of a force to preserve motion were still accepted theory in Galileo's time. In fact, in 1590, when Galileo was 26, he wrote a short work called *De Motu* (*On Motion*) that deals with the proper places of objects and their natural motions.

In Galileo's time and for the two preceding millennia, scholars had commonly tried to resolve problems of science by referring to authority. To analyze the flight of a cannonball, for instance, they would turn to the writings of Aristotle and other classical philosophers and try to deduce what those philosophers would have said on the subject. This generated a great deal of discussion but little real progress. Galileo broke with this tradition and conducted his own experiments.

He began by studying the motions of falling bodies, but he quickly discovered that the velocities were so great and the times so short that he could not measure them accurately. Consequently, he began using polished bronze balls rolling down gently sloping inclines. In that instance, the velocity is lower and the time longer. Using an ingenious water clock, he was able to measure the time the balls took to roll given distances down the incline, and he correctly recognized that these times are proportional to the times taken by falling bodies.

He found that falling bodies do not fall at constant rates, as Aristotle had said, but are accelerated. That is, they move faster with each passing second. Near Earth's surface, a falling object will have a velocity of 9.8 m/s (32 ft/s) at the end of 1 second, 19.6 m/s (64 ft/s) after 2 seconds, 29.4 m/s (96 ft/s) after 3 seconds, and so on. Each passing second adds 9.8 m/s (32 ft/s) to the object's velocity (❚ Figure 5-3). In modern terms, this is called the **acceleration of gravity** at Earth's surface.

Galileo also discovered that the acceleration does not depend on the weight of the object. This, too, is contrary to the teachings of Aristotle, who believed that heavy objects, containing more earth and water, fall with higher velocity. Galileo found that the acceleration of a falling body is the same whether it is heavy or light. According to some accounts, he demonstrated this by dropping balls of iron and wood from the top of the Leaning Tower of Pisa to show that they would fall together and hit the ground at the same time (❚ Figure 5-4a). In fact, he probably didn't perform this experiment. It would not have been conclusive anyway because of air resistance. More than 300 years later, Apollo 15 astronaut David Scott, standing on the airless moon, demonstrated Galileo's discovery by dropping a feather and a steel geologist's hammer. They hit the lunar surface at the same time (Figure 5-4b).

Having described natural motion, Galileo turned his attention to violent motion—that is, motion directed other than toward an object's proper place in the cosmos. He pointed out that an object rolling down an incline is accelerated toward Earth and that an object rolling up the same incline is decelerated. If the surface

FIGURE 5-3
Galileo found that a falling object is accelerated downward. Each second, its velocity increases by 9.8 m/s (32 ft/s).

were perfectly horizontal and frictionless, he reasoned, there could be no acceleration or deceleration to change the object's velocity, and, in the absence of friction, the object would continue to move forever. In his own words, "any velocity once imparted to a moving body will be rigidly maintained as long as the external causes of acceleration or retardation are removed."

a

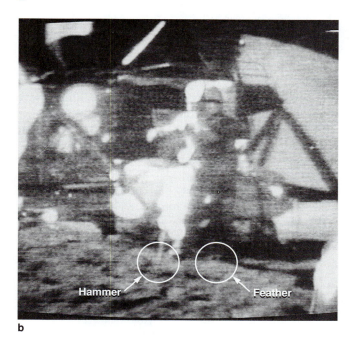

b

FIGURE 5-4
(a) According to tradition, Galileo demonstrated that the acceleration of a falling body is independent of its weight by dropping balls of iron and wood from the Leaning Tower of Pisa. In fact, air resistance would have confused the result. (b) In a historic television broadcast from the moon on August 2, 1971, David Scott dropped a hammer and a feather at the same instant. In the vacuum of the lunar surface, they fell together. *(NASA)*

This is contrary to Aristotle's belief that motion can continue only if a force is present to maintain it. In fact, Galileo's statement is a perfectly valid summary of the law of inertia, which became Newton's first law of motion.

Galileo published his work on motion in 1638, two years after he had become entirely blind and only four years before his death. The book was called *Mathematical Discourses and Demonstrations Concerning Two New Sciences, Relating to Mechanics and to Local Motion*. It is known today as *Two New Sciences*.

The book is a brilliant achievement for a number of reasons. To understand motion, Galileo had to abandon the authority of the ancients, devise his own experiments, and draw his own conclusions. In a sense, this was the first example of experimental science. But Galileo also had to generalize his experiments to discover how nature worked. Though his apparatus was finite and plagued by friction, he was able to imagine an infinite, totally frictionless plane on which a body moves at constant velocity. In his workshop, the law of inertia was obscure; but, in his imagination, it was clear and precise.

NEWTON AND THE LAWS OF MOTION

Newton's three laws of motion (▌Table 5-1) are critical to our understanding of orbital motion. They apply to any moving object, from an automobile driving along a highway to galaxies colliding with each other.

The first law is really a restatement of Galileo's law of inertia. An object continues at rest or in uniform motion in a straight line unless acted upon by some force. Astronauts drifting in space will travel at constant rates in straight lines forever if no forces act on them (▌Figure 5-5a).

Newton's first law also explains why a projectile continues to move after all forces have been removed—for instance, why an arrow continues to move after leaving the bowstring. The object continues to move

because it has momentum. We can think of an object's **momentum** as a measure of the amount of motion.

Momentum depends on velocity and mass.* A low-velocity object such as a paper clip tossed across a room has little momentum, and we could easily catch it in our hand. But the same paper clip fired at the speed of a rifle bullet would have tremendous momentum, and we would not dare try to catch it.

Momentum also depends on the mass of an object (Parameters of Science 5-1). To see how, imagine that, instead of tossing us a paper clip, someone tosses us a bowling ball. A bowling ball contains much more mass than a paper clip and therefore has much greater momentum at the same velocity.

Newton's first law explains the consequences of the conservation of momentum. When we say that momentum is conserved, we mean that it remains constant until something acts to change it. A moving object has a given amount of momentum. To change that momentum, we must exert some force on the object to change either the speed or the direction. Newton's first law and the concept of momentum came from the work of Galileo.

*Mathematically, momentum is the product of mass and velocity.

F a
m
$F = ma$

Action

Reaction

FIGURE 5-5
Newton's three laws of motion.   a   b   c

Mass

One of the most fundamental parameters in science is **mass,** a measure of the amount of matter in an object. A bowling ball, for example, contains a large amount of mass; but a child's rubber ball contains less matter than the bowling ball, and we say it is less massive.

Mass is not the same as weight. Our weight is the force that Earth's gravity exerts on the mass of our bodies. Because gravity pulls us downward, we press against the bathroom scale and we can measure our weight. Floating in space, we would have no weight at all; a bathroom scale would be useless. But our bodies would still contain the same amount of matter, so we would still have mass.

Sports analogies illustrate the importance of mass in dramatic ways. A bowling ball, for example, must be massive in order to have a large effect on the pins it strikes. Imagine trying to knock down all the pins with a bowling ball no more massive than a balloon. Even in space, where the bowling ball would be weightless, a low-mass bowling ball would have little effect on the pins. On the other hand, runners want track shoes that have low mass and thus are easy to move. Imagine trying to run a 100-meter dash wearing track shoes that were as massive as bowling balls. They would be very hard to move, and it would be difficult to accelerate away from the starting block. The shot put takes muscle because the shot is massive, not because it is heavy. Imagine throwing the shot in space where it would have no weight. It would still be massive, and it would take great effort to start it moving.

Mass is a unique measure of the amount of material in an object. Using the metric system (Appendix), we will measure mass in kilograms.

Newton's second law of motion discusses forces, and Galileo did not talk about forces. He spoke instead of accelerations. Newton saw that an acceleration is the result of a force acting on a mass (Figure 5-5b). Newton's second law is commonly written as:

$$F = ma$$

Once again, we must carefully define terms. An **acceleration** is a change in velocity, and a **velocity** is a directed rate of motion. By rate of motion, we mean, of course, a speed, but the word *directed* has a special meaning. Speed itself does not have any direction associated with it, but velocity does. If you drive a car in a circle at 55 mph, your speed is constant, but your velocity is changing. Thus, an object experiences an acceleration if its speed changes or if its direction of motion changes. Every automobile has three accelerators—the gas pedal, the brake pedal, and the steering wheel. All three change the auto's velocity.

The acceleration of a body is proportional to the force applied. This is reasonable. If we push gently against a grocery cart, we do not expect a large acceleration. The second law of motion also says that the acceleration is inversely proportional to the mass of the body. This, too, is reasonable. If the cart were filled with bricks and we pushed it gently, we would expect very little result. If it were full of inflated balloons, however, it would move easily in response to a gentle push. Finally, the second law says that the resulting acceleration is in the direction of the force. This is also what we would expect. If we push on a cart that is not moving, we expect it to begin moving in the direction we push.

The second law of motion is important because it establishes a precise relationship between cause and effect (Window on Science 5-1). Objects do not just move. They accelerate due to the action of a force. Moving objects do not just stop. They decelerate due to a force. Also, moving objects don't just change direction for no reason. Any change in direction is a change in velocity and requires the presence of a force. Aristotle said that objects move because they have a tendency to move. Newton said that objects move due to a specific cause, a force.

Newton's third law of motion specifies that for every action there is an equal and opposite reaction. In other words, forces must occur in pairs directed in opposite directions. For example, if you stand on a skateboard and jump forward, the skateboard will shoot away backward. As you jump, your feet exert a force against the skateboard, which accelerates it toward the rear. But forces must occur in pairs, so the skateboard must exert an equal but opposite force on your feet that accelerates your body forward (Figure 5-5c).

MUTUAL GRAVITATION

The three laws of motion led Newton to consider the force that causes objects to fall. The first and second laws tell us that falling bodies accelerate downward because some force must be pulling downward on them. Newton wondered what that force could be.

Newton was also aware that some force has to act on the moon. The moon follows a curved path around Earth, and motion along a curved path is accelerated motion. The second law says that a force is required in order to make the moon follow that curved path.

Newton wondered if the force that holds the moon in its orbit could be the same force that causes rocks to roll downhill—gravity. He was aware that gravity extends at least as high as the tops of mountains, but he

The Unstated Assumption of Science: Cause and Effect

One of the most often used and least often stated principles of science is cause and effect, and we could argue that Newton's second law of motion was the first clear statement of the principle. Ancient philosophers such as Aristotle argued that objects moved because of tendencies. They said that earth and water, and objects made of earth and water, had a natural tendency to move toward the center of the universe. This natural motion had no cause but was inherent in the nature of the objects. But Newton's second law says $F = ma$. If an object (of mass m) changes its motion (a in the equation), then it must be acted on by a force (F in the equation). Any effect (a) must be the result of a cause (F).

The principle of cause and effect goes far beyond motion. The principle of cause and effect gives scientists confidence that every effect has a cause. Hearing loss in certain laboratory rats, color changes in certain chemical dyes, and explosions on certain stars are all effects that must have causes. All of science is focused on understanding the causes of the effects we see. If the universe were not rational, then we could never expect to discover causes. Newton's second law of motion was arguably the first statement that the behavior of the universe depends rationally on causes.

---

did not know if it could extend all the way to the moon. He believed that it could, but he thought it would be weaker at greater distances. He also guessed that its strength would decrease as the square of the distance increased.

This relationship, the **inverse square law,** was familiar to Newton from his work on optics, where it applied to the intensity of light. A screen set up 1 unit from a candle flame receives a certain amount of energy on each square meter. However, if that screen is moved to a distance of 2 units, the light that originally illuminated 1 m² must cover 4 m² (❚ Figure 5-6). Thus, the intensity of the light is inversely proportional to the square of the distance to the screen.

Newton made two assumptions that enabled him to predict the strength of Earth's gravity at the distance of the moon. He assumed that the strength of gravity follows the inverse square law and that the critical distance is not the distance from Earth's surface but the distance from Earth's center. Because the moon is about 60 Earth radii away, Earth's gravity at the distance of the moon should be about 60^2 times less than at Earth's surface. Instead of being 9.8 m/s² at Earth's surface, it should be about 0.0027 m/s² at the distance of the moon.

Now, Newton wondered, could this acceleration keep the moon in orbit? He knew the moon's distance and its orbital period, so he could calculate the actual acceleration needed to keep it in its curved path. The answer is 0.0027 m/s². To the accuracy of Newton's data for Earth's radius, it was exactly what his assumptions predicted. The moon is held in its orbit by gravity, and gravity obeys the inverse square law.

Newton's third law says that forces always occur in pairs, and this leads to the conclusion that gravity is mutual. If Earth pulls on the moon, then the moon must pull on Earth. Gravitation is a general property of the universe. The sun, the planets, and all their moons must also attract each other by mutual gravitation. In fact, every particle in the universe must attract every other particle, and Newtonian gravity is often called universal mutual gravitation.

Clearly the force of gravity depends on mass. Although our bodies are made of mass, our personal gravitational fields do not attract personal satellites. Larger masses have stronger gravity. From an analysis of the third law of motion, Newton realized that the mass that resists acceleration in the first law must be the same as the mass associated with gravity. Then when two bodies attract one another and accelerate toward each other, the two masses must be equally involved. Newton performed precise experiments with pendulums and confirmed this equivalence between the mass that resists acceleration and the mass that causes gravity. From this, combined with the inverse square law, he

FIGURE 5-6

The inverse square law. A light source is surrounded by spheres with radii of 1 unit and 2 units. The light falling on an area of 1 m² on the inner sphere spreads to illuminate an area of 4 m² on the outer sphere. Thus, the brightness of the light source is inversely proportional to the square of the distance.

was able to write the famous formula for the gravitational force between two masses, M and m.

$$F = -G\frac{Mm}{r^2}$$

The constant G is the gravitational constant, and r is the distance between the masses. The negative sign tells us the force is attractive, pulling the masses together and making r decrease. In plain English, Newton's law of gravitation states: The force of gravity between two masses M and m is proportional to the product of the masses and inversely proportional to the square of the distance between them.

Newton's description of gravity was a difficult idea for physicists of his time to accept because it is an example of action at a distance. Earth and moon exert forces on each other although there is no physical connection between them. Modern scientists resolve this problem by referring to gravity as a **field**. Earth's presence produces a gravitational field directed toward Earth's center. The strength of the field decreases according to the inverse square law. Any particle of mass in that field experiences a force that depends on the mass of the particle and the strength of the field at the particle's location. The resulting force is directed toward the center of the field.

The field is an elegant way to describe gravity, but it does not tell us what gravity is. For that we must wait until later in this chapter, when we discuss Einstein's theory of curved space-time.

REVIEW CRITICAL INQUIRY

What do the words *universal* and *mutual* mean when we say "universal mutual gravitation"?

Newton realized that the force that makes an apple accelerate downward is the same as the force that accelerates the moon and holds it in its orbit. We can learn more by thinking about Newton's third law of motion, which says that forces always occur in pairs. If Earth attracts the moon, then the moon must attract Earth. That is, gravitation is *mutual* between any two objects.

Furthermore, if Earth's gravity attracts the apple and the moon, then it must attract the sun, and the third law tells us that the sun must attract Earth. But if the sun attracts Earth, then it must also attract the other planets and even distant stars, which, in turn, must attract the sun and each other. Step by step, Newton's third law of motion leads us to conclude that gravitation must apply to all masses in the universe. That is, gravitation must be *universal*.

Aristotle explained gravity in a totally different way. How did Aristotle account for a falling apple? Could that explanation account for a hammer falling on the surface of the moon?

Newton first came to understand gravity by thinking about the orbital motion of the moon. Gravitation is tremendously important in astronomy, and its most important manifestation is orbital motion.

5-2 ORBITAL MOTION

Newton's laws of motion and gravitation make it possible to understand why the moon orbits Earth and how the planets move along their orbits. We can even discover why Kepler's laws work.

ORBITS

To understand how an object can orbit another object, we must see orbital motion as Newton did. Objects in orbit are falling. We will explore Newton's insight by analyzing the motion of objects orbiting Earth.

Study "❚ Orbiting Earth" on pages 84 and 85, and notice three important ideas. First, an object orbiting Earth is actually falling (being accelerated) toward Earth's center. The object continuously misses Earth because of its orbital velocity. Second, notice that objects orbiting each other actually revolve around their center of mass. Finally, notice the difference between closed orbits and open orbits. If you want to leave Earth never to return, you must give your spaceship a high enough velocity so it will follow an open orbit.

ORBITAL VELOCITY

If we were about to ride a rocket into orbit, we would have a critical question. How fast do we have to travel to stay in orbit? An object's circular velocity is the lateral velocity the object must have to remain in a circular orbit. If we assume the mass of our spaceship is small compared with the mass of the object we expect to orbit, Earth in this case, then the circular velocity is:

$$V_c = \sqrt{\frac{GM}{r}}$$

In this formula, M is the mass of the central body in kilograms, r is the radius of the orbit in meters, and G is the gravitational constant, 6.67×10^{-11} m³/s²kg. This formula is all we need to calculate how fast an object must travel to stay in a circular orbit.

For example, how fast does the moon travel in its orbit? Earth's mass is 5.98×10^{24} kg, and the radius of the moon's orbit is 3.84×10^8 m. Then the moon's velocity is:

$$V_c = \sqrt{\frac{6.67 \times 10^{-11} \times 5.98 \times 10^{24}}{3.84 \times 10^8}} = \sqrt{\frac{39.9 \times 10^{13}}{3.84 \times 10^8}}$$

$$= \sqrt{1.04 \times 10^6} = 1020 \text{ m/s} = 1.02 \text{ km/s}$$

This calculation shows that the moon travels 1.02 km along its orbit each second. That is the circular velocity at the distance of the moon.

Orbiting Earth

We can understand orbital motion by thinking of a cannon ball falling around Earth in a circular path.

Imagine a cannon on a high mountain aimed horizontally. A little gunpowder gives the cannonball a low velocity, and it doesn't travel very far before falling to Earth.

More gunpowder gives the cannonball a higher velocity, and it travels farther.

With enough gunpowder, the cannonball travels so fast it never strikes the ground. Earth's gravity pulls it toward Earth's center, but Earth's surface curves away from it at the same rate it falls. It is in orbit.

North Pole

A satellite above Earth's atmosphere feels no friction and will fall around Earth indefinitely.

The velocity needed to stay in a circular orbit is called the **circular velocity.** Just above Earth's atmosphere, circular velocity is 7790 m/s or about 17,400 miles per hour, and the orbital period is about 90 minutes.

Earth satellites eventually fall back to Earth if they orbit too low and experience friction with the upper atmosphere.

Ace Astronomy™
Go to AceAstronomy and click Active Figures to see "Newton's Cannon" and fire your own version of Newton's cannon.

A Geosynchronous Satellite

At a distance of 42,250 km (26,260 miles) from Earth's center, a satellite orbits with a period of 24 hours.

The satellite orbits eastward, and Earth rotates eastward under the moving satellite.

The satellite remains fixed above a spot on Earth's equator.

A **geosynchronous satellite** orbits eastward with the rotation of Earth and remains above a fixed spot — ideal for communications and weather satellites.

Ace Astronomy™
Go to AceAstronomy and click Active Figures to see "Geosynchronous Orbit" and place your own satellite in geosynchronous orbit.

According to Newton's first law of motion, the moon should follow a straight line and leave Earth forever. Because it follows a curve, Newton knew that some force must continuously accelerate it toward Earth — gravity. Each second the moon moves 1020 m (3350 ft) eastward and falls about 1.6 mm (about 1/16 inch) toward Earth. The combination of these motions produces the moon's curved orbit. The moon is falling.

Straight line motion of the moon

Motion toward Earth

Curved path of moon's orbit

Earth

Astronauts in orbit around Earth feel weightless, but they are not "beyond Earth's gravity," to use a term from old science fiction movies. Like the moon, the astronauts are accelerated toward Earth by Earth's gravity, but they travel fast enough along their orbits that they continually "miss the Earth." They are literally falling around Earth. Inside or outside a spacecraft, astronauts feel weightless because they and their spacecraft are falling at the same rate. Rather than saying they are weightless, we should more accurately say they are in free fall.

To be precise we should not say that an object orbits Earth. Rather the two objects orbit each other. Gravitation is mutual, and if Earth pulls on the moon, the moon pulls on Earth. The two bodies revolve around their common **center of mass,** the balance point of the system.

Two bodies of different mass balance at the center of mass, which is located closer to the more massive object. As the two objects orbit each other, they revolve around their common center of mass.

Center of mass

The center of mass of the Earth–moon system lies only 4708 km (2926 miles) from the center of Earth — inside the Earth. As the moon orbits the center of mass on one side, the Earth swings around the center of mass on the opposite side.

Ace Astronomy™ Go to AceAstronomy and click Active Figures to see "Center of Mass" and change the mass ratio in this diagram.

Closed orbits return the orbiting object to its starting point. The moon and artificial satellites orbit Earth in closed orbits. Below, the cannonball could follow an elliptical or a circular closed orbit.

If the cannonball travels as fast as **escape velocity,** the velocity needed to leave a body, it will enter an open orbit. An **open orbit** does not return the cannonball to Earth. It will escape.

Hyberbola

A cannonball with a velocity greater than escape velocity will follow a hyperbola and escape from Earth.

Parabola

North Pole

A cannonball with escape velocity will follow a parabola and escape.

As described by Kepler's Second Law, an object in an elliptical orbit has its lowest velocity when it is farthest from Earth (apogee), and its highest velocity when it is closest to Earth (perigee). Perigee must be above Earth's atmosphere, or friction will rob the satellite of energy and it will eventually fall back to Earth.

Ellipse

Circle

Ellipse

A satellite just above Earth's atmosphere is only about 200 km above Earth's surface, or 6578 km from Earth's center, so Earth's gravity is much stronger and the satellite must travel much faster to stay in a circular orbit. We can use the formula above to find that the circular velocity just above Earth's atmosphere is about 7790 m/s, or 7.79 km/s. This is about 17,400 miles per hour and shows why putting satellites into Earth orbit takes such large rockets. Not only must the rocket lift the satellite above Earth's atmosphere, but the rocket must tip over and accelerate the satellite to circular velocity.

CALCULATING ESCAPE VELOCITY

If we launch a rocket upward, it will consume its fuel in a few moments and reach its maximum speed. From that point on, it will coast upward. How fast must a rocket travel to coast away from Earth and escape? We know, of course, that no matter how far it travels, it can never escape from Earth's gravity. The effects of Earth's gravity extend to infinity. It is possible, however, for a rocket to travel so fast initially that gravity can never slow it to a stop. Thus, it could leave Earth.

The escape velocity is the velocity required to escape from the surface of an astronomical body. Here we are interested in escaping from Earth or a planet; in later chapters, we will consider the escape velocity from stars, galaxies, and even a black hole.

The escape velocity, V_e, is given by a simple formula:

$$V_e = \sqrt{\frac{2GM}{r}}$$

Here G is the gravitational constant 6.67×10^{-11} m^3/s^{2}kg, M is the mass of the astronomical body in kilograms, and r is its radius in meters. (This formula is very similar to the formula for circular velocity.)

We can find the escape velocity from Earth by looking up its mass, 5.98×10^{24} kg, and its radius, 6.38×10^6 m. Then the escape velocity is:

$$V_e = \sqrt{\frac{2 \times 6.67 \times 10^{-11} \times 5.98 \times 10^{24}}{6.38 \times 10^6}} = \sqrt{\frac{7.98 \times 10^{14}}{6.38 \times 10^6}}$$

$$= \sqrt{1.25 \times 10^8} = 11,200 \text{ m/s} = 11.2 \text{ km/s}$$

This is equal to about 25,000 miles per hour.

Notice that the formula tells us that escape velocity depends on both mass and radius. A massive body might have a low escape velocity if it has a very large radius. We will meet such objects when we discuss giant stars. On the other hand, a rather low-mass body could have a very large escape velocity if it had a very small radius, a condition we will discuss when we meet black holes.

Circular velocity and escape velocity are two aspects of Newton's laws of gravity and motion. Once Newton understood gravity and motion, he could do what Kepler had failed to do—he could explain why the planets obey Kepler's laws of planetary motion.

KEPLER'S LAWS REEXAMINED

Now that we understand Newton's laws, gravity, and orbital motion, we can understand Kepler's laws of planetary motion in a new way.

Kepler's first law says that the orbits of the planets are ellipses with the sun at one focus. Kepler wondered why the planets keep moving along these orbits, and now we know the answer. They move because there is nothing to slow them down. Newton's first law says that a body in motion stays in motion unless acted on by some force. The gravity of the sun accelerates the planets inward toward the sun and holds them in their orbits, but it doesn't pull backward on the planets, so they don't slow to a stop. With no friction, they must continue to move.

The orbits of the planets are ellipses because gravity follows the inverse square law. In one of his most famous problems, Newton proved that if a planet moves in a closed orbit under the influence of an attractive force that follows the inverse square law, then the planet must follow an elliptical path.

Kepler's second law says that a planet moves faster when it is near the sun and slower when it is farther away. Once again, Newton's discoveries explain why. Earlier we saw that a body moving on a frictionless surface will continue to move in a straight line until it is acted on by some force; that is, the object has momentum. In a similar way, an object set rotating on a frictionless surface will continue rotating until something acts to speed it up or slow it down. Such an object has **angular momentum,** a measure of the rotation of the body about some point. A planet circling the sun has a given amount of angular momentum; and, with no outside influences to speed it up or slow it down, it must conserve its angular momentum. That is, its angular momentum must remain constant.

Mathematically, a planet's angular momentum is the product of its mass, velocity, and distance from the sun. This explains why a planet must speed up as it comes closer to the sun along an elliptical orbit. Its angular momentum is conserved, so as its distance from the sun decreases, its velocity must increase. In the same way, the planet's velocity must decrease as its distance from the sun increases.

This conservation of angular momentum is actually a common human experience. Skaters spinning slowly can draw their arms and legs closer to their axis of rotation and, through conservation of angular momentum, spin faster (❚ Figure 5-7). To slow their rotation, they again extend their arms. Similarly, divers can spin rapidly in the tuck position and slow their rotation by stretching into the extended position.

Energy

One of the most fundamental parameters in science is **energy.** Physicists define energy as the ability to do work, but we might paraphrase that definition as the ability to produce a change. Certainly a moving body has energy. A planet moving along its orbit, a cement truck rolling down the highway, and a golf ball sailing down the fairway all have the ability to produce a change. Imagine colliding with any of these objects! But energy need not be represented by motion. Sunlight falling on a green plant, on photographic film, or on unprotected skin can produce chemical changes, and thus light is a form of energy.

Energy can take many forms. Batteries and gasoline are examples of chemical energy, and uranium fuel rods con-

tain nuclear energy. A tank of hot water contains thermal energy. Even a weight on a high shelf can represent stored energy. Imagine a bowling ball falling off a high shelf onto your desk. That would produce significant change.

Much of science is the study of how energy flows from one place to another place, producing changes. A biologist might study the way a nerve cell transmits energy along its length to a muscle, while a geologist might study how energy flows as heat from Earth's interior and deforms Earth's surface. In such processes, we see energy being transformed from one state to another. Sunlight (energy) is absorbed by ocean plants and stored as sugars and starches (energy). When the plant dies, it and

other ocean life are buried and become oil (energy), which we pump to the surface and burn in automobile engines to produce motion (energy).

Aristotle believed that all change originated in the motion of the starry sphere and flowed down to Earth. Modern science has found a more sophisticated description of the continual change we see around us, but we are still interested in the way energy flows through the world and produces change. Energy is the pulse of the natural world.

Using the metric system (Appendix), we will measure energy in **joules** (abbreviated **J**). One joule is about as much energy as that given up when an apple falls from a table to the floor.

Kepler's third law is also explained by a conservation law, but in this case it is the law of conservation of energy (Parameters of Science 5-2). A planet orbiting the sun has a specific amount of energy that depends only on its average distance from the sun. That energy can be divided between energy of motion and energy stored in the gravitational attraction between the planet and the sun. The energy of motion depends on how fast the planet moves, and the stored energy depends on the size of its orbit. The relation between these two kinds

of energy is fixed by Newton's laws. That means there has to be a fixed relationship between the rate at which a planet moves around its orbit and the size of the orbit—between its orbital period P and the orbit's semi-major axis a. This is just Kepler's third law.

NEWTON'S VERSION OF KEPLER'S THIRD LAW

The equation for circular velocity is actually a version of Kepler's third law, as we can prove with three lines of simple algebra. The result is one of the most useful formulas in astronomy.

The equation for circular velocity, as we have seen, is:

$$V_c = \sqrt{\frac{GM}{r}}$$

The orbital velocity of a planet is simply the circumference of its orbit divided by the orbital period:

$$V = \frac{2\pi r}{P}$$

If we substitute this for V in the equation for circular velocity and solve for P^2, we get:

$$P^2 = \frac{4\pi^2}{GM} r^3$$

Here M is just the total mass of the system in kilograms. For a planet orbiting the sun, we can use the mass of the sun for M, because the mass of the planet is negligible compared to the mass of the sun. In a later chapter,

FIGURE 5-7
Skaters demonstrate conservation of angular momentum when they spin faster by drawing their arms and legs closer to their axis of rotation.

we will apply this formula to two stars orbiting each other, and then the mass M will be the sum of the two masses. For a circular orbit, r equals the semimajor axis a, so this formula is a general version of Kepler's third law, $P^2 = a^3$. In Kepler's version, we use the units AU and years, but in this formula, known as Newton's version of Kepler's third law, we use units of meters, seconds, and kilograms. G, of course, is the gravitational constant.

This is a very powerful formula. Astronomers use it to find the masses of bodies by observing orbital motion. If, for example, we observe a moon orbiting a planet and we can determine the size of its orbit, r, and the orbital period, P, we can use this formula to solve for M, the total mass of the planet plus the moon. There is no other way to find masses in astronomy, and we will use this formula in later chapters to find the masses of stars, galaxies, and planets.

This discussion is a good illustration of the power of Newton's work. By carefully defining motion and gravity and by giving them mathematical expression, Newton was able to derive new truths, among them Newton's version of Kepler's third law. His work changed science and made it into something new, precise, and exciting.

ASTRONOMY AFTER NEWTON

Newton published his work in July 1687 in a book called *Philosophiae Naturalis Principia Mathematica* (*Mathematical Principles of Natural Philosophy*), now known simply as *Principia* (❙ Figure 5-8). It is one of the most important books ever written. The principles changed astronomy, changed science, and changed the way we think about nature.

Principia changed astronomy and ushered in the age of gravitational astronomy. No longer did astronomers appeal to the whim of the gods to explain things in the heavens. No longer did they speculate on why the planets move. They now knew that the motions of the heavenly bodies are governed by simple, universal rules that describe the motions of everything from planets to falling apples. Suddenly the universe was understandable in simple terms.

Newton's laws of motion and gravity made it possible for astronomers to calculate the orbits of planets and moons. Not only could they explain how the heavenly bodies move, they could predict future motions (Window on Science 5-2). This subject, known as gravitational astronomy, dominated astronomy for almost 200 years and is still important. It included the calculation of the orbits of comets and asteroids and the theoretical prediction of the existence of two planets, Neptune and Pluto.

Principia also changed science in general. The works of Copernicus and Kepler had been mathematical, but no book before had so clearly demonstrated the power of mathematics as a language of precision. Newton's arguments in *Principia* were so powerful an illustration of the quantitative study of nature that scientists around the world adopted mathematics as their most powerful tool.

Also, *Principia* changed the way we think about nature. Newton showed that the rules that govern the universe are simple. Particles move according to three rules of motion and attract each other with a force called gravity. These motions are predictable, and that makes the universe a vast machine based on a few simple rules. It is complex only in that it contains a vast number of particles. In Newton's view, if he knew the location and motion of every particle in the universe, he could, in principle, derive the past and future of the universe in every detail. This mechanical determinism has been undermined by modern quantum mechanics, but it dominated science for more than two centuries during which scientists thought of nature as a beautiful clockwork that would be perfectly predictable if we knew how all the gears meshed.

FIGURE 5-8

Newton, working from the discoveries of Galileo and Kepler, derived three laws of motion and the principle of mutual gravitation. He and some of his discoveries are honored on this English pound note. Notice the diagram of orbital motion in the background and the open copy of *Principia* in Newton's hands.

Testing a Theory by Prediction

When you read about any science, you should notice that scientific theories face in two directions. They look back into the past and explain phenomena we have previously observed. For example, Newton's laws of motion and gravity explained how the planets moved. But theories also face forward in that they enable us to make predictions about what we should find as we explore further. Thus, Newton's laws allowed astronomers to calculate the orbits of comets, predict their return, and eventually understand their origin.

Scientific predictions are important in two ways. First, if a theory leads to a prediction and scientists later discover the prediction was true, the theory is confirmed, and scientists gain confidence that it is a true description of nature. But predictions are important in science in a second way. Using an existing theory to make a prediction may lead us into an unexplored avenue of knowledge. Thus, the first theories of genetics made predictions that confirmed the genetic theory of inheritance, but those predictions also created a new understanding of how living creatures evolve.

As you read about any scientific theory, think about both what it can explain and what it can predict.

Most of all, Newton's work broke the last bonds between science and formal philosophy. Newton did not speculate on the good or evil of gravity. He did not debate its meaning. Not more than a hundred years before, scientists would have argued over the "reality" of gravity. Newton didn't care for these debates. He wrote, "It is enough that gravity exists and suffices to explain the phenomena of the heavens."

REVIEW CRITICAL INQUIRY

How do Newton's laws of motion explain the orbital motion of the moon?

If Earth and the moon did not attract each other, the moon would move in a straight line in accord with Newton's first law of motion and vanish into space. Instead, gravity pulls the moon toward Earth's center, and the moon accelerates toward Earth. This acceleration is just enough to pull the moon away from its straight-line motion and cause it to follow a curve around Earth. In fact, it is correct to say that the moon is falling, but because of its lateral motion it continuously misses Earth.

Every orbiting object is falling toward the center of its orbit but is moving laterally fast enough to compensate for the inward motion, and it follows a curved orbit. If this is true, then how can astronauts float inside spacecraft in a "weightless" state? Why might "free fall" be a more accurate term?

Newton's laws of motion and gravitation are critical in astronomy not only because they describe orbital motion but also because they describe the interaction of astronomical bodies. Newton made other discoveries, but we reserve these for later chapters. Now we move forward two centuries to see how Einstein described gravity in a new and powerful way.

5-3 EINSTEIN AND RELATIVITY

In the early years of the last century, Albert Einstein (1879–1955) (∎ Figure 5-9) began thinking about how motion and gravity interact. He soon gained international fame by showing that Newton's laws of motion and gravity were only partially correct. The revised theory became known as the theory of relativity. As we will see, there are really two theories of relativity.

SPECIAL RELATIVITY

Einstein began by thinking about how moving observers see events around them. His analysis led him to the first postulate of relativity, also known as the principle of relativity:

> **First postulate** (the principle of relativity): Observers can never detect their *uniform* motion except relative to other objects.

FIGURE 5-9

Einstein's greatest accomplishments, the special theory of relativity and the general theory of relativity, were developed as he analyzed the nature of motion and time.

You may have experienced the first postulate while sitting on a train in a station. You suddenly notice that the train on the next track has begun to creep out of the station. However, after several moments you realize that it is your own train that is moving and that the other train is still motionless on its track.

Consider another example. Suppose you are floating in a spaceship in interstellar space and another spaceship comes coasting by (Figure 5-10a). You might conclude that it is moving and you are not, but someone in the other ship might be equally sure that you are moving and it is not. The principle of relativity says that there is no experiment you can perform to decide which ship is moving and which is not. This means that there is no such thing as absolute rest—all motion is relative.

Because neither you nor the people in the other spaceship could perform any experiment to detect your absolute motion through space, the laws of physics must have the same form in both spaceships. Otherwise, experiments would give different results in the two ships, and you could decide who was moving. Thus, a more general way of stating the first postulate refers to these laws of physics:

> **First postulate** (alternate version): The laws of physics are the same for all observers, no matter what their motion, so long as they are not *accelerated*.

The words *uniform* and *accelerated* are important. If either spaceship were to fire its rockets, then its velocity would change. The crew of that ship would know it because they would feel the acceleration pressing them into their couches. Accelerated motion, therefore, is different—we can always tell which ship is accelerating and which is not. The postulates of relativity discussed here apply only to observers in uniform motion. That is why the theory is called **special relativity.**

The first postulate fit with Einstein's conclusion that the speed of light must be constant for all observers. No matter how you move, your measurement of the speed of light has to give the same result (Figure 5-10b). This became the second postulate of special relativity:

> **Second postulate:** The velocity of light is constant and will be the same for all observers independent of their motion relative to the light source.

Once Einstein had accepted the basic postulates of relativity, he was led to some startling discoveries. Newton's laws of motion and gravity

FIGURE 5-10

(a) The principle of relativity says that observers can never detect their uniform motion, except relative to other observers. Thus, neither of these travelers can decide who is moving and who is not. (b) If the principle of relativity is correct, the velocity of light must be a constant for all observers. If the velocity of light depended on the motion of the observer through space, then these travelers could decide who was moving and who was not.

worked well as long as distances were small and velocities were low. But when we begin to think of very large distances or very high velocities, Newton's laws are no longer adequate to describe what happens. Instead, we must use relativistic physics. For example, special relativity shows that the observed mass of a moving particle depends on its velocity. The higher the velocity, the greater the mass of the particle. This is not significant at low velocities, but it becomes very important as the velocity approaches the velocity of light. Such increases in mass are observed whenever physicists accelerate atomic particles to high velocities (▌ Figure 5-11).

This discovery led to yet another insight. The relativistic equations that describe the energy of a moving particle predict that the energy of a motionless particle is not zero. Rather, its energy at rest is m_0c^2. This is of course the famous equation:

$$E = m_0c^2$$

The c is the speed of light and the m_0 is the mass of the particle when it is at rest. (We must specify mass this way because one of the consequences of relativity is that a particle's mass depends on its velocity.) This simple formula suggests that mass and energy are related, and we will see in later chapters how nature can convert one into the other inside stars.

For example, suppose that we convert 1 kg of matter into energy. We must express the velocity of light as 3×10^8 m/s, and our result is 9×10^{16} joules (J) (approximately equal to a 20-megaton nuclear bomb). (Recall that a joule is a unit of energy roughly equivalent to the energy given up when an apple falls from a table to the floor.) Our simple calculation shows that the energy equivalent of even a small mass is very large.

Other relativistic effects include the slowing of moving clocks and the shrinkage of lengths measured in the direction of motion. A detailed discussion of the major consequences of the special theory of relativity is beyond the scope of this book. Instead, we must consider Einstein's second advance, the general theory.

THE GENERAL THEORY OF RELATIVITY

In 1916, Einstein published a more general version of the theory of relativity that dealt with accelerated as well as uniform motion. This **general theory of relativity** contained a new description of gravity.

Einstein began by thinking about observers in accelerated motion. Imagine an observer sitting in a windowless spaceship. Such an observer cannot distinguish between the force of gravity and the inertial forces produced by the acceleration of the spaceship (▌ Figure 5-12). This led Einstein to conclude that gravity and acceleration through space-time are related, a conclusion now known as the equivalence principle:

FIGURE 5-11
The observed mass of moving electrons depends on their velocity. As the ratio of their velocity to the velocity of light, v/c, gets larger, the mass of the electrons in terms of their mass at rest, m/m_0, increases. Such relativistic effects are quite evident in particle accelerators, which accelerate atomic particles to very high velocities.

Equivalence principle: Observers cannot distinguish locally between inertial forces due to acceleration and uniform gravitational forces due to the presence of a massive body.

The importance of the general theory of relativity lies in its description of gravity. Einstein concluded that gravity, inertia, and acceleration are all associated with the way space and time are related. This relation is often referred to as curvature, and a one-line description of general relativity explains a gravitational field as a curved region of space-time:

Gravity according to general relativity: Mass tells space-time how to curve, and the curvature of space-time (gravity) tells mass how to accelerate.

Thus, we feel gravity because Earth's mass causes a curvature of space-time. The mass of our bodies responds to that curvature by accelerating toward Earth's center. According to general relativity, all masses cause curvature, and the larger the mass, the more severe the curvature.

CONFIRMATION OF THE CURVATURE OF SPACE-TIME

Einstein's general theory of relativity has been confirmed by a number of experiments, but two are worth mentioning here because they were among the first tests of the theory. One involves Mercury's orbit, and the other involves eclipses of the sun.

Johannes Kepler understood that the orbit of Mercury is elliptical, but only since 1859 have astronomers known that the long axis of the orbit sweeps around the sun in a motion called precession (▌ Figure 5-13). The total observed precession is 5600.73 seconds of arc per

FIGURE 5-12

(a) An observer in a closed spaceship on the surface of a planet feels gravity. (b) In space, with the rockets smoothly firing and accelerating the spaceship, the observer feels inertial forces that are equivalent to gravitational forces.

century (as seen from Earth), or about 1.5° per century. This precession is produced by the gravitation of Venus, Earth, and the other planets. However, when astronomers used Newton's description of gravity, they calculated that the precession should amount to only 5557.62 seconds of arc per century. Thus, Mercury's orbit is advancing 43.11 seconds of arc per century faster than Newton's law predicted.

This is a tiny effect. Each time Mercury returns to perihelion, its closest point to the sun, it is about 29 km (18 miles) past the position predicted by Newton's laws. This is such a small distance compared with the planet's diameter of 4850 km that it could never have been detected had it not been cumulative. Each orbit, Mercury gains 29 km, and in a century it gains over 12,000 km—more than twice its own diameter. Thus,

this tiny effect, called the advance of perihelion of Mercury's orbit, accumulated into a serious discrepancy in the Newtonian description of the universe.

The advance of perihelion of Mercury's orbit was one of the first problems to which Einstein applied the principles of general relativity. First he calculated how much the sun's mass curves space-time in the region of Mercury's orbit, and then he calculated how Mercury moves through the space-time. The theory predicted that the curved space-time should cause Mercury's orbit to advance by 43.03 seconds of arc per century, well within the observational accuracy of the excess (Figure 5-13b).

Einstein was elated with this result, and he would be even happier with modern studies that have shown that Mercury, Venus, Earth, and even Icarus, an asteroid that comes close to the sun, have orbits observed to

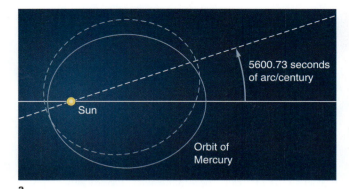

5600.73 seconds
of arc/century

Sun

Orbit of
Mercury

a

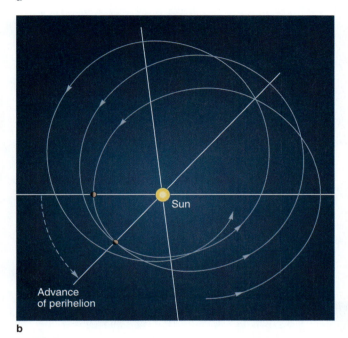

Sun

Advance
of perihelion

b

FIGURE 5-13
(a) Mercury's orbit precesses 5600.73 seconds of arc per century—43.11 seconds of arc per century faster than predicted by Newton's laws. (b) Even when we ignore the influences of the other planets, Mercury's orbit is not a perfect ellipse. Curved space-time near the sun distorts the orbit from an ellipse into a rosette. The advance of Mercury's perihelion is exaggerated about a million times in this figure.

TABLE 5-2		
Precession in Excess of Newtonian Physics		
Planet	Observed Excess Precession (seconds of arc per century)	Relativistic Prediction (seconds of arc per century)
Mercury	43.11 ± 0.45	43.03
Venus	8.4 ± 0.48	8.6
Earth	5.0 ± 1.2	3.8
Icarus	9.8 ± 0.8	10.3

by curved space-time just as a rolling golf ball is deflected by undulations in a putting green. Einstein predicted that starlight grazing the sun's surface would be deflected by 1.75 seconds of arc (▌ Figure 5-14). Starlight passing near the sun is normally lost in the sun's glare, but during a total solar eclipse stars beyond the sun could be seen. As soon as Einstein published his theory, astronomers rushed to observe such stars and thus test the curvature of space-time.

The first solar eclipse following Einstein's announcement in 1916 was June 8, 1918. It was cloudy at some observing sites, and results from other sites were inconclusive. The next occurred on May 29, 1919, only months after the end of World War I, and was visible from Africa and South America. British teams went to both Brazil and Príncipe, an island off the coast of Africa. First, they photographed that part of the sky where the sun would be located during the eclipse and measured the positions of the stars on the plates. Then, during the eclipse, they photographed the same star field with the eclipsed sun located in the middle. After measuring the plates, they found slight changes in the

be slipping forward due to the curvature of space-time near the sun (▌ Table 5-2).

This same effect has been detected in pairs of stars that orbit each other. In some cases, the advance of perihelion agrees with general relativity; in many cases, the sizes and masses of the stars are not well enough known for us to be certain of the theoretical rate of advance we should expect. But in a few cases, the stars' orbits appear to be changing faster than predicted. This may be a critical test for Einstein's theory. This shows how science continues to test theories over and over even after they are widely accepted.

A second test of the curvature of space-time was directly related to the motion of light through the curved space-time near the sun. The equations of general relativity predicted that light would be deflected

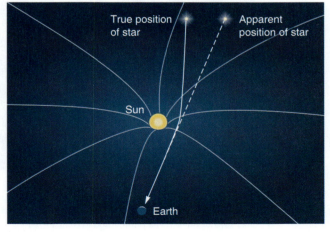

True position
of star

Apparent
position of star

Sun

Earth

FIGURE 5-14
Like a depression in a putting green, the curved space-time near the sun deflects light from distant stars and makes them appear to lie slightly farther from the sun than their true positions.

positions of the stars. During the eclipse, the positions of the stars on the plates were shifted outward, away from the sun (❙ Figure 5-15). If a star had been located at the edge of the solar disk, it would have been shifted outward by about 1.8 seconds of arc. This represents good agreement with the theory's prediction.

This test has been repeated at many total solar eclipses since 1919, with similar results. The most accurate results were obtained in 1973 when a Texas–Princeton team measured a deflection of 1.66 ± 0.18 seconds of arc—good agreement with Einstein's theory.

The general theory of relativity is critically important in modern astronomy. We will discuss it again when we meet black holes, distant galaxies, and the big bang universe. The theory revolutionized modern physics by providing a theory of gravity based on the geometry of curved space-time. Thus, Galileo's inertia and Newton's mutual gravitation are shown to be fundamental properties of space and time.

no longer be in a closed spaceship. As long as we make no outside observations, we can't tell whether our spaceship is firing its rockets and accelerating through space or resting on the surface of a planet where gravity gives us weight.

Einstein took the equivalence principle to mean that gravity and acceleration through space-time are somehow related. The general theory of relativity gives that relationship mathematical form and shows that gravity is really a disturbance in space-time that physicists refer to as curvature. Thus, we say "mass tells space-time how to curve, and space-time tells mass how to move." The equivalence principle led Einstein to an explanation for gravity.

But what about the second postulate of special relativity? Why does it have to be true if the first postulate is true? And what does the second postulate tell us about the nature of uniform motion?

REVIEW CRITICAL INQUIRY

What does the equivalence principle tell us?

The equivalence principle says that there is no observation we can make inside a closed spaceship to distinguish between uniform acceleration and gravitation. Of course, we could open a window and look outside, but then we would

Our discussion of the origin of astronomy began with the builders of Stonehenge and reaches the modern day with Einstein's general theory of relativity. Now that we have seen where astronomy came from, we are ready to see how it helps us understand the nature of the universe. Our first question should be "How do astronomers get information?" The answer involves the astronomer's most basic tool, the telescope; that is the subject of the next chapter.

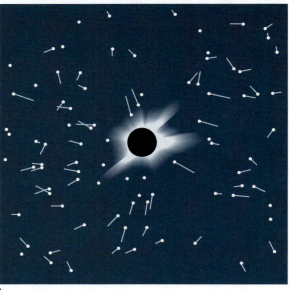

a   b

FIGURE 5-15

(a) Schematic drawing of the deflection of starlight by the sun's gravity. Dots show the true positions of the stars as photographed months before. Lines point toward the positions of the stars during the eclipse. (b) Actual data from the eclipse of 1922. Random uncertainties of observation cause some scatter in the data, but in general the stars appear to move away from the sun by 1.77 seconds of arc at the edge of the sun's disk. The deflection of stars is magnified by a factor of 2300 in both (a) and (b).

SUMMARY

Galileo took the first step toward understanding motion and gravity when he began to study falling bodies. He found that a falling object is accelerated; that is, it falls faster and faster with each passing second. The rate at which it accelerates, termed the acceleration of gravity, is 9.8 m/s² (32 ft/s²) at Earth's surface and does not depend on the weight of the object. According to tradition, Galileo demonstrated this by dropping balls of iron and wood from the Leaning Tower of Pisa to show that they would fall together. Finally, Galileo stated the law of inertia. In the absence of friction, a moving body on a horizontal plane will continue moving forever.

Newton adopted Galileo's law of inertia as his first law of motion. The second law of motion establishes the relationship between the force acting on a body, its mass, and the resulting acceleration. The third law says that forces occur in pairs acting in opposite directions.

Newton also developed an explanation for the accelerations that Galileo had discovered—gravity. By considering the motion of the moon, Newton was able to show that objects attract each other with a gravitational force that is proportional to the product of their masses and inversely proportional to the square of the distance between them.

If we understand Newton's laws of motion and gravity, we can better understand orbital motion. An object in space near Earth would move along a straight line and quickly leave Earth were it not for Earth's gravity accelerating the object toward Earth's center and forcing it to follow a curved path, an orbit. If there is no friction, the object will fall around its orbit forever.

Newton's laws also illuminate the meaning of Kepler's three laws of planetary motion. The planets follow elliptical orbits because gravity follows the inverse square law. The planets move faster when closer to the sun and slower when farther away because they conserve angular momentum. The same law makes ice skaters spin faster when they draw their arms and legs nearer their bodies. A planet's orbital period squared is proportional to its orbital radius cubed because the moving planet conserves energy.

In fact, Newtonian gravity and motion show that an ellipse is only one of a number of orbits that a body can follow. The circle and ellipse are closed orbits that return to their starting points. If an object moves at circular velocity, V_c, it will follow a circular orbit. If its velocity equals or exceeds the escape velocity, V_e, it will follow a parabola or hyperbola. These orbits are termed "open" because the object never returns to its starting place.

Newton's laws changed astronomy and our view of nature. They made it possible for astronomers to predict the motions of the heavenly bodies using the analytic power of mathematics. Newton's laws also show that the apparent complexity of the universe is based on a few simple principles, or natural laws.

Einstein published two theories that extended Newton's laws of motion and gravity. The special theory of relativity, published in 1905, applies to observers in uniform motion. The theory holds that the speed of light is a constant for all observers and that mass and energy are related by the expression $E = m_0c^2$.

The general theory of relativity, published in 1916, holds that a gravitational field is a curvature of space-time caused by the presence of a mass. Thus, Earth's mass curves space-time, and the mass of our bodies responds to that curvature by accelerating toward Earth's center. This curvature of space-time was confirmed by the slow advance in perihelion of the orbit of Mercury and by the deflection of starlight observed during a 1919 total solar eclipse.

NEW TERMS

natural motion	geosynchronous satellite
violent motion	center of mass
acceleration of gravity	closed orbit
momentum	escape velocity
mass	open orbit
acceleration	angular momentum
velocity	energy
inverse square law	joule (J)
field	special relativity
circular velocity	general theory of relativity

REVIEW QUESTIONS

Ace⊜Astronomy™ Assess your understanding of this chapter's topics with additional quizzing and animations at **http:// astronomy.brookscole.com/seeds8e**

1. Why wouldn't Aristotle's explanation of gravity work if Earth was not the center of the universe?

2. According to the principles of Aristotle, what part of the motion of a baseball pitched across the home plate is natural motion? What part is violent motion?

3. If we drop a feather and a steel hammer at the same moment, they should hit the ground at the same instant. Why doesn't this work on Earth, and why does it work on the moon?

4. What is the difference between mass and weight? between speed and velocity?

5. Why did Newton conclude that some force had to pull the moon toward Earth?

6. Why do we conclude that gravity has to be mutual and universal?

7. How does the concept of a field explain action at a distance? Name another kind of field also associated with action at a distance.

8. Why can't a spacecraft go "beyond Earth's gravity"?

9. What is the center of mass of the Earth–moon system? Where is it?

10. How do planets orbiting the sun and skaters conserve angular momentum?

11. Why is the period of an open orbit undefined?

12. How does the first postulate of special relativity imply the second?

13. When we ride a fast elevator upward, we feel slightly heavier as the trip begins and slightly lighter as the trip ends. How is this phenomenon related to the equivalence principle?

14. From your knowledge of general relativity, would you expect radio waves from distant galaxies to be deflected as they pass near the sun? Why or why not?

DISCUSSION QUESTIONS

1. How did Galileo idealize his inclines to conclude that an object in motion stays in motion until it is acted on by some force?

2. Give an example from everyday life to illustrate each of Newton's laws.

PROBLEMS

1. Compared with the strength of Earth's gravity at its surface, how much weaker is gravity at a distance of 10 Earth radii from Earth's center? at 20 Earth radii?

2. Compare the force of gravity on the surface of the moon with the force of gravity at Earth's surface.

3. If a small lead ball falls from a high tower on Earth, what will be its velocity after 2 seconds? after 4 seconds?

4. What is the circular velocity of an Earth satellite 1000 km above Earth's surface? (*Hint:* Earth's radius is 6380 km.)

5. What is the circular velocity of an Earth satellite 36,000 km above Earth's surface? What is its orbital period? (*Hint:* Earth's radius is 6380 km.)

6. What is the orbital period of an imaginary satellite orbiting just above Earth's surface? Ignore friction with the atmosphere.

7. Repeat the previous problem for Mercury, Venus, the moon, and Mars.

8. Describe the orbit followed by the slowest cannonball on page 84 on the assumption that the cannonball could pass freely through Earth. (Newton got this problem wrong the first time he tried to solve it.)

9. If you visited an asteroid 30 km in radius with a mass of 4×10^{17} kg, what would be the circular velocity at its surface? A major league fastball travels 90 mph. Could a good pitcher throw a baseball into orbit around the asteroid?

10. What is the orbital period of a satellite orbiting just above the surface of the asteroid in Problem 9?

11. What would be the escape velocity at the surface of the asteroid in Problem 9? Could a major league pitcher throw a baseball off of the asteroid?

CRITICAL INQUIRIES FOR THE WEB

1. Einstein's general theory of relativity predicts the curvature of space-time, but here on Earth we have little opportunity to observe such effects. Find an astronomical situation in which space-time curvature is evident from our observations, and describe the effect of the curvature on what we see when we view these objects.

2. Communications satellites are obvious uses of the geosynchronous orbit, but can you think of other uses for such orbits? Find an Internet site that uses or displays information gleaned from geosynchronous orbit that provides a useful service.

 Visit the Seeds *Foundations of Astronomy* companion Web site for critical thinking exercises, articles, and additional readings from InfoTrac College Edition, Brooks/Cole's online student library.

LIGHT AND TELESCOPES

He burned his house down
for the fire insurance
And spent the proceeds
on a telescope.

Robert Frost, The Star-Splitter

Courtesy William Keel

Chapter 6

GUIDEPOST

Previous chapters have described the sky as it appears to our unaided eyes, but modern astronomers turn powerful telescopes on the sky. Chapter 6 introduces us to the modern astronomical telescope and its delicate instruments.

The study of the universe is so challenging, astronomers cannot ignore any source of information; that is why they use the entire spectrum, from gamma rays to radio waves. This chapter shows how critical it is for astronomers to understand the nature of light.

In each of the chapters that follow, we will study the universe using information gathered by the telescopes and instruments described in this chapter.

What do fleas living on rats have to do with modern astronomy? That may sound like the beginning of a bad joke, but it is actually related to the subject of this chapter. We will examine the tools that modern astronomers use, and those tools are connected by an interesting sequence of events to rats and their fleas.

The most horrible disease in history—the black plague—was spread by fleabites, and the fleas lived on the rats that infested the cities. Plague broke out in London in 1665, and, although no one knew how the disease spread, people who lived in the country were less likely to get the plague than city dwellers. All who could left the cities. When the plague reached Cambridge, the colleges were closed, and both students and faculty fled to the English countryside. One who fled was the young Isaac Newton. From the summer of 1665 to 1667, he spent most of his time in his mother's cottage in the small village of Woolsthorpe. While he was there, he conducted an experiment that changed the history of science.

Boring a hole in a shutter, he admitted a thin beam of sunlight into his darkened room. A glass prism placed in the beam threw a rainbow of color—a spectrum—across the wall. When he used a second prism to recombine the colors, they produced white light. From this and other experiments conducted in his bedroom, he concluded that white light was made up of a mixture of all the colors of the rainbow.

When the plague abated, Newton returned to the university, where he began experimenting with telescopes. He discovered that telescopes made of lenses produced colored fringes around bright objects in the field of view because the glass lenses broke the light into colors, just as his prism broke up the sunlight. To solve the problem, Newton designed and built a telescope containing a mirror instead of a lens. Although his first model hardly exceeded 1 inch in diameter, when Newton presented it to the Royal Society in 1671, it established his reputation as a scientist.

For a century after Newton's first telescope, astronomers did little with such devices, but as instrument makers grew more skilled, large telescopes became the principal tool of the astronomer. Telescopes are important in astronomy because they gather light and concentrate it for study. The larger the telescope, the more light it gathers. Thus astronomers are still striving to build bigger telescopes to gather more light from the objects in the sky (❙ Figure 6-1). Like Newton's original telescope, almost all modern telescopes use mirrors rather than lenses to avoid spreading the light into its component colors.

6-1 RADIATION: INFORMATION FROM SPACE

Just as a book on bread baking might begin with a discussion of flour, our chapter on telescopes begins with a discussion of light—not just visible light, but the entire range of radiation from the sky.

LIGHT AS A WAVE AND A PARTICLE

Light is merely one form of radiation, called **electromagnetic radiation** because it is associated with changing electric and magnetic fields that travel through

 Ace✸Astronomy™ The AceAstronomy icon throughout the text indicates an opportunity for you to test yourself on key concepts and to explore animations and interactions on the AceAstronomy Web site at **http://astronomy.brookscole.com/seeds8e**

FIGURE 6-1
Astronomers build giant telescopes on high mountains where the air is clear and steady. (a) In this time exposure, star trails are visible behind the dome of the northern Gemini Telescope. The tail lights of a vehicle illuminate the road and a shooting star (a meteor) is visible near the horizon. (b) Inside the dome, the telescope holds a mirror 8.2 meters in diameter to focus light. Note the figure at right for scale. *(NOAO/AURA/NSF)*

a   b

space and transfer energy from one place to another. When light enters our eye, the fluctuating electric and magnetic fields carry energy that stimulates nerve endings, and we see what we call light.

The oscillating electric and magnetic fields that constitute electromagnetic radiation move through space at about 300,000 km/s (186,000 mi/s). This speed is commonly referred to as the speed of light, *c,* but it is in fact the speed of all such radiation in a vacuum.

It may seem odd to use the word "radiation" when we speak of light. The word can be used to refer to high-energy particles emitted from radioactive atoms, and we have all leaned to be a little bit concerned when we see the word "radiation." But it really refers to anything that spreads outward from a source. Light radiates from a source, so we can correctly refer to light as a form of radiation.

Electromagnetic radiation is a wave phenomenon; that is, it is associated with a periodically repeating disturbance, or wave. We are familiar with waves in water. If we disturb a quiet pool of water, waves spread across the surface. Imagine that we use a meter stick to measure the distance between the successive peaks of a wave. This distance is the **wavelength,** usually represented by the Greek letter lambda (λ). If we were measuring ripples in a pond, we might find that the wavelength is a few centimeters, whereas the wavelength of ocean waves might be a hundred meters or more. There is no restriction on the wavelength of electromagnetic radiation. Wavelengths can range from smaller than the diameter of an atom to larger than that of Earth.

Because all electromagnetic radiation travels at the speed of light, wavelength is related to **frequency,** the number of cycles that pass in one second. Short-wavelength radiation has a high frequency; long-wavelength radiation has a low frequency. To understand this, imagine watching an electromagnetic wave race past us while we count its peaks (■ Figure 6-2). If the wavelength is short, we will count many peaks in one second; if the wavelength is long, we will count few peaks per second. The dials on radios are marked in frequency, but they could just as easily be marked in wavelength. The relation between wavelength and frequency is a simple one:

$$\lambda = \frac{c}{f}$$

That is, the wavelength equals the speed of light *c* divided by the frequency *f*. Notice that the larger (higher) the frequency, the smaller (shorter) the wavelength. In most cases, astronomers use wavelength rather than frequency.

Radio waves can have wavelengths from a few millimeters for microwaves to kilometers. In contrast, the wavelength of light is so short that we must use more convenient units. In this book, we will use **nanometers (nm)** because this unit is consistent with the International System of units. One nanometer is 10^{-9} meter,

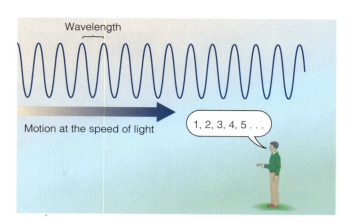

FIGURE 6-2

All electromagnetic waves travel at the speed of light. The wavelength is the distance between successive peaks. The frequency of the wave is the number of peaks that pass us in one second.

and visible light has wavelengths that range from about 400 nm to about 700 nm. Another unit that astronomers commonly use, and a unit that you will see in many references on astronomy, is the **Angstrom (Å).** One Angstrom is 10^{-10} meter, and visible light has wavelengths between 4000 Å and 7000 Å.

You may find radio astronomers describing wavelengths in centimeters or millimeters, and infrared astronomers often refer to wavelengths in micrometers (or microns). One micrometer (μm) is 10^{-6} meter. Whatever unit is used to describe the wavelength, we must keep in mind that all electromagnetic radiation is the same phenomenon.

What exactly is electromagnetic radiation? Although we have been discussing its wavelength, it is incorrect, or at least incomplete, to say that electromagnetic radiation is a wave. It sometimes has the properties of a wave and sometimes has the properties of a particle. For instance, the beautiful colors in a soap bubble arise from the wave nature of light. On the other hand, when light strikes the photoelectric cell in a camera's light meter, it behaves like a stream of particles carrying specific amounts of energy. Throughout his life, Newton believed that light was made up of particles, but we now recognize that light can behave as both particle and wave. Our model of light is thus more complete than Newton's. We will refer to "a particle of light" as a **photon,** and we can recognize its dual nature by thinking of it as a bundle of waves.

The amount of energy a photon carries depends on its wavelength. The shorter the wavelength, the more energy the photon carries; the longer the wavelength, the less energy it contains. This is easy to remember because short wavelengths have high frequencies, and we expect rapid fluctuations to be more energetic. We can express this relationship in a simple formula:

$$E = \frac{hc}{\lambda}$$

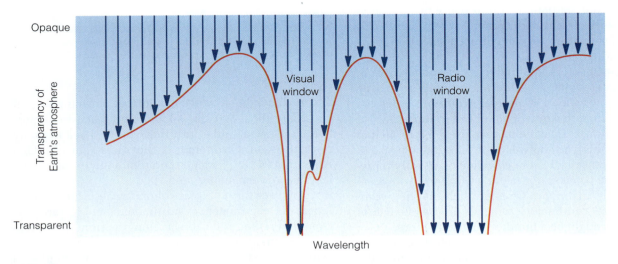

FIGURE 6-3

The spectrum of visible light, extending from red to violet, is only part of the electromagnetic spectrum. Most radiation is absorbed in Earth's atmosphere, and only radiation in the visual window and the radio window can reach Earth's surface

Here h is Planck's constant (6.6262×10^{-34} joule s), c is the speed of light (3×10^8 m/s), and λ is the wavelength in meters. A photon of visible light carries a very small amount of energy, but a photon with a very short wavelength can carry much more.

THE ELECTROMAGNETIC SPECTRUM

A spectrum is an array of electromagnetic radiation in order of wavelength. We are most familiar with the spectrum of visible light, which we see in rainbows. The colors of the spectrum differ in wavelength, with red having the longest wavelength and violet the shortest. The visible spectrum is shown at the top of ▌Figure 6-3.

The average wavelength of visible light is about 0.00005 cm. We could put 50 light waves end to end across the thickness of a sheet of household plastic wrap. Measured in nanometers, the wavelength of visible light ranges from about 400 to 700 nm. Just as we sense the wavelength of sound as pitch, we sense the wavelength of light as color. Light near the short-wavelength end of the visible spectrum (400 nm) looks violet to our

eyes, and light near the long-wavelength end (700 nm) looks red.

Figure 6-3 shows how the visible spectrum makes up only a small part of the entire electromagnetic spectrum. Beyond the red end of the visible spectrum lies **infrared radiation,** where wavelengths range from 700 nm to about 0.1 cm. Our eyes are not sensitive to this radiation, but our skin senses it as heat. A "heat lamp" is just a bulb that gives off principally infrared radiation.

Beyond the infrared part of the electromagnetic spectrum lie radio waves. Microwaves have wavelengths of a millimeter to a few centimeters and are used for radar and long-distance telephone communication. Longer wavelengths are used for UHF and VHF television transmissions. FM, military, governmental, and ham radio signals have wavelengths up to a few meters, and AM radio waves can have wavelengths of kilometers.

The distinction between the wavelength ranges is not sharp. Long-wavelength infrared radiation and the shortest microwave radio waves are the same. Similarly, there is no clear division between the short-wavelength infrared and the long-wavelength part of the visible spectrum. It is all electromagnetic radiation.

Look once again at the electromagnetic spectrum in Figure 6-3, and notice that electromagnetic waves shorter than violet are called **ultraviolet.** Electromagnetic waves even shorter are called X rays, and the shortest are gamma rays. Again, the boundaries between these wavelength ranges are not clearly defined.

X rays and gamma rays can be dangerous, and even ultraviolet photons have enough energy to do us harm. Small doses produce a suntan and larger doses sunburn and skin cancers. Contrast this to the lower-energy infrared photons. Individually they have too little energy to affect skin pigment, a fact that explains why you can't get a tan from a heat lamp. Only by concentrating many low-energy photons in a small area, as in a microwave oven, can we transfer significant amounts of energy.

We are interested in electromagnetic radiation because it brings us clues to the nature of stars, planets, and other celestial objects. Earth's atmosphere is opaque to most electromagnetic radiation, as shown by the graph at the bottom of Figure 6-3. Gamma rays, X rays, and some radio waves are absorbed high in Earth's atmosphere, and a layer of ozone (O_3) at an altitude of about 30 km absorbs ultraviolet radiation. Water vapor in the lower atmosphere absorbs the longer wavelength infrared radiation. Only visible light, some shorter wavelength infrared, and some radio waves reach Earth's surface through two wavelength regions called **atmospheric windows.** Obviously, if we wish to study the sky from Earth's surface, we must look out through one of these windows.

REVIEW CRITICAL INQUIRY

What could we see if our eyes were sensitive only to X rays?

Sometimes the critical analysis of an idea is easier if we try to imagine a totally new situation. In this case, we might at first expect to be able to see through walls, but remember that our eyes detect only light that already exists. There are almost no X rays bouncing around at Earth's surface, so if we had X-ray eyes, we would be in the dark and would be unable to see anything. Even when we look up at the sky, we would see nothing, because Earth's atmosphere is not transparent to X rays. If Superman can see through walls, it is not because his eyes can detect X rays.

But suppose our eyes were sensitive only to infrared waves or to radio waves. Would we be in the dark?

Now that we know something about electromagnetic radiation, we can study the tools astronomers use to gather and analyze that radiation.

6-2 OPTICAL TELESCOPES

Astronomers build optical telescopes to gather light and focus it into sharp images. This requires sophisticated optical and mechanical designs, and it leads astronomers to build gigantic telescopes on the tops of high mountains. To begin, we need to understand the terminology of telescopes, but it is more important to understand how different kinds of telescopes work and why some are better than others.

TWO KINDS OF TELESCOPES

Astronomical telescopes focus light into an image in one of two ways, as shown in ▌Figure 6-4. A lens bends (refracts) the light as it passes through the glass and brings it to a focus to form a small inverted image. A mirror—a concave piece of glass with a reflective surface—forms an image by reflecting the light. In either case, the **focal length** is the distance from the lens or mirror to the image formed of a distant light source, such as a star (▌Figure 6-5). Short-focal-length lenses and mirrors must be strongly curved, and long-focal-length lenses and mirrors are less strongly curved. Grinding the proper shape on a lens or mirror is a delicate, time-consuming, and expensive process.

Because there are two ways to focus light, there are two kinds of astronomical telescopes. **Refracting telescopes** use a large lens to gather and focus the light, whereas **reflecting telescopes** use a concave mirror. The advantages of the reflecting telescope have made it the preferred design for modern observatories.

The main lens in a refracting telescope is called the **primary lens,** and the main mirror in a reflecting telescope is called the **primary mirror.** These are also called the **objective lens** and **mirror.** Both kinds of telescopes form a very small, inverted image that is difficult to observe directly, so astronomers use a small lens called the **eyepiece** to magnify the image and make it convenient to view (▌Figure 6-6).

Refracting telescopes suffer from a serious optical distortion that limits their usefulness. When light is refracted through glass, shorter wavelengths bend more than longer wavelengths, and blue light comes to a focus closer to the lens than does red light (▌Figure 6-7a). If we focus the eyepiece on the blue image, the red light is out of focus, and we see a red blur around the image. If we focus on the red image, the blue light blurs. The color separation is called **chromatic aberration.** Telescope designers can grind a telescope lens of two components made of different kinds of glass and so bring two different wavelengths to the same focus (Figure 6-7b). This does improve the image, but these **achromatic lenses** are not totally free of chromatic aberration, because other wavelengths still blur. Telescopes made with such lenses were popular until the end of the 19th century.

FIGURE 6-4

(a) To see how a lens can focus light, we trace four light rays from the flame and base of a candle through a lens, where they are refracted to form an inverted image.
(b) A mirror forms an image by reflection from a concave surface. Notice that the light is reflected from the aluminized front surface of the mirror and does not enter the glass.

The primary lens of a refracting telescope is very expensive to make because it must be achromatic, and the glass must be pure and flawless because the light passes through the lens. The four surfaces must be ground precisely, and the lens can be supported only along its edge. The largest refracting telescope in the world was completed in 1897 at Yerkes Observatory in Wisconsin. Its lens is 1 m (40 in.) in diameter and weighs half a ton. Larger refracting telescopes are prohibitively expensive.

Reflecting telescopes are much less expensive because the light reflects from the front surface of the mirror. Consequently only the front surface need be ground to precise shape. Also, the glass of the mirror need not be perfectly transparent, and the mirror can be supported over its back surface to reduce sagging.

Most important, reflecting telescopes do not suffer from chromatic aberration because the light is reflected toward the focus before it can enter the glass. Thus, every large astronomical telescope built since the beginning of the 20th century has been a reflecting telescope.

THE POWERS OF A TELESCOPE

A telescope can aid our eyes in only three ways—the three powers of a telescope. They make images brighter, more detailed, and larger.

Most interesting celestial objects are faint sources of light, so we need a telescope that can gather large amounts of light to produce a bright image. **Light-gathering power** refers to the ability of a telescope to collect light. Catching light in a telescope is like catch-

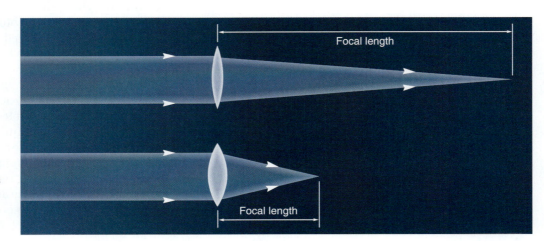

FIGURE 6-5
The focal length of a lens is the distance from the lens to the point where parallel rays of light come to a focus. The lens at the top has a longer focal length than does the lens at the bottom.

ACTIVE FIGURE 6-6

(a) A refracting telescope uses a primary lens to focus starlight into an image that is magnified by a lens called an eyepiece. The primary lens has a long focal length, and the eyepiece has a short focal length. (b) A reflecting telescope uses a primary mirror to focus the light by reflection. A small secondary mirror reflects the starlight back down through a hole in the middle of the primary mirror to the eyepiece.

Ace Astronomy™ Go to AceAstronomy and click Active Figures to see "Refractors and Reflectors." You can watch light flowing through the telescopes.

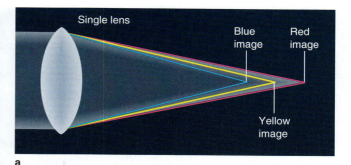

FIGURE 6-7

(a) A normal lens suffers from chromatic aberration because short wavelengths bend more than long wavelengths. (b) An achromatic lens, made in two parts, can bring any two colors to the same focus, but other colors remain slightly out of focus.

ing rain in a bucket—the bigger the bucket, the more rain it catches (❚ Figure 6-8). Light-gathering power is proportional to the area of the telescope objective. A lens or mirror with a large area gathers a large amount of light. The area of a circular lens or mirror of diameter D is $\pi\left(\frac{D}{2}\right)^2$. To compare the relative light-gathering powers (*LGP*) of two telescopes *A* and *B*, we can calculate the ratio of the areas of their objectives, which reduces to the ratio of their diameters (*D*) squared.

$$\frac{LGP_A}{LGP_B} = \left(\frac{D_A}{D_B}\right)^2$$

For example, suppose we compare a telescope 24 cm in diameter with a telescope 4 cm in diameter. The ratio of the diameters is 24/4, or 6, but the larger telescope does not gather 6 times as much light. Light-gathering power increases as the ratio of diameters squared, so it gathers 36 times more light than the smaller telescope. This example shows the importance of diameter in astronomical telescopes. Even a small increase in diameter produces a large increase in light-gathering power and allows astronomers to study much fainter objects.

FIGURE 6-8

Gathering light is like catching rain in a bucket. A large-diameter telescope gathers more light and has a brighter image than a smaller telescope of the same focal length.

(a) Stars are so far away that their images are points, but the wave nature of light surrounds each star image with diffraction fringes (much magnified in this computer model). (b) Two stars close to each other have overlapping diffraction fringes and become impossible to detect separately. *(Computer model by M. A. Seeds)*

Ace ⊙ **Astronomy**™ Go to AceAstronomy and click Active Figures to see "Resolution and Telescopes." Control mirror diameter and watch resolution change.

a

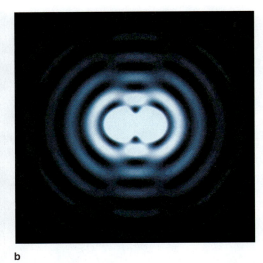

b

The second power, **resolving power,** refers to the ability of the telescope to reveal fine detail. Because light acts as a wave, it produces a small **diffraction fringe** around every point of light in the image, and we cannot see any detail smaller than the fringe (▌ Figure 6-9). Astronomers can't eliminate diffraction fringes, but the larger a telescope is in diameter, the smaller the diffraction fringes are. Thus the larger the telescope, the better its resolving power.

For optical telescopes, we estimate the resolving power by calculating the angular distance between two stars that are just barely visible through the telescope as two separate images. The resolving power, α, in seconds of arc, equals 11.6 divided by the diameter of the telescope in centimeters:

$$\alpha = \frac{11.6}{D}$$

For example, the resolving power of a 25-cm telescope is 11.6 divided by 25, or 0.46 second of arc. No matter how perfect the telescope optics, this is the smallest detail we can see through that telescope.

In addition to resolving power, two other factors—lens quality and atmospheric conditions—limit the detail we can see through a telescope. A telescope must contain high-quality optics to achieve its full potential resolving power. Even a large telescope shows us little detail if its optics are marred with imperfections. Also, when we look through a telescope, we are looking through miles of turbulent air in Earth's atmosphere, which makes the image dance and blur, a condition called **seeing.** On a night when the atmosphere is unsteady and the images are blurred, the seeing is bad (▌ Figure 6-10). Even under good seeing conditions, the detail visible through a large telescope is limited, not by its diffraction fringes, but by the air through which the telescope must look. A telescope performs better on a high mountaintop where the air is thin and steady,

but even there Earth's atmosphere limits the detail the best telescopes can reveal to about 0.5 second of arc.

This limitation on the amount of information in an image is related to the limitation on the accuracy of a measurement. All measurements have some built-in uncertainty (Window on Science 6-1), and scientists must learn to work within those limitations.

The third and least important power of a telescope is **magnifying power,** the ability to make the image bigger. Because the amount of detail we can see is limited

Visual-wavelength image

FIGURE 6-10

The left half of this photograph of a galaxy is from an image recorded on a night of poor seeing. Small details are blurred. The right half of the figure is from an image recorded on a night when Earth's atmosphere above the telescope was steady and the seeing was better. Much more detail is visible under good seeing conditions. *(Courtesy William Keel)*

Resolving Power and the Resolution of a Measurement

Have you ever seen a movie in which the hero magnifies a newspaper photo and reads some tiny detail? It isn't really possible, because newspaper photos are made up of tiny dots of ink, and no detail smaller than a single dot will be visible no matter how much you magnify the photo. In fact, all images are made up of elements of some sort, and that means there is a limit to the amount of detail you can see in an image. In an astronomical image, the resolution is often set by seeing. It is foolish to

attempt to see a detail in the image that is smaller than the resolution.

This limitation is true of all measurements in science. A zoologist might be trying to measure the length of a live snake, or a sociologist might be trying to measure the attitudes of people toward drunk driving, but both face limits to the resolution of their measurements. The zoologist might specify that the snake was 43.28932 cm long, and the sociologist might say that 98.2491 percent of people oppose drunk driving, but a critic

might point out that it isn't possible to make these measurements that accurately. The resolution of the techniques does not justify the accuracy implied.

Science is based on measurement, and whenever we make a measurement we should ask ourselves how accurate that measurement can be. The accuracy of the measurement is limited by the resolution of the measurement technique, just as the amount of detail in a photograph is limited by the resolution of the photo

---

by the seeing conditions and the resolving power, very high magnification does not necessarily show us more detail. Also, we can change the magnification by changing the eyepiece, but we cannot alter the telescope's light-gathering power or resolving power.

We calculate the magnification of a telescope by dividing the focal length of the objective by the focal length of the eyepiece:

$$M = \frac{F_o}{F_e}$$

For example, if a telescope has an objective with a focal length of 80 cm and we use an eyepiece whose focal length is 0.5 cm, the magnification is 80/0.5, or 160 times.

The search for light-gathering power and high resolution explains why nearly all major observatories are located far from major cities and usually on high mountains. Astronomers avoid cities because **light pollution,** the brightening of the night sky by light scattered from artificial outdoor lighting, can make it impossible to see faint objects (❙ Figure 6-11). In fact, many residents of cities are unfamiliar with the beauty of the night sky because they can see only the brightest stars. Nevertheless, nature's own light pollution, the moon, is so bright it drowns out fainter objects, and astronomers are often unable to observe on the nights near full moon when faint objects cannot be observed even with the largest telescopes on high mountains.

Astronomers prefer to place their telescopes on carefully selected high mountains. The air there is thin and more transparent, but, most important, astronomers select mountains where the air flows smoothly and is not turbulent. This produces the best seeing. Building an observatory on top of a high mountain far from civilization is difficult and expensive, but the dark sky and steady seeing make it worth the effort (❙ Figure 6-12).

BUYING A TELESCOPE

Thinking about how we should shop for a new telescope will not only help us if we decide to buy one but will also illustrate some important points about astronomical telescopes.

Assuming we have a fixed budget, we should buy the highest-quality optics and the largest-diameter telescope we can afford. Of the two things that limit what we see, optical quality is under our control. We can't make the atmosphere less turbulent, but we should buy good optics. If we buy a telescope from a toy store and it has plastic lenses, we shouldn't expect to see very much. Also, we want to maximize the light-gathering power of our telescope, so we want to purchase the largest-diameter telescope we can afford. Given a fixed budget, that means we should buy a reflecting telescope rather than a refracting telescope. Not only will we get more diameter per dollar, but our telescope will not suffer from chromatic aberration.

We can safely ignore magnification. Department stores and camera stores may advertise telescopes by quoting their magnification, but it is not an important number. What we can see is fixed by light-gathering power, optical quality, and Earth's atmosphere. Besides, we can change the magnification by changing eyepieces.

Other things being equal, we should choose a telescope with a solid mounting that will hold the telescope steady and allow us to point at objects easily. Computer-controlled pointing systems are available for a price on many small telescopes. A good telescope on a poor mounting is almost useless.

We might be buying a telescope to put in our backyard, but we must think about the same issues astronomers consider when they design giant telescopes to go on mountaintops. In fact, some of the newest telescopes solve these traditional problems in new ways.

Visual-wavelength image

FIGURE 6-11

This satellite view of the continental United States at night shows the light pollution produced by outdoor lighting. Not only does the glare drown out the fainter stars and interfere with astronomy, but it wastes electrical power. Many astronomers work with city governments to enact laws that improve lighting on the ground and reduce light scattered into the night sky. *(NOAA)*

FIGURE 6-13

Astronomers have begun building multiple telescopes to solve different problems. Two Gemini telescopes have been built, one in Hawaii and one in Chile, to observe the entire sky. The Large Binocular Telescope (LBT) carries two 8.4-m mirrors that combine their light. The four telescopes of the VLT are housed in separate domes, but they can combine to work as a very large telescope. *(Gemini: NOAO/ AURA/NSF; LBT: Large Binocular Telescope Project and European Industrial Engineer; VLT: ESO)*

The LBT has the light-gathering power of a 11.8-m telescope and the resolving power of a 22.8-m telescope.

The mirrors in the VLT telescopes are 8.2 m in diameter.

The two Gemini telescopes have 8.1-m mirrors with active optics.

NEW-GENERATION TELESCOPES

For most of the 20th century, astronomers faced a serious limitation on the size of astronomical telescopes. Traditional telescope mirrors were made thick to avoid sagging that would distort the reflecting surface, but those thick mirrors were heavy. The 5-m (200-in.) mirror on Mount Palomar weighs 14.5 tons. These traditional telescopes were big, heavy, and expensive.

Modern astronomers have solved these problems in a number of ways. Study "▌Modern Astronomical Telescopes" (pages 108 and 109) and notice three important advances in telescope design made possible by high-speed computers. First, astronomers can now build simpler, stronger telescope mountings and depend on computers to move the telescope and follow the westward motion of the stars as Earth rotates.

Second, notice that computer control of the shape of telescope mirrors allows the use of thin, lightweight

FIGURE 6-12

The domes of four giant telescopes are visible at upper left at Paranal Observatory, built by the European Southern Observatory at an altitude of 2635 m (8660 ft) atop the mountain Cerro Paranal in the Atacama desert of northern Chile. Located 120 km (75 mi) from the nearest city, the mountaintop was chosen for its steady air and isolation from lights. It is believed to be the driest pace on Earth. *(ESO)*

mirrors—either "floppy" mirrors or segmented mirrors. Lowering the weight of the mirror lowers the weight of the rest of the telescope and makes it stronger and less expensive. Also, thin mirrors cool faster at nightfall and produce better images.

The third advance to notice is the way astronomers use high-speed computers to reduce seeing distortion caused by Earth's atmosphere. Only a few decades ago, many astronomers argued that it wasn't worth building more large telescopes on Earth's surface because of the limitations set by seeing. Now a number of new giant telescopes have been built, and more are in development that can partially overcome the seeing problem.

An international collaboration of astronomers have built the Gemini telescopes with 8.1-m thin mirrors (▌Figure 6-13). One is located in the northern hemisphere and one in the southern hemisphere to cover the entire sky. The European Southern Observatory has built the Very Large Telescope (VLT) high in the remote Andes mountains of northern Chile (Figure 6-13). The VLT consists of four telescopes with computer-controlled mirrors 8.2 m in diameter and only 17.5 cm (6.9 in.) thick. The four telescopes can work singly or can combine their light to work as one large telescope. Italian and American astronomers are building the Large Binocular Telescope, which carries a pair of

Modern Astronomical Telescopes

The traditional telescopes described on this page are limited by complexity, weight, and Earth's atmosphere. Modern solutions are shown on the opposite page.

In larger telescopes the light can be focused to a **prime focus** *position high in the telescope tube. Although it is a good place to image faint objects, the prime focus is inconvenient for large instruments. A* **secondary mirror** *can reflect the light through a hole in the primary mirror to a* **Cassegrain focus.** *This focal arrangement may be the most common form of astronomical telescope.*

Secondary mirror

With the secondary mirror removed, the light converges at the prime focus. In large telescopes, astronomers can ride inside the prime-focus cage, although most observations are now made by instruments connected to computers in a separate control room.

The Cassegrain focus is convenient and has room for large instruments.

Traditional mirrors are thick to prevent the optical surface from sagging and distorting the image as the telescope is moved around the sky. Large mirrors can weigh many tons and are expensive to make and difficult to support. Also, they cool slowly at nightfall. Expansion and contraction in the cooling mirror causes distortion in the images.

Smaller telescopes are often found with a **Newtonian focus,** the arrangement that Isaac Newton used in his first reflecting telescope. The Newtonian focus is inconvenient for large telescopes.

Newtonian focus

The 4-meter Mayall Telescope at Kitt Peak National Observatory in Arizona can be used at either the prime focus or the Cassegrain focus. Note the human figure at lower right.

Many small telescopes use a **Schmidt-Cassegrain focus.** A thin correcting plate improves the image but is too slightly curved to introduce serious chromatic aberration.

Thin correcting lens

Schmidt-Cassegrain telescope

Prime focus cage

Secondary mirror

Cassegrain focus

Primary mirror (inside)

Equatorial mounting

Westward rotation about polar axis follows stars.

To north celestial pole

Polar axis

North Pole

Eastward rotation of Earth

Alt-azimuth mounting

Computer control of motion about both axes follows stars.

To north celestial pole

To north celestial pole

North Pole

Eastward rotation of Earth

Telescope mountings must contain a **sidereal drive** to move smoothly westward and counter the eastward rotation of Earth. The traditional **equatorial mounting** has a **polar axis** parallel to Earth's axis, but the modern **alt-azimuth mounting** moves like a cannon — up and down and left to right. Such mountings are simpler to build but need computer control to follow the stars.

Unlike traditional thick mirrors, thin mirrors, sometimes called floppy mirrors, weigh less and require less massive support structures. Also, they cool rapidly at nightfall and there is less distortion from uneven expansion and contraction.

Floppy mirror

Computer-controlled thrusters   Support structure

Grinding a large mirror may remove tons of glass and take months, but new techniques speed the process. Some large mirrors are cast in a rotating oven that causes the molten glass to flow to form a concave upper surface. Grinding and polishing such a preformed mirror is much less time consuming.

Mirrors made of segments are economical because the segments can be made separately. The resulting mirror weighs less and cools rapidly.

Segmented mirror

Computer-controlled thrusters   Support structure

The two largest telescopes in the world, the Keck I and Keck II telescopes in Hawaii, contain segmented mirrors 10 m in diameter.

The thrusters are located behind the mirror segments in this face-on photo of the Keck I mirror. The technician kneels in the light baffle in front of the Cassegrain hole in the center of the mirror.

Both floppy mirrors and segmented mirrors sag under their own weight. Their optical shape must be controlled by computer-driven thrusters under the mirror in what is called **active optics.**

Adaptive optics off

Adaptive optics on

Object revealed as a pair of stars.

1 second of arc

Adaptive optics uses high-speed computers to monitor the image distortion caused by Earth's atmosphere and adjust the optics many times a second to compensate. This can reduce the blurring due to seeing and dramatically improve image quality in Earth-based telescopes.

8.4-m mirrors on a single mounting (Figure 6-13). Around the world astronomers are drawing plans for large telescopes, including truly gigantic instruments with segmented mirrors 30 m and even 100 m in diameter. Computer control of the optics makes such huge telescopes worth considering.

Computer control has made massive surveys of the sky practical. We have mentioned earlier the Hipparchos satellite that surveyed the entire sky, measuring the parallax of over a million years. Other surveys are being made. The Sloan Digital Sky Survey is mapping the sky, measuring the position and brightness of 100 million stars and galaxies at a number of wavelengths. The Two-Micron All Sky Survey (2MASS) has mapped the entire sky at three wavelengths in the infrared. Other surveys are being made at many other wavelengths. Every night large telescopes scan the sky, and billions of bytes of data are compiled automatically in immense sky atlases. Astronomers will study those data banks for decades to come.

The days when astronomers worked beside their telescopes through long, dark, cold nights are nearly gone. The complexity and sophistication of telescopes require a battery of computers, and almost all research telescopes are run from control rooms that astronomers call warm rooms. Astronomers don't need to be kept warm, but computers demand comfortable working conditions (❚ Figure 6-14).

INTERFEROMETRY

One of the reasons astronomers build big telescopes is to increase resolving power, and astronomers have been able to achieve very high resolution by connecting multiple telescopes together to work as if they were

FIGURE 6-15
In an astronomical interferometer, smaller telescopes can combine their light through specially designed optical tunnels to simulate a larger telescope with a resolution set by the separation of the smaller telescopes

a single telescope. This method of synthesizing a larger telescope is known as **interferometry** (❚ Figure 6-15).

To work as an interferometer, the separate telescopes must combine their light through a network of mirrors, and the path that each light beam travels must be controlled so that it does not vary more than some small fraction of the wavelength. Turbulence in Earth's atmosphere constantly distorts the light, and the high-speed computers must continuously adjust the light paths. Recall that the wavelength of light is very short, roughly 0.0005 mm, so building optical interferometers is one of the most difficult technical problems that astronomers face. Infrared- and radio-wavelength interferometers are slightly easier to build because the wavelengths are longer. In fact, as we will discover later in this chapter, the first astronomical interferometers worked at radio wavelengths.

The VLT shown in Figure 6-13 consists of four 8.2-m telescopes that can operate separately, but they can be linked together through underground tunnels with three 1.8-m telescopes on the same mountaintop. The resulting optical interferometer provides the resolution of a telescope 200 meters in diameter.

Other telescopes can work as interferometers. The two Keck 10-m telescopes can be used as an interferometer. The Navy Prototype Optical Interferometer built near Flagstaff, Arizona has small telescopes located along three arms up to 250 m in length. It is being used to study technical aspects of optical interferometry and to make high-precision measures of star positions. The CHARA array on Mt. Wilson combines six 1-meter telescopes to create the equivalent of a tele-

FIGURE 6-14
In a telescope control room, computers monitor everything from the shape of the optics to the weather forecast. Furthermore, most instruments such as cameras and spectrographs are controlled by computers and store their data in computer memory. Astronomers work through the night controlling the computers that control the telescope and its instruments. *(AAO)*

scope one-fifth of a mile in diameter. The Large Binocular Telescope shown in Figure 6-13 can be used as an interferometer.

Although turbulence in Earth's atmosphere can be partially averaged out in an interferometer, plans are being made to put interferometers in space. The Space Interferometry Mission, for example, will work at visual wavelengths and study everything from the cores of erupting galaxies to planets orbiting nearby stars.

REVIEW CRITICAL INQUIRY

Why do astronomers build observatories at the tops of mountains?

Astronomers have joked that the hardest part of building a new observatory is constructing the road to the top of the mountain. It certainly isn't easy to build a large, delicate telescope at the top of a high mountain, but it is worth the effort. A telescope on top of a high mountain is above the thickest part of Earth's atmosphere. There is less air to dim the light, and there is less water vapor to absorb infrared radiation. Even more important, the thin air on a mountaintop causes less disturbance to the image, and thus the seeing is better. A large telescope on Earth's surface has a resolving power much better than the distortion caused by Earth's atmosphere. Thus, it is limited by seeing, not by its own diffraction. It really is worth the trouble to build telescopes atop high mountains.

Astronomers not only build telescopes on mountaintops; they also build gigantic telescopes many meters in diameter. What are the problems and advantages in building such giant telescopes?

Astronomers sometimes refer to a telescope that produces distorted images as a "light bucket." In a sense, all astronomical telescopes are light buckets, because the light they focus into images tells us very little until it is recorded and analyzed by special instruments attached to the telescopes.

6-3 SPECIAL INSTRUMENTS

Looking through a telescope doesn't tell us much. To use an astronomical telescope to learn about stars, we must be able to analyze the light the telescope gathers. Special instruments attached to the telescope make that possible.

IMAGING SYSTEMS

The original imaging device in astronomy was the photographic plate. It could record faint objects in long exposures and could be stored for later analysis. But photographic plates have been almost entirely replaced in astronomy by electronic imaging systems.

Most modern astronomers use **charge-coupled devices (CCDs)** to record images. A CCD is a specialized computer chip containing roughly a million microscopic light detectors arranged in an array about the size of a postage stamp. These devices can be used like a small photographic plate, but they have dramatic advantages. They can detect both bright and faint objects in a single exposure, are much more sensitive than a photographic plate, and can be read directly into computer memory for later analysis. Although CCDs for astronomy are extremely sensitive and therefore expensive, less sophisticated CCDs are used in most video cameras and digital cameras.

The image from a CCD is stored as numbers in computer memory, so it is easy to manipulate the image to bring out details that would not otherwise be visible. For example, astronomical images are often reproduced as negatives with the sky white and the stars dark. This makes the faint parts of the image easier to see (∎ Figure 6-16a). Astronomers also manipulate images to produce **false-color images** in which the colors represent different levels of intensity and are not related to the true colors of the object (Figure 6-16b).

Measurements of intensity and color were made in the past using a photometer, a highly sensitive light meter attached to a telescope. Today, however, most

Visual-wavelength image

a

Visual-wavelength image

b

FIGURE 6-16

(a) This CCD image of the star cluster known as M3 has been reproduced as a negative to reveal detail such as faint stars in the cluster visible in the inset. *(A. Saha, NOAO/WIYN/NOAO/NSF)* (b) This CCD image of two galaxies has been given false color according to brightness. The faintest regions are blue and green, and the brighter regions are red and white. The colors in such false-color images are not usually related to the actual colors of the object. *(WIYN/NOAO/NSF)*

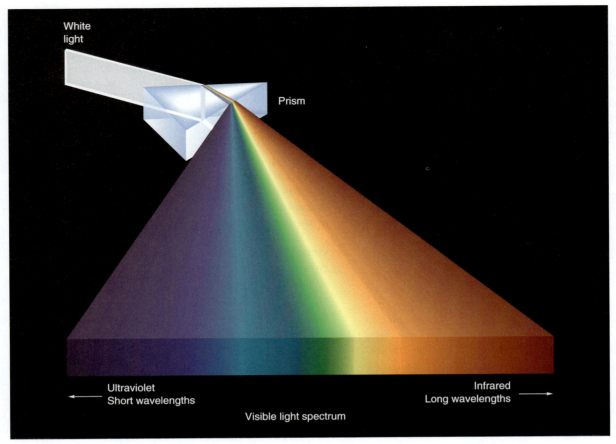

White
light

Prism

Ultraviolet
Short wavelengths

Infrared
Long wavelengths

Visible light spectrum

FIGURE 6-17
A prism bends light by an angle that depends on the wavelength of the light. Short wavelengths bend most, and long wavelengths least. Thus, white light passing through a prism is spread into a spectrum.

such measurements are made on CCD images. Because the CCD image is easily digitized, brightness and color can be measured to high precision.

THE SPECTROGRAPH

To analyze light in detail, we need to spread the light out according to wavelength into a spectrum, a task performed by a **spectrograph.** We can understand how this works if we reproduce the experiment performed by Isaac Newton in 1666. Placing a prism in a beam of sunlight spreads it into a beautiful spectrum. From this Newton concluded that white light was made of a mixture of all the colors.

Newton didn't think in terms of wavelength, but we can use that modern concept to see that the light passing through the prism is bent at an angle that depends on the wavelength. Violet (short wavelength) bends most, and red (long wavelength) least. Thus, the white light entering the prism is spread into a spectrum (▌Figure 6-17). A typical prism spectrograph contains more than one prism to spread the light farther and lenses to guide the light into the prism and to focus the light into a camera.

Nearly all modern spectrographs use a grating in place of a prism. A **grating** is a piece of glass with thousands of microscopic parallel lines scribed onto its surface. Different wavelengths of light reflect from the grating at slightly different angles, so white light is spread into a spectrum and can be recorded, often by a CCD camera.

Because astronomers must measure the wavelengths of a spectrum, they use a **comparison spectrum** as a calibration of their spectrograph. Special bulbs built into the spectrograph produce emission spectra of such atoms as thorium and argon, or neon. The wavelengths of these spectral lines have been measured to high precision in the laboratory, so astronomers can use the spectra of these light sources like a road map to measure the wavelengths of spectral lines in the spectrum of a planet, star, or galaxy to high accuracy.

One of the newest advances in astronomical instrumentation is called the S-CAM. Developed by the European Space Agency, the S-CAM is a CCD camera in which each pixel records the energy of every photon. That is enough information to create a spectrum for any pixel in the image. In that way the S-CAM combines the advantages of the CCD camera and the spectrograph.

Because astronomers understand how light interacts with matter, a spectrum carries a tremendous amount of information (as we will see in the next chapter), and that makes a spectrograph the astronomer's most powerful instrument. An astronomer recently remarked, "We don't know anything about an object till we get a spectrum," and that is only a slight exaggeration.

REVIEW CRITICAL INQUIRY

What is the difference between light going through a lens and light passing through a prism?

A refracting telescope producing chromatic aberration and a prism dispersing light into a spectrum are two examples of the same thing, but one is bad and one is good. When light passes through the curved surfaces of a lens, different wavelengths are bent by slightly different amounts, and the different colors of light come to focus at different focal lengths. This produces the color fringes in an image called chromatic aberration, and that's bad. But the surfaces of a prism are made to be precisely flat, so all of the light enters the prism at the same angle, and any given wavelength is bent by the same amount wherever it meets the prism. Thus, white light is dispersed into a spectrum. We could call the dispersion of light by a prism "controlled chromatic aberration," and that's good.

CCDs have been very good for astronomy, and they are now widely used. Explain why they are more useful than photographic plates.

So far, our discussion has been limited to visual wavelengths. Now it is time to consider the rest of the electromagnetic spectrum.

6-4 RADIO TELESCOPES

All the telescopes and instruments we have discussed look out through the visible light window in Earth's atmosphere, but there is another window running from a wavelength of 1 cm to about 1 m (see Figure 6-3). By building the proper kinds of instruments, we can study the universe through this radio window.

OPERATION OF A RADIO TELESCOPE

A radio telescope usually consists of four parts: a dish reflector, an antenna, an amplifier, and a recorder (❙ Figure 6-18). The components, working together, make it possible for astronomers to detect radio radiation from celestial objects.

The dish reflector of a radio telescope, like the mirror of a reflecting telescope, collects and focuses radiation. Because radio waves are much longer than light waves, the dish need not be as smooth as a mirror. In some radio telescopes, the reflector may not even be dish-shaped, or the telescope may contain no reflector at all.

FIGURE 6-18

In most radio telescopes, a dish reflector concentrates the radio signal on the antenna. The signal is then amplified and recorded. For all but the shortest radio waves, wire mesh is an adequate reflector (photo). *(Courtesy Seth Shostak/SETI Institute)*

Though a radio telescope's dish may be many meters in diameter, the antenna may be as small as your hand. Like the antenna on a TV set, its only function is to absorb the radio energy and direct it along a cable to an amplifier. After amplification, the signal goes to some kind of recording instrument. Most radio observatories record data directly into computer memory. However it is recorded, an observation with a radio telescope measures the amount of radio energy coming from a specific point on the sky.

Because humans can't see radio waves, astronomers must convert them into something perceptible. One way is to measure the strength of the radio signal at various places in the sky and draw a map in which contours mark areas of uniform radio intensity. We might compare such a map to a seating diagram for a baseball stadium in which the contours mark areas in which the seats have the same price (■ Figure 6-19a). Contour maps are very common in radio astronomy and are often reproduced using false colors (Figure 6-19b).

LIMITATIONS OF THE RADIO TELESCOPE

A radio astronomer works under three handicaps: poor resolution, low intensity, and interference. We saw that the resolving power of an optical telescope depends on the diameter of the objective lens or mirror. It also depends on the wavelength of the radiation. At very long wavelengths, like those of radio waves, images become fuzzy because of the large diffraction fringes. As with an optical telescope, the only way to improve the resolving power is to build a bigger telescope. Consequently, radio telescopes must be quite large.

Even so, the resolving power of a radio telescope is not good. A dish 30 m in diameter receiving radiation with a wavelength of 21 cm has a resolving power of about 0.5°. Such a radio telescope would be unable to show us any details in the sky smaller than the moon. Fortunately, radio astronomers can combine two or more radio telescopes to form a **radio interferometer** capable of much higher resolution. For example, the Very Large Array (VLA) consists of 27 dish antennas spread in a Y-shape across the New Mexico desert (■ Figure 6-20). In combination, they have the resolving power of a radio telescope 36 km (22 mi) in diameter. The VLA can resolve details smaller than 1 second of arc. Eight new dish antennas being added across New Mexico will give the VLA 10 times better resolving power. Another large radio interferometer, the Very Long Baseline Array (VLBA), consists of matched radio dishes spread from Hawaii to the Virgin Islands and has an effective diameter almost as large as Earth.

The second handicap radio astronomers face is the low intensity of the radio signals. We saw earlier that the energy of a photon depends on its wavelength. Photons of radio energy have such long wavelengths that their individual energies are quite low. In order to get strong signals focused on the antenna, the radio astronomer must build large collecting dishes.

The largest fully steerable radio telescope in the world is at the National Radio Astronomy Observatory in Green Bank, West Virginia (■ Figure 6-21a). The telescope has a reflecting surface 100 meters in diameter, big enough to hold an entire football field, and can be pointed anywhere in the sky. Its surface consists of 2004 computer-controlled panels that adjust to maintain the shape of the reflecting surface.

The largest radio dish in the world is 300 m (1000 ft) in diameter. So large a dish can't be supported in the usual way, so it is built into a mountain valley in Arecibo, Puerto Rico. The reflecting dish is a thin metallic surface supported above the valley floor by ca-

FIGURE 6-19
(a) A contour map of a baseball stadium shows regions of similar admission prices. The most expensive seats are those behind home plate. (b) A false-color-image radio map of Tycho's supernova remnant, the expanding shell of gas produced by the explosion of a star in 1572. The radio contour map has been color-coded to show intensity. Red is the strongest radio intensity, and violet the weakest. *(NRAO)*

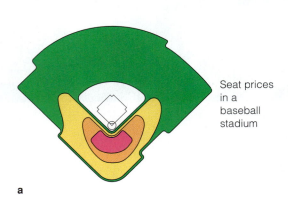

Seat prices in a baseball stadium

a

b

FIGURE 6-20

The Very Large Array uses 27 radio dishes, which can be moved to different positions along a Y-shaped set of tracks across the New Mexico desert. They are shown here in the most compact arrangement. Signals from the dishes are combined to create very-high-resolution radio maps of celestial objects. *(NRAO)*

bles attached near the rim, and the antenna hangs above the dish on cables from three towers built on three mountain peaks that surround the valley (Figure 6-21b). Although this telescope can look only overhead, the operators can change its aim slightly by moving the antenna and by waiting for Earth's rotation to point the telescope in the proper direction. This may sound clumsy, but the telescope's ability to detect weak radio sources, together with its good resolution, makes it one of the most important radio observatories in the world.

The third handicap the radio astronomer faces is interference. A radio telescope is an extremely sensitive radio receiver listening to radio signals thousands of times weaker than artificial radio and TV transmissions. Such weak signals are easily drowned out by interference. Sources of such interference include everything from poorly designed transmitters in Earth satellites to automobiles with faulty ignition systems. To avoid this kind of interference, radio astronomers locate their telescopes as far from civilization as possible. Hidden deep in mountain valleys, they are able to listen to the sky protected from human-made radio noise.

FIGURE 6-21

(a) The largest steerable radio telescope in the world is the GBT located in Green Bank, West Virginia. With a diameter of 100 m, it stands higher than the Statue of Liberty. *(Mike Bailey: NRAO/AUII)* (b) The 300-m (1000-ft) radio telescope in Arecibo, Puerto Rico, hangs from cables over a mountain valley. The Arecibo Observatory is part of the National Astronomy and Ionosphere Foundation. *(David Parker/Science Photo Library)*

ADVANTAGES OF RADIO TELESCOPES

Building large radio telescopes in isolated locations is expensive, but three factors make it all worthwhile. First, and most important, a radio telescope can show us where clouds of cool hydrogen are located between the stars. Because 90 percent of the atoms in the universe are hydrogen, that is important information. Large clouds of cool hydrogen are completely invisible to normal telescopes, because they produce no visible light of their own and reflect too little to be detected on

a

b

photographs. However, cool hydrogen emits a radio signal at the specific wavelength of 21 cm. (We will see how the hydrogen produces this radiation when we discuss the gas clouds in space in Chapter 10.) The only way we can detect these clouds of gas is with a radio telescope that receives the 21-cm radiation. These hydrogen clouds are the places where stars are born, and that is one reason that radio telescopes are important.

Nevertheless, there is a second reason. Because radio signals have relatively long wavelengths, they can penetrate the vast clouds of dust that obscure our view at visual wavelengths. Light waves are short, and they interact with tiny dust grains floating in space; thus, the light is scattered and never penetrates the dust to reach optical telescopes on Earth. However, radio signals from far across the galaxy pass unhindered through the dust, giving us an unobscured view.

Finally, a radio telescope can detect objects that are more luminous at radio wavelengths than at visible wavelengths. This includes everything from the coldest clouds of gas to the hottest stars. Some of the most distant objects in the universe, for instance, are detectable only at radio wavelengths.

REVIEW CRITICAL INQUIRY

Why do optical astronomers build big telescopes, while radio astronomers build groups of widely separated smaller telescopes?

Optical astronomers build large telescopes to maximize light-gathering power, but the problem for radio telescopes is resolving power. Because radio waves are so much longer than light waves, a single radio telescope can't see details in the sky much smaller than the moon. By linking radio telescopes miles apart, radio astronomers build a radio interferometer that can simulate a radio telescope miles in diameter and thus increase the resolving power.

The difference between the wavelengths of light and radio waves makes a big difference in building the best telescopes. But why don't radio astronomers want to build their telescopes on mountaintops as optical astronomers do?

Our atmosphere causes trouble for Earth's astronomers in two ways. It distorts images, and it absorbs many wavelengths. The only way to avoid these limitations completely is to send telescopes above the atmosphere, into space.

6-5 SPACE ASTRONOMY

Ground-based telescopes can operate only at wavelengths in the visual and radio windows of the atmosphere. Most of the rest of the electromagnetic radiation—infrared, ultraviolet, X ray, and gamma ray—never reaches Earth's surface. To observe at these wavelengths, telescopes must fly above the atmosphere in high-flying aircraft, rockets, balloons, and satellites. The only exception is some observations of the shorter infrared wavelengths that can be made from high mountains.

INFRARED ASTRONOMY

Some infrared radiation does leak through our atmosphere. This radiation enters narrow, partially open atmospheric windows scattered from 1200 nm to about 40,000 nm. Infrared astronomers usually measure wavelength in micrometers (10^{-6} meters), so they refer to this wavelength range as 1.2 to 40 microns. In this range, called the near infrared, much of the radiation is absorbed by water vapor, carbon dioxide, and oxygen molecules in Earth's atmosphere, so it is an advantage to place telescopes on mountains where the air is thin and dry. A number of important infrared telescopes, for example, observe from the 4150-m (13,600-ft) summit of Mauna Kea in Hawaii (❙ Figure 6-22). At this altitude, they are above much of the water vapor, which is the main absorber of infrared.

The far-infrared range, which includes wavelengths longer than 40 microns, can tell us about planets, comets, forming stars, and other cool objects, but these wavelengths are absorbed high in the atmosphere. To observe in the far infrared, telescopes must venture to high altitudes. Remotely operated infrared telescopes suspended under balloons have reached altitudes as high as 41 km (25 mi). For many years, a NASA jet transport carried a 91-cm infrared telescope and a crew of astronomers to altitudes of 12,000 m (40,000 ft) to get above 99 percent of the water vapor in Earth's atmosphere. Now retired from service, that airborne observatory will soon be replaced with the Stratospheric Observatory for Infrared Astronomy (SOFIA), a Boeing 747 that will carry a 2.5-m telescope to the fringes of the atmosphere.

The ultimate solution is to place infrared telescopes in space above the atmosphere. In the early 1980s, the Infrared Astronomy Satellite (IRAS) mapped the sky at infrared wavelengths (❙ Figure 6-23). In the middle 1990s, European astronomers launched the Infrared Space Observatory, which carried detectors more sensitive than IRAS.

The newest space telescope observing in the infrared is the Spitzer Space Telescope Facility launched in 2003. It carries a mirror of 85 cm in diameter that is cooled by liquid helium to less than 5.5 K and can observe from 3 to 180 microns. The telescope is so sensitive to heat, it cannot be placed in orbit around Earth (❙ Figure 6-24).

We can understand why a telescope observing in the infrared must have its optics cooled if we note that

FIGURE 6-22

Comet Hale-Bopp hangs in the sky over the 3-meter NASA infrared telescope atop Mauna Kea. The air at high altitudes is steady and so dry that it is transparent to shorter infrared photons. Infrared astronomers can observe with the lights on in the telescope dome. Their instruments are usually insensitive to visible light. *(Courtesy William Keel)*

infrared radiation is emitted by warm objects. If the telescope is warm it will emit many times more infrared radiation than that coming from a distant object. Imagine trying to look at a dim, moonlit scene through binoculars that are glowing brightly. In a telescope observing near-infrared wavelengths, only the detector, the element on which the infrared radiation is focused, must be cooled. To observe in the far infrared, however, as IRAS did, the entire telescope must be cooled.

ULTRAVIOLET ASTRONOMY

Ultraviolet radiation with wavelengths shorter than about 290 nm, the far ultraviolet, is completely absorbed by the ozone layer in our atmosphere. The ozone layer extends from 20 km to about 40 km above Earth's surface. Telescopes to observe in the far ultraviolet must get above the ozone layer, and that means they must go into space.

One of the most successful ultraviolet observatories was the International Ultraviolet Explorer (IUE). Launched in 1978, it carried a 45-cm (18-in.) telescope and used TV systems to record spectra from 320 nm to 115 nm. Although IUE was expected to last only a year or two, it survived and was widely used by astronomers from all around the world until, partially shut down because of budget cutbacks, it finally failed in 1996.

FIGURE 6-23

Space telescopes observing at many wavelengths. The Infrared Astronomy Satellite (IRAS) surveyed the sky at a number of infrared wavelengths, detecting clouds of cold gas in space. The Extreme Ultraviolet Explorer (EUVE) observed in the short-wavelength ultraviolet, and Chandra observes at X-ray wavelengths. None of these observations could be made from below Earth's atmosphere. *(NASA and Chandra X-Ray Center, Smithsonian Astrophysical Observatory)*

FIGURE 6-24
Because the Spitzer Space Telescope observes at long-infrared wavelengths, it is very sensitive to heat, so the main telescope is shielded from direct sunlight by a sunscreen. Also, the telescope does not orbit Earth, a source of infrared radiation. Rather, the telescope is trailing Earth as it orbits the sun. *(NASA/JPL/Caltech)*

The Extreme Ultraviolet Explorer (EUVE) telescope was launched in 1992 to observe at wavelengths from 100 nm to 10 nm. With four telescopes and more modern detectors than IUE, EUVE surveyed the entire sky and studied specific targets at its shorter wavelength range, as shown in Figure 6-23.

Observations in the infrared reveal cool material, but observations in the ultraviolet tend to show hot, excited regions. Hot stars and hot gas are mapped by such telescopes.

X-RAY ASTRONOMY

Beyond the UV, at wavelengths from 10 nm to 0.01 nm, lie the X rays. These photons can be produced only by high-energy processes, so we see them coming from very hot regions in stars and from violent events such as matter smashing onto a neutron star—a subject we will discuss in Chapter 14. X-ray images can give us information about the heavens that we can get in no other way.

Although early X-ray observations of the sky were made from balloons and small rockets in the 1960s, the age of X-ray astronomy did not really begin until 1970, when an X-ray telescope named Uhuru (Swahili for "freedom") was put into orbit. Uhuru detected nearly 170 separate sources of celestial X rays. In the late 1970s, the three High Energy Astronomy Observatories (HEAO) satellites, carrying more sensitive and more sophisticated equipment, pushed the total to many hundreds. The second HEAO satellite, named the Einstein Observatory, used special optics to produce X-ray images.

For many years, astronomers have worked to create a major X-ray observatory in space. In 1999, the Advanced X-Ray Astrophysics Facility (AXAF) was carried into orbit by the Space Shuttle and renamed Chandra in honor of the late Indian-American Nobel Laureate Subrahmanyan Chandrasekhar, who worked on theoretical astrophysics in many areas of astronomy which the X-ray observatory Chandra studies (Figure 6-23).

Focusing X rays is difficult because the high-energy X-ray photons do not reflect off conventional mirrors but rather penetrate into the mirrors. The optics in Chandra are specially designed mirrors in which the X-ray photons graze the surface of the mirrors at very small angles. Under these circumstances, the photons are reflected to form an image. Just as a CCD chip records visible-light photons, the detector in Chandra has been designed to absorb and record X-ray photons to produce an X-ray image. Chandra can detect X-ray-emitting objects 50 times fainter than any previous X-ray telescope and can resolve details 10 times smaller.

GAMMA-RAY TELESCOPES

Gamma rays have wavelengths even shorter than X rays, and that means they have even higher energies. They are not understood very well, but they appear to be produced by the hottest and most violent objects in the universe: exploding stars, erupting galaxies, neutron stars, and black holes.

We don't understand gamma rays well because they are so difficult to detect. First, they cannot be focused or detected as X rays can. Thus, gamma-ray telescopes are more complex and can see less detail than similar X-ray telescopes. Also, gamma rays are such high-energy photons that natural processes produce only a few. Gamma-ray telescopes must count gamma rays one at a time. For example, a gamma-ray telescope observed an intense source for a month and counted only 3000 gamma-ray photons—one every 15 minutes.

In addition, gamma rays are almost totally absorbed by our atmosphere, so gamma-ray telescopes must observe from orbit. NASA has operated three important gamma-ray telescopes aboard satellites. The 17-ton Gamma Ray Observatory was launched in April 1991. At 10 to 50 times the sensitivity of any previous gamma-ray telescope, it first mapped the entire sky and then observed selected targets.

COSMIC RAYS

All of the radiation we have discussed in this chapter has been electromagnetic radiation. **Cosmic rays,** however, are not really rays; they are subatomic particles traveling at tremendous velocities that strike our atmosphere from space. Almost no cosmic rays reach the ground, but they do smash into gas atoms in the upper atmosphere, and fragments of these collisions shower down on us day and night over our entire lives. These secondary cosmic rays are passing through you as you read this sentence.

Some cosmic-ray research can be done from high mountains or high-flying aircraft; but, to study cosmic rays in detail, detectors must go into space. A number of cosmic-ray detectors have been carried into orbit, but this area of astronomical research is just beginning to bear fruit.

We can't be sure where cosmic rays come from. Because they are atomic particles with electric charges, they are deflected by the magnetic fields spread through our galaxy, and that means we can't tell where they are coming from. The space between the stars is a glowing fog of cosmic rays. Some lower-energy cosmic rays come from the sun, but many cosmic rays are probably produced by the violent explosions of dying stars. At present, cosmic rays largely remain a mystery. We will discuss them again in future chapters.

THE HUBBLE SPACE TELESCOPE

Astronomers dreamed of having a giant telescope in orbit from the 1920s on. Such a telescope would not only allow observation at nonvisible wavelengths but would also avoid the turbulence in Earth's atmosphere. That dream was fulfilled on April 24, 1990, when the Hubble Space Telescope (HST) was released into orbit (❙ Figure 6-25a). Named after the man who discovered the expansion of the universe (Chapter 19), it is the largest orbiting telescope ever built. As big as a large bus, the HST carries a 2.4-m (96-in.) mirror. The light can be directed to any of its spectrographs and cameras.

Although many space telescopes observe at wavelengths that are absorbed in Earth's atmosphere, the Hubble Space Telescope is designed to observe mainly at visual wavelengths. It was placed in orbit to avoid the blurring (seeing) and dimming of starlight caused by Earth's atmosphere. The telescope has far exceeded its original design criteria. It can detect objects so faint it can look 50 times farther into space than can conventional Earth-based telescopes. With its precise optics and its location above Earth's distorting atmosphere, it can resolve details 10 times smaller than can be resolved by any conventional telescope on Earth. Every few years, astronauts visit the telescope in orbit and install new instruments.

HST is controlled from the Space Telescope Science Institute located at Johns Hopkins University in Baltimore, Maryland. Astronomers there plan the telescope's observing schedule to maximize the scientific return while engineers study the operation of the telescope. HST transmits its data to the Institute, where it is analyzed. Because the telescope is a national facility, anyone may propose research projects for the telescope. Competition is fierce, and only the most worthy projects win approval. Nevertheless, even some amateur astronomers have had projects approved for the Space Telescope.

The Hubble Space Telescope has been a phenomenal success, exploring such mysteries as the age of the universe, the birth and death of stars, and weather on nearby planets (Figure 6-25b). Visits by astronauts will

Visual-wavelength image

a

b

FIGURE 6-25

The Hubble Space Telescope is the largest orbiting telescope ever launched. It was carried into orbit by the space shuttle in 1990. (b) This image of Mars recorded by the Hubble Space Telescope reveals thin clouds drifting around high volcanoes at the left and details of the polar caps at the top. *(NASA)* keep it in operation for years to come, but astronomers are already planning an even larger space telescope that will someday orbit above Earth's atmosphere.

REVIEW CRITICAL INQUIRY

Why can infrared astronomers observe from high mountaintops, while X-ray astronomers must observe from space? Infrared radiation is absorbed by water vapor in Earth's atmosphere. If we built our infrared telescope on top of a high mountain, we would be above most of the water vapor in the atmosphere, and we could collect some infrared radiation from the stars. The longer-wavelength infrared radiation is absorbed much higher in the atmosphere, so we couldn't observe it from our mountaintop. Similarly, X rays are absorbed in the uppermost layers of the atmosphere, and we would not be able to find any mountain high enough to get an X-ray telescope above those absorbing layers. To observe the stars at X-ray wavelengths, we would need to put our telescope in space, above Earth's atmosphere.

X-ray and far-infrared telescopes must observe from space, but the Hubble Space Telescope observes in the visual wavelength range. Then why must the Hubble Space Telescope observe from orbit?

The tools of the astronomer are designed to gather radiation from the sky and extract information. Perhaps no tool is as important as the spectrograph, because no form of observation is as loaded with information as a spectrum. In the next chapter, we will see how we can harvest the information in a star's spectrum.

SUMMARY

Electromagnetic radiation is an electric and magnetic disturbance that transports energy at the speed of light. The electromagnetic spectrum includes radio waves, infrared radiation, visible light, ultraviolet radiation, X rays, and gamma rays.

We can think of "a particle of light," a photon, as a bundle of waves that sometimes acts as a particle and sometimes as a wave. The energy a photon carries depends on its wavelength. The wavelength of visible light, usually measured in nanometers (10^{-9} m), ranges from 400 nm to 700 nm. Infrared and radio photons have longer wavelengths and carry less energy. Ultraviolet, X-ray, and gamma-ray photons have shorter wavelengths and carry more energy.

Astronomical telescopes are of two types, refractor and reflector. A refractor uses a lens to bend the light and focus it into an image. Because of chromatic aberration, refracting telescopes cannot bring all colors to the same focus, resulting in color fringes around the images. An achromatic lens partially corrects for this, but such lenses are expensive and cannot be made larger than about 1 m in diameter.

Reflecting telescopes use a mirror to focus the light and are less expensive than refracting telescopes of the same diameter. In addition, reflecting telescopes do not suffer from chromatic aberration. Thus, most recently built telescopes are reflectors.

The largest traditional telescopes in the world have thick, solid mirrors of great weight. New telescopes use thin mirrors controlled by computers or mirrors made up of segments. The two Keck Telescopes in Hawaii use segments to make up mirrors 10 m in diameter.

The powers of a telescope are light-gathering power, resolving power, and magnifying power. The first two of these depend on the telescope's diameter; thus, astronomical telescopes often have large diameters.

Special instruments attached to a telescope analyze the light it gathers. The photographic plate records vast amounts of detail for later analysis, but electronic imaging systems such as charge-coupled devices have largely replaced photographic plates. CCDs are more sensitive, record both faint and bright objects, and can be read directly into computer memory. Spectrographs, using prisms or gratings, break the starlight into a spectrum, which then can be photographed or electronically recorded.

To observe radio signals from celestial objects, we need a radio telescope, which usually consists of a dish reflector, an antenna, an amplifier, and a data recorder such as a computer. Such an instrument measures the intensity of radio signals over the sky and constructs radio maps. The poor resolution of the radio telescope can be improved by combining it with another radio telescope to make a radio interferometer. The three principal advantages of radio telescopes are that they can detect the very cold gas clouds in space, they can detect regions of very hot gas produced by exploding stars or erupting galaxies, and they can look through the dust clouds that block our view at optical wavelengths.

Observations at some wavelengths in the near infrared are possible from high mountaintops or from high-flying aircraft. At these altitudes, the air is thin and dry, and the infrared radiation can reach the telescope. At longer infrared wavelengths, the telescope must observe from above Earth's atmosphere—from orbit. To observe at the short ultraviolet wavelengths, a telescope must be above the ozone layers in Earth's atmosphere. That means it must be in orbit. X-ray and gamma-ray telescopes must also be placed in orbit above Earth's absorbing atmosphere.

The largest orbiting telescope is the Hubble Space Telescope, launched in the spring of 1990. The Chandra X-ray telescope was put into orbit in 1999, and the Spitzer Space Telescope, a major infrared observatory in space, began observing in 2003.

NEW TERMS

electromagnetic radiation	photon
wavelength	infrared radiation
frequency	ultraviolet radiation
nanometer (nm)	atmospheric window
Angstrom (Å)	focal length

refracting telescope

reflecting telescope

primary lens, mirror

objective lens, mirror

eyepiece

chromatic aberration

achromatic lens

light-gathering power

resolving power

diffraction fringe

seeing

magnifying power

light pollution

prime focus

secondary mirror

Cassegrain focus

Newtonian focus

Schmidt-Cassegrain focus

sidereal drive

equatorial mounting

polar axis

alt-azimuth mounting

active optics

adaptive optics

interferometry

charge-coupled device (CCD)

false-color image

spectrograph

grating

comparison spectrum

radio interferometer

cosmic ray

REVIEW QUESTIONS

Ace Astronomy™ Assess your understanding of this chapter's topics with additional quizzing and animations at **http:// astronomy.brookscole.com/seeds8e**

1. Why do nocturnal animals usually have large pupils in their eyes? How is that related to astronomical telescopes?

2. Astronomers have not constructed a single major refracting telescope since about 1900. Why not?

3. How have computers and new technology changed the design of astronomical telescopes and their mountings?

4. Small telescopes are often advertised as "200 power" or "magnifies 200 times." As someone knowledgeable about astronomical telescopes, how would you improve such advertisements?

5. An astronomer recently said, "Some people think I should give up photographic plates." Why might she change to something else?

6. What purpose do the colors in a false-color image or false-color radio map serve?

7. Why can we observe in the infrared from high mountains and aircraft but must go into space to observe in the far ultraviolet?

8. The moon has no atmosphere at all. What advantages would we have if we had an observatory on the lunar surface?

DISCUSSION QUESTIONS

1. Why does the wavelength response of the human eye match so well the visual window of Earth's atmosphere?

2. Most people like beautiful sunsets with brightly glowing clouds, bright moonlit nights, and twinkling stars. Most astronomers don't. Why?

PROBLEMS

1. The thickness of the plastic in plastic bags is about 0.001 mm. How many wavelengths of red light is this?

2. What is the wavelength of radio waves transmitted by a radio station with a frequency of 100 million cycles per second?

3. Compare the light-gathering powers of one of the Keck telescopes and a 0.5-m telescope.

4. How does the light-gathering power of one of the Gemini telescopes compare with that of the human eye? (*Hint:* Assume that the pupil of your eye can open to about 0.8 cm.)

5. What is the resolving power of a 25-cm telescope? What do two stars 1.5 seconds of arc apart look like through this telescope?

6. Most of Galileo's telescopes were only about 2 cm in diameter. Should he have been able to resolve the two stars mentioned in Problem 5?

7. How does the resolving power of the 5-m telescope compare with that of the Hubble Space Telescope? Why does the HST outperform the 5-m telescope?

8. If we build a telescope with a focal length of 1.3 m, what focal length should the eyepiece have to give a magnification of 100 times?

9. Astronauts observing from a space station need a telescope with a light-gathering power 15,000 times that of the human eye, capable of resolving detail as small as 0.1 second of arc, and having a magnifying power of 250. Design a telescope to meet their needs. Could you test your design by observing stars from Earth?

10. A spy satellite orbiting 400 km above Earth is supposedly capable of counting individual people in a crowd. What minimum-diameter telescope must the satellite carry? (*Hint:* Use the small-angle formula.)

CRITICAL INQUIRIES FOR THE WEB

1. Research in chemistry, physics, and biology is supported in part by industry. Because astronomy has few industrial applications, it is not well supported by industry. Visit Web sites for major observatories and find out who pays their bills.

2. Visit Web sites for the major observatories in space, such as Hubble Space Telescope and Chandra, and check on their latest status. What kinds of observations are they making, and what kinds of discoveries are they making?

3. Astronomers are leaders in efforts to reduce electric power wasted on outdoor lighting because it produces

light pollution. Find Web sites on light pollution and see if your town is making any effort to preserve the beauty of the night sky and simultaneously save electrical power.

EXPLORING *THESKY*

1. Astronomical telescopes using equatorial mountings must be aligned precisely with the north celestial pole. Locate Polaris and determine how far it is from the north celestial pole. (*Hint:* Use **Reference Lines** under the **View** menu and check **Grid** under Equatorial. Be sure the spacing is set to auto/fine. Then locate the Little Dipper and zoom in on Polaris.)

 Visit the Seeds *Foundations of Astronomy* companion Web site for critical thinking exercises, articles, and additional readings from InfoTrac College Edition, Brooks/Cole's online student library.

ATOMS AND STARLIGHT

Awake! for Morning in the Bowl of Night
Has flung the Stone that puts the Stars to Flight:
And Lo! the Hunter of the East has caught
The Sultan's Turret in a Noose of Light.

The Rubáiyát of Omar Khayyám, *Trans. Edward FitzGerald*

AURA/NOAO/NSF

Chapter 7

GUIDEPOST

Some chapters in textbooks do little more than present facts. The chapters in this book attempt to present astronomy as organized understanding. But this chapter is special. It presents us with a tool. The interaction of light with matter gives astronomers clues about the nature of the heavens, but the clues are meaningless unless astronomers understand how atoms leave their traces on starlight. Thus, we dedicate an entire chapter to understanding how atoms interact with light.

This chapter marks a transition in the way we look at nature. Earlier chapters described what we see with our eyes and explained those observations using models and theories. With this chapter, we turn to modern astrophysics, the application of physics to the study of the sky. Now we can search out secrets of the stars that lie beyond the grasp of our eyes.

If this chapter presents us with a tool, then we should use it immediately. The next chapter will apply our new tool to understanding the sun.

VIRTUAL LABORATORIES

PROPERTIES OF LIGHT AND ITS INTERACTION WITH MATTER

THE DOPPLER EFFECT

THE SPECTRAL SEQUENCE AND THE H–R DIAGRAM

No laboratory jar on Earth holds a sample labeled "star stuff," and no instrument has ever probed inside a star. The stars are far beyond our reach, and the only information we can obtain about them comes to us hidden in light (▋ Figure 7-1). Whatever we want to know about the stars we must catch in a noose of light.

We Earthbound humans knew almost nothing about stars until the early 19th century, when the Munich optician Joseph von Fraunhofer studied the solar spectrum and found it interrupted by some 600 dark lines. As scientists realized that the lines were related to the various atoms in the sun and found that stellar spectra had similar patterns of lines, the door to an understanding of stars finally opened.

In this chapter, we will go through that door by considering how atoms interact with light to produce spectral lines. We begin with the hydrogen atom because it is the most common atom in the universe as well as the simplest. Other atoms are larger and more complicated, but in many ways their properties resemble those of hydrogen.

Once we understand how an atom's structure can interact with light to produce spectral lines, we will recognize certain patterns in stellar spectra. By classifying the spectra according to these patterns, we can arrange the stars in a sequence according to temperature. One of the most important pieces of information revealed in a star's spectrum is its temperature.

But, properly analyzed, a stellar spectrum can tell us much more. The spectrum gives us information about the chemical composition of the star and the star's motion relative to Earth.

7-1 STARLIGHT

If you look at the stars in the constellation Orion, you will notice that they are not all the same color (see Figure 2-4). Betelgeuse, in the upper left corner, is quite red; Rigel, in the lower right corner, is blue. These differences in color arise from the way the stars produce light and give us our first clue to the temperatures of stars.

TEMPERATURE AND HEAT

Temperature is one of the defining characteristics of a star. That is, if we know a star's temperature and a few other properties, such as size, we can understand the star. But if we don't know a star's temperature, we can know almost nothing about it.

Ace◐**Astronomy**™ The AceAstronomy icon throughout the text indicates an opportunity for you to test yourself on key concepts and to explore animations and interactions on the AceAstronomy Web site at **http://astronomy.brookscole .com/seeds8e**

Visual-wavelength image

FIGURE 7-1
What's going on here? The sky is filled with beautiful and mysterious objects that lie far beyond our reach, in the case of the nebula NGC 6751 about 6500 ly beyond our reach. The only way to understand such objects is by analyzing the light we receive from them. Such an analysis reveals that this object is a dying star surrounded by the expanding shell of gas it ejected a few thousand years ago. We will discuss this phenomenon in Chapter 10. *(NASA Hubble Heritage Team/STScI/AURA)*

A gas is made up of particles—atoms and molecules—that are in constant motion, colliding with one another millions of times a second. The **temperature** of a gas is a measure of the average velocity of the particles. If a gas is hot, the particles are moving very rapidly; if it is cool, the particles are moving more slowly. The motion of the particles represents stored energy, and we feel that energy as heat. Read Parameters of Science 7-1 and notice the distinction we make between the temperature, the stored energy, and the heat we feel.

We will measure the temperature of stars on the **Kelvin temperature scale.** The Kelvin scale sets its zero point at **absolute zero,** the temperature at which the particles of a gas have no remaining motion that we can extract as heat. Absolute zero is −273.2°C, which is −459.7°F. (See Appendix A.) The temperatures of the stars range from 2000 K to 40,000 K or more and can be deduced from an analysis of starlight.

THE ORIGIN OF STARLIGHT

The starlight we see comes from gases in the outer surface of the star—the photosphere. The gases deep inside the star also emit light, but it is absorbed before it can escape, and the low-density gas above the photosphere is too thin to emit significant amounts of light. Thus, the photosphere of a star is that layer of gases dense enough to emit significant amounts of light but thin enough to allow the photons to escape. (Recall that we met the photosphere of the sun in Chapter 3.)

To see how the photosphere of a star can produce light, we must consider two things: how photons are produced and how the temperature of a material is related to motion among its atoms.

First, a photon can be produced by a changing electric field. An **electron** is a negatively charged subatomic particle, and if we disturb the motion of an electron, the sudden change in the electric field around it can produce an electromagnetic wave. For example, if you run a comb through your hair while standing near a radio tuned to an AM station, you can produce popping noises on the radio. The moving comb disturbs the electrons in both the comb and your hair, building static electricity. The sudden sparks of static electricity produce electromagnetic waves that the radio picks up as noise. This illustrates an important principle: When we change the motion of an electron, we generate an electromagnetic wave.

Second, recall that temperature is a measure of the average motion among the particles in a material. In the gases of a hot star, the atoms move faster, on average, than the atoms in a cool star.

Now we can understand why a hot object glows. The hotter an object is, the more violent the motion among its particles. The agitated particles collide with electrons, and when electrons are accelerated, part of the energy is carried away as a photon. The radiation emitted by a heated object is called **black body radiation,** a name that refers to the way a perfect emitter of

radiation would behave. A perfect emitter would also be a perfect absorber and at room temperature would look black. Thus we often use the term *black body radiation* to refer to objects that glow brightly.

Black body radiation is quite common. In fact, it is responsible for the light emitted by an incandescent lightbulb. Electricity flowing through the filament of the lightbulb heats it to high temperature, and it glows. We also recognize the light emitted by a heated horseshoe in the blacksmith's forge as black body radiation. Many objects in astronomy, including stars, emit radiation approximately as if they were black bodies.

Hot objects emit black body radiation, but so do cold objects. Ice cubes are cold, but their temperature is higher than absolute zero, so they contain some thermal energy and must emit some black body radiation. The coldest gas drifting in space has a temperature only a few degrees above absolute zero, but it too emits black body radiation.

We must discuss two important features of black body radiation. First, the hotter an object is, the more black body radiation it emits. Hot objects emit more radiation because their agitated particles travel faster and collide more often. Thus we expect a glowing coal from a fire to emit more total energy than an ice cube of the same size.

The second feature we must discuss is the relationship between the temperature of the object and the wavelengths of the photons it emits. The wavelength of a photon emitted when a particle collides with an electron depends on the violence of the collision. Only a violent collision can produce a short wavelength (high-energy) photon. Because extremely violent collisions don't occur very often, short-wavelength photons are rare. Similarly, most collisions are not extremely gentle, so long-wavelength (low-energy) photons are also rare. Consequently, black body radiation is made up of photons with a distribution of wavelengths, and very short and very long wavelengths are rare. The **wavelength of maximum intensity (λ_{max})**, the wavelength at which the object emits the most radiation, occurs at some intermediate wavelength.

▍Figure 7-2 shows the intensity of radiation versus wavelength for three objects of different temperatures. The curves are high in the middle and low at either end, telling us that these objects emit most intensely at intermediate wavelengths. The total area under each curve is proportional to the total energy emitted, and we can see that the hotter object emits more total energy than the cooler objects. Closer examination of the curves reveals that the wavelength of maximum intensity depends on temperature. The hotter the object, the shorter the wavelength of maximum intensity. Notice in this figure how temperature determines the color of a glowing black body. The hotter object emits more blue light than red and thus looks blue, and the cooler object emits more red than blue and consequently looks red. We will see this among stars. Hot stars look blue and cool stars red.

Notice that cool objects may emit little visible radiation but are still producing black body radiation. For example, the human body has a temperature of 310 K and emits black body radiation mostly in the infrared part of the spectrum. Infrared security cameras can detect burglars by the radiation they emit, and mosquitoes can track you down in total darkness by homing in on your infrared radiation. Although we humans emit lots of infrared radiation, we rarely emit higher-energy photons; we almost never emit X-ray or gamma-ray photons. Our wavelength of maximum intensity lies in the infrared part of the spectrum.

TWO RADIATION LAWS

Black body radiation can be described by two simple laws that will help us understand the nature of starlight. One law is related to energy and one to color.

As we saw in the previous section, a hot surface emits more black body radiation than a cool surface. That is, it emits more energy. Recall from Chapter 5 that we measure energy in units called joules (J); 1 joule is about the energy of an apple falling from a table to the floor. The total radiation given off by 1 square meter of the object in joules per second equals a constant number, represented by σ, times the temperature raised to the fourth power.* This relationship is often called the Stefan–Boltzmann law:

*For the sake of completeness, we should note that the constant σ equals 5.67×10^{-8} J/m²s degree⁴.

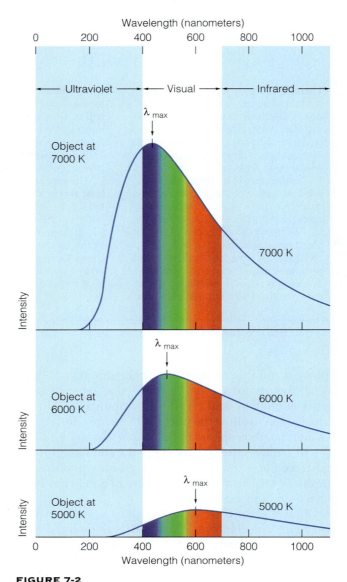

FIGURE 7-2
Black body radiation from three bodies at different temperatures demonstrates that a hot body radiates more total energy and that the wavelength of maximum intensity is shorter for hotter objects. The hotter object here will look blue to our eyes, while the cooler object will look red.

$$E = \sigma\, T^4 \ (\text{J/s/m}^2)$$

How does this help us understand stars? Suppose a star the same size as the sun had a surface temperature that was twice as hot as the sun's surface. Then each square meter of that star would radiate not twice as much energy but 2^4 or 16 times as much energy. From this law we see that a small difference in temperature can produce a very large difference in the amount of energy emitted.

The second radiation law is related to the color of stars. In the previous section, we saw that hot stars look blue and cool stars look red. Wien's law tells us that the wavelength at which a star radiates the most energy, the wavelength of maximum (λ_{max}), depends only on the star's temperature:

$$\lambda_{max} = \frac{3,000,000}{T}$$

That is, the wavelength of maximum radiation in nanometers equals 3 million divided by the temperature on the Kelvin scale.

This is a powerful tool in astronomy, because it means we can relate the temperature of a star and its wavelength of maximum intensity. For example, we might find a star that has a surface temperature of 3000 K. Then its wavelength of maximum intensity would be 3,000,000/3000, or 1000 nm—in the near infrared. Later we will meet objects much hotter than most stars; such objects radiate most of their energy at very short wavelengths. The hottest stars, for instance, radiate most of their energy in the ultraviolet.

THE COLOR INDEX

Hot stars look blue, but if we want to know the temperature of the star's surface, we must have a way of measuring color. In astronomy, the **color index** is a measure of color.

To measure the color index of a star, we need to measure brightness through a set of standard filters. The most commonly used filters are the blue (B) and visual (V). Each filter isolates a specific part of the spectrum. The B filter, for example, will not allow any light through except those photons with wavelengths between about 400 nm and 480 nm. The V filter is transparent to photons with wavelengths between about 500 nm and 600 nm, roughly approximating the sensitivity of the human eye.

If we measure the magnitude of a star through each of these filters, the difference between the two magnitudes is a number related to the color of the star. This is commonly written B-V and is referred to as the B-V color index.

For example, a hot star radiates much more blue light than red and will be brighter through the B filter (Figure 7-2). That means that the B magnitude will be smaller, and the B-V color index will be negative. The bluest stars have a B-V color index of about −0.4 and have surface temperatures of about 50,000 K. Red stars radiate much more red light than blue, so they will be fainter through the blue filter. Then B-V for a red star will be positive. The reddest stars have a B-V color index of about 2 and surface temperatures of about 2000 K.

| REVIEW CRITICAL INQUIRY |

How could a doctor measure someone's temperature without touching him or her?

Doctors and nurses use a handheld device to measure body temperature by observing the infrared radiation emerging from a patient's ear. We might suspect the device depends on the Stephan–Boltzmann law and measures the intensity of the infrared radiation. A person with a fever will emit more energy than a healthy person. However, a healthy person with a large ear canal would emit more energy than a person with a small ear canal, so measuring intensity won't be accurate. The device actually depends on Wien's law in that it measures the "color index" of the infrared radiation. A patient with a fever will emit at a shorter wavelength of maximum intensity, and the infrared radiation emerging from that person's ear will be a tiny bit "bluer" than normal.

Astronomers can measure the temperatures of stars the same way. Use Figure 7-2 to explain how that works.

The spectrum of a star is bursting with information. To interpret that information, astronomers must think carefully about how atoms and light interact at the surfaces of stars.

7-2 ATOMS

The atoms in the surface layers of stars leave their marks on the light the stars emit. By understanding what atoms are and how they interact with light, we can decode the spectra of the stars. We begin this section by constructing a working model of an atom.

A MODEL ATOM

To think about atoms and how they can interact with light, we must create a working model of an atom. In Chapter 2, we created a working model of the sky, the celestial sphere. We identified and named the important parts and described how they were located and how they interacted. We begin our study of atoms by creating a model of an atom.

In our model atom, we identify a positively charged **nucleus** at the center; this nucleus consists of two kinds of particles. **Protons** carry a positive electrical charge, and **neutrons** have no charge. Thus, the nucleus has a net positive charge.

The nucleus in our model atom is surrounded by a whirling cloud of orbiting electrons, low-mass particles with a negative charge. In a normal atom, the number of electrons equals the number of protons, and the positive and negative charges balance to produce a neutral atom. Because protons and neutrons each have a mass 1836 times greater than that of an electron, most of the mass of the atom lies in the nucleus. The hydrogen atom is the simplest of all atoms. The nucleus is a single proton orbited by a single electron, with a total mass of only 1.67×10^{-27} kg, about a trillionth of a trillionth of a gram.

An atom is mostly empty space. To see this we can construct a simple scale model. The nucleus of a hydrogen atom is a proton with a diameter of about 0.0000016 nm, or 1.6×10^{-15} m. If we multiply this by one trillion (10^{12}) we can represent the nucleus of our model atom with a grape seed, which is about 0.16 cm in diameter. The region of a hydrogen atom containing the whirling electron has a diameter of about 0.4 nm or 4×10^{-10} m. Multiplying by a trillion magnifies the diameter to about 400 m, or about 4.5 football fields laid end to end (▌Figure 7-3). When you imagine a grape seed in the midst of a sphere 4.5 football fields in diameter, you can see that an atom is mostly empty space.

DIFFERENT KINDS OF ATOMS

There are over a hundred kinds of atoms called chemical elements. Which element an atom is depends only on the number of protons in the nucleus. For example, carbon has six protons in its nucleus. An atom with one more proton than this is nitrogen, and an atom with one fewer proton is boron.

Although the number of protons in an atom of an element is fixed, we can change the number of neutrons in an atom's nucleus without changing the atom significantly. For instance, if we add a neutron to the carbon nucleus, we still have carbon, but it is slightly heavier than normal carbon. Atoms that have the same number of protons but a different number of neutrons are **isotopes.** Carbon has two stable isotopes. One form contains six protons and six neutrons for a total of 12 particles and is thus called carbon-12. Carbon-13 has six protons and seven neutrons in its nucleus (▌Figure 7-4).

Protons and neutrons are bound tightly into the nucleus, but the electrons are held loosely in the electron cloud. Running a comb through your hair creates a static charge by removing a few electrons from their atoms. This process is called **ionization,** and the atom that has lost one or more electrons is an **ion.** A carbon atom is neutral if it has six electrons to balance the positive charge of the six protons in its nucleus. If we ionize the atom by removing one or more electrons, the atom is left with a net positive charge. Under some circumstances, an atom may capture one or more extra electrons, giving it more negative charges than positive. Such a negatively charged atom is also considered an ion.

Atoms that collide may form bonds with each other by exchanging or sharing electrons. Two or more atoms bonded together form a **molecule.** Atoms do collide in

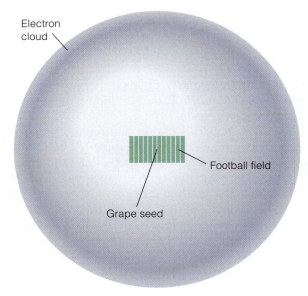

FIGURE 7-3
Magnifying a hydrogen atom by 10^{12} makes the nucleus the size of a grape seed and the diameter of the electron cloud about 4.5 times longer than a football field. The electron itself is still too small to see.

stars, but the high temperatures cause violent collisions that are unfavorable for chemical bonding. Only in the coolest stars are the collisions gentle enough to permit the formation of chemical bonds. We will see later that the presence of molecules such as titanium oxide (TiO) in a star is a clue that the star is cool. In later chapters,

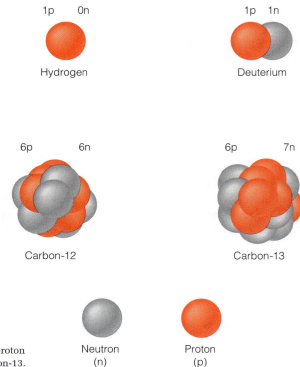

FIGURE 7-4
Some common isotopes. A rare isotope of hydrogen, deuterium, contains a proton and a neutron in its nucleus. Two isotopes of carbon are carbon-12 and carbon-13.

Quantum Mechanics: The World of the Very Small

Quantum mechanics is the set of rules that describe how atoms and subatomic particles behave. When we think about large objects such as stars, planets, aircraft carriers, and hummingbirds, we don't have to think about quantum mechanics, but on the atomic scale, particles behave in ways that seem unfamiliar to us.

One of the principles of quantum mechanics specifies that we can not know simultaneously the exact location and motion of a particle. This is why physicists refer to the electrons in an atom as if they were a cloud of negative charge surrounding the nucleus. Because we can't know the position and

motion of the electron, we can't really describe it as a small particle following an orbit. We can use that image as a model to help our imaginations, but the reality is much more interesting, and describing the electrons as a charge cloud gives us a better and more sophisticated model of an atom.

This raises some serious questions about reality. Is an electron really a particle at all? Quantum mechanics describes particles as waves, and waves as particles. If we can't know simultaneously the position and motion of a specific particle, how can we know how it will react to a collision with a photon or another particle? The answer is that we can't know,

and that seems to violate the principle of cause and effect (Window on Science 5-1).

Needless to say, we can't explore these discrepancies here. Scientists and philosophers of science continue to struggle with the meaning of reality on the quantum-mechanical level. Here we should note that the reality we see on the scale of stars and hummingbirds is only part of nature. We have constructed some models to help us think about nature on the scale of atoms, but the truth is much more interesting and much more exciting than anything we see on larger scales. Although we use models of atoms to study stars, there is still much to learn about the atoms themselves.

---

we will see that molecules can form in cool gas clouds in space and in the atmospheres of planets.

ELECTRON SHELLS

We mentioned the whirling cloud of orbiting electrons in a general way, but the specific way electrons behave within the cloud is very important in astronomy.

The electrons are bound to the atom by the attraction between their negative charge and the positive charge on the nucleus. This attraction is known as the **Coulomb force,** after the French physicist Charles-Augustin de Coulomb (1736–1806). If we wish to ionize an atom, we need a certain amount of energy to pull an electron away from its nucleus. This energy is the electron's **binding energy,** the energy that holds it to the atom.

An electron may orbit the nucleus at various distances. If the orbit is small, the electron is close to the nucleus, and a large amount of energy is needed to pull it away. Therefore, its binding energy is large. An electron orbiting farther from the nucleus is held more loosely, and less energy will pull it away. It therefore has less binding energy. The size of an electron's orbit is related to the energy that binds it to the atom.

Nature permits atoms only certain amounts (quanta) of binding energy, and the laws that describe how atoms behave are called the laws of **quantum mechanics** (Window on Science 7-1). Much of our discussion of atoms is based on the laws of quantum mechanics.

Because atoms can have only certain amounts of binding energy, our model atoms can have orbits of only certain sizes, called **permitted orbits.** These are like steps in a staircase: You can stand on the number-one step or the number-two step, but not on the number-

one-and-one-quarter step. The electron can occupy any permitted orbit but not orbits in between.

The arrangement of permitted orbits depends primarily on the charge of the nucleus, which in turn depends on the number of protons. Thus, each kind of element has its own pattern of permitted orbits (█ Figure 7-5). Isotopes of the same elements have nearly the same pattern because they have the same number of protons. However, ionized atoms have orbital patterns that differ from their un-ionized forms. Thus the arrangement of permitted orbits differs for every kind of atom and ion.

REVIEW CRITICAL INQUIRY

How many hydrogen atoms would it take to cross the head of a pin?

This is not a frivolous question. In answering it, we will discover how small atoms really are, and we will see how powerful physics can be as a way to understand nature. First, we will assume that the head of a pin is about 1 mm in diameter. That is 0.001 m. The size of a hydrogen atom is represented by the diameter of the electron cloud, roughly 0.4 nm. Because 1 nm equals 10^{-9} m, we multiply and discover that 0.4 nm equals 4×10^{-10} m. To find out how many atoms would stretch 0.001 m, we divide the diameter of the pinhead by the diameter of an atom. That is, we divide 0.001 m by 4×10^{-10} m, and we get 2.5×10^6. It would thus take 2.5 million hydrogen atoms lined up side by side to cross the head of a pin.

This shows how tiny an atom is and also how powerful basic physics is. A bit of arithmetic gives us a view of nature beyond the capability of our eyes. Now use another bit of arithmetic to calculate how many hydrogen atoms you would need to add up to the mass of a paper clip (1 g).

FIGURE 7-5

The electron in an atom may occupy only certain, permitted orbits. Because different elements have different charges on their nuclei, the elements have different patterns of permitted orbits.

The electron orbits in atoms are familiar territory to astronomers because the electrons in those orbits can interact with light. Such interactions fill starlight with clues to the nature of the stars.

7-3 THE INTERACTION OF LIGHT AND MATTER

If light did not interact with matter, you would not be able to see these words. In fact, you would not exist, because, among other problems, photosynthesis would be impossible and there would be no grass, wheat, bread, beef, cheeseburgers, or any other kind of food. The interaction of light and matter makes our lives possible, and it also makes it possible for us to understand our universe.

We begin our study of light and matter by considering the hydrogen atom. As we noted earlier, hydrogen is both simple and common. Roughly 90 percent of all atoms in the universe are hydrogen.

THE EXCITATION OF ATOMS

Each orbit in an atom represents a specific amount of binding energy, so physicists commonly refer to the orbits as **energy levels.** Using this terminology, we can say that an electron in its smallest and most tightly bound orbit is in its lowest permitted energy level. We can move the electron from one energy level to another by supplying enough energy to make up the difference between the two energy levels. It is like moving a flowerpot from a low shelf to a high shelf; the greater the distance between the shelves, the more energy we need to raise the pot. The amount of energy needed to move the electron is the energy difference between the two energy levels.

If we move the electron from a low energy level to a higher energy level, we say the atom is an **excited atom.** That is, we have added energy to the atom in moving its electron. If the electron falls back to the lower energy level, that energy is released.

An atom can become excited by collision. If two atoms collide, one or both may have electrons knocked into a higher energy level. This happens very commonly in hot gas, where the atoms move rapidly and collide often.

Another way an atom can get the energy that moves an electron to a higher energy level is to absorb a photon. Only a photon with exactly the right amount of energy can move the electron from one level to another. If the photon has too much or too little energy, the atom cannot absorb it. Because the energy of a photon depends on its wavelength, only photons of certain wavelengths can be absorbed by a given kind of atom. ▌Figure 7-6 shows the lowest four energy levels of the hydrogen atom along with three photons the atom could absorb. The longest-wavelength photon only has enough energy to excite the electron to the second energy level, but the shorter-wavelength photons can excite the electron to higher levels. A photon with too much or too little energy cannot be absorbed. Because the hydrogen atom has many more energy levels than shown in Figure 7-6, it can absorb photons of many different wavelengths.

Atoms, like humans, cannot exist in an excited state forever. The excited atom is unstable and must eventually (usually within 10^{-6} to 10^{-9} seconds) give up the energy it has absorbed and return its electron to the lowest energy level. Because the electrons eventually tumble down to this bottom level, physicists call it the **ground state.**

When the electron drops from a higher to a lower energy level, it moves from a loosely bound level to one more tightly bound. The atom then has a surplus of energy—the energy difference between the levels—that it can emit as a photon. Study the sequence of events in ▌ Figure 7-7 to see how an atom can absorb and emit photons. Because each type of atom or ion has its

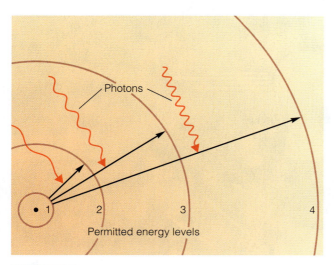

FIGURE 7-6

A hydrogen atom can absorb only those photons that move the atom's electron to one of the higher-energy orbits. Here three different photons are shown along with the change they would produce if they were absorbed.

unique set of energy levels, each type absorbs and emits photons with a unique set of wavelengths. Thus we can identify the elements in a gas by studying the characteristic wavelengths of light absorbed or emitted.

The process of excitation and emission is a common sight in urban areas at night. A neon sign glows when atoms of the neon gas in the tube are excited by electricity flowing through the tube. As the electrons in the electric current flow through the gas, they collide with the neon atoms and excite them. As we have seen, immediately after an atom is excited, its electron drops back to a lower energy level, emitting the surplus energy as a photon of a certain wavelength. The photons emitted by excited neon produce a reddish-orange glow. Signs of other colors, erroneously called "neon," contain other gases or mixtures of gases instead of pure neon.

THE FORMATION OF A SPECTRUM

The spectrum of a star is formed as light passes outward through the gases near its surface. Study ▍ "Atomic Spectra" on pages 132 and 133. Notice three important properties of spectra.

First, there are three kinds of spectra. When we see one of these types of spectra, we can recognize the kind of matter that emitted the light.

Notice also that the wavelengths of the photons that are absorbed or emitted in a spectrum are determined by the atomic energy levels in the atoms. The emitted photons coming from a hot cloud of hydrogen gas have the same wavelengths as the photons absorbed by hydrogen atoms in the gases of a star. Different atoms have different energy levels, so their spectra contain different patterns of spectral lines.

Finally, notice that the hydrogen atom produces many spectral lines from the ultraviolet to the infrared, but only three are visible to our eyes. These hydrogen lines are prominent in the spectra of stars and can tell us the temperature of the stars.

REVIEW **CRITICAL INQUIRY**

What spectrum would we see if we observed molten iron?

Molten iron is a dense liquid, and the atoms and molecules collide so often they emit all wavelengths and we would see a continuous spectrum. That is Kirchhoff's first law. But in order to see the molten iron, we would have to look through the hot vapors rising from it. Photons on their way to our spectrograph would pass through these gases, and atoms in the gases would absorb certain wavelengths. So what we would really see would be the continuous spectrum of the molten iron with weak absorption lines caused by the gases above the iron. This is what Kirchhoff's third law describes.

But suppose we had a very sensitive spectrograph that could look at the hot gases above the molten iron from the side so as to avoid looking directly at the molten iron. What kind of spectrum would the hot gases emit?

Whatever kind of spectrum astronomers look at, the most common spectral lines are the Balmer lines of hydrogen, the only hydrogen lines we can study from Earth's surface. In the next section, we will see how Balmer lines can tell us a star's temperature.

7-4 STELLAR SPECTRA

In later chapters, we use spectra to study galaxies and planets, but we begin here by studying the spectra of stars. Such spectra are the easiest to understand, and the nature of stars is central to our study of all celestial objects.

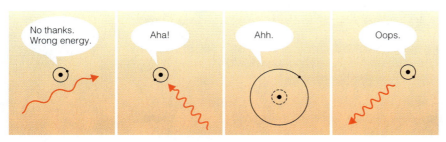

FIGURE 7-7

An atom can absorb a photon only if the photon has the correct amount of energy. The excited atom is unstable and within a fraction of a second returns to a lower energy level, reradiating the photon in a random direction.

Atomic Spectra

To understand how we can analyze a spectrum, we begin with a simple incandescent lightbulb.

The hot filament emits black body radiation, which forms a **continuous spectrum**.

An **absorption spectrum** results when radiation passes through a cool gas. In this case we can imagine that our lightbulb is surrounded by a cool cloud of gas. Atoms in the gas absorb photons of certain wavelengths, which are missing from the spectrum, and we see their positions as dark **absorption lines**. Such spectra are sometimes called **dark-line spectra**.

An **emission spectrum** is produced by photons emitted by an excited gas. We could see **emission lines** by turning our telescope aside so that photons from the bright bulb did not enter our telescope. The photons we would see would be those emitted by the excited atoms near the bulb. Such spectra are also called **bright-line spectra**.

The spectrum of a star is an absorption spectrum. The denser layers of the photosphere emit black body radiation. Gases in the atmosphere of the star absorb their specific wavelengths and form dark absorption lines in the spectrum.

In 1859, long before scientists understood atoms and energy levels, the German scientist Gustav Kirchhoff formulated three rules, now known as **Kirchhoff's laws**, that describe the three types of spectra.

Spectrograph

Telescope

Continuous spectrum

Gas atoms

Absorption spectrum

Emission spectrum

Absorption spectrum

KIRCHHOFF'S LAWS

Law I: **The Continuous Spectrum**

A solid, liquid, or dense gas excited to emit light will radiate at all wavelengths and thus produce a continuous spectrum.

Law II: **The Emission Spectrum**

A low-density gas excited to emit light will do so at specific wavelengths and thus produce an emission spectrum.

Law III: **The Absorption Spectrum**

If light comprising a continuous spectrum passes through a cool, low-density gas, the result will be an absorption spectrum.

Ace Astronomy™ Go to AceAstronomy and click Active Figures to see "Kirchhoff's Laws" and the three kinds of spectra formed.

954.6 nm
1005.0 nm
1093.8 nm
1281.8 nm
1875.1 nm

388.9 nm
397.0 nm
410.2 nm
434.0 nm
486.1 nm
656.3 nm

Paschen series
(IR)

H_ζ

H_β
H_α

Balmer series
(Visible-UV)

Lyman
series
(UV)

Nucleus

The electron orbits in the hydrogen atom are shown here as energy levels. When an electron makes a **transition** from one orbit to another, it changes the energy stored in the atom. In this diagram, arrows pointed outward represent transitions that result from the absorption of a photon. Arrows pointed inward represent transitions that result in the emission of a photon. Long arrows represent large amounts of energy and correspondingly short-wavelength photons.

In this drawing of the hydrogen spectrum, emission lines in the infrared and ultraviolet are shown as gray. Only the first three lines of the Balmer series are visible to our eyes.

ons in the hydrogen atom can be grouped into the **Lyman series, Balmer series, Paschen** etc. Transitions and the resulting spectral lines are d by Greek letters. Only the first few transitions in three series are shown here.

Excited clouds of gas in space emit light at all of the Balmer wavelengths, but we see only the red, blue, and violet photons blending to create the pink color typical of ionized hydrogen.

AURA/NOAO/NSF

The shorter-wavelength lines in each series blend together.

2000 nm

1500 nm

Infrared

Paschen lines

1000 nm

500 nm

Visible

H_α

Balmer lines

H_β

H_γ

100 nm

Ultraviolet

Lyman lines

dern
mers
work
spectra as
ds of light.
pectra are
recorded
lly, so it is
represent
as graphs
f intensity

Intensity

H_γ

H_β

H_α

500

600

700

Wavelength (nm)

wavelength. Here the artwork above suggests the appearance of a stellar spectrum. The below reveals details not otherwise visible and allows comparison of relative intensities. Notice that dark absorption lines in the spectrum appear as dips in the curve of intensity.

THE BALMER THERMOMETER

We can use the Balmer absorption lines as a thermometer to find the temperatures of stars. Earlier we saw how to estimate temperature from color, but the strengths of the Balmer lines in a star's spectrum give a much more accurate estimate of the star's temperature.

The Balmer thermometer works because the Balmer absorption lines are produced only by atoms whose electrons are in the second energy level. If the star is cool, there are few violent collisions between atoms to excite the electrons, and most atoms have their electron in the ground state. If most electrons are in the ground state, they can't absorb photons in the Balmer series. As a result, we should expect to find weak Balmer absorption lines in the spectra of cool stars.

In hot stars, on the other hand, there are many violent collisions between atoms, exciting electrons to high energy levels or knocking the electron clear out of some atoms; that is, some atoms are ionized. Thus, few hydrogen atoms have their electron in the second orbit to form Balmer absorption lines, and we should expect hot stars, like cool stars, to have weak Balmer absorption lines.

At some intermediate temperature, the collisions are just right to excite large numbers of electrons into the second energy level. With many atoms excited to the second level, the gas absorbs Balmer-wavelength photons strongly and thus produces strong Balmer lines.

To summarize, the strength of the Balmer lines depends on the temperature of the star's surface layers. Both hot and cool stars have weak Balmer lines, but medium-temperature stars have strong Balmer lines.

Theoretical calculations can predict just how strong the Balmer lines should be for stars of various temperatures. The details of these calculations are not important to us, but the results are. The curve in ▌Figure 7-8a shows the strength of the Balmer lines for various stellar temperatures. We could use this as a temperature indicator except that the curve gives us two answers. A star with Balmer lines of a certain strength might have either of two temperatures, one high and one low. We must examine other spectral lines to choose the correct temperature.

We have seen how the strength of the Balmer lines depends on temperature. Temperature has a similar ef-

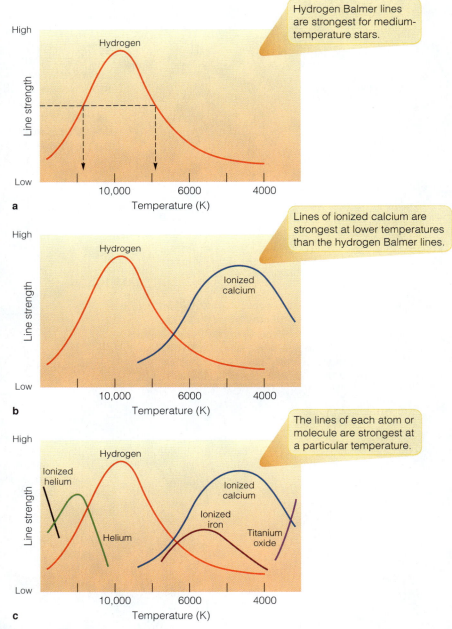

Hydrogen Balmer lines are strongest for medium-temperature stars.

Lines of ionized calcium are strongest at lower temperatures than the hydrogen Balmer lines.

The lines of each atom or molecule are strongest at a particular temperature.

FIGURE 7-8

The strength of spectral lines can tell us the temperature of a star. (a) Balmer hydrogen lines alone are not enough because Balmer lines of a certain strength could be produced by a hotter star or a cooler star. (b) Adding another atom to the diagram helps, and (c) adding many atoms and molecules to the diagram gives us a precise tool to find the temperatures of stars.

fect on the spectral lines of other elements, but the temperature at which they reach maximum strength differs for each element (Figure 7-8b). If we add these elements to our graph, we get a handy tool for finding the stars' temperatures (Figure 7-8c).

How do we use this tool? We determine a star's temperature by comparing the strengths of its spectral lines with our graph. For instance, if we recorded a spectrum of a star and found medium-strength Balmer lines and strong helium lines, we could conclude it had a temperature of about 20,000 K. But if the star had weak hydrogen lines and strong lines of ionized iron, we would assign it a temperature of about 5800 K, similar to that of the sun.

The spectra of stars cooler than about 3000 K contain dark bands produced by molecules such as titanium oxide (TiO). Because of their structure, molecules can absorb photons at many wavelengths, producing numerous, closely spaced spectral lines that blend together to form bands. These molecular bands appear only in the spectra of the coolest stars because, as mentioned before, molecules in cool stars are not subject to the violent collisions that would break up molecules in hotter stars. Thus the presence of dark bands in a star's spectrum indicates that the star is very cool.

From stellar spectra, astronomers have found that the hottest stars have surface temperatures above 40,000 K and the coolest about 2000 K. Compare these with the surface temperature of the sun, which is about 5800 K.

SPECTRAL CLASSIFICATION

We have seen that the strengths of spectral lines depend on the surface temperature of the star. From this we can predict that all stars of a given temperature should have similar spectra. If we learn to recognize the pattern of spectral lines produced by a 6000-K star, for instance, we need not use Figure 7-8c every time we see that kind of spectrum. We can save time by classifying stellar spectra rather than analyzing each one individually.

The first widely used classification system was devised by astronomers at Harvard during the 1890s and 1900s. One of them, Annie J. Cannon, personally inspected and classified the spectra of over 250,000 stars. The spectra were first classified in groups labeled A through Q, but some groups were later dropped, merged with others, or reordered. The final classification includes the seven **spectral classes**, or **types**, still used today: O, B, A, F, G, K, M.*

This sequence of spectral types, called the **spectral sequence,** is important because it is a temperature sequence. The O stars are the hottest, and the temperature continues to decrease down to the M stars, the coolest. For maximum precision, astronomers divide each spectral class into 10 subclasses. For example, spectral class A consists of the subclasses A0, A1, A2, . . . A8, A9. Next comes F0, F1, F2, and so on. This finer division gives us a star's temperature to an accuracy within about 5 percent. The sun, for example, is not just a G star, but a G2 star, with a temperature of 5800 K.

We classify a star by the lines and bands in its spectrum, as shown in ▌ Table 7-1. For example, if it has weak Balmer lines and lines of ionized helium, it must be an O star. This table is based on the same information we used in Figure 7-8c.

▌ Figure 7-9 shows 13 stellar spectra ranging from the hottest at the top to the coolest at the bottom. Notice how special features change gradually from hot to cool stars. Although these spectra are attractive, astronomers rarely work with spectra as color images. Rather, they display spectra as graphs of intensity versus wavelength with dark absorption lines as dips in the graph (▌ Figure 7-10). Such graphs show more detail than photographs. Notice also that the overall curves are similar to black body curves. The wavelength of maximum is in the infrared for the coolest stars and in the ultraviolet for the hottest stars.

Compare Figures 7-9 and 7-10 and notice how the strength of spectral lines depends on temperature. Note that the Balmer lines are strongest in A stars, where the temperature is moderate but still high enough to excite the electrons in hydrogen atoms to the second energy level, where they can absorb Balmer wavelength

*Generations of astronomy students have remembered the spectral sequence using the mnemonic "Oh, Be A Fine Girl (Guy), Kiss Me." More recent suggestions from students include, "Oh Boy, An F Grade Kills Me," and "Only Bad Astronomers Forget Generally Known Mnemonics."

TABLE 7-1
Spectral Classes

Spectral Class	Approximate Temperature (K)	Hydrogen Balmer Lines	Other Spectral Features	Naked-Eye Example
O	40,000	Weak	Ionized helium	Meissa (O8)
B	20,000	Medium	Neutral helium	Achernar (B3)
A	10,000	Strong	Ionized calcium weak	Sirius (A1)
F	7,500	Medium	Ionized calcium weak	Canopus (F0)
G	5,500	Weak	Ionized calcium medium	Sun (G2)
K	4,500	Very weak	Ionized calcium strong	Arcturus (K2)
M	3,000	Very weak	TiO strong	Betelgeuse (M2)

FIGURE 7-9
These spectra show stars from hot O stars at the top to cool M stars at the bottom. The Balmer lines of hydrogen are strongest about A0, but the two closely spaced lines of sodium in the yellow are strongest for very cool stars. Helium lines appear only in the spectra of the hottest stars. Bands produced by the molecule titanium oxide are strong in the spectra of the coolest stars. *(AURA/NOAO/NSF)*

photons. In the hotter stars (O and B), the Balmer lines are weak because the higher temperature excites the electrons to energy levels above the second or ionizes the atoms. The Balmer lines in cooler stars (F through M) are also weak, but for a different reason. The lower temperature cannot excite many electrons to the second energy level, so few hydrogen atoms are capable of absorbing Balmer wavelength photons.

The spectral lines of other atoms also change from class to class. Helium is visible only in the spectra of the hottest classes, and titanium oxide bands only in the coolest. Two lines of ionized calcium increase in strength from A to K and then decrease from K to M. Because the strength of these spectral lines depends on temperature, it requires only a few moments to study a star's spectrum and determine its temperature.

The study of spectral types is a century old, but astronomers continue to discover new types of stars. The **L dwarfs,** found in 1998, are cooler and fainter than M stars. The L dwarfs are clearly a different type of star. The spectra of M stars contain bands produced by metal oxides such as titanium oxide, but L dwarf spectra contain bands produced by chromium hydride and iron hydride (❙ Figure 7-11). The **T dwarfs,** discovered

in 2000, are even cooler and fainter than L dwarfs. The development of giant telescopes and highly sensitive infrared cameras and spectrographs is allowing astronomers to find and study these coolest of stars.

THE COMPOSITION OF THE STARS

It seems as though it should be easy to find the composition of the sun and stars just by looking at their spectra, but it turns out to be a difficult task that wasn't well understood until the 1920s. The story of how astronomers first discovered the composition of the stars is worth telling, not only because it is the story of an important American astronomer who never got proper credit, but also because the story illustrates the temperature dependence of spectral features. The story begins in England.

As a child in England, Cecilia Payne (1900–1979) excelled in classics, languages, mathematics, and literature, but her first love was astronomy. After finishing Newnham College in Cambridge, she left England, sensing that there were no opportunities in England for a woman of science. In 1922, Payne arrived at Harvard, where she eventually earned her Ph.D., although the

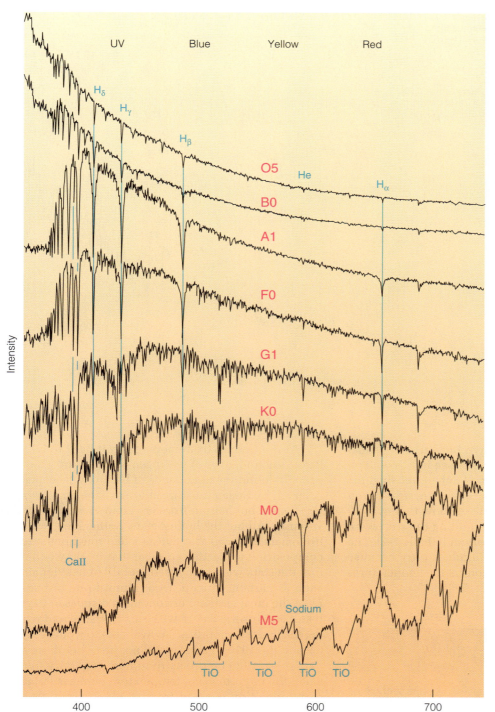

FIGURE 7-10
Modern digital spectra are often represented by graphs of intensity versus wavelength. Dark absorption lines are dips in intensity. The hottest stars are at the top and the coolest at the bottom. Hydrogen Balmer lines are strongest at about A0, while lines of ionized calcium (CaII) are strong in K stars. Titanium oxide (TiO) bands are strongest in the coolest stars. Compare these spectra with Figures 7-8c and 7-9. *(Courtesy NOAO, G. Jacoby, D. Hunter, and C. Christian)*

degree was awarded by Radcliffe because Harvard did not then admit women.

In her thesis, Payne attempted to relate the strength of the absorption lines in stellar spectra to the physical conditions in the atmospheres of the stars. This was not easy because, as we have seen in this chapter, a given spectral line can be weak because the atom is rare or because the temperature is too high or too low for that atom to be able to absorb efficiently. If we see sodium lines in a star's spectrum, we can be sure that the star contains sodium atoms, but if we see no sodium lines, we must consider the possibility that the star is too hot or too cool for sodium to produce spectral lines.

Payne's problem was to untangle these two factors and find the true temperatures of the stars and the true abundance of the atoms in their atmospheres. Recent

FIGURE 7-11

The spectrum of an M star and an L star compared: M-star spectra contain strong bands produced by the titanium oxide molecule (TiO), but L-star spectra contain weak or absent TiO bands. Notice the missing TiO band near 710 nm in the L-star spectrum above. Instead, the spectra of L stars contain bands produced by chromium hydride (CrH) and iron hydride (FeH). *(Adapted from data by J. Davy Kirkpatrick)*

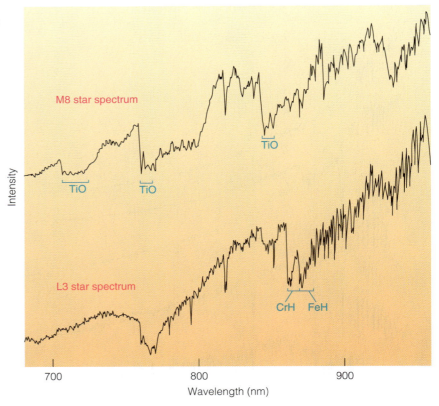

advances in atomic physics gave her the theoretical tools she needed. About the time Payne left Newnham College, Indian physicist Meghnad Saha published his work on the ionization of atoms. Drawing from such theoretical work, Payne was able to show that over 90 percent of the atoms in stars (including the sun) were hydrogen and most of the rest helium (■ Table 7-2). The heavier atoms seemed more abundant only because they are better at absorbing photons at the temperatures of stars.

TABLE 7-2
The Most Abundant Elements in the Sun

Element	Percentage by Number of Atoms	Percentage by Mass
Hydrogen	91.0	70.9
Helium	8.9	27.4
Carbon	0.03	0.3
Nitrogen	0.008	0.1
Oxygen	0.07	0.8
Neon	0.01	0.2
Magnesium	0.003	0.06
Silicon	0.003	0.07
Sulfur	0.002	0.04
Iron	0.003	0.1

At the time, astronomers found it hard to believe Payne's abundances of hydrogen and helium. They especially found the abundance of helium unacceptable. After all, hydrogen lines are at least visible in most stellar spectra, but helium lines are almost invisible in the spectra of all but the hottest stars. Rather, nearly all astronomers assumed that the stars had roughly the same composition as Earth's surface; that is, they believed that the stars were composed mainly of heavier atoms such as carbon, silicon, iron, aluminum, and so on. Even the most eminent astronomers dismissed Payne's result as illusory. Faced with this pressure and realizing the limited opportunities available to women in science in the 1920s, Payne could not press her discovery.

It was 1929 before astronomers generally understood the importance of temperature on measurements of composition derived from stellar spectra. At that point, astronomers recognized that stars are mostly hydrogen and helium, but Payne received no credit.

Payne worked for many years as a staff astronomer at the Harvard College Observatory with no formal position on the faculty. She married Russian astronomer Sergei Gaposchkin in 1934 and was afterward known as Cecilia Payne-Gaposchkin. In 1956, when Harvard accepted women to its faculty, she was appointed a full professor and chair of the Harvard astronomy department.

Cecilia Payne-Gaposchkin's work on the chemical composition of the stars illustrates the importance of fully understanding the interaction between light and

— Approaching

— Receding

← Red   Blue →

a

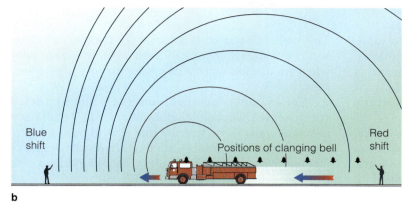

Blue shift

Positions of clanging bell

Red shift

b

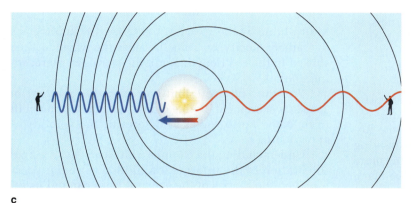

c

FIGURE 7-12

The Doppler effect. (a) A blue shift appears in the spectrum of a star approaching Earth (top spectrum). A red shift appears in the spectrum of a star moving away from Earth (bottom spectrum). *(The Observatories of the Carnegie Institution of Washington)* (b) The clanging bell on a moving fire truck produces sounds that move outward (black circles). An observer ahead of the truck hears the clangs closer together, while an observer behind the truck hears them farther apart. (c) A moving source of light emits waves that move outward (black circles). An observer in front of the light source observes a shorter wavelength (a blue shift), and an observer behind the light source observes a longer wavelength (a red shift).

matter. Only a detailed understanding of the physics could lead her to the correct composition. As we turn our attention to other information that can be derived from stellar spectra, we again discover the importance of understanding light.

THE DOPPLER EFFECT

Surprisingly, one of the pieces of information hidden in a spectrum is the velocity of the light source. Astronomers can measure the wavelengths of lines in a star's spectrum and find the velocity of the star. The **Doppler effect** is the apparent change in the wavelength of radiation caused by the motion of the source.

If a star is moving toward Earth, the lines in its spectrum will be shifted slightly toward shorter wavelengths. That is, they are shifted toward the blue end of the spectrum—a **blue shift.** If a star is moving away from Earth, the lines are shifted slightly toward the red end of the spectrum—a **red shift.** These Doppler shifts are small and don't change the color of the star. But it is easy to detect these changes in wavelength in a star's spectrum.

❙ Figure 7-12a shows the Doppler effect in two spectra of the star Arcturus. The lines in the top spectrum are slightly blue shifted because the spectrum was recorded when Earth, following its orbit, was moving toward Arcturus. The lines in the bottom spectrum are red shifted because it was recorded six months later, when Earth was moving away from Arcturus.

The Doppler effect tells us how rapidly the distance between us and the source of light is increasing or decreasing. It does not matter whether we are moving or the star is moving. Only the relative velocity is important. Also, the Doppler shift is only sensitive to the part of the velocity directed away from us or toward us. This part of the velocity is called the **radial velocity (V_r).** The **transverse velocity** is the part of the star's velocity perpendicular to our line of sight (❙ Figure 7-13). The Doppler effect cannot be used to measure the transverse velocity.

The Doppler effect occurs for any kind of radiation, not just light. Radio astronomers, for instance, observe the Doppler effect when they measure the wavelength of radio waves coming from objects moving toward or away from Earth. Whatever wavelength range we consider—radio, X rays, gamma rays—we will still refer to

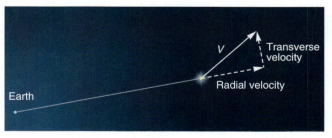

Earth   V   Transverse velocity   Radial velocity

FIGURE 7-13

The radial velocity of a star is the part of the star's velocity V that is directed away from or toward Earth. The Doppler effect can tell us a star's radial velocity but cannot tell us the transverse velocity—the part of the velocity perpendicular to the radial direction.

a lengthening of the observed wavelength as a red shift and a shortening of the observed wavelength as a blue shift.

We can even detect the Doppler shift in sound. Sound is not electromagnetic radiation, of course; it is a mechanical wave transmitted through air, but because it is a wave, it is subject to the Doppler effect. Sounds with long wavelengths have low pitches, and sounds with short wavelengths have higher pitches. When we hear a truck pass by, we hear a Doppler shift. Its sound is shifted to shorter wavelengths and higher pitches (a "blue" shift) while it is approaching and is shifted to longer wavelengths and lower pitches (a "red" shift) after it passes. As the truck passes, we hear the pitch drop.

We have described how the Doppler shift works, but we haven't explained why it works. To see why the Doppler shift occurs, we can think about a fire truck approaching us with a bell clanging once a second. When the bell clangs, the sound travels ahead of the truck to reach our ears. One second later, the bell clangs again, but not at the same place. During that one second, the fire truck moved closer to us, so the bell is closer at its second clang. Now the sound has a shorter distance to travel and reaches our ears a little sooner than it would have if the fire truck were not approaching. The third time the bell clangs, it is even closer. By timing the bell, we could observe that the clangs are slightly less than 1 second apart, all because the fire truck is approaching. If the fire truck were moving away from us, we would hear the clangs sounding more than one second apart, because each successive clang of the bell occurs farther from us.

Figure 7-12b shows a fire truck moving toward one observer and away from another observer. The position of the bell at each clang is shown by the small black bells with the sound spreading outward as black circles. We can see that the clangs are squeezed together ahead of the fire truck and stretched apart behind.

Now we can substitute a source of light waves for the clanging of the bell. If the source is approaching, then each time the source emits the peak of a wave it will be slightly closer to us, and we will observe a shorter wavelength. If it is moving away, we will observe a longer wavelength. This is shown in Figure 7-12c, where the peaks of the waves appear compressed in front of the moving light source and stretched out behind the source.

Of course, how great the change in wavelength is depends on the velocity. Just as a slowly moving car has a smaller Doppler effect in sound than does a high-speed airplane, a slowly moving star has a smaller Doppler effect in light than does a high-speed star. Thus, we can measure the velocity by measuring the amount by which the spectral lines are shifted in wavelength. A large Doppler shift means a high radial velocity.

CALCULATING THE DOPPLER VELOCITY

It is easy to calculate the radial velocity of an object from its Doppler shift. The formula is a simple ratio relating the radial velocity V_r divided by the speed of light c to the change in wavelength $\Delta\lambda$ divided by the unshifted wavelength λ_0:

$$\frac{V_r}{c} = \frac{\Delta\lambda}{\lambda_0}$$

This expression is quite accurate for the low radial velocities of stars, but we will need a better version later when we discuss objects moving with very high velocities.

For example, we might observe a line in a star's spectrum with a wavelength of 600.1 nm. Laboratory measurements show that the line should have a wavelength of 600 nm. That is, its unshifted wavelength is 600 nm. What is the star's radial velocity? First we note that the change in wavelength is 0.1 nm:

$$\frac{V_r}{c} = \frac{0.1}{600} = 0.000167$$

Multiplying by the speed of light, 3×10^5 km/s, gives the radial velocity, 50 km/s. Because the wavelength is shifted to the red (lengthened), the star must be receding from us.

You may be quite familiar with this method of speed measurement. Police radar uses the Doppler shift in reflected radio waves to determine the velocity of approaching cars.

Understanding the Doppler shift leads us to a final illustration of the information hidden in stellar spectra. Even the shapes of the spectral lines can tell us secrets about the stars.

THE SHAPES OF SPECTRAL LINES

When astronomers refer to the shape of a spectral line, they mean the variation of intensity across the line. An absorption line, for instance, is darkest in the center and grows brighter to each side. This shape is represented by a **line profile,** a graph of brightness as a function of wavelength across a spectral line (❚ Figure 7-14).

The exact shape of a line profile can tell us a great deal about a star, but the most important characteristic is the width of the line. Spectral lines are not perfectly narrow; if they were, we could not see them. They have a natural width because nature allows an atom some leeway in the energy it may absorb or emit. The quantum mechanics behind this effect is beyond the scope of our discussion, but the result is very simple. In the absence of all other effects, spectral lines have a natural width of about 0.001 to 0.00001 nm—very narrow indeed.

The natural widths of spectral lines are not important in most branches of astronomy because other effects smear out the lines and make them much broader. For example, if a star spins rapidly, the Doppler effect

FIGURE 7-14

A line profile is a graph that shows the intensity of starlight as a function of wavelength across a spectral line. In these two examples, the spectra show the same spectral line from the spectra of two A1 stars. Thus, both stars have the same temperature. Differences in the density of the gas in the atmospheres of the stars cause the differences in the widths of the spectral lines. Such differences in the width of the spectral lines are easily measured from such line profiles. *(Courtesy NOAO, G. Jacoby, D. Hunter, and C. Christian)*

will broaden the spectral lines. As the star rotates, one side will recede from us, and the other side will approach. Light from the receding side will be red shifted, and light from the approaching side will be blue shifted, so any spectral lines will be broadened.

Another important process is called **Doppler broadening.** To consider this process, let us imagine that we photograph the spectrum of a jar full of hydrogen atoms (█ Figure 7-15). Because the gas has some thermal energy (it is not at absolute zero), the gas atoms are in motion. Some will be coming toward our spectrograph, and some will be receding. Most, of course, will not be traveling very fast, but some will be moving very quickly. The photons emitted by the atoms approaching us will have slightly shorter wavelengths because of the Doppler effect, and photons emitted by atoms receding from us will have slightly longer wavelengths. Thus, the Doppler shifts due to the motions of the individual atoms will smear the spectral line out and make it broader. We have described the Doppler broadening of an emission line, but the effect is the same for absorption lines.

The extent of Doppler broadening depends on the temperature of the gas. If the gas is cold, the atoms travel at low velocities, and the Doppler shifts are small (Figure 7-15a). If the gas is hot, however, the atoms travel faster, Doppler shifts are larger, and the lines will be wider (Figure 7-15b).

Another form of broadening, **collisional broadening,** is caused by collisions between atoms, and thus it depends on the density of the gas (Parameters of Science 7-2). Densities in astronomy cover an enormous range, from one atom per cubic centimeter in space to millions of tons of atoms per cubic centimeter inside dead stars. Clearly, we should expect such densities to affect the way atoms collide with one another and thus how they absorb and emit photons.

Collisional broadening spreads out spectral lines when the atoms absorb or emit photons while they are colliding with other atoms, ions, or electrons. The collisions disturb the energy levels in the atoms, making it possible for the atoms to absorb a slightly wider range

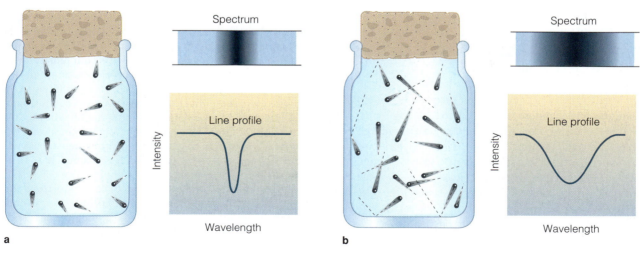

FIGURE 7-15

Doppler broadening. The atoms of a gas are in constant motion. Photons emitted by atoms moving toward the observer will have slightly shorter wavelengths, and those emitted by atoms moving away will have slightly longer wavelengths. This broadens the spectral line. If the gas is cool (a), the atoms do not move very fast, the Doppler shifts are small, and the line is narrow. If the gas is hot (b), the atoms move faster, the Doppler shifts are larger, and the line is broader.

Density

Density is a measure of the amount of matter in a given volume. Density is expressed as mass per volume, such as grams per cubic centimeter. The density of water, for example, is about 1 g/cm³.

To get a feel for density, imagine holding a brick in one hand and a similar-sized block of styrofoam in the other hand. We can easily tell that the brick contains more matter than the styrofoam block even though both are the same size. The brick weighs more than the styrofoam, but it isn't really the weight that we should consider. Rather, we should think about the mass of the two objects. In space, where they have no weight, the brick and the styrofoam would still have mass, and we could tell by moving them about that the brick

contains more mass than the styrofoam. For example, imagine tapping each object gently against your ear. The massive brick would be easy to distinguish from the low-mass styrofoam block even in weightlessness.

When we think of density, we divide mass by volume, and our minds make that comparison at an instinctive level. We can sense the density of an object just by handling it. Gift shops sometimes sell imitation rocks made of styrofoam as humorous gifts. "Rocks" made of styrofoam seem odd when we handle them because our brain expects rocks to be dense.

Density is a common parameter in science because it is a general property of materials. Lead, for example, has a

density of about 7 g/cm³, and rock has a density of 3 to 4 g/cm³. Water and ice have densities of about 1 g/cm³. If we knew that a small moon had a density of 1.5 g/cm³, we could immediately draw some conclusions about what kinds of materials it might be made of—ice and a little rock, but not much lead. The density of an object is a basic clue to its composition.

Density also determines how materials behave. The low-density gases in a nebula behave in one way, but the same gases under high density inside a star behave in a different way. Thus, astronomers must consider the density of materials when they describe astronomical objects.

---

of wavelengths. Thus, the spectral lines are wider. Because atoms in a dense gas collide more often than atoms in a low-density gas, collisional broadening depends on the density of the gas, but temperature is also a factor. Atoms in a hot gas travel faster and collide more often and more violently than atoms in a cool gas. Once again, the physics of the interaction of light and matter gives us a tool to understand starlight. In later chapters, we will see how collisional broadening affects the widths of spectral lines that are formed in normal stars like the sun, in giant stars, and in gas in interstellar space.

The spectra of the stars are filled with clues about the gas emitting the light. Much of astronomy is based on unraveling these clues. We will begin in the following chapter by studying the nearest star—the sun.

SUMMARY

Stars emit black body radiation from the dense gases of their photospheres, and as that radiation passes through the less dense gas of the star's atmosphere, the atoms absorb photons of certain wavelengths to produce dark lines in the spectrum.

A heated solid, liquid, or dense gas emits black body radiation that contains all wavelengths and thus produces a continuous spectrum. Black body radiation is most intense at the wavelength of maximum, λ_{max}, which depends on the temperature of the radiating body. Hot objects emit mostly short-wavelength radiation, whereas cool objects emit mostly long-wavelength radiation. This effect gives us clues to the temperatures of stars—hot stars appear blue and cool stars red.

A low-density gas excited to emit radiation will produce an emission (or bright-line) spectrum. Light from a source of a continuous spectrum passing through a low-density gas will produce an absorption (or dark-line) spectrum. These three kinds of spectra are described by Kirchhoff's laws.

The lines in spectra are produced by the electrons that surround the nucleus of the atom. The electrons may occupy only certain permitted energy levels, and photons may be absorbed or emitted when electrons move from one energy level to another. Because each kind of atom and ion has a different set of energy levels, each kind of atom and ion can absorb or emit only photons of a certain wavelength. The hydrogen atom can produce the Lyman series lines in the ultraviolet, the Balmer series in the visual and near ultraviolet, the Paschen series in the infrared, and many more.

REVIEW **CRITICAL INQUIRY**

Why are helium lines weak and calcium lines strong in the visible spectrum of the sun?

To analyze this problem, we must recall that the ability of an atom or ion to absorb light depends on temperature. Helium is quite abundant in the sun, but the surface of the sun is too cool to excite helium atoms and enable them to easily absorb visible-wavelength photons. On the other hand, calcium is easily ionized, and calcium atoms that have lost an electron are very good absorbers of photons at temperatures like those at the sun's surface. Thus, calcium lines are strong in the solar spectrum even though calcium ions are rare, and helium lines are weak even though helium atoms are common.

If we are to analyze a star's spectrum and consider the strength of spectral lines, we must take into account the temperature of the star. Why don't we need to take temperature into account when we use the Doppler effect to measure a star's radial velocity?

In cool stars, the Balmer lines are weak because most atoms are not excited out of the ground state. In hot stars, the Balmer lines are weak because most atoms are excited to higher orbits or ionized. Only at medium temperatures are the Balmer lines strong. We can use this effect as a thermometer for determining the temperature of a star. In its simplest form, this amounts to classifying the stars' spectra in the spectral sequence O, B, A, F, G, K, M.

Stellar spectra can tell us the chemical compositions of the stars, but we must be careful to consider the temperature of the star in our analysis. In general, 90 percent of the atoms in a star are hydrogen.

When a source of radiation is approaching us, we observe shorter wavelengths, a blue shift, and when it is receding, we observe longer wavelengths, a red shift. This Doppler effect makes it possible for the astronomer to measure a star's radial velocity, that part of its velocity directed toward or away from the earth.

The widths of spectral lines are affected by a number of processes. With no outside influences, a spectral line has a natural width that is very small. Doppler broadening, due to the thermal motion of the gas atoms, depends on the temperature of the gas. Collisional broadening occurs when the atoms emitting or absorbing photons collide with other atoms, ions, or electrons. The collisions disturb the atomic energy levels and slightly alter the wavelengths the atoms can absorb or emit. In a dense, hot gas, where the gas atoms collide often, collisional broadening is important.

NEW TERMS

temperature	absorption spectrum (dark-line spectrum)
Kelvin temperature scale	
absolute zero	absorption line
thermal energy	emission spectrum (bright-line spectrum)
electron	
black body radiation	emission line
wavelength of maximum intensity (λ_{max})	Kirchhoff's laws
	transition
color index	Lyman series
nucleus	Balmer series
proton	Paschen series
neutron	spectral class or type
isotope	spectral sequence
ionization	L dwarf
ion	T dwarf
molecule	Doppler effect
Coulomb force	blue shift
binding energy	red shift
quantum mechanics	radial velocity (V_r)
permitted orbit	transverse velocity
energy level	line profile
excited atom	Doppler broadening
ground state	collisional broadening
continuous spectrum	density

REVIEW QUESTIONS

1. Why can a good blacksmith judge the temperature of a piece of heated iron by its color?

2. Why do hot stars look bluer than cool stars?

3. Use black body radiation to explain why you could hold a cigarette between your lips and light it with a match, but you couldn't hold the cigarette in your lips while you light it by sticking the tip into a bonfire. The match and bonfire have about the same temperature. (This is just one of the hazards of smoking.)

4. What is the difference between a neutral atom, an ion, and an excited atom?

5. How do the energy levels in an atom determine which wavelength photons it can absorb or emit?

6. What kind of spectrum would you expect to record if you observed molten lava? In practice, to view molten lava you must look through gases boiling out of the lava. What kind of spectrum might you see in that case?

7. Why do the strengths of the Balmer lines depend on the temperature of the star?

8. Why does a stellar spectrum tell us about the surface layers but not the deeper layers of the star?

9. Explain the similarities among Figure 7-8, Figure 7-9, Figure 7-10, and Table 7-1.

10. Why do we not see TiO bands in the spectra of hot stars?

11. In observing a star's spectrum, we note that the lines of a certain element are not present. Are we safe in concluding that the star does not contain this element? Why or why not?

12. If a star moves exactly perpendicular to a line connecting it to Earth, will the Doppler effect change its spectrum? Why or why not?

13. Why would we expect a star that rotates very rapidly to have broad spectral lines? What else could broaden the lines?

DISCUSSION QUESTIONS

1. In what ways is our model of an atom a scientific model? How can we use it when it is not a completely correct description of an atom?

2. Can you think of classification systems we commonly use to simplify what would otherwise be very complex measurements? Consider foods, movies, cars, grades, clothes, and so on.

PROBLEMS

1. Human body temperature is about 310 K (98.6°F). At what wavelength do humans radiate the most energy? What kind of radiation do we emit?

2. If a star has a surface temperature of 20,000 K, at what wavelength will it radiate the most energy?

3. Infrared observations of a star show that it is most intense at a wavelength of 2000 nm. What is the temperature of the star's surface?

4. If we double the temperature of a black body, by what factor will the total energy radiated per second per square meter increase?

5. If one star has a temperature of 6000 K and another star has a temperature of 7000 K, how much more energy per second will the hotter star radiate from each square meter of its surface?

6. Transition A produces light with a wavelength of 500 nm. Transition B involves twice as much energy as A. What wavelength light does it produce?

7. Determine the temperatures of the following stars based on their spectra. Use Figure 7-8.
   a. medium-strength Balmer lines, strong helium lines
   b. medium-strength Balmer lines, weak ionized calcium lines
   c. TiO bands very strong
   d. very weak Balmer lines, strong ionized calcium lines

8. To which spectral classes do the stars in Problem 7 belong?

9. In a laboratory, the Balmer beta line has a wavelength of 486.1 nm. If the line appears in a star's spectrum at 486.3 nm, what is the star's radial velocity? Is it approaching or receding?

10. The highest-velocity stars an astronomer might observe have velocities of about 400 km/s. What change in wavelength would this cause in the Balmer gamma line? (*Hint:* Wavelengths are given on page 133).

CRITICAL INQUIRIES FOR THE WEB

1. To what extent is it possible to tell the spectral type of a star based on observation of its color? Search for color index data on the stars in Table A-9 (Appendix A) to see how B-V indices correlate with spectral type. Observational follow-up: Is it possible to rate star colors in such a way as to determine spectral type through visual observation? (Search the Web for indications that others have tried this.)

2. Is it possible to tell the direction of motion of the sun in space by observing radial velocities of other stars? Search the Internet for stellar radial velocity data for stars in a variety of directions on the celestial sphere and look for trends that might indicate our motion in space.

EXPLORING *THESKY*

1. Locate the following stars, click on them, and determine their spectral types: Antares in Scorpius, Betelgeuse in Orion, Aldebaran in Taurus, Sirius in Canis Major, Rigel in Orion.

2. How are spectral types correlated with the colors of stars? (*Hint:* Locate Orion and choose **Spectral Colors** under the **View** menu.)

 Visit the Seeds *Foundations of Astronomy* companion Web site for critical thinking exercises, articles, and additional readings from InfoTrac College Edition, Brooks/Cole's online student library.

THE SUN—OUR STAR

All cannot live on the piazza, but everyone may enjoy the sun.

Italian Proverb

Royal Swedish Academy of Sciences

GUIDEPOST

The preceding chapter described how we can get information from a spectrum. In this chapter, we apply these techniques to the sun, to learn about its complexities.

This chapter gives us our first close look at how scientists work, how they use evidence and hypothesis to understand nature. Here we will follow carefully developed logical arguments to understand our sun.

Most important, this chapter gives us our first detailed look at a star. The chapters that follow will discuss the many kinds of stars that fill the heavens, but this chapter shows us that each of them is both complex and beautiful; each is a sun.

VIRTUAL LABORATORIES

SOLAR WIND AND COSMIC RAYS

HELIOSEISMOLOGY

Visual-wavelength image

THE SUN

Data File One

An image of the sun in visible light shows a few sunspots. The Earth–moon system is added for scale. *(Daniel Good)*

Average distance from Earth	1.00 AU (1.495979×10^8 km)
Maximum distance from Earth	1.0167 AU (1.5210×10^8 km)
Minimum distance from Earth	0.9833 AU (1.4710×10^8 km)
Average angular diameter seen from Earth	0.53° (32 minutes of arc)
Period of rotation	25.38 days at equator
Radius	6.9599×10^5 km (109 $R_\odot$)*
Mass	1.989×10^{30} kg (333,000 $M_\oplus$)*
Average density	1.409 g/cm³
Escape velocity at surface	617.7 km/s
Luminosity	3.826×10^{26} J/s
Surface temperature	5800 K
Central temperature	15×10^6 K
Spectral type	G2 V
Apparent visual magnitude	−26.74
Absolute visual magnitude	4.83

*In astronomy the symbols ☉ and ⊕ represent the sun and Earth, respectively.

A wit once remarked that solar astronomers would know a lot more about the sun if it were farther away. This comment contains a grain of truth; the sun is only a humdrum star, and there are billions like it in the sky, but the sun is the only one close enough to show surface detail. Solar astronomers can see so much detail in the swirling currents of gas and arching bridges of magnetic force that present theories seem inadequate to describe it. Yet the sun is not a complicated object. It is just a star.

In their general properties, stars are very simple. They are great balls of hot gas held together by their own gravity. Their gravity would make them collapse into small, dense bodies were they not so hot. The tremendously hot gas inside stars has such a high pressure that the stars would surely explode were it not for their own confining gravity. Like soap bubbles, stars are simple structures balanced between opposing forces that individually would destroy them. Thus, we study the sun as a close-up example of a star.

Another reason to study the sun is that life on Earth depends critically on the sun. Very small changes in the sun's luminosity can alter Earth's climate, and a slightly larger change might make Earth uninhabitable. In addition, we get nearly all our energy from the sun— oil and coal are merely stored sunlight—and our pleasant climate is maintained by energy from the sun. In fact, the sun's atmosphere of very thin gas reaches out past Earth's orbit, and thus any change in the sun, such as an eruption or a magnetic storm, can have a direct effect on Earth.

Finally, we study the sun because it is beautiful. Our analysis of sunlight will reveal that the sun is both powerful and delicate. Thus, we study the sun not only because it is *a star*, not only because it is *our star*, but because it is the sun.

8-1 THE SOLAR ATMOSPHERE

The sun is 109 times Earth's diameter and 333,000 times Earth's mass. This seems dramatic, but look at ❙ Data File One and notice the sun's density. It is only a little bit more dense than water. So, although the sun is very large and very massive, it must be a gas from its surface to its center. When we look at the sun we see only the outer layers of this vast sphere of gas. In fact, these outer layers, the solar atmosphere, extend high above the visible surface of the sun.

Ace Astronomy™ The AceAstronomy icon throughout the text indicates an opportunity for you to test yourself on key concepts and to explore animations and interactions on the AceAstronomy Web site at **http://astronomy.brookscole .com/seeds8e**

HEAT FLOW IN THE SUN

Simple logic tells us that energy in the form of heat is flowing outward from the sun's interior. The solar spectrum reveals that the temperature of the sun's surface is about 5800 K. At that temperature, every square centimeter of the sun's surface must be radiating more energy than a 6000-watt lightbulb. With all that energy radiating into space, the sun's surface would cool rapidly if energy did not flow up from the interior to keep the surface hot.

Not until the 1930s did astronomers understand how the sun makes its energy. Nuclear reactions occur in the core of the sun and generate energy, which flows upward as heat and keeps the surface hot. We will discuss these nuclear reactions in detail later in this chapter, but here we must recognize the importance of the energy flowing outward through the sun's surface.

As we study the atmosphere of the sun, we will find many phenomena that are driven by the energy flowing outward. Like a pot of boiling soup on a hot stove, the surface of the sun is in constant activity as the heat flows up from below.

THE PHOTOSPHERE

The photosphere, the sun's visible surface, seems smooth and featureless, marked occasionally by **sunspots,** dark spots on the solar disk. Recall that Galileo was the first to see sunspots. We will discuss sunspots later in this chapter; but, for now, we must consider the sun's smooth face, the photosphere.

The photosphere is not a solid surface. Remember that the sun is gaseous from its outer atmosphere right down to its center. The photosphere is the thin layer of gas from which we receive most of the sun's light. It is less than 500 km deep and has an average temperature of about 5800 K. If the sun magically shrank to the size of a bowling ball, the photosphere would be no thicker than a layer of tissue paper wrapped around the ball (❙ Figure 8-1a). For comparison, the chromosphere lies above the photosphere and is only a few times thicker in extent, but the corona, beginning above the chromosphere, extends far above the visible surface (Figure 8-1b).

Below the photosphere, the gas is denser and hotter and therefore radiates plenty of light, but that light cannot escape from the sun because of the outer layers of gas. Thus, we cannot detect light from these deeper layers. Above the photosphere, the gas is less dense and so is unable to radiate much light. The photosphere is the layer in the sun's atmosphere that is dense enough to emit plenty of light but not so dense that the light can't escape.

One reason the photosphere is so shallow is related to the hydrogen atom. Because the temperature of the photosphere is sufficient to ionize some atoms, there are a large number of free electrons in the gas. Neutral hydrogen atoms can add an extra electron and become

a

b

FIGURE 8-1

(a) A cross section at the edge of the sun shows the relative thickness of the photosphere and chromosphere. Earth is shown for scale. On this scale, the disk of the sun would be more than 1.5 m (5 feet) in diameter. (b) The corona extends from the top of the chromosphere to great height above the photosphere. This photograph, made during a total solar eclipse, shows only the inner part of the corona. *(Daniel Good)*

an H⁻ (H-minus) ion, but this extra electron is held so loosely that almost any photon has energy enough to free it. In the process, of course, the photon is absorbed. Thus, the H⁻ ions are very good absorbers of photons and make the gas of the photosphere very opaque. Light from below cannot escape easily, and we see a well-defined surface—the thin photosphere.

Although the photosphere appears to be substantial, it is really a very-low-density gas. Even in the deepest and densest layers visible, the photosphere is 3400 times less dense than the air we breathe. To find gases as dense as the air we breathe, we would have to descend about 7×10^4 km below the photosphere, about 10 percent of the way to the sun's center. With a fantastically efficient insulation system, we could fly a spaceship right through the photosphere.

The spectrum of the sun is an absorption spectrum, and that can tell us a great deal about the photosphere. We know from Kirchhoff's third law that an absorption spectrum is produced when a source of a continuous spectrum is viewed through a gas. In the case of the photosphere, the deeper layers are dense enough to produce a continuous spectrum, but atoms in the photosphere absorb photons of specific wavelengths, producing the absorption lines we see.

In good photographs, the photosphere has a mottled appearance because it is made up of dark-edged regions. The regions are called granules, and the visual pattern is called **granulation** (▌Figure 8-2a). Each granule is about the size of Texas and lasts for only 10 to 20 minutes before fading away. Faded granules are continuously replaced by new granules. Spectra of these granules show that the centers are a few hundred degrees hotter than the edges, and Doppler shifts reveal that the centers are rising and the edges are sinking at speeds of about 0.4 km/second.

From this evidence, astronomers recognize granulation as the surface effects of convection just below the photosphere. **Convection** occurs when hot fluid rises and cool fluid sinks, as when, for example, a convection current of hot gas rises above a candle flame. You can create convection in a liquid by adding a bit of cool nondairy creamer to an unstirred cup of hot coffee. The cool creamer sinks, warms, rises, cools, sinks again, and so on, creating small regions on the surface of the coffee that mark the tops of convection currents. Viewed from above, these regions look much like solar granules. Rising currents of hot gas heat small regions of the photosphere, which, being slightly hotter, emit more black body radiation and look brighter. The cool sinking gas of the edges emits less light and thus looks darker (Figure 8-2b). Thus the granulation reveals that energy flows upward from below by convection and heats the photosphere.

Spectroscopic studies of the solar surface have revealed another kind of granulation. **Supergranules** are regions about 30,000 km in diameter (about 2.3 times Earth's diameter) and include about 300 granules. These supergranules are regions of very slowly rising currents that last a day or two. They may be the surface traces of larger currents of rising gas deeper under the photosphere.

The edge, or **limb,** of the solar disk is dimmer than the center (see the figure in Data File One). This **limb darkening** is caused by the absorption of light in the photosphere. When we look at the center of the solar disk, we are looking directly down into the sun, and we see deep, hot, bright layers in the photosphere. But when we look near the limb of the solar disk, we are looking at a steep angle and cannot see as deeply. The photons we see come from shallower, cooler, dimmer layers in the photosphere. Limb darkening proves that the temperature in the photosphere decreases with height, as we would expect if energy is flowing up from below.

THE CHROMOSPHERE

Above the photosphere lies the chromosphere. Solar astronomers define the lower edge of the chromosphere as lying just above the visible surface of the sun with its upper regions blending gradually with the corona. We can think of the chromosphere as being an irregular layer with a depth on average less than Earth's diameter (see Figure 8-1). Because the chromosphere is roughly 1000 times fainter than the photosphere, we can see it with our unaided eyes only during a total solar eclipse when the moon covers the brilliant photosphere. Then, the chromosphere flashes into view as a thin line of pink just above the photosphere. The word *chromosphere* comes from the Greek word *chroma,* meaning "color." The pink color is produced by the combined light of three bright emission lines—the red, blue, and violet Balmer lines of hydrogen.

Astronomers know a great deal about the chromosphere from its spectrum. The chromosphere produces an emission spectrum, and Kirchhoff's second law tells us the chromosphere must be an excited, low-density gas. The density is about 10^8 times less dense than the air we breathe.

From spectra, astronomers can tell that atoms in the lower chromosphere are ionized, and atoms in the

a

b

Granule

Sinking gas   Rising gas

FIGURE 8-2

(a) A visible-light photo of the sun's surface shows granulation. *(AURA/NOAO/NSF)* (b) This model explains granulation as the tops of rising convection currents just below the photosphere. Heat flows upward as rising currents of hot gas and downward as sinking currents of cool gas. The rising currents heat the solar surface in small regions that we see as granules.

higher layers of the chromosphere are even more highly ionized. That is, they have lost more electrons. From this, astronomers can find the temperature in different parts of the chromosphere. Just above the photosphere the temperature falls to a minimum of about 4500 K, and then rises rapidly (▮ Figure 8-3). The region where the temperature increases fastest is called the **transition region** because it makes the transition from the lower temperatures of the photosphere and chromosphere to the extremely high temperatures of the corona.

Solar astronomers often image the sun at specific wavelengths. For example, hydrogen atoms in the layers of gas above the photosphere absorb photons at the wavelength of the Balmer alpha line very efficiently, and these photons cannot escape easily from the photosphere. An image formed using only these wavelengths, often called a **filtergram,** reveals structure above the photosphere in the chromosphere. In a similar way, an image recorded in the far ultraviolet or in the X-ray part of the spectrum reveals other structures in the solar atmosphere. This chapter contains a number of such solar images.

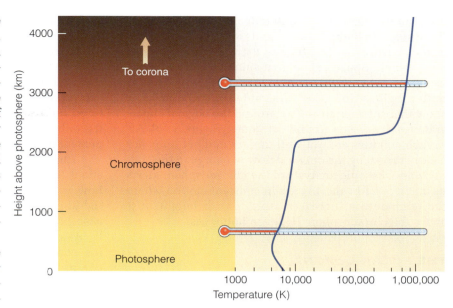

FIGURE 8-3

If we could place thermometers in the sun's atmosphere, we could construct a graph of temperature versus height as shown at the right. The curve shows that the temperature at the photosphere is about 5800 K, falls slightly in the lower chromosphere, and then rises rapidly.

The most common features of the chromosphere are shown in ▮ Figure 8-4. **Filaments** are long dark features silhouetted against the brighter surface. We will see later in this chapter that filaments are related to solar activity. The **spicules** are flamelike jets of gas rising upward into the chromosphere and lasting 5 to

FIGURE 8-4

H_α filtergrams reveal complex structure in the chromosphere including long, dark filaments and spicules springing from the edges of supergranuals twice the diameter of Earth. *(NOAA/SEL/USAF; © 1971 NOAO/NSO)*

15 minutes. Seen at the limb (edge) of the sun's disk, these spicules blend together and look like flames covering a burning prairie (Figure 8-1a), but they are not flames at all. Spectra show that spicules are cooler gas from the lower chromosphere extending upward into hotter regions. Images at the center of the solar disk show that spicules spring up around the edge of supergranules like weeds around flagstones (Figure 8-4b). Although spicules are not well understood yet, they are clearly driven by the outward flow of energy in the sun.

Spectroscopic analysis of the chromosphere alerts us that it is a low-density gas in constant motion where the temperature increases rapidly with height. Just above the chromosphere lies even hotter gas.

THE SOLAR CORONA

The outermost part of the sun's atmosphere is called the corona, after the Greek word for crown. The corona is so dim that it is not visible in Earth's daytime sky because of the glare of scattered light from the brilliant photosphere. During a total solar eclipse, however, when the moon covers the photosphere, we can see the innermost parts of the corona, as shown in Figure 8-1b.

Specialized telescopes called **coronagraphs** can block the light of the photosphere with a mask and image the corona. Such telescopes have traditionally been located on high mountaintops where the air is thin and scatters little sunlight. Now coronagraphs are carried by satellites in the vacuum of space where scattered light is not a problem. Such coronagraphs can trace the corona out beyond 20 solar radii, almost 10 percent of the way to Earth (▌ Figure 8-5).

The spectrum of the corona can tell us a great deal about the coronal gases and simultaneously illustrate how astronomers can analyze a spectrum. Some of the light from the outer corona produces a spectrum with absorption lines the same as the sun's spectrum. This light is just sunlight reflected from dust particles in the corona. In contrast, some of the light from the corona produces a continuous spectrum that lacks absorption lines, and that happens when sunlight from the photosphere is scattered off of free electrons in the ionized coronal gas. We might expect this light to reproduce the absorption spectrum of the photosphere, but it does not because the coronal gas has a temperature over 1 million K, and the electrons travel very fast. When photons scatter off of these high-speed elec-

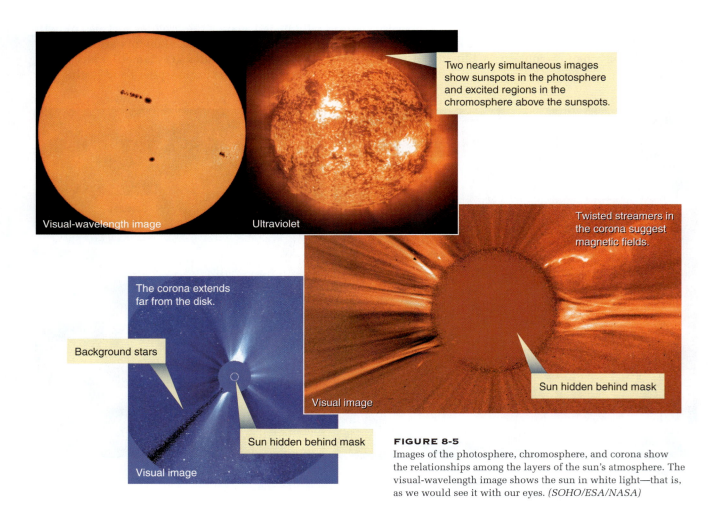

Two nearly simultaneous images show sunspots in the photosphere and excited regions in the chromosphere above the sunspots.

Visual-wavelength image

Ultraviolet

Twisted streamers in the corona suggest magnetic fields.

The corona extends far from the disk.

Background stars

Sun hidden behind mask

Visual image

Sun hidden behind mask

Visual image

FIGURE 8-5

Images of the photosphere, chromosphere, and corona show the relationships among the layers of the sun's atmosphere. The visual-wavelength image shows the sun in white light—that is, as we would see it with our eyes. *(SOHO/ESA/NASA)* trons, the photons suffer random Doppler shifts that smear out any absorption lines to produce a continuous spectrum.

Superimposed on the corona's continuous spectrum are emission lines of highly ionized gases. In the lower corona, the atoms are not as highly ionized as they are at higher altitudes, and this tells us that the temperature of the corona rises with altitude. Just above the chromosphere, the temperature is about 500,000 K; but in the outer corona the temperature can be as high as 2 million K or more.

The spectrum of the corona tells us that it is exceedingly hot gas, but it is not very bright. Its density is very low, only 10^6 atoms/cm^3 in its lower regions. That is about a trillion times less dense than the air we breath. In its outer layers the corona contains only 1 to 10 atoms/cm^3, better than the best vacuum on Earth. Because of this low density, the hot gas does not emit much radiation.

Astronomers have wondered for years how the corona and chromosphere can be so hot. Heat flows from hot regions to cool regions, so how can the heat from the photosphere, with a temperature of only 5800 K, flow out into the much hotter chromosphere and corona? Observations made by the SOHO satellite have mapped a **magnetic carpet** of looped magnetic fields extending up through the photosphere (▌Figure 8-6). Turbulence below the surface may be whipping these fields about and churning the low-density gases of the chromosphere and corona. That could heat the gas. In this instance, energy appears to flow outward as the agitation of the magnetic fields.

Gas from the solar atmosphere follows the magnetic fields pointing outward and flows away from the sun in a breeze called the **solar wind.** Like an extension of the corona, the low-density gases of the solar wind blow past Earth at 300 to 800 km/s with gusts as high as 1000 km/s. Thus Earth is bathed in the corona's hot breath.

As the solar wind carries gas away from the sun, the sun loses mass. This is only a minor loss for an object as massive as the sun. The sun loses about 10^7 tons per year, but that is only 10^{-14} of a solar mass per year. Even in its entire life as a stable star, the sun will lose only a tiny fraction of its mass through the solar wind.

Do other stars have chromospheres, coronae, and stellar winds like the sun? Ultraviolet and X-ray observations suggest the answer is yes. The spectra of many stars contain emission lines in the far ultraviolet that could only have formed in the low-density, high-temperature gases of a chromosphere and corona. Also, many stars are sources of X rays, which appear to have been produced by coronae. Thus, observational evidence gives us good reason to believe that the sun, for all its complexity, is a typical star.

FIGURE 8-6

This extreme ultraviolet image of a section of the sun's lower corona has been given a green color. White and black areas are regions of opposite magnetic polarity. Computer models show the location of the magnetic fields that connect these regions. The largest loops shown here could encircle Earth. The entire surface of the sun is covered by this magnetic carpet. *(Stanford-Lockheed Institute for Space Research, Palo Alto, CA, and NASA GSFC)*

HELIOSEISMOLOGY

It seems that the light we receive from the sun can tell us nothing about the layers below the photosphere, but solar astronomers have found a way to explore the sun's interior by **helioseismology,** the study of the modes of vibration of the sun. Just as geologists can study Earth's interior by observing how sound waves produced by earthquakes are reflected and transmitted by the layers of Earth's interior, so too can solar astronomers explore the sun's interior by studying how vibrations travel through the sun.

The waves are believed to arise as mechanical vibrations in the convective zone just below the photosphere. Helioseismologists refer to these waves as acoustic vibrations, but we might call them sound. As the waves travel down into the sun, they are bent back up toward the surface by the increasing temperature. Short-wavelength waves penetrate less deeply and travel shorter distances than longer-wavelength waves, as shown in ▌Figure 8-7a. When the waves reach the surface, they cause sections of the photosphere to move up and down by small amounts—roughly plus or minus 15 km. This covers the surface of the sun with a pattern of rising and falling regions that can be mapped using the Doppler effect (Figure 8-7b).

The sun can oscillate in about 10 million different ways, and each mode of oscillation has its own characteristic wavelength and its own unique pattern on the solar surface. The waves producing different modes penetrate to different depths where conditions can weaken or strengthen a wave. By discovering which wavelength waves are actually present, solar astronomers can determine the temperature, density, pressure, composition, and motion at different depths inside the sun.

a

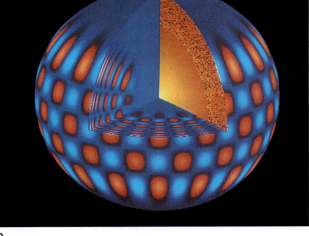

b

FIGURE 8-7

Helioseismology: (a) This cross section through the sun's equator shows how waves with short wavelengths do not penetrate as deeply as longer-wavelength waves. Where waves reflect off the underside of the solar surface, they cause a pattern of rising and falling regions (red and blue in this diagram) that can be observed by the Doppler effect. (b) This model shows only one of the 10 million possible patterns that are superimposed on the sun's surface. *(AURA/NOAO/NSF)*

Of course, with 10 million possible wavelengths, the observations and analysis are difficult. A single wave produces a complicated pattern of motion on the solar surface (Figure 8-7b). Large amounts of data are necessary, so helioseismologists have used a network of telescopes around the world operated by the Global Oscillation Network Group (GONG). The sun never sets on GONG. The network can observe the sun continuously for weeks at a time. The SOHO satellite in space has also observed solar oscillations continuously and can detect motions as slow as 1 mm/s (0.002 mph). Solar astronomers can then use high-speed computers to separate the different patterns on the solar surface and measure the strength of the waves at many different wavelengths.

Helioseismology sounds almost magical, but we can understand it better if we think of a duck pond. If we stood at the shore of a duck pond and looked down at the water, we would see ripples arriving from all parts of the pond. Because every duck on the pond contributes to the ripples, we could, in principle, study the ripples near the shore and draw a map showing the position and velocity of every duck on the pond. Of course, it would be difficult to untangle the different ripples, but all of the information would be there, lapping at the rocks at our feet.

Helioseismology has allowed astronomers to map the temperature, density, and rate of rotation inside the sun. They have been able to detect great currents of gas flowing below the photosphere and the emergence of sunspots before they appear in the photosphere. Helio-

seismology can even locate sunspots on the back side of the sun, sunspots that are not yet visible from Earth.

REVIEW CRITICAL INQUIRY

How deeply into the sun can we see?

This is a simple question, but it has a very interesting answer. When we look into the layers of the sun, our sight does not really penetrate into the sun. Rather, our eyes record photons that have escaped from the sun and traveled outward through the layers of the sun's atmosphere. If we observe at a wavelength at the center of a dark absorption line, then the photosphere and lower chromosphere are opaque, photons can't escape to our eyes, and the only photons we can see come from the upper chromosphere. What we see are the details of the upper chromosphere. On the other hand, if we observe at a wavelength that is not easily absorbed (a wavelength between spectral lines), the atmosphere is more transparent, and photons from deep inside the photosphere can escape to our eyes. There is a limit, however, set by the H^- ion, a hydrogen atom with an extra electron. At a certain depth, there is so much of this ion that the sun's atmosphere is opaque for almost all wavelengths, few photons can escape, and we can't see deeper.

By choosing the proper wavelength, solar astronomers can observe to different depths. But the corona is so thin and the gas below the photosphere so dense that this method doesn't work in these regions. How can we observe the corona and the deeper layers of the sun?

So far we have thought of the sun as a static, unchanging ball of gas with energy flowing outward from the interior and through the atmospheric layers. In fact, the sun is a highly variable body whose appearance is constantly changing. It is now time to think of the active sun.

8-2 SOLAR ACTIVITY

The sun is unquiet. It is home to slowly changing spots larger than Earth and vast eruptions that dwarf our imaginations. All of these seemingly different forms of solar activity have one thing in common—magnetic fields. The weather on the sun is magnetic.

SUNSPOTS AND ACTIVE REGIONS

Solar activity is often visible with even a small telescope, but we should exercise great caution in observing the sun. Sunlight is very intense, and when it enters our eye it is absorbed and converted into heat. Equally dangerous is the infrared radiation in sunlight. Our eyes can't detect the infrared, but it is converted to heat in our eyes and can burn and scar the retina. Looking at the sun with the unaided eye can cause permanent damage, and it is even more dangerous to look at the sun through a telescope, which concentrates the sunlight. ▌Figure 8-8a shows a safe way to observe the sun with a small telescope.

When we observe the sun at visual wavelengths, we see the gases of the photosphere. Dark spots on the photosphere are called sunspots, and observations over a period of days will show the sun rotating and the sunspots growing larger or shrinking away (Figure 8-8b).

If we look carefully at sunspots, we see that they consist of a dark center called the umbra and an outer border called the penumbra (▌Figure 8-9). Sunspots come in a wide range of sizes, but we might observe that the average spot is about 40 seconds of arc in diameter. The small-angle formula (Chapter 3) tells us that an average sunspot is about twice the diameter of Earth. Such spots last a week or so and tend to occur in pairs or in groups of one or two major spots accompanied by smaller spots. A large group of sunspots is called an **active region,** may contain up to 100 spots, and could last as long as two months.

FIGURE 8-8
Looking through a telescope at the sun is dangerous, but you can always view the sun safely with a small telescope by projecting its image on a white screen (a). If you sketch the location and structure of sunspots on successive days (b), you will see the rotation of the sun and gradual changes in the size and structure of sunspots.

a   b

FIGURE 8-9

(a) Sunspots often occur in pairs or in complex groups. This large sunspot group is a number of times larger than Earth's diameter. By analogy with the parts of a shadow, the dark center of the sunspot is called an umbra, and the brighter outer border is called the penumbra. Of course, sunspots are not shadows but are slightly cooler regions of the solar surface. *(SOHO/ ESA/NASA)* (b) This extremely high-resolution image shows detailed flow patterns in the penumbra as well as granulation on the solar surace. *(Royal Swedish Academy of Sciences)*

a

Sunspots look dark because they are cooler than the photosphere. By noting which atoms are excited in the spectra of sunspots, astronomers can tell that the umbra has a temperature of about 4240 K compared with 5800 K for the photosphere. Because the amount of black body radiation emitted depends on the temperature to the fourth power (the Stefan–Boltzmann law in Chapter 7), this small difference in temperature makes a big difference in brightness, and the spot looks quite dark compared to the photosphere. In fact, a sunspot umbra emits quite a bit of light. If the sun were removed and only an average-size sunspot were left behind, it would glow brighter than the full moon.

A clue to the origin of sunspots appeared in 1908 when American astronomer George Ellery Hale discovered magnetic fields in sunspots. A magnetic field affects the permitted energy levels in an atom. With no magnetic field present, a particular atom might absorb

Visual

Slit allows light from sunspot to enter spectrograph.

Spectral line split by Zeeman effect

Ultraviolet filtergram

Sunspot groups

Magnetic image

Magnetic fields around sunspot groups

Infrared image

Sunspot

Brighter photosphere around sunspot

FIGURE 8-10

The Zeeman effect splits spectral lines into three or more components. The amount of Zeeman splitting allows astronomers to map the north and south magnetic fields on the sun and show that the sunspot groups are highly magnetic. Infrared images show that energy that would have flowed upward through a sunspot is deflected and emerges around the sunspot. *(Zeeman: AURA/NOAO/NSF; Magnetic map: J. Harvey/NSO and HAO/NCAR; IR image: Dan Gezari, NASA/Goddard)*

photons of a certain wavelength, producing a spectral line. If the atom is located in a magnetic field, however, the energy levels are split into multiple levels, and the single spectral line could appear as three or more spectral lines at slightly different wavelengths. This is known as the **Zeeman effect** (❚ Figure 8-10). The separation of the spectral lines depends on the strength of the magnetic field, so Hale could measure the strength of magnetic fields on the sun by looking for this effect. He found that the field in a sunspot is about 1000 times stronger than the sun's average field. Modern solar telescopes can use the Zeeman effect to map the magnetic field across the sun's surface. The magnetic map in Figure 8-10 shows that the magnetic field is strongest at the location of active regions. This suggests that the powerful magnetic field causes a sunspot by inhibiting circulation. Ionized gas is made up of electrically charged particles, and such particles cannot move freely in a magnetic field. Astronomers often say the magnetic field is "frozen into" the ionized gas. This simply means that the gas and magnetic field are locked together. Rising currents of hot gas just under the photosphere might be slowed by the strong magnetic fields in sunspots, causing a decrease in the temperature and producing a dark spot. The energy unable to emerge through the sunspot is evidently deflected and emerges in the photosphere around the sunspot.

The magnetic field that emerges through sunspots extends up into the chromosphere and corona above active regions, and high-temperature gas becomes trapped in these magnetic fields. At far-ultraviolet wavelengths, this trapped gas is much brighter than the cool photosphere and outlines the arched magnetic fields, giving us further evidence that active regions are dominated by magnetic fields (❚ Figure 8-11).

Do other stars have sunspots, or rather "starspots," on their surfaces? This is a difficult question, because, except for the sun, the stars are so far away that no surface detail is visible. Some stars, however, vary in brightness in ways that suggest they are mottled by randomly placed, dark spots. As the star rotates, its total brightness changes slightly, depending on the number of spots facing in our direction. This explains the variation of the RS Canum Venaticorum stars, whose spots may cover as much as 25 percent of the surface.

Also, some stars show spectral features that suggest the presence of magnetic fields and starspots. Ultraviolet observations reveal stars whose spectra contain emission lines commonly produced by the regions

FIGURE 8-11

(a) At visible wavelengths we see a sunspot group. (b) At far-ultraviolet wavelengths, the TRACE satellite can detect the hot gas trapped in the magnetic fields arching above the sunspot group. *(NASA/TRACE)* Notice the resemblance in shape to the magnetic field around a bar magnet as revealed by iron fillings. *(M. Seeds)*

Visual-wavelength image

a

Far-UV image

b

around spots on the sun. This suggests that these stars, too, have spots. No star except the sun can be resolved into a disk, so spots cannot be seen directly. Nevertheless, spectroscopic analysis has allowed astronomers to detect spots on the surfaces of certain stars (❚ Figure 8-12). Such results tell us that the sunspots we see on our sun are not unusual.

THE SUNSPOT CYCLE

The total number of sunspots visible on our sun is not constant. In 1843, the German amateur astronomer Heinrich Schwabe noticed that the number of sunspots

FIGURE 8-12

A very large, dark star spot lies near the axis of rotation of the star known as HD 12545. A bright spot on the star was located roughly opposite the dark spot, suggesting a magnetic field produced the spots. The size and shape of the spots on this star change over a period of a month or so. This map was constructed by observing the changing Doppler shifts in the shapes of spectral lines as the star rotated over a period of 24 days. The sun is shown for scale. *(K. Strassmeier, Vienna, AURA/NOAO/NSF)*

varies in a period of about 11 years. This is now known as the sunspot cycle (❚ Figure 8-13). At sunspot maximum, there are often as many as 100 spots visible at any one time, but at sunspot minimum there are only a few small spots. A sunspot minimum occurred in the mid-1990s with a maximum in early 2001.

At the beginning of each sunspot cycle, the spots begin to appear in the sun's middle latitudes about 35° above and below the sun's equator. As the cycle proceeds, the spots appear at lower latitudes until, near the end of the cycle, they are appearing within 5° of the sun's equator. If we plot the latitude of the appearance of sunspots versus time, the diagram takes on the shape of butterfly wings as shown in Figure 8-13. Such diagrams are known as **Maunder butterfly diagrams,** named after E. Walter Maunder of the Royal Greenwich Observatory, who first published such a diagram in 1922.

THE SUN'S MAGNETIC CYCLE

Sunspots are magnetic phenomena, so the 11-year cycle of sunspots must be caused by cyclical changes in the sun's magnetic field. To explore that idea, we must begin with the sun's rotation.

The sun does not rotate as a rigid body. It is a gas from its outermost layers down to its center, so some parts of the sun rotate faster than other parts. The equatorial region of the photosphere rotates faster than do regions at higher latitudes (❚ Figure 8-14a). At the equator, the photosphere rotates once every 25 days, but at a latitude of 45° one rotation takes 27.8 days. Helioseismology can map the rotation throughout the interior (Figure 8-14b). This phenomenon is called **differential rotation,** and it is clearly linked with the magnetic cycle.

Astronomers believe that the sun's overall magnetic field is produced by its rotation and the outward

FIGURE 8-13

The number of sunspots varies in an 11-year cycle. At times of sunspot minimum, very few spots are visible on the sun, but at times of sunspot maximum, there are many spots. Early in each cycle, the spots appear at higher latitudes on the sun (farther from the sun's equator) as is shown in the Maunder butterfly diagram at the bottom of this figure. Later in the cycle, the spots appear closer to the sun's equator, producing a pattern in the graph much like butterfly wings. *(Magnetic maps: AURA/NOAO/ NSF; Butterfly diagram: HAO/SMM/NASA/MSFC and D. Hathaway)*

flow of energy. The gases of the sun are highly ionized, so they are very good conductors of electricity. When an electrical conductor rotates rapidly and is stirred by convection, it can convert some of the energy flowing outward as convection into a magnetic field. This process is called the **dynamo effect,** and it is believed to produce Earth's magnetic field as well. Helioseismologists have found evidence that the sun's magnetic field is generated in convection currents deep under the photosphere. Once the magnetic field is created, the convection and differential rotation stretch, twist, and tangle the field to produce a magnetic cycle.

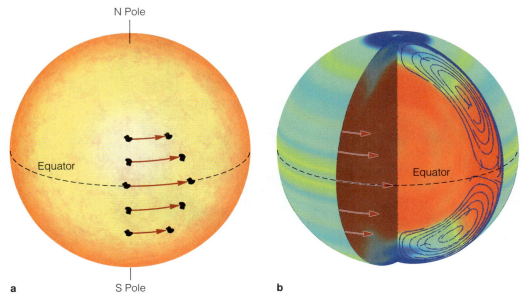

FIGURE 8-14

(a) In general, the photosphere of the sun rotates faster at the equator than at higher latitudes. If we started five sunspots in a row, they would not stay lined up as the sun rotates. (b) Detailed analysis of the sun's rotation from helioseismology reveals regions of slow rotation (blue) and rapid rotation (red). Such studies show that the interior of the sun rotates differentially and that currents similar to the trade winds in Earth's atmosphere flow through the sun. *(NASA/SOI)*

The magnetic behavior of sunspots gives us an insight into how the magnetic cycle works. Sunspots tend to occur in pairs, and the magnetic field around the pair resembles that around a bar magnet with one end magnetic north and the other end magnetic south. At any one time, sunspot pairs south of the sun's equator have reversed polarity compared to those north of the sun's equator. ▌Figure 8-15 illustrates this by showing sunspot pairs south of the sun's equator with magnetic south poles leading and sunspots north of the sun's equator with magnetic north poles leading. At the end of an 11-year sunspot cycle, the new spots appear with reversed magnetic polarity.

This magnetic cycle is not fully understood, but the **Babcock model** (named for its inventor) explains the magnetic cycle as a progressive tangling of the solar magnetic field. Because the electrons in an ionized gas are free to move, the gas is a very good conductor of electricity, and any magnetic field in the gas is "frozen" into the gas. If the gas moves, the magnetic field must move with it. Thus the sun's magnetic field is frozen into its gases, and the differential rotation wraps this field around the sun like a long string caught on a hubcap. Rising and sinking gas currents twist the field into ropelike tubes, which tend to float upward. Where these magnetic tubes burst through the sun's surface, sunspot pairs occur (▌Figure 8-16).

The Babcock model explains the reversal of the sun's magnetic field from cycle to cycle. As the magnetic field becomes tangled, adjacent regions of the sun's surface are dominated by magnetic fields that point in different directions. After about 11 years of tangling, the field becomes so complex that adjacent regions of the

solar surface begin changing their magnetic field to agree with neighboring regions. Quickly the entire field rearranges itself into a simpler pattern, and differential rotation begins winding it up to start a new cycle. But the

The Solar Magnetic Cycle

For simplicity, a single line of the solar magnetic field is shown.

Differential rotation drags the equatorial part of the magnetic field ahead.

As the sun rotates, the magnetic field is eventually dragged all the way around.

Differential rotation wraps the sun in many turns of its magnetic field.

Where loops of tangled magnetic field rise through the surface, sunspots occur.

FIGURE 8-15

The magnetic polarity of sunspot groups in the sun's southern hemisphere is the reverse of that in the northern hemisphere. When a new sunspot cycle begins, the magnetic polarity is reversed—magnetic north becomes magnetic south and vice versa. Compare this diagram with the magnetic maps in Figures 8-10 and 8-13.

FIGURE 8-16

The Babcock model of the solar magnetic cycle explains the sunspot cycle as a consequence of the sun's differential rotation gradually winding up the magnetic field.

newly organized field is reversed, and the next sunspot cycle begins with magnetic north replaced by magnetic south. Thus the complete magnetic cycle is 22 years long, and the sunspot cycle is 11 years long.

This magnetic cycle may even explain the Maunder butterfly diagrams. As a sunspot cycle begins, the twisted tubes of magnetic force first begin to float upward and produce sunspot pairs at higher latitude. Consequently the first spots in a cycle appear further north and south of the equator. Later in the cycle, when the field is more tightly wound, the tubes of magnetic force arch up through the surface closer to the equator. Thus the later sunspot pairs in a cycle appear closer to the equator.

Notice the power of a scientific model. The Babcock model may in fact be incorrect in some or all details, but it gives us a framework on which to organize all of the complex solar activity. Even though our models of the sky (see Chapter 2) and the atom (see Chapter 7) were only partially correct, they served as organizing themes to guide our thinking. Similarly, although the precise details of the solar magnetic cycle are not yet understood, the Babcock model gives us a general picture of the behavior of the sun's magnetic field (Window on Science 8-1).

MAGNETIC CYCLES ON OTHER STARS

Because we believe that the sun is a representative star, we should expect other stars to have similar cycles of starspots. We can't see individual spots from Earth, of course, but certain features in stellar spectra are associated with magnetic fields. Regions of strong magnetic

fields on the solar surface emit strongly at the central wavelengths of the two strongest lines of ionized calcium. This calcium emission appears in the spectra of other sunlike stars and tells us that these stars, too, have strong magnetic fields on their surfaces. These stars presumably have starspots as well.

In 1966, astronomers began measuring the strength of this H and K emission in the spectra of stars. The stars to be observed were selected to be similar to the sun. With temperatures ranging from 1000 K hotter than the sun to 3000 K cooler, these stars were considered most likely to have sunlike magnetic activity on their surfaces.

The observations show that the strength of the emissions in the spectra of these stars varies from year to year. The calcium emission averaged over the sun's disk varies with the sunspot cycle, and similar periodic variations can be seen in the spectra of the stars studied (❙ Figure 8-17). The star 107 Piscium, for instance, appears to have a starspot cycle lasting 9 years. Thus, we can be sure that stars like the sun do have magnetic fields and are subject to magnetic cycles.

The rotation periods of these stars are also apparent from the observations. If a star has a major region of magnetic activity and rotates every 20 days, we see the emission appear for 10 days and then disappear for 10 days as the rotation of the star carries the region to the far side of the star.

These observations confirm our belief that the sun is an average sort of star, that is, not peculiar. Most other stars like our sun have magnetic fields and starspots and go through magnetic cycles. More important, these studies are helping us understand our sun better.

FIGURE 8-17

The average amount of emission in the lines of calcium (calcium flux) is related to magnetic activity. In the sun, this emission is stronger when sunspot activity is higher. The sunlike star 107 Piscium appears to have a magnetic cycle, while the star Tau Ceti does not. *(Adapted from data by Baliunas and Saar)*

CHROMOSPHERIC AND CORONAL ACTIVITY

The solar magnetic fields extend high into the chromosphere and corona where they produce beautiful and powerful phenomena. Study ▌ "Magnetic Solar Phenomena" on pages 162 and 163 and notice three important points.

First, notice that all solar activity is magnetic. We do not experience such events on Earth because Earth's magnetic field is weak, and our atmosphere is not ionized and so is free to move independent of the magnetic field. On the sun, however, the weather is a magnetic phenomenon.

Second, tremendous energy can be stored in arches of magnetic field. We see these near the limb of the sun as prominences, and, seen from above in filtergrams, we seem them as filaments (Figure 8-4). When that stored energy is released, it can trigger powerful eruptions, and, although these eruptions occur far from Earth, they can affect us in dramatic ways.

The third thing to notice is how solar astronomers use every part of the electromagnetic spectrum to study the sun. Some things are easily visible to our eyes, but many phenomena on the sun can only be studied at nonvisible wavelengths. To understand our sun, we need every piece of evidence we can find.

THE SOLAR CONSTANT

Even a small change in the sun's energy output could produce dramatic changes in Earth's climate. The con-

tinued existence of our civilization and our species depends on the constancy of our sun, but we know very little about the variation of the sun's energy output.

The energy production of the sun can be measured by adding up all of the energy falling on 1 square meter of Earth's surface during 1 second. Of course, some correction for the absorption of Earth's atmosphere is necessary, and we must count all wavelengths from X rays to radio waves. The result, which is called the **solar constant,** amounts to about 1360 joules per square meter per second. A change in the solar constant of only 1 percent could change Earth's average temperature by 1 to 2°C (about 1.8 to 3.6°F). For comparison, during the last ice age Earth's average temperature was about 5°C cooler than it is now.

Some of the best measurements of the solar constant were made by instruments aboard the Solar Maximum Mission satellite. These have shown variations in the energy received from the sun of about 0.1 percent that lasted for days or weeks. Superimposed on that random variation is a long-term decrease of about 0.018 percent per year that has been confirmed by observations made by sounding rockets and balloons and by the NIMBUS 7 satellite. This long-term decrease may be related to a cycle of activity on the sun with a period longer than the 22-year magnetic cycle.

Small, random fluctuations will not affect our climate, but a long-term decrease over a decade or more could cause worldwide cooling. History contains some evidence that the solar constant may have varied in the past. The "Little Ice Age" was a period of unusually cool weather in Europe and America that lasted from about 1500 to 1850. The average temperature worldwide was about 1 K cooler than it is now. Although many things affect climate, this cooling could have been caused by a decrease in the solar constant of only 1 percent.

This period of cool weather included an era of few sunspots now known as the **Maunder minimum*** (▌ Figure 8-18). Between 1645 and 1715, the sun was unusually quiet. Very large sunspots can sometimes be seen at sunrise or sunset when the sun is dimmed, but there is no record of any naked-eye sighting of sunspots during this period. Historical records show that astronomers were using the newly invented telescope to observe the sun, but they only observed a few sunspots. Reports of total solar eclipses during this period make no mention of the corona or chromosphere, and there is almost no record of auroral displays. The Maunder minimum seems to have been a period of reduced solar activity that resulted in a cooling of Earth's climate.

In contrast, a period called the Grand Maximum, lasting from about 1100 to about 1250 AD, saw a warming of Earth's climate. The Vikings explored Greenland

---

*Ironically, the Maunder minimum coincides with the reign of Louis XIV of France, the "Sun King."

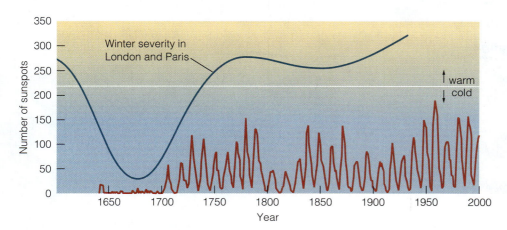

FIGURE 8-18

Historical records show that there were very few sunspots during the late 17th century and that Earth's climate was significantly cooler then. Evidence such as this suggests a link between solar activity and the solar constant, a link that has been confirmed by precise measurements of the solar constant made by instruments on satellites above Earth's atmosphere.

and parts of North America but were forced to abandon their settlements when the climate grew colder.

Other minima and maxima have been found in climate data taken from studies of tree rings. Evidently, solar activity can increase the solar constant very slightly and affect Earth's climate in dramatic ways. The future of our civilization on Earth may depend on our learning to understand the solar constant.

REVIEW CRITICAL INQUIRY

What kind of activity would the sun have if it didn't rotate differentially?

This is a really difficult question because we can see only one star close up and thus have no other examples. Nevertheless, we can make an educated guess by thinking about the Babcock model. If the sun didn't rotate differentially, with its equator traveling faster than higher latitudes, then the magnetic field might not get twisted up, and there might not be a solar cycle. Twisted tubes of magnetic field might not form and rise through the photosphere to produce prominences and flares, although convection might tangle the magnetic field and produce some activity. Is the magnetic activity that heats the chromosphere and corona driven by differential rotation or by convection? It is hard to guess, but without differential rotation, the sun might not have a strong magnetic field and high-temperature gas above its photosphere.

This is very speculative, but sometimes in the critical analysis of ideas it helps to imagine a change in a single important factor and try to understand what might happen. For example, what do you think the sun would be like if it had no convection inside?

The sun is beautiful and complex, but the evidence tells us that all of the powerful activity on its surface is driven by the energy flowing up from its interior. It is time to descend into the sun and ask how that energy is made.

8-3 NUCLEAR FUSION IN THE SUN

Astronomers often use the wrong words to describe energy generation in the sun and stars. Astronomers will say, "The star ignites hydrogen burning." We use the word *ignite* to mean *catch on fire,* and we use *burn* to mean *on fire.* What goes on inside stars isn't really burning in the usual sense.

We must also use the word *atom* with care. The interior of the sun is so hot the gas is totally ionized. That is, the electrons are not attached to the atomic nuclei, and the gas is an atomic soup of rapidly moving particles colliding with each other at high velocity. When we discuss nuclear reactions inside stars, we refer to atomic nuclei and not to atoms.

How exactly can the nucleus of an atom yield energy? The answer lies in the forces that hold the nuclei together.

NUCLEAR BINDING ENERGY

The sun generates its energy by breaking and reconnecting the bonds between the particles *inside* atomic nuclei. This is quite different from the way we generate energy by burning wood in a fireplace. The process of burning wood extracts energy by breaking and reconnecting chemical bonds between atoms in the wood. Chemical bonds are formed by the electrons in atoms, and we saw in Chapter 7 that the electrons are bound to the atoms by the electromagnetic force. Thus chemical energy originates in the electromagnetic force.

There are only four known forces in nature: the force of gravity, the electromagnetic force, the **weak force,** and the **strong force.** The weak force is involved in the radioactive decay of certain kinds of nuclear particles, and the strong force binds together atomic nuclei. Thus nuclear energy comes from the strong force.

Nuclear power plants on Earth generate energy through **nuclear fission** reactions that split uranium nuclei into less massive fragments. A uranium nucleus contains a total of 235 protons and neutrons, and it splits into a range of fragments containing roughly half

Magnetic Solar Phenomena

Magnetic phenomena in the chromosphere and corona, like magnetic weather, result as constantly changing magnetic fields on the sun trap ionized gas to produce beautiful arches and powerful outbursts. Some of this solar activity can affect Earth's magnetic field and atmosphere.

A **prominence** is composed of ionized gas trapped in a magnetic arch rising up through the photosphere and chromosphere into the lower corona. Seen during total solar eclipses at the edge of the solar disk, prominences look pink because of the three Balmer emission lines. The image below shows the arch shape suggestive of magnetic fields. Seen from above against the sun's bright surface, prominences form dark filaments, as in Figure 8-4.

Sacramento Peak Observatory

H-alpha filtergram

This ultraviolet image of the solar surface was made by the NASA TRACE spacecraft. It shows hot gas trapped in magnetic arches extending above active regions. At visual wavelengths, we would see sunspot groups in these active regions.

Quiescent prominences may hang in the lower corona for many days, whereas eruptive prominences burst upward in hours. The eruptive prominences below are many Earth diameters long.

Far-UV image

Trace/NASA

The gas in prominences may be 60,000 to 80,000 K, quite cold compared with the low-density gas in the corona, which may be as hot as a million Kelvin.

SOHO, EIT, ESA and NASA

Solar **flares** rise to maximum in minutes and decay in an hour. They occur in active regions where oppositely directed magnetic fields meet and cancel each other out in what astronomers call **reconnections.** Energy stored in the magnetic fields is released as short-wavelength photons and as high-energy protons and electrons. X-ray and ultraviolet photons reach Earth in 8 minutes and increase ionization in our atmosphere, which can interfere with radio communications. Particles from flares reach Earth hours or days later as gusts in the solar wind, which can distort Earth's magnetic field and disrupt navigation systems. Solar flares can also cause surges in electrical power lines and damage to Earth satellites.

Far-UV image

An ultraviolet image shows an active region experiencing a flare.

Auroras occur about 130 km above Earth's surface when energy in the solar wind guided by Earth's magnetic field excites gases in the upper atmosphere.

...oseismology ...ge

This sequence of three images of the photosphere taken 5 minutes apart shows sound waves rushing away from a solar flare at 50 km/s. This flare was equivalent to an earthquake 40,000 times stronger than the quake that destroyed San Francisco in 1906. The most powerful flares can be a billion times more powerful than a hydrogen bomb.

Much of the solar wind comes from **coronal holes,** where the magnetic field does not loop back into the sun. These open magnetic fields allow ionized gas in the corona to flow away as the solar wind. The dark areas in this X-ray image are coronal holes.

X-ray image

Sun

CME

Visual-wavelength image

Magnetic reconnections can release enough energy to blow large amounts of ionized gas outward from the corona in **coronal mass ejections (CMEs).** These produce violent gusts in the solar wind, and if one blows past Earth it can create electrical currents up to a million megawatts, which flow down into Earth's magnetic poles and excite atoms in Earth's upper atmosphere to emit photons, which we see as an aurora, as shown above.

Coronal hole

as many particles. Because the fragments produced are more tightly bound than the uranium nuclei, binding energy is released during uranium fission.

Stars make energy in **nuclear fusion** reactions that combine light nuclei into heavier nuclei. The most common reaction, including that in the sun, fuses hydrogen nuclei (single protons) into helium nuclei (two protons and two neutrons). Because the nuclei produced are more tightly bound than the original nuclei, energy is released. Notice in ▌ Figure 8-19 that both fusion and fission reactions move downward in the diagram toward more tightly bound nuclei. Thus both produce energy by releasing the binding energy of atomic nuclei.

HYDROGEN FUSION

Not until the 1930s did astronomers realize how the sun generates energy. The sun fuses together four hydrogen nuclei to make one helium nucleus. Because one helium nucleus has 0.7 percent less mass than four hydrogen nuclei, it seems that some mass vanishes in the process:

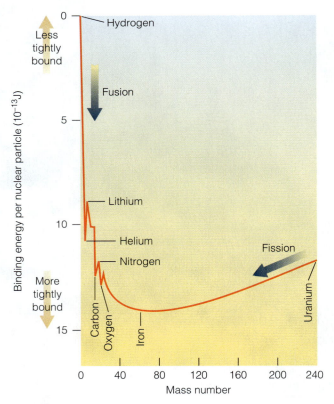

FIGURE 8-19

The red line in this graph shows the binding energy per particle, the energy that holds particles inside an atomic nucleus. The horizontal axis shows the atomic mass number of each element, the number of protons and neutrons in the nucleus. Both fission and fusion nuclear reactions move downward in the diagram (arrows) toward more tightly bound nuclei. Iron has the most tightly bound nucleus, so no nuclear reactions can begin with iron and release energy.

4 hydrogen nuclei = 6.693×10^{-27} kg
1 helium nucleus = 6.645×10^{-27} kg
difference in mass = 0.048×10^{-27} kg

However, this mass does not actually vanish; it merely changes form. The equation $E = m_0c^2$ reminds us that mass and energy are related, and under certain circumstances mass may become energy and vice versa. Thus, the 0.048×10^{-27} kg does not vanish but merely becomes energy. To see how much, we use Einstein's equation:

$$E = m_0c^2$$
$$= (0.048 \times 10^{-27} \text{ kg})(3 \times 10^8 \text{ m/s})^2$$
$$= 0.43 \times 10^{-11} \text{ J}$$

This is a very small amount of energy, hardly enough to raise a housefly one-thousandth of an inch. Because one reaction produces such a small amount of energy, it is obvious that many reactions are necessary to supply the energy needs of a star. The sun, for example, needs 10^{38} reactions per second, transforming 5 million tons of mass into energy every second, just to stay hot enough to resist its own gravity.

It seems from this that nuclear fusion is very powerful, especially when we calculate that the fusion of a milligram of hydrogen (roughly the mass of a match head) produces as much energy as burning 30 gallons of gasoline. However, the nuclear reactions in the sun are spread through a large volume in its core, and any single gram of matter produces little energy. A person of normal mass eating a normal diet produces about 4000 times more heat per gram than the matter in the core of the sun. The sun produces a lot of energy because it contains a lot of grams of matter in its core.

Two nuclei in the sun's core can fuse only if they come close together, but atomic nuclei have positive charges and repel each other with an electrostatic force. This repulsion between the positively charged nuclei has been called the **Coulomb barrier.** To overcome this barrier, the atomic nuclei must collide at high velocity. If the particles in a material are moving at high velocities, we say that the material is hot. Thus, atomic fusion reactions can only occur if the gas is very hot—at least 10^7 K.

Fusion reactions in the sun also require that the gas be very dense—denser than solid lead. We know that the sun requires 10^{38} reactions per second to manufacture sufficient energy. But fusion occurs in only a small percentage of all collisions, so the sun requires many collisions between nuclei each second. Only where the gas is very dense are there enough collisions to meet the sun's energy needs.

Nuclear fusion requires high temperature and high density. Only in the central regions of the sun is the gas hot enough and dense enough for hyrodgen to fuse. Thus all of the energy that flows outward through the sun's surface is created in the sun's core. Sunlight is nuclear power.

We can symbolize the fusion reactions in the sun with a simple nuclear reaction:

$$4\ {}^1H \rightarrow {}^4He + energy$$

In this equation, ${}^1H$ represents a proton, the nucleus of the hydrogen atom, and ${}^4He$ represents the nucleus of a helium atom. The superscripts indicate the approximate weight of the nuclei (the number of protons plus the number of neutrons). The actual steps in the process are more complicated than this convenient summary suggests. Instead of waiting for four hydrogen nuclei to collide simultaneously, a highly unlikely event, the process can proceed step by step in a chain of reactions—the proton–proton chain.

The **proton–proton chain** is a series of three nuclear reactions that builds a helium nucleus by adding together protons. This process is efficient at temperatures above 10,000,000 K. The sun, for example, manufactures over 90 percent of its energy in this way.

The three steps in the proton–proton chain entail these reactions:

$$\begin{aligned}{}^1H + {}^1H &\rightarrow {}^2H + e^+ + \nu \\ {}^2H + {}^1H &\rightarrow {}^3He + \gamma \\ {}^3He + {}^3He &\rightarrow {}^4He + {}^1H + {}^1H\end{aligned}$$

In the first step, two hydrogen nuclei (two protons) combine to form a heavy hydrogen nucleus called **deuterium,** emitting a particle called a positron, e^+ (a positively charged electron), and a **neutrino,** ν (a subatomic particle having an extremely low mass and a velocity nearly equal to the velocity of light). In the second reaction, the heavy hydrogen nucleus absorbs another proton and, with the emission of a gamma ray, γ, becomes a lightweight helium nucleus. Finally, two light helium nuclei combine to form a common helium nucleus and two hydrogen nuclei. Because the last reaction needs two ${}^3He$ nuclei, the first and second reactions must occur twice (Figure 8-20). The net result of this chain reaction is the transformation of four hydrogen nuclei into one helium nucleus plus energy.

The energy appears in the form of gamma rays, positrons, the energy of motion of the particles, and neutrinos. The gamma rays are photons that are absorbed by the surrounding gas before they can travel more than a fraction of a millimeter. This heats the gas and helps maintain the pressure. The positrons produced in the first reaction combine with free electrons, and both particles vanish, converting their mass into gamma rays. Thus, the positrons

also help keep the center of the star hot. In addition, when fusion produces new nuclei, they fly apart at high velocity. This energy of motion helps raise the temperature of the gas. The neutrinos, however, resemble photons except that they almost never interact with other particles. The average neutrino could pass unhindered through a lead wall 1 ly thick. Thus, the neutrinos do not help heat the gas but race out of the star at the speed of light, carrying away roughly 2 percent of the energy produced.

It is time to ask the critical question that lies at the heart of science. What is the evidence to support our theoretical explanation of how the sun makes its energy? The search for that evidence introduces us to one of the great problems of modern astronomy.

THE SOLAR NEUTRINO PROBLEM

The center of a star seems forever hidden from us, but the sun is transparent to neutrinos because these subatomic particles almost never interact with normal matter. Nuclear reactions in the sun's core produce floods of neutrinos that rush out of the sun and off into space. If we could detect these neutrinos, we could probe the sun's interior.

Because neutrinos almost never interact with atoms, we never feel the flood of over 10^{12} solar neutrinos that flow through our bodies every second. Even at night, neutrinos from the sun rush through Earth as if it weren't there, up through our beds, through us, and onward into space. Obviously we are lucky to be transparent to neutrinos, but it means that neutrinos are extremely hard to detect. Certain nuclear reactions, however, can be triggered by a neutrino of the

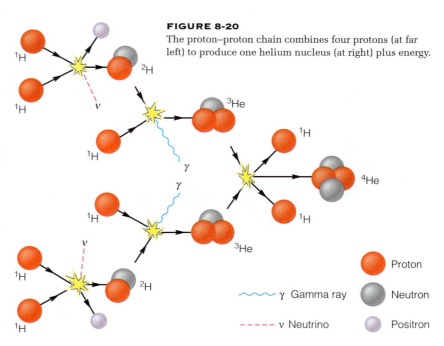

FIGURE 8-20
The proton–proton chain combines four protons (at far left) to produce one helium nucleus (at right) plus energy.

right energy; and in the late 1960s, chemist Raymond Davis, Jr., began using such a reaction to detect solar neutrinos.

Davis filled a 100,000-gallon tank with the cleaning fluid perchloroethylene (C_2Cl_4). Theory predicts that about once a day, a solar neutrino will convert a chlorine atom in the tank into radioactive argon, which can be detected later by its radioactive decay. To protect the detector from cosmic rays from space, the tank was buried nearly a mile deep in a South Dakota gold mine (❙ Figure 8-21a). Of course, the mile of rock overhead has no effect on the neutrinos.

The result of the Davis experiment startled astronomers. The cleaning fluid detected too few neutrinos—not one neutrino per day as predicted by models of the sun, but about one every three days. The experiment was refined, tested, and calibrated for three decades; but it did not find the missing neutrinos. Other detectors were built, and they too counted too few neutrinos coming from the sun.

The missing solar neutrinos were one of the great mysteries of modern astronomy. Some scientists argued that astronomers didn't correctly understand how the sun and stars make their energy, but other scientists wondered if there was something about neutrinos that could explain the problem. Astronomers have great confidence in their theories of the sun's interior and helioseismology confirmed those theories, so astronomers did not abandon their theories immediately (Window on Science 8-2).

As the 21st century began, scientists were able to solve the mystery. Physicists know of three kinds of neutrinos, which they call flavors. The Davis experiment could detect (or taste) only one flavor, electron-neutrinos. Theory hinted that the electron-neutrinos produced in the core of the sun might oscillate among the three flavors as they rushed out through the sun. Observations begun in 2000 confirm this theory (Figure 8-21b). Some of the electron-neutrinos produced in the sun transform into tau- and muon-neutrinos, which most detectors cannot count.

This solution to the solar neutrino problem is exciting because neutrinos can't oscillate unless they have mass. Neutrinos were long thought to be massless, but if they have even a small mass, they are so common their gravity could affect the evolution of the universe

a

b

FIGURE 8-21

(a) The Davis solar neutrino experiment used cleaning fluid and could detect only one of the three flavors of neutrinos. *(Brookhaven National Laboratory)* (b) The Sudbury Neutrino Observatory is a 12-meter-diameter globe containing water rich in deuterium in place of hydrogen. Buried 6800 feet deep in an Ontario mine, it can detect all three flavors of neutrinos and confirms that neutrinos oscillate. *(Photo courtesy of SNO)*

Avoiding Hasty Judgments: Scientific Faith

Scientists like to claim that every scientific belief is based on evidence, that every theory has been tested, and that the moment a theory fails a test, it is discarded or revised. The truth is much more complicated than that, and the solar neutrino problem is a good illustration. If the detection of solar neutrinos contradicted the theory of stellar structure, why wasn't the theory abandoned?

While scientists do indeed have tremendous respect for evidence, they also have a faith in theories that have been tested successfully many times. If a theory has been tested and confirmed over and over, they may even begin to call it a natural law, and that means scientists have great faith in its truth. That is, they have confidence that the theory or law is a good description of how nature works.

Nevertheless, it is not unusual for an experiment or an observation to contradict well-established theories. In many cases, the experiments and observations are simple mistakes, or they have not been interpreted correctly. Scientists resist abandoning a well-tested theory even when an observation continues to contradict it. If confidence in the theory is stronger than the evidence, scientists begin by testing the evidence. Can it be right? Do we understand it correctly? Of course, if the evidence cannot be impeached and it continues to contradict the theory, scientists must eventually abandon or modify the theory no matter how many times it has previously been tested and confirmed. This ultimate reliance on evidence is the distinguishing characteristic of science.

It is only human nature to hang on to the principles you have come to trust, and that confidence in well-tested scientific principles helps scientists avoid rushing to faulty judgments. For example, claims for perpetual motion machines occasionally crop up in the news, but the world's scientists don't instantly abandon the laws of energy and motion pending an analysis of the latest claim. Of course, if such a claim did prove true, the entire structure of scientific knowledge would come crashing down, but because the known laws of energy and motion have been well tested and no perpetual motion machine has ever been successful, scientists know which way to bet. Like the keel on a ship, confidence in well-tested theories and laws keeps the scientific boat from rocking before every little breeze.

Some forms of faith must be absolute and unshakable—religious faith, for example. But scientific faith may be better described as scientific confidence because it must be open to change. If, ultimately, a single experiment or observation conclusively contradicts our most cherished law of nature, we as scientists must abandon that law and find a new way to understand nature. We can see this scientific confidence at work in many controversies, from the origin of the human race, to the meaning of IQ measurements; but one of the best examples is the solar neutrino problem. While astronomers struggled to understand the origin of solar neutrinos, they continued to have confidence that they did indeed understand how stars make energy.

---

as a whole—something we will discuss in Chapter 18. The detection of neutrino oscillation excites astronomers for another reason. It confirms the theories that describe the interior of the sun and stars.

REVIEW CRITICAL INQUIRY

Why does nuclear fusion require that the gas be very hot?
Only under certain conditions do the nuclei of atoms fuse together to form a new nucleus. Inside a star, the gas is ionized, which means the electrons have been stripped off the atoms, and the nuclei are bare and have a positive charge. For hydrogen fusion, the nuclei are single protons. These atomic nuclei repel each other because of their positive charges, so they must collide with each other violently to overcome that repulsion and get close enough together to fuse. If the atoms in a gas are moving rapidly, we say it has a high temperature, and so nuclear fusion requires that the gas have a very high temperature. If the gas is cooler than about 10 million K, hydrogen can't fuse because the protons don't collide violently enough to overcome the repulsion of their positive charges.

It is easy to see why nuclear fusion in the sun requires high temperature, but why does it require high density?

The sun is beautiful and complex, with great eruptions, prominences, active regions, and dark spots sweeping across its surface like magnetic weather. All of this activity is driven by the outward flow of energy generated by hydrogen fusion in the core. Presumably other stars are like the sun, but other stars are so far away we can't see their surfaces. Through the largest telescopes, they look like nothing more than points of light. How can we learn about the stars? We begin to answer that question in the next chapter.

SUMMARY

The atmosphere of the sun consists of three layers: photosphere, chromosphere, and corona. The photosphere, or visible surface, is a thin layer of low-density gas from which visible photons most easily escape. It is marked by granulation, a pattern produced by circulation below the photosphere.

The chromosphere is visible to the unaided eye during total solar eclipses when the moon covers the bright photosphere and the chromosphere flashes into view as a bright pink layer of gas. The spectrum shows that the upper chromosphere is composed of ionized gas much hotter than the photosphere. Images recorded at wavelengths absorbed by specific atoms reveal that the chromosphere is filled with large jets called spicules.

The corona is the sun's outermost layer. The inner corona is visible during solar eclipses, but it is best studied with special telescopes called coronagraphs located on high mountains or in space. Spectra show that it is composed of extremely hot gas extending very far from the sun. The corona and chromosphere appear to be heated by interactions with the sun's magnetic field.

Sunspots are the most prominent example of solar activity. A sunspot appears to have a dark center, called the umbra, and a slightly lighter border, called the penumbra. A sunspot seems dark because it is slightly cooler than the rest of the photosphere. The average sunspot is about twice the size of Earth and contains magnetic fields about 1000 times stronger than the sun's average field. Sunspots are thought to form because the magnetic field inhibits convection. The average number of sunspots visible varies with a period of about 11 years and appears to be related to the solar magnetic cycle.

Alternate sunspot cycles have reversed magnetic polarity, and this has been explained by the Babcock model of the magnetic cycle. In this model, the differential rotation of the sun winds up the magnetic field. Tangles in the field rise to the surface and cause active regions containing sunspots. When the field becomes strongly tangled, it reorders itself into a simpler but reversed field, and the cycle starts over.

Prominences and flares are other examples of solar activity. Prominences occur in the chromosphere; their arched shape shows that they are formed of ionized gas trapped in the magnetic field. In filtergrams of the chromosphere, we see prominences as filaments silhouetted against the brighter surface. Flares, too, seem to be related to the magnetic field. They are sudden eruptions of X-ray, ultraviolet, and visible radiation and high-energy particles that occur among the twisted magnetic fields around sunspot groups.

Activity in the corona is also guided by the magnetic field. The corona seems to be composed of streamers of thin, hot gas escaping from the magnetic field. In some regions of the corona, the magnetic field does not loop back to the sun, and the gas escapes unimpeded. These regions are called coronal holes and are believed to be the source of the solar wind.

The sun generates its energy near its center where the temperature and density are high enough for nuclear fusion reactions that combine hydrogen to make helium. Observations of too few neutrinos coming from the sun's core have been difficult to understand, but neutrino oscillation now explains the missing neutrinos.

NEW TERMS

sunspot	transition region
granulation	filtergram
convection	filament
supergranule	spicule
limb	coronagraph
limb darkening	magnetic carpet

solar wind	coronal hole
helioseismology	coronal mass ejection (CME)
active region	solar constant
Zeeman effect	Maunder minimum
Maunder butterfly diagram	weak force
differential rotation	strong force
dynamo effect	nuclear fission
Babcock model	nuclear fusion
prominence	Coulomb barrier
flare	proton–proton chain
reconnection	deuterium
aurora	neutrino

REVIEW QUESTIONS

Ace ⬤Astronomy™ Assess your understanding of this chapter's topics with additional quizzing and animations at **http://astronomy.brookscole.com/seeds8e**

1. Why can't we see deeper than the photosphere?

2. What evidence do we have that granulation is caused by convection?

3. How are granules and supergranules related? How do they differ?

4. How can astronomers detect structure in the chromosphere?

5. What evidence do we have that the corona has a very high temperature?

6. What heats the chromosphere and corona to high temperature?

7. How are astronomers able to explore the layers of the sun below the photosphere?

8. What evidence do we have that sunspots are magnetic?

9. How does the Babcock model explain the sunspot cycle?

10. What does the spectrum of a prominence tell us? What does its shape tell us?

11. How can solar flares affect Earth?

12. Why does nuclear fusion require high temperatures?

13. Why does nuclear fusion in the sun occur only near the center?

14. How can astronomers detect neutrinos from the sun?

15. How can neutrino oscillation explain the solar neutrino problem?

DISCUSSION QUESTIONS

1. What energy sources on Earth cannot be thought of as stored sunlight?

2. What would the spectrum of an auroral display look like? Why?

3. What observations would you make if you were ordered to set up a system that could warn astronauts in orbit of dangerous solar flares? Such a warning system exists.

PROBLEMS

1. The radius of the sun is 0.7 million km. What percentage of the radius is taken up by the chromosphere?

2. The smallest detail visible with ground-based solar telescopes is about 1 second of arc. How large a region does this represent on the sun? (*Hint:* Use the small-angle formula.)

3. What is the angular diameter of a star like the sun located 5 ly from Earth? Is the Hubble Space telescope able to detect detail on the surface of such a star?

4. If a sunspot has a temperature of 4240 K and the solar surface has a temperature of 5800 K, how many times brighter is the surface compared to the sunspot? (*Hint:* Use the Stefan–Boltzmann law, Chapter 7.)

5. A solar flare can release 10^{25} J. How many megatons of TNT would be equivalent? (*Hint:* A 1-megaton bomb produces about 4×10^{15} J.)

6. The United States consumes about 2.5×10^{19} J of energy in all forms in a year. How many years could we run the United States on the energy released by the solar flare in Problem 5?

7. Neglecting energy absorbed or reflected by our atmosphere, the solar energy hitting 1 square meter of Earth's surface is 1360 J/s (the solar constant). How long does it take a baseball diamond (90 ft on a side) to receive 1 megaton of solar energy? (*Hint:* See Problem 5.)

8. How much energy is produced when the sun converts 1 kg of mass into energy?

9. How much energy is produced when the sun converts 1 kg of hydrogen into helium? (*Hint:* How does this problem differ from Problem 8?)

10. A 1-megaton nuclear weapon produces about 4×10^{15} J of energy. How much mass must vanish when a 5-megaton weapon explodes?

CRITICAL INQUIRIES FOR THE WEB

1. Do disturbances in one layer of the solar atmosphere produce effects in other layers? We have seen that filtergrams are useful in identifying the layers of the solar atmosphere and the structures within them. Visit a Web site that provides daily solar images. Then, choose today's date (or one near it) and examine the sun in several wavelengths to explore the relation between disturbances in various layers.

2. Explore the Web to find out how auroral activity is affected as solar activity rises and falls through the solar cycle. What changes in auroral visibility occur during this cycle? In what other ways can the increased activity associated with a solar maximum affect Earth?

3. Explore the Web to find photos and observations of auroras. From what places on Earth are auroras most often seen?

4. What can you find on the Web about Earth-based efforts to generate energy through nuclear fusion? How do nuclear fusion power experiments attempt to trigger and control nuclear fusion? So-called "cold fusion" has been abandoned as a false trail. How did it resemble nuclear fusion?

EXPLORING *THESKY*

1. Locate the six photos of the sun provided in *TheSky,* and attempt to draw in the sun's equator in each photo. (*Hint:* In the sun's information box, choose **More Information** and then **Multimedia.** What features are visible in these images that help us recognize the orientation of the sun's equator?)

 Visit the Seeds *Foundations of Astronomy* companion Web site for critical thinking exercises, articles, and additional readings from InfoTrac College Edition, Brooks/Cole's online student library.

THE FAMILY OF STARS

Ice is the silent language of the peak;
and fire the silent language of the star.

Conrad Aiken, And in the Human Heart

Chapter 9

GUIDEPOST

Science is based on measurement, but measurement in astronomy is very difficult. Even with the powerful modern telescopes described in Chapter 6, it is impossible to measure directly simple parameters such as the diameter of a star. This chapter shows how we can use the simple observations that are possible, combined with the basic laws of physics, to discover the properties of stars.

With this chapter, we leave our sun behind and begin our study of the billions of stars that dot the sky. In a sense, the star is the basic building block of the universe. If we hope to understand what the universe is, what our sun is, what our Earth is, and what we are, we must understand the stars.

In this chapter we will find out what stars are like. In the chapters that follow, we will trace the life stories of the stars from their births to their deaths.

VIRTUAL LABORATORIES

BINARY STARS

ASTRONOMICAL DISTANCE SCALES

THE SPECTRAL SEQUENCE AND THE H–R DIAGRAM

If the silent language of the stars is fire, then the silent language of fire is light. To understand the family relations among the stars, we must learn to speak the language of light.

Unfortunately, the stars are unimaginably remote. The nearest star is the sun, only 150 million km (93 million miles) away, so close that light takes only about 8 minutes to reach Earth. The next nearest star is nearly 300,000 times farther away, a distance so great that the starlight takes 4 years to reach Earth. The distances from Earth to the stars are so large we must express them using the light-year, the distance light travels in a year—about 9.5 trillion kilometers or about 5.9 trillion miles.

Finding out what a star is like is quite difficult. When we look at the stars, we look across vast distances and see only bright points of light. To understand stars, we must analyze starlight with great care (❚ Figure 9-1). We will concentrate in this chapter on finding out three things about stars—how much energy they emit, how big they are, and how much mass they contain. These three properties, combined with stellar temperatures—already discussed in Chapter 7—will give us an insight into the family relations among stars.

Although we begin with three things to learn about stars, we immediately meet a detour. To find out how much energy a star emits, we must know how far away it is. A fantastically luminous star might look very dim if it were far away. Only when we know the distances to the stars can we decide how much light they are actually emitting. Our short detour will provide us with a method of measuring stellar distances.

After we reach our goals, we will put our data together to find out what the average star is like. Which types of stars are most common? Which are rare? By the time we finish this chapter, we will have a good idea what the family of the stars is like.

9-1 MEASURING THE DISTANCES TO STARS

Distance is the most important and the most difficult measurement in astronomy, and astronomers have found many different ways to estimate the distance to stars. Yet each of those ways depends on a direct geometrical method that is much like the method surveyors would use to measure the distance across a river they cannot cross. We will begin by reviewing this method, and then we will apply it to stars.

Ace⚫Astronomy™ The AceAstronomy icon throughout the text indicates an opportunity for you to test yourself on key concepts and to explore animations and interactions on the AceAstronomy Web site at **http://astronomy.brookscole .com/seeds8e**

Infrared image

FIGURE 9-1

The modern quest to understand the universe is manifest in this photo of the Eta Carinae Nebula, roughly 7500 light-years from Earth. Only by the careful analysis of starlight have astronomers learned that the stars in the nebula are only 3 million years old and that the brightest star at lower left contains 100 times more matter than the sun. (*2MASS Sky Survey and IPAC*)

THE SURVEYOR'S METHOD

To measure the distance across a river, a team of surveyors begins by driving two stakes into the ground. The distance between the stakes is the baseline of the measurement. The surveyors then choose a landmark on the opposite side of the river, a tree perhaps, thus establishing a large triangle marked by the two stakes and the tree. Using their surveyor's instruments, they sight the tree from the two ends of the baseline and measure the two angles on their side of the river.

Knowing two angles of this large triangle and the length of the side between them, the surveyors can then find the distance across the river by simple trigonometry. Another way to find the distance is to construct a scale drawing. For example, if the baseline is 50 m and the angles are 66° and 71°, we can draw a line 50 mm long to represent the baseline. Using a protractor, we can construct angles of 66° and 71° at each end of the baseline, and then, as shown in ❚ Figure 9-2, extend the two sides until they meet at *C*. Point *C* on our drawing is the location of the tree. Measuring the height of our triangle, we find it to be 64 mm and thus conclude that the distance from the baseline to the tree is 64 m.

The more distant an object is, the longer the baseline we must use to measure the distance to the object. We could use a baseline 50 m long to find the distance across a river, but to measure the distance to a mountain on the horizon, we might need a baseline 1000 m long.

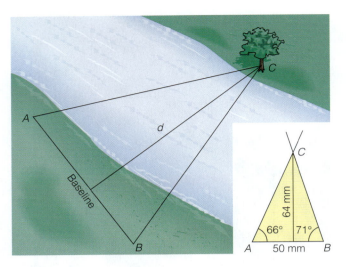

FIGURE 9-2
You can find the distance *d* across a river by measuring the baseline and the angles *A* and *B* and then constructing a scale drawing of the triangle.

Modern surveyors use computers for these problems and not simple drawings. The point is not how the problem is solved, but that it can be solved. Surveyors use simple triangulation to find the distance across a river.

THE ASTRONOMER'S METHOD

To find the distance to a star, we must use a very long baseline, the diameter of Earth's orbit. If we take a photograph of a nearby star and then wait 6 months, Earth will have moved halfway around its orbit. We can then take another photograph of the star. This second photograph is taken at a point in space 2 AU (astronomical units) from the point where the first photograph was taken. Thus our baseline equals the diameter of Earth's orbit, or 2 AU.

We then have two photographs of the same part of the sky taken from slightly different locations in space. When we examine the photographs, we will discover that the star is not in exactly the same place in the two photographs. This apparent shift in the position of the star is called *parallax*.

Parallax is the apparent change in the position of an object due to a change in the location of the observer. We saw in Chapter 4 that parallax is an everyday experience. Our thumb, held at arm's length, appears to shift position against a distant background when we look with first one eye and then with the other. In this case, the baseline is the distance between our eyes, and the parallax is the angle through which our thumb appears to move when we change eyes. The farther away we hold our thumb, the smaller the parallax. If we know the length of the baseline and measure the parallax, we can calculate the distance from our eyes to our thumb.

Because the stars are so distant, their parallaxes are very small angles, usually expressed in seconds of arc. The quantity that astronomers call **stellar parallax (*p*)** is half the total shift of the star, as shown in ▌Figure 9-3. Astronomers measure the parallax, and surveyors measure the angles at the ends of the baseline, but both measurements tell us the same thing—the shape of the triangle and thus the distance to the object in question.

The distance to a star with parallax *p* is given by the simple formula

$$d = \frac{1}{p}$$

where the parallax is measured in seconds of arc and the distance is measured in a unit of distance invented by astronomers, the **parsec (pc)**,* defined as the distance to a star with a parallax of 1 second of arc. One parsec turns out to equal 206,265 AU, or 3.26 ly. This makes it very easy to calculate the distance to a star given the parallax. For example, the star Altair has a parallax of 0.20 second of arc. Then the distance to Altair in parsecs is 1 divided by 0.20, which equals 5 pc. To convert to light-years, we multiply 5 pc by 3.26 and discover that Altair is 16.3 ly away.

Measuring the small angle *p* is very difficult. The nearest star to the sun, α Centauri, has a parallax of

*The parsec is used throughout astronomy because it simplifies the calculation of distance. However, there are instances when the light-year is also convenient. Consequently, the chapters that follow use either parsecs or light-years as convenience and custom dictate.

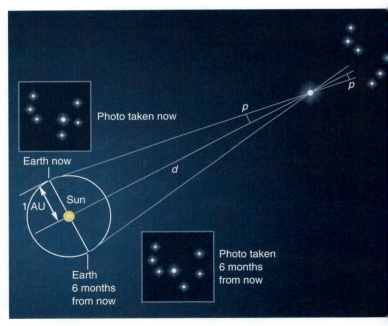

FIGURE 9-3
We can measure the parallax of a nearby star by photographing it from two points along Earth's orbit. For example, we might photograph it now and again 6 months from now. Half of the star's total change in position from one photograph to the other is its stellar parallax *p*.

only 0.76 second of arc, and more distant stars have even smaller parallaxes. To see how small these angles are, hold a piece of paper edgewise at arm's length. The thickness of the paper covers an angle of about 30 seconds of arc. You can see that the parallax of a star, smaller than 1 second of arc, must be very difficult to measure accurately.

The blurring caused by Earth's atmosphere smears star images into tiny disks about one second of arc in diameter, and that makes it difficult to measure parallax. Even averaging many observations, astronomers on Earth cannot measure parallax with an uncertainty smaller than about 0.002 seconds of arc. To limit their uncertainty to 10 percent or better, ground-based astronomers can't measure parallaxes smaller than about 0.02 seconds of arc, which corresponds to a distance of 50 pc. That means ground-based parallax measurements are limited to only the closest stars. Since the first stellar parallax was measured in 1838, ground-based astronomers have been able to measure accurate parallaxes for only about 10,000 stars.

In 1989, the European Space Agency launched the satellite Hipparcos to measure stellar parallaxes from orbit above the blurring effects of Earth's atmosphere. The little satellite observed for 4 years, and the data were reduced by the most sophisticated computers to produce two parallax catalogs in 1997. One catalog contains 120,000 stars with parallaxes 20 times more accurate than ground-based measurements. The other catalog contains over a million stars with parallaxes as accurate as ground-based parallaxes. The Hipparcos data have given astronmers new insights into the nature of stars.

Before we drop the subject of parallax measurement, we should mention a related observation that reveals the motions of the stars through space.

PROPER MOTION

All the stars in the sky, including our sun, are moving along orbits around the center of our galaxy. We don't notice that motion over periods of years, but over the centuries it can significantly distort the shape of constellations (❚ Figure 9-4).

If we photograph a small area of the sky on two dates separated by 10 years or more, we can notice that some of the stars in the photograph have moved very slightly against the background stars. This motion, expressed in units of seconds of arc per year, is the **proper motion** of the stars. For example, the star Altair (α Aquilae) has a proper motion of 0.662 second of arc per year, but Albireo (β Cygni) has a proper motion of only 0.002 second of arc per year.

Why should one star have a larger proper motion than another star? For one thing, a star might be moving almost directly toward or away from us, and thus its position on the sky would change very slowly. That

The Changing Shape of the Big Dipper

100,000 years ago the Big Dipper had a different shape.

Proper motion is moving the stars of the Big Dipper across the sky.

100,000 years in the future, the Big Dipper will have a distorted shape.

FIGURE 9-4
"Proper motion" refers to the slow movement of the stars across the sky.

is, its transverse velocity (Chapter 7) could be quite low, resulting in a small proper motion.

Another reason why a star might have a small proper motion is that it could be quite far away from us. Then even a large transverse velocity would not produce a large proper motion. Thus Albireo, at a distance of 116 pc, has a smaller proper motion, and Altair, at a distance of only 4.9 pc, has a larger proper motion.

We can use proper motion to look for nearby stars. If we see a star with a small (or zero) proper motion, it is probably a distant star. We cannot be sure, because it could be a nearby star traveling almost directly toward or away from us. But usually stars with small proper motions are distant. You have probably seen this effect if you watch birds. Distant geese move slowly across the sky, but a nearby bird flits quickly across our field of view. Similarly, stars with large proper motions are usually nearby stars. In fact, stars with proper motions above a certain limit must be nearby. A distant star traveling fast enough to have a proper motion above this limit would have such a high velocity that it would have escaped from the galaxy long ago.

Thus, proper motions can give us statistical clues to distance. We can locate stars that are probably nearby by looking for stars with large proper motions.

Having found a way to locate nearby stars and measure their distances, we are ready to discuss the first of the three stellar properties—brightness. Our goal is to find out how much energy stars emit.

9-2 INTRINSIC BRIGHTNESS

Our eyes tells us that some stars look brighter than others, and in Chapter 2 we used the scale of apparent magnitudes to refer to stellar brightness. The faintest stars we can see with the naked eye are about sixth magnitude. Brighter stars have magnitudes represented by smaller numbers and the brightest stars we see in the sky have negative magnitudes, such as Sirius, whose apparent magnitude is −1.47.

The scale of apparent magnitudes only tells us how bright stars look, however, and we want to know their true, or intrinsic, brightness. *Intrinsic* means "belonging to the thing." When we refer to the intrinsic brightness of a star, we mean a measure of the total amount of light the star emits. Apparent magnitudes are deceptive, however. An intrinsically very bright star might appear faint if it were far away. To find the true brightness of stars, we must correct the apparent magnitudes for the influence of distance.

BRIGHTNESS AND DISTANCE

If we see lights at night, it is difficult to determine which are less powerful but nearby and which are highly luminous but farther away (❚ Figure 9-5). We face the same problem when we look at stars, and to resolve that problem, we must think carefully about how brightness depends on distance.

When we look at a bright light, our eyes respond to the visual-wavelength energy falling on the eye's retina, which tells us how bright the object looks. Thus, the brightness is related to the flux of energy entering our eye. **Flux** is the energy in joules (J) per second falling on one square meter. Recall that a joule is about as much energy as is released when an apple falls from a table onto the floor. We commonly refer to 1 joule per second as 1 watt. The light flux entering our eye is directly related to the intensity of the light. The more flux entering our eye, the brighter the light looks.

If we placed a screen 1 meter square near a light-bulb, a certain amount of flux would fall on the screen. If we moved the screen twice as far from the bulb, the light that previously fell on the screen would be spread

ACTIVE FIGURE 9-5

To judge the true brightness of a light source, we need to know how far away it is. With no clues to distance, the distant headlight on a truck might look as bright as the nearby headlight on a bicycle.

Ace⊙Astronomy™ Go to AceAstronomy and click Active Figures to find "Brightness and Distance." You can take control of this figure.

Observer

to cover an area four times larger, and the screen would receive only one-fourth as much light. If we tripled the distance to the screen, it would receive only one-ninth as much light. Thus, the flux we receive from a light source is inversely proportional to the square of the distance to the source. This is known as the inverse square relation (Figure 5-6). (We first encountered the inverse square relation in Chapter 5, where it was applied to the strength of gravity.)

Now we understand how the brightness of a star depends on its distance. If we know the apparent magnitude of a star and its distance from Earth, we can use the inverse square law to correct for distance and learn the intrinsic brightness of the star. Astronomers do that using a special kind of magnitude scale as described in the next section.

ABSOLUTE VISUAL MAGNITUDE

Judging the intrinsic brightness of stars would be easier if they were all at the same distance. Although astronomers can't move stars about and line them up at some standard distance, they can use the inverse square law to calculate how bright a star of known distance would appear at any other distance. Using this method, they refer to the intrinsic brightness of a star as its **absolute visual magnitude (M_v)**—the apparent visual magnitude the star would have if it were 10 pc away.

The symbol for absolute visual magnitude is an uppercase M with a subscript v. The symbol for apparent visual magnitude is a lowercase m with a subscript v. The subscript tells us that the visual magnitude system is based only on the wavelengths of light we can see. Other magnitude systems are based on other parts of the electromagnetic spectrum such as the infrared, ultraviolet, and so on. Yet another magnitude system refers to the total energy emitted at all wavelengths. We will limit our discussion to visual magnitudes.

The intrinsically brightest stars known have absolute magnitudes of about −8, and the faintest about +19. The nearest star to the sun, α Centauri, is only 1.4 pc away, and its apparent magnitude is 0.0, indicating that it looks bright in the sky. However, its absolute magnitude is 4.39, telling us it is not intrinsically very bright. Because we know the distance to the sun and can measure its apparent magnitude, we can find its absolute magnitude—about 4.78. If the sun were only 10 pc away from us, it would look no brighter than the faintest star in the handle of the Little Dipper.

CALCULATING ABSOLUTE VISUAL MAGNITUDE

How can astronomers find the absolute visual magnitude of a star? This question leads us to one of the most common formulas in astronomy, a formula that relates a star's magnitude to its distance.

The **magnitude–distance formula** relates the apparent magnitude m_v, the absolute magnitude M_v, and the distance d in parsecs:

$$m_v - M_v = -5 + 5 \log_{10}(d)$$

If we know any two of the parameters in this formula, we can easily calculate the third. If we are interested in knowing how we might find the absolute magnitude of a star, then we must consider an example in which we know the distance and apparent magnitude of a star. Suppose a star has a distance of 50 pc and an apparent magnitude of 4.5. A pocket calculator tells us that the log of 50 is 1.70, and −5 + 5 × 1.70 equals 3.5, so we know that the absolute magnitude is 3.5 magnitudes brighter than the apparent magnitude. The absolute magnitude is thus 1.0, since 4.5 minus 3.5 is 1.0. (Remember that smaller numbers mean brighter magnitudes.) If this star were 10 pc away, it would be a first-magnitude star.

Astronomers also use the magnitude–distance formula to calculate the distance to a star if the apparent and absolute magnitudes are known. For that purpose, it is handy to rewrite the formula in the following form:

$$d = 10^{(m_v - M_v + 5)/5}$$

If we knew that a star had an apparent magnitude of 7 and an absolute magnitude of 2, then $m_v - M_v$ is 5 magnitudes, and the distance would be 10^2 or 100 parsecs.

The magnitude difference $m_v - M_v$ is known as the **distance modulus,** a measure of how far away the star is. The larger the distance modulus, the more distant the star. We can use the magnitude–distance formula to construct a table of distance and distance modulus (❚ Table 9-1).

TABLE 9-1	
Distance Moduli	
$m_v - M_v$	d(pc)
0	10
1	16
2	25
3	40
4	63
5	100
6	160
7	250
8	400
9	630
10	1000
⋮	⋮
15	10,000
⋮	⋮
20	100,000
⋮	⋮

The magnitude–distance formula may seem awkward at first, but a pocket calculator makes it easy to use. It is important because it performs a critical function in astronomy: It allows us to convert observations of distance and apparent magnitude into absolute magnitude, a measure of the true brightness of the star. Once we know the absolute magnitude, we can go one step further and figure out the total amount of energy a star is radiating into space.

LUMINOSITY

The **luminosity (L)** of a star is the total amount of energy the star radiates in 1 second—not just visible light, but all wavelengths. To find a star's luminosity, we begin with its absolute visual magnitude, make a small correction, and compare the star with the sun.

The correction we must make adjusts for the radiation emitted at wavelengths we cannot see. Absolute visual magnitude includes only visible light. The absolute magnitudes of hot stars and cool stars will underestimate their total luminosities because those stars radiate significant amounts of radiation in the ultraviolet or infrared parts of the spectrum. We can correct for the missing radiation because the amount of missing energy depends only on the star's temperature. For hot and cool stars, the correction can be large, but for medium-temperature stars like the sun, the correction is small. Adding the proper correction to the absolute visual magnitude changes it into the **absolute bolometric magnitude**—the absolute magnitude the star would have if we could see all wavelengths.

Once we know a star's absolute bolometric magnitude, we can find its luminosity by comparing it with the sun. The absolute bolometric magnitude of the sun is +4.7. For every magnitude a star is brighter than 4.7, it is 2.512 times more luminous than the sun. (Recall from Chapter 2 that a difference of 1 magnitude corresponds to an intensity ratio of 2.512.) Thus, a star with an absolute bolometric magnitude of 2.7 is 2 magnitudes brighter than the sun, which means it is 6.3 times more luminous (6.3 is approximately 2.512 × 2.512).

Arcturus, for example, has an absolute bolometric magnitude of −0.3. That makes it 5 magnitudes brighter than the sun. A difference of 5 magnitudes is defined to be a factor of 100 in brightness, so the luminosity of Arcturus is 100 times the sun's luminosity, or 100 $L_\odot$.

The symbol $L_\odot$ represents the luminosity of the sun, a number we can calculate in a direct way. Because we know how much solar energy hits 1 square meter in 1 second just above Earth's atmosphere (the solar constant defined in the previous chapter) and how far it is from Earth to the sun, it is a simple matter to calculate how much energy the sun must radiate in all directions to provide Earth with the energy it receives per second (see Problem 9 at the end of this chapter). We find that the luminosity of the sun is about 4×10^{26} joules/s.

Let's review: If we can measure the parallax of a star, we can find its distance, calculate its absolute visual magnitude, correct for the light we can't see to find the absolute bolometric magnitude, and then find the luminosity in terms of the sun. If we want to know the luminosity in joules per second, we need only multiply by the sun's luminosity.

When we look at the night sky, the stars look much the same; yet our study of distances and luminosities has revealed an astonishing fact. Some stars are almost a million times more luminous than the sun, and some are nearly a million times fainter. Clearly, the family of stars is filled with interesting characters.

REVIEW CRITICAL INQUIRY

Why do hot stars have an absolute bolometric magnitude that is quite different from their absolute visual magnitude? Recall that the absolute visual magnitude, M_v, is the magnitude we would see if the star were 10 pc away. But our eyes can't see all wavelengths, so a very hot star, which radiates much of its energy in the ultraviolet, would not look very bright to our eyes. The absolute bolometric magnitude includes all wavelengths, so hot stars have absolute bolometric magnitudes much brighter (smaller numbers) than their absolute visual magnitudes because of the ultraviolet energy they radiate. If we want to calculate the luminosities of hot stars, we must use absolute bolometric magnitudes. Otherwise, we'll leave out the ultraviolet radiation they emit, and the luminosities we calculate will be too small.

Stars like the sun radiate most of their energy at visible wavelengths, and we can see that energy, so there isn't a big difference between the absolute bolometric and absolute visual magnitudes for these medium-temperature stars. But what about the cool stars? Explain their absolute bolometric magnitudes.

Although we had to make a detour in order to find the distances to the stars, we have reached our first goal. We have found a way to discover the energy emitted by stars. The range of stellar luminosities is very large. The most luminous stars have luminosities of about 10^5 $L_\odot$, and the least luminous are roughly 10^{-4} $L_\odot$. Knowing the luminosities of stars helps us toward our second goal, finding their diameters.

9-3 THE DIAMETERS OF STARS

One fundamental property of stars is their diameter. Are they all the same size, or are some larger and some smaller? We know little about stars until we know their diameters. We certainly can't see their diameters through a telescope; the stars appear much too small for us to resolve their disks and measure their diameters. But there is a way to find out how big stars really are. If we know their temperature and luminosity, we can find their diameter. That relationship will introduce us to the most important diagram in astronomy, where we will discover the family relations among the stars.

LUMINOSITY, RADIUS, AND TEMPERATURE

The luminosity and temperature of a star can tell us its diameter if we understand the two factors that affect a star's luminosity: surface area and temperature. For example, you can eat dinner by candlelight because the candle flame has a small surface area, and although it is very hot, it cannot radiate much heat; it has a low luminosity. However, if the candle flame were 12 ft tall, it would have a very large surface area from which to radiate, and although it might be no hotter than a normal candle flame, its luminosity would drive you from the table.

In a similar way, a hot star may not be very luminous if it has a small surface area. It could be highly luminous if it were larger, and even a cool star could be very luminous if it were very large and so had a large surface area from which to radiate. Thus, we see that both temperature and surface area help determine the luminosity of a star.

Stars are spheres, and the surface area of a sphere is $4\pi R^2$. If we measure the radius in meters, this is the number of square meters on the surface of a star. Each square meter radiates like a black body, and the total energy given off each second is σT^4. The luminosity of the star is the surface area multiplied by the energy radiated per square meter:

$$L = 4\pi R^2 \sigma T^4$$

If we divide by the same quantities for the sun, we can cancel out the constants and get a simple formula for the luminosity of a star in terms of its radius and temperature:

$$\frac{L}{L_\odot} = \left(\frac{R}{R_\odot}\right)^2 \left(\frac{T}{T_\odot}\right)^4$$

Here the symbol $\odot$ stands for the sun, and the formula tells us that the luminosity in terms of the sun equals the radius in terms of the sun squared times the temperature in terms of the sun raised to the fourth power.

Suppose a star is 10 times the sun's radius but only half as hot. How luminous would it be?

$$\frac{L}{L_\odot} = \left(\frac{10}{1}\right)^2 \left(\frac{1}{2}\right)^4 = \frac{100}{1}\frac{1}{16} = 6.25$$

The formula tells us that it would be 6.25 times the sun's luminosity.

How can we use this to tell us the diameters of the stars? If we see a cool star that is very luminous, we know it must be very large, and if we see a hot star that is not very luminous, we know it must be very small.

The formula allows us to calculate these sizes. For instance, suppose that a star is 40 times the luminosity of the sun and twice as hot. If we put these numbers into our formula, we get:

$$\frac{40}{1} = \left(\frac{R}{R_\odot}\right)^2 \left(\frac{2}{1}\right)^4$$

Solving for the radius, we get:

$$\left(\frac{R}{R_\odot}\right)^2 = \frac{40}{2^4} = \frac{40}{16} = 2.5$$

So the radius is:

$$\frac{R}{R_\odot} = \sqrt{2.5} = 1.58$$

The star is 1.58 times larger in radius than the sun.

Because a star's luminosity depends on its surface area and its temperature, we can sort the stars by diameter, and thus discover which are large and which are small, if we can sort them by luminosity and temperature. Astronomers use a special diagram for that sorting.

THE H–R DIAGRAM

The **Hertzsprung–Russell (H–R) diagram,** named after its originators, Ejnar Hertzsprung and Henry Norris Russell, is a graph that separates the effects of temperature and surface area on stellar luminosities and enables us to sort the stars according to their diameters. Before we discuss the details of the H–R diagram (as it is often called), let us look at a similar diagram we might use to sort automobiles.

We can plot a diagram such as ▮ Figure 9-6 to show horsepower versus weight for various makes of cars. And in so doing, we find that in general the more a car weighs, the more horsepower it has. Most cars fall somewhere along the sequence of cars, running from heavy, high-powered cars to light, low-powered models. We might call this the main sequence of cars. But some cars have much more horsepower than normal for their weight—the sport or racing models—and the economy models have less power than normal for cars of the same weight. Just as this diagram helps us understand the different kinds of autos, the H–R diagram helps us understand the kinds of stars.

Now we are ready to plot an H–R diagram. We put the luminosity of the stars on the vertical axis and the temperature on the horizontal axis as in the H–R diagram in ▮ Figure 9-7. A star is represented by a point in the graph that tells us the luminosity of the star and its surface temperature (Window on Science 9-1). A point near the top represents a very luminous star, and a point near the bottom represents a low-luminosity star. A point toward the right represents a cool star, and a point toward the left represents a hot star.

Just as in our plot of cars, our graph separates the stars into groups. Roughly 90 percent of normal stars fall somewhere along the **main sequence** running from

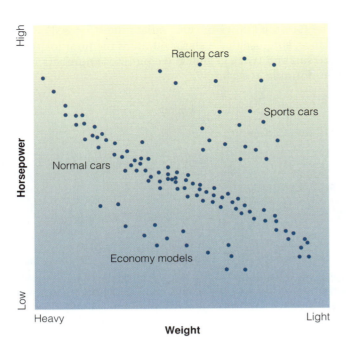

FIGURE 9-6

We could analyze automobiles by plotting their horsepower versus their weight and thus reveal relationships between various models. Most would lie somewhere along the main sequence of "normal" cars.

upper left to lower right. The sun is a main-sequence star. Stars not on the main sequence fall into a few separate groups on the H–R diagram, and we will see in the next section that stars in these groups have similar sizes. That is why the H–R diagram is so important. It sorts stars by size.

Before we discuss the different groups of stars in the H–R diagram, we should note that astronomers plot H–R diagrams in a number of different ways. However they are plotted, the diagrams always show the same information—brightness versus temperature. ❙ Figure 9-8 shows two common forms of the H–R diagram.

Astronomers use H–R diagrams so often that they usually skip the words "the point that represents the star." Rather they will say that a star is located in a certain place in the diagram. Of course, they mean the point that represents the luminosity and temperature of the star and not the star itself.

Notice that the location of a star in the H–R diagram has nothing to do with the location of the star in space. Furthermore, a star may move in the H–R diagram as it ages and its luminosity and temperature change, but such motion in the diagram has nothing to do with the star's motion in space.

GIANTS, SUPERGIANTS, AND DWARFS

The dominant feature of an H–R diagram is the main sequence running from upper left to lower right. It makes sense that the hottest stars are the most luminous, but we have learned that both temperature and

size determine a star's luminosity. That alerts us that the H–R diagram contains information about the diameters of stars. In general, stars at the top are larger and stars at the bottom are smaller.

The relation among luminosity, radius, and temperature is a precise mathematical relationship that can be used to draw lines of constant radius across an H–R diagram. ❙ Figure 9-9 is an H–R diagram on which slanting dashed lines show the location of stars of certain radii. For example, locate the dashed line labeled $1\ R_\odot$. That line passes through the point marked "Sun" and represents the location of any star whose radius equals that of the sun. Of course, the line slants down to the right because cooler stars are always fainter than hotter stars of the same size.

Now we can identify certain relationships within the family of stars. The stars called **giants** lie at the right above the main sequence. Although these stars are cool, they are luminous because they are 10 to 100 times larger than the sun. The **supergiants** lie near the top of the H–R diagram and are 10 to 1000 times the size of the sun (❙ Figure 9-10). Betelgeuse in Orion is a supergiant. If it magically replaced the sun at the center of our solar system, it would swallow up Mercury, Venus, Earth, Mars, and would just reach Jupiter. μ Cephei is believed to be over three times larger than Betelgeuse. It would reach nearly to the orbit of Uranus.

At the bottom of the H–R diagram lie the economy models, stars that are very low in luminosity because they are very small. At the bottom end of the main sequence, the **red dwarfs** are not only small, but they are

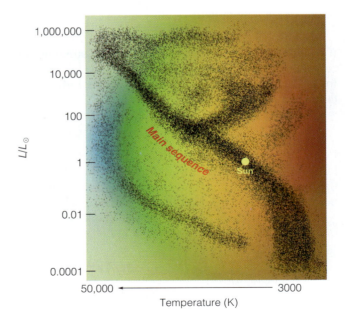

FIGURE 9-7

In this H–R diagram, the luminosity of stars relative to the sun is plotted on the vertical axis, and the temperature of their surface on the horizontal axis. Notice how the background colors are shaded to illustrate temperature. Most stars fall along a main sequence running from hot, luminous stars, to cool, faint stars. *(Adapted from an ESO graphic)*

Exponential Scales

Graphs are important in all branches of science, and many graphs have scales that are exponential (related to the power of some base, such as 10^2, 10^3, 10^4, and so on). Because we are accustomed to thinking about graphs with linear scales, exponential scales can be misleading. H–R diagrams are good examples of graphs with exponential scales, but such scales are common in all the sciences, including chemistry, biology, and geology. They also appear in economics, government, and other fields that deal with quantitative data.

In an exponential scale, each step (tick mark) along the axis corresponds to a constant factor. For example, suppose we plotted a diagram in which the vertical axis showed weight on an exponential scale. At the very bottom of the graph, we could plot the weight of a cat, and above that our own weight. Near the top of the graph, we could plot the weight of a bull elephant, and just above that the weight of a whale. In fact, just above that we could plot the weight of all the whales on Earth plus all the elephants, people, and cats. An exponential scale compresses the high numbers and spreads out the low numbers. Thus, our weight on the graph would seem much greater than that of a cat and surprisingly close to that of an elephant. Exponential scales are very helpful for compressing a wide range of numbers into a single graph, but they can be misleading.

For an astronomical example, notice that the vertical axis of the H–R diagram in Figure 9-7 has tick marks separated by a constant factor of 100 in stellar luminosity. If a star is just one tick mark higher than another star, it is *100 times* more luminous.

To see how misleading this can be, try drawing an H–R diagram with a linear scale on the vertical axis. Put zero luminosity at the bottom, and make each tick mark equal to an increase of 100,000 $L_\odot$, with 1,000,000 $L_\odot$ at the top. You will discover that it is difficult to plot all of the different kinds of stars, but their relative luminosities will be much more clearly illustrated.

---

also cool, and that means they can't radiate much energy; they have very low luminosities. In contrast, the **white dwarfs** lie in the lower left of the H–R diagram, and, although some are very hot, they are so small they can't be very bright. They are all roughly the size of Earth and can't radiate much energy from their small surface areas.

The H–R diagram tells us that there is a great range in the sizes of stars. The largest are 100,000 times larger than the smallest. The distribution of stars in the H–R

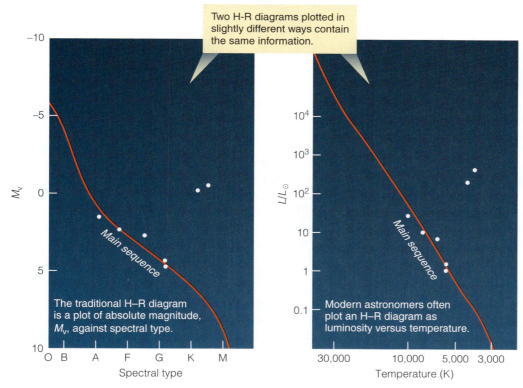

Two H-R diagrams plotted in slightly different ways contain the same information.

The traditional H-R diagram is a plot of absolute magnitude, M_v, against spectral type.

Modern astronomers often plot an H-R diagram as luminosity versus temperature.

FIGURE 9-8

Here the astronomer has plotted the same data in two different ways. Both graphs are H–R diagrams, and they reveal the same information about the stars. Of the seven stars, five are main-sequence stars and two are not.

diagram according to size is a clue to how stars are born and how they die. We will follow those clues in the chapters that follow, but, before we can do that, we must gather more information about the diverse family of stars.

LUMINOSITY CLASSIFICATION

We can tell from a star's spectrum what kind of star it is. Recall from Chapter 7 that collisional broadening can make spectral lines wider when the density of a gas is relatively high and the atoms collide often.

Main-sequence stars are relatively small and have dense atmospheres in which the gas atoms collide often and distort their electron energy levels. Thus, lines in the spectra of main-sequence stars are broad. On the other hand, giant stars are larger, their atmospheres are less dense, and the atoms disturb one another relatively little. The widths of spectral lines are partially determined by the density of the gas.

Some spectral lines are particularly sensitive to density. If the atoms collide often in a dense gas, their energy levels become distorted and the spectral lines are broadened. Hydrogen Balmer lines are an example. In the spectrum of a main-sequence star, the Balmer lines are broad because the star's atmosphere is dense and the hydrogen atoms collide often. In the spectrum of a giant star, the lines are narrower (❙ Figure 9-11), because the giant star's atmosphere is less dense and the hydrogen atoms collide less often. In the spectrum of a supergiant star, the Balmer lines are very narrow.

Thus we can look at a star's spectrum and tell roughly how big it is. These are called **luminosity classes,** because the size of the star is the dominating factor in determining luminosity. Supergiants, for example, are very luminous because they are very large.

The luminosity classes are represented by the Roman numerals I through V, with supergiants further subdivided into types Ia and Ib, as follows:

Luminosity Classes

Ia	Bright supergiant
Ib	Supergiant
II	Bright giant
III	Giant
IV	Subgiant
V	Main-sequence star

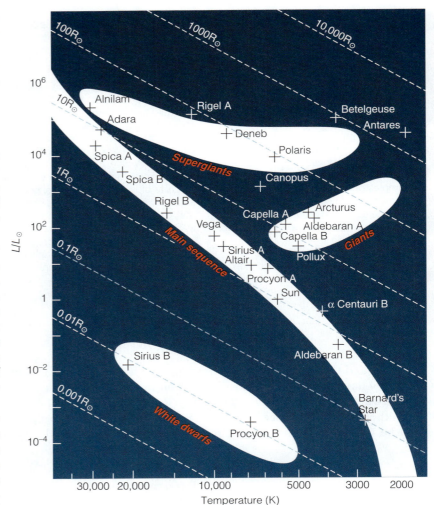

FIGURE 9-9

An H–R diagram showing the luminosity and temperature of many well-known stars. The dashed lines are lines of constant radius. (Individual stars that orbit each other are designated A and B, as in Spica A and Spica B.)

We can distinguish between the bright supergiants (Ia) such as Rigel (β Orionis) and the regular supergiants (Ib) such as Polaris, the North Star. The star Adhara (ε Canis Majoris) is a bright giant (II), Capella (α Aurigae) is a giant (III), and Altair (α Aquilae) is a subgiant (IV). The sun is a main-sequence star (V). The luminosity class appears after the spectral type, as in G2 V for the sun. White dwarfs don't enter into this classification, because their spectra are peculiar.

We can plot the positions of the luminosity classes on the H–R diagram (❙ Figure 9-12). Remember that these are rather broad classifications and that the lines on the diagram are only approximate. A star of luminosity class III may lie slightly above or below the line labeled III.

Luminosity classification is subtle and not too accurate, but it is an important tool in modern astronomy. As we will see in the next section, luminosity classification, combined with the H–R diagram, gives us a way

FIGURE 9-10

(a) This H–R diagram shows the relative sizes of stars. Giant stars are 10 to 100 times larger than the sun, and white dwarfs are about the size of Earth. (The dots representing white dwarfs here are much too large.) Supergiants are too large for this diagram. (b) To visualize the size of the largest stars, imagine that the sun is the size of one of your eyeballs. Then the largest supergiants would be the size of a hot air balloon.

to find the distance to stars that are too far away to have measurable parallaxes.

SPECTROSCOPIC PARALLAX

We can measure the stellar parallax of nearby stars, but most stars are too distant to have measurable parallaxes. We can find the distances to these stars if we can photograph their spectra and determine their luminosity classes in a process called **spectroscopic parallax**—the estimation of the distance to a star from its spectral type, luminosity class, and apparent magnitude. Spectroscopic parallax is not an actual measure of parallax, but it does tell us the distance to the star.

The method of spectroscopic parallax depends on the H–R diagram. If we photograph the spectrum of a star, we can determine its spectral class, and that tells us its horizontal location in the H–R diagram. We can also determine its luminosity class by looking at the widths of its spectral lines, and that tells us the star's vertical location in the diagram. Once we plot the point that represents the star in the H–R diagram, we can read off its absolute magnitude. As we have seen earlier in this chapter, we can find the distance to a star by comparing its apparent and absolute magnitudes.

For example, Spica is classified B1 V, and its apparent magnitude is +1. We can plot this star in an H–R diagram such as that in Figure 9-12, where we would find that it should have an absolute magnitude of about −3. Therefore, its distance modulus is 1 minus (−3), or 4, and the distance (from Table 9-1) is about 63 pc.

FIGURE 9-11

Differences in widths and strengths of spectral lines distinguish the spectra of supergiants, giants, and main-sequence stars. *(Adapted from H. A. Abt, A. B. Meinel, W. W. Morgan, and J. W. Tapscott,* An Atlas of Low-Dispersion Grating Stellar Spectra, *Kitt Peak National Observatory, 1968)*

The first two goals of this chapter have been achieved; we know how to find the luminosities and diameters of stars. We have discovered that there are four different kinds of stars—main-sequence stars, giants, supergiants, and white dwarfs—and that these types differ in their intrinsic properties, luminosity and diameter. But this discovery raises an entirely new question. How massive are these stars?

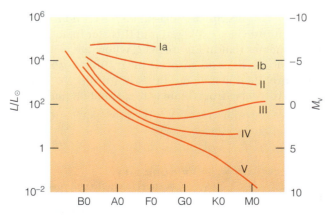

FIGURE 9-12
The approximate location of the luminosity classes on the H–R diagram.

9-4 THE MASSES OF STARS

Our third goal is to find out how much matter stars contain, that is, to know their mass. Do they all contain about the same mass as our sun, or are some more massive and others less? Unfortunately, it's difficult to determine the mass of a star. Looking through a telescope at a star, we see only a point of light that tells nothing about the star. Gravity is the key. Matter produces a gravitational field, and we can figure out how much matter a star contains if we watch an object move through the star's gravitational field. To find the masses of stars, we must study **binary stars,** pairs of stars that orbit each other.

BINARY STARS IN GENERAL

The key to finding the mass of a binary star is understanding orbital motion, which provides clues to mass.

In Chapter 5, we illustrated orbital motion by imagining a cannonball fired from a high mountain (see page 84). If Earth's gravity didn't act on the cannonball, it would follow a straight-line path and leave Earth forever. Because Earth's gravity pulls it away from its straight-line path, the cannonball follows a curved path around Earth—an orbit. When two stars orbit each other, their mutual gravitation pulls them away from straight-line paths and makes them follow closed orbits around a point between the stars.

Each star in a binary system moves in its own orbit around the system's center of mass, the balance point of the system. If the stars were connected by a massless rod and placed in a uniform gravitational field such as that near Earth's surface, the system would balance at its center of mass like a child's seesaw (see page 85). If one star were more massive than its companion, then the massive star would be closer to the center of mass and would travel in a smaller orbit, while the lower-mass star would whip around in a larger orbit (Figure 9-13). The ratio of the masses of the stars M_A/M_B equals r_B/r_A, the inverse of the ratio of the radii of the orbits. If one star has an orbit twice as large as the other star's orbit, then it must be half as massive. Getting the ratio of the masses is easy, but that doesn't tell us the individual masses of the stars, which is what we really want to know. That takes further analysis.

To find the mass of a binary star system, we must know the size of the orbits and the orbital period—the length of time the stars take to complete one orbit. The smaller the orbits are and the shorter the orbital period is, the stronger the stars' gravity must be to hold each other in orbit. For example, if two stars whirl rapidly around each other in small orbits, then their gravity must be very strong to prevent their flying apart. Such stars would have to be very massive. From the size of the orbits and the orbital period, we can figure out the masses of the stars from Newton's laws.

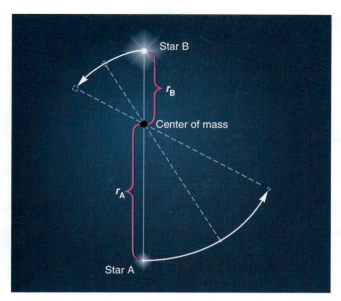

ACTIVE FIGURE 9-13

As stars in a binary star system revolve around each other, the line connecting them always passes through the center of mass, and the more massive star is always closer to the center of mass.

Ace⟲Astronomy™ Go to AceAstronomy and click Active Figures to see "Center of Mass." Experiment to see how mass ratio affects the size of the orbits.

CALCULATING THE MASSES OF BINARY STARS

According to Newton's laws of motion and gravity, the total mass of two stars orbiting each other is related to the average distance a between them and their orbital period P. If the masses are M_A and M_B, then:

$$M_A + M_B = \frac{a^3}{P^2}$$

In this formula, we will measure a in AU, P in years, and the mass in solar masses.

Notice that this formula is related to Kepler's third law of planetary motion (see Table 4-1). Almost all of the mass of the solar system is in the sun. If we apply this formula to any planet in our solar system, the total mass is 1 solar mass. Then the formula becomes $P^2 = a^3$, which is Kepler's third law.

In other star systems, the total mass is not necessarily 1 solar mass, and this gives us a way to find the masses of binary stars. If we can find the average distance in AU between the two stars and their orbital period in years, the sum of the masses of the two stars is just a^3/P^2.

Example A: If we observe a binary system with a period of 32 years and an average separation of 16 AU, what is the total mass? *Solution:* The total mass equals $16^3/32^2$, which equals 4 solar masses.

Example B: Let's call the two stars in the previous example A and B. Suppose star A is 12 AU away from the center of mass, and star B is 4 AU away. What are the individual masses? *Solution:* The ratio of the masses must be 12:4, which is the same as 3:1. What two numbers add up to 4 and have the ratio of 3:1? In this case, the answer is easy. Star B must be 3 solar masses and Star A must be 1 solar mass.

Actually, figuring out the mass of a binary star system is not as easy as it might seem from our discussion. The orbits of the two stars may be elliptical, and, although the orbits lie in the same plane, that plane can be tipped at an unknown angle to our line of sight, further distorting the shapes of the orbits. Astronomers must find ways to correct for these distortions. In addition, astronomers analyzing binary systems must find the distances to the stars so they can estimate the true size of the orbits in astronomical units. Notice that finding the masses of binary stars requires a number of steps to get from what we can observe to what we really want to know, the masses. Constructing such sequences of steps is an important part of science (Window on Science 9-2).

Although there are many different kinds of binary stars, three types are especially important for determining stellar masses. We will discuss these separately in the next sections.

VISUAL BINARY SYSTEMS

In a **visual binary system,** the two stars are separately visible in the telescope. Only a pair of stars with large orbits can be separated visually, for if the orbits are small, the star images blend together in the telescope and we see only a single point of light. Because visual binary systems must have large orbits, they also have long orbital periods. Some take hundreds or even thousands of years to complete a single orbit.

Astronomers study visual binary systems by measuring the position of the two stars directly at the telescope or in images. In either case, the astronomers need measurements over many years to map the orbits. The first frame of ▌Figure 9-14 shows a photograph of the bright star Sirius, which is a visual binary system made up of the bright star Sirius A and its white dwarf companion Sirius B. The photo was taken in 1960. Successive frames in Figure 9-14 show the motion of the two stars as observed since 1960 and the orbits the stars follow. The orbital period is 50 years.

Binary systems are common; more than half of all stars are members of binary star systems. Few, however, can be analyzed completely. Many visual binaries are so far apart that their periods are much too long for practical mapping of their orbits. Other binary systems are so close together they are not visible as separate stars.

Learning About Nature Through Chains of Inference

Scientists can rarely observe the things they really want to know, so they must construct chains of inference. We can't observe the mass of stars directly, so we must find a way to use what we can observe, orbital period and angular separation, and step by step figure out the parameters we need to reach our goal. In this case, the goal is the masses of the two stars. We follow the chain of inference from the observable parameters to the unobservable quantities we want to know.

Chains of inference can be mathematical, as when geologists use earthquakes to calculate the temperature and density of Earth's interior. There is no way to drill a hole to Earth's center and lower a thermometer or recover a sample, so the internal temperature and density are not directly observable. Nevertheless, the speed of vibrations from distant earthquakes depends on the temperature and density of the rock they pass through. Geologists can't measure the speed of the vibrations deep inside Earth; but they can measure the delay in the arrival time at different locations on the surface, and that allows them to work their way back to the speed and, finally, the temperature and density.

Chains of inference can also be non-mathematical. Biologists studying the migration of whales can't follow individual whales for years at a time, but they can observe feeding and mating in different locations; take into consideration food sources, ocean currents, and water temperatures; and construct a chain of inference that leads back to the seasonal migration pattern for whales.

This chapter contains a number of chains of inference because it describes how we know what stars are like. Almost all sciences use chains of inference. If you can link the observable parameters step by step to the final conclusions, you have gained a strong insight into the nature of science.

SPECTROSCOPIC BINARY SYSTEMS

If the stars in a binary system are close together, the telescope, limited by diffraction and by seeing, shows us a single point of light. Only by looking at a spectrum, which is formed by light from both stars and contains spectral lines from both, can we tell that there are two stars present and not one. Such a system is a **spectroscopic binary system.**

▌Figure 9-15 shows a pair of stars orbiting each other in identical circular orbits. From the diagram we can guess that the two stars have equal masses, and we notice that the circular orbit appears elliptical because we see it nearly edge-on. Of course, if this were a spectroscopic binary system, we would not see the separate stars. Nevertheless, the Doppler shift would tell us there were two stars orbiting each other. In the first frame of the figure, star A is approaching us while star B recedes from us. In the spectrum, we see a spectral line from star A blue shifted while the same spectral line from star B is red shifted. As we watch the two stars revolve around their orbits, they alternately approach and recede from us, and we see their spectral lines Doppler shifted first toward the blue and then toward the red. In a real spectroscopic binary, we can't see the individual stars, but the sight of pairs of spectral lines moving back and forth across each other would alert us that we were observing a spectroscopic binary system (▌Figure 9-16).

We can find the orbital period of a spectroscopic binary system by waiting to see how long it takes for the spectral lines to return to their starting positions. We can measure the size of the Doppler shifts to find the orbital velocities of the two stars. If we multiply velocity times orbital period, we can find the circumference of the orbit, and from that we can find the radius of the orbit. Of course, if we know the orbital period and the size of the orbit we can calculate the mass. One important detail is missing, however. We don't know how much the orbits are inclined to our line of sight.

We can find the inclination of a visual binary system because we can see the shape of the orbits. In a spectroscopic binary system, however, we cannot see the individual stars, so we can't find the inclination or untip the orbits. Recall that the Doppler effect only tells us the radial velocity, the part of the velocity directed toward or away from us. Because we cannot find the inclination, we cannot correct these radial velocities to find the true orbital velocities. Consequently, we cannot find the true masses. All we can find from a spectroscopic binary system is a lower limit to the masses.

More than half of all stars are in binary systems, and most of those are spectroscopic binary systems. Many of the familiar stars in the sky are actually pairs of stars orbiting each other (▌Figure 9-17).

We might wonder what happens when the orbits of a spectroscopic binary system lie exactly edge-on to Earth. The result is the most useful kind of binary system.

ECLIPSING BINARY SYSTEMS

As we said earlier, the orbits of the two stars in a binary system always lie in a single plane. If that plane is nearly edge-on to Earth, then the stars can cross in front of each other as seen from Earth. Imagine a model of a binary star system in which a cardboard disk represents the orbital plane and balls represent the stars, as in ▌Figure 9-18. If we see the model from the edge, then the balls that represent the stars can move in front of

A Visual Binary Star System

The bright star Sirius A has a faint companion Sirius B (arrow), a white dwarf.

Visual

1960

Over the years astronomers can watch the two move and map their orbits.

1970

Center of mass

A line between the stars always passes through the center of mass of the system.

1980

The star closest to the center of mass is the most massive.

1990

Orbit of white dwarf

The elliptical orbits are tipped at an angle to our line of sight.

Orbit of Sirius A

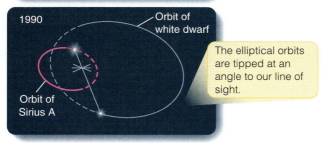

FIGURE 9-14
The orbital motion of Sirius A and Sirius B can reveal their individual masses. *(Photo © UC Regents/Lick Observatory)*

A Spectroscopic Binary Star System

Approaching   Receding

A   B

Stars orbiting each other produce spectral lines with Doppler shifts.

⟵ Blue shift   A   B   Red shift ⟶

B

A

As the stars circle their orbits, the spectral lines move together.

⟵ Blue shift   A B   Red shift ⟶

B

A

When the stars move perpendicular to our line of sight, there are no Doppler shifts.

⟵ Blue shift   A + B   Red shift ⟶

B

A

Spectral lines shifting apart and then merging are a sign of a spectroscopic binary.

⟵ Blue shift   B A   Red shift ⟶

B   A

The size of the Doppler shifts contains clues to the masses of the stars.

⟵ Blue shift   B   A   Red shift ⟶

FIGURE 9-15
From Earth, a spectroscopic binary looks like a single point of light, but the Doppler shifts in its spectrum reveal the orbital motion of the two stars.

each other as they follow their orbits. The small star crosses in front of the large star, and then, half an orbit later, the large star crosses in front of the small star. When one star moves in front of the other, it blocks some of the light, and we say the star is eclipsed. Such a system is called an **eclipsing binary system.**

Seen from Earth, the two stars are not visible separately. The system looks like a single point of light. But when one star moves in front of the other star, part of the light is blocked, and the total brightness of the point of light decreases. ▌ Figure 9-19 shows a smaller

Time in days

0.061

0.334

1.019

1.152

1.338

1.886

2.038

2.148

2.821

2.859

3.145

3.559

3.654

3.677

654.0   655.0
Wavelength (nm)

FIGURE 9-16

Fourteen spectra of the star HD80715 are shown here as graphs of intensity versus wavelength. A single spectral line (arrow in top spectrum) splits into a pair of spectral lines (arrows in third spectrum), which then merge and split apart again. These changing Doppler shifts reveal that HD80715 is a spectroscopic binary. *(Adapted from data courtesy of Samuel C. Barden and Harold L. Nations)*

star moving in an orbit around a larger star, first eclipsing the larger star and then being eclipsed as it moves behind. The resulting variation in the brightness of the system is shown as a graph of brightness versus time, a **light curve.**

The light curves of eclipsing binary systems contain tremendous amounts of information about the stars, but the curves can be difficult to analyze. Figure 9-19 shows

a

b

FIGURE 9-17

(a) At the bend of the handle of the Big Dipper lies a pair of stars, Mizar and Alcor. Through a telescope we discover that Mizar has a fainter companion and so is a member of a visual binary system. (b) Spectra of Mizar recorded at different times show that it is itself a spectroscopic binary system rather than a single star. In fact, both the faint companion to Mizar and the nearby star Alcor are also spectroscopic binary systems. *(The Observatories of the Carnegie Institution of Washington)*

an idealized system. ▌Figure 9-20 shows the light curve of a real system in which the stars have dark spots on their surfaces and are so close to each other that their shapes are distorted.

Once the light curve of an eclipsing binary system has been accurately observed, we can construct a chain of inference that leads to the masses of the two stars. We can find the orbital period easily, and we can get spectra showing the Doppler shifts of the two stars. We can find the orbital velocity because we don't have to untip the orbits; we know they are nearly edge-on or there would not be eclipses. Then we can find the size of the orbits and the masses of the stars.

Earlier we used luminosity and temperature to calculate the radii of stars, but eclipsing binary systems give us a way to measure the sizes of stars directly. From the light curve we can tell how long it took for the small star to cross the large star. Multiplying this time interval by the orbital velocity of the small star gives us the diameter of the larger star. We can also determine the diameter of the small star by noting how long it took to disappear behind the edge of the large star. For

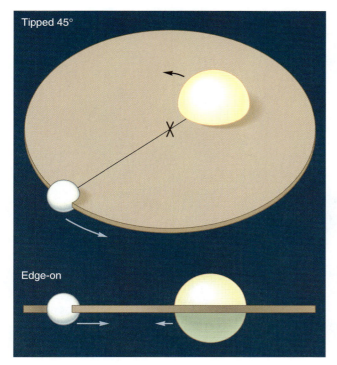

FIGURE 9-18
Imagine a model of a binary system with balls for stars and a disk of cardboard for the plane of the orbits. Only if we view the system edge-on do we see the stars cross in front of each other.

example, if it took 300 seconds for the small star to disappear while traveling 500 km/s relative to the large star, then it must be 150,000 km in diameter.

Of course, there are complications due to the inclination and eccentricity of orbits, but often these effects can be taken into account, and the system can tell us not only the masses of its stars but also their diameters.

Algol (β Persei) is one of the best-known eclipsing binaries, because its eclipses are visible to the naked eye. Normally, its magnitude is about 2.15, but its brightness drops to 3.4 in eclipses that occur every 68.8 hours. Although the nature of the star was not recognized until 1783, its periodic dimming was probably known to the ancients. *Algol* comes from the Arabic for "the demon's head," and it is associated in constellation mythology with the severed head of Medusa, the sight of whose serpentine locks turned mortals to stone (❙ Figure 9-21). Indeed, in some accounts, Algol is the winking eye of the demon.

From the study of binary stars, astronomers have found that the masses of stars range from roughly 0.1 solar mass at the low end to somewhere between 60 and 100 solar masses at the high end. The most massive star ever found in a binary system is an O3 star with a mass of 56.9 solar masses. A few other stars are believed to be more massive, but they do not lie in binary systems, so astronomers can only estimate their masses.

An Eclipsing Binary Star System

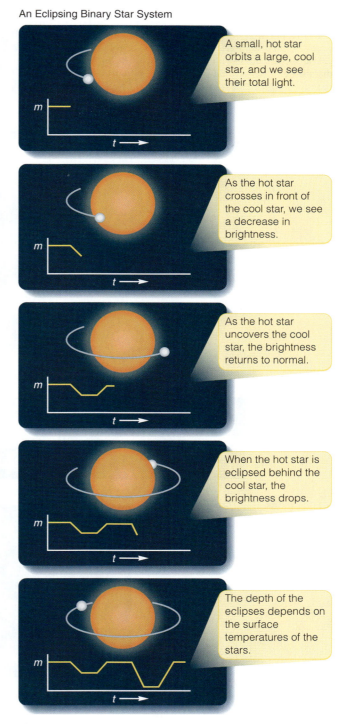

A small, hot star orbits a large, cool star, and we see their total light.

As the hot star crosses in front of the cool star, we see a decrease in brightness.

As the hot star uncovers the cool star, the brightness returns to normal.

When the hot star is eclipsed behind the cool star, the brightness drops.

The depth of the eclipses depends on the surface temperatures of the stars.

FIGURE 9-19
From Earth, an eclipsing binary looks like a single point of light, but changes in brightness reveal that two stars are eclipsing each other. Doppler shifts in the spectrum combined with the light curve, shown here as magnitude versus time, can reveal the size and mass of the individual stars.

FIGURE 9-20

The observed light curve of the binary star VW Cephei (lower curve) shows that the two stars are so close together their gravity distorts their shapes. Slight distortions in the light curve reveal the presence of dark spots at specific places on the star's surface. The upper curve shows what the light curve would look like if there were no spots. *(Graphics created with Binary Maker 2.0)*

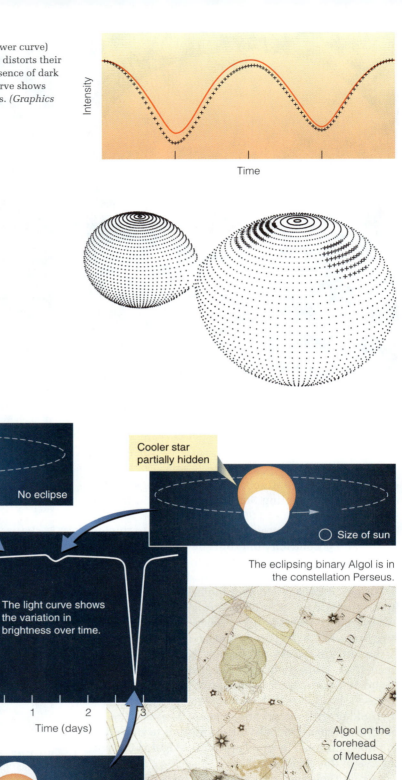

FIGURE 9-21

The eclipsing binary Algol consists of a hot B star and a cooler G or K star. The eclipses are partial, meaning that neither star is completely hidden during eclipses. The orbit here is drawn as if the cooler star were stationary.

WINDOW ON SCIENCE 9-3

Information About Nature: Basic Scientific Data

In a simple sense, science is the process by which we look at data and search for relationships that tell us how nature works. But, as a result, science sometimes requires large amounts of data. For example, astronomers need to know the masses and luminosities of many stars before they can begin looking for relationships.

Compiling basic data is one of the common forms of scientific study. This work may not seem very exciting to an outsider, but scientists often love their work not so much because they want to know nature's secrets but because they love the process of studying nature. Using a microscope or a telescope is fun. Gathering plants in the rain forest or geological samples from a cliff face can

be tremendously exciting and satisfying. It is sometimes the process of science that is most rewarding, and enjoyment of the process can lead scientists to gather significant amounts of information.

Solving a single binary star system to find the masses of the stars does not tell an astronomer a great deal about nature, but solving a binary is like solving a puzzle. It is fun and it is satisfying. Over the years, many astronomers have added their results to the growing data file on stellar masses. We can now analyze that data to search for relationships between the masses of stars and other parameters such as diameter and luminosity.

The history of science is filled with stories of hardworking scientists who compiled large amounts of data that

later scientists used to make important discoveries. For example, Tycho Brahe spent 20 years recording the positions of the stars and planets (Chapter 4). He must have loved his work to spend 20 years on his windy island, but he did not live to use his data. It was his successor, Johannes Kepler, who used Tycho Brahe's data to discover the laws of planetary motion.

Whatever science you study, you will encounter accumulations of measurements and observations that have been compiled over the years, including everything from the hardness of rocks to the attention span of infants. Determining these basic scientific data is as much a part of science as is testing a hypothesis.

REVIEW CRITICAL INQUIRY

When we look at the light curve for an eclipsing binary system with total eclipses, how can we tell which star is hotter? If we assume that the two stars in an eclipsing binary system are not the same size, then we can refer to them as the larger star and the smaller star. When the smaller star moves behind the larger star, we lose the light coming from the total area of the small star. And when the smaller star moves in front of the larger star, it blocks off light from the same amount of area on the larger star. In both cases, the same amount of area, the same number of square meters, is hidden from our sight. Then the amount of light lost during an eclipse depends only on the temperature of the hidden surface because temperature is what determines how much a single square meter can radiate per second. When the surface of the hotter star is hidden, the brightness will fall dramatically, but when the surface of the cooler star is hidden, the brightness will not fall as much. So we can look at the light curve and point to the deeper of the two eclipses and say, "That is where the hotter star is behind the cooler star."

But eclipsing binaries also tell us the diameters of the stars. How could we look at the light curve of an eclipsing binary with total eclipses and find the ratio of the diameters?

Binary stars give us a powerful tool with which to discover the masses of the stars. With that information, we are now ready to assemble our data and begin trying to understand the secrets of the family of stars.

9-5 A SURVEY OF THE STARS

We have achieved the three goals we set for ourselves at the start of this chapter. We know how to find the luminosities, diameters, and masses of stars. Now we can put that data (Window on Science 9-3) together to paint a family portrait of the stars; like most family portraits, both similarities and differences are important clues to the history of the family. As we begin trying to understand how stars are born and how they die, we ask a simple question: What is the average star like? Answering that question is both challenging and illuminating.

MASS, LUMINOSITY, AND DENSITY

The H–R diagram is filled with patterns that give us clues as to how stars are born, how they age, and how they die. When we add our data, we see traces of those patterns.

If we label an H–R diagram with the masses of the plotted stars, as is done in ❙ Figure 9-22, we discover that the main-sequence stars are ordered by mass. The most massive main-sequence stars are the hot stars. As we run our eye down the main sequence, we find lower-mass stars; and the lowest-mass stars are the coolest, faintest main-sequence stars.

Stars that do not lie on the main sequence are not in order according to mass. Some giants and supergiants are massive, while others are no more massive than our sun. All white dwarfs have about the same

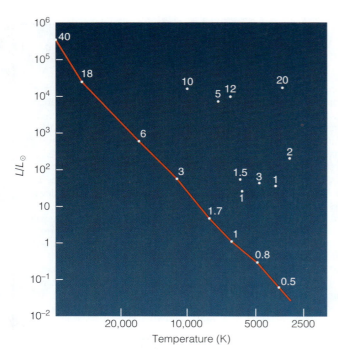

FIGURE 9-22

The masses of the plotted stars are labeled on this H–R diagram. Notice that the masses of main-sequence stars decrease from top to bottom but that masses of giants and supergiants are not arranged in any ordered pattern.

mass, somewhere in the narrow range of 0.5 to about 1 solar mass.

Because of the systematic ordering of mass along the main sequence, these main-sequence stars obey a **mass–luminosity relation**—the more massive a star is, the more luminous it is (▮ Figure 9-23). In fact, the mass–luminosity relation can be expressed as a simple formula:

$$L = M^{3.5}$$

That is, a star's luminosity (in terms of the sun's luminosity) equals its mass (in solar masses) raised to the 3.5 power. For example, a star of 4 solar masses has a luminosity of approximately $4^{3.5}$, or $4 \times 4 \times 4 \times \sqrt{4}$. This equals 64×2, or 128. So a 4-solar-mass star will have a luminosity about 128 times the luminosity of the sun. This is only an approximate equation, as shown by the yellow line in Figure 9-23.

Notice how large the range in luminosity is. The range of masses extends from about 0.08 solar mass to about 50 solar masses—a factor of 600. But the range of luminosities extends from about 10^{-6} to about 10^{6} solar luminosities—a factor of 10^{12}. Clearly, a small difference in mass causes a large difference in luminosity.

Giants and supergiants do not follow the mass–luminosity relation very closely, and white dwarfs not at all. In the next chapters, the mass–luminosity relation will help us understand how stars generate their energy.

Though mass alone does not reveal any pattern among giants, supergiants, and white dwarfs, density does. Once we know a star's mass and diameter, we can calculate its average density by dividing its mass by its volume. Stars are not uniform in density but are most dense at their centers and least dense near their surface. The center of the sun, for instance, is about 100 times as dense as water; its density near the visible surface is about 3400 times less dense than Earth's atmosphere at sea level. A star's average density is intermediate between its central and surface densities. The sun's average density is approximately 1 g/cm³—about the density of water.

Main-sequence stars have average densities similar to the sun's, but giant stars, being large, have low average densities, ranging from 0.1 to 0.01 g/cm³. The enormous supergiants have still lower densities, ranging from 0.001 to 0.000001 g/cm³. These densities are thinner than the air we breathe; and, if we could insulate ourselves from the heat, we could fly an airplane through these stars. Only near the center would we be in any danger, for there the material is very dense—about 3,000,000 g/cm³.

The white dwarfs have masses about equal to the sun's, but are very small, only about the size of Earth. Thus, the matter is compressed to densities of 3,000,000 g/cm³ or more. On Earth, a teaspoonful of this material would weigh about 15 tons.

Density divides stars into three groups. Most stars are main-sequence stars with densities like the sun's. Giants and supergiants are very-low-density stars, and white dwarfs are high-density stars. We will see in later chapters that these densities reflect different stages in the evolution of stars.

FIGURE 9-23

The mass–luminosity relation shows that the more massive a main-sequence star is, the more luminous it is. The open circles represent white dwarfs, which do not obey the relation. The yellow line represents the equation $L = M^{3.5}$.

SURVEYING THE STARS

If we want to know what the average person thinks about a certain subject, we take a survey. If we want to know what the average star is like, we must survey the stars. Such surveys reveal important relationships among the family of stars.

Not many decades ago, surveying large numbers of stars was an exhausting task, but modern computers have changed that. Specially designed telescopes controlled by computers can make millions of observations per night, and high-speed computers can compile and analyze these vast surveys and create easy-to-use data bases. Chapter 6 mentioned the Sloan Digital Sky Survey and the 2Mass infrared survey. Those and other surveys produce mountains of data that astronomers can mine searching for relationships within the family of stars.

What could we learn about stars from a survey of the stars near the sun? Because we believe the sun is in a typical place in the universe, such a survey could reveal general characteristics of the stars and might reveal unexpected processes in the formation and evolution of stars. Study "❚ The Family of Stars" on pages 192 and 193 and notice three important points.

First, taking a survey is difficult because we must be sure we get an honest sample. If we don't survey enough stars or if we don't notice the faintest stars, we can get biased results. The second important point is that most stars are faint, and the most luminous stars are rare. The most common kinds of stars are the lower-main-sequence red dwarfs and the white dwarfs. Also notice that luminous stars, although they are rare, are easily visible even at great distances. Many of the brighter stars in the sky are highly luminous stars that we see even though they lie far away.

The night sky is a beautiful carpet of stars, but they are not all the same. Some are giants and supergiants, and some are dwarfs. The family of the stars is rich in its diversity.

REVIEW CRITICAL INQUIRY

Why is it difficult to be sure how common O and B stars are?

In our survey of the stars, the number of O and B stars per million cubic parsecs is difficult to determine because these stars are quite rare. We know that O and B stars are not common because there are none near us in space. If we take our survey by counting stars within 63 pc of the sun, a distance that would include a million cubic parsecs, we find no O or B stars that close. To find these stars we must look to great distances, and at such distances we can't measure their distance accurately. Thus, the chain of inference we use to find the number of sunlike stars in space does not work as well for these more luminous stars.

Now consider the chain of inference for the most common stars. Why is it difficult to be sure just how common these stars are?

In this chapter we set out to measure the basic properties of stars. Once we found the distance to the stars, we were able to find their luminosities, diameters, and masses—rather mundane data. But we have now discovered a puzzling situation. The largest and most luminous stars are so rare we might joke that they hardly exist, and the average stars are such dinky low-mass things they are hard to see even if they are near us in space. Why does nature make stars in this peculiar way? To answer that question, we must explore the birth, life, and death of stars. We begin that quest in the next chapter.

SUMMARY

Our goal in this chapter was to characterize the stars by finding their luminosities, diameters, and masses. Before we could begin, we needed to find the distances to stars. Only by first knowing the distance could we find the other properties of the stars.

We can measure the distance to the nearer stars by observing their parallaxes. The more distant stars are so far away that their parallaxes are unmeasurably small. To find the distances to these stars, we must use spectroscopic parallax. Stellar distances are commonly expressed in parsecs. One parsec is 206,265 AU—the distance to an imaginary star whose parallax is 1 second of arc.

Once we know the distance to a star, we can find its intrinsic brightness, expressed as its absolute magnitude or its luminosity. A star's absolute magnitude is the apparent magnitude we would see if the star were only 10 pc away. The luminosity of the star is the total energy radiated in 1 second, usually expressed in terms of the luminosity of the sun.

The H–R diagram plots stars according to their intrinsic brightness and their surface temperature. In the diagram, roughly 90 percent of all normal stars fall on the main sequence, the more massive being hotter, larger, and more luminous. The giants and supergiants, however, are much larger and lie above the main sequence. They are more luminous than main-sequence stars of the same temperature. Some of the white dwarfs are hot stars, but they fall below the main sequence because they are so small.

The large size of the giants and supergiants means their atmospheres have low densities and their spectra have sharper spectral lines than the spectra of main-sequence stars. In fact, it is possible to assign stars to luminosity classes by the widths of their spectral lines. Class V stars are main-sequence stars with broad spectral lines. Giant stars (III) have sharper lines, and supergiants (I) have extremely sharp spectral lines.

The only direct way we can find the mass of a star is by studying binary stars. When two stars orbit a common center of mass, we can find their masses by observing the period and sizes of their orbits.

Given the mass and diameter of a star, we can find its average density. On the main sequence, the stars are about as dense as the sun, but the giants and supergiants are very low-density stars. Some are much thinner than air. The white dwarfs, lying below the main sequence, are tremendously dense.

The Family of Stars

What is the most common kind of star? Are some rare? Are some common? To answer those questions we must survey the stars. To do so we must know their spectral class, their luminosity class, and their distance. Our census of the family of stars produces some surprising demographic results.

A sphere 62 pc in radius encloses a million cubic parsecs.

62 pc

Earth

We could survey the stars by observing every star within 62 pc of Earth. Such a survey would tell us how many stars of each type are found within a volume of a million cubic parsecs.

Our survey faces two problems.

1. The most luminous stars are so rare we find few in our survey region. There are no O stars at all within 62 pc of Earth.

2. Lower-main-sequence M stars, called red dwarfs, and white dwarfs are so faint they are hard to locate even when they are only a few parsecs from Earth. Finding every one of these stars in our survey sphere is a difficult task.

In this chart showing most of the constellation Canis Major, stars are represented as dots with colors assigned according to spectral class. The brightest stars in the sky tend to be the rare, highly luminous stars, which look bright even though they are far away. Most stars are of very low luminosity, so nearby stars tend to be very faint red dwarfs.

●	O and B
●	A
●	F
●	G
●	K
●	M

Red dwarf
15 pc

o² Canis Majoris
B3Ia   790 pc

Red dwarf
17 pc

δ Canis Majoris
F8Ia   550 pc

σ Canis Majoris
M0Iab   370 pc

η Canis Majoris
B5Ia   980 pc

ε Canis Majoris
B2II   130 pc

In this histogram, bars rise from an H-R diagram to represent the frequency of stars in space.

Stars per 10^6 pc^3

Red dwarfs and white dwarfs are the most common kinds of stars.

Red dwarfs

White dwarfs

O and B stars, super-giants, and giants are so rare their bars are not visible in this graph.

Supergiants

Giants

Main sequence

M_v

Sirius A (α Canis Majoris) is the brightest star in the sky. With a spectral type of A1V, it is not a very luminous star. It looks bright because it is only 2.6 pc away.

Sirius B is a white dwarf that orbits Sirius A. Although Sirius B is not very far away, it is much too faint to see with the unaided eye.

The universe is filled with uncountable stars, but their family relationships tell us how they were born and how they will die.

Brightest stars

Supergiants

Giants

Sun

Main sequence

M_v

The brightest stars in the sky tend to be highly luminous stars — upper-main-sequence stars, giants, or supergiants

B0 A0 F0 G0 K0   M0

Nearest stars

The nearest stars in space tend to be very faint stars — lower-main-sequence red dwarfs or white dwarfs.

Main sequence

Sun

White dwarfs

Red dwarfs

M_v

B0 A0 F0 G0 K0   M0

Hubble Heritage Team

Visual-wavelength image

Although white dwarfs and red dwarfs are very common, not a single one is bright enough to be visible to the unaided eye.

The mass–luminosity relation says that the more massive a star is, the more luminous it is. Main-sequence stars follow this rule closely, the most massive being the upper-main-sequence stars and the least massive the lower-main-sequence stars. Giants and supergiants do not follow the relation precisely, and white dwarfs not at all.

A survey in the neighborhood of the sun shows us that lower-main-sequence stars are the most common type. The hot stars of the upper main sequence are very rare. Giants and supergiants are also rare, but white dwarfs are quite common, although they are faint and hard to find.

NEW TERMS

stellar parallax (*p*)

parsec (pc)

proper motion

flux

absolute visual magnitude (M_v)

magnitude–distance formula

distance modulus ($m_v - M_v$)

luminosity (*L*)

absolute bolometric magnitude

H–R (Hertzsprung–Russell) diagram

main sequence

giants

supergiants

red dwarf

white dwarf

luminosity class

spectroscopic parallax

binary stars

visual binary system

spectroscopic binary system

eclipsing binary system

light curve

mass–luminosity relation

REVIEW QUESTIONS

Ace⊙Astronomy™ Assess your understanding of this chapter's topics with additional quizzing and animations at http://astronomy.brookscole.com/seeds8e

1. Why are parallax measurements made from Earth-based telescopes limited to the nearest stars?

2. How would having an observatory on Mars help astronomers measure parallax more accurately?

3. How did having an observatory in orbit above Earth's atmosphere help astronomers measure parallax more accurately?

4. For which stars does absolute visual magnitude differ least from absolute bolometric magnitude? Why?

5. How can a cool star be more luminous than a hot star? Give some examples.

6. How can we be certain that the giant stars are actually larger than the sun?

7. Describe the steps in using the methods of spectroscopic parallax. Do we really measure a parallax?

8. Give the approximate radii and intrinsic brightnesses of stars in the following classes: G2 V, G2 III, G2 Ia.

9. Why can we find the masses of the stars in visual binary systems and in eclipsing binary systems, but not in spectroscopic binary systems?

10. What is the mass–luminosity relation?

11. What measurements must we make to find out how common different kinds of stars are?

DISCUSSION QUESTIONS

1. If someone asked you to compile a list of the nearest stars to the sun based on your own observations, what measurements would you make, and how would you analyze them to detect nearby stars?

2. The sun is sometimes described as an average star. What is the average star really like?

PROBLEMS

1. If a star has a parallax of 0.050 second of arc, what is its distance in parsecs? in light-years? in AU?

2. If a star has a parallax of 0.016 second of arc and an apparent magnitude of 6, how far away is it, and what is its absolute magnitude?

3. Complete the following table:

m_v	M_v	*d* (pc)	*P* (seconds of arc)
—	7	10	—
11	—	1000	—
—	−2	—	0.025
4	—	—	0.040

4. If a main-sequence star has a luminosity of 400 $L_\odot$, what is its spectral type? (*Hint:* See Figure 9-12).

5. If a star has an apparent magnitude equal to its absolute magnitude, how far away is it in parsecs? in light-years?

6. If a star has an absolute bolometric magnitude that is eight magnitudes brighter than the sun, what is the star's luminosity?

7. If a star has an absolute bolometric magnitude that is one magnitude fainter than the sun, what is the star's luminosity?

8. An O8 V star has an apparent magnitude of +1. Use the method of spectroscopic parallax to find the distance to the star.

9. Find the luminosity of the sun, given the radius of Earth's orbit and the solar constant (Chapter 8). Make your calculation in two steps. First, use $4\pi R^2$ to calculate the surface area in square meters of a sphere surrounding the sun with a radius of 1 AU. Second, multiply by the solar constant to find the total solar energy passing through the

sphere in 1 second. That is the luminosity of the sun. Compare your result with that in Data File One.

10. In the following table, which star is brightest in apparent magnitude? most luminous in absolute magnitude? largest? farthest away?

Star	Spectral Type	m_v
a	G2 V	5
b	B1 V	8
c	G2 Ib	10
d	M5 III	19
e	White dwarf	15

11. What is the total mass of a visual binary system if its average separation is 8 AU and its period is 20 years?

12. Measure the wavelengths of the iron lines in Figure 9-16 and plot a graph showing the radial velocities of the two stars in this system. What is the period? Assuming that the orbit is circular and edge-on, what is the total mass? What are the individual masses?

13. If the eclipsing binary in Figure 9-19 has a period of 32 days, an orbital velocity of 153 km/s, and an orbit that is nearly edge-on, what is the circumference of the orbit? the radius of the orbit? the mass of the system?

14. If the orbital velocity of the eclipsing binary in Figure 9-19 is 153 km/s and the smaller star becomes completely eclipsed in $2\frac{1}{2}$ hours, what is its diameter?

15. What is the luminosity of a 4-solar-mass main-sequence star? of a 9-solar-mass main-sequence star? of a 7-solar-mass main-sequence star?

CRITICAL INQUIRIES FOR THE WEB

1. The Hertzsprung–Russell diagram was named for two famous astronomers. Who were they? What did they do to earn such an honor?

2. Algol is a famous binary star. What can you find out about the mythology of Algol?

3. An entire class of binary stars is called the Algol binary stars. How would you characterize such stars?

EXPLORING *THESKY*

1. Locate the following stars and determine their apparent magnitude, parallax, distance in parsecs and in light-years, and spectral classification. (*Hint:* To center on an object, use **Find** under the **Edit** menu and type the object's name followed by a period.)
   Sirius
   Aldebaran
   Vega
   Deneb
   Betelgeuse
   Antares
   Altair

2. Use the spectral type and parallax of the stars above to estimate their distance. Compare with the distances given in *TheSky*.

3. Take a survey of the stars. Center on Orion and adjust the field until it is about 100° wide. Click on the 10 brightest stars and record the spectral types. Now zoom in until only a few dozen stars are in the frame. Click on the 10 faintest stars and record their spectral types. Is there a difference between the brightest and faintest stars? Is the result what you expected? Are certain kinds of stars missing from this computer data base?

4. Repeat Activity 3 for a region centered on the Big Dipper.

 Visit the Seeds *Foundations of Astronomy* companion Web site for critical thinking exercises, articles, and additional readings from InfoTrac College Edition, Brooks/Cole's online student library.

Chapter 10

when he shall die,
Take him and cut him out in little stars,
And he will make the face of heaven so fine
That all the world will be in love with night,
And pay no worship to the garish sun.

Shakespeare, Romeo and Juliet *III, ii, 21*

The Hubble Heritage Team STScI/AURA/NASA

GUIDEPOST

In a discussion of bread baking, we might begin with a chapter on wheat and flour. In our discussion of the birth and death of stars, the theme of the next five chapters, we begin with a chapter about the gas and dust between the stars. It is the flour from which nature bakes stars.

This chapter clearly illustrates how astronomers use the interaction of light and matter to learn about nature on the astronomical scale. That tool, which we developed in Chapter 7, "Starlight and Atoms," is powerfully employed here, especially when we include observations at many different wavelengths.

We also see in this chapter the interplay of observation and theory. Neither is useful alone, but together they are a powerful method for studying nature, a method generally known as science.

Juliet loved Romeo so much she compared him to the beauty of the stars, but had she known what was between the stars, she might have compared him to that instead. True, the gas and dust between the stars are mostly dark and cold, but where they are illuminated by stars they create beautiful nebulae, and where they are densest they give birth to beautiful stars. If there is beauty in vast extent and sweeping power, then the **interstellar medium,** the gas and dust between the stars, could steal worship from the garish stars.

We are interested in the interstellar medium because it is the matter from which stars are born. We will discover both large and small clouds of gas and dust, and both hot and cold gas, and we will find that the densest clouds can contract under the influence of their own gravity and create new stars. Clearly, before we can understand how stars are born, we must understand what fills the spaces between the stars.

Another reason we are interested in studying the interstellar medium is that it further illustrates the importance of the interaction of light and matter. Our study of starlight and atoms in Chapter 7 will help us understand the vast but nearly invisible clouds of gas and dust between the stars.

In Chapter 1, we saw that we live in a disk-shaped galaxy, the Milky Way (Figure 1-11). The interstellar medium is confined mostly within a few hundred parsecs of the plane of the disk of the galaxy; if we journeyed out of the galaxy, we would find conditions to be quite different. In our study of the interstellar medium, we confine ourselves to the disk of our galaxy in the neighborhood of the sun—probably a typical part of interstellar space.

To detect the interstellar medium, we must use all the tools of the astronomer. Observations at radio, infrared, visual, ultraviolet, and X-ray wavelengths reveal the properties of the interstellar medium. We will analyze images and spectra, consider the nature of atoms, and utilize the physics of light in order to unlock the secrets hidden in the material between the stars.

10-1 VISIBLE-WAVELENGTH OBSERVATIONS

A quick glance at the night sky does not reveal any matter between the stars, but evidence of its existence is clear when we examine photographs and spectra. In fact, if we look carefully, one important piece of evidence is visible to the naked eye in a familiar constellation.

Ace**Astronomy**™ The AceAstronomy icon throughout the text indicates an opportunity for you to test yourself on key concepts and to explore animations and interactions on the AceAstronomy Web site at **http://astronomy.brookscole .com/seeds8e**

NEBULAE

On a cold, clear winter night, Orion hangs high in the southern sky, a large constellation composed of brilliant stars. If you look carefully at Orion's sword, you will see that one of the stars is a hazy cloud (Figure 2-4). A small telescope reveals even more such clouds of gas and dust. Astronomers refer to these clouds as **nebulae** (singular, nebula), from the Latin word for mist or cloud.

Study ▐ "Three Kinds of Nebulae" on pages 198 and 199 and notice three important points. First, very hot stars can excite clouds of gas and dust to emit light, and that tells us the clouds contain mostly hydrogen gas at very low densities. Second, where slightly cooler stars illuminate clouds we get evidence that the dust in the clouds is made up of very small particles. The third thing to notice is that some dense clouds of gas and dust are only detectable where they are silhouetted against background regions filled with stars or bright nebulae.

A detailed analysis of the spectra of nebulae can tell us even more. Certain lines in the spectra of emission nebulae are called **forbidden lines** because they are almost never seen in excited gas on Earth. Two good examples are the strong green lines at 495.9 nm and 500.7 nm produced by oxygen atoms that have lost two electrons. (Following our convention for naming ions, twice-ionized oxygen is OIII.) The oxygen ions can become excited by collision with a high-energy photon or a rapidly moving ion or electron, and the atom can emit various-wavelength photons as its electron cascades back down to lower energy levels. However, transitions between certain energy levels are so unlikely they are called "forbidden." If an electron enters such an energy level, it will remain there for a relatively long time before it can decay further and emit the appropriate photon. For a normal transition, the electron might wait 10^{-8} to 10^{-7} second. If it enters one of these **metastable levels,** the electron may be stuck there for as long as an hour before it becomes likely to fall to a lower level and emit the proper photon.

This explains why we don't see these forbidden lines in laboratories on Earth. The atoms in a dense gas collide with each other so often that there isn't time for an electron in a metastable level to decay and emit a photon. Such electrons get excited back up to higher levels before they can drop downward and emit a photon. In an emission nebula, the gas has a very low density, and an atom could go for an hour or more between collisions, giving an electron in a metastable level time to decay to a lower level and emit a photon at a supposedly forbidden wavelength. The forbidden lines clearly show that nebulae have very low densities. This is a dramatic example of how astronomers can use knowledge of atomic physics to understand astronomical objects.

Three Kinds of Nebulae

At visual wavelengths, astronomers see three kinds of nebulae. The differences between these clouds of gas and dust give us insights into the interstellar medium.

Emission nebulae are also called **HII regions,** following the custom of naming gas with a roman numeral to show its state of ionization. HI is neutral hydrogen, and HII is ionized.

In an HII region, the ionized nuclei and free electrons are mixed. When a nucleus captures an electron, the electron falls down through the atomic energy levels, emitting photons at specific wavelengths. Spectra tell us that the nebulae have compositions much that like of the sun – mostly hydrogen.

Emission nebulae have densities of 100 to 1000 atoms per cubic centimeter, equivalent to the best vacuums produced in laboratories on Earth.

Emission nebulae are produced when a hot star excites the gas near it to produce an emission spectrum. The star must be hotter than about B1 (25,000 K). Cooler stars do not emit enough ultraviolet radiation to ionize the gas. Emission nebulae have a distinctive pink color produced by the blending of the red, blue, and violet Balmer lines.

ESO

Visual-wavelength image

A **reflection nebula** is produced when starlight scatters from a dusty nebula. Consequently, the spectrum of a reflection nebula is just the reflected absorption spectrum of starlight. Gas is surely present in a reflection nebula, but it is not excited to emit photons.

Reflection nebulae NGC 1973, 1975, and 1977 lie just north of the Orion nebula. The pink tints are produced by ionized gases deep in the nebulae.

Reflection nebulae look blue for the same reason the sky looks blue. Short wavelengths scatter more easily than long wavelengths.

Sunlight enters Earth's atmosphere

Blue photons are scattered more easily than longer wavelengths.

Blue photons enter our eyes from all directions, making the sky look blue.

Visual-wavelength image   AATB

The blue color of reflection nebulae tells us that the dust particles must be very small in order to preferentially scatter the blue photons. Interstellar dust grains must have diameters ranging from 0.001 mm down to 1 nm or so.

Ace Astronomy™

Go to AceAstronomy and click Active Figures to see "Scattering in Earth's Atmosphere." Take control of this diagram.

The hottest star in the Pleiades star cluster is Merope, a B3 star. It is not hot enough to ionize the gas so we see a reflection nebula rather than an emission nebula.

Merope

Visual-wavelength image

A dusty reflection nebula is located very close to the star Merope just out of the image to upper right. The glare from the star is caused by internal reflections in the telescope, but the wispy nature of the nebula is real. The intense light from the star is pushing the dust particles away and may destroy the little nebula over the next few thousand years.

Merope

Visual   NASA

Reflection   **Emission**
              Trifid Nebula

The Milky Way in Sagittarius contains two nebulae that dramatically demonstrate the difference between emission and reflection nebulae.

Emission

Visual   **Lagoon Nebula**

Dark Nebula Barnard 86

Star Cluster NGC6520

Visual

Dark nebulae are dense clouds of gas and dust that obstruct our view of more distant stars. Some are generally round, but others are twisted and distorted, suggesting that even when there are no nearby stars to ionize the gas or produce a reflection nebula, there are breezes and currents pushing through the interstellar medium.

Northern Coalsack

Cygnus

Milky Way

Great Rift

The Horsehead Nebula in Orion is a dark nebula silhouetted against a more distant emission nebula.

Visual

Large dark nebulae obstruct our view of more distant stars and form holes and rifts along the Milky Way. The Great Rift extends from Cygnus to Sagittarius.

The three kinds of nebulae introduce us to the interstellar medium. To learn more, we have to think carefully about light passing through this gas and dust.

EXTINCTION AND REDDENING

The dust in space is called **interstellar dust.** Its presence is dramatically evident in dark nebulae, but simple observations tell us that the interstellar dust is spread throughout space, making up roughly 1 percent of the mass of the interstellar medium.

One way we know that dust is present in the interstellar medium is that it makes distant stars appear fainter than they would if space were perfectly transparent. This phenomenon is called **interstellar extinction,** and in the neighborhood of the sun it amounts to about 2 magnitudes per thousand parsecs. That is, if a star lies 1000 pc from Earth, it will look about 2 magnitudes fainter than it would if space were perfectly empty. If it were 2000 pc away, it would look about 4 magnitudes dimmer, and so on. This is a dramatic effect, and it shows that the interstellar medium is not confined to a few nebulae scattered here and there. The spaces between the stars are far from empty.

Another way we can detect the presence of dust is through the effect it has on the colors of stars. An O star should be blue, but some stars with the spectrum of an O star seem much redder than they should be. Termed **interstellar reddening,** this effect is caused by dust particles scattering light. As we saw in the case of the reflection nebulae, the dust particles are small, with diameters roughly equal to the wavelength of light, and they scatter shorter wavelengths better than longer wavelengths. That means they scatter blue photons more often than red photons. Thus, the light from a distant star has lost some of its blue photons because of scattering, and consequently the star looks redder (❚ Figure 10-1).

As we discussed earlier, this scattering of blue light is what makes the sky blue, but it is also what makes distant city lights look yellow. If you view the lights of a city at night from a high-flying aircraft or a distant mountaintop, the lights will look yellow. As you descend toward the city, the lights will become bluer. The light from the city is reddened by microscopic particles in the air. If the particles are especially dense, we call them smog.

Astronomers can measure the amount of reddening by comparing two stars of the same spectral type, one of which is dimmed more than the other. The more obscured star will look redder. If we plot the difference in brightness between the two stars as a function of wavelength, we get a curve that shows the reddening. That is, it shows how the starlight is dimmed at different wavelengths (❚ Figure 10-2). In general, the light is dimmed in proportion to the reciprocal of the wavelength, a pattern typical of scattering from small dust particles. Laboratory measurements show that the high extinction at about 220 nm is caused by a form of carbon, so we must conclude that some of the dust particles are carbon. Other evidence suggests that some grains contain silicates and metals and may have coatings of water ice, frozen ammonia, or carbon-based molecules.

INTERSTELLAR ABSORPTION LINES

If we look at the spectra of distant stars, we can see dramatic evidence of an interstellar medium. Of course, we see spectral lines produced by the gas in the atmospheres of the stars, but we also see sharp spectral lines produced by the gas in the interstellar medium. These **interstellar absorption lines** give us a new way to study the gas between the stars.

Astronomers can recognize interstellar absorption lines in three ways: by their ionization, by their widths, and by their multiple components. Some stellar spectra contain absorption lines that just don't belong because they represent the wrong ionization state. For example, if we look at the spectrum of a very hot star such as an O star, we would expect to see no lines of once-ionized calcium (CaII) because that ion cannot exist at the high temperatures in the atmosphere of such a hot star. But many O-star spectra contain lines of CaII, so we must conclude that these lines were produced not in the star but in the interstellar medium.

In addition, the widths of the interstellar lines give away their identity. In the hot atmosphere of a star, the atoms move rapidly, and the Doppler effect broadens spectral lines. Also, the atoms collide with each other often enough to blur the electron energy levels, and that also broadens the spectral lines. We called these two effects Doppler broadening and collisional broadening in Chapter 7. These effects make the lines in the spectrum of a main-sequence star quite broad, but even in the atmosphere of a giant or supergiant star, the gas atoms move rapidly and collide often enough to broaden the spectral lines. Interstellar lines, on the other hand, are exceedingly sharp. This tells us that the interstellar matter is cold and has a low density. If it were hot, Doppler broadening would smear out the lines due to the motions of individual atoms. If the gas were dense, collisional broadening would produce wider lines. The exceedingly narrow widths of the interstellar absorption lines are typical of cold, low-density gas.

Another revealing characteristic of the interstellar lines is that they are often split into two or more components. These multiple components have slightly different wavelengths and appear to have been produced when the light from the star passed through different clouds of gas on its way to Earth. Because the clouds of gas have slightly different radial velocities, they produce absorption lines with slightly Doppler-shifted wavelengths. ❚ Figure 10-3 illustrates all three of these characteristics of interstellar absorption lines—the wrong ionization, narrow widths, and multiple components.

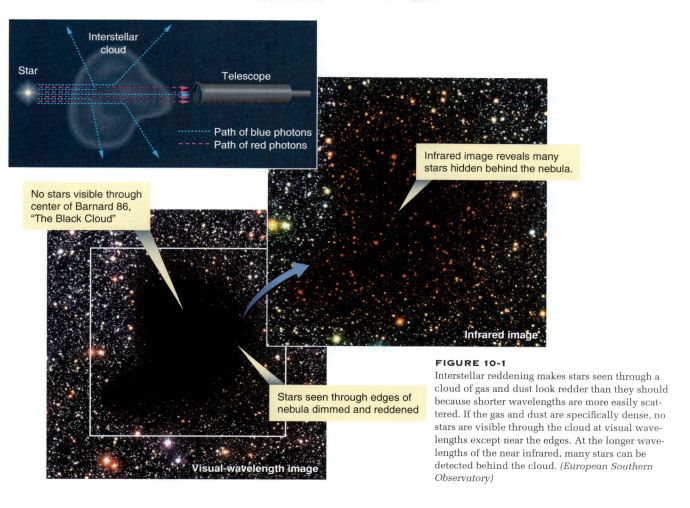

Interstellar cloud

Star

Telescope

Path of blue photons
Path of red photons

No stars visible through
center of Barnard 86,
"The Black Cloud"

Infrared image reveals many
stars hidden behind the nebula.

Stars seen through edges of
nebula dimmed and reddened

Infrared image

Visual-wavelength image

FIGURE 10-1

Interstellar reddening makes stars seen through a
cloud of gas and dust look redder than they should
because shorter wavelengths are more easily scattered. If the gas and dust are specifically dense, no
stars are visible through the cloud at visual wavelengths except near the edges. At the longer wavelengths of the near infrared, many stars can be
detected behind the cloud. *(European Southern
Observatory)*

Astronomers disagree as to the structure of the interstellar medium, and the boundaries and characteristics of clouds are ill defined. Nevertheless, astronomers find it convenient to categorize clouds into a few
main varieties. Studies of interstellar absorption lines
reveal clouds of neutral gas (and presumably dust) with
densities of 10 to a few hundred atoms/cm^3. Because
these clouds are not ionized, they are called **HI clouds.**
These clouds must be 50 to 150 pc in diameter and have
masses of a few solar masses. The gas temperature is
only about 100 K. Starlight seems to pass through six to
ten of these clouds for every 1000 pc near the plane of
our galaxy. While it is easy to think of these clouds as
more or less spherical blobs of gas, observations show
that they are usually twisted into long filaments, flattened into thin sheets, or tangled into chaotic shapes—
further evidence that the interstellar medium is not static and motionless.

Between these HI clouds of neutral gas lies a hot
intercloud medium with a temperature of a few thousand K and a density of only about 0.1 atom/cm^3. This
intercloud medium consists of ionized hydrogen (HII).
The gas is partially ionized by the ultraviolet photons in
starlight. If it is to be in equilibrium with the HI clouds
(if the HI clouds are not expanding or contracting), the
pressure in both regions should be about the same.

High

Low

Extinction

IR   Visible   UV

1000 500   300   200   125
Wavelength λ [nm]

FIGURE 10-2

Interstellar extinction, the dimming of starlight by dust between
the stars, depends strongly on wavelength. Infrared radiation is
only slightly affected, but ultraviolet light is strongly scattered.
The strong extinction at about 220 nm is caused by a certain form
of carbon dust in the interstellar medium.

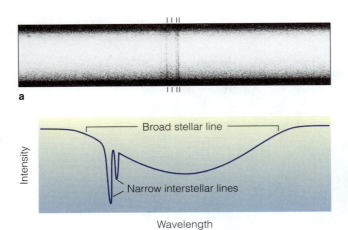

a

b

FIGURE 10-3

Interstellar absorption lines can be recognized in three ways. (a) The B0 supergiant ∈ Orionis is much too hot to show spectral lines of once-ionized calcium (CaII), yet this short segment of its spectrum reveals narrow, multiple lines of CaII (tick marks) that must have been produced in the interstellar medium. *(The Observatories of the Carnegie Institution of Washington)* (b) Spectral lines produced in the atmospheres of stars are much broader than the spectral lines produced in the interstellar medium. *(Adapted from a diagram by Binnendijk)* In both (a) and (b) the multiple interstellar lines are produced by separate interstellar clouds with slightly different radial velocities.

The pressure in a gas depends on its density and its temperature (Parameters of Science 10-1). In the interstellar medium, the HI clouds are cool but dense, and the HII clouds of the intercloud medium are hot but low in density. Thus the two regions could have similar pressures even though their temperatures are quite different. Nevertheless, observations reveal some regions where the pressures are not equal, and that is poorly understood.

How can the intercloud medium be ionized when it is not close to hot stars? To understand how, imagine that we are atoms floating in the intercloud medium. Ultraviolet photons from distant stars are not common, but they do whiz by now and then. Soon we become ionized by absorbing one of these photons and losing an electron. In a denser gas, we would quickly find another electron, capture it, and become neutral again, but the interstellar medium has such a low density that we must wait a very long time to find an electron. Thus, the atoms of the intercloud medium spend almost all their time in an ionized state because of the low density.

Studies of interstellar absorption lines give us further clues about the composition of the interstellar gas. It is much like the composition of the sun. Hydrogen is most abundant, with helium second. Light elements such as carbon, nitrogen, and oxygen are as common as in the sun, but elements such as iron, calcium, and titanium are less abundant than they are in the sun. Most likely, these elements are missing from the gas because they have condensed to form the dust.

Observations at visible wavelengths can give us important information about the interstellar medium, but in order to paint a complete picture of the matter between the stars, we must observe at other wavelengths.

10-2 LONG- AND SHORT-WAVELENGTH OBSERVATIONS

The interstellar medium is mostly invisible to our eyes, so we must observe at the longer wavelengths of infrared and radio radiation and at the shorter wavelengths of ultraviolet and X-ray radiation to further explore the interstellar medium.

21-CM OBSERVATIONS

Cold, neutral hydrogen floating in space can emit electromagnetic radiation with a wavelength of 21 cm. This **21-cm radiation** allows radio astronomers to map the distribution of neutral hydrogen (HI) throughout our galaxy.

The existence of the 21-cm radiation was predicted theoretically in the 1940s by H. C. van de Hulst, but it was not detected until 1951. We can understand how a theoretical prediction could be made by thinking of the structure of a hydrogen atom. A hydrogen atom consists of a proton and an electron, and physicists know that both of these particles must spin like tiny tops. They can never stop. Because these particles have an electrostatic charge, their rapid spin is the same as the circulation of an electric current through a coil of wire.

Pressure

One of the most fundamental parameters in science is **pressure,** a measure of force per unit area. Doctors measure blood pressure and astronomers measure gas pressure, but they are really measuring the same fundamental parameter.

Pressure is expressed in the units of force per unit area. When we inflate the tires on a car, we use the unit pounds per square inch. A typical pressure might be 34 lb/in.[2]. It is important to note that this is not the total force pushing out on the inside of the tire but only the force exerted on a single square inch. When you stand, your weight exerts a force on the floor, and the pressure under your shoes is your weight divided by the surface area of your shoes' soles. A typical pressure might be only 4 lb/in.[2]. If you step on someone's toe, that is the pressure you exert. Of

course, if you were wearing ice skates, your weight would be spread over a much smaller area, the area of the bottom of the blade, and you might exert a pressure of 150 lb/in.[2] or more. We must be careful not to step on someone's toe while we are wearing ice skates. The pressure would be dangerously high.

Astronomers are most commonly interested in the behavior of matter when it is a gas, and pressure in a gas arises when atoms or molecules collide. Consider, for example, how the gas molecules colliding with the inside of a balloon exert an outward force on the rubber and keep the balloon inflated. If the gas is hot, the atoms or molecules move rapidly, and the resulting pressure is higher than for a cooler gas. If the gas is dense, there will be many gas parti-

cles colliding with the inside of the balloon, and the pressure will be higher than for a lower-density gas. Thus, pressure depends on both the temperature and the density of the gas.

Notice that pressure and density are related, but they are not at all the same thing. Density is a measure of the amount of matter in a given volume, and pressure is a measure of the force that matter exerts on its surroundings. A very-low-density gas and a very-high-density gas might have the same pressure if they had different temperatures.

In daily life, we think of pressure when we inflate an automobile tire, but pressure is common in nearly all of the sciences. Astronomers must consider pressure in thinking about the gas inside stars and the thin gas between the stars.

It creates a magnetic field. Because the charge on the proton is positive and the charge on the electron is negative, the magnetic fields are reversed when the particles spin in the same direction.

The magnetic fields of the spinning proton and electron affect the binding energy that holds the electron to the proton in a hydrogen atom. We have all played with small magnets and noticed that they repel each other in one orientation and attract each other if we turn one around. In the same way, the small magnetic fields produced by the spinning proton and electron can repel or attract each other. In one orientation the electron is slightly less tightly bound, and in the other orientation it is slightly more tightly bound (❚ Figure 10-4). The energy difference is very small, but it makes a big difference in astronomy.

Now we can follow van de Hulst's logic and predict the existence of 21-cm radiation. Because an electron in the ground state of a hydrogen atom could spin in either of two directions—the same way as the proton or the opposite way—the ground state must really be two energy levels separated by the tiny amount of energy produced by the spinning particles. If an electron is spinning such that it is in the higher of the two states, it can spontaneously flip over and spin in the other direction. Then the atom must drop to the lower of the two energy levels, and the excess energy is radiated away as a photon. The energy difference is small, so the photon has a long wavelength—21 cm (❚ Figure 10-5).

The 21-cm radiation was predicted theoretically but it could not be detected in laboratory experiments. Of the two closely spaced energy levels, the upper one

Same spins

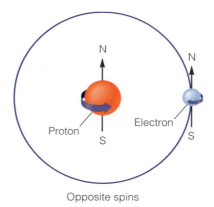

Opposite spins

FIGURE 10-4
Both the proton and electron in a neutral hydrogen atom spin and consequently have small magnetic fields. When they spin in the same direction, their magnetic fields are reversed, and when they spin in opposite directions, their magnetic fields are aligned. As explained in the text, this allows cold, neutral hydrogen in space to emit radio photons with a wavelength of 21 cm.

FIGURE 10-5

The lowest energy level of the hydrogen atom, the ground state, is actually two closely spaced energy levels that differ because the proton and electon spin. In this diagram, we would need a magnifying glass to distinguish the energy levels that make up the ground state. When an atom decays from the upper energy level to the lower energy level, it emits a photon with a wavelength of 21 cm.

is metastable. Once an electron gets caught in the upper energy level, it will, on average, stay there for 11 million years before spontaneously dropping to the lower energy level and emitting a photon of 21-cm wavelength. The atoms in a gas in a laboratory jar will collide with each other millions of times a second, so none of those atoms can remain undisturbed with its electron in the upper energy level long enough to produce a photon of 21-cm radiation. Atoms in space, however, collide much less often, so a few do manage to emit 21-cm radiation. Thus, the existence of 21-cm radiation was confirmed observationally by radio astronomers who detected it coming from clouds of neutral hydrogen in space.

The 21-cm observations give us a way to map the cold, neutral hydrogen that fills much of our galaxy. Astronomers can locate individual clouds and measure their motion by observing the Doppler shifts in the 21-cm radiation (▮ Figure 10-6).

Of course, ionized hydrogen lacks an electron, so it can't emit 21-cm radiation. Further, hydrogen atoms locked in molecules are also unable to emit 21-cm wavelength photons. We must find other ways to study these parts of the interstellar medium.

MOLECULES IN SPACE

Radio telescopes can also detect radiation from various molecules in the interstellar medium. A molecule can store energy in a number of different ways. For example, it can rotate at different rates, or the atoms in a molecule can vibrate as if they were linked together by small springs. If a molecule suffers a collision or absorbs a photon, it can be excited to vibrate and rotate in some higher energy state. Quickly, however, it will return to a lower energy state and radiate the excess energy as a photon. Because these energy levels are closely spaced, the emitted photons typically have low energies, and we detect them in the radio or far-infrared part of the electromagnetic spectrum. Just as neutral hydrogen radiates at a specific wavelength of 21 cm, many natural molecules radiate at their own unique wavelengths.

Unfortunately for astronomers, molecules of hydrogen (H_2) are difficult to detect. In the far ultraviolet, molecular hydrogen can be detected by the photons it absorbs, but these observations must be made by specialized telescopes in space. Mapping the location of molecular hydrogen would be easier if the molecules emitted radio-wavelength photons, but a molecule containing two identical atoms does not radiate in the radio part of the spectrum. However, clouds of gas dense enough to form molecular hydrogen also form tiny amounts of other molecules, and many of them are good emitters of radio energy. Nearly 100 different molecules have been detected (▮ Table 10-1). Some are quite complex, and it is not clear how they form. Most astronomers believe that the atoms meet and bond to form molecules on the surfaces of dust grains. Some of these molecules have not yet been synthesized on Earth, but others are common, such as N_2O (nitrous oxide), also known as laughing gas. Ethyl alcohol, which some humans drink, has also been detected. Although this is a very rare molecule compared to molecular hydrogen, an interstellar cloud can contain ethyl alcohol in amounts equivalent to 10^{28} fifths of whisky (about 100 Earth masses).

One of the most important of the interstellar molecules may be carbon monoxide (CO). It is one of the poisonous gases that comes out of the tailpipes of cars, but it is also a very good emitter of radio energy at a wavelength of 2.6 mm. An interstellar cloud may contain only 1 CO molecule for every 10,000 molecules of hydrogen, but the CO can be detected while the molecular hydrogen cannot. Thus, radio astronomers can map the interstellar clouds by searching for the CO radio emission; where they detect CO, they can be certain that molecular hydrogen is common.

Another important constituent of the interstellar medium is OH. Because this combination of an oxygen

a

b

FIGURE 10-6

(a) This radio map of the entire sky shows the distribution of neutral hydrogen. The Milky Way is the bright band running from left to right. Notice how wispy and irregular the hydrogen is. *(J. Dickey, UMn, F. Lockman, NRAO, SkyView)* (b) These 21-cm radio observations contain many peaks, each produced by a separate cloud of neutral hydrogen with its own radial velocity. *(Adapted from observations by Burton)*

atom and a hydrogen atom is not electrically neutral, it is called a radical instead of a molecule. The OH radical forms in interstellar clouds and is a good emitter of radio energy.

These molecules are very fragile, and a high-energy photon such as an ultraviolet photon has enough energy to break the molecule into separate atoms. Thus, the molecules cannot exist outside the dense clouds. Only deep inside the densest clouds, where dust absorbs and scatters the short-wavelength, high-energy photons, can the molecules survive. The very fact that these molecules are detected tells us that some of these **molecular clouds** must be very dense.

The molecules, especially CO, are such good radiators of energy that they cool the clouds to low temperatures. Thermal energy is present in the cloud as motion among the atoms and molecules. When a molecule collides with an atom or another molecule, some of the energy of motion can be stored in the rotation and vibration of a molecule. When the molecule emits that energy as a radio or infrared photon, the energy escapes from the cloud. In this way, molecular radiation can cool the interior of the cloud and keep it very cold.

The largest of these cool, dense clouds are called **giant molecular clouds.** They are 15 to 60 pc across and may contain 100 to 1,000,000 solar masses. The internal temperature is a frigid 10 K. Although we detect these clouds by their carbon monoxide molecular emission, remember that the gas is mostly hydrogen.

Giant molecular clouds are the nests of star birth. Deep inside these great clouds, gravity can pull the matter inward and create new stars. That is a story we will discuss in detail in the next chapter. For now, we must study the dirty part of the interstellar medium—the dust. To do this, we must trade our radio telescope for an infrared telescope.

TABLE 10-1
Selected Molecules Detected in the Interstellar Medium

H_2	molecular hydrogen	H_2S	hydrogen sulfide
C_2	diatomic carbon	N_2O	nitrous oxide
CN	cyanogen	H_2CO	formaldehyde
CO	carbon monoxide	C_2H_2	acetylene
NO	nitric oxide	NH_3	ammonia
OH	hydroxyl	HCO_2H	formic acid
NaCl	common table salt	CH_4	methane
HCN	hydrogen cyanide	CH_3OH	methyl alcohol
H_2O	water	CH_3CH_2OH	ethyl alcohol

INFRARED RADIATION FROM DUST

The dust in the interstellar medium makes up only about 1 percent of the mass, and at a temperature of 100 K (−143°C) or less it is very cold. Nevertheless, it is easy to detect at infrared wavelengths. To see how such cold dust can radiate a lot of energy, consider a simple experiment with the dust in an imaginary giant molecular cloud.

For the sake of quick calculation, let us assume that our giant molecular cloud has a mass of about 10^5 solar masses. Only 1 percent of that mass is dust, so all of the dust in the cloud amounts to a mass of 10^3 solar masses. Imagine that we could collect all of the dust into a single sphere. It would be only about 10 times the diameter of the sun, and its surface area would be about 100 times that of the sun. But suppose we left the dust as separate specks, each 0.0005 mm in diameter, a typical size. Then the cloud would contain about 10^{43} dust specks, and the total surface area of the dust would be about 10^{29} times that of the sun. When matter is finely divided into dust, it has a very large surface area, and even though it is much colder than the sun, its vast surface area can radiate tremendous amounts of infrared radiation.

In 1983, the Infrared Astronomy Satellite mapped the sky at far-infrared wavelengths and found the galaxy filled with infrared radiation from dust. Most of this dust is confined to the region near the plane of our galaxy, and, as we would expect, the dust is thickest where the gas is thickest. Much of the dust is distributed in wispy clouds that became known as the **infrared cirrus** because of their overall resemblance to cirrus clouds in Earth's atmosphere. The infrared cirrus consists of dusty clouds of interstellar matter with temperatures of about 30 K (▌Figure 10-7a). Studies of CO emission from molecular clouds show that at least some of the infrared cirrus is associated with molecular clouds within a few hundred parsecs of the sun. This suggests that the gas and dust in the molecular clouds is not uniform but rather very patchy.

As we mentioned before, the interstellar dust appears to be composed of carbon, silicates, and iron, elements that are underabundant in the gas. Some grains appear to contain water ice contaminated with ammonia, methane, and other compounds.

Infrared observations can help us understand the interstellar medium, but X-ray and ultraviolet observations complete the picture by allowing us to detect a part of the gas between the stars that is otherwise invisible.

X RAYS FROM THE INTERSTELLAR MEDIUM

The interstellar medium seems very cold, so we would hardly expect to detect X rays. These high-energy photons are commonly produced by high temperatures. Nevertheless, X-ray telescopes can help us understand the interstellar medium.

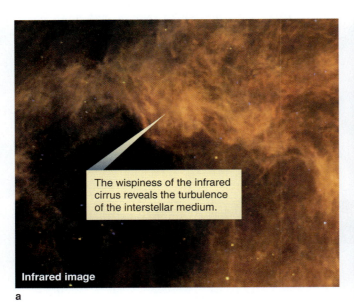

The wispiness of the infrared cirrus reveals the turbulence of the interstellar medium.

Infrared image

a

About 450 ly in diameter, the cavity has been inflated by many supernova explosions.

Cygnus

Cygnus Superbubble

A nearby cavity only 110 ly in diameter formed by the explosion of a single star

Cygnus Loop

X-ray image

b

FIGURE 10-7

Infrared and X-ray observations tell us about different components of the interstellar medium. (a) This far-infrared image shows the infrared cirrus, wispy clouds of very cold dusty gas spread all across the sky. *(NASA/IPAC, Courtesy Deborah Levine)* (b) This ROSAT X-ray image shows the Cygnus Superbubble, a cavity filled with gas from exploding stars. The gas is heated and emits X rays where the expanding bubble pushes into the surrounding gas. *(Courtesy Steve Snowden and Max-Planck Institute for Extraterrestrial Physics, Germany)*

X-ray telescopes above Earth's atmosphere have detected X rays from a part of the interstellar medium with a very high temperature. This gas has been called the **coronal gas** because it has temperatures of 10^6 K or higher, as does the sun's corona. Of course, the coronal gas in the interstellar medium is not related in any way to the actual corona of the sun.

We might expect the coronal gas to have a very high pressure because it is so hot, but it has a very low density. It contains only 0.0004 to 0.003 particles/cm^3. That is, we would have to search through a few hundred to a few thousand cubic centimeters of coronal gas to find a single particle—an ionized atom or an electron. Because of its very low density, the gas pressure in the coronal gas can be roughly the same as in the HI clouds and the intercloud medium. Nevertheless, exceptions are known. The cloud of interstellar matter near the sun has a pressure that is more than 20 times less than the pressure of neighboring coronal gas. Such differences in pressure are not well understood.

The coronal gas appears to originate when a massive star explodes violently in what astronomers call a supernova. Although rare, such explosions blast large amounts of very hot gas outward in expanding shells. Gas flowing away from very hot, young stars must add to the coronal gas. In some cases, neighboring shells of coronal gas may expand into each other and merge to form larger volumes, but an earlier theory that coronal gas fills a large network of tunnels and shells throughout interstellar space seems to be contradicted by evidence. Probably about 20 percent of the space between the stars is filled with isolated pockets of coronal gas.

The Cygnus superbubble appears to be related to the coronal gas. Located in Cygnus, it is a very large shell of hot gas about 450 pc in diameter (Figure 10-7b). The energy needed to create such a shell is equivalent to hundreds of supernova explosions. The bubble may have developed as a large cluster of stars was born and grew old and the most massive stars died in supernova explosions. Several of these large bubbles are known.

ULTRAVIOLET OBSERVATIONS OF THE INTERSTELLAR MEDIUM

We can divide the ultraviolet spectrum into the near ultraviolet, with wavelengths only slightly shorter than visible light, and the far ultraviolet, with much shorter wavelengths. Only a few decades ago, astronomers believed that far-ultraviolet photons could not travel far through the interstellar medium because they would be absorbed so easily by neutral hydrogen atoms. These atoms of neutral hydrogen are good absorbers of far-ultraviolet photons because the photons have enough energy to ionize the atoms. In fact, even one atom of HI per cubic centimeter would make the interstellar medium opaque to far-ultraviolet photons, and we would be unable to see more than a light-year into space with a far-ultraviolet telescope.

When the Extreme Ultraviolet Explorer (EUVE) satellite was put into Earth orbit in 1992, it discovered that the interstellar medium was only partly cloudy. While some regions were filled with clouds of neutral hydrogen and so were opaque, other regions were filled with hot, ionized hydrogen that was transparent to far-ultraviolet photons. This confirms the description of the interstellar medium produced by X-ray observations.

The far-ultraviolet observations tell us that the sun is located just inside a large region of hot, ionized hydrogen, while only a few light-years from the sun lies a cool, neutral hydrogen cloud that is opaque to far-ultraviolet photons. This **local bubble or void** of gas in which the sun is located appears to be linked to other hot, transparent regions. Thus, ultraviolet observations, combined with X-ray observations, suggest that the apparently cold and empty regions between the stars are filled with a complex, evolving mixture of hot and cold gas.

REVIEW CRITICAL INQUIRY

If hydrogen is the most common molecule, why do astronomers depend on the CO molecule to map molecular clouds?

Although hydrogen is the most common atom in the universe and molecular hydrogen the most common molecule, a molecule of hydrogen does not radiate in the radio part of the spectrum. Consequently, radio astronomers cannot detect it. But the much less common CO (carbon monoxide) molecule is a very efficient radiator of radio energy, so radio astronomers use it as a tracer of molecular clouds. Wherever radio telescopes reveal a great cloud of CO, we can be confident that most of the gas is molecular hydrogen.

We can also be confident that the molecular clouds contain dust, because it is the dust that protects the molecules in the cloud from the ultraviolet radiation that would otherwise break down the molecules into atoms.

Dust doesn't radiate longer-wavelength radio energy, so we must study it in the infrared. Although the dust in a molecular cloud is very cold and makes up a small percentage of the total mass, it is a very good radiator of infrared radiation. How can a small amount of cold dust radiate vast amounts of infrared radiation?

From one end of the spectrum to the other, we have used every wavelength to study the interstellar medium. Now we will try to put our data in order and create a model of the gas and dust between the stars. That is the first step toward understanding how stars are born.

TABLE 10-2
Four Components of the Interstellar Medium

Component	Temperature (K)	Density (atoms/cm^3)	Gas
HI clouds	50–150	1–1000	Neutral hydrogen; other atoms ionized
Intercloud medium (HII)	10^3–10^4	0.01	Partially ionized
Coronal gas	10^5–10^6	10^{-4}–10^{-3}	Highly ionized
Molecular clouds	20–50	10^3–10^5	Molecules

10-3 A MODEL OF THE INTERSTELLAR MEDIUM

When we look at bright nebulae like the Great Nebula in Orion, we see a part of the interstellar medium, but what we see is only a very special region that happens to be close enough to hot, bright stars to become excited. Most of the interstellar medium is invisible to our eyes, so we must gather the evidence—the observational facts (Window on Science 10-1)—and use them to develop a model of the interstellar medium.

We can divide the interstellar medium into four basic components (■ Table 10-2). Our problem is to describe these components and explain how they interact and evolve. We will discover that their evolution is intimately connected to the process of star formation and death.

FOUR COMPONENTS OF THE INTERSTELLAR MEDIUM

The interstellar medium is not at all uniform. Rather, it is lumpy, and the lumps differ dramatically in temperature and density.

HI clouds are cool, with temperatures of 50 to 150 K and densities of ten to a few hundreds of atoms per cubic centimeter. These clouds are only a few parsecs in diameter and contain a few solar masses.

Between the cool HI clouds lies the warm intercloud medium of HII, with temperatures of a few thousand Kelvin and densities of 0.01 atoms/cm^3. The intercloud medium is in approximate equilibrium with the HI clouds in that the hot, low-density gas has about the same pressure as the colder, denser gas in the HI clouds.

Molecular clouds are especially dense. Molecules cannot survive if they are exposed to ultraviolet photons in starlight, so they can form only in the densest clouds, where dust absorbs and scatters ultraviolet photons. The molecular clouds can be very large, with diameters up to 60 pc and masses up to a million solar masses, but they are also very cold. Strong evidence suggests that stars are born when giant molecular clouds

contract under the influence of their own gravity, a subject we will explore in detail in the next chapter.

Pushing through this stew of interstellar clouds are regions of coronal gas. Most of this very hot gas is probably produced in supernova explosions, although some may be gas flowing away from very hot stars. With temperatures up to a million degrees and densities as low as 10^{-4} atoms/cm^3, the coronal gas occupies a large part of the interstellar medium.

Astronomers believe that the HI clouds make up about 25 percent of the interstellar mass, and the intercloud medium about 50 percent. The coronal gas contributes only 5 percent of the mass, although it seems to occupy about one-fifth of the volume. The giant molecular clouds amount to about 25 percent of the mass. If we add up these percentages, we get slightly more than 100 percent, which illustrates the uncertainty inherent in our model of the interstellar medium.

As the stars move through the interstellar medium, they meet no resistance, but they do have a dramatic effect on the gas and dust. The Pleiades, for example, is a relatively young star cluster that is moving rapidly through space. As it moves through the interstellar medium, it is leaving behind a trail like that left by a boat in water (■ Figure 10-8). This wake has been detected in IRAS images and is apparently produced by the ultraviolet radiation from the stars in the cluster. None of the stars is hot enough to ionize the gas, but there are a number of relatively hot stars; the ultraviolet radiation from those stars heats, but does not quite ionize, the gas. The heated gas then expands and forms a wake showing the path of the cluster through the interstellar medium.

The Pleiades clearly illustrates the close relationship between the stars and the interstellar medium. In fact, we can now outline a cycle that links the stars to the gas and dust between them.

THE INTERSTELLAR CYCLE

The story of the interstellar medium is closely linked to star formation. We will see in the next chapter that stars form in giant molecular clouds. As soon as a group

Understanding Science: Separating Facts from Theories

The fundamental work of science is testing theories by comparing them with facts. As we think about science, we need to distinguish clearly between facts and theories. The facts are the evidence against which we test the theories.

Scientific facts are those observations or experimental results of which we are confident. An astronomer makes observations of stars, and a botanist collects samples of related plants. A fact could be a precise measurement, such as the mass of a star expressed as a specific number, or it could be a simple observation, such as that a certain butterfly no longer visits a certain mountain valley. In each case, the scientist is gathering facts.

A theory, however, is a conjecture as to how nature works. If we are uncertain of the theory, we might call it a hypothesis. In any case, these conjectures are not facts; they are attempts to explain how nature works. In a sense, a theory or hypothesis is a story that scientists have made up to explain how nature works in some specific case. These stories can be wonderfully detailed and ingenious, but without evidence they are nothing more than hunches.

When one of these stories is tested against the facts and confirmed, we have more confidence that the story is more or less right. The more a story is tested successfully, the more confidence we have in it. The facts represent reality, and every theory or hypothesis must be repeatedly tested against reality.

We can't test one theory against another theory. Theories are not evidence; they are conjectures. If we were allowed to test one theory against another theory, we might fall into the trap of circular reasoning. "Elves make the flowers bloom. I know that elves exist because the flowers bloom." That is using two theories to confirm each other, and it leads us to nonsense. Only facts can be evidence.

Nor can we use a theory to deduce facts. We can use a theory to make predictions, which we can then test against facts, but the predictions themselves can never be certainties, so they can't be facts. The only way to arrive at facts is to consult nature and make direct measurements or observations.

As we study different problems in astronomy, we must carefully distinguish between facts and theories. Facts are the basic building blocks of all science, and they can come only from the careful study of nature. As Galileo would say, we must read the book of nature.

---

of stars forms, the hottest stars begin ionizing the gas to produce emission nebulae. The pressure of the starlight and the gas flowing away from the hot, young stars pushes the interstellar cloud outward and may disrupt the cloud entirely.

The composition of the interstellar dust suggests that it is formed mostly in the atmospheres of cool stars. There the temperatures are low enough for some atoms to condense into specks of solid matter, much as soot can condense in a candle flame. The pressure of the starlight can push these dust specks out of the star and thus replenish the interstellar medium. Other stars that eject mass into space, such as supernovae, probably also add to the supply of interstellar dust.

The most massive stars die quickly in supernova explosions, and those tremendously violent events blast high-temperature gas outward that further disrupts the interstellar medium (Figure 10-9). Much of the motion we see in interstellar clouds and their twisted shapes are probably produced by these supernova explosions and the hot coronal gas they produce. It seems that our galaxy produces about two supernova explosions per century (although most are not visible from Earth because of obscuration by dust) and that these explosions keep the interstellar medium stirred and create the vast regions filled with coronal gas. In fact, the sun lies inside such a region of high-temperature, low-density gas called the local bubble or void. With a typical diameter

FIGURE 10-8

This infrared image shows the brighter stars of the Pleiades as small crosses. (Compare with photo on page 199.) The motion of the cluster from left to right has left a wake in the interstellar medium. *(Courtesy Richard E. White; image rendered by Duncan Chesley of American Image, Inc.)*

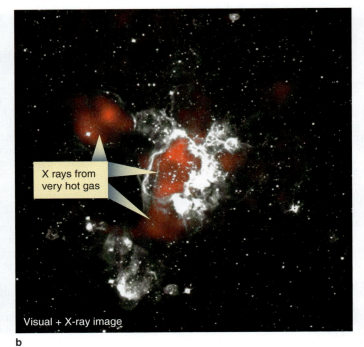

X rays from very hot gas

a

b

FIGURE 10-9

(a) The nebula N44C is part of a larger region where young, hot, massive stars have formed and where at least one supernova has blown a bubble filled with hot gas. *(The Hubble Heritage Team STScI/ AURA/NASA)* (b) This image of the entire N44 region combines a visual-wavelength image (white) with an X-ray image (red). X rays reveal regions of very high temperature gas that has been ejected from supernova explosions. *(R. Chris Smith, CTIO and Y.-H. Chu, UICU)*

of a few hundred parsecs, the local bubble may be a cavity in the interstellar medium inflated by a supernova explosion within the last million years or so.

While supernovae and hot stars keep the interstellar medium in motion, the natural gravitation of gas clouds and collisions between clouds may gradually build more massive clouds. In the most massive clouds, the dust protects the interior from ultraviolet photons, and molecules can form. These giant molecular clouds eventually give birth to new stars, and the cycle begins all over again.

The Trifid Nebula (▌ Figure 10-10) is a dramatic illustration of this cycle. Measuring over 12 pc in diameter, the nebula is illuminated by a number of hot, young stars that have apparently formed recently from the gas. Near the stars, the gas is ionized and glows as a pink-red HII region, but farther from the stars, where the ultraviolet radiation is weaker, the gas is not ionized. Nevertheless, dust in the nebula scatters blue light, so this un-ionized part of the nebula is visible as a blue reflection nebula. Dark lanes of obscuring dust cross the face of the nebula as if to remind us again of the importance of dust in the interstellar cycle.

Our study of the interstellar medium is incomplete for two reasons. First, astronomers don't yet understand all of its components or how those components interact. Second, we can't fully understand the interstellar medium until we understand how stars are born and how they die. We begin that story in the next chapter.

FIGURE 10-10
The Trifid Nebula embodies an important part of the interstellar cycle. A dense cloud of gas has given birth to stars, and the hottest are ionizing the gas to produce the pink emission nebula, while dust scatters light to produce the blue reflection nebula. As the nebula is disrupted by the newborn stars, the gas and dust merge with the interstellar medium. *(AURA/NOAO/NSF and Nigel Sharp)*

SUMMARY

The interstellar medium, the gas and dust between the stars, is confined near the plane of our Milky Way Galaxy. We see clear evidence of an interstellar medium when we look at an emission nebula, a cloud of gas near one or more hot stars whose ultraviolet radiation ionizes the hydrogen and makes the nebula glow like a giant neon sign. The red, blue, and violet Balmer lines blend together to produce the characteristic pink-red glow of ionized hydrogen.

A reflection nebula is produced by gas and dust illuminated by a star that is not hot enough to ionize the gas. Rather, the dust scatters the starlight to produce a reflection of the stellar absorption spectrum. Because shorter-wavelength photons scatter more than longer-wavelength photons, reflection nebulae look blue. The daytime sky looks blue for the same reason.

A dark nebula is a cloud of gas and dust that is visible because it blocks the light of distant stars. We see such nebulae as dark shapes.

Further evidence of an interstellar medium is the extinction, or dimming, of the light of distant stars, and interstellar reddening. Light from distant stars suffers scattering by dust particles in the interstellar medium, and blue light is scattered more than red light. This makes distant stars look redder than their spectral types suggest. The dependence of this extinction on wavelength tells us that the scattering dust particles are very small. The dust is made of carbon, silicates, iron, and ice.

Interstellar absorption lines in the spectra of distant stars are very narrow. The interstellar gas is cold and has a very low density, and this makes the interstellar lines much narrower than the spectral lines produced in stars. Multiple interstellar lines tell us that the light has passed through more than one interstellar cloud on its way to Earth. Radio observations at a wavelength of 21 cm also reveal multiple clouds of neutral hydrogen. These neutral clouds drift through a warmer but lower-density intercloud medium. Radio telescopes tuned to other wavelengths have detected nearly 100 different molecules in the interstellar medium, most of them found in giant molecular clouds.

The Infrared Astronomy Satellite has detected dust in the interstellar medium in the form of the infrared cirrus. X-ray and far-ultraviolet observations have detected very hot coronal gas produced by supernova explosions.

The four main components of the interstellar medium are the small neutral HI clouds, the warm intercloud medium, coronal gas, and molecular clouds. Stars are born in the dense molecular clouds, and the energy from hot stars and supernova explosions causes currents in the interstellar medium and creates the coronal gas.

NEW TERMS

interstellar medium

nebula

emission nebula

HII region

reflection nebula

dark nebula

forbidden line

metastable level

interstellar dust

interstellar extinction

interstellar reddening

interstellar absorption lines

HI clouds

intercloud medium

pressure

21-cm radiation

molecular cloud

giant molecular clouds

infrared cirrus

coronal gas

local bubble or void

REVIEW QUESTIONS

Ace⟲Astronomy™ Assess your understanding of this chapter's topics with additional quizzing and animations at **http:// astronomy.brookscole.com/seeds8e**

1. What evidence do we have that the spaces between the stars are not totally empty?

2. What evidence do we have that the interstellar medium contains both gas and dust?

3. How do the spectra of HII regions differ from the spectra of reflection nebulae? Why?

4. Why are interstellar lines so narrow? Why do some spectral lines forbidden in spectra on Earth appear in spectra of interstellar clouds and nebulae? What does that tell us?

5. How is the blue color of a reflection nebula related to the blue color of the daytime sky?

6. Why do distant stars look redder than their spectral types suggest?

7. If starlight on its way to Earth passed through a cloud of interstellar gas that was hot instead of very cold, would you expect the interstellar absorption lines to be broader or narrower than usual? Why?

8. How can the HI clouds and the intercloud medium have similar pressures when their temperatures are so different?

9. What does the shape of the 21-cm radio emission line of neutral hydrogen tell us about the interstellar medium?

10. What produces the coronal gas?

DISCUSSION QUESTIONS

1. When we see distant streetlights through smog, they look dimmer and redder than they do normally. But when we see the same streetlights through fog or falling snow, they look dimmer but not redder. Use your knowledge of the interstellar medium to discuss the relative sizes of the particles in smog, fog, and snowstorms compared to the wavelength of light.

2. If you could see a few stars through a dark nebula, how would you expect their spectra and colors to differ from similar stars just in front of the dark nebula?

PROBLEMS

1. A small nebula has a diameter of 20 seconds of arc and a distance of 1000 pc from Earth. What is the diameter of the nebula in parsecs? in meters?

2. The dust in a molecular cloud has a temperature of about 50 K. At what wavelength does it emit the maximum energy? (*Hint:* Consider black body radiation, Chapter 7.)

3. Extinction dims starlight by about 1.9 magnitudes per 1000 pc. What fraction of photons survives a trip of 1000 pc? (*Hint:* Consider the definition of the magnitude scale in Chapter 2.)

4. If the total extinction through a dark nebula is 10 magnitudes, what fraction of photons makes it through the cloud? (*Hint:* See Problem 3.)

5. The density of air in a child's balloon 20 cm in diameter is roughly the same as the density of air at sea level, 10^{19} particles/cm³. To how large a diameter would you have to expand the balloon to make the gas inside the same density as the interstellar medium, about 1 particle/cm³? (*Hint:* The volume of a sphere is $\frac{4}{3}\pi R^3$.)

6. If a giant molecular cloud has a diameter of 30 pc and drifts relative to neighboring clouds at 20 km/s, how long will it take to travel its own diameter?

7. An HI cloud is 4 pc in diameter and has a density of 100 hydrogen atoms/cm³. What is its total mass in kilograms? (*Hints:* The volume of a sphere is $\frac{4}{3}\pi R^3$, and the mass of a hydrogen atom is 1.67×10^{-27} kg.)

8. Find the mass in kilograms of a giant molecular cloud that is 30 pc in diameter and has a density of 300 hydrogen molecules/cm³. (*Hint:* See Problem 7.)

9. At what wavelength does the coronal gas radiate most strongly? (*Hint:* Consider black body radiation, Chapter 7.)

CRITICAL INQUIRIES FOR THE WEB

1. What causes the varied colors in images of gaseous nebulae that grace textbooks and Web sites? Search the Web for a color image of a nebula in the Messier catalog. Describe the structure and colors you see in terms of the concepts discussed in this chapter. (Be careful to distinguish real color from false color when answering this question. See Window on Science 13-1 for more information about false-color images.)

2. An interesting highlight in the history of astronomy is the discovery of "Nebulium." Search the Internet for information on this once-mysterious source of nebular spectral lines. How can astronomers today explain these lines in terms of known elements?

EXPLORING *THESKY*

1. The following nebulae are all star-formation regions. What kind of nebulae are they? (*Hint:* To center on an object, use **Find** under the **Edit** menu. Choose **Messier Objects** and pick from the list.)

    M42, M20, M8, M17

2. Locate M8 in *TheSky,* zoom in, and identify other nebulae in the region. Study the photo of NGC6559. What kind of nebula is it?

3. Locate M42, M20, and M8. Zoom in and identify the spectral type of the central star(s). What common characteristic do these stars share? Why?

 Visit the Seeds *Foundations of Astronomy* companion Web site for critical thinking exercises, articles, and additional readings from InfoTrac College Edition, Brooks/Cole's online student library.

THE FORMATION OF STARS

Jim he allowed [the stars] was made, but I allowed they happened.
Jim said the moon could'a laid them; well, that looked kind of reasonable,
so I didn't say nothing against it, because I've seen a frog lay most as many,
so of course it could be done.

Mark Twain, The Adventures of Huckleberry Finn

N. Walborn and J. Maiz-Apellániz, STScI, R. Barbá, La Plata Observatory, and NASA

Chapter 11

GUIDEPOST

Previous chapters have used the basic principles of physics as a way to deduce things about stars and the interstellar medium. All of the data we have amassed will now help us understand the life stories of the stars in this chapter and those that follow.

In this chapter, we use the laws of physics in a new way. We develop theories and models based on physics that help us understand how stars work. For instance, what stops a contracting star and gives it stability? We can understand this phenomenon because we understand some of the basic laws of physics.

Throughout this chapter and the chapters that follow, we search for evidence. What observational facts confirm or contradict our theories? That is the basis of all science, and it must be part of any critical analysis of what we know and how we know it.

Stars exist because of gravity. They form because gravity makes clouds of gas contract, and they generate nuclear energy because gravity squeezes their cores to unearthly densities and temperatures. In the end, stars die because they exhaust their fuel supply and can no longer withstand the force of their own gravity.

A star can remain stable only by maintaining great pressure in its interior. Gravity tries to make it contract. However, if the internal temperature is high enough, pressure pushes outward just enough to balance gravity. Thus, a star is a battlefield where pressure and gravity struggle for dominance.

Only by generating tremendous amounts of energy can a star maintain the gravity–pressure balance. The sun, for example, generates 6×10^{13} times more energy per second than all of the coal, oil, natural gas, and nuclear power plants on Earth. Like the sun, most stars generate their power by fusion reactions that consume hydrogen and produce helium. However, we will discover that not all stars fuse their hydrogen in the same way as does the sun.

In this chapter, we will see how gravity creates stars from the thin gas of space and how nuclear reactions inside stars generate energy. We will see how the flow of that energy outward toward the surface of the star balances gravity and makes the stars stable. In the next two chapters, we will follow the life story of stars from the beginning of their stable lives to their ultimate deaths.

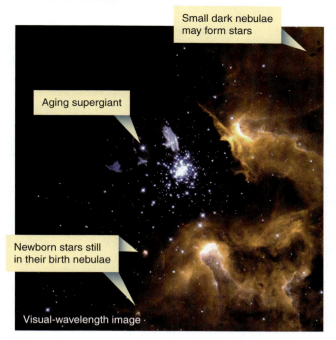

Small dark nebulae may form stars

Aging supergiant

Newborn stars still in their birth nebulae

Visual-wavelength image

FIGURE 11-1

This young star cluster contains hot, luminous O stars, which are blowing away the nebula from which the cluster formed. Yet star formation is continuing in the nebula as it is compressed. Young in years, the supergiant at upper left has already begun its death throes as it has ejected a ring of gas around its equator and clouds of gas from its poles. The entire cycle of star birth and star death is illustrated in this true color image. (*W. Brandner, JPL/IPAC, E.K. Grebel, Univ. of Washington, Y. Chu, Univ. of Illinois, and NASA*)

11-1 MAKING STARS FROM THE INTERSTELLAR MEDIUM

Stars have been forming continuously since our galaxy took shape over 10 billion years ago. We know this for two reasons. First, the sun is only about 5 billion years old, a relative newcomer compared to the older stars in our galaxy. Second, we can see hot, blue stars such as Spica (α Virginis), a B1 main-sequence star. As we will see in the next chapter, such massive stars have very short lives. In fact, a star like Spica can last only 10 million years and thus must have formed recently.

The key to understanding star formation is the correlation between young stars and clouds of gas. Where we find the youngest groups of stars, we also find large clouds of gas illuminated by the hottest and brightest of the new stars (▌Figure 11-1). This leads us to suspect that stars form from such clouds, much as raindrops condense from the water vapor in a thundercloud. Indeed,

the giant molecular clouds discussed in the preceding chapter can give birth to entire clusters of new stars.

The central problem for our discussion of star formation is how these large, low-density, cold clouds of gas become comparatively small, high-density, hot stars. Gravity is the key.

STAR BIRTH IN GIANT MOLECULAR CLOUDS

The giant molecular clouds are the sites of active star formation, yet they are very unlike stars. With a typical diameter of 50 pc and a typical mass exceeding 10^5 solar masses, a giant molecular cloud is vastly larger than a star. Also, the gas in a giant molecular cloud is about 10^{20} times less dense than a star and has temperatures of only a few degrees Kelvin. These clouds can form stars because gravity can force some small regions of the clouds to contract to high density and high temperature.

Radio observations show that at least some giant molecular clouds develop dense cores that are only 0.1 pc in radius and that contain roughly 1 solar mass. A single giant molecular cloud may contain many of these dense cores and thus can give birth to star clusters containing hundreds of stars. However, both theory and observations suggest that many giant molecular clouds cannot begin the formation of dense cores spontaneously. At least four factors resist the contraction of a gas cloud, and gravity must overcome those four factors before star formation can begin.

First, thermal energy in the gas is present as motion among the atoms and molecules. Even at temperatures of 10 K, the average hydrogen molecule moves at about 0.35 km/s (almost 800 mph). This thermal motion would make the cloud drift apart if gravity were too weak to hold it together.

The interstellar magnetic field is the second factor that gravity must overcome to make the cloud contract. Neutral atoms and molecules are unaffected by a magnetic field, but ions, having an electric charge, cannot move freely through a magnetic field. Although the gas in a molecular cloud is mostly neutral, there are some ions, and thus a magnetic field can exert a force on the gas. The magnetic field present all through our galaxy averages only about 10^{-4} times as strong as that on Earth, but it can act like an internal spring to resist the contraction of the gas cloud.

The third factor is rotation. Everything in the universe rotates to some extent. As the gas cloud begins to contract, it spins more and more rapidly as it conserves angular momentum, just as ice skaters spin faster as they pull in their arms (Figure 5-7). This rotation can become so rapid that it resists further contraction of the cloud.

The turbulence in the interstellar medium is the fourth thing that could prevent a cloud from contracting. Our discussion of the interstellar medium in the previous chapter revealed that nebulae are often twisted and distorted by moving currents in the interstellar medium. This turbulence could make it difficult for a large molecular cloud to contract.

Given these four resistive factors, it seems surprising that any giant molecular clouds can contract, form dense cores, and eventually form stars. In some cases, the gas in giant molecular clouds can gradually recombine with free electrons and become less ionized. Neutral gas is free to "slip past" the magnetic field and contract. This gradual process has been observed to form dense cores inside isolated molecular clouds.

In contrast, both theory and observation suggest that many giant molecular clouds are triggered to form stars by a passing **shock wave,** the astronomical equivalent of a sonic boom (■ Figure 11-2). During such a triggering event, a few regions of the large cloud can be compressed to such high densities that the resistive factors can no longer oppose gravity, and star formation begins.

At least four different processes can produce shock waves that trigger star formation. Supernova explosions (Chapter 13) can produce powerful shock waves that rush through the interstellar medium. Also, the ignition of very hot stars can ionize nearby gas and drive it away to produce a shock wave where it pushes into

Shock Wave Triggers Star Formation

A shock wave (red) approaches an interstellar gas cloud.

The shock wave passes through and compresses the cloud.

Motions in the cloud continue after the shock wave passes.

The densest parts of the cloud become gravitationally unstable.

Contracting regions of gas give birth to stars.

FIGURE 11-2

In this summary of a computer model, an interstellar gas cloud is triggered into star formation by a passing shock wave. The events summarized here might span about 6 million years.

the colder, denser interstellar matter. A third trigger is the collision of molecular clouds. Because the clouds are large, they are likely to run into each other occasionally; and because they contain magnetic fields, they cannot pass through each other. A collision between such clouds can compress parts of the clouds and trigger star formation. The fourth trigger is the spiral pattern of our Milky Way Galaxy (see Figure 1-11). One theory suggests that the spiral arms are shock waves that travel around the galaxy like the moving hands of a clock (Chapter 15). As a cloud passes through a spiral arm, the cloud could be compressed, and star formation could begin.

Of course, a single giant molecular cloud containing a million solar masses does not contract to form a single humongous star. The cloud fragments and the densest parts form a number of dense cores. Exactly why a cloud fragments isn't fully understood, but the rotation, magnetic field, and turbulence probably play important roles. In any case, a giant cloud of gas typically contracts to form a number of newborn stars.

HEATING BY CONTRACTION

We have explained how clouds of interstellar gas can become dense enough to make stars, but we have not explained how the gas can become hot enough. The answer, once again, is gravity.

To consider the formation of a star, we now shift our attention to a single dense core of gas destined to become a star. Once a small cloud of gas begins to contract, gravity draws the atoms toward the center, and the atoms gather speed as they fall. In fact, astronomers refer to this early stage in the formation of a star as **free-fall contraction.** Whereas the atoms may have had low velocities to start with, by the time they have fallen most of the way to the center of the cloud, they are traveling at high velocities. We have defined thermal energy as the agitation of the particles in a gas, so this increase in velocity is a step toward heating the gas. But we can't say that the gas is hot simply because all of the atoms are moving rapidly. The air in the cabin of a jet airplane is traveling rapidly, but it isn't hot because all of the atoms are moving in generally the same direction with the plane. To convert the high velocity of the infalling atoms into thermal energy, the motion must become randomized, and that happens when the atoms begin to collide with one another as they fall into the central region of the cloud. The jumbled, random motion of the atoms represents thermal energy, and the temperature of the gas increases.

This is an important principle in astronomy. Whenever a cloud of gas contracts, gravitational energy is converted into thermal energy, and the gas grows hotter. Whenever a gas cloud expands, thermal energy is converted into gravitational energy, and the gas cools. This principle applies not only to clouds of interstellar gas, but also to contracting and expanding stars, as we will see in the following chapters.

Our study of gas clouds has shown us how nature can begin the contraction of dense cores in giant molecular clouds and how contraction can heat the gas. Now we will construct a detailed story of the transformation from gas cloud to star.

PROTOSTARS

To follow the story of star formation farther, we must concentrate on a single fragment of a collapsing cloud as it contracts, heats up, and begins to behave like a star. Although the term **protostar** is used rather loosely by astronomers, we will define it here to be a prestellar object that is hot enough to radiate infrared radiation but not hot enough to generate energy by nuclear fusion.

A protostar begins life as a contracting cloud of gas and dust falling inward under the influence of gravity (❙ Figure 11-3). As the cloud contracts, it develops a higher-density region at the center and a low-density envelope. Mass continues to flow inward from the outer parts of the cloud. That is, the cloud contracts from the inside out, with the protostar taking shape deep inside an enveloping cloud of cold, dusty gas. These clouds have been called **cocoons** because they hide the forming protostar from our view as it takes shape.

The contraction of the cloud cannot lead directly to the formation of a small protostar, because the cloud has some net rotation. Everything rotates one way or another. The rotation of the cloud may be undetectable at first, but its rotation must grow more and more pronounced as it contracts and conserves its angular momentum. The rapidly spinning core of the cloud must

FIGURE 11-3

A contracting protostar begins as a very cold, very faint fragment of a gas cloud as shown in this H–R diagram extended to low temperatures. Contraction heats the protostar, but it remains hidden inside its dust cloud. The behavior of dust in the cloud makes the exact behavior of a protostar difficult to model as it approaches the main sequence.

Formation of a Protostellar Disk

A slowly rotating cloud of gas begins to contract.

Conservation of angular momemtum spins the cloud faster and it flattens...

into a growing protostar at the center of a rotating disk of gas and dust.

FIGURE 11-4
The rotation of a contracting gas cloud forces it to flatten into a disk, and the protostar grows at the center.

flatten into a spinning disk like a blob of pizza dough spun into the air. Gas that has lost its angular momentum through collisions can sink directly to the center of the cloud, where the protostar begins to grow, surrounded by the disk. As more gas falls inward, it passes through the disk, giving up much of its angular momentum before it sinks into the protostar (❚ Figure 11-4).

Basic physics predicts that protostars should form at the centers of spinning disks of gas, which are called **protostellar disks.** That is important to us because astronomers believe that planets form within these disks. Our Earth formed in a disk around the protosun 4.6 billion years ago. As we will see in Chapter 19, the evidence is very strong that planetary systems form in these protostellar disks.

If we could see a protostar without its cocoon, it would be a very luminous and cool red star in the upper right part of the H–R diagram. The dust cocoon, however, absorbs almost all of the visible radiation and, growing warm, reradiates the energy as infrared radiation. Thus we should not expect to see protostars at visible wavelengths, but they should be emitting infrared radiation.

When the protostar becomes hot enough, it can drive away the surrounding gas and dust and become

visible. The location in the H–R diagram where protostars first emerge from their cocoons and become visible is called the **birth line.** Once a star crosses the birth line, it continues to contract toward the main sequence with a speed that depends on its mass. More massive stars have stronger gravity and contract more rapidly (❚ Figure 11-5). The sun took about 30 million years to reach the main sequence, but a 30-solar-mass star takes only 30,000 years. A 0.2-solar-mass star needs about 1 billion years to contract from a gas cloud to the main sequence.

The theory of star formation takes us into an unearthly realm filled with unfamiliar processes and objects. We might think it was nothing more than a fairy tale if observations did not support the theory.

EVIDENCE OF STAR FORMATION

In astronomy, evidence means observations. We must ask what observations confirm our theories of star formation. Unfortunately, a protostar is not easy to observe. The protostar stage is less than 0.1 percent of a star's lifetime, so although that is a long time in human terms, we cannot expect to find many stars in the protostar stage. Furthermore, protostars form deep inside clouds of dusty gas that absorb any light the protostar might emit. We must depend on observations at infrared wavelengths to search for hidden protostars.

Study ❚ "Observational Evidence of Star Formation" on pages 220 and 221 and notice three important points.

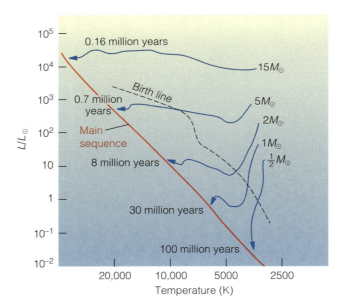

FIGURE 11-5
The more massive a protostar is, the faster it contracts. A 1-$M_\odot$ star requires 30 million years to reach the main sequence. (Recall that $M_\odot$ means "solar mass.") The dashed line is the birth line, where contracting protostars first become visible as they dissipate their surrounding clouds of gas and dust. Compare with Figure 11-3, which shows the evolution of a protostar of about 1-$M_\odot$ as a dashed line up to the birth line and as a solid line from the birth line to the main sequence. (*Illustration design by author*)

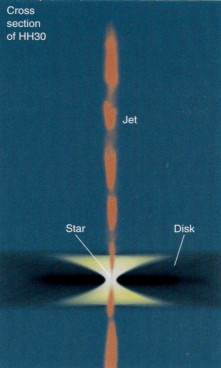

Cross section of HH30

Jet

Star   Disk

FIGURE 11-6

In the object HH30, a newly formed star lies at the center of a dense disk of dusty gas that is narrow near the star and thicker farther away. Although the star is hidden from us by the edge-on dusty disk, the star illuminates the inner surface of the disk. Interactions between the infalling material in the disk and the spinning star eject jets of gas along the axis of rotation. *(C. Burrows, STScI & ESA, WFPC 2 Investigation Definition Team, NASA)*

0.75 to 3 solar masses, but the stars in **O associations,** extended groups of O stars, are more massive. It is not clear why some gas clouds give birth to compact star clusters held together by their own gravity and others give birth to larger associations not bound together by gravity.

Not only can we locate evidence of star formation, but we can find evidence that star formation can stimulate more star formation. If a gas cloud produces massive stars, those massive stars can ionize the gas nearby and drive it away. Where that gas pushes into surrounding gas, it can compress the gas and trigger more star formation (Figure 11-7). Another way massive stars can trigger star formation is through the explosion of a massive star in a supernova, a subject we will explore in detail in Chapter 13. Such an explosion can drive a shock wave through surrounding gas and trigger more star formation. Like a grass fire spreading through the interstellar medium, star formation can reignite itself if it creates massive stars (Figure 11-8). Of course, lower-mass stars also form from such star formation, but they are not hot enough, nor do they explode as supernovae, so they alone cannot trigger further star formation.

First, we can be sure that star formation is going on right now because we can find regions containing stars so young they must have formed recently. They lie between the birth line and the main sequence. Second, notice that regions of star formation are rich in gas and dust and contain infrared protostars and stars still contracting toward the main sequence. The third thing to notice is how observations give us insight into the process by which stars form. For example, the observation of jets coming from hidden protostars alerts us that protostars form surrounded by disks of gas and dust.

In some cases, these dark disks of gas and dust are clearly visible around newborn stars (Figure 11-6). Such disks are exciting not only because they are evidence of star formation, but because they are the regions where planets form. We will discuss these disks later in this chapter and again when we discuss the origin of our solar system.

An **association** is a widely distributed star cluster that is not held together by its own gravity—its stars wander away as the association ages. These associations consist of young stars, because the stars wander apart so quickly. The constellation Orion, a known region of star formation, is filled with T Tauri stars in a **T association.** T Tauri stars are relatively low-mass objects ranging from

Star-formation pillar

Star-formation pillar

Star-formation pillar

UV + visual + IR image

FIGURE 11-7

The hot, massive stars in the star cluster 30 Doradus are pouring out intense ultraviolet radiation and powerful winds of hot gas that are pushing back and compressing the surrounding nebula. Slightly denser regions in the nebula protect the gas behind them to form pillars like those in the Eagle Nebula that point back at the star cluster. These pillars are each a few light-years long. The dense gas at the tips of the pillars is being compressed and may form more stars within the next few million years. *(N. Walborn and J. Maiz-Apellániz, STScI, R. Barbá, La Plata Observatory, and NASA)*

Observational Evidence of Star Formation

At visible, infrared, and radio wavelengths, astronomers find many objects in the sky that tell us how stars form.

The nebula around the star S Monocerotis is bright with hot stars. Such stars live short lives of only a few million years, so they must have formed recently. Such regions of young stars are common. The entire constellation of Orion is filled with young stars and clouds of gas and dust.

Nebulae containing young stars usually contain **T Tauri** stars. These stars fluctuate irregularly in brightness, and many are bright in the infrared, suggesting they are surrounded by dust clouds and in some cases by dust disks. Doppler shifts show that gas is flowing away from many T Tauri stars. Located near the birth line in the H–R diagram, the T Tauri stars appear to be newborn stars just blowing away their dust cocoons. T Tauri stars appear to have ages ranging from 100,000 years to 100,000,000 years.

Spectra of T Tauri stars show signs of an active chromosphere as we might expect from young, rapidly rotating stars with powerful dynamos and strong magnetic fields.

Visual-wavelength image

AATB

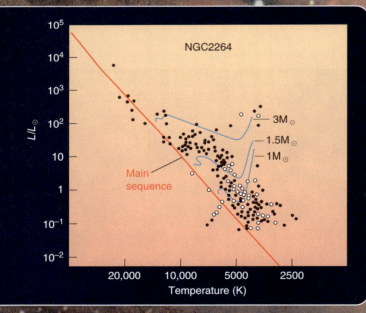

The star cluster NGC2264, imbedded in the nebula on this page, is only a few million years old. Lower-mass stars have not yet reached the main sequence, and the cluster contains many T Tauri stars (open circles), which are found above and to the right of the main sequence, near the birth line.

Bok globules, named after astronomer Bart Bok, are dense, dusty clouds seen silhouetted against glowing nebulae. Only a light-year or so in diameter, they contain from 10 to 1000 solar masses, and infrared observations show that most are very cold at their centers. Only a few have warm interiors and appear to be contracting to form protostars.

Visual

NASA

HH34S

Gas jet

Young star

Visual

Visual

HH34N

At the center of this image, a newborn star is emitting powerful jets to left and right. Where the jets strike the interstellar medium, they produce Herbig–Haro objects. The inset shows how irregular the jet is. Such jets can be over a light year long and contain gas traveling at 100 km/s or more.

Herbig–Haro object

Herbig–Haro objects, named after the two astronomers who first described them, are small nebulae that fluctuate in brightness. They appear to be produced by flickering jets from newborn stars exciting the interstellar medium.

Dusty disk

Jet

Jet

Herbig–Haro object

Matter flowing into a protostar swirls through a thick disk and, by a process believed to involve magnetic fields, ejects high-energy jets in opposite directions. Observation of these **bipolar flows** is evidence that protostars are surrounded by disks because only disks could focus the flows into jets.

The Eagle Nebula, shown at the right in a ground-based photograph, is a region of known star formation. Turn the page a quarter turn to the right to see the eagle flying with a salmon in its talons.

Visual

AURA/NOAO/NSF

EGGs

EGGs

Visual

This Hubble Space Telescope image of part of the Eagle Nebula shows that radiation from bright stars out of the field to the upper right is evaporating the dust and driving away the gas to expose small globules of denser gas and dust. Infrared observations show that about 15 percent of these objects have formed protostars, and at least one seems to contain a newborn star (arrow). In part because these objects were first found in the Eagle Nebula, astronomers have enjoyed calling them EGGs — evaporating gaseous globules. They are evidently stars exposed in the act of forming.

The study of star formation is an exciting part of modern astronomy, and astronomers are using every wavelength to search for more information about the birth of stars. Although much remains to be discovered, the general outline is clear. Gravity creates stars by forcing interstellar clouds to contract and grow hotter.

REVIEW CRITICAL INQUIRY

What evidence do we have that stars are forming right now?

First, we should note that some extremely luminous stars can't live very long, so when we see such stars, such as the hot, blue stars in Orion, we know they must have formed in the last few million years. Star formation must be a continuous process, or we would not see any of these short-lived stars. But we have more direct evidence when we look at T Tauri stars, which lie above the main sequence and are often associated with gas and dust. In fact, we see entire associations of T Tauri stars as well as other associations of short-lived O stars. Many regions of gas and dust contain objects visible in the infrared that appear to be actual protostars. Furthermore, we can find cases where jets flowing outward from protostars create Herbig–Haro objects. In a few cases, such as the Eagle Nebula, the dense cores of gas and dust that are the forerunners of star formation are exposed by the evaporation of the nebula.

There seems to be no doubt that star formation is an ongoing process, and we can even discuss how it happens. For example, what evidence do we have that protostars are often surrounded by disks of gas and dust?

FIGURE 11-8

Generations of stars appear in the two star clusters in this Hubble Space Telescope image mosaic. The large cluster of yellow stars is about 50 million years old, but the bright white stars scattered through the image are massive young stars in a cluster only about 4 million years old. The younger cluster lies 200 ly beyond the older cluster, suggesting that star formation began 50 million years ago and created the nearer cluster. Supernova explosions in that cluster could have compressed gas clouds nearby and triggered the formation of the slightly more distant younger cluster. *(R. Gilmozzi, STScI/ESA, Shawn Ewald, JPL, and NASA)*

Gravity pulls the gas and dust together to make new stars, but this raises a natural question. If all this matter is falling inward to make a star, why does the contracting star stop? How do contracting protostars turn themselves into stable main-sequence stars? The answer is that the contracting stars begin to generate their own energy.

11-2 THE SOURCE OF STELLAR ENERGY

Stars are born when gravity pulls matter together, but what stops the contraction? Contracting stars somehow begin to make large amounts of energy, and somehow that energy stops the contraction and stabilizes the stars. In this section, we see how stars make energy; in the next section, we will see how contracting stars reach stability. This story will lead our imaginations into a region where our bodies can never go—the heart of a star.

A REVIEW OF THE PROTON–PROTON CHAIN

In Chapter 8 we visited the center of the sun and discovered that it manufactured energy through hydrogen fusion using a series of nuclear reactions called the proton–proton chain. The reaction must begin with the fusion of two protons. Protons are the nuclei of hydrogen atoms and have positive charges and repel each other with a force called the Coulomb barrier. Consequently, the proton–proton chain cannot occur if the gas temperature is lower than about 10 million Kelvin. High-velocity collisions are required to overcome the Coulomb barrier, and high velocity means high temperature.

We also discovered that the gas must be dense if the proton–proton chain is to produce significant energy. The fusion of two protons is unlikely, so many collisions are necessary to produce a few fusion reactions. Also, a single cycle through the proton–proton chain produces only a tiny amount of energy, so a vast number of fusion reactions are needed to supply the energy the sun needs. There will be a large number of fusion reactions only if the gas density is high.

Our study of the sun revealed that the proton–proton chain can produce energy only near the sun's center, where the temperature and density are high. In fact, only about 30 percent of the sun's mass is actually hot enough to fuse.

We might expect other stars to fuse hydrogen the same way the sun does, and we would be right for most stars. Some stars, however, can fuse hydrogen using a different recipe, and that makes a big difference to the structure of those stars.

THE CNO CYCLE

Main-sequence stars more massive than 1.1 solar mass fuse hydrogen into helium using the **CNO cycle,** a hydrogen fusion process that uses carbon, nitrogen, and oxygen as steppingstones.

Look carefully at ▍Figure 11-9 and notice the steps in the CNO cycle. The cycle begins with a carbon-12 nucleus absorbing a proton and becoming nitrogen-13, which decays to become carbon-13. The carbon-13 nucleus absorbs a second proton and becomes nitrogen-14, which absorbs a third proton and becomes oxygen-15. The oxygen-15 decays to become nitrogen-15, which absorbs a fourth proton, ejects a helium nucleus, and becomes carbon-12. Notice that carbon-12 begins the cycle and ends the cycle, and, along the way, four protons are combined to make a helium nucleus. This CNO cycle has the same outcome as the proton–proton chain, but it is different in an important way.

The CNO cycle begins with a carbon nucleus combining with a proton, a bare hydrogen nucleus. Because a carbon nucleus has a positive charge six times higher than a proton, the Coulomb barrier is high, and temperatures higher than 16 million Kelvin are required to make the CNO cycle work.

The difference between the proton–proton chain and the CNO cycle is their dependence on temperature. The CNO cycle is so critically sensitive to temperature that it produces almost no energy at all at temperatures below 16 million Kelvin. The proton–proton chain can make energy at these lower temperatures, so the sun makes nearly all of its energy using the proton–proton chain. At slightly higher temperatures, the proton–proton chain makes slightly more energy, but the CNO cycle produces a flood of energy. Thus stars with higher-temperature cores make nearly all of their energy using the CNO cycle, while stars with cooler cores make their energy using the proton–proton chain.

The CNO cycle and the proton–proton chain both fuse hydrogen to make helium, but their sensitivity to temperature is dramatic. We will see later in this chapter how this affects the internal structure of the stars.

$$^{12}C + {}^{1}H \rightarrow {}^{13}N + \gamma$$
$$^{13}N \rightarrow {}^{13}C + e^{+} + \nu$$
$$^{13}C + {}^{1}H \rightarrow {}^{14}N + \gamma$$
$$^{14}N + {}^{1}H \rightarrow {}^{15}O + \gamma$$
$$^{15}O \rightarrow {}^{15}N + e^{+} + \nu$$
$$^{15}N + {}^{1}H \rightarrow {}^{4}He + {}^{12}C$$

$\sim\sim\sim$ γ Gamma ray
$----$ ν Neutrino
⬤ Positron

FIGURE 11-9

The CNO cycle uses ^{12}C as a catalyst to combine four hydrogen nuclei (^{1}H) to make one helium nucleus (^{4}He) plus energy. The carbon nucleus reappears at the end of the process, ready to start the cycle over.

---| **REVIEW** CRITICAL INQUIRY |---

Why does the CNO cycle require higher temperatures than the proton–proton cycle?

Nuclear fusion happens when atomic nuclei collide and fuse to form a new nucleus. Inside a star, the gas is ionized, so the nuclei have no orbiting electrons, and thus the nuclei have positive charges. Two objects with positive charges will repel each other in what we have called the Coulomb barrier. To start fusion, the nuclei must collide at high enough velocity to overcome that barrier. Heavier atomic nuclei such as carbon have higher positive charges, so the Coulomb barrier is higher. That requires that the gas be hotter so the collisions will be violent enough to overcome the barrier.

The CNO cycle and the proton–proton chain are strikingly similar in some ways but different in other ways. Describe these similarities and differences.

Because the law of gravity and the rules of nuclear fusion determine how stars work, we can understand what the inside of a star is like. That is, we can describe the internal structure of a star.

11-3 STELLAR STRUCTURE

Gravity makes stars contract, and when the density and temperature at the centers of the stars are high enough, nuclear fusion begins making energy. Why do the stars stop contracting? Somehow, the energy generated at the center of the star flows outward to the surface, and that flow of energy stops the contraction. To understand the stable structure of a star, we must first understand how energy flows through the star and how that energy flow affects the entire star.

ENERGY TRANSPORT

The sun and other stars generate nuclear energy in their deep interiors, and that energy must flow outward to their surfaces to replace the energy radiated into space as light and heat. If this outward flow of energy in the sun were suddenly shut off, the sun's surface would gradually cool and dim, and the sun would begin to shrink. Thus, stars can exist only as long as energy can move from their cores to their surfaces. In the material of which stars are made, energy can move by conduction, radiation, or convection.

Conduction is the most familiar form of heat flow. If you hold the bowl of a spoon in a candle flame, the handle of the spoon grows warmer. Heat, in the form of the motion of the particles in the metal, is conducted from particle to particle up the handle, until the particles under your fingers begin to move faster and you sense heat (▮ Figure 11-10). Thus, conduction requires close contact between the particles. Because matter in most stars is gaseous, conduction is an unimportant means of energy flow. Conduction is significant only in peculiar stars that have tremendous internal densities.

The transport of energy by radiation is another familiar experience. Put your hand beside a candle flame, and you can feel the heat. What you actually feel are infrared photons radiated by the flame (Figure 11-10). Because photons are packets of energy, your hand grows warm as it absorbs them. Radiation is the principal means of energy transport in the sun's interior. Photons are absorbed and reemitted in random directions over and over as they work their way outward. The process of absorption and reemission breaks the high-energy photons common near the sun's center into large numbers of low-energy photons in the cooler outer layers. The energy of a single high-energy photon at the center of the sun takes about 1 million years to reach the sun's surface, where it emerges as roughly 2500 photons of visible light.

The flow of energy by radiation depends on how difficult it is for the photons to move through the gas. If the gas is cool and dense, the photons are more likely to be absorbed or scattered, and thus the radiation does not penetrate the gas very well. We call such a gas opaque. In a hotter, lower-density gas, the photons can

ACTIVE FIGURE 11-10
The three modes by which energy may be transported from the flame of a candle, as shown here, are the three modes of energy transport within a star.

Ace⊛Astronomy™ Go to AceAstronomy and click Active Figures to see "Conduction, Convection, and Radiation." Take control of this diagram.

penetrate more easily; such a gas is less opaque. The **opacity** of a gas—its resistance to the flow of radiation—depends strongly on its temperature.

If the opacity of a gas is high, radiation cannot flow through it easily. Energy moving outward from the center of the sun first moves through the hotter gas of the deep interior. Because the gas is so hot, it is transparent, and the energy moves as radiation. But the outer portions of the sun are cooler and therefore more opaque. Like water behind a dam, energy builds up, raising the temperature until the gas begins to churn. Hot gas, being less dense, rises, and cool gas, being denser, sinks. This is convection, the third way energy can move in a star.

Convection is a common experience; the wisp of smoke rising above a candle flame travels upward in a small convection current (Figure 11-10). If you hold your hand above the flame, you can feel the rising current of hot gas. In stars, energy may be carried upward by rising currents of hot gas hundreds or thousands of miles in diameter.

Convection is important in stars because it both carries energy and mixes the gas. Convection currents flowing through the layers of a star tend to homogenize the gas, giving it a uniform composition throughout the convective zone. As you might expect, this mixing affects the fuel supply of the nuclear reactions, just as the stirring of a campfire makes it burn more efficiently.

The convection in the sun stirs a zone just below the visible surface (▮ Figure 11-11). We see the tops of these rising currents of hot gas as granulation in the solar photosphere (Figure 8-2). That is, we can see the upper layers of the solar convection zone. Below that, the energy moves as radiation, and the gas does not get mixed. Deep in the interior of the sun, there is no convection. Hydrogen nuclei are fusing to form helium, and nothing removes the helium ashes or brings fresh

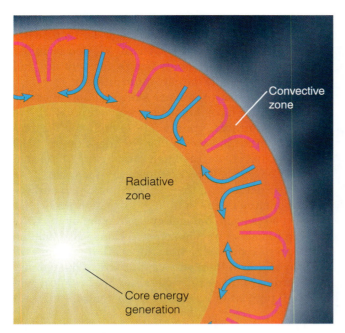

ACTIVE FIGURE 11-11

A cross section of the sun. Near the center, nuclear fusion reactions generate high temperatures. Energy flows outward through the radiation zone as photons. In the cooler, more opaque outer layers, the energy is carried outward by rising convection currents of hot gas (red) and sinking currents of cooler gas (blue).

Ace ◈ Astronomy™ Go to AceAstronomy and click Active Figures to see "Inside the Sun." Watch energy flow outward from the core to the surface.

hydrogen down into the core. Thus, the sun is like a great pot of mashed potatoes burning at the bottom and stirred only slightly at the top. We will see later that most stars have interiors like the sun's.

WHAT SUPPORTS THE SUN?

From its surface to its interior, the sun is gaseous. On Earth, a puff of gas, vapor from a smokestack perhaps, dissipates rapidly, driven by the motion of the air and by the random motions of the atoms in the gas. However, the sun differs from a puff of gas in a very important characteristic—its mass. The sun is over 300,000 times more massive than Earth, and that mass produces a tremendous gravitational field that draws the sun's gases into a sphere.

With such a strong gravitational field, it might seem as if the gases of the sun should compress into a tiny ball, but a second force balances gravity and prevents the sun from shrinking. The gas of which the sun is made is quite hot; and, because it is hot, it has a high pressure. The pressure pushes outward; indeed, without the restraint of the sun's gravity, it would blow the sun apart. Thus, the sun is balanced between two forces—gravity trying to squeeze it tighter and gas pressure trying to make it expand.

To discuss the forces inside the sun, we can imagine that the sun's interior is divided into concentric shells like those in an onion (❙ Figure 11-12). We can then discuss the temperature, density, pressure, and so on in each shell. Keep in mind, however, that these helpful shells do not really exist. The sun and stars are not composed of separable layers. The layers are only a convenience in our discussion.

The gravity–pressure balance that supports the sun is a fundamental part of stellar structure known as the law of **hydrostatic equilibrium.** It says that, in a stable star like the sun, the weight of the material pressing downward on a layer must be balanced by the pressure of the gas in that layer. *Hydro* implies that we are discussing a fluid—the gases of the star. *Static* implies that the fluid is stable—neither expanding nor contracting.

The law of hydrostatic equilibrium can prove to us that the interior of the sun must be very hot. Near the sun's surface, there is little weight pressing down on the gas, so the pressure must be low, implying a low temperature. But as we go deeper into the sun, the weight becomes larger, so the pressure, and therefore the temperature, must also increase (❙ Figure 11-13). Near the sun's surface, the pressure is only a few times Earth's atmospheric pressure, and the temperature is only about 5800 K. At the center of the sun, the weight pressing down equals a pressure of about 3×10^{14} times Earth's atmospheric pressure. To support that weight, the gas temperature has to be almost 15 million K.

Of course, the interior of the sun is kept hot by the nuclear reactions occurring at the core, and the outward flow of energy keeps each layer in the sun hot enough to support the weight pressing down from above. In a sense, the sun is supported by the flow of energy from its center to its surface. Turn off that energy, and

FIGURE 11-12

To discuss the structure of a star, it is helpful to divide its interior into concentric shells much like the layers in an onion. This model is, of course, only an aid to our imaginations. Stars are not really divided into separable layers.

Surface

Center

Weight

Pressure

FIGURE 11-13
The law of hydrostatic equilibrium says the pressure in each layer must balance the weight on that layer. As a result, pressure and temperature must increase from the surface of a star to its center.

gravity would gradually force the sun to collapse into its center.

INSIDE STARS

In Chapter 9, we discovered that the stars on the main sequence are ordered according to mass, with the upper-main-sequence stars being most massive and the lower-main-sequence stars being least massive. Combining this with what we know about hydrogen fusion tells us that there are two kinds of main-sequence stars: upper-main-sequence stars, which fuse hydrogen on the CNO cycle, and lower-main-sequence stars, which fuse hydrogen on the proton–proton chain. Viewed from the outside, these stars differ only in size, temperature, and luminosity, but inside they are quite different.

The upper-main-sequence stars are more massive and thus must have higher central temperatures to withstand their own gravity. These high central temperatures permit the star to fuse hydrogen on the CNO cycle, and that affects the internal structure of the star.

The CNO cycle is very temperature-sensitive. If the central temperature of the sun rose by 10 percent, energy production by the proton–proton chain would rise by about 46 percent, but energy production by the CNO cycle would shoot up 350 percent. This means that the more massive stars generate almost all of their energy in a tiny region at their very centers where the temperature is highest. A 10-solar-mass star, for instance, generates 50 percent of its energy in its central 2 percent of mass.

This concentration of energy production at the very center of the star causes a "traffic jam" as the energy tries to flow away from the center. Transport of energy by radiation can't drain away the energy fast enough, and the central core of the star churns in convection as hot gas rises upward and cooler gas sinks downward. Farther from the center, the traffic jam is less severe, and the energy can flow outward as radiation. Thus, massive stars have convective cores at their centers and radiative envelopes extending from their cores to their surfaces (❚ Figure 11-14).

Main-sequence stars less massive than about 1.1 solar masses cannot get hot enough to fuse much hydrogen on the CNO cycle. They generate nearly all of their energy by the proton–proton chain, which is not as sensitive to temperature, and thus the energy generation occurs in a larger region in the star's core. The sun, for example, generates 50 percent of its energy in a region that contains 11 percent of its mass. Because the energy generation is not concentrated at the very center of the star, no traffic jam develops, and the energy flows outward as radiation. Only near the surface, where the gas is cooler and therefore more opaque, does convection stir the gas. Consequently, the less massive stars on the main sequence have radiative cores and convective envelopes.

The lowest-mass stars have a slightly different kind of structure. For stars less than about 0.4 solar mass, the gas is relatively cool compared to the inside of more massive stars, and the radiation cannot flow outward easily. Thus, the entire bulk of these low-mass stars is stirred by convection.

We have answered the questions of how the energy in stars is made and how it gets from the core to the surface. We are now ready to answer the question: What makes stars stable?

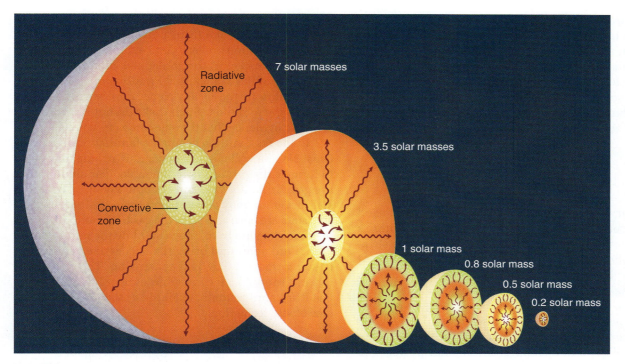

FIGURE 11-14

Inside stars. The more massive stars have small convective cores and radiative envelopes. Stars like the sun have radiative cores and convective envelopes. The lowest-mass stars are convective throughout. *(Illustration design by author)*

THE PRESSURE–TEMPERATURE THERMOSTAT

We can now understand what stops a newborn star from contracting into a tiny ball. Once nuclear fusion begins, energy flows outward, heats the layers of gas, and stops the contraction. This raises an interesting question. How does the star manage to make just the right amount of energy? The answer involves the relationship between pressure and temperature.

In a star, the nuclear reactions generate just enough energy to balance the inward pull of gravity. Consider what would happen if the reactions began to produce too much energy. The star balances gravity by generating energy, so the extra energy flowing out of the star would force it to expand. The expansion would lower the central temperature and density and slow the nuclear reactions until the star regained stability. Thus the star has a built-in regulator that keeps the nuclear reactions from occurring too rapidly.

The same thermostat keeps the reactions from dying down. Suppose the nuclear reactions began making too little energy. Then the star would contract slightly, increasing the central temperature and density and increasing the nuclear energy generation.

The stability of a star depends on this relation between pressure and temperature. If an increase or decrease in temperature produces a corresponding change in pressure, the thermostat functions correctly, and the star is stable. We will see in the next chapter how the thermostat accounts for the mass–luminosity relation. In Chapter 13, we will see what happens to a star when the thermostat breaks down completely and the nuclear fires burn unregulated.

REVIEW **CRITICAL INQUIRY**

What would happen if the sun stopped generating energy?
Stars are supported by the outward flow of energy generated by nuclear fusion in their interiors. That energy keeps each layer of the star just hot enough for the gas pressure to support the weight of the layers above. Each layer in the star must be in hydrostatic equilibrium; that is, the inward weight must be balanced by outward pressure. If the sun stopped making energy in its interior, nothing would happen at first, but over many thousands of years the loss of energy from its surface would reduce the sun's ability to withstand its own gravity, and it would begin to contract. We wouldn't notice much for 100,000 years or so, but eventually the sun would lose its battle with gravity.

Stars are elegant in their simplicity. Nothing more than a cloud of gas held together by gravity and warmed by nuclear fusion, a star can achieve stability by balancing its weight through the generation of nuclear energy. But how does the star manage to make exactly the right amount of energy to support its weight?

We have traced the birth of stars from the first instability in the interstellar medium to the final equilibrium of nuclear fusion and gravity. We have constructed a complete story of star formation, but we should demand evidence to support our theories. One of the best places to search for evidence of star formation is in the Great Nebula in Orion. It is just part of a great storm of star birth sweeping through Orion.

11-4 THE ORION NEBULA

On a clear winter night, you can see with your naked eye the Great Nebula of Orion as a fuzzy wisp in Orion's sword. With binoculars or a small telescope it is striking, and through a large telescope it is breathtaking. At the center lie four brilliant blue-white stars known as the Trapezium, the brightest of a cluster of a few hundred stars. Surrounding the stars are the glowing filaments of a nebula more than 8 pc across. Like a great thundercloud illuminated from within, the churning currents of gas and dust suggest immense power. The significance of the Orion Nebula lies hidden, figuratively and literally, beyond the visible nebula. The region is ripe with star formation.

EVIDENCE OF YOUNG STARS

We should not be surprised to find star formation in Orion. The constellation is a brilliant landmark in the winter sky because it is marked by hot, blue stars. These stars are bright in our sky, not because they are nearby, but because they are tremendously luminous. These O and B stars cannot live more than a few million years, so we can conclude that they must have been born recently. Furthermore, the constellation contains large numbers of T Tauri stars, which are known to be young. Orion is rich with young stars.

The history of star formation in the constellation of Orion is written in its stars. The stars at Orion's west shoulder are about 12 million years old, while the stars of Orion's belt are about 8 million years old. The stars of the Trapezium at the center of the Great Nebula are no older than 2 million years. Apparently, star formation began near the west shoulder, and the massive stars that formed there triggered the formation of the stars we see in Orion's belt. That star formation may have triggered the formation of the stars we see in the Great Nebula. Like a grass fire, star formation has swept across Orion from northwest to southeast.

Study ▌ "Star Formation in the Orion Nebula" on pages 230 and 231 and notice four points. First, the nebula we see is only a small part of a vast, dusty molecular cloud. We see the nebula because the stars born within it have ionized the gas and driven it outward, breaking out of the molecular cloud. Also notice that a single very hot star is almost entirely responsible for ionizing the gas. The third point to notice is that infrared observations reveal clear evidence of active star formation deeper in the molecular cloud just to the northwest of the Trapezium. Finally, notice that many stars visible in the Orion Nebula are surrounded by disks of gas and dust. Such disks do not last long and are clear evidence that the stars are very young.

In the next million years, the familiar outline of the Great Nebula will change, and a new nebula may begin to form as the protostars in the molecular cloud ionize the gas, drive it away, and become visible. Centers of star formation may develop and then dissipate as massive stars are born and force the gas to expand. If enough massive stars are born, they can blow the entire molecular cloud apart and bring the successive generations of star formation to a final conclusion. The Great Nebula in Orion and its invisible molecular cloud are a beautiful and dramatic example of the continuing cycles of star formation.

| REVIEW CRITICAL INQUIRY |

What did Orion look like to the ancient Egyptians, to the first humans, and to the dinosaurs?

The Egyptian civilization had its beginning only a few thousand years ago, and that is not very long in terms of the history of Orion. The stars we see in the constellation are hot and young, but they are a few million years old, so the Egyptians saw the same constellation we see. (They called it Osiris.) Even the Orion Nebula hasn't changed very much in a few thousand years, and Egyptians may have admired it in the dark skies along the Nile.

Our oldest human ancestors lived about 3 million years ago, and that was about the time that the youngest stars in Orion were forming. Thus, they may have looked up and seen some of the stars we see, but some stars have formed since that time. Also, the Great Nebula is excited by the Trapezium stars, and they are not more than a few million years old, so our early ancestors probably didn't see the Great Nebula.

The dinosaurs saw something quite different. The last of the dinosaurs died about 65 million years ago, long before the birth of the brightest stars in the constellation we see. The dinosaurs, had they had the brains to appreciate the view, might have seen bright stars along the Milky Way, but they didn't see Orion. All of the stars in the sky are moving through space, and the sun is orbiting the center of our galaxy. Over many millions of years, the stars move appreciable distances across the sky. The night sky above the dinosaurs contained totally different star patterns.

The Orion Nebula is the product of a giant molecular cloud, but such a cloud can't continue spawning new stars forever. What processes limit star formation in a molecular cloud?

The ancient Aztecs of central Mexico told the story of how the stars, known as the Four Hundred Southerners, were scattered across the sky when they lost a cosmic battle with their brother, the great war god Huitzilopochtli. Modern astronomy tells a less colorful story of the birth of stars, but the modern story is supported by evidence and leads us to ask further questions. If stars are born, then how do they die? We will begin that story in the next chapter.

SUMMARY

The largest and densest clouds in the interstellar medium are the giant molecular clouds. Triggered by the compression of a passing shock wave, the densest regions of such a cloud may begin to contract under the pull of their own gravity and eventually fragment to form a cluster or association of stars.

A contracting protostar begins as a very large, cool object, but the acceleration of the atoms as they fall inward causes the temperature of the gas to rise. Rotation in the gas causes some of the material to settle into a rapidly rotating disk around the growing protostar. Although the protostar is hot and luminous, it is hidden inside the dusty remains of the cloud from which it formed and thus remains invisible until the cloud is driven away by the increasing luminosity of the protostar.

Protostars can be detected at infrared wavelengths because the longer-infrared-wavelength photons can escape from the dusty clouds. Bipolar flows are ejected along the axis of rotation of the protostar and its disk. Where these jets strike the surrounding gas, they can excite small nebulae called Herbig–Haro objects. T Tauri stars appear to be very young stars just shedding their cocoons of gas and dust. Very young star clusters contain T Tauri stars, and some are born in very large, loosely bound groups called T associations.

As a protostar's center grows hot enough to fuse hydrogen into helium, the star settles onto the main sequence to begin its long, stable, hydrogen-fusion life. The sun is typical of these stars. It generates energy in its interior through the proton–proton chain, and that energy flows outward from the core as radiation and, in the outermost layers, as convection. All lower-mass stars such as the sun generate energy by the proton–proton chain, but more massive stars have hotter interiors and use the CNO cycle.

Because of the temperature sensitivity of the CNO cycle, most of the energy generated by stars more massive than about 1.1 solar masses is produced very near the star's center. The resulting traffic jam as this energy flows outward causes the core to be convective, whereas the outer part of the star transports energy by radiation. In stars less massive than about 1.1 solar masses, the energy is generated on the proton—proton chain in a larger volume of the core. Thus, there is no traffic jam, and these stars have radiative cores and convective envelopes.

In all stable stars, the nuclear reactions are regulated by the pressure–temperature thermostat. Thus, stars cannot generate less energy or more energy than that needed to support them against their own gravity.

The Orion Nebula is an example of a star-forming region. The bright stars we now see are ionizing part of the gas and forcing it to expand, and this gas is pushing into cooler regions of the cloud. Radio and infrared observations show that the compressed regions of the gas cloud are giving birth to a new generation of stars.

NEW TERMS

shock wave	bipolar flow
free-fall contraction	association
protostar	T association
cocoon	O association
protostellar disk	CNO (carbon–nitrogen– oxygen) cycle
birth line	opacity
T Tauri star	hydrostatic equilibrium
Bok globule	
Herbig–Haro object	

REVIEW QUESTIONS

Ace ◓Astronomy™ Assess your understanding of this chapter's topics with additional quizzing and animations at **http:// astronomy.brookscole.com/seeds8e**

1. What factors resist the contraction of a cloud of interstellar matter?

2. Explain four different ways a giant molecular cloud can be triggered to contract.

3. What evidence do we have that (a) star formation is a continuing process? (b) protostars really exist? (c) the Orion region is actively forming stars?

4. How does a contracting protostar convert gravitational energy into thermal energy?

5. How does the geometry of bipolar flows and Herbig–Haro objects support our hypothesis that protostars are surrounded by rotating disks?

6. How does the CNO cycle differ from the proton–proton chain? How is it similar?

7. How does the extreme temperature sensitivity of the CNO cycle affect the structure of stars?

8. How does energy get from the core of a star, where it is generated, to the surface, where it is radiated into space?

9. Describe the principle of hydrostatic equilibrium as it relates to the internal structure of a star.

10. How does the pressure–temperature thermostat control the nuclear reactions inside stars?

Star Formation in the Orion Nebula

The visible Orion Nebula shown below is a pocket of ionized gas on the near side of a vast, dusty molecular cloud that fills much of the southern part of the constellation Orion. The molecular cloud can be mapped by radio telescopes. To scale, the cloud would be many times larger than this page. As the stars of the Trapezium were born in the cloud, their radiation has ionized the gas and pushed it away. Where the expanding nebula pushes into the larger molecular cloud, it is compressing the gas and may be triggering the formation of the protostars that can be detected at infrared wavelengths within the molecular cloud.

Side view of Orion Nebula

Hot Trapezium stars

Protostars

To Earth

Expanding ionized hydrogen

Molecular cloud

Daniel Good

A few hundred stars lie within the nebula, but only the four brightest, those in the Trapezium, are easy to see with a small telescope. A fifth star, at the narrow end of the Trapezium, may be visible on nights of good seeing.

The cluster of stars in the nebula is less than 2 million years old. This must mean the nebula is similarly young.

Trapezium

Visual-wavelength image

The near-infrared image below reveals about 50 low-mass, very cool stars that must have formed recently.

Infrared

NASA

X-ray

NASA/CXC/SAO

Roughly 1000 young stars with hot chromospheres appear in this X-ray image of the Orion Nebula.

Photons with enough energy to ionize H

Energy radiated by O6 star

Energy radiated by B1 star

Energy

0   100   200   300
Wavelength (nanometers)

Of all the stars in the Orion Nebula, only one is hot enough to ionize the gas. Only photons with wavelengths shorter than 91.2 nm can ionize hydrogen. The second-hottest stars in the nebula are B1 stars, and they emit little of this ionizing radiation. The hottest star, however, is an O6 star 30 times the mass of the sun. At a temperature of 40,000 K, it emits plenty of photons with wavelengths short enough to ionize hydrogen. Remove that one star, and the nebula would turn off its emission.

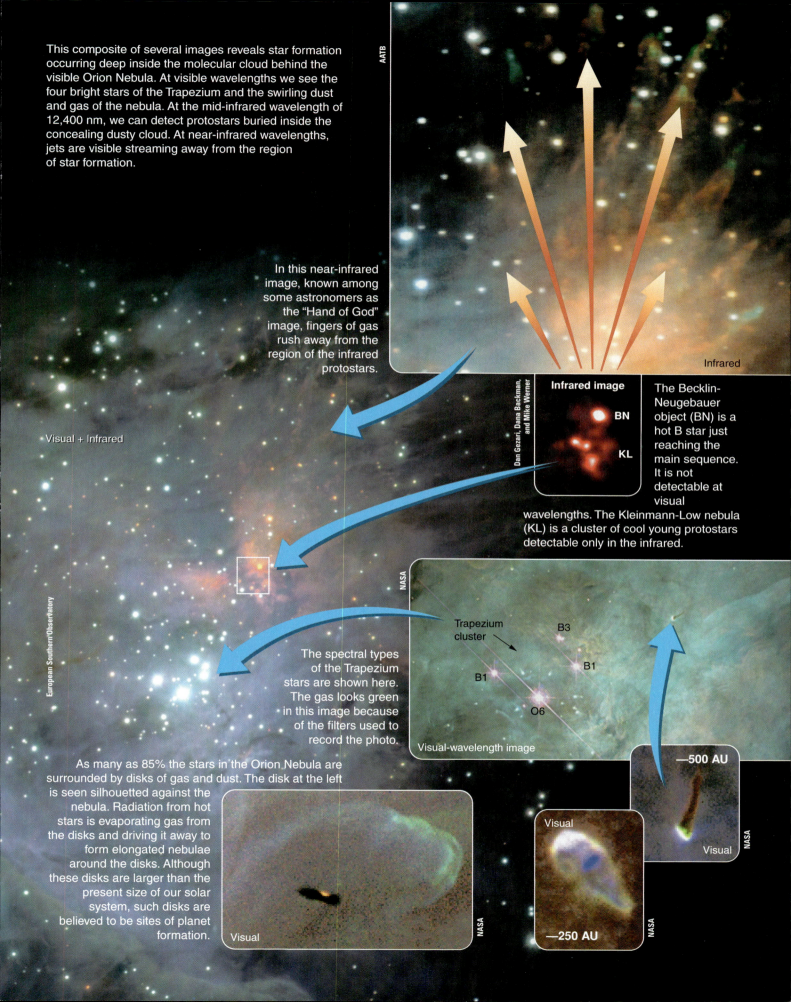

This composite of several images reveals star formation occurring deep inside the molecular cloud behind the visible Orion Nebula. At visible wavelengths we see the four bright stars of the Trapezium and the swirling dust and gas of the nebula. At the mid-infrared wavelength of 12,400 nm, we can detect protostars buried inside the concealing dusty cloud. At near-infrared wavelengths, jets are visible streaming away from the region of star formation.

AATB

Infrared

In this near-infrared image, known among some astronomers as the "Hand of God" image, fingers of gas rush away from the region of the infrared protostars.

Visual + Infrared

European Southern Observatory

Dan Gezari, Dana Backman, and Mike Werner

Infrared image

BN

KL

The Becklin-Neugebauer object (BN) is a hot B star just reaching the main sequence. It is not detectable at visual wavelengths. The Kleinmann-Low nebula (KL) is a cluster of cool young protostars detectable only in the infrared.

NASA

Trapezium cluster

B3

B1

B1

O6

The spectral types of the Trapezium stars are shown here. The gas looks green in this image because of the filters used to record the photo.

Visual-wavelength image

As many as 85% the stars in the Orion Nebula are surrounded by disks of gas and dust. The disk at the left is seen silhouetted against the nebula. Radiation from hot stars is evaporating gas from the disks and driving it away to form elongated nebulae around the disks. Although these disks are larger than the present size of our solar system, such disks are believed to be sites of planet formation.

Visual

—500 AU

Visual

Visual

—250 AU

NASA

NASA

NASA

DISCUSSION QUESTIONS

1. Ancient astronomers, philosophers, and poets assumed that the stars were eternal and unchanging. Is there any observation they could have made or any line of reasoning that could have led them to conclude that stars don't live forever?

2. How does hydrostatic equilibrium relate to hot-air ballooning?

PROBLEMS

1. The ring around the bright star in Figure 11-1 has a radius of about 7 seconds of arc. If the cluster is 20,000 ly from Earth, what is the radius of the ring in light-years?

2. If a giant molecular cloud is 50 pc in diameter and a shock wave can sweep through it in 2 million years, how fast is the shock wave going in kilometers per second?

3. If a giant molecular cloud has a mass of 10^{35} kg, and it converts 1 percent of its mass into stars during a single encounter with a shock wave, how many stars can it make? Assume the stars each contain 1 solar mass.

4. If a protostellar disk is 200 AU in radius, and the disk plus the forming star contain 2 solar masses, what is the orbital velocity at the outer edge of the disk in kilometers per second?

5. If a contracting protostar is five times the radius of the sun and has a temperature of only 2000 K, how luminous will it be? (*Hint:* See Chapter 9.)

6. The gas in a bipolar flow can travel as fast as 100 km/s. If the length of the jet is 1 ly, how long does it take for a blob of gas to travel from the protostar to the end of the jet?

7. If a T Tauri star is the same temperature as the sun but is ten times more luminous, what is its radius? (*Hint:* See Chapter 9.)

8. Circle all of the ^{1}H and ^{4}He nuclei in Figure 11-9 and explain how the CNO cycle can be summarized by $4\ ^1\text{H} \rightarrow\ ^4\text{He} + \text{energy}$.

9. How much energy is produced when the CNO cycle converts 1 kg of mass into energy? Is your answer different if the mass is fused by the proton–proton chain?

10. If the Orion Nebula is 8 pc in diameter and has a density of about 600 hydrogen atoms/cm^3, what is its total mass? (*Hint:* The volume of a sphere is $\frac{4}{3}\pi R^3$.)

11. The hottest star in the Orion Nebula has a surface temperature of 40,000 K. At what wavelength does it radiate the most energy? (*Hint:* See Chapter 7.)

CRITICAL INQUIRIES FOR THE WEB

1. Use the Web to supply additional details concerning the evolution of protostars and T Tauri stars.

2. Astronomers continue to study the Orion Nebula and star formation in the molecular cloud behind the nebula. What is the latest news from Orion?

3. What is BM Orionis? Where is it located, and what does it do? Why might it be interesting to observe with even a small telescope?

EXPLORING *THESKY*

1. The following nebulae are all star formation regions. What kind of nebulae are they? (*Hint:* To center on an object, use **Find** under **Edit.** Choose **Messier Objects** and pick from the list.)

   M42, M20, M8, M17

2. Locate M8 in *TheSky*, zoom in, and identify other nebulae in the region. Study the photo of NGC6559. What kind of nebula is it?

 Visit the Seeds *Foundations of Astronomy* companion Web site for critical thinking exercises, articles, and additional readings from InfoTrac College Edition, Brooks/Cole's online student library.

STELLAR EVOLUTION

*We should be unwise to trust scientific inference very far
when it becomes divorced from opportunity for observational test.*

Sir Arthur Eddington, The Internal Constitution of the Stars

NASA/CXC/SAO/HST; Jon Morse and J. Hester

Chapter 12

GUIDEPOST

This chapter is the heart of any discussion of astronomy. Previous chapters showed how astronomers make observations with telescopes and how they analyze their observations to find the luminosity, diameter, and mass of stars. All of that aims at understanding what stars are.

This is the middle of three chapters that tell the story of stars. The preceding chapter told us how stars form, and the next chapter tells us how stars die. This chapter is the heart of the story—how stars live.

As always, we accept nothing at face value. We expect theory to be supported by evidence. We expect carefully constructed models to help us understand the structure inside stars. In short, we exercise our critical faculties and analyze the story of stellar evolution rather than merely accepting it.

After this chapter, we will know how stars work, and we will be ready to study the rest of the universe, from galaxies that contain billions of stars to the planets that form around individual stars.

The stars are going out. Although stars enjoy long, stable lives on the main sequence, they suffer the first pangs of stellar age as they gradually consume their hydrogen fuel. Once hydrogen is exhausted at their centers, they swell rapidly into giant stars 10 to 1000 times the size of the sun. For a short time, a few million to a billion years, they are majestic beacons visible across thousands of parsecs.

Although we say a star swells rapidly and spends a short time as a giant, these changes happen slowly compared with a human life. How can we be sure stars really do age? Stellar evolution and the formation of giant stars might be just an astronomer's daydream.

In this chapter, we see the full interplay of theory and evidence. We use the basic laws of physics, combined with our knowledge of the nature of stars, to create a theory to describe how main-sequence stars maintain their equilibrium and how such stars change when they exhaust their nuclear fuels. Theory alone, however, is never enough. At each step, we compare these theories with the evidence. We will discover that clusters of stars can make the slow evolution of stars visible even during our short lifetimes, and we will find that some evolving stars can become unstable and pulsate like beating hearts. Those pulsating stars tell us more about how stars evolve.

From beginning to end, this chapter tells the story of the evolution of stars from main sequence to giant. We will trace the passage of stars through this phase of their existence and see how they generate their energy and why they swell so large. We begin with the main-sequence stars.

12-1 MAIN-SEQUENCE STARS

If Shakespeare were alive today, he would probably have something sarcastic to say about modern astronomers. "How do these stargazers pretend to know the hearts of the eternal stars?" he might ask. In fact, one of the greatest triumphs of modern astronomy is the discovery that the stars are not eternal and that mere humans can indeed probe their centers. We can know the conditions at the center of a star because stars are fundamentally very simple objects. We begin by considering how we can use stellar models to understand the interior of main-sequence stars; later we will use these same models to understand what happens to stars when they grow old.

STELLAR MODELS

Astronomers describe the insides of stars by imagining that the stars are divided into concentric shells as shown in Figure 11-12. Of course, stars are not made like onions, but defining these shells mathematically helps astronomers discuss the conditions at different levels inside the star—what astronomers call the "structure" of the star.

The structure of a star is described by four simple laws of physics, two of which we have already discussed. In Chapter 11, we met the law of hydrostatic equilibrium, which says that the weight pressing down on a shell in a star must be balanced by the pressure pushing outward in that shell. That is, at every level in a star, the weight of the overlying matter must be supported by the pressure. Of course, that means that there can be no empty shells; an empty gap could exert no pressure and would be unable to support the weight of the material above. Chapter 11 also introduced us to the law of energy transport, which describes how energy flows from hot to cool regions by radiation, convection, or conduction.

To these two laws we add two basic laws of nature. The **conservation of mass law** says that the total mass of the star must equal the sum of the masses in its shells. Of course, no shell can be empty, and there is no such thing as negative mass. The **conservation of energy law** says that the amount of energy flowing out the top of a shell in the star must be equal to the amount of energy coming in at the bottom of the shell plus whatever energy is generated within the shell. Of course, that means that the energy leaving the surface of the star, its luminosity, must equal the sum of the energy generated in all of the shells inside the star. This is like saying that the total number of new cars driving out of a factory must equal the sum of cars manufactured on each assembly line. No car can vanish into nothing or appear from nothing. Energy in a star may not vanish without a trace or appear out of nowhere.

The laws of stellar structure, described in general terms in ▌ Table 12-1, can be written as mathematical equations. By solving those equations in a special way, astronomers can build a mathematical model of the inside of a star.

If we wanted to build a model of a star, we would have to divide the star into about 100 concentric shells and then write down the four equations of stellar structure for each shell. We would then have 400 equations that would have 400 unknowns, namely, the temperature, density, mass, and energy flow in each shell. Solving 400 equations simultaneously is not easy, and the first such solutions, done by hand before the invention of electronic computers, took months of work. Now a properly programmed computer can solve the equations in a few seconds and print a table of numbers that represent the conditions in each shell of the star. Such a table is a **stellar model.**

TABLE 12-1
The Four Laws of Stellar Structure

1. Hydrostatic equilibrium — The weight on each layer is balanced by the pressure in that layer.
2. Energy transport — Energy moves from hot to cool by radiation, convection, or conduction.
3. Conservation of mass — Total mass equals the sum of the shell masses. No gaps are allowed.
4. Conservation of energy — Total luminosity equals the sum of the energies generated in each shell.

our model of the sun's magnetic cycle, for instance. Both kinds of models are useful, but the quantitative model gives us deeper insights into how nature works, reflecting the power of mathematics as a precise way of thinking (Window on Science 12-1).

Stellar models also let us look into a star's past and future. In fact, we can use models as time machines to follow the evolution of stars over billions of years. To look into a star's future, for instance, we use a stellar model to determine how fast the star uses its fuel in each shell. As the fuel is consumed, the chemical composition of the gas changes and the amount of energy generated declines. By calculating the rate of these changes, we can predict what the star will look like at any point in the future.

Although this sounds simple, it is actually a highly challenging problem involving nuclear and atomic physics, thermodynamics, and sophisticated computational methods. Only since the 1950s have electronic computers made the rapid calculation of stellar models possible, and the advance of astronomy since then has been heavily influenced by the use of such models to study the structure and evolution of stars. Our summary of star formation in this chapter is based on thousands of stellar models. We will continue to rely on theoretical models as we study main-sequence stars in the next section and the deaths of stars in the next chapter.

WHY THERE IS A MAIN SEQUENCE

We can now understand why there is a main sequence. The clue is the law of hydrostatic equilibrium, and we can use models of stars to analyze this clue.

There is a main sequence because the centers of contracting protostars eventually grow hot enough to

The table shown in ▌Figure 12-1 is a model of the sun. The bottom line, for radius equal to 0.00, represents the center of the sun, and the top line, for radius equal to 1.00, represents the surface. The other lines tell us the temperature and density in each shell, the mass inside each shell, and the fraction of the sun's luminosity flowing outward through the shell. We can use the table to understand the sun. The bottom line tells us the temperature at the center of the sun is about 15 million Kelvin. At such a high temperature, the gas is highly transparent, and energy flows as radiation. Nearer the surface, the temperature is lower, the gas is more opaque, and the energy is carried by convection.

Notice that stellar models are quantitative; that is, properties have specific numerical values. Earlier in this book, we examined models that were qualitative—

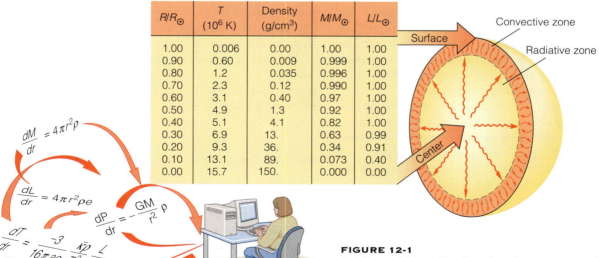

$R/R_\odot$	T (10^6 K)	Density (g/cm³)	$M/M_\odot$	$L/L_\odot$
1.00	0.006	0.00	1.00	1.00
0.90	0.60	0.009	0.999	1.00
0.80	1.2	0.035	0.996	1.00
0.70	2.3	0.12	0.990	1.00
0.60	3.1	0.40	0.97	1.00
0.50	4.9	1.3	0.92	1.00
0.40	5.1	4.1	0.82	1.00
0.30	6.9	13.	0.63	0.99
0.20	9.3	36.	0.34	0.91
0.10	13.1	89.	0.073	0.40
0.00	15.7	150.	0.000	0.00

$$\frac{dM}{dr} = 4\pi r^2 \rho$$

$$\frac{dL}{dr} = 4\pi r^2 \rho e$$

$$\frac{dP}{dr} = -\frac{GM}{r^2}\rho$$

$$\frac{dT}{dr} = \frac{-3}{16\pi ac}\frac{\bar{\kappa}\rho}{T^3}\frac{L}{r^2}$$

FIGURE 12-1
A stellar model is a table of numbers that represent conditions inside a star. Such tables can be computed using the four laws of stellar structure, shown here in mathematical form. The table in this figure describes the sun. (*Illustration design by author*)

WINDOW ON SCIENCE 12-1

Quantitative Thinking with Mathematical Models

One of the most powerful tools in science is the mathematical model, a group of equations carefully designed to mimic the behavior of the object scientists want to study. Astronomers build mathematical models of stars using only four equations, but other systems are much more complicated and may require many more equations.

Many sciences use mathematical models. Medical scientists have built mathematical models of the nerves that control the human heart, and physicists have built mathematical models of the inside of an atomic nucleus. Economists have built mathematical models of certain aspects of economic systems, such as the municipal bond market, and Earth scientists have built mathematical models of Earth's atmosphere. In each case, the mathematical model allows the scientists to study something that is too

difficult to study in the real world. The model can reveal regions we cannot observe, speed up a slow process, slow down a fast process, and allow scientists to perform experiments that would be impossible in reality. Astronomers, for example, can change the abundance of certain chemical elements in a model star to see how its structure depends on its composition.

Many of these mathematical models require very large and fast computers, in some cases supercomputers. A modern computer takes only seconds to compute a stellar model, but models such as the motions of the planets in our solar system millions of years into the future require the largest and fastest computers in the world and highly sophisticated mathematical methods. Most astronomers are apt computer programmers, and some are experts. A few astrono-

mers have built their own specialized computers to calculate specific kinds of models. Thus, the mathematical model is one of the most important tools in astronomy.

As is true for any scientific model, a mathematical model is only as reliable as the assumptions that go into its creation. In Window on Science 2-1, we saw that the celestial sphere was an adequate model of the sky for some purposes but breaks down if we extend it too far. So, too, with mathematical models. We can think of a mathematical model as a numerical expression of one or more theories. While such models can be very helpful, they are always based on theory and so must be compared with the real world at every opportunity. Models must always be tested against experiment and observation. Otherwise, they might lead us astray.

---

begin hydrogen fusion, and that stops their contraction. They reach equilibrium somewhere along a line in the H–R diagram that we call the main sequence. Hot stars are more luminous and cool stars are less luminous, as we would expect. There is, however, a mystery about the main sequence that we can now solve. Why does the luminosity of a star depend on its mass?

In Chapter 9, we used binary stars to find the masses of stars, and we discovered that the masses of main-sequence stars are ordered along the main sequence. The least massive stars are at the bottom, and the most massive stars are at the top. Further, we discovered a direct relationship between the mass of a star and its luminosity—the mass–luminosity relation. This is one of the most fundamental observations in astronomy, and stellar models can help us understand why there must be a mass–luminosity relation, and that tells us why there has to be a main sequence.

The keys to the mass–luminosity relation are the law of hydrostatic equilibrium, which says that pressure must balance weight, and the pressure–temperature thermostat, which regulates energy production. We have seen that a star's internal pressure stays high because the generation of thermonuclear energy keeps its interior hot. Because more massive stars have more weight pressing down on the inner layers, their interiors must have high pressures and thus must be hot. For example, the temperature at the center of a 15-solar-mass star is about 34,000,000 K, more than twice the central temperature of the sun.

Because massive stars have hotter cores, their nuclear reactions burn more fiercely. That is, their pressure–temperature thermostat is set higher. The nuclear fuel at the center of a 15-solar-mass star fuses over 3000 times more rapidly than the fuel at the center of the sun. The rapid reactions in the cores of massive stars produce more energy, but that energy cannot remain in the core. The energy must flow outward toward the cooler surface, and in doing so, the flowing energy heats each level in the star and enables it to support the weight pressing inward. When all that energy reaches the surface, it radiates into space and makes the star highly luminous. Thus there must be a mass–luminosity relation because each star must support its weight by generating nuclear energy.

The main sequence is elegant in its simplicity. It exists because stars balance their weight by fusing hydrogen in their cores. To understand the main-sequence stars even better, we can look at the ends of the main sequence.

THE ENDS OF THE MAIN SEQUENCE

Stellar structure gives us a way to study the extreme ends of the main sequence, the most massive and least massive stars. The first are rare, but the latter are common. Nevertheless, both are very difficult to study.

We can identify two reasons why there is an upper limit to the mass of stars. First, it appears that gas clouds contracting to form a star can fragment and form two or more stars. If a gas cloud contains more mass, it is more likely to fragment, so we don't find very massive stars because those gas clouds broke into smaller fragments and formed multiple-star systems.

But second, stellar models tell us that stars of roughly 100 solar masses are unstable at formation. To support the tremendous weight in such stars, the internal gas must be very hot, and that means it must emit floods of radiation that flow outward through the star and blow gas away from the star's surface in powerful stellar winds. This mass loss from very massive stars could reduce a 60-solar-mass star to less than 30 solar masses in a million years.

When we compare our theory with observation, we discover that it is difficult to find truly massive stars. Most of the O and B stars we see in the sky have masses of 10 to 25 solar masses. Our survey of stars at the end of Chapter 9 revealed that the stars at the upper end of the main sequence are very rare, so we must search to great distances to find just a few. Nevertheless, a few stars are known that are thought to be very massive, and their spectra contain blue-shifted emission lines. Kirchhoff's laws tell us that emission lines come from excited low-density gas, and the blue shift must be caused by a Doppler shift in gas coming toward us. These stars are losing mass.

▌Figure 12-2 shows a famous star, Eta Carinae, that is believed to be a massive binary in which the stars contain 60 solar masses and 70 solar masses. They may have formed with about 100 solar masses each, but mass loss is reducing their mass. An eruption 150 years ago made Eta Carinae the second brightest star in the sky and ejected the two expanding lobes of dusty gas. Although the star has faded, it is still very active, and

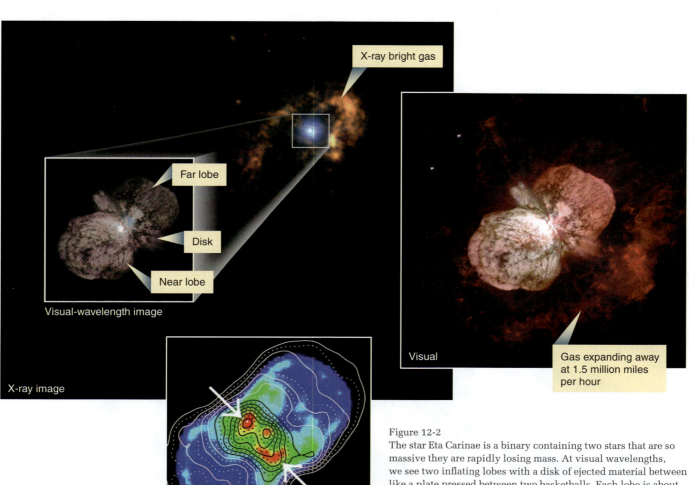

Figure 12-2
The star Eta Carinae is a binary containing two stars that are so massive they are rapidly losing mass. At visual wavelengths, we see two inflating lobes with a disk of ejected material between like a plate pressed between two basketballs. Each lobe is about half a light-year in diameter. At X-ray wavelengths, we see very hot gas excited by collision with high-speed gas ejected from the stars. An infrared image reveals a 15-solar-mass torus (doughnut) of gas and dust squeezing the outflowing gas into the two lobes. (*NASA/CXC/SAO/HST; Jon Morse and J. Hester; IR image: ISO, Courtesy ESA*)

more recent eruptions have ejected jets and an equatorial disk of gas and dust. Whatever their status, these massive stars are clearly unstable.

The lower end of the main sequence is difficult to study, not because the stars are rare but because they are dim. If a red dwarf from the lower end of the main sequence replaced the sun, it would shine only a few times brighter than the full moon. Such stars are difficult to find even when they are only a few light-years away.

Stellar models predict that stars less massive than 0.08 solar masses cannot get hot enough to ignite hydrogen fusion. These **brown dwarfs** should cool as they convert their gravitational energy into heat and radiate it away. A star of 0.08 solar masses fusing hydrogen has a surface temperature of about 2500 K, but brown dwarfs should have temperatures of 1000 K or so, giving them a color even ruddier than red dwarfs—thus the term "brown dwarf."

Brown dwarfs were difficult to find because they are so faint, but large surveys and infrared studies have turned up lots. Some are located in binary systems with normal stars (❚ Figure 12-3a), but large numbers are free-floating objects without stellar companions (Figure 12-3b).

Brown dwarfs are clearly different from normal stars. Some brown dwarfs such as Gliese 229B in Figure 12-3 have methane bands in their spectra, and that means they must be quite cool. Methane molecules would be broken up at the temperatures of true stars. Color variations suggest that some brown dwarfs may be cool enough to have weather patterns.

The discovery of brown dwarfs has created a controversy over what we call them. Are they failed stars or are they planets? To discuss this we use a new unit of mass related to Jupiter, the giant planet in our solar system. Jupiter is one thousand times less massive than the sun, and astronomers refer to the mass of brown dwarfs in Jupiter masses. Thus a brown dwarf must be less than about 80 Jupiter masses. Generally, astronomers think of stars as bodies that generate energy by nuclear fusion, and a star less than 80 Jupiter masses can't heat its center to the 2.7 million Kelvin temperature needed to start hydrogen fusion. Then brown dwarfs aren't stars. But astronomers tend to think of a planet as a nonluminous body that orbits a star, so can a brown dwarf floating free in space be called a planet?

Brown dwarfs more massive than 13 Jupiter masses can fuse deuterium, a heavy isotope of hydrogen, and some astronomers draw the line between stars and planets at 13 Jupiter masses. However, deuterium fusion does not generate much energy and does not stop the star from cooling rapidly. Not everyone agrees that this is a good dividing line.

Another approach to the controversy is to note how these objects form. A star forms from the contraction of a gas cloud, but astronomers believe planets form from the accumulation of solid bits of matter in disk-shaped nebulae around stars (a subject we will discuss in Chapter 19). In that case, free-floating brown dwarfs can't be planets.

This is a controversy over the definition of words, but it illustrates our uncertainty concerning the lower end of the main sequence. Perhaps the study of brown dwarfs will eventually help us better understand what stars and planets are and how they form.

THE LIFE OF A MAIN-SEQUENCE STAR

A normal main-sequence star supports its weight by fusing hydrogen into helium, but its supply of hydrogen is limited. As it consumes its hydrogen, the chemical composition in its core changes, and the star evolves. Mathematical models of stars allow astronomers to follow that evolution.

a
b

FIGURE 12-3

(a) Gliese 229B (arrow) is a brown dwarf that is a binary companion to the bright star at the left. The bright ray of light is caused by scattered light in the telescope. *(Kulkarni, Caltech, Golimowski, JHU, NASA)* (b) Too dim to see at visual wavelengths, about 50 free-floating brown dwarfs appear in an infrared image of the center of the Orion nebula. The bright stars of the Trapezium are located at the center. *(STScI and NASA)*

Hydrogen fusion combines four nuclei into one. Thus, as a main-sequence star consumes its hydrogen, the total number of nuclei in its interior decreases. Each newly made helium nucleus can exert the same pressure as a hydrogen nucleus, but because the gas has fewer nuclei, its total pressure is less. This unbalances the gravity–pressure stability, and gravity squeezes the core of the star more tightly. As the core contracts, its temperature and density increase, and the nuclear reactions burn faster, releasing more energy and making the star more luminous. This additional energy flowing outward through the envelope forces the outer layers to expand and cool, so the star becomes slightly larger, brighter, and cooler.

As a result of these gradual changes in main-sequence stars, the main sequence is not a sharp line across the H–R diagram but rather a band (blue in ▌Figure 12-4). A star begins its stable life fusing hydrogen and falls on the lower edge of this band, the **zero-age main sequence (ZAMS).** As it combines hydrogen nuclei to make helium nuclei, the star slowly changes. In the H–R diagram, the point that represents the star's luminosity and surface temperature moves upward and to the right, eventually reaching the upper edge of the main sequence, just as the star exhausts nearly all of the hydrogen in its center. Thus, we find main-sequence stars plotted throughout this band at various stages of their main-sequence lives.

These gradual changes in the sun will spell trouble for Earth. When the sun began its main-sequence life about 5 billion years ago, it was only about 75 percent of its present luminosity. This, by the way, makes it difficult to explain how Earth has remained at roughly its present temperature for at least 3 billion years. Some experts suggest that Earth's atmosphere has gradually changed and thus compensated for the increasing luminosity of the sun.

By the time the sun leaves the main sequence in a few billion years, it will have twice its present luminosity. This will raise the average temperature on Earth by at least 19°C (34°F). As this happens over the next billion years, the polar caps will melt, the oceans will evaporate, and much of the atmosphere will vanish into space. Clearly, the future of Earth as the home of life is limited by the evolution of the sun.

Once a star leaves the main sequence, it evolves rapidly and dies. The average star spends 90 percent of its life fusing hydrogen on the main sequence. This explains why 90 percent of all normal stars are main-sequence stars. We are most likely to see a star during that long, stable period when it is on the main sequence.

The number of years a star spends on the main sequence depends on its mass (▌Table 12-2). Massive stars use fuel rapidly and live short lives, but low-mass stars conserve their fuel and shine for billions of years. For example, a 25-solar-mass star will exhaust its hydrogen and die in only about 7 million years. This means, for one thing, that life is very unlikely to develop on planets orbiting massive stars. These stars do not live long enough for life to get started and evolve into complex creatures. We will discuss this problem in detail in Chapter 26.

Very-low-mass stars, the red dwarfs, use their fuel so slowly they should survive for 200 to 300 billion years. Because the universe seems to be only 10 to 20 billion years old, red dwarfs must still be in their infancy. None of them should have exhausted their hydrogen fuel yet.

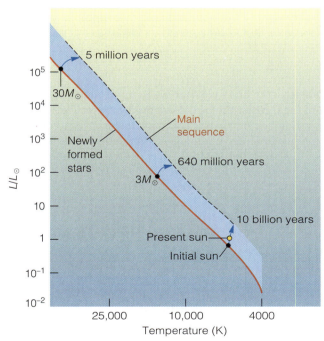

FIGURE 12-4
The main sequence is not a line but a band (shaded). Newly formed stars begin fusing hydrogen on the lower edge, the zero-age main sequence (ZAMS). As hydrogen changes their composition, the stars move across the band. Notice that lower-mass stars take longer to cross the band of the main sequence than stars of higher mass.

TABLE 12-2 Main-Sequence Stars			
Spectral Type	Mass (sun = 1)	Luminosity (sun = 1)	Approximate Years on Main Sequence
O5	40	405,000	1×10^6
B0	15	13,000	11×10^6
A0	3.5	80	440×10^6
F0	1.7	6.4	3×10^9
G0	1.1	1.4	8×10^9
K0	0.8	0.46	17×10^9
M0	0.5	0.08	56×10^9

Nature makes more low-mass stars than massive stars, but this fact is not sufficient to explain the vast numbers of low-mass stars that fill the sky. An additional factor is the stellar lifetimes. Because low-mass stars live long lives, there are more of them in the sky than massive stars. Look at page 193 and notice how much more common the lower-main-sequence stars are than the massive O and B stars. The main-sequence K and M stars are so faint they are difficult to locate, but they are very common. The O and B stars are luminous and easy to locate, but because of their fleeting lives, there are never more than a few on the main sequence at any one time.

THE LIFE EXPECTANCIES OF STARS

To understand how nature makes stars and how stars evolve, we must be able to estimate how long they can survive. In general, massive stars live short lives and lower-mass stars live long lives, but we can make more accurate estimates by using simple stellar models. In fact, we can calculate the approximate life expectancy of a star from its mass.

Because main-sequence stars consume their fuel at a constant rate, we can estimate the amount of time a star spends on the main sequence—its life expectancy T—by dividing the amount of fuel by the rate of fuel consumption. This is a common calculation. If we drive a truck that carries 20 gallons of fuel and uses 5 gallons of fuel per hour, we know the truck can run for 4 hours.

The amount of fuel a star has is proportional to its mass, and the rate at which it burns its fuel is proportional to its luminosity; thus, we could make a first estimate of the star's life expectancy by dividing mass by luminosity. A 2-solar-mass star is about 11 times more luminous than the sun and should live about 2/11, or 18 percent as long as the sun. We can, however, make the calculation even easier if we remember that the mass–luminosity relation tells us that the luminosity of a star equals $M^{3.5}$. The life expectancy then is:

$$T = \frac{M}{L} = \frac{M}{M^{3.5}} \quad \text{or} \quad T = \frac{1}{M^{2.5}}$$

This means that we can estimate the life expectancy of a star by dividing 1 by the star's mass raised to the 2.5 power. If we express the mass in solar masses, the life expectancy will be in solar lifetimes.

For example, how long can a 4-solar-mass star live?

$$T = \frac{1}{4^{2.5}} = \frac{1}{4 \cdot 4 \cdot \sqrt{4}} = \frac{1}{32} \text{ solar lifetimes}$$

Detailed studies of models of the sun show that the sun, presently 5 billion years old, can last another 5 billion years. Thus, a solar lifetime is approximately 10 billion years, and a 4-solar-mass star will last about (10 billion)/32 or about 310 million years.

Our estimation of stellar life expectancies is very approximate. For example, our model ignores mass loss, which may affect the life expectancies of very luminous and very faint stars. Nevertheless, it serves to illustrate a very important point. Stars that are only slightly more massive than the sun have dramatically shorter lifetimes on the main sequence.

REVIEW CRITICAL INQUIRY

Why are main-sequence O stars so rare compared with stars like the sun (G stars)?

In Chapter 9, we discussed the family of stars and noticed that space contains only about 0.036 main-sequence O stars per million cubic parsecs, but that there are about 2000 main-sequence G stars per million cubic parsecs (see page 193). Dividing, we discover that main-sequence G stars are over 50,000 times more common than O stars. Our discussion of the lives of main-sequence stars gives us a way to analyze these numbers. O stars live very short lifetimes. Our simple formula predicts a 20-solar-mass O star has a life expectancy of about 6 million years. The sun, however, will live a total of about 10 billion years or about 1600 times longer than an O star. Assuming nature makes equal numbers of different kinds of stars, we would expect to see about 1600 times more main-sequence G stars than O stars. But we see over 50,000 times more, so there must be another factor involved. Nature must be making vastly more lower-mass stars than massive stars. By comparing theory and observation, our analysis has revealed something new about how nature makes stars.

Now we can understand why O stars are so rare. What factors must you take into consideration to explain why main-sequence M stars are even more common than G stars?

When a star finally exhausts its hydrogen fuel, it can no longer resist the pull of its own gravity. The contraction that began when it was a protostar resumes and begins the process that leads to its death. It can delay its end by fusing other fuels, but, as we will discover in Chapter 13, nothing can steal gravity's final victory.

12-2 POST-MAIN-SEQUENCE EVOLUTION

In earlier chapters, we asked three questions: Why are giant stars so large? Why are they so uncommon? And why do they have such low densities? Now we are ready to answer those questions by discussing the evolution of stars after they leave the main sequence.

EXPANSION INTO A GIANT

To understand how stars evolve, we must recognize that they are not well mixed; that is, their interiors are not stirred. The centers of stars like the sun are radiative, meaning the energy moves as radiation and not as circulating currents of heated gas. The gas does not move in such stars, and thus they are not mixed at all. More massive stars have convective cores that mix the central regions (Figure 11-14), but these regions are not very large, and thus, for the most part, these stars, too, are not mixed. (The lowest-mass stars are an exception that we will examine in the next chapter.)

In this respect, stars are like a campfire that is not stirred; the ashes accumulate at the center, and the fuel in the outer parts never gets used. Nuclear fusion consumes hydrogen nuclei and produces helium nuclei, the "ashes," at the star's center. Nothing mixes the interior of the star, so the helium nuclei remain where they are in the center of the star, and the hydrogen in the outer parts of the star is not mixed down to the center where it can be fused.

The helium ashes that accumulate in the star's core cannot fuse into heavier elements because the temperature is too low. As a result, the helium accumulates, the hydrogen is used up, and the core becomes an inert ball of helium. As this happens, the energy production in the core falls, and the weight of the outer layers forces the core to contract.

Although the contracting helium core cannot generate nuclear energy, it does grow hotter because it converts gravitational energy into thermal energy (see Chapter 11). The rising temperature heats the unprocessed hydrogen just outside the core, hydrogen that was never before hot enough to fuse. When the temperature of the surrounding hydrogen becomes high enough, it ignites in a shell of fusing hydrogen. Like a grass fire burning outward from an exhausted campfire, the hydrogen-fusing shell burns outward, leaving helium ash behind and increasing the mass of the helium core.

At this stage in its evolution, the star overproduces energy; that is, it produces more energy than it needs to balance its own gravity. The helium core, having no nuclear energy sources, must contract, and that contraction converts gravitational energy into thermal energy—the contraction heats the helium core. Some of that heat leaks outward through the star. At the same time, the hydrogen-fusing shell produces energy as the contracting core brings fresh hydrogen closer to the center of the star and heats it to high temperature. The result is a flood of energy flowing outward through the outer layers of the star, forcing them to puff up and swelling the star into a giant (❚ Figure 12-5).

The expansion of the envelope dramatically changes the star's location in the H–R diagram. Just as contraction heats a star, expansion cools it. As the outer layers of gas expand, energy is absorbed in lifting and expanding the gas. The loss of that energy lowers the temperature of the gas. Consequently the point that represents the star in the H–R diagram moves quickly to the right (in less than a million years for a star of 5 solar

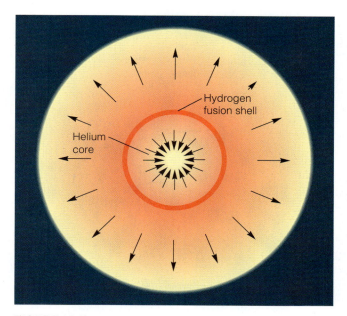

FIGURE 12-5
When a star runs out of hydrogen at its center, it ignites a hydrogen-fusing shell. The helium core contracts and heats, while the envelope expands and cools. (For a scale drawing, see Figure 12-8.)

masses). A massive star moves to the right across the top of the H–R diagram and becomes a supergiant, while a medium-mass star like the sun becomes a red giant (❚ Figure 12-6). As the radius of a giant star continues to increase, the enlarging surface area makes the star more luminous, moving its point upward in the H–R diagram. Aldebaran, the glowing red eye of Taurus the bull, is such a red giant, with a diameter 25 times that of the sun, but with a surface temperature only half that of the sun.

More massive stars cross higher in the H–R diagram and become supergiants, stars even larger than giants. Betelgeuse (α Orionis) is a very cool, red supergiant over 800 times the diameter of the sun. Rigel (β Orionis) is a supergiant 50 times larger than the sun. It may seem odd to say that Rigel, a blue star, has expanded and cooled. At a temperature of 12,000 K, Rigel looks quite blue to our eyes. When it was on the main sequence, Rigel had a much hotter surface, but it would not have looked much bluer because we cannot see the ultraviolet radiation that the hottest stars emit. Any star as hot or hotter than Rigel will look blue to our eyes, as shown in Figure 7-2. Thus, blue Rigel is indeed a star that has expanded and cooled.

Our discussion of post-main-sequence evolution explains a number of things we noticed in earlier chapters. Giants and supergiants are large because they have expanded, and they have very low densities for the same reason. Also, the giant and supergiant region of the H–R diagram contains a jumble of different mass stars because all main-sequence stars funnel through this part of the diagram as they die. (See Figure 9-22.)

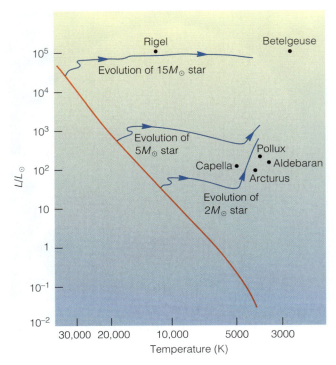

ACTIVE FIGURE 12-6

The evolution of a massive star moves the point that represents it in the H–R diagram to the right of the main sequence into the region of the supergiants such as Rigel and Betelgeuse. The evolution of a medium-mass star moves its point in the H–R diagram into the region of the giants such as those shown here.

Ace✪Astronomy™ Go to AceAstronomy and click Active Figures to see "Evolution of Stars" and compare the evolution of a massive star and a low-mass star.

DEGENERATE MATTER

Although the hydrogen-fusing shell can force the envelope of the star to expand, it cannot stop the contraction of the helium core. Because the core has no energy source, gravity squeezes it tighter, and it becomes very small. If we represent the helium core of a 5-solar-mass star with a quarter, the outer envelope of the star would be about the size of a baseball diamond. Yet the core would contain about 12 percent of the star's mass compressed to very high density. When gas is compressed to such extreme densities, it begins to behave in astonishing ways that can alter the evolution of the star. Thus, to follow the story of stellar evolution, we must consider the behavior of gas at extremely high densities.

Normally, the pressure in a gas depends on its temperature. The hotter the gas is, the faster its particles move and the more pressure it exerts. But the gas inside a star is ionized, so there are two kinds of particles: atomic nuclei and free electrons. If the gas is compressed to very high densities, for example, in the core of a giant star, the difference between these two kinds of particles is important.

If the density is very high, the particles of the gas are forced close together, and two laws of quantum me-

chanics become important. First, quantum mechanics says that the moving electrons confined in the star's core can have only certain amounts of energy, just as the electron in an atom can occupy only certain energy levels (see Chapter 7). We can think of these permitted energies as the rungs of a ladder. An electron can occupy any rung, but not the spaces between.

The second quantum-mechanical law (called the Pauli Exclusion Principle) says that two identical electrons cannot occupy the same energy level. Because electrons spin in one direction or the other, two electrons can occupy an energy level if they spin in opposite directions. That level is then completely filled, and a third electron cannot enter because, whichever way it spins, it will be identical to one or the other of the two electrons already in the level. Thus, no more than two electrons can occupy the same energy level.

A low-density gas has few electrons per cubic centimeter, so there are plenty of energy levels available. If a gas becomes very dense, however, nearly all of the lower energy levels may be occupied (▌Figure 12-7). In such a gas, a moving electron cannot slow down, because slowing down would decrease its energy, and there are no open energy levels for it to drop down to. It can speed up only if it can absorb enough energy to leap to the top of the energy ladder, where there are empty energy levels.

When a gas is so dense that the electrons are not free to change their energy, astronomers call it **degenerate matter**. Although it is a gas, it has two peculiar properties that can affect the star. First, the degenerate gas resists compression. To compress the gas, we must push against the moving electrons, and changing their motion means changing their energy. That requires tremendous effort, because we must boost them to the

FIGURE 12-7

Electron energy levels are arranged like rungs on a ladder. In a low-density gas, many levels are open, but in a degenerate gas, all lower energy levels are filled.

top of the energy ladder. Thus, degenerate matter, though still a gas, takes on the consistency of hardened steel.

Second, the pressure of degenerate gas does not depend on temperature but rather on the speed of the electrons, which cannot be changed without tremendous effort. The temperature, however, depends on the motion of all the particles in the gas, both electrons and nuclei. If we add heat to the gas, most of that energy goes to speed up the motions of the nuclei, and only a few electrons can absorb enough energy to reach the empty energy levels at the top of the energy ladder. Thus, changing the temperature of the gas has almost no effect on the pressure.

These two properties of degenerate matter become important when stars end their main-sequence lives and approach their final collapse (Window on Science 12-2). Eventually, many stars collapse into white dwarfs, and we will discover that these tiny stars are made of degenerate matter. But long before that, the cores of many giant stars become so dense that they are degenerate, a situation that can produce a cosmic bomb when helium begins to fuse.

HELIUM FUSION

Hydrogen fusion leaves behind helium ash, which cannot fuse at the relatively low temperatures of hydrogen fusion. Helium nuclei have a positive charge twice that of a hydrogen nucleus (a proton), which means the Coulomb barrier is higher. At the temperature of hydrogen fusion, the particles move too slowly and collide too gently to produce helium fusion. Thus the helium that accumulates is an ash that cannot fuse until the temperature rises significantly.

As the star becomes a giant star, fusing hydrogen in a shell, the inner core of helium ash contracts and grows hotter. It may even become degenerate. Finally, as the temperature approaches 100,000,000 K, helium nuclei begin to fuse together to make carbon.

We can summarize the helium-fusing process in two steps:

$$^4\text{He} + {}^4\text{He} \rightarrow {}^8\text{Be} + \gamma$$
$$^8\text{Be} + {}^4\text{He} \rightarrow {}^{12}\text{C} + \gamma$$

This process is complicated by the fact that beryllium-8 is very unstable and may break up into two helium nuclei before it can absorb another helium nucleus. Three helium nuclei can also form carbon directly, but such a triple collision is unlikely. Because a helium nucleus is called an alpha particle, astronomers often refer to helium fusion as the **triple alpha process.**

Some stars begin helium fusion gradually, but stars in a certain mass range begin helium fusion with an explosion called the **helium flash.** This explosion is caused by the density of the helium, which can reach 1,000,000 g/cm^3. On Earth, a teaspoon of this material would weigh as much as a large truck. At these densities, the gas is degenerate, and its pressure no longer depends on temperature. Thus, the pressure–temperature thermostat that controls the nuclear fusion reactions no longer works.

When the helium ignites, it generates energy, which raises the temperature. Because the pressure–temperature thermostat is not working in the degenerate gas, the core does not respond to the higher temperature by expanding. Rather, the higher temperature forces the reactions to go faster, which makes more energy, which raises the temperature, which makes the reactions go faster, and so on. The ignition of helium fusion in a degenerate gas results in a runaway explosion so violent that for a few minutes the helium core generates more energy than an entire galaxy. At its

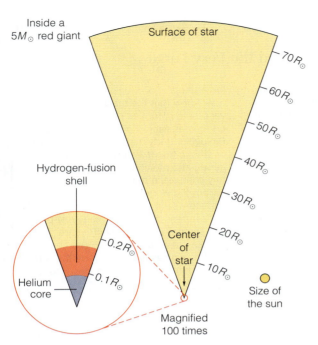

FIGURE 12-8
As a star becomes a giant, its envelope (yellow) expands to enormous size, while its core contracts and becomes very small. *(Illustration design by the author)*

peak, the core generates 10^{14} times more energy per second than the sun.

Although the helium flash is sudden and powerful, it does not destroy the star. In fact, if you were observing a giant star as it experienced a helium flash, you would see no outward evidence of an eruption. The helium core is quite small (■ Figure 12-8), and all of the energy of the explosion is absorbed by the distended envelope. Also, the helium flash is a very short-lived event. In a matter of minutes, the core of the star becomes so hot it is no longer degenerate, the pressure–temperature thermostat brings the helium fusion under control, and the star proceeds to fuse helium steadily in its core. We will discuss this post-helium-flash evolution later.

Not all stars experience a helium flash. Stars less massive than about 0.4 solar mass can never get hot enough to ignite helium, and stars more massive than about 3 solar masses ignite helium before their cores become degenerate (■ Figure 12-9). In such stars, pressure depends on temperature, so the pressure–temperature thermostat keeps the helium fusion under control.

If the helium flash occurs only in some stars and is a very short-lived event that is not visible from outside the star, why should we worry about it? The answer is that it limits the reliability of the mathematical models astronomers use to study stellar evolution. The medium-mass stars, which experience the helium flash, are those that we must consider most carefully in studying stellar evolution. Massive stars are not very common; low-mass stars evolve so slowly that we cannot see much ev-

idence of low-mass stellar evolution. Thus, studies of stellar evolution must concentrate on medium-mass stars, which do experience the helium flash. But the helium flash occurs so rapidly and so violently that computer programs cannot follow the changes in the star's internal structure in detail. To follow the evolution of medium-mass stars like the sun past the helium flash, astronomers must make assumptions about the way the helium flash affects stars' internal structures.

Post-helium-flash evolution is generally understood. As the core temperature rises, the gas ceases to be degenerate, and helium fusion proceeds under the control of the pressure–temperature thermostat. Throughout these events, the hydrogen-fusion shell continues to produce energy, but the new helium-fusion energy produced in the core makes the core expand. That expansion absorbs energy previously used to support the outer layers of the star. As a result, the outer layers contract, and the surface of the star grows slightly hotter. In the H–R diagram, the point that represents the star initially moves down toward lower luminosity and to the left toward higher temperature. Later, as the star stabilizes fusing helium in its core, its luminosity recovers, but its surface grows hotter.

Helium fusion produces carbon and oxygen as "ash," and these accumulate at the center of the star. They do not fuse because the core is not hot enough. Thus, the helium in the core is gradually converted into carbon–oxygen ashes. As this happens, the core contracts, grows hotter, and ignites a helium-fusion shell. At this stage, the star has two shells producing energy. A hydrogen-fusion shell eats its way outward, leaving behind helium ash; a helium-fusion shell eats its way

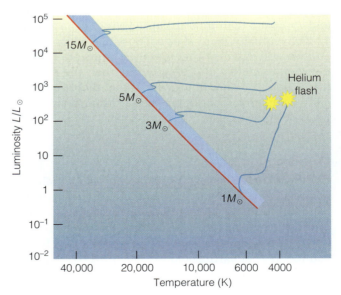

FIGURE 12-9
Stars like the sun suffer a helium flash, but more massive stars begin helium fusion without a helium flash. The lower-mass stars cannot get hot enough to ignite helium. The zero-age main sequence is shown here in red, with the band of the main sequence in blue.

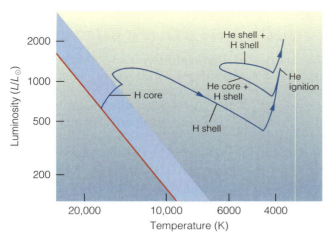

FIGURE 12-10

The evolution of a 5-$M_\odot$ star illustrates stages in energy generation. From the zero-age main sequence (red) across the band of the main sequence (blue), the star makes energy in a hydrogen-fusion core. Hydrogen-shell fusion begins as the star leaves the main sequence and continues after helium fusion ignites in the core. When the core exhausts its helium, the star makes energy with a helium-fusion shell around the core and a hydrogen-fusion shell outside the helium-fusion shell.

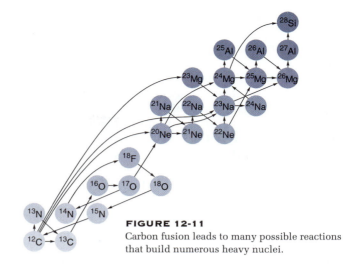

FIGURE 12-11
Carbon fusion leads to many possible reactions that build numerous heavy nuclei.

outward into the helium, leaving behind carbon–oxygen ash. Unable to generate energy in its carbon–oxygen core, the star expands its outer layers and its point in the H–R diagram moves back toward the right, completing a loop (▌Figure 12-10).

FUSING ELEMENTS HEAVIER THAN HELIUM

Modeling the evolution of stars after helium exhaustion requires great sophistication. The inert core of carbon–oxygen ash contracts and becomes hotter. Stars more massive than about 4 solar masses can heat their cores to temperatures as high as 600,000,000 K, hot enough to fuse carbon. Subsequently, the stars can fuse oxygen, silicon, and other heavy elements.

Carbon fusion in a star begins a complex network of reactions illustrated in ▌Figure 12-11, where each circle represents a possible nucleus and each arrow represents a different nuclear reaction. Nuclei can react by capturing a proton, capturing a neutron, capturing a helium nucleus, or by combining directly with other nuclei. Also, unstable nuclei can decay by ejecting an electron, ejecting a positron, ejecting a helium nucleus or by splitting into fragments. Astronomers who model the structure and evolution of these stars must use sophisticated nuclear physics to compute the energy production and changes in chemical composition as the stars fuse heavier elements.

These stars develop complex internal structure as they fuse heavier and heavier elements in concentric shells. Like great spherical cakes with many layers, the stars may be fusing a number of different elements simultaneously in concentric shells.

Eventually, all stars face collapse. Less massive stars cannot get hot enough to ignite the heavier nuclei. The more massive stars can ignite heavy nuclei, but these stars consume their fuels at a tremendous rate, so they run through their available fuel quickly (▌Table 12-3). No matter what the star's mass, it will eventually run out of usable fuels, and gravity will win the struggle with pressure. We will explore the ultimate deaths of stars in the next chapter.

TABLE 12-3
Nuclear Reactions in Massive Stars

Nuclear Fuel	Nuclear Products	Minimum Ignition Temperature	Main-Sequence Mass Needed to Ignite Fusion	Duration of Fusion in a 25-$M_\odot$ Star
H	He	4×10^6 K	0.1 $M_\odot$	7×10^6 yr
He	C, O	120×10^6 K	0.4 $M_\odot$	0.5×10^6 yr
C	Ne, Na, Mg, O	600×10^6 K	4 $M_\odot$	600 yr
Ne	O, Mg	1.2×10^9 K	~ 8 $M_\odot$	1 yr
O	Si, S, P	1.5×10^9 K	~ 8 $M_\odot$	~ 0.5 yr
Si	Ni to Fe	2.7×10^9 K	~ 8 $M_\odot$	~ 1 day

The time a star spends as a giant or supergiant is small compared with its life on the main sequence. The sun, for example, will spend about 10 billion years as a main-sequence star but only about a billion years as a giant. The more massive stars pass through the giant stage even more rapidly. Because of the short time a star spends as a giant, we are unlikely to see many such stars. This illustrates an important principle in astronomy: the shorter the time a given evolutionary stage takes, the less likely we are to see stars in that particular stage. Thus, we see a great many main-sequence stars, but few giants (see page 193).

REVIEW CRITICAL INQUIRY

What uncertainties limit the accuracy of our story of stellar evolution?

Mathematical models of stars apply four basic laws of physics to describe the inside of a star. Modern astronomers have used such models to study stellar evolution, a process that takes billions of years and thus is unobservable for humans. But two things limit the accuracy of the models. First, some events in the history of a star happen so fast that computers can't predict the results in a reasonable amount of time. In fact, some astronomers estimate that a computer following the evolution of a model star undergoing the helium flash would have to recompute the model in such short time steps that the computer program would take millions of years to predict events that occur inside the star in just seconds. Thus, truly rapid changes such as explosions in stars make the models uncertain.

Second, we have seen how the general properties of a giant star can depend on the subatomic properties of matter. Not all of those subatomic properties are well understood. For example, the solar neutrino problem hints that physicists might not understand exactly how neutrinos and other subatomic particles behave, and if that is true, then stellar models may be slightly different from the real stars they are supposed to represent.

Can you think of any other physical processes that have been omitted from stellar models? What features of the sun's surface suggest motions and forces that are ignored in most stellar models?

We cannot see the stars evolving, so we must test our theoretical story of stellar evolution against observation. To paraphrase the quotation at the beginning of this chapter, we shouldn't believe any theory until we compare it with reality—seeing is believing. Thus, in the rest of this chapter, we will search for evidence to support the theory of stellar evolution. At the same time, we will discover some interesting consequences of the fact that the stars are not eternal.

12-3 EVIDENCE OF EVOLUTION: STAR CLUSTERS

The theory of stellar evolution is so complex and involves so many assumptions that astronomers would have little confidence in it were it not for H–R diagrams of clusters of stars. By observing the properties of star clusters of different ages, we can see clear evidence of the evolution of stars.

To grasp the difficulty of understanding stellar evolution, consider an analogy suggested by William Herschel. Suppose a visitor who had never seen a tree wandered through a forest for an hour looking at the falling seeds, rising sprouts, young saplings, mature trees, and fallen logs. Could such an observer understand the life cycle of the trees? Astronomers face the same problem when they try to understand the life story of the stars.

We do not live long enough to see stars evolve. We see only a momentary glimpse of the universe, a snapshot in which all the stages in the life cycle of the stars are represented. Unscrambling these stages and putting them in order is an almost impossible task. Only by looking at selected groups of stars—star clusters—can we see the pattern.

OBSERVING STAR CLUSTERS

Just as Sherlock Holmes studies peculiar dust on a lamp shade as a clue to a crime, astronomers look at star clusters and say, "Aha!" Each star cluster freezes a moment in the evolution of the cluster and makes the evolution of the stars visible.

Study ❚ "Star Cluster H–R Diagrams" on pages 248 and 249 and notice three important points. First, there are two kinds of star clusters, but they are similar in the way their stars evolve. We will discuss these clusters in more detail in a later chapter. Second, notice that we can estimate the age of a star cluster by observing the distribution of the points that represent its stars in the H–R diagram. Finally, we notice that the shape of a star cluster's H–R diagram is governed by the evolutionary path the stars take. By comparing clusters of different ages, we can visualize how stars evolve almost as if we were watching a film of a star cluster evolving over billions of years.

Were it not for star clusters, astronomers would have little confidence in the theories of stellar evolution. Star clusters make that evolution visible and assure astronomers that they really do understand how stars are born, live, and die.

THE EVOLUTION OF STAR CLUSTERS

What we know about star formation and stellar evolution can help us understand star clusters and how they evolve.

A star cluster is formed when a cloud of gas contracts, fragments, and forms a group of stars. Some groups of stars, which we have called associations, are so widely scattered the group is not held together by its own gravity. The stars in an association wander away from each other rather quickly. Some groups of stars are more compact, and even after the stars become luminous and the gas and dust are blown away, gravity holds the group together as a star cluster.

Open clusters are not as old or as massive as the globular clusters, and that helps us understand their appearance. Close encounters between stars are rare in an open cluster, and such clusters have an irregular appearance. The globular clusters appear to be nearly perfect globes because encounters between their stars are more common and because the globular clusters have had time to evenly distribute the energy of motion among all of the stars.

As a star cluster ages, some stars traveling a bit faster than others can escape. Globular clusters are compact and massive and have survived for 11 billion years or more. A star cluster with only a few stars widely distributed may evaporate completely as its stars escape one by one. Our sun probably formed in a star cluster 5 billion years ago, but there is no way to trace it back to its family home even if that star cluster still exists.

REVIEW CRITICAL INQUIRY

Why do open clusters contain only a small number of giant stars but many main-sequence stars?

When we look at a star cluster, we see a snapshot of the cluster that freezes it in time. In the H–R diagram of an open cluster, we see a main sequence containing many stars, but typically we see significantly fewer giant stars. Those giant stars were once main-sequence stars in the cluster, but they exhausted the hydrogen fuel in their cores and expanded to become giants. A star like the sun can fuse hydrogen on the main sequence for many billions of years but can remain a giant for only a billion years or less. More massive stars evolve even faster. Thus, at any given moment, we see most of the stars in a star cluster on the main sequence and only a few, caught in the snapshot, as giant stars.

Star clusters are important in astronomy because their H–R diagrams reveal the progress of stellar evolution. How do they reveal the age of the star cluster?

Star clusters give us dramatic confirmation that stars evolve and also reveal some of the details of stellar structure and evolution. Certain pulsating giant stars give us further information about how stars work.

12-4 EVIDENCE OF EVOLUTION: VARIABLE STARS

For millennia, poets and astronomers believed that the stars were constant and unchanging. Modern theory tells us that the stars are slowly evolving as they consume their fuels. But can we really be sure the stars are changing and evolving? Certain stars that pulsate and change their brightness give us evidence of stellar evolution and illustrate once more the importance of energy flowing outward through the layers of the stars.

A **variable star** is any star that changes its brightness in a periodic way. Variable stars include many different kinds of stars. Some variable stars are eclipsing binaries (Chapter 9), but many others are stars that expand and contract, heat and cool, and grow more luminous and less luminous because of internal characteristics. These stars are often called **intrinsic variables** to distinguish them from eclipsing binaries. Of these intrinsic variables, one particular kind is centrally important to modern astronomy even though the first star of that class was discovered centuries ago.

CEPHEID AND RR LYRAE VARIABLE STARS

In 1784, the deaf and mute English astronomer John Goodricke, then 19 years old, discovered that the star δ Cephei is variable, changing its brightness by almost a magnitude over a period of 5.37 days (❙ Figure 12-12a). Goodricke died at the age of 21, but his discovery ensures his place in the history of astronomy. Since 1784, hundreds of stars like δ Cephei have been found, and they are now known as **Cepheid variable stars.**

Cepheid variables are supergiant or bright giant stars of spectral type F or G. The fastest complete a cycle—bright to faint to bright—in about 2 days, whereas the slowest take as long as 60 days. A plot of a variable's magnitude versus time, a light curve, shows a typical rapid rise to maximum brightness and a slower decline (Figure 12-12b). Some Cepheids change their brightness by only 0.5 magnitude (about 10 percent), while others change by more than a magnitude.

Supergiants and bright giants are rare, and not all are Cepheids. Nevertheless, some familiar stars are Cepheids. Polaris, the North Star, is a Cepheid with a period of 3.9696 days and a very low amplitude.

The **RR Lyrae variable stars** are related to the Cepheids. The RR Lyrae stars pulsate with periods of less than a day and are fainter than the faintest Cepheids.

Studies of Cepheids and RR Lyrae stars reveal two related facts of great importance. First, there is a **period–luminosity relation** that connects the period of pulsation of a Cepheid to its luminosity. The longer-period Cepheids are about 40,000 times more luminous than the sun, while the shortest-period Cepheids are only a few hundred times more luminous than the sun. Second, there are two types of Cepheids. Type I Cepheids,

Star Cluster H-R Diagrams

Star clusters are beautiful, but they are also keys to understanding the evolution of stars. All of the stars in a star cluster formed at about the same time from the same cloud of gas. That means the stars have the same age and the same initial chemical composition. The differences we see among stars in a star cluster must arise from differences in mass, and that makes stellar evolution visible.

An **open cluster** is a collection of 10 to 1000 stars in a region about 25 pc in diameter. Some open clusters are quite small and some are large, but they all have an open, transparent appearance because the stars are not crowded together.

**Open Cluster
The Jewel Box**

In a star cluster each star follows its orbit around the center of mass of the cluster.

Visual-wavelength image

A **globular cluster** can contain 10^5 to 10^6 stars in a region only 10 to 30 pc in diameter. The term "globular cluster" comes from the word "globe," although globular cluster is pronounced like "glob of butter." These clusters are nearly spherical, and the stars are much closer together than the stars in an open cluster. Globular clusters are discussed in more detail in Chapter 15.

**Globular Cluster
47 Tucanae**

Visual-wavelength image

Astronomers can construct an H–R diagram for a star cluster by plotting a point to represent the luminosity and temperature of each star.

The most massive stars have died.

Only a few stars are in the giant stage.

Main sequence

The lowest-mass stars are still on the main sequence.

$L/L_\odot$

10^4
10^3
10^2
10
1
10^{-1}

20,000   10,000   5000   3000

The H–R diagram of a star cluster can make the evolution of stars visible. The key is to remember that all of the stars in the star cluster have the same age but differ in mass. The H–R diagram of a star cluster gives us a snapshot of the evolutionary state of the stars at the time we happen to be alive. The diagram here represents a medium-age star cluster in which the most massive stars have died and a few medium mass stars are in the act of evolving off of the main sequence to become giants.

As a star cluster ages, its main sequence grows shorter like a candle burning down. We can judge the age of a star cluster by looking at the turnoff point, the point on the main sequence where stars evolve to the right to become giants. Stars at the **turnoff point** have lived out their lives and are about to die. Consequently, the life expectancy of the stars at the turnoff point equals the age of the cluster.

Ace ⊛ Astronomy™

Go to AceAstronomy and click Active Figures to see "Cluster Turnoff" and notice how the shape of

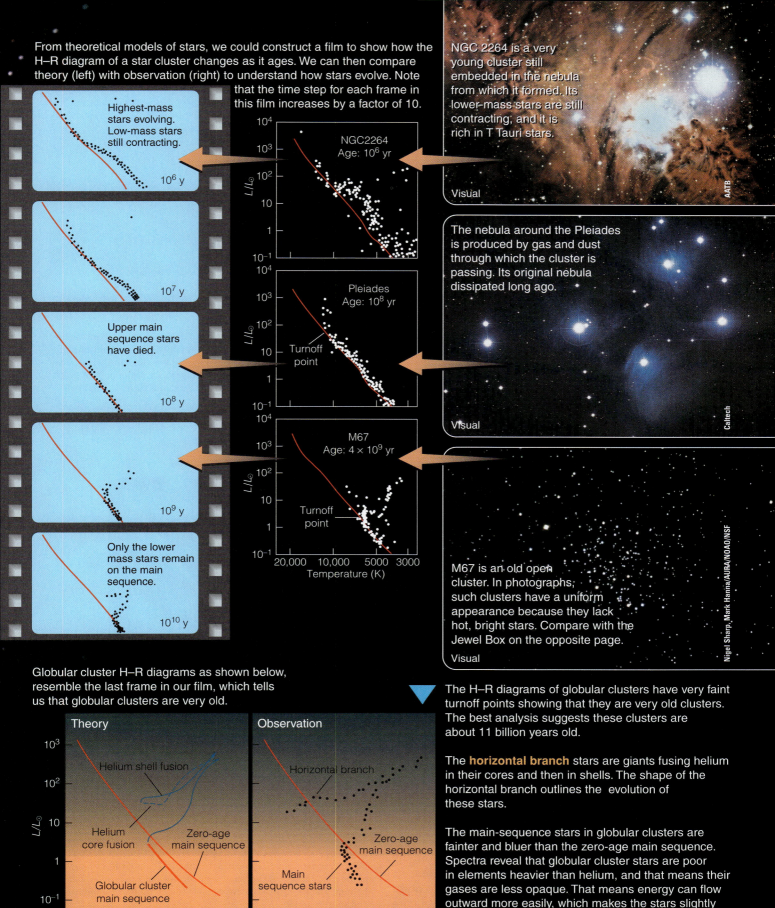

From theoretical models of stars, we could construct a film to show how the H–R diagram of a star cluster changes as it ages. We can then compare theory (left) with observation (right) to understand how stars evolve. Note that the time step for each frame in this film increases by a factor of 10.

Highest-mass stars evolving. Low-mass stars still contracting.

10^6 y

10^7 y

Upper main sequence stars have died.

10^8 y

10^9 y

Only the lower mass stars remain on the main sequence.

10^{10} y

NGC2264
Age: 10^6 yr

Pleiades
Age: 10^8 yr

Turnoff point

M67
Age: 4×10^9 yr

Turnoff point

$L/L_\odot$

Temperature (K)

NGC 2264 is a very young cluster still embedded in the nebula from which it formed. Its lower-mass stars are still contracting, and it is rich in T Tauri stars.

Visual
AATB

The nebula around the Pleiades is produced by gas and dust through which the cluster is passing. Its original nebula dissipated long ago.

Visual
Caltech

M67 is an old open cluster. In photographs, such clusters have a uniform appearance because they lack hot, bright stars. Compare with the Jewel Box on the opposite page.

Visual
Nigel Sharp, Mark Hanna/AURA/NOAO/NSF

Globular cluster H–R diagrams as shown below, resemble the last frame in our film, which tells us that globular clusters are very old.

Theory

Helium shell fusion

Helium core fusion

Zero-age main sequence

Globular cluster main sequence

$L/L_\odot$

Temperature (K)

Observation

Horizontal branch

Zero-age main sequence

Main sequence stars

Temperature (K)

The H–R diagrams of globular clusters have very faint turnoff points showing that they are very old clusters. The best analysis suggests these clusters are about 11 billion years old.

The **horizontal branch** stars are giants fusing helium in their cores and then in shells. The shape of the horizontal branch outlines the evolution of these stars.

The main-sequence stars in globular clusters are fainter and bluer than the zero-age main sequence. Spectra reveal that globular cluster stars are poor in elements heavier than helium, and that means their gases are less opaque. That means energy can flow outward more easily, which makes the stars slightly smaller and hotter. Again the shape of star cluster H–R diagrams illustrates principles of stellar evolution.

a

b

FIGURE 12-12

(a) The star δ Cephei (upper right) changes its brightness from about magnitude 3.6 at brightest to about magnitude 4.5 at faintest. The magnitudes of a few stars in the constellation Cepheus are given here for comparison. The constellation is shown as it appears on evenings in the autumn. (b) A graph of the brightness of δ Cephei versus time shows that it varies in brightness with a period slightly longer than 5 days.

including δ Cephei itself, have chemical compositions like that of the sun, but type II Cepheids and the RR Lyrae stars are poor in elements heavier than helium. These two facts can be summarized in a graph of absolute magnitude versus period (Figure 12-13). The two types of Cepheids are clearly separated, and the fainter RR Lyrae stars lie at the lower left.

Cepheid and RR Lyrae variable stars are critically important in astronomy because they can be used to find the distance to star clusters and galaxies. We will discuss this use of variable stars in later chapters; here we are interested in them as peculiar giant stars that are unstable.

PULSATING STARS

Why should a star pulsate? Why should there be a period–luminosity relation? Why are there two types of Cepheids? The answers to these questions help us understand the insides of stars. To find the answers, we start with the evolution of stars in the H–R diagram.

After a star leaves the main sequence, it fuses hydrogen in a shell, and the point that represents it in the H–R diagram moves to the right as the star becomes a cool giant. When it ignites helium in its core, the point moves back to the left. Soon, however, the helium core

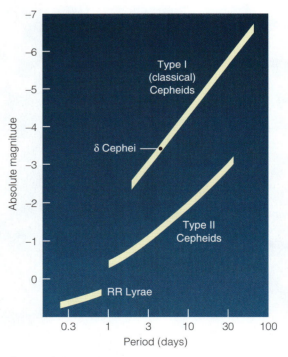

FIGURE 12-13

The Cepheid variable stars obey a period–luminosity relation: The longer-period stars are more luminous. The Cepheids are divided into two types. Type I Cepheids have chemical compositions like that of the sun, but type II Cepheids are poor in elements heavier than helium. The RR Lyrae stars are variable stars with lower luminosity and shorter periods.

is exhausted, and helium fusion continues only in a shell, forcing the star to expand again, and the point in the H–R diagram moves back to the right. Because giant stars evolve through these stages, their size and temperature change in complicated ways. Certain combinations of size and temperature make the star unstable, and it pulsates as an intrinsic variable star. In the H–R diagram, the region where size and temperature lead to pulsation is called the **instability strip** (❚ Figure 12-14).

Why do some stars pulsate? The answer has to do with the flow of energy out of the star's interior. We have seen that a star supports its own weight by generating energy and that the flow of that energy outward prevents the star from contracting. The star is balanced between the outward flow of energy and the inward force of gravity. We could make a stable star pulsate if we squeezed it and then released it. The star would rebound outward, and it might oscillate for a few cycles, but the oscillation would eventually run down because of friction. The energy of the moving gas would be converted to heat and radiated away, and the star would return to stability. To make a star pulsate continuously, some process must drive the oscillation, just as a spring in a wind-up clock is needed to keep it ticking.

Studies using stellar models show that variable stars in the instability strip pulsate like beating hearts because of an energy-absorbing layer in their outer envelopes. This layer is the region where helium is partially ionized. Above this layer the temperature is too low to ionize helium, and below the layer it is hot enough to ionize all of the helium. Like a spring, the helium ionization zone can absorb energy when it is compressed and release it when the zone expands. That is enough to keep the star pulsating.

We can follow this pulsation in our imaginations and see how the helium ionization zone makes a star pulsate. As the outer layers of a pulsating star expand, the ionization zone expands, and the ionized helium becomes less ionized and releases stored energy, which forces the expansion to go even faster. The surface of the star overshoots its equilibrium position—it expands too far—and eventually falls back. As the surface layers contract, the ionization zone is compressed and becomes more ionized, which absorbs energy. Robbed of some of their energy, the layers of the star can't support the weight, and they compress further. This allows the contraction to go even faster, and the in-falling layers overshoot the equilibrium point until the pressure inside the star slows them to a stop and makes them expand again. Thus, the surface of these stars is not in hydrostatic equilibrium but rather pulsates in and out.

Cepheids change their radius by 5 to 10 percent as they pulsate, and this motion can be observed as Doppler shifts in their spectra. When they expand, the surface layers approach, and we see a blue shift. When they contract, the surface layers recede, and we see a

FIGURE 12-14
Both the Cepheid and RR Lyrae variable stars lie in the instability strip (dashed lines) because their variability is caused by the same mechanism.

red shift. Although a 10 percent change in radius seems like a large change, it affects only the outer layers of the star. The center of the star is much too dense to be affected.

Now we can understand why there is an instability strip. Only stars in the instability strip have their helium ionization zone at the right depth to act as a spring and drive the pulsation. A star will pulsate if the zone lies at the right depth below the surface. In cool stars, the zone is too deep and cannot overcome the weight of the layers above it. That is, the spring is overloaded and can't expand. In hot stars, the helium ionization zone lies near the surface, and there is little weight above it to compress it. That is, the spring never gets squeezed. The stars in the instability strip have the right temperature and radius for the helium ionization zone to fall exactly where it is most effective, and thus those stars pulsate.

We can also understand why there is a period–luminosity relation. The more massive stars are larger and more luminous. When they cross the instability strip, they cross higher in the H–R diagram. But the larger a star is, the more slowly it pulsates—just as large bells go "bong" and little bells go "ding." Pulsation period depends on size, which depends on mass, and mass also determines luminosity. So the most luminous Cepheids at the top of the instability strip are also the largest and pulsate the slowest (❚ Figure 12-15). There has to be a period–luminosity relation because period and luminosity both depend on mass.

FIGURE 12-15
The instability strip (shaded) is a region of the H–R diagram in which stars are unstable and pulsate as variable stars. Massive stars cross the strip at high luminosities, and because of their large radius they pulsate with longer periods, such as 50 days. Lower-mass stars cross the instability strip at lower luminosities, and because of their smaller radius they pulsate with shorter periods, such as 3 days. This accounts for the period–luminosity relation.

We also know enough about stars to understand why there are two types of Cepheid variable stars. Type I Cepheids have chemical abundances roughly like those of the sun, but type II Cepheids are poor in elements heavier than helium, and that means the gases in type II Cepheids are not as opaque as the gases in type I Cepheids. The all-important energy flowing outward can escape more easily in type II Cepheids, and they reach a slightly different equilibrium. For stars of the same period of pulsation, type II Cepheids are less luminous (Figure 12-13).

PERIOD CHANGES IN VARIABLE STARS

Some Cepheids have been observed to change their periods of pulsation, and those changes seem to arise from the evolution of the stars.

The evolution of a star may carry it through the instability strip a number of times, and each time it can become a variable star. If we could watch a star as it entered the instability strip, we could see it begin to pulsate. Similarly, if we could watch a star leave the instability strip, we could watch it stop pulsating. Such an observation would confirm our belief that the stars are evolving. Unfortunately, stars evolve so slowly that these events are surely rare.

One famous Cepheid, RU Camelopardalis, was observed to stop pulsating temporarily. Its pulsations died away in 1966, and many astronomers assumed that it had stopped pulsating because it was leaving the instability strip. But its pulsations resumed in 1967. Polaris may be going through a similar change now, but it does not lie near the edge of the instability strip in the H–R diagram. Thus, it may not be an example of a star evolving out of the instability strip.

Even more dramatic evidence of stellar evolution can be found in the slowly changing periods of Cepheid variables. Even a tiny change in period can become easily observable because it accumulates, just as a clock that gains a second each day becomes obviously fast over a time as long as a year. Some Cepheids have periods that are growing shorter, and others have periods that are growing longer. These changes occur because the stars are evolving and changing their radii; contraction shortens the period, and expansion lengthens the period. The star X Cygni, for example, has a period that is gradually growing shorter, and it has been "gaining" since observations began in the late 19th century (∎ Figure 12-16). These changes in the periods of pulsating stars give us dramatic evidence that the stars are evolving.

Some kinds of variable stars lie outside the instability strip because their pulsation is driven by different mechanisms. It is sufficient here for us to discuss the Cepheids. They are common, they illustrate important ideas about stellar structure, and their changing periods make stellar evolution clearly evident. Also, we will use Cepheids in later chapters to study the distances to galaxies and the fate of the universe.

REVIEW CRITICAL INQUIRY

How do Cepheid variable stars give us evidence that stars are evolving?

If we observe Cepheids for a few years, we may notice that some pulsations are slowing down and others are speeding up. This is a very small change in the period of pulsation, but if we observe the stars for long periods of time, the change in their pulsation causes them to gain or lose just as a clock can gain or lose time over many weeks. These changes in period are caused by the expansion or contraction of the stars as they evolve. The period of pulsation depends on the size of the star, so pulsations of a star that is expanding should slow down slightly, and those of a star that is contracting should speed up slightly. When we observe these small changes in the pulsation periods of Cepheid variable stars, we are seeing direct evidence that the stars are evolving.

If we wanted to conduct this research project, what observations would we need to make, and how would we analyze them to detect changes in period? Think about a clock. How would we determine whether a clock was ticking a tiny bit too fast or too slow?

FIGURE 12-16

Like a clock running slow, the Cepheid variable star X Cygni has been reaching maximum light late for most of this century. As it evolves to the left across the instability strip, it is slowly contracting, and its period is growing shorter.

This chapter has discussed the evolution of stars, a process that occurs so slowly we cannot expect to see much happen in a single human lifetime. Thus, the study of stellar evolution depends heavily on theory and its verification by evidence from observations of real stars. We have seen that star clusters make the evolution of stars visible in the H–R diagram and that the pulsation of the intrinsic variable stars can give us insight into how giant stars evolve through the instability strip. We have not yet considered, however, the most interesting question of stellar evolution. What happens to a star when it uses up the last of its nuclear fuel? Clearly, stars must die—the question is how. We will explore that question in the next chapter.

SUMMARY

Almost everything we know about the internal structure of stars comes from mathematical stellar models. The models are based on four simple laws of stellar structure. Two laws say that mass and energy must be conserved and spread smoothly through the star. Another, the law of hydrostatic equilibrium, says that the star must balance the weight of its layers by its internal pressure. Yet another says that energy can flow outward only by conduction, convection, or radiation.

The mass–luminosity relation is explained by the requirement that a star support the weight of its layers by its internal pressure. The more massive a star is, the more weight it must support and the higher its internal pressure must be. To keep its pressure high, it must be hot and generate large amounts of energy. Thus, the mass of a star determines its luminosity. The massive stars are very luminous and lie along the upper main sequence. The less-massive stars are fainter and lie lower on the main sequence.

How long a star can stay on the main sequence depends on its mass. The more massive a star is, the faster it uses up its hydrogen fuel. A 25-solar-mass star will exhaust its hydrogen and die in only about 7 million years, but the sun can last for 10 billion years.

When a main-sequence star exhausts its hydrogen, it does not just wink out. Its core contracts, and it begins to fuse hydrogen in a shell around its core. The outer parts of the star—its envelope—swell, and the star becomes a giant. Because of this expansion, the surface of the star cools, and it moves toward the right in the H–R diagram. The most massive stars move across the top of the diagram as supergiants.

As the core of the star continues to contract, it finally ignites helium fusion, a process that converts helium into carbon. If the core becomes degenerate before helium ignites, the pressure of the gas does not depend on its temperature, and when helium ignites, it explodes in the helium flash. Although the helium flash is violent, the star absorbs the extra energy and quickly brings the helium-fusing reactions under control.

After helium is exhausted in the core, it can fuse in a shell around the core. Then, if the star is massive enough, it can contract and ignite other fuels, such as carbon.

Confirmation of stellar evolution comes from star clusters. Because all the stars in a cluster have about the same distance, composition, and age, we can see the effects of stellar evolution in the H–R diagram of a cluster. Massive stars evolve faster than low-mass stars, so in a given cluster the most massive stars leave the main sequence first. We can judge the age of such a cluster by looking at the turnoff point, the location on the main sequence where the stars turn off to the right and become giants. The life expectancy of a star at the turnoff point equals the age of the cluster.

There are two types of star clusters. Open clusters contain 10 to 1000 stars and have an open, transparent appearance. Globular clusters contain 10^5 to 10^6 stars densely packed into a spherical shape. The open clusters tend to be young to middle-aged, but globular clusters tend to be very old—11 billion years or more. Also, globular clusters tend to be poor in elements heavier than helium.

A simple understanding of stellar structure explains many of the properties of pulsating stars such as the Cepheid and RR Lyrae variables. These intrinsic variable stars lie in an instability strip in the H–R diagram because they contain a layer in their envelopes that stores and releases energy as they expand and contract. Stars outside the instability strip do not pulsate, because the layer is too deep or too shallow to make the stars unstable. The Cepheids obey a period–luminosity relation because more massive stars, which are more luminous, become larger and pulsate more slowly.

Some Cepheids have periods that are slowly changing, showing that the evolution of the star is changing the star's radius and thus its period of pulsation.

NEW TERMS

conservation of mass law	globular cluster
conservation of energy law	turnoff point
stellar model	horizontal branch
brown dwarf	variable star
zero-age main sequence (ZAMS)	intrinsic variable
degenerate matter	Cepheid variable star
triple alpha process	RR Lyrae variable star
helium flash	period–luminosity relation
open cluster	instability strip

REVIEW QUESTIONS

1. Why is there a lower end to the main sequence? Why is there an upper end?

2. What is a brown dwarf?

3. Why is there a mass–luminosity relation?

4. Why does a star's life expectancy depend on mass?

5. Why do expanding stars become cooler and more luminous?

6. What causes the helium flash? Why does it make it difficult for astronomers to understand the later stages of stellar evolution?

7. How do some stars avoid the helium flash?

8. Why are giant stars so low in density?

9. Why are lower-mass stars unable to ignite more massive nuclear fuels such as carbon?

10. How can we estimate the age of a star cluster?

11. How do star clusters confirm that stars evolve?

12. How do some variable stars prove that stars are evolving?

DISCUSSION QUESTIONS

1. How do we know that the helium flash occurs if it cannot be observed? Can we accept an event as real if we can never observe it?

2. Can you think of ways that chemical differences could arise in stars in a single star cluster? Consider the mechanism that triggered their formation.

PROBLEMS

1. In the model shown in Figure 12-1, how much of the sun's mass is hotter than 13,000,000 K?

2. What is the life expectancy of a 16-solar-mass star? of a 50-solar-mass star?

3. How massive could a star be and still survive for 5 billion years?

4. If the sun expanded to a radius 100 times its present radius, what would its density be? (*Hint:* The volume of a sphere is $\frac{4}{3}\pi R^3$.)

5. If a giant star 100 times the diameter of the sun were 1 pc from us, what would its angular diameter be? (*Hint:* Use the small-angle formula, in Chapter 3.)

6. What fraction of the volume of a 5-solar-mass giant star is occupied by its helium core? (*Hints:* See Figure 12-8. The volume of a sphere is $\frac{4}{3}\pi R^3$.)

7. If the stars at the turnoff point in a star cluster have masses of about 4 solar masses, how old is the cluster?

8. If an open cluster contains 500 stars and is 25 pc in diameter, what is the average distance between the stars? (*Hints:* What share of the volume of the cluster surrounds the average star? The volume of a sphere is $\frac{4}{3}\pi R^3$.)

9. Repeat Problem 8 for a typical globular cluster containing a million stars in a sphere 25 pc in diameter.

10. If a Cepheid variable star has a period of pulsation of 2 days and its period increases by 1 second, how late will it be in reaching maximum light after 1 year? after 10 years? (*Hint:* How many cycles will it complete in a year?)

CRITICAL INQUIRIES FOR THE WEB

1. Find Hertzsprung–Russell (or color-magnitude) diagrams for five star clusters. Based on the distributions of stars on the diagrams, rank the clusters in order from oldest to youngest. At what value (spectral type, temperature, or B-V color index) does each cluster have its turnoff?

2. What is the latest news about Eta Carinae? It is a complex system and is constantly under study as astronomers try to understand what is happening to its stars.

3. Both professional and amateur astronomers study variable stars. Search the Web for the American Association of Variable Star Observers and see what information you can find. Can you find recent data and light curves for known variables like δ Cephei and X Cyg?

4. What is the latest news about brown dwarfs?

EXPLORING *THESKY*

1. Locate the open clusters below and describe their location with respect to the Milky Way Galaxy.

    M7, M41, M44, M45

2. Locate the clusters in Activity 1 and zoom in until you can see individual stars. Plot an H–R diagram for each cluster using the spectral type and apparent magnitude of the stars you see. Discuss the ages of the clusters. (*Hint:* Remember you can see only the brightest stars in each cluster.)

3. Use the H–R diagram of the Pleiades star cluster (M45) from Activity 2 to find the distance to the cluster. (*Hint:* How bright should the main-sequence stars be in absolute magnitude?)

 Visit the Seeds *Foundations of Astronomy* companion Web site for critical thinking exercises, articles, and additional readings from InfoTrac College Edition, Brooks/Cole's online student library.

THE DEATHS OF STARS

Natural laws have no pity.

Robert Heinlein, The Notebooks of Lazarus Long

Hubble Heritage Team, AURA/STScI/NASA

GUIDEPOST

As you read this chapter, take a moment to be astonished and proud that the human race knows how stars die. Finding the masses of stars is difficult, and understanding the invisible matter between the stars takes ingenuity, but we humans have used what we know about stars and what we know about the physics of energy and matter to solve one of nature's deepest mysteries. We know how stars die.

The previous chapter made heavy use of theory in describing the internal structure of stars. We tested that theory against observations of clusters of stars and variable stars. In this chapter, we have more direct observational evidence to help us understand how stars die. We still need theory, but images and spectra of dying stars give us landmarks to guide our theoretical explorations.

This chapter leads us onward, not only to discuss the remains of dead stars—neutron stars and black holes—but it also leads us into discussions of the formation of our galaxy, the origin of the universe, the formation of planets, and the origin of life. Our world is made of atoms that were born in the deaths of stars. This chapter of endings is also a chapter of beginnings.

VIRTUAL LABORATORY

STELLAR EXPLOSIONS, NOVAE, AND SUPERNOVAE

Gravity is patient—so patient it can kill stars. Occasionally astronomers see a new star appear in the sky, grow brighter, fade away after a few weeks or a year, and vanish. We will discover that a **nova,** what seems to be a new star in the sky, is in fact produced by a very old, dying star, and that a **supernova,** a particularly luminous and long-lasting nova, is caused by the explosive death of a star. Modern astronomers find a few novae (plural of *nova*) each year, but supernovae (plural) are so rare that there are only one or two supernovae each century in our galaxy. Astronomers know that supernovae occur because they occasionally see them flare in other galaxies, and because we see in our own galaxy stars about to explode as supernovae and the shattered remains of these violent explosions (▌ Figure 13-1).

In the previous chapter, we saw that stars resist their own gravity by generating energy through nuclear fusion. The energy keeps their interiors hot, and the resulting high pressure balances gravity and prevents the star from collapsing. Stars, however, have limited fuel, and gravity is patient. When the stars exhaust their fuel, gravity wins, and the stars die.

The mass of a star is critical in determining its fate. Massive stars use up their nuclear fuel at a furious rate and die after only a few million years, while the lowest-mass stars use their fuels sparingly and may be able to live hundreds of billions of years. Massive stars can die in violent explosions that we see as supernovae, but low-mass stars die quiet deaths. To follow the evolution of stars to their graves, we will sort the stars into three categories according to their masses: low-mass red dwarfs, medium-mass sunlike stars, and massive upper-main-sequence stars. These stars lead dramatically different lives and die in different ways.

13-1 LOWER-MAIN-SEQUENCE STARS

The stars of the lower main sequence share a common characteristic: They have relatively low masses. That means that they face similar fates as they exhaust their nuclear fuels.

When a star exhausts one nuclear fuel, its interior contracts and grows hotter until the next nuclear fuel ignites. The contracting star heats up by converting gravitational energy into thermal energy, so low-mass stars cannot get very hot, and this limits the fuels they can ignite. The lowest-mass stars, for example, cannot get hot enough to ignite helium fusion.

Ace◐Astronomy™ The AceAstronomy icon throughout the text indicates an opportunity for you to test yourself on key concepts and to explore animations and interactions on the AceAstronomy Web site at **http://astronomy.brookscole .com/seeds8e**

Structural differences divide the low-mass stars into two subgroups—very-low-mass stars and medium-mass stars such as the sun. The critical difference between the two groups is the extent of interior convection (see Figure 11-14). If the star is convective, fuel is constantly mixed, and the resulting evolution of the star is drastically altered.

RED DWARFS

Stars less massive than about 0.4 solar mass—the red dwarfs—have two advantages over more massive stars. First, they have very small masses, and thus they have very little weight to support. Their pressure–temperature thermostats are set low, and they consume their hydrogen fuel very slowly. Our discussion of the life expectancies of stars in Chapter 12 predicted that the red dwarfs would live very long lives.

The red dwarfs have a second advantage because they are totally convective. That is, they are stirred by circulating currents of hot gas rising from their interiors and cool gas sinking inward from their surfaces. This mixes the stars like pots of soup that are constantly stirred. Hydrogen is consumed, and helium accumulates uniformly throughout the stars; and that has two important consequences for red dwarfs.

Because they use up their hydrogen uniformly, they can never develop a helium core and a hydrogen fusion shell. That means they can never become giant stars. They are doomed to convert their hydrogen into helium, a fuel they are not massive enough to fuse, and then die slow, unremarkable deaths.

Because red dwarf stars are completely mixed and can fuse all of their hydrogen, their lives are prolonged even further. Models of a 0.1-solar-mass red dwarf predicted it would take 2 billion years to contract to the main sequence and 6 trillion years to use up its hydrogen fuel. Because the universe is only about 14 billion years old, none of these stars should have exhausted its fuel yet.

SUNLIKE STARS

Stars with masses between roughly 0.4 and 4 solar masses,* including the sun, ignite hydrogen and then helium and become giants, but they cannot get hot enough to ignite carbon, the next fuel in the sequence (see Table 12-3). When they reach that impasse, they can no longer maintain their stability, and their interiors contract while their envelopes expand.

To understand the fate of these stars, we must consider two concepts: mixing and expansion. The interiors of these sunlike stars are not well mixed (see Figure 11-14). As we learned in Chapter 11, stars of 1.1 solar masses or less, including the sun, have no convection

---

*This mass limit is uncertain, as are many of the masses stated here. The evolution of stars is highly complex, and such parameters are not precisely known.

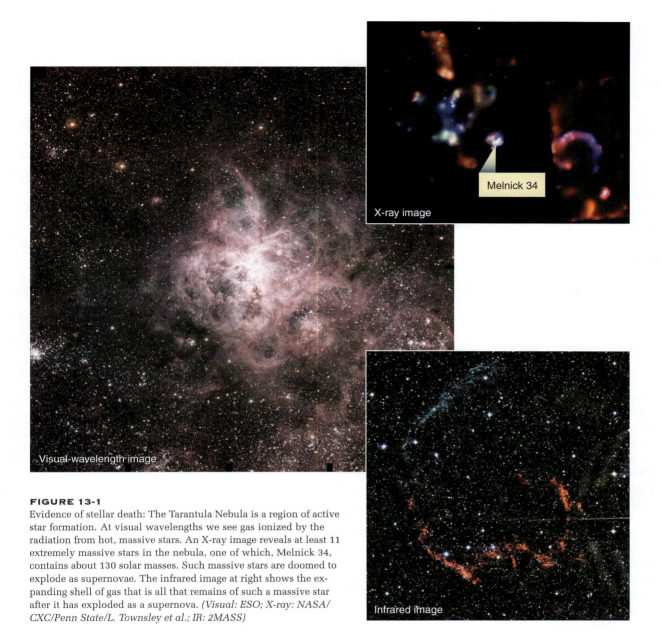

FIGURE 13-1

Evidence of stellar death: The Tarantula Nebula is a region of active star formation. At visual wavelengths we see gas ionized by the radiation from hot, massive stars. An X-ray image reveals at least 11 extremely massive stars in the nebula, one of which, Melnick 34, contains about 130 solar masses. Such massive stars are doomed to explode as supernovae. The infrared image at right shows the expanding shell of gas that is all that remains of such a massive star after it has exploded as a supernova. *(Visual: ESO; X-ray: NASA/ CXC/Penn State/L. Townsley et al.; IR: 2MASS)*

near their centers, so they are not mixed at all. Stars more massive than 1.1 solar masses have small zones of convection at their centers, but this mixes no more than about 12 percent of the star's mass. Thus, medium-mass stars, whether they have convective cores or not, are not mixed, and the helium accumulates in an inert helium core surrounded by unprocessed hydrogen. When this core contracts, the unprocessed hydrogen just outside the core ignites in a shell and swells the star into a giant.

As a giant, the star fuses helium in its core and then in a shell surrounding a core of carbon and oxygen. This core contracts and grows hotter, but it cannot become

hot enough to ignite the carbon. Thus, the carbon–oxygen core is a dead end for these medium-mass stars.

Because no nuclear reactions can begin in the carbon–oxygen core, it cannot resist the weight pressing down on it. In addition, just outside the carbon–oxygen core, the helium-fusion shell converts helium into carbon and thus increases the mass of the carbon–oxygen core, so the core must contract. The energy released by the contracting core, plus the energy generated in the helium- and hydrogen-fusing shells, flows outward and makes the envelope of the star expand.

This forces the star to become a very large giant. Its radius may become as large as the radius of Earth's orbit, and its surface becomes as cool as 2000 K. Such a star can lose large amounts of mass from its surface.

MASS LOSS FROM SUNLIKE STARS

We have clear evidence that stars like the sun lose mass. The sun itself is losing mass. The solar wind is a gentle breeze of gas blowing outward from the solar corona and carrying mass into space.

Other stars like the sun are losing mass, and the evidence lies in their spectra. Ultraviolet and X-ray spectra of many stars reveal strong emission lines that must originate in hot chromospheres and coronas like the sun's. If these stars have atmospheres like the sun's, they presumably have similar winds of hot gas. Also, the spectra of some giant stars contain blue-shifted absorption lines. The blue shifts must be Doppler shifts produced as the gas flowing out of the star comes toward us.

The solar wind is not very strong; the sun loses about 0.001 solar mass per billion years. Even over the entire lifetime of the sun, that would not significantly alter the sun's mass. However, once the sun becomes a giant star, it may lose mass more rapidly. Because the spectra of some of these stars do not show the characteristic emission, we must assume that the stars do not have hot coronas, but other processes could drive mass loss. The stars are so large that gravity is weak at their surfaces, and convection in the cool gas can drive shock waves outward and power mass loss. In addition, some giants are so cool that specks of carbon dust condense in their atmospheres, just as soot can condense in a fireplace. The pressure of the star's radiation can push this dust and any atoms that collide with the dust completely out of the star.

Another factor that can cause mass loss is periodic eruptions in the helium-fusion shell. The triple-alpha process that fuses helium into carbon is extremely sensitive to temperature, and that can make the helium-fusion shell so unstable that it may experience eruptions called **thermal pulses.** Every 200,000 years or so, the shell can erupt and suddenly produce energy equivalent to a million times the luminosity of the sun. Almost all of this energy goes into lifting the outer layers of the star and driving gas away from the surface.

Whatever drives this mass loss from giants, it can affect the mass of the star appreciably in a short time. A star expanding as its carbon–oxygen core contracts could lose an entire solar mass in only 10^5 years, which is not a long time in the evolution of a star. Thus, a star that began its existence on the main sequence with a mass of 8 solar masses might reduce its mass to only 3 solar masses in half a million years.

Stellar mass loss confuses our story of stellar evolution. We would like to say that stars more massive than a certain limit will evolve one way, and stars less massive will evolve another way. But stars may lose enough mass to alter their own evolution. Thus, we must consider both the initial mass a star has on the main sequence and the mass it retains after mass loss. Because we don't know exactly how effective mass loss is, it is difficult for us to be exact about the mass limits in our discussion of stellar evolution.

The mass loss we have discussed so far has been a gradual process, but a sunlike star with a contracting carbon–oxygen core expands so far that it can expel all of its outer layers. This surface expulsion is helped by the growing pressure of the radiation inside the star and by instabilities in the helium-fusion shell. The expanding layers are visible as small nebulae.

PLANETARY NEBULAE

When a medium-mass star like the sun becomes a distended giant, it can expel its outer atmosphere to form one of the most interesting objects in astronomy, a **planetary nebula,** so called because it looks like the greenish-blue disk of a planet such as Uranus or Neptune. In fact, a planetary nebula has nothing to do with a planet. It is composed of ionized gases expelled by a dying star.

Study ▌ "The Formation of Planetary Nebulae" on pages 260 and 261 and notice four things. First, we can understand what planetary nebulae are like by using simple observational methods we have discussed before. Second, notice the model astronomers have developed to explain planetary nebulae. The real nebulae are more complex than our simple model of a slow wind and a fast wind, but the model gives us a way to organize the observed phenomena. The third thing to notice is how oppositely directed jets (much like bipolar flows) produce many of the asymmetries seen in planetary nebulae. Finally, notice the fate of the star itself; it must contract into a white dwarf.

The medium-mass stars, including the sun, must die by forming planetary nebulae and contracting into white dwarfs. This suggests another way we can search for evidence of the deaths of medium-mass stars: We can study white dwarfs.

WHITE DWARFS

White dwarfs are among the most common stars. The survey we took in Chapter 9 revealed that white dwarfs and red dwarfs, although very faint, are the dominant kinds of stars (page 193). Now we can recognize white dwarfs as the remains of medium-mass stars that fused hydrogen and helium, failed to ignite carbon, drove away their outer layers to form planetary nebulae, and collapsed and cooled to form white dwarfs. The billions of white dwarfs in our galaxy must be the remains of medium-mass stars.

The first white dwarf discovered was the faint companion to Sirius. In that visual binary system, the bright star is Sirius A. The white dwarf, Sirius B, is 10,000 times fainter than Sirius A. The orbital motions of the stars (shown in Figure 9-14) tell us that the white dwarf's mass is 0.98 solar mass, and its blue-white color tells us that its surface is hot, about 25,000 K. Although it is very hot, it has a very low luminosity, so it must have a small surface area—in fact, its diameter is about twice Earth's. Dividing its mass by its volume reveals that it is very dense—over 3×10^6 g/cm^3. On Earth, a teaspoonful of Sirius B material would weigh more than 15 tons (■ Figure 13-2). Thus, basic observations and simple physics lead us to the conclusion that white dwarfs are astonishingly dense.

A normal star is supported by energy flowing outward from its core, but a white dwarf cannot generate energy by nuclear fusion. It has exhausted its hydrogen and helium fuel and produced carbon. As the star collapses into a white dwarf it converts gravitational energy into thermal energy, and its interior becomes very hot, but it cannot get hot enough to ignite carbon fusion. The star contracts until it becomes degenerate. Although a tremendous amount of energy flows out of the hot interior, it is not the energy flow that supports the star. The white dwarf is supported against its own gravity by the inability of its degenerate electrons to pack into a smaller volume.

The interior of a white dwarf is mostly carbon and oxygen ions floating among a whirling storm of degenerate electrons. It is the degenerate electrons that exert the pressure to support the star's weight, but most of its mass is represented by the carbon and oxygen ions. Theory predicts that as the star cools these ions will lock together to form a crystal lattice, so there may be some truth in thinking of aging white dwarfs as great crystals of carbon and oxygen. Near the surface, where the pressure is lower, a layer of ionized gases makes up a hot atmosphere.

The tremendous surface gravity of white dwarfs—100,000 times that of Earth—affects their atmospheres in strange ways. The heavier atoms in the atmosphere tend to sink, leaving the lightest gases at the surface. We see some white dwarfs with atmospheres of almost pure hydrogen, whereas others have atmospheres of nearly pure helium. Still others, for reasons not well understood, have atmospheres that contain traces of heavier atoms. In addition, the powerful surface gravity pulls the white dwarf's atmosphere down into a very shallow layer. If Earth's atmosphere were equally shallow, people on the top floors of skyscrapers would have to wear oxygen masks.

Clearly, a white dwarf is not a normal star. It generates no nuclear energy, is almost totally degenerate, and, except for a thin layer at its surface, contains no gas. Instead of calling a white dwarf a "star," we can

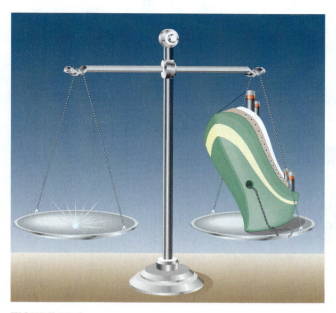

FIGURE 13-2
The degenerate matter from inside a white dwarf is so dense that a lump the size of a beach ball would, transported to Earth, weigh as much as an ocean liner.

call it a **compact object.** In the next chapter, we will discuss two other compact objects—neutron stars and black holes.

A white dwarf's future is bleak. As it radiates energy into space, its temperature gradually falls, but it cannot shrink any smaller because its degenerate electrons cannot get closer together. This degenerate matter is a very good thermal conductor, so heat flows to the surface and escapes into space, and the white dwarf gets fainter and cooler, moving downward and to the right in the H–R diagram. Because the white dwarf contains a tremendous amount of heat, it needs billions of years to radiate that heat through its small surface area. Eventually, such objects may become cold and dark, so-called **black dwarfs.** Our galaxy is not old enough to contain black dwarfs. The coolest white dwarfs in our galaxy are about the temperature of the sun.

Perhaps the most interesting thing about white dwarfs has come from mathematical models. The equations predict that if we added mass to a white dwarf, its radius would *shrink,* because added mass would increase its gravity and squeeze it tighter. If we added enough to raise its total mass to about 1.4 solar masses, its radius would shrink to zero (■ Figure 13-3). This is called the **Chandrasekhar limit** after Subrahmanyan Chandrasekhar, the astronomer who discovered it. It seems to imply that a star more massive than 1.4 solar masses could not become a white dwarf unless it shed mass in some way.

Stars do lose mass. Young stars have strong stellar winds, and aging giants also lose mass (■ Figure 13-4).

The Formation of Planetary Nebulae

Simple observations tell us what planetary nebulae are like. Their angular size and their distance tell us that their radii range from 0.2 to 3 light-years. The presence of emission lines in their spectra assures us they are excited, low-density gas. Doppler shifts show they are expanding at 10 to 20 km/s. If we divide radius by velocity, we find planetary nebulae are no more than 10,000 years old. Older planetary nebulae evidently become mixed into the interstellar medium.

We see about 1500 planetary nebulae in our sky. Because planetary nebulae are short-lived formations, we can conclude that they must be a common part of stellar evolution. All stars like the sun, up to a mass of about 8 solar masses, are destined to die by forming planetary nebulae.

IC3568   Visual

This nearly spherical planetary nebula has a low-luminosity outer envelope and a highly excited inner region.

Visual

The Ring Nebula in the constellation Lyra is visible even in small telescopes. Note the hot blue star at its center and the radial texture in the gas, suggesting outward motion.

The process that produces planetary nebulae involves two stellar winds. First, as an aging giant, the star gradually blows away its outer layers in a slow breeze of low-excitation gas that is not easily visible. Once the hot interior of the star is exposed, it ejects a high-speed wind that overtakes and compresses the gas of the slow wind like a snowplow, while ultraviolet radiation from the hot remains of the central star excites the gases to glow like a giant neon sign.

Slow stellar wind from a red giant

The gases of the slow wind are not easily detectable.

Fast wind from exposed interior

We see a planetary nebula where the fast wind compresses the slow wind.

Images from the Hubble Space Telescope reveal that asymmetry is the rule in planetary nebulae rather than the exception. A number of causes have been suggested. A disk of gas around a star's equator might form during the slow-wind stage and then deflect the fast wind into oppositely directed flows. Another star or planets orbiting the dying star, rapid rotation, or magnetic fields might cause these peculiar shapes. The Hour Glass Nebula seems to have formed when a fast wind overtook an equatorial disk (white in the image). The nebula Menzel 3, as do many planetary nebulae, shows evidence of multiple ejections.

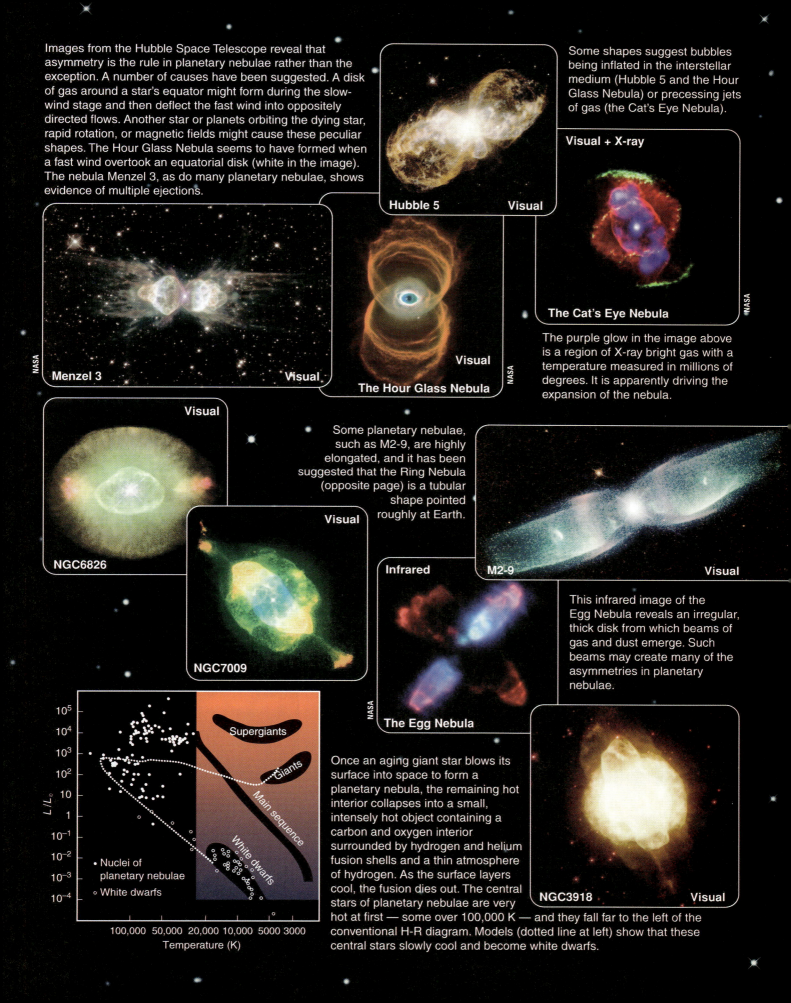

Hubble 5 — Visual

Some shapes suggest bubbles being inflated in the interstellar medium (Hubble 5 and the Hour Glass Nebula) or precessing jets of gas (the Cat's Eye Nebula).

Visual + X-ray

NASA

The Cat's Eye Nebula

The purple glow in the image above is a region of X-ray bright gas with a temperature measured in millions of degrees. It is apparently driving the expansion of the nebula.

NASA

Menzel 3 — Visual

Visual

NASA

The Hour Glass Nebula

Visual

NGC6826

Some planetary nebulae, such as M2-9, are highly elongated, and it has been suggested that the Ring Nebula (opposite page) is a tubular shape pointed roughly at Earth.

Visual

NGC7009

Infrared

M2-9 — Visual

NASA

The Egg Nebula

This infrared image of the Egg Nebula reveals an irregular, thick disk from which beams of gas and dust emerge. Such beams may create many of the asymmetries in planetary nebulae.

- Nuclei of planetary nebulae
- White dwarfs

$L/L_\odot$

10^5
10^4
10^3
10^2
10
1
10^{-1}
10^{-2}
10^{-3}
10^{-4}

Supergiants

Giants

Main sequence

White dwarfs

100,000 50,000 20,000 10,000 5000 3000
Temperature (K)

NGC3918 — Visual

Once an aging giant star blows its surface into space to form a planetary nebula, the remaining hot interior collapses into a small, intensely hot object containing a carbon and oxygen interior surrounded by hydrogen and helium fusion shells and a thin atmosphere of hydrogen. As the surface layers cool, the fusion dies out. The central stars of planetary nebulae are very hot at first — some over 100,000 K — and they fall far to the left of the conventional H-R diagram. Models (dotted line at left) show that these central stars slowly cool and become white dwarfs.

FIGURE 13-3

The more massive a white dwarf, the smaller its radius. Stars more massive than the Chandrasekhar limit of 1.4 solar masses cannot be white dwarfs.

This suggests that stars more massive than the Chandrasekhar limit can eventually die as white dwarfs if they reduce their mass. Theoretical models show an 8-solar-mass star should be able to reduce its mass to 1.4 solar masses before it collapses, and slightly more massive stars may also be able to get under the limit. Thus, a wide range of medium-mass stars eventually die as white dwarfs.

REVIEW CRITICAL INQUIRY

How will the sun die?

We have seen that lower-main-sequence stars like the sun die by producing planetary nebulae and then becoming white dwarfs. For the sun, this process will begin in a few billion years, when it exhausts the hydrogen in its core. It will then expand to become a giant star. Although the surface of the sun will become cooler, it will expand until it has such a large surface that it will be about 100 times more luminous than it is now. As a giant star, the sun will have a strong solar wind carrying gas into space and pushing back the interstellar medium. Eventually, the loss of mass will expose deeper layers in the sun. These extremely hot layers will ionize the gas around the sun and propel it outward to scoop up previously expelled gases and produce a planetary nebula. Soon the last remains of the sun will collapse inward and slowly cool to form a small, dense white dwarf. The surface will be very hot but so small that the sun will be about 100 times fainter than it is now.

The story of stellar evolution tells us that our sun will someday create a planetary nebula and become a white dwarf. At that point, what will have become of all the hydrogen the sun once contained?

a

b

FIGURE 13-4

Many stars lose mass as they age. (a) This far-infrared map of the constellation Orion shows clouds of gas and dust heated and pushed away by hot, young stars. The Orion Nebula is the brightest area at lower right, and the large ring at upper right is expanding away from the O8 giant star λ Orionis. *(NASA/IPAC, Courtesy Deborah Levine)* (b) The massive star WR124 is ejecting mass in a violent stellar wind, forming chaotic arcs and blobs. *(Grosdidier and Moffat, University of Montreal; Joncas, Université Laval; Acker, Observatorie de Strasbourg; and NASA)*

Lower-main-sequence stars die by producing beautiful planetary nebulae and shrinking to become white dwarfs. Now we are ready to see what happens when one of these dying stars has a stellar companion.

13-2 THE EVOLUTION OF BINARY STARS

So far we have discussed the deaths of stars as if they were all single objects that never interact. But more than half of all stars are members of binary star systems. Most such binaries are far apart, and one of the stars can swell into a giant and eventually collapse without affecting the companion star. The two stars in some systems, however, are close together. When the more massive star begins to expand, it interacts with its companion star in peculiar ways.

These interacting binary stars are interesting objects themselves, but they are also important because they help us explain observed phenomena such as nova explosions. In the next chapter, we will use them to help us find black holes.

MASS TRANSFER

Binary stars can sometimes interact by transferring mass from one star to the other. To understand this process, we must understand how the gravity of the two stars controls the matter in the binary system.

Each star in a binary system controls a region of space near it, and if we could make these regions visible they would look like two teardrop-shaped lobes enclosing the stars and meeting tip to tip between the stars. These two volumes are called the **Roche lobes**, and the surface of the lobes is called the **Roche surface**. Because of the interaction of the gravity of the two stars and their rotation around their center of mass, there are five points of stability in the orbital plane called **Lagrangian points**.* ▌Figure 13-5 illustrates the arrangement of the Roche surface around the stars and the Lagrangian points.

The Lagrangian points are important because they are locations where gas is stable in the revolving binary system. Gas at the L_2 or L_3 point tends to leak away, but gas at the L_4 or L_5 point can be trapped. (We will use the L_4 and L_5 points when we discuss planetary satellites and the orbits of asteroids in later chapters.) For our purposes here, the inner Lagrangian point L_1 located between the two stars is critical. If matter from one star can reach the inner Lagrangian point, it can

*The Lagrangian points are named after French mathematician Joseph Louis Lagrange, who solved this famous mathematical problem around the time of the French Revolution.

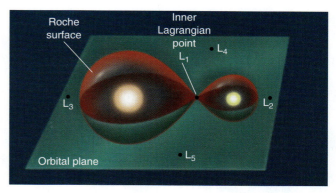

FIGURE 13-5

A pair of binary stars controls the region of space located inside the Roche surface. The Lagrangian points are locations of stability, with the inner Lagrangian point making a connection through which the two stars can transfer matter.

flow onto the other star. That is, the stars can transfer mass through the inner Lagrangian point.

In general, there are only two ways matter can escape from a star and reach the inner Lagrangian point. First, if a star has a strong stellar wind, some of the gas blowing away from the star can pass through the inner Lagrangian point and be captured by the other star. Second, if an evolving star expands so far that it fills its Roche surface, it will be forced to take on the teardrop shape of the Roche surface. We have seen that effect in our study of binary stars (Figure 9-20). If the star expands farther, matter will flow through the inner Lagrangian point (like water in a pond flowing over a dam) and will fall into the other star. Mass transfer driven by a stellar wind tends to be slow, but mass transfer driven by an expanding star can occur rapidly.

RECYCLED STELLAR EVOLUTION

Mass transfer between stars can affect their evolution in surprising ways. In fact, it provides the explanation for a problem that puzzled astronomers for many years.

In some binary systems, the less massive star has become a giant, while the more massive star is still on the main sequence. If more massive stars evolve faster than lower-mass stars, how does the low-mass star in such binaries manage to leave the main sequence first? This is called the Algol paradox, after the binary system Algol (see Figure 9-21).

Mass transfer explains how this could happen. Imagine a binary system that contains a 5-solar-mass star and a 1-solar-mass companion (▌Figure 13-6). The two stars formed at the same time, so the more massive star will evolve faster and leave the main sequence first. When it expands into a giant, it can fill its Roche surface and transfer matter to the low-mass companion. Thus, the massive star could evolve into a lower-

The Evolution of a Binary System

Star B is more massive than Star A.

Star B becomes a giant and loses mass to Star A.

Star B loses mass, and Star A gains mass.

Star A is a massive main-sequence star with a lower-mass giant companion—an Algol system.

Star A has now become a giant and loses mass back to the white dwarf that remains of Star B.

FIGURE 13-6
A pair of stars orbiting close to each other can exchange mass and modify their evolution.

mass star, and the companion could gain mass and become a massive star still on the main sequence. We might find a system such as Algol containing a 5-solar-mass main-sequence star and a 1-solar-mass giant.

The evolution of close binary stars could result in one of the stars having its outer layers "peeled" away. A massive star expanding to become a giant could lose its outer layers to its companion and then collapse to form a lower-mass peculiar star. A few such peculiar stars are known.

Another exotic result of the evolution of close binary systems is the merging of the stars. We see many binaries in which both stars have expanded to fill their Roche surfaces and spill mass out into space. If the stars are close enough together and expand rapidly enough, theorists believe, the two stars could merge into a single, rapidly rotating giant star. Most giants rotate slowly because they conserved angular momentum as they expanded and consequently slowed. Examples of rapidly rotating giant stars are known, however, and they are believed to be the results of merged binary stars. Inside the distended envelope of such a star, the cores of the two stars could continue to orbit each other until friction slowed them down and they sank to the center.

Mass transfer can lead to dramatic violence. The first four frames of Figure 13-6 show mass transfer producing a system like Algol. The last frame shows an additional stage in which the giant star has expelled its outer layers and has collapsed to form a white dwarf. The more massive companion has expanded, transferring matter back onto the white dwarf. Such systems can become the site of tremendous explosions. To see how this can happen, we must consider in detail how mass falls into a star.

ACCRETION DISKS

We say that matter flows through the inner Lagrangian point and falls onto the white dwarf, but the matter cannot fall directly into a white dwarf because of the conservation of angular momentum (as described in Chapter 5). Instead, it falls into a whirlpool around the white dwarf. For a common example, consider a bathtub full of water. Gentle currents in the water give it some angular momentum, but its slow circulation is not apparent until we pull the stopper. Then, as the water rushes toward the drain, conservation of angular momentum forces it to form a whirlpool. This same effect forces gas falling into a white dwarf to form a whirling disk of gas called an **accretion disk** (❙ Figure 13-7).

Two important things happen in an accretion disk. First, the gas in the disk grows very hot due to friction and tidal forces. The disk also acts as a brake, ridding the gas of its angular momentum and allowing it to fall into the white dwarf. The temperature of the gas in the inner parts of an accretion disk can exceed 1,000,000 K, and the gas can emit X rays. In addition, the matter falling inward from the accretion disk can cause a violent explosion if it accumulates on the white dwarf.

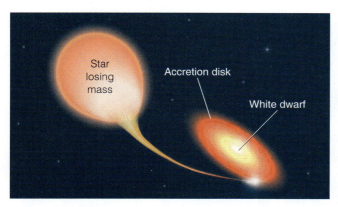

FIGURE 13-7
Matter falling into a compact object forms a whirling accretion disk. Friction and tidal forces can make the disk very hot.

NOVA EXPLOSIONS

At the beginning of this chapter we saw that the word "nova" refers to a new star that appears in the sky for a while and then fades away (Figure 13-8). The evidence tells us that it is not a new star at all. After a nova fades, astronomers can photograph the spectrum of the object, and they invariably find a closely spaced binary star containing a normal star and a white dwarf. A nova is evidently an explosion involving a white dwarf in a binary system.

Observational evidence can tell us how nova explosions occur. As the explosion begins, spectra show blue-shifted absorption lines, which tells us the gas is dense and coming toward us at a few thousand kilometers per second. After a few days, the spectral lines change to emission lines, which tells us the gas has thinned, but the blue shifts remain, showing that a cloud of debris had been ejected into space.

Theory gives us a way to understand nova explosions as the result of mass transfer from a normal star, through the inner Lagrangian point, through an accretion disk, onto the surface of the white dwarf. Because the matter comes from the surface of a normal star, it is rich in unfused fuel, mostly hydrogen, and when it accumulates on the surface of the white dwarf, it forms a layer of unprocessed fuel.

As the layer of fuel grows deeper, it becomes hotter and denser and eventually becomes degenerate, much like the core of a sunlike star approaching the helium flash. In such a gas, the pressure–temperature thermostat does not work, so the layer of unfused hydrogen is a thermonuclear bomb waiting to explode. By the time the white dwarf has accumulated about 100 Earth masses of hydrogen, the temperature at the base of the hydrogen layer reaches millions of degrees, and the density is 10,000 times the density of water. Suddenly, the hydrogen begins to fuse on the proton–proton chain, and the energy released drives the temperature

so high that the CNO cycle begins. With no pressure–temperature thermostat to control the fusion, in seconds the temperature shoots to 100 million degrees. Quickly the temperature rises high enough to force the gas to expand, and it stops being degenerate. But it is too late. By the time the gas begins to expand, the amount of energy released is enough to blow the surface layers of the star into space as a violently expanding shell of hot gas, which we see from Earth as a nova.

The explosion of its surface hardly disturbs the white dwarf or its companion star. Mass transfer quickly resumes, and a new layer of fuel begins to accumulate. How fast the fuel builds up depends on the rate of mass transfer. Accordingly, we can expect novae to repeat each time an explosive layer accumulates. Many novae take thousands of years to build an explosive layer, but some take only a few years (Figure 13-9).

THE END OF EARTH

Astronomy is about us. Although we have been discussing the deaths of medium-mass stars, white dwarfs, and novae explosions, we have also been discussing the future of our planet. The sun is a medium-mass star

FIGURE 13-8
Nova Cygni 1975 near maximum at about second magnitude (top) and later when it had declined to about eleventh magnitude (bottom). *(Photo © UC Regents)*

FIGURE 13-9

Nova T Pyxidis erupts about every two decades, expelling shells of gas into space. The shells of gas are visible from ground-based telescopes, but the Hubble Space Telescope reveals much more detail. The shell consists of knots of excited gas that presumably form when a new shell collides with a previous shell. *(M. Shara and R. Williams, STScI; R. Gilmozzi, ESO; and NASA)*

Ground-based image

Visual

Hubble Space Telescope image

Visual

and must eventually die by becoming a giant, possibly producing a planetary nebula, and collapsing into a white dwarf. That will spell the end of Earth.

Mathematical models of the sun suggest that it may survive for 5 billion years or so, but it is already growing more luminous as it fuses hydrogen into helium. In a few billion years, it will exhaust hydrogen in its core and swell into a giant star about 100 times its present radius. That giant sun will be about as large as the orbit of Earth, so that will mark the end of our world. Whether or not the expanding sun becomes large enough to totally engulf Earth, its growing luminosity will certainly evaporate our oceans, drive away our atmosphere, and even vaporize most of Earth's crust.

Ace Astronomy™ Go to AceAstronomy and click Active Figures to see "The Future of the Sun" and how Earth's temperature will change as the sun evolves.

While it is a giant star, the sun will lose mass into space. This mass loss is a relatively gentle process, so any cinder that might remain of Earth would not be disturbed. Of course, when the sun finally collapses into a white dwarf, the solar system will become a much colder place, but not much of Earth will remain by then.

If the collapsing sun becomes hot enough, it will ionize the expelled gas to form a planetary nebula. Some models, however, suggest that the sun is not quite massive enough. It will collapse into a hot object that will cool to become a white dwarf, but it may not become hot enough to ionize the expelled gas and "turn on" a planetary nebula. An astronomer recently expressed disappointment that the dying sun might not leave behind a beautiful planetary nebula, but that embarrassment lies a few billion years in the future.

There is no danger that the sun will explode as a nova; it has no binary companion. And, as we will see, the sun is not massive enough to die the violent death of the massive stars.

The most important lesson of astronomy is that we are part of the universe and not just observers. The atoms we are made of are destined to return to the interstellar medium in just a few billion years. That's a long time, and it is possible that the human race will migrate to other planetary systems. That might save the human race, but our planet is star dust.

REVIEW CRITICAL INQUIRY

How does modern astronomy explain the Algol paradox?

When we encounter a paradox in nature, it is usually a warning that we don't understand things as well as we thought we did. The Algol paradox is a good example. The binary star Algol contains a lower-mass giant star and a more massive main-sequence star. Because the two stars must have formed together, they are the same age; and the more massive star should have evolved first and left the main sequence. But in binary systems such as Algol, the lower-mass star has left the main sequence, or so it seems.

We can understand this paradox if we add mass transfer to our theory of stellar evolution. The first star to leave the main sequence must be the more massive star; but if the stars are close together, the star that is initially more massive can fill its Roche surface and transfer mass back to its companion. The companion can grow more massive, and the giant can grow less massive. Thus, the lower-mass giant star we see in the system may have originally been more massive and may have evolved away from the main sequence, leaving its companion behind to increase in mass as it decreased.

Mass transfer can explain the Algol paradox, and it can also explain the violent explosions called novae. How does mass transfer explain why novae can explode over and over in the same binary system?

Our study of the deaths of low- and medium-mass stars has led us to discuss planetary nebulae, white dwarfs, and the cause of novae. But there are more violent fish in the cosmic sea. What causes supernovae? To find out, we must consider how massive stars die.

13-3 THE DEATHS OF MASSIVE STARS

We have seen that low- and medium-mass stars die relatively quietly as they exhaust their hydrogen and helium and then eject their surface layers to form planetary nebulae. In contrast, massive stars live spectacular lives and destroy themselves in violent explosions.

NUCLEAR FUSION IN MASSIVE STARS

Stars on the upper main sequence have too much mass to die as white dwarfs, but their evolution begins much like that of their lower-mass cousins. They consume the hydrogen in their cores and ignite hydrogen shells; as a result, they expand and become giants or, for the most massive stars, supergiants. Their cores contract and fuse helium first in the core and then in a shell, producing a carbon–oxygen core.

A massive star can lose significant mass as it ages, but if it still has a mass over 4 solar masses when its carbon–oxygen core contracts, it can reach a temperature of 600 million Kelvin and ignite carbon fusion. The fusion of carbon produces heavier nuclei such as oxygen and neon. As soon as the carbon is exhausted in the core, the core contracts, and carbon ignites in a shell. This pattern of core ignition followed by shell ignition continues with fuel after fuel, and the star develops a layered structure at its center (❙ Figure 13-10), with a hydrogen-fusion shell above a helium-fusion shell above a carbon-fusion shell and so on. After carbon fusion, oxygen, neon, magnesium, and heavier elements fuse right up to iron.

The fusion of these nuclear fuels goes faster and faster as the massive star evolves rapidly. Recall that massive stars must consume their fuels rapidly to support their great weight, but other factors also cause the heavier fuels like carbon, oxygen, and silicon to fuse at increasing speed. For one thing, the amount of energy released per fusion reaction decreases as the mass of the fusing atom increases. To support its weight, a star

Infrared image

a

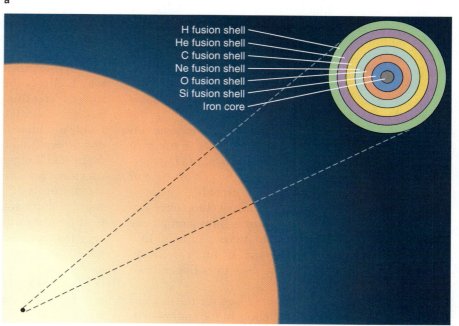

H fusion shell
He fusion shell
C fusion shell
Ne fusion shell
O fusion shell
Si fusion shell
Iron core

b

ACTIVE FIGURE 13-10

Massive stars live fast and die young. (a) Hidden deep inside the molecular cloud where it was born, the massive star AFGL2591 is rapidly expelling rings of gas to the right. Similar clouds of gas being expelled to the left are hidden behind a dusty disk surrounding the star. *(Gemini Observatory/NSF/C. Aspin)* (b) Inside an aging massive star lies an Earth-size core containing most of the mass. As nuclear fuels fuse one after the other, the core develops layers surrounding an inert iron core at the center.

Ace ◐Astronomy™ Go to AceAstronomy and click Active Figures to see "Inside Stars" and compare the interior of a sunlike star with that of a massive star.

must fuse oxygen much faster than it fused hydrogen. Also there are fewer atoms in the core of the star by the time heavy atoms begin to fuse. Four hydrogens made a helium atom, and three heliums made a carbon, so there are 12 times fewer atoms of carbon available for fusion than there were hydrogen. This means the heavy elements are used up and fusion goes very quickly in massive stars (▌Table 13-1). Hydrogen fusion can last 7 million years in a 25-solar-mass star, but that same star will fuse its oxygen in 6 months and its silicon in a day.

THE IRON CORE

Heavy-element fusion ends with iron, because nuclear reactions that use iron as a fuel cannot produce energy. Nuclear reactions can produce energy if they proceed from less tightly bound nuclei to more tightly bound nuclei. As shown in Figure 8-19, both nuclear fission and nuclear fusion produce nuclei that are more tightly bound than the starting fuel. Look again at Figure 8-19, and notice that iron is the most tightly bound nucleus of all. No nuclear reaction, fission or fusion, that starts with iron can produce a more tightly bound nucleus, and that means that iron is a dead end.

When a massive star develops an iron core, nuclear fusion cannot produce energy, and the core contracts and grows hotter. The shells around the core burn outward, fusing lighter elements and leaving behind more iron, which further increases the mass of the core. When the mass of the iron core exceeds 1.3 to 2 solar masses, the core must collapse.

As the core begins to contract, two processes can make it contract even faster. Heavy nuclei in the core can capture high-energy electrons, thus removing thermal energy from the gas. This robs the gas of some of the pressure it needs to support the crushing weight of the outer layers. Also, in more massive stars, temperatures are so high that many photons have gamma-ray wavelengths and can break more massive nuclei into less-massive nuclei. This reversal of nuclear fusion absorbs energy and allows the core to collapse even faster.

Although a massive star may live for millions of years, its iron core—about 500 km in diameter—collapses in only a few thousandths of a second. This collapse happens so rapidly that our most powerful computers are not capable of following the details. One thing is clear, however. The collapse of the iron core in a massive star triggers a star-destroying explosion—a supernova.

THE SUPERNOVA DEATHS
OF MASSIVE STARS

Modern theory predicts that the collapse of a massive star can eject the outer layers of the star to produce a supernova explosion while the core of the star col-

TABLE 13-1
Heavy-Element Fusion in a 25-$M_\odot$ Star

Fuel	Time	Percentage of Lifetime
H	7,000,000 years	93.3
He	500,000 years	6.7
C	600 years	0.008
O	0.5 years	0.000007
Si	1 day	0.00000004

lapses to form a neutron star or a black hole. We will examine neutron stars and black holes in detail in the next chapter; here we concentrate on the process that triggers the supernova explosion.

A supernova explosion is rare, remote, rapid, and violent. It is an event in nature that is extremely difficult to study in person, which is why astronomers have used powerful mathematical modeling techniques and high-speed supercomputers to explore the inside of a star exploding as a supernova. Such models allow astronomers to experiment on an exploding star as if the star were in a laboratory beaker.

Mathematical models reveal that the key to the supernova explosion is the collapse of the iron core, which allows the rest of the interior of the star to fall inward, creating a tremendous "traffic jam" as all the nuclei try to fall toward the center. It is as if all the residents of Indiana suddenly tried to drive their cars into downtown Indianapolis. There would be a tremendous traffic jam, not only downtown but also in the suburbs. As more cars arrived, the traffic jam would spread outward. Similarly, as the inner core of the star falls inward, a shock wave (a traffic jam) develops and begins to move outward. Containing about 100 times more energy than necessary to destroy the star, such a shock wave was first thought to be the cause of the supernova explosion.

Computer models revealed, however, that the shock wave spreading outward through the collapsing star stalls within a few hundredths of a second. Matter flowing inward smothers the shock wave and pushes it back into the star. If this happened in a real star, it presumably would collapse without any visible explosion.

However, theory predicts that 99 percent of the energy released in the collapse will appear as neutrinos. In the sun, neutrinos zip outward unimpeded by the gas of the solar layers, but in a collapsing star the density is as high as 10^{12} g/cm³, nearly as dense as an atomic nucleus. This gas is opaque to neutrinos, and they are partially absorbed in the gas. Not only does the tremendous burst of neutrinos remove energy from the core and allow the core to collapse even faster, but the neutrinos are absorbed outside the core and heat those layers. For some years, astronomers thought that these neutrinos could spread energy across the shock wave and, within a quarter of second or so, reaccelerate the stalled shock wave. The computer models, however, refused to explode.

The final ingredient that the models needed to make them explode was turbulent convection. When the collapse begins, the very center of the star forms a highly dense core, the beginnings of a neutron star, and the infalling material bounces off of that core. With the temperature shooting up, the bouncing material produces highly turbulent convection currents that give the stalled shock wave an outward boost (❚ Figure 13-11). Within a second or so, the shock wave begins to push outward, and after just a few hours it bursts out through the surface, blasting the star apart in a supernova explosion.

The supernova we see is the brightening of the star as its distended outer layers are blasted outward. As the months pass, the cloud of gas expands, thins, and begins to fade. But the way it fades in some cases tells astronomers about the death throes of the star. Essentially all of the iron in the core of the star is destroyed when the core collapses, but the violence in the outer layers can produce densities and temperatures high enough to trigger nuclear fusion reactions that produce as much as half a solar mass of radioactive nickel-56. The nickel gradually decays to form radioactive cobalt, which decays to form normal iron. The rate at which the supernova fades matches the rate at which these radioactive elements decay. Thus the destruction of iron in the core of the star is matched by the production of iron through nuclear fusion in the expanding outer layers.

The presence of nuclear fusion in the outer layers of the supernova testifies to the violence of the explosion. A typical supernova is equivalent to the explosion of 10^{28} megatons of TNT—about 3 million solar masses of high explosive.

Collapsing massive stars can trigger cosmic violence, but astronomers have observed more than one kind of supernova.

TYPES OF SUPERNOVAE

Supernovae are rare and only a few have been seen in our galaxy, but astronomers have been able to observe supernovae occurring in other galaxies (❚ Figure 13-12). From these data accumulated over decades, astronomers have noticed that there are two main types. **Type I supernovae** have spectra that contain no hydrogen lines. They reach a maximum brightness about 4 billions times more luminous than the sun, decline rapidly at first and then more slowly. **Type II supernovae** have spectra containing hydrogen lines. They reach a maximum brightness up to about 0.6 billion times more luminous than the sun, decline to a standstill and then fade rapidly. The light curves in ❚ Figure 13-13 summarize the behavior of the two types of supernovae.

FIGURE 13-11

The inside of an exploding star is shown in this mathematical model of a supernova explosion. The image here represents a region 150 km on a side with the center of the star at the lower left corner. Black arrows indicate the motion of the gas. Turbulent material (blue) has bounced off the core and is rushing outward to meet the rest of the star, which is falling inward (red at upper right). The innermost part of the star is forming a neutron star (red at lower left). To show the entire star at this scale, this figure would have to be 140 km in diameter. *(Courtesy Adam Burrows, John Hayes, and Bruce Fryxell)*

The evidence is clear that type II supernovae occur when a massive star develops an iron core and collapses. They occur in regions of active star formation where there is plenty of gas and dust. These are the

FIGURE 13-12

This supernova (arrow) was seen in the outskirts of the galaxy NGC4526. From its spectrum and from the way its brightness faded with time, astronomers recognize it as a type Ia supernova produced by the collapse and total destruction of a white dwarf. *(High-Z Supernova Search Team, HST, NASA)*

FIGURE 13-13
Type I supernovae decline rapidly at first and then more slowly, but type II supernovae pause for about 100 days before beginning a steep decline. Supernova 1987A was odd in that it did not rise directly to maximum brightness. These light curves have been adjusted to the same maximum brightness. Generally, type II supernovae are about two magnitudes fainter than type I.

regions where we would expect to find massive stars. The spectra of type II supernovae contain hydrogen lines, as we would expect from the explosion of a massive star that contains large amounts of hydrogen in its outer layers.

Type I supernovae are further classified into type Ia and type Ib according to their spectra and light curves. Type Ia supernovae are often found in regions where star formation ended long ago, and they lack hydrogen lines in their spectra, so they can't be caused by the deaths of massive stars. Rather, the evidence shows that a type Ia occurs when a white dwarf gaining mass in a binary star system exceeds the Chandrasekhar limit and collapses.

The collapse of a white dwarf is different from the collapse of a massive star because the core of the white dwarf contains usable fuel. As the collapse begins, the temperature shoots up, but the gas cannot halt the collapse because it is degenerate, and the temperature–pressure thermostat is turned off in the core. Even as carbon fusion begins, the increased temperature cannot increase the pressure and make the gas expand and slow the reactions. The core of the white dwarf is a bomb. The carbon–oxygen core fuses suddenly in violent nuclear reactions called **carbon deflagration.** The word *deflagration* means to be totally destroyed by fire, in this case by nuclear fusion. In a flicker of a stellar lifetime the entire star is consumed, and the outermost layers are blasted away in a violent explosion that at its brightest is 3 to 6 times more luminous than a type II supernova. The white dwarf is entirely destroyed. No neutron star or black hole is left behind. Of course, we see no hydrogen lines in the spectrum of a type Ia supernova because white dwarfs contain very little hydrogen.

The less common type Ib supernovae are believed to occur when a massive star in a binary system loses its hydrogen-rich outer layers to its companion star.

The remains of the massive star could develop an iron core and collapse, producing a supernova explosion that lacks hydrogen lines in the spectrum.

To summarize, a type II supernova is caused by the collapse of a massive star. A type Ia is caused by the collapse of a white dwarf. A type Ib is caused by the collapse of a massive star that has lost its outer envelope of hydrogen.

Much of our discussion so far has been based on theory, so it is time to compare our theory with observation of real supernova explosions.

OBSERVATIONS OF SUPERNOVAE

In 1054, Chinese astronomers saw a "guest star" appear in the constellation we know as Taurus, the Bull. The star quickly became so bright it was visible in the daytime. After a month's time, it slowly faded, taking almost two years to vanish from sight. When modern astronomers turned their telescopes to the location of the guest star, they found a cloud of gas about 1.35 pc in radius, expanding at 1400 km/s. Projecting the expansion back in time, they concluded that the expansion must have begun about nine centuries ago, just when the guest star made its visit. Thus, we think the nebula, now called the Crab Nebula because of its shape (❙ Figure 13-14), marks the site of the 1054 supernova.

Supernovae are rare. Only a few have been seen with the naked eye in recorded history. Arab astrono-

FIGURE 13-14
The Crab Nebula, the remains of the supernova of AD 1054, is located in the constellation Taurus and is bright enough to see with a small telescope. At visual wavelengths, the nebula consists of a network of expanding filaments containing a hazy glow. *(European Southern Observatory)*

False-Color Images and Reality

Astronomers use computers to add false color to images and reveal things our eyes cannot see. Radio, infrared, ultraviolet, and X-ray radiation are not visible to our eyes, so data collected at these wavelengths can be displayed in images that are given colors visible to our eyes. Of course, the colors used are entirely arbitrary. Radio astronomers might choose to use reds and yellows to color a radio map of a cloud of gas, or they might choose greens and blues. The colors make the radio energy visible to our eyes, but the choice of colors is up to the astronomer analyzing the image.

The variation in color across an image does have meaning, however.

Radio astronomers might decide to use reds for the regions of strongest radio signal, yellows for fainter regions, and blues for the faintest. Thus, we could look at the radio map of a gas cloud and immediately pick out the parts of the cloud that were emitting the strongest radio energy.

In addition to using false color in images recorded in wavelengths our eyes cannot see, astronomers often use false color in photographs recorded at visible wavelengths. These images can reveal to our eyes subtle variations in brightness not visible in the original photographs.

False-color images are common in astronomy, but they are also used in other sciences. Doctors often analyze medical X-ray images, CAT scans, and so on by converting the images to false color. Biologists use false color to analyze microscopic photographs, and geologists use false color to study photographs of Earth recorded by a satellite at various wavelengths.

We might think of false-color images as deceptive, but they actually help us see parts of the universe that would otherwise never register on our limited eyes.

---

mers saw one in 1006, and the Chinese saw one in 1054. European astronomers observed two—one in 1572 (Tycho's supernova) and one in 1604 (Kepler's supernova). Also, the guest stars of 185, 386, 393, and 1181 may have been supernovae. For 383 years following 1604, no naked-eye supernova appeared, and then in February 1987 a star in the southern sky exploded. We will examine the great supernova of 1987 later in this section. Here we will explore supernovae in general.

Most supernovae are discovered in distant galaxies, and searching for these supernovae was once a tedious job. Robotic telescopes now observe long lists of galaxies each night searching for the appearance of a new star. Detecting supernovae early is important because astronomers need to study the early stages of the explosion in order to understand the complex processes that occur in the rapidly expanding cloud of gas.

The supernova explosion fades to obscurity in a year or two, but an expanding shell of gas marks the explosion site. The gas, originally expelled at 10,000 to 20,000 km/s, may carry away one-fifth of the mass of the star. The collision of that expanding gas with the surrounding interstellar medium can sweep up even more gas and excite it to produce a **supernova remnant,** the nebulous remains of a supernova explosion.

Supernova remnants look quite delicate and do not survive very long—a few tens of thousands of years—before they gradually mix with the interstellar medium and vanish. The Crab Nebula is a young remnant, only about 950 years old, and isn't very large, only a few parsecs in diameter (Figure 13-14). Older remnants can be larger. Some supernova remnants are detectable only at radio and X-ray wavelengths. They have become too tenuous to emit detectable light, but the collision of the expanding hot gas with the interstellar medium can

generate radio and X-ray radiation and allows us to create images of them at these nonvisible wavelengths (Window on Science 13-1). In general, supernova remnants are tenuous spheres of gas expanding into the interstellar medium (■ Figure 13-15). We saw in Chapter 11 that the compression of the interstellar medium by expanding supernova remnants can trigger star formation.

Radio images of supernova remnants reveal that they are emitting **synchrotron radiation**—electromagnetic energy radiated by high-speed electrons spiraling through a magnetic field (■ Figure 13-16). Low-speed electrons radiate at longer wavelengths, and high-speed electrons radiate at shorter wavelengths, so synchrotron radiation is spread over a wide range of wavelengths. The presence of synchrotron radiation coming from a nebula tells us that the nebula contains magnetic fields and energetic electrons.

Some of the energy coming from the Crab Nebula is synchrotron radiation, and it is so energetic that some of it has wavelengths as short as light. The general hazy or foggy glow of the nebula is produced by synchrotron radiation. As the electrons in a nebula emit synchrotron radiation, they slow. To emit synchrotron radiation as light, the electrons in the Crab Nebula must be traveling very fast, but over the 950-year history of the nebula, those electrons should have radiated their energy away and slowed down. That is, the synchrotron glow of the nebula can't have lasted since the supernova explosion. The observation of synchrotron radiation coming from the Crab Nebula today tells us that something must be continuously supplying energy to keep the electrons moving rapidly. We will solve this mystery in the next chapter when we locate a neutron star in the center of the Crab Nebula.

The supernova remnant called the Cygnus Loop is about 10,000 years old and 80 ly in diameter.

Visible light produced by gas expanding into surrounding interstellar medium

Supernova remnant N132D is 3000 years old and 80 ly in diameter. It is 180,000 ly from Earth.

Shock waves in expanding gas heat it to millions of degrees, and it emits X-rays.

Cassiopeia A (Cas A) is a supernova remnant about 300 years old and 10 ly in diameter.

Visual-wavelength image

X-ray image

Cas A is also bright at radio wavelengths.

Infrared radiation from dust condensing out of the gas

Infrared image

Radio image

FIGURE 13-15

A supernova remnant is an expanding bubble of hot gas created by a supernova explosion. As the remnant expands and pushes into neighboring gas, it can emit radiation at many wavelengths. *(Cygnus Loop: Caltech; N132D: NASA/SAO/CXC; IR: 2MASS; Radio: NRAO/AUI)*

THE GREAT SUPERNOVA OF 1987

Until 1987, astronomers had never looked at a bright supernova through a telescope. Nearly all supernovae are in distant galaxies and thus are very faint. The last supernova visible to the naked eye, Kepler's supernova of 1604, predated the invention of the telescope. Then, in late February 1987, the news raced around the world. Astronomers in Chile had discovered a naked-eye supernova in the Large Magellanic Cloud, a small galaxy very near our Milky Way Galaxy (∎ Figure 13-17). Because the supernova was only 20 degrees from the south celestial pole, it could be studied only from southern latitudes. It was named SN1987A to denote the first supernova discovered in 1987.

The hydrogen-rich spectrum suggested that the supernova was a type II, caused by the collapse of the core of a massive star. As the months passed, however, the light curve proved to be odd (Figure 13-13) in that it paused for a few weeks before rising to its final maximum. From photographs of the area made some years before, astronomers were able to determine that the star that exploded, cataloged as Sanduleak −69°202, was not the expected red supergiant but rather a hot, blue supergiant of only 15 solar masses and 50 solar radii, not extreme for a supergiant. Theorists now believe that the star was chemically poor in elements heavier than helium and had consequently contracted and heated up after a phase as a cool, red supergiant, during which it lost mass into space. The relatively small

FIGURE 13-16

Synchrotron radiation is emitted when an electron spirals around a magnetic field line and radiates its energy away as photons. Only a few photons are shown here, for clarity.

FIGURE 13-17

Supernova 1987A eventually reached a peak brightness of magnitude 2.9, nearly as bright as the stars in the Big Dipper, and was easily visible to the unaided eye. An earlier photograph of the region, left, allows astronomers to identify the star that exploded (arrow). *(Anglo-Australian Observatory)*

size of the supergiant may explain the pause in the light curve. Much of the energy of the explosion went into blowing apart the smaller, denser-than-usual star and making it expand.

The brightening of the supernova after the first few weeks seems to have been caused by the decay of radioactive nickel into cobalt. Theory predicts the production of such nickel atoms in the explosion, and their decay into cobalt would release gamma rays that would heat the expanding shell of gas and make it brighter. About 0.07 solar mass of nickel was produced, about 20,000 times the mass of Earth.

The cobalt atoms are also unstable, but they decay more slowly, so it was not until sometime later, after much of the nickel had decayed, that the decay of cobalt into iron began providing energy to keep the expanding gas hot and luminous. Although these processes had been predicted, they were clearly observed in SN1987A. Gamma rays from the decay of cobalt to iron were detected, and cobalt and iron are clearly visible in the infrared spectra of the supernova.

As the supernova dimmed, photographs revealed bright rings of gas (❙ Figure 13-18). Comparison with mathematical models shows that the rings were produced by two stellar winds. When the star was a red supergiant, it expelled a slow stellar wind. Later when it became a blue supergiant it expelled a fast stellar wind. The interaction of these two winds shaped by a magnetic field and illuminated by the supernova explosion produces the rings. Compare this with the rings seen in some of the planetary nebulae shown on page 261.

Two independent observations confirm that SN1987A probably gave birth to a neutron star. Theory predicts that the collapse of a massive star's core should liberate a tremendous blast of neutrinos that leave the star hours before the shock wave from the

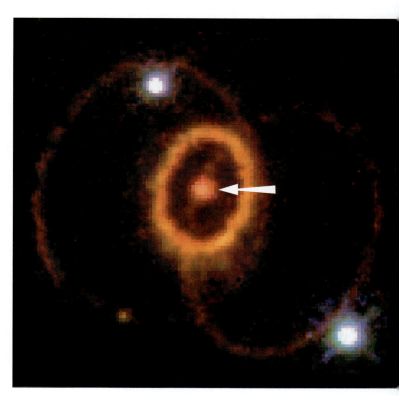

FIGURE 13-18

Supernova 1987A (arrow) is surrounded by three rings. (The bright stars are foreground stars not related to the supernova.) The rings are caused by the interaction of a fast stellar wind, an older slow stellar wind, and a magnetic field. The rings resemble the rings seen in some planetary nebulae such as the Hour Glass Nebula on page 261. *(Christopher Burrows, ESA/STScI, NASA)* interior blows the star apart. Two detectors, one in Ohio and one in Japan, recorded a burst of neutrinos passing through Earth at 2:35:41 AM EST on February 23, 1987, about 18 hours before the supernova was seen. The detectors caught only 19 neutrinos during a 12-second interval, but recall that neutrinos hardly ever react with normal matter. The full flood of neutrinos was immense. Within a few seconds of that time, roughly 20 trillion neutrinos passed harmlessly through each human body on Earth. The detection of the neutrino blast confirms that the collapsing core gave birth to a neutron star.

The expanding gas shell of the supernova will continue to thin and cool, and eventually astronomers on Earth will be able to peer inside and see what is left of Sanduleak −69°202. Will we see the expected neutron star, or will theories need further revision? For the first time in almost 400 years, astronomers can observe a bright supernova to test their theories.

LOCAL SUPERNOVAE AND LIFE ON EARTH

Although supernovae are rare events, they are very powerful and could affect life on planets orbiting nearby stars. In fact, supernovae explosions long ago may have affected Earth's climate and the evolution of life.

If a supernova occurred within about 50 ly of Earth, the human race would have to abandon the surface and live below ground for at least a few decades. The burst of gamma rays and high-energy particles from the supernova explosion could kill many life forms and cause serious genetic damage in others. The only way we could avoid this radiation would be to move our population into tunnels below Earth's surface. Of course, if a supernova did occur, we would not have time to dig enough tunnels.

Even if we could survive in tunnels long enough for the radioactivity on Earth's surface to subside, we might not like Earth when we emerged. Genetic mutation induced by radioactivity could alter plant and animal life so seriously that we might not be able to support our population. After all, we humans depend almost totally on grass for food. The basic human foods—milk, butter, eggs, wheat, tomatoes, lettuce, and meat (Big Macs, in other words)—are grass and similar vegetation processed into different forms. Seafood is merely processed ocean plankton and plant life, which might also be altered or damaged by a local supernova explosion. Even if surface life survived the radiation, damage to the delicate upper layers of our atmosphere might alter the climate dramatically.

Local supernovae have been suggested as a possible cause of occasional climate changes and extinctions in Earth's past. It is quite possible that, as the sun moves through space, supernovae explode near Earth every few hundred million years. A really close call could trigger an extinction.

But do supernovae really affect Earth? In 1979, scientists studying Antarctic ice cores containing very old ice noticed that ice from the years 1181, 1320, 1572, and 1604 contains high abundances of nitrates. Bright supernovae were seen in the years 1181, 1572, and 1604. Ultraviolet radiation from those supernovae might have altered Earth's upper atmosphere and produced acid rain that was recorded in the ice as nitrates. But what of the ice from 1320? In 1999, X-ray astronomers noticed a supernova remnant in the southern sky. When they calculated its age, they found the supernova should have occurred about 1320. It isn't clear why there is no historical record of this supernova or why the nitrates appear in Antarctic ice cores but not in Greenland ice cores. In any case, this evidence may alert us that even distant supernova explosions can affect Earth in subtle ways.

A study of cores from the seafloor has revealed a deposit of the isotope iron-60 in a layer that was laid down at the time of the Pliocene-Pleistocene extinction about two million years ago. Iron-60 is produced in supernova explosions and has a half-life of only 1.5 million years. To reach earth before it decayed, this iron must have been produced in a supernova explosion no more than 100 ly away. Thus, the Pliocene-Pleistocene extinction may have been caused by a nearby supernova explosion.

Earth seems to be fairly safe at the moment. No star within 50 ly is known to be a massive giant capable of exploding as a supernova. Massive stars are rare, and most are far away. The most massive star known is 25,000 ly from Earth and poses no threat (❙ Figure 13-19). We only need to worry about nearby supernova explosions, and massive stars are so rare and so easy to locate that we can be sure that no type II supernova explosion will occur near us for the time being.

Type Ia supernovae are caused by collapsing white dwarfs, and white dwarfs are both very common and very hard to locate. There is no way to be sure that a white dwarf teetering on the edge of the Chandrasekhar limit does not lurk near us in space. We can take confidence, however, in the extreme rarity of supernova explosions. Earth faces greater dangers from human ignorance than from exploding stars.

REVIEW CRITICAL INQUIRY

Why does a type II supernova explode?

A type II supernova occurs when a massive star reaches the end of its usable fuel and develops an iron core. The iron is the final ash produced by nuclear fusion, and it cannot fuse because iron is the most tightly bound nucleus. When energy generation begins to fall, the star contracts, but because iron can't ignite there is no new energy source to stop the contraction. In seconds, the core of the star falls inward and a shock wave moves outward. Aided by a flood of neutrinos and sudden convective turbulence, the shock wave blasts the star apart, and we see it brighten as its surface gases expand into space.

Type II supernovae are easy to recognize because their spectra contain hydrogen lines. Use what you know about type Ia supernovae to explain why the spectra of these supernovae do not contain visible hydrogen lines.

The study of the deaths of stars has led us to discover astonishing objects of unbelievable density, temperature, and violence—all consequences of the victory of gravity over matter. But we have not considered the strangest circumstance of all. What happens when degenerate matter can't support the weight of the dying star, that is, when the mass of the compact object exceeds the Chandrasekhar limit? The answer is in the title of the next chapter.

FIGURE 13-19

The most massive star known is about 100 times more massive than the sun and 10 million times more luminous. Originally about 200 solar masses, it is rapidly expelling mass. At a distance of 25,000 ly, it will not be dangerous when it explodes as a supernova within the next few million years. *(Don F. Figer, UCLA, NASA)*

SUMMARY

When a star's central hydrogen-fusing reactions cease, its core contracts and heats up, igniting a hydrogen-fusing shell and swelling the star into a cool giant. The contraction of the star's core ignites helium first in the core and later in a shell. If the star is massive enough, it can eventually fuse carbon and other elements.

How a star evolves depends on its mass. Stars less massive than about 0.4 solar mass are completely mixed and have very little hydrogen left when they die. They cannot ignite a hydrogen shell or a helium core, so they cannot become giants. Their fuel will last for trillions of years. Our universe is not old enough for any of these stars to have died.

Medium-mass stars between about 0.4 and 4 solar masses become giants and fuse helium but cannot fuse carbon. They swell very large as their carbon–oxygen cores contract and eventually eject their surface layers to form planetary nebulae. Such nebulae are expanding gas ionized by the high-temperature core of the star left at the center. These planetary nebula nuclei eventually collapse to form white dwarfs.

Because stars lose mass through stellar winds, the mass of a star can decrease as it evolves. Thus, it seems possible for a star with a main-sequence mass as great as eight solar masses to lose enough mass to produce a planetary nebula and a white dwarf. No white dwarf can exist if it is more massive than the Chandrasekhar limit of 1.4 solar masses, so the most massive stars cannot collapse into white dwarfs. Presumably, they collapse to form other compact objects—neutron stars or black holes.

Evolving stars in close binary systems can transfer matter from one star to the other, producing an accretion disk around the star gaining mass. This can dramatically alter the evolution of the stars. If the star gaining mass is a white dwarf, the added fuel on its surface can explode as a nova.

The most massive stars can get hot enough to ignite carbon and other nuclear fuels. Eventually, they develop iron cores, which cannot release energy through fusion reactions, and collapse to produce a supernova explosion as a powerful shock wave, aided by a blast of neutrinos and violent turbulence, rushes out of the core. The expanding shell of gas produces a supernova remnant, and the remains of the core of the star collapse to form a neutron star or black hole. Supernovae caused by the collapse of a massive star are called type II supernovae.

A type Ia supernova is produced by the collapse of a white dwarf that has gained enough matter from a binary companion to exceed the Chandrasekhar limit. The collapse of the white dwarf causes rapid nuclear fusion, called carbon deflagration, and the entire star is destroyed. Unlike the spectra of type II supernovae, the spectrum of a type Ia supernova contains no hydrogen lines.

NEW TERMS

nova

supernova

thermal pulse

planetary nebula

compact object

black dwarf

Chandrasekhar limit

Roche lobe

Roche surface

Lagrangian point

accretion disk

supernova (type I)

supernova (type II)

carbon deflagration

supernova remnant

synchrotron radiation

REVIEW QUESTIONS

Ace☉Astronomy™ Assess your understanding of this chapter's topics with additional quizzing and animations at **http:// astronomy.brookscole.com/seeds8e**

1. Why can't the lowest-mass stars become giants?

2. Presumably, all the white dwarfs we see in our galaxy were produced by sunlike stars of medium mass. Why couldn't any of the white dwarfs we see have been produced by the deaths of the lowest-mass stars?

3. What leads us to believe that stars can lose mass?

4. What kind of spectrum does the gas in a planetary nebula produce? Where does it get the energy to radiate?

5. The coolest stars we see at the center of planetary nebulae are about 25,000 K. Why don't we see planetary nebulae containing cooler central stars? (*Hint:* What kind of photons excite the gas in a planetary nebula?)

6. As white dwarfs cool, they move toward the lower right in the H–R diagram (Figure 9-9 and page 261), maintaining constant radius. Why don't they contract as they cool?

7. All white dwarfs are about the same mass. Why?

8. Why have no white dwarfs cooled to form black dwarfs in our galaxy?

9. What happens to a star when it becomes a giant if it has a close binary companion?

10. Why do novae repeat but supernovae do not?

11. Why can't massive stars generate energy from iron fusion?

12. How can we use the spectrum to tell the difference between a type I supernova and a type II supernova? Why does this difference arise?

13. What processes produce type I and type II supernovae?

14. Why do supernova remnants emit X rays?

DISCUSSION QUESTIONS

1. How certain can we be that the sun will never explode as a supernova? What does it mean when a scientist uses the word "certainty"?

2. If we lived for a billion years, we might notice a star that exploded occasionally as a nova and finally destroyed itself in a supernova explosion. What could account for this evolution?

PROBLEMS

1. Use the formula in Chapter 12 to compute the life expectancy of a 0.4-solar-mass star. Why might this be an underestimate if the star is fully mixed?

2. The Ring Nebula in Lyra is a planetary nebula with an angular diameter of 72 seconds of arc and a distance of 5000 ly. What is its linear diameter? (*Hint:* Use the small-angle formula.)

3. If the Ring Nebula is a light-year in diameter and is expanding at a velocity of 15 km/s, typical of planetary nebulae, how old is it? (*Hint:* 1 ly = 9×10^{12} km, and 1 y = 3.15×10^7 s.)

4. Suppose that a planetary nebula is 1 pc in diameter and the Doppler shifts in its spectrum show that it is expanding at 30 km/s. How old is it? (*Hint:* See Problem 3.)

5. A planetary nebula photographed 20 years ago and photographed today has increased its radius by 0.6 second of arc. If Doppler shifts in its spectrum show that it is expanding at a velocity of 20 km/s, how far away is it? (*Hint:* First figure out how many parsecs it has increased in radius in 20 years. Then use the small-angle formula.)

6. The Crab Nebula is now 1.35 pc in radius and is expanding at 1400 km/s. About when did the supernova occur?

7. Doppler shifts in the spectra of the Cygnus Loop show that it is now expanding at 70 km/s. If it is increasing its radius by 0.03 second of arc per year, how far away is it? (*Hint:* How many kilometers will its radius increase in 1 year? Use the small-angle formula.)

8. The Cygnus Loop is now 2.6° in diameter and lies about 500 pc distant. If it is 10,000 years old, what was its average velocity of expansion? (*Hint:* Find its radius in parsecs first.)

9. The supernova remnant Cassiopeia A is expanding in radius at a rate of about 0.5 second of arc per year. Doppler shifts show that the velocity of expansion is about 5700 km/s. How far away is the nebula?

10. Cassiopeia A has a radius of about 2.5 minutes of arc. If it is expanding at 0.5 second of arc per year, when did the supernova explosion occur? (There is no record of a supernova being seen at that time.)

CRITICAL INQUIRIES FOR THE WEB

1. As seen on pages 260 and 261, planetary nebulae show an incredible diversity in appearance. Browse the Web for images and information on these dying stars, and discuss why we see such a range of shapes of planetary nebulae.

2. Naked-eye supernovae in our galaxy are rare, but astronomers have noted supernovae in other galaxies for years. Search the Web for summaries of observations of recent supernovae. Are similar numbers of type I and type II being seen? Compare the number of supernovae seen during the last few years with that of two decades ago. Why are we finding so many more supernovae in recent years than in the past?

3. The word "nova" is used in many ways, so you must be thoughtful to search out photos and observations of recently discovered novae. Limit your search to astronomy categories, and also try pages about variable stars.

4. What is U Scorpii?

EXPLORING *THESKY*

1. Locate the planetary nebulae M57, M97, and M27. How does their shape distinguish them from the star-formation nebulae such as M42 and M8? (*Hint:* To find an object, use **Find** under **Edit.** Choose **Messier Objects** and pick from the list.)

2. The Crab Nebula is M1. Locate it, zoom in, measure its angular size in seconds of arc, and compute its diameter assuming it is about 6000 ly from Earth.

3. Locate the supernova remnant called the Cygnus Loop just south of ε Cygni. How big is this object in angular diameter compared to the full moon? (*Hint:* Under the **View** menu, choose **Labels** and **Setup.** Check **Bayer Designation,** go to Cygnus, and zoom in on ε Cygni until the Cygnus Loop appears.)

 Visit the Seeds *Foundations of Astronomy* companion Web site for critical thinking exercises, articles, and additional readings from InfoTrac College Edition, Brooks/Cole's online student library.

NEUTRON STARS AND BLACK HOLES

Almost anything is easier to get into than out of.

Agnes Allen

Chapter 14

ESO and NASA

GUIDEPOST

The preceding chapters have traced the story of stars from their birth as clouds of gas in the interstellar medium to their final collapse. This chapter finishes the story by discussing the kinds of objects that remain after a massive star dies.

How strange and wonderful that we humans can talk about places in the universe where gravity is so strong it bends space, slows time, and curves light back on itself! To carry on these discussions, astronomers have learned to use the language of relativity.

Throughout this chapter, remember that our generalized discussions are made possible by astronomers studying general relativity in all its mathematical sophistication. That is, our understanding rests on a rich foundation of theory.

This chapter ends the story of individual stars. The next three chapters, however, extend that story to include the giant communities in which stars live— the galaxies.

VIRTUAL LABORATORIES

 NEUTRON STARS AND PULSARS

 GENERAL RELATIVITY AND BLACK HOLES

Gravity always wins. In a star's struggle to withstand its own gravity, the star must lose and gravity must win. Gravity ensures that this star's last remains must eventually reach one of three final states—white dwarf, neutron star, or black hole. These objects, often called compact objects, are small, high-density monuments to the power of gravity. Every star faces an ultimate collapse into such an object.

We discussed white dwarfs in the preceding chapter. In this chapter, we complete our story of stellar death by discussing the other two forms a dead star can take—neutron stars and black holes. In many ways, we learn more about normal stars by studying the remains of dead stars.

In this chapter, we must compare evidence and hypothesis with great care. Theory predicts the existence of these objects; but, by their nature, they are difficult to detect. Is theory right? Do these objects really exist? To confirm the theories, astronomers have searched for real objects that could be identified as having the properties predicted by theory. That is, they have tried to find real neutron stars and real black holes. No medieval knight ever rode off on a more difficult quest, but modern astronomers have found objects that do seem to be neutron stars and black holes. We can judge their success by critically analyzing the evidence to see if it does confirm the theory.

14-1 NEUTRON STARS

Theory predicts that the death of a massive star could leave behind a **neutron star,** an object containing a little over one solar mass compressed to a radius of about 10 km and heated to high temperatures (Figure 14-1). Modern astronomers have found many objects that appear to be neutron stars, but we must tell their story with care in order to separate theory from observation.

Theory predicts that a neutron star should spin a number of times a second, be nearly as hot at its surface as the inside of the sun, and have a magnetic field a trillion times stronger than Earth's. Two questions should occur to us immediately. First, how could any theory predict such a wondrously unbelievable star? And second, do such neutron stars really exist?

THEORETICAL PREDICTION OF NEUTRON STARS

The neutron was discovered in the laboratory in 1932, and its properties suggested something fantastic. Neutrons spin in much the way that electrons do, which

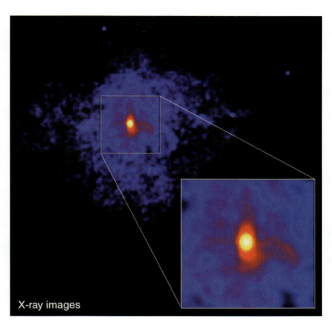

FIGURE 14-1

A supernova explosion seen in 1181 AD left behind an expanding supernova remnant. The Chandra X-Ray Observatory has imaged the nebula in X rays and finds a tiny hot object within—a neutron star. (*NASA/SAO/CXC/P. Slane et al.*)

means that neutrons must obey the Pauli exclusion principle. In that case, if neutrons are packed together tightly enough, they can become degenerate just as electrons do. White dwarfs are supported by degenerate electrons, and, in 1932, the Russian physicist Lev Landau predicted that neutron stars might exist supported by degenerate neutrons. Of course, the inside of a neutron star would have to be much denser than the inside of a white dwarf.

Only two years later, in 1934, Walter Baade and Fritz Zwicky provided a pedigree for neutron stars. They suggested that some of the most luminous novae in the historical record were not true novae but were caused by the collapse of a massive star in an even larger explosion they called a supernova. What was left of the core of the star, they proposed, was a small, high-density neutron star.

Atomic physics gives us an explanation of how the collapsing core of a massive star could form a neutron star. If the collapsing core is more massive than the Chandrasekhar limit of 1.4 solar masses, then it cannot reach stability as a white dwarf. The weight is too great to be supported by degenerate electrons. The collapse of the core continues, and the atomic nuclei are broken apart by gamma rays. Almost instantly, the increasing density forces the freed protons to combine with electrons and become neutrons by the emission of neutrinos:

$$e + p \rightarrow n + \nu$$

The burst of neutrinos, as we saw in the previous chapter, helps blow the star apart; and the core of the star, in

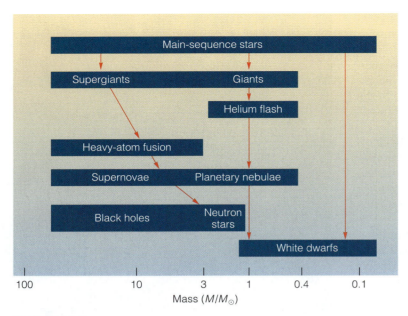

FIGURE 14-2

How a star evolves depends on its mass. The lowest-mass stars will someday become white dwarfs, while stars like the sun become giants, produce planetary nebulae, and then collapse into white dwarfs. Mass loss can move stars to the right in this diagram, so a star of 8 or less solar masses may be able to shed enough mass to die as a planetary nebula and a white dwarf. A more massive supergiant may be able to eject enough mass to leave behind a neutron star instead of a black hole. *(Illustration design by author)*

a fraction of a second, is transformed into a contracting ball of neutrons left behind by the supernova explosion as a neutron star.

Our discussion of stellar evolution in the previous chapter revealed that a star of 8 solar masses or less could lose enough mass to die as a planetary nebula leaving behind a white dwarf. More massive stars up to a limit of about 15 solar masses will lose mass rapidly, but they cannot reduce their mass fast enough and will apparently face death as a supernova explosion leaving behind a neutron star (❚ Figure 14-2). (We will see later in this chapter that the most massive stars probably leave behind black holes.)

Theoretical calculations predict that a neutron star will be only 10 or so kilometers in radius (❚ Figure 14-3) and will have a density of about 10^{14} g/cm^3. On Earth, a sugar-cube-sized lump of this material would

FIGURE 14-3

A tennis ball and a road map illustrate the relative size of a neutron star. Such an object, containing slightly more than the mass of the sun, would fit with room to spare inside the beltway around Washington, DC. *(Photo by author)*

weigh 100 million tons. This is roughly the density of the atomic nucleus, and we can think of a neutron star as matter with all of the empty space squeezed out of it.

How massive can a neutron star be? That is a critical question, and a difficult one to answer because we don't know the strength of pure neutron material. Scientists can't make such matter in the laboratory, so its properties must be predicted theoretically. The most widely accepted calculations suggest that a neutron star cannot be more massive than 2 to 3 solar masses. If a neutron star were more massive than that, the degenerate neutrons would not be able to support the weight, and the object would collapse (presumably into a black hole).

Simple physics, the physics we have used in previous chapters to discuss normal stars, predicts that neutron stars should be hot, spin rapidly, and have strong magnetic fields. We have seen that contraction heats the gas in a star. As the gas particles fall inward, they pick up speed, and when they collide, their high speeds become thermal energy. The sudden collapse of the core of a massive star to a radius of 10 km should heat it to millions of degrees. Furthermore, neutron stars should cool slowly because the heat can escape only from the surface, and neutron stars are so small they have little surface from which to radiate. Thus, basic theory predicts that neutron stars should be very hot.

The conservation of angular momentum predicts that neutron stars should spin rapidly. All stars rotate to some extent because they form from swirling clouds of interstellar matter. As such a star collapses, it must rotate faster because it conserves angular momentum.

Recall that we see this happen when ice skaters spin slowly with their arms extended and then speed up as they pull their arms closer to their bodies (see Figure 5-7). In the same way, a collapsing star must spin faster as it pulls its matter closer to its axis of rotation. If the sun collapsed to a radius of 10 km, its period of rotation would decrease from 25 days to about 0.001 second. We might expect the collapsed core of a massive star to rotate 10 or 100 times a second.

Basic theory also predicts that a neutron star should have a powerful magnetic field. Whatever magnetic field a star has is frozen into the star. The gas of the star is ionized, and that means the magnetic field cannot move easily through the gas. When the star collapses, the magnetic field is carried along and squeezed into a smaller area, which could make the field a billion times stronger. Because some stars have magnetic fields over 1000 times stronger than the sun's, we might expect a neutron star to have a magnetic field as much as a trillion times stronger than the sun's. For comparison, that is about 10 million times stronger than any magnetic field ever produced in the laboratory.

Basic physics predicted the properties of neutron stars, but it also predicted that such objects should be difficult to observe. Neutron stars are very hot, so from our understanding of black body radiation we can predict they will radiate most of their energy in the X-ray part of the spectrum, radiation that could not be observed in the 1940s and 1950s because astronomers could not put their telescopes above Earth's atmosphere. Also, the small surface areas of neutron stars mean that they will be faint objects. Thus astronomers of the mid-20th century were not surprised that none of the newly predicted neutron stars were found. Neutron stars were, at that point, entirely theoretical objects. Progress came not from theory but from observation.

THE DISCOVERY OF PULSARS

In November 1967, Jocelyn Bell, a graduate student at Cambridge University in England, found a peculiar pattern on the paper chart from a radio telescope. Unlike other radio signals from celestial bodies, this was a series of regular pulses (▌Figure 14-4). At first she and the leader of the project, Anthony Hewish, thought the signal was interference, but they found it day after day in the same place in the sky. Clearly, it was celestial in origin.

Another possibility, that it came from a distant civilization, led them to consider naming it LGM for Little Green Men. But within a few weeks, the team found three more objects in other parts of the sky pulsing with different periods. The objects were clearly natural, and the team dropped the name LGM in favor of **pulsar**—a contraction of *pulsing star*. The pulsing radio source Bell had observed with her radio telescope was the first known pulsar.

FIGURE 14-4

The 1967 detection of regularly spaced pulses in the output of a radio telescope led to the discovery of pulsars. This record of the radio signal from the first pulsar, CP1919, contains regularly spaced pulses (marked by ticks). The period is 1.33730119 seconds.

As more pulsars were found, astronomers argued over their nature. Periods ranged from 0.033 to 3.75 seconds and were nearly as exact as an atomic clock. Months of observation showed that many of the periods were slowly growing longer by a few billionths of a second per day. Thus, whatever produced the regular pulses had to be highly precise, nearly as exact as an atomic clock, but it had to gradually slow down.

It was easy to eliminate possibilities. Pulsars could not be stars. A normal star, even a small white dwarf, is too big to pulse that fast. Nor could a star with a hot spot on its surface spin fast enough to produce the pulses. Even a small white dwarf would fly apart if it spun 30 times a second.

The pulses themselves gave the astronomers a clue. The pulses lasted only about 0.001 second. This places an upper limit on the size of the object producing the pulse. If a white dwarf blinked on and then off in that interval, we would not see a 0.001-second pulse. The near side of the white dwarf would be about 6000 km closer to us, and light from the near side would arrive 0.022 seconds before the light from the bulk of the white dwarf. Thus its short blink would be smeared out into a longer pulse. This is an important principle in astronomy—an object cannot change its brightness appreciably in an interval shorter than the time light takes to cross its diameter. If pulses from pulsars are no longer than 0.001 second, then the objects cannot be larger than 300 km (190 miles) in diameter.

Only a neutron star is small enough to be a pulsar. In fact, a neutron star is so small, it can't vibrate slowly enough, but it can spin as fast as 1000 times a second without flying apart. The missing link between pulsars and neutron stars was found in late 1968, when astronomers discovered a pulsar at the heart of the Crab Nebula (▌Figure 14-5). The Crab Nebula is a supernova remnant, and theory predicts that some supernovae leave behind a neutron star.

The short pulses and the discovery of the pulsar in the Crab Nebula are strong hints that pulsars are neutron stars. If we combine theory and observation, we can devise a model of a pulsar.

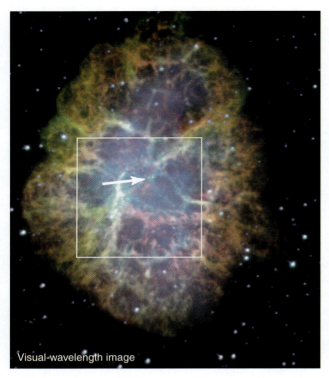

FIGURE 14-5

The pulsar at the center of the Crab Nebula (arrow) is detectable in visual-wavelength photographs. The star to the left of the pulsar lies much closer to Earth and is not in the Crab Nebula. The white box outlines the area imaged on page 285. *(Caltech)*

A MODEL PULSAR

Scientists often work by building a model of a natural phenomenon—not a physical model made of plastic and glue, but an intellectual conception of how nature works in a specific instance. The model may be limited and incomplete, but it helps us organize our understanding of pulsars.

The modern model of a pulsar has been called the **lighthouse model** and is shown in ❚ "The Lighthouse Model of a Pulsar" on pages 284 and 285. Notice four important points. First, a pulsar does not pulse but rather emits beams of radiation that sweep around the sky as the neutron star rotates. Second, notice that the mechanism that produces the beams involves extremely high energies and is not fully understood. The third thing to notice is that we tend to see only those pulsars whose beams sweep over Earth. Finally, notice how modern space telescopes observing at various wavelengths can help us confirm and refine our model.

Neutron stars are not simple objects, and modern astronomers need both general relativity and quantum mechanics to try to understand them. Nevertheless, the life story of pulsars can be understood in terms of the lighthouse model.

The model of a pulsar as a spinning neutron star won the support of many astronomers because of two

properties of pulsars. First, many pulsars are slowing down. Their periods are increasing by a few billionths of a second each day—a change radio astronomers can measure using atomic clocks. Evidently, the spinning neutron star is converting some of its energy of rotation into various kinds of electromagnetic energy and a powerful outflow of high-speed particles called a **pulsar wind** (❚ Figure 14-6). About 99.9 percent of the energy released by the slowing of the neutron star is carried away by the pulsar wind, and only 0.1 percent goes into producing the radio beams we detect. The energy that keeps the Crab Nebula glowing nearly 1000 years after the explosion is coming from the rotational energy of the neutron star and the pulsar wind that the neutron star produces.

The second property of pulsars that supports the neutron-star model of a pulsar is the **glitch**—a sudden increase in the pulse rate seen in some pulsars (❚ Figure 14-7). Two theories have been proposed to explain these changes, and both depend on the internal structure of spinning neutron stars.

One theory suggests that "starquakes" occur on the surface of a neutron star and produce glitches. To understand this theory we must note that theoretical models of neutron stars suggest that their interiors are fluids composed mostly of neutrons. Near the surface, where the pressure is lower, matter can exist as a rigid crust composed mostly of iron nuclei about 10^{16} times stronger than steel. This crust might be a few hundred meters thick. The rapidly spinning neutron star would be slightly flattened, but as it slowed, its gravity would squeeze it until the crust broke in a "starquake"—a neutron-star earthquake. When the crust breaks, the neutron star could become slightly less flattened, and the conservation of angular momentum would cause the neutron star to spin slightly faster. From Earth we would see a sudden increase in the pulse rate.

The starquake theory has a drawback; starquakes should occur every few thousand years on any given neutron star, but glitches are more common. The Vela pulsar has glitched 13 times in 25 years. A newer theory may explain most glitches. Because the liquid in the core of a neutron star can circulate with almost no friction, vortexes should develop in the fluid. Like swarms of frictionless tornadoes, they should store large amounts of angular momentum. Calculations applying quantum mechanics to these vortices show that as the neutron star slows, swarms of vortices should suddenly transfer their angular momentum to the crust. That would make the crust spin faster and increase the pulse rate. That is, it would produce a glitch. This theory has been confirmed in laboratory experiments with rotating fluids.

A rare glitch may be caused by a starquake, but most are probably caused by changes in internal vortices. Notice, in any case, that both explanations depend on the pulsar being a spinning neutron star. Con-

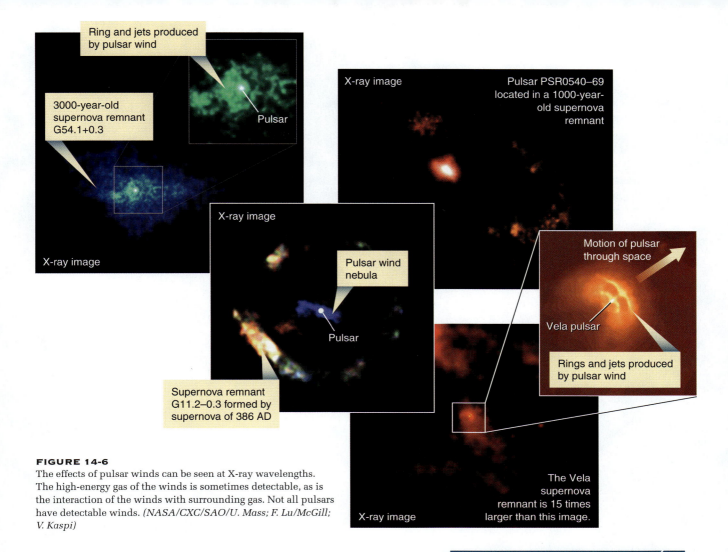

FIGURE 14-6

The effects of pulsar winds can be seen at X-ray wavelengths. The high-energy gas of the winds is sometimes detectable, as is the interaction of the winds with surrounding gas. Not all pulsars have detectable winds. *(NASA/CXC/SAO/U. Mass; F. Lu/McGill; V. Kaspi)*

sequently, glitches give astronomers further confidence that pulsars are spinning neutron stars.

Since the discovery of the first pulsar, over 1000 have been found, and there are probably about 100 million in our galaxy. Of course, we see only those pulsars whose beams sweep over Earth. Most neutron stars are invisible from Earth. Nevertheless, astronomers can describe how neutron stars are born, how they age, and how they die.

THE EVOLUTION OF PULSARS

When a pulsar first forms, it is spinning quickly, perhaps nearly 100 times a second. The energy it radiates into space comes from its energy of rotation, so as it blows away its pulsar wind and blasts beams of radiation outward, its rotation slows. The average pulsar is apparently only a few million years old, and the oldest is about 10 million years old. Presumably, older neutron stars rotate too slowly to generate detectable radio beams.

FIGURE 14-7

Soon after pulsars were discovered, radio astronomers accumulated enough data to show that pulsars were very gradually slowing down. That is, their pulses were growing longer. Some pulsars, such as the Vela pulsar whose data are shown here, experience glitches in which the pulsar suddenly speeds up only to resume its more leisurely decline.

The Lighthouse Model of a Pulsar

Astronomers think of pulsars not as pulsing objects, but rather as objects emitting beams. As they spin, the beams sweep around the sky; when a beam sweeps over us, we detect a pulse of radiation. Understanding the details of this lighthouse model is a challenge, but the implications are clear. Although a neutron star is only a few kilometers in radius, it can produce powerful beams. Also, we tend to notice only those pulsars whose beams happen to sweep over us.

In this artist's conception, gas trapped in the neutron star's magnetic field is excited to emit light and outline the otherwise invisible magnetic field.

Beams of electromagnetic radiation would probably be invisible unless they excited local gas to glow.

What color should an artist use to paint a neutron star? With a temperature of a million degrees, the surface emits most of its electromagnetic radiation at X-ray wavelengths. Nevertheless, it would probably look blue-white to our eyes.

How a neutron star can emit beams is one of the challenging problems of modern astronomy, but astronomers have a general idea. A neutron star contains a powerful magnetic field and spins very rapidly. The spinning magnetic field generates a tremendously powerful electric field, and the field causes the production of electron–positron pairs. As these charged particles are accelerated through the magnetic field, they emit photons in the direction of their motion, which produce powerful beams of electromagnetic radiation emerging from the magnetic poles.

Neutron Star Rotation with Beams

As in the case of Earth, the magnetic axis of a neutron star could be inclined to its rotational axis.

The rotation of the neutron star will sweep its beams around like beams from a lighthouse.

While a beam points roughly toward Earth, we detect a pulse.

While neither beam is pointed toward us, we detect no energy.

Beams may not be as exactly symmetric as in this model.

Ace✪Astronomy™

Go to AceAstronomy and click Active Figures to see "Neutron Star" and adjust the inclination of the neutron star's magnetic field to produce pulses.

Crab pulsar

Visual-wavelength image

The hazy glow of the Crab Nebula is produced by synchrotron radiation. In the nearly 10 centuries since the supernova, the high-speed electrons should have radiated their energy away, and the synchrotron radiation should have faded. Evidently, the Crab pulsar powers the nebula. The Hubble Space Telescope image above shows the region boxed in Figure 14-5. Circular wisps excited by the pulsar change and flicker from day to day.

Red and yellow in this image show the radio and visual extent of the Crab Nebula. Blue, an X-ray image, reveals that the central pulsar has flung off rings of highly ionized gas in a disk up to a light-year in radius. The pulsar is ejecting jets of high-energy particles perpendicular to the disk.

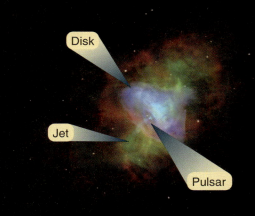

Disk

Jet

Pulsar

Composite image
Radio + visual + X-ray

X-ray image of Puppis supernova remnant

Neutron star

Hubble Space Telescope visual-wavelength image

Neutron star

If a pulsar's beams do not sweep over us, we detect no pulses and the neutron star is difficult to find. A few such objects are known, however. The Puppis A supernova remnant is about 4000 years old and contains a point source of X rays believed to be a neutron star. The isolated neutron star in the right-hand image has a temperature of 700,000 K.

NASA

NASA

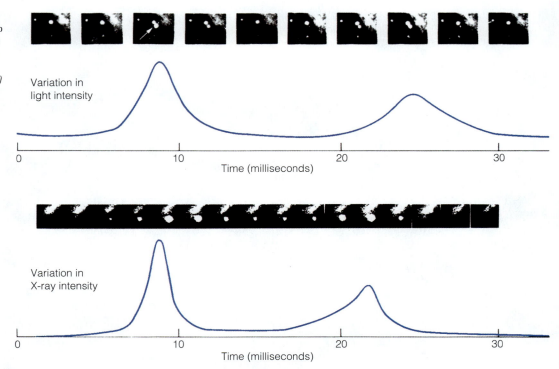

FIGURE 14-8

High-speed images of the Crab Nebula pulsar (arrow) show it pulsing at visual wavelengths and at X-ray wavelengths. *(© AURA, Inc., NOAO, KPNO)* The period of pulsation is 33 milliseconds, and each cycle includes two pulses as its two beams of unequal intensity sweep over Earth. *(Courtesy F. R. Harnden, Jr., from* The Astrophysical Journal, *published by the University of Chicago Press; © 1984 The American Astronomical Society)*

From this we can expect that a young neutron star should emit powerful beams of radiation. The Crab Nebula gives us an example of such a system. Only about 950 years old, the Crab pulsar is so powerful it emits photons of radio, infrared, visible, X-ray, and gamma-ray wavelengths (▌ Figure 14-8). Careful measurements of its brightness with high-speed instruments show that it blinks twice for every rotation. One beam sweeps almost directly over us, and we detect a strong pulse. Half a rotation later, the edge of the other beam sweeps past us, and we detect a weaker pulse.

We would expect only the most energetic pulsars to produce short-wavelength photons and thus pulse at visible wavelengths. The Crab Nebula pulsar is young and powerful, and it produces visible pulses, and so does a pulsar called the Vela pulsar (located in the Southern Hemisphere constellation Vela). The Vela pulsar is fast, pulsing about 11 times a second, and, like the Crab Nebula pulsar, is located inside a supernova remnant. Its age is estimated at about 20,000 to 30,000 years, young in terms of the average pulsar. Thus we suspect that pulsars are capable of producing optical pulses only when they are young.

We might expect to find all pulsars inside supernova remnants, but the statistics must be examined with care. Not every supernova remnant contains a pulsar, and not every pulsar is located inside a supernova remnant. Many supernova remnants probably contain pulsars whose beams never sweep over Earth, and it is difficult to detect such pulsars. Also, some pulsars move through space at high velocity, which suggests that supernova explosions can occur asymmetrically, perhaps because of the violent turbulence in the exploding core.

Such a violent, off-center explosion could give the neutron star a high velocity through space (▌ Figure 14-9). Also, some supernovae probably occur in binary systems and fling the two stars apart at high velocity. In any case, pulsars are known to have such high velocities that many probably escape the disk of our galaxy. We should not be surprised that many neutron stars quickly leave their supernova remnant behind. And finally, we must remember that a pulsar can remain detectable for 10 million years or so, but a supernova remnant cannot survive more than about 50,000 years before it is mixed into the interstellar medium. Consequently, we might expect that most pulsars are not located in supernova remnants, and that most supernova remnants do not contain pulsars.

Careful analysis, however, shows that there are more empty supernova remnants than we should expect. Where are these missing pulsars? Searches suggest that many neutron stars may not be emitting powerful radio beacons. If they are born with very strong magnetic fields, they may have slowed down rapidly and may be rotating so slowly they do not emit beams of radio energy. These objects would have magnetic fields 1000 times stronger than those of a typical pulsar and have been dubbed **magnetars.** We will see more evidence that such magnetars do exist. This line of research is ongoing, but it suggests that we have more to learn about the evolution of pulsars.

The explosion of Supernova 1987A in February 1987 apparently formed a neutron star. We can draw this conclusion because a burst of neutrinos was detected passing through Earth, and theory predicts that the collapse of a massive star's core into a neutron star would

FIGURE 14-9
Many neutron stars have high velocities through space. Here the neutron star known as RX J185635-3754 was photographed on three different dates as it rushes past background stars. *(NASA and F. M. Walter)*

produce such a burst of neutrinos. At first the neutron star would be hidden at the center of the expanding shells of gas ejected into space, but as the gas expands and thins we might be able to see it. Or, if its beams don't sweep over Earth, we might be able to detect it from its X-ray and gamma-ray emission. Although, years after the explosion, no neutron star has been detected, astronomers continue to watch the site, hoping to see the youngest pulsar known.

One reason pulsars are so fascinating is the extreme conditions we find in spinning neutron stars. To see natural processes of even greater violence, we have only to look at pulsars in binary systems.

BINARY PULSARS

Over a thousand pulsars are now known, and some are located in binary systems. These pulsars are of special interest because we can learn more about the neutron star by studying the orbital motions of the binary. Also, in some cases, mass can flow from the companion star onto the neutron star, and that produces high temperatures and X rays.

The first binary pulsar was discovered in 1974 when astronomers Joseph Taylor and Russell Hulse noticed that the pulse period of the pulsar PSR 1913+16 was changing. The period first grew longer and then grew shorter in a cycle that took 7.75 hours. Thinking of the Doppler shifts seen in spectroscopic binaries, the radio astronomers realized that the pulsar had to be in a binary system with an orbital period of 7.75 hours. When the orbital motion of the pulsar carries it away from Earth, we see the pulse period slightly lengthened, just as the wavelength of light emitted by a receding source is lengthened. That is, we see a red shift. Then, when the pulsar rounds its orbit and approaches

Earth, we see the pulse period slightly shortened—a blue shift. From these changing Doppler shifts, the astronomers could calculate the radial velocity of the pulsar around its orbit just as if it were a spectroscopic binary star (Chapter 9). The resulting graph of radial velocity versus time could be analyzed to find the shape of the pulsar's orbit (❙ Figure 14-10). When Taylor and Hulse analyzed PSR 1913+16, they discovered that the binary system consisted of two neutron stars separated by a distance roughly equal to the radius of our sun.

Yet another surprise was hidden in the motion of PSR 1913+16. In 1916, Einstein's general theory of relativity described gravity as a curvature of space-time. Einstein realized that any rapid change in a gravitational field should spread outward at the speed of light as **gravitational radiation.** Gravity waves have not been detected, but Taylor and Hulse were able to show that the orbital period of the binary pulsar was slowly grow-

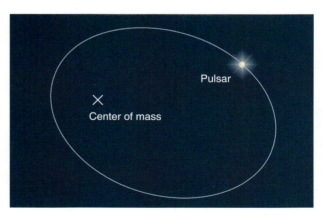

FIGURE 14-10
The radial velocity of pulsar PSR 1913+16 can be found from the Doppler shifts in its period of pulsation. The shape of the radial velocity curve is typical for a spectroscopic binary with an elliptical orbit and allows astronomers to determine the pulsar's orbit. In the plotted orbit, the center of mass does not appear to be at a focus of the elliptical orbit, because the orbit is inclined. *(Adapted from data by Joseph Taylor and Russell Hulse)*

ing shorter because the stars are gradually spiraling toward each other. They are radiating orbital energy away as gravitational radiation. Taylor and Hulse won the Nobel prize in 1993 for their work with binary pulsars.

Dozens of binary pulsars have been found; by analyzing the Doppler shifts in their pulse periods, astronomers can estimate the mass of the neutron stars. Typical masses are about 1.35 solar masses, in good agreement with models of neutron stars.

Binary pulsars can emit strong gravitational waves because the neutron stars contain large amounts of mass in a small volume. This also means that binary pulsars can be sites of tremendous violence because of the strength of gravity at the surface of a neutron star. An astronaut stepping onto the surface of a neutron star would be instantly smushed into a layer of matter only 1 atom thick. Matter falling onto a neutron star can release titanic amounts of energy. If you dropped a single marshmallow onto the surface of a neutron star from a distance of 1 AU, it would hit with an impact equivalent to a 3-megaton nuclear warhead. In general, a particle falling from a large distance to the surface of a neutron star will release energy equivalent to $0.2 \, mc^2$, where m is the particle's mass at rest. Even a small amount of matter flowing from a companion star to a neutron star can generate high temperatures and release X rays and gamma rays.

As an example of such an active system, we can examine Hercules X-1. It emits pulses of X rays with a period of about 1.2 seconds, but every 1.7 days the pulses vanish for a few hours (■ Figure 14-11). Astronomers can understand this system by comparing it to an eclipsing binary star. Hercules X-1 seems to contain a 2-solar-mass star with a temperature of 7000 K and a neutron star that orbit each other with a period of 1.7 days. Matter flowing from the normal star into an accretion disk around the neutron star can reach temperatures of millions of degrees and emit a powerful X-ray glow. Interactions with the neutron star's magnetic field can produce beams of X rays that sweep around with the rotating neuron star. We receive a pulse of X rays every time a beam points our way. The X rays shut off every 1.7 days when the neutron star is eclipsed behind the normal star. The X rays from the neutron star and its accretion disk heat the near side of the normal star to about 20,000 K. As the system rotates, we alternately see the hot side of the star and then the cool side, and its brightness at visible wavelengths varies. Hercules X-1 is a complex system and is still not well understood, but our quick analysis serves to illustrate how complex and powerful such binary systems are during mass transfer.

THE FASTEST PULSARS

Our discussion of pulsars suggests that newborn pulsars should blink rapidly and old pulsars should blink

a

b

FIGURE 14-11
Sometimes the X-ray pulses from Hercules X-1 are on and sometimes they are off. A graph of X-ray intensity versus time looks like the light curve of an eclipsing binary. *(Insets: J. Trümper, Max-Planck Institute)* (b) In Hercules X-1, matter flows from a star into an accretion disk around a neutron star producing X rays, which heat the near side of the star to 20,000 K compared with only 7000 K on the far side. X rays turn off when the neutron star is eclipsed behind the star.

slowly, but the handful that blink the fastest may be quite old. One of the fastest known pulsars is cataloged as PSR 1937+21 in the constellation Vulpecula. It pulses 642 times a second and is slowing down only slightly. The energy stored in the rotation of a neutron star at this rate is equal to the total energy of a supernova explosion, so it seemed difficult at first to explain this pulsar. It now appears that PSR 1937+21 is an old neutron star that has gained mass and rotational energy from a companion in the binary system. Like water hitting a mill wheel, the matter falling on the neutron star

No scientific theory or hypothesis can be proved correct. We can test a theory over and over by performing experiments or making observations, but we can never prove that the theory is absolutely true. It is always possible that we have misunderstood the theory or the evidence, and the next observation we make might disprove the theory.

For example, we might propose the theory that the sun is mostly iron. We might test the theory by looking at the iron lines in the solar spectrum, and the strength of the iron lines would suggest that our theory is right. Although our observation has confirmed our theory, it has not proven the theory is right. We might confirm the theory many times before we realized that iron absorbs photons much more efficiently than hydrogen. Although the hydrogen lines are weak in the sun's spectrum, they tell us that most of the atoms in the sun are hydrogen and not iron.

The nature of scientific thinking can lead to two kinds of mistakes. Sometimes nonscientists will say, "You scientists just want to tear everything down—you don't believe in anything." Scientists test a theory over and over to test its worth. If a theory survives many tests, scientists begin to have confidence it is true.

The second error nonscientists make is to say, "You scientists are never sure of anything." Again, the scientist knows that no theory can be proven correct. That the sun will rise tomorrow is very likely, and scientists have great confidence in that theory. But in the end it is still a theory.

People will say of an idea they dislike, "That is only a theory," as if a theory were simply a random guess. In fact, a theory can be a well-tested truth in which all scientists have great confidence. Yet we can never prove that any theory is absolutely true.

---

has spun it up to 642 rotations per second. With its weak magnetic field, it slows down very slowly and will continue to spin for a very long time.

A number of other very fast pulsars have been found. They are known generally as **millisecond pulsars** because their pulse periods are almost as short as a millisecond (0.001 s). This produces some fascinating physics because the pulse period of a pulsar equals the rotation period of a neutron star. If a neutron star 10 km in radius spins 642 times a second, as does PSR 1937+21, then the period is 0.0016 second, and the equator of the neutron star must be traveling about 40,000 km/s. That is fast enough to flatten the neutron star into an ellipsoidal shape and is nearly fast enough to break it up.

All scientists should be made honorary citizens of Missouri, the "Show Me" state, because scientists demand evidence. The hypothesis that the millisecond pulsars are spun up by mass transfer from a companion star is quite reasonable, but astronomers demand evidence, and evidence has been found. For example, the pulsar PSR J1740-5340 has a period of 42 milliseconds and is orbiting with a bloated red star that is losing mass to the neutron star. This appears to be a pulsar in the act of being spun up to high speed. For another example, consider the X-ray source XTE J1751-305, a pulsar with a period of only 2.3 milliseconds. X-ray observations show that it is in the act of gaining mass from a companion star. The orbital period is only 42 minutes, and the mass of the companion star is only 0.014 solar masses. The evidence suggests this neutron star has devoured all but the last morsel of its binary partner.

Although some millisecond pulsars have binary companions, some are solitary neutron stars. A pulsar known as the Black Widow may explain how a fast pulsar can lack a companion. The Black Widow has a period of 1.6 milliseconds, meaning it is spinning 622 times per second, and it orbits with a low-mass compan-

ion. Presumably the neutron star was spun up by mass flowing from the companion, but spectra show that the blast of radiation and high-energy particles from the neutron star is now boiling away the surface of the companion. The Black Widow has eaten its fill and is now evaporating the remains of its companion. It will soon be a solitary millisecond pulsar.

"Show me," say scientists, and in the case of neutron stars, the evidence seems very strong. Of course, we can never prove a theory is absolutely true (Window on Science 14-1), but the evidence for neutron stars is so strong that astronomers have great confidence that they really do exist. Other theories that describe how they emit beams of radiation and how they form and evolve are less certain, but continuing observations at many wavelengths are expanding our understanding of these last embers of massive stars. In fact, observations of one pulsar have turned up objects no one predicted.

PULSAR PLANETS

Finding planets orbiting stars other than the sun is very difficult, and only about a hundred are known. Oddly, the first such planets were found orbiting a neutron star.

Because a pulsar's period is so precise, astronomers can detect tiny variations by comparison with atomic clocks. When astronomers checked pulsar PSR 1257+12, they found variations in the period of pulsation much like those caused by the orbital motion of the binary pulsar (❙ Figure 14-12a). However, in the case of PSR 1257+12, the variations were much smaller; and, when they were interpreted as Doppler shifts, it became evident that the pulsar was being orbited by at least two objects with planetlike masses of 4.3 and 3.9 Earth masses. The gravitational tugs of the planets make the pulsar wobble about the center of mass of the system by no more than 800 km, and that produces the tiny changes in period (Figure 14-12b).

Astronomers greeted this discovery with both enthusiasm and skepticism. As usual, they looked for ways to test the hypothesis. Simple gravitational theory predicts that the planets should interact and slightly modify each other's orbit. When the data were analyzed, that interaction was found, further confirming the hypothesis that the variations in the period of the pulsar are caused by planets. In fact, further data revealed the presence of a third planet of about the mass of Earth's moon, and a fourth planet with a mass of about 100 Earth masses is now believed to follow a much larger orbit. This illustrates the astonishing precision of studies based on pulsar timing. At least one other pulsar is known where small shifts in pulse timing imply the presence of an orbiting planet.

Astronomers wonder how a neutron star can have planets. The planets that orbit PSR 1257+12 are very close to the pulsar; the inner three orbit at 0.19 AU, 0.36 AU, and 0.47 AU—closer to the pulsar than Venus is to the sun. Any planets that orbit a star would be lost or vaporized when the star exploded. Furthermore, a star about to explode as a supernova would be a large giant or a supergiant, and planets only a few AU distant would be inside such a large star and could not survive. It seems more likely that these planets are the remains of a stellar companion that was devoured by the neutron star. In fact, the pulsar is very fast (162 pulses per second), suggesting that it was spun up in a binary system.

Another pulsar planet has been found orbiting in a binary system containing a neutron star and a white dwarf. Because this system is located in a very old star cluster and contains a white dwarf, astronomers suspect that the planet may be very old. Planets probably orbit other neutron stars, and small shifts in the timing of the pulses may eventually reveal their presence.

We can imagine what these worlds might be like. Planets formed from the remains of dying stars might be rich in heavy elements, but truly ancient planets might be poor in such elements. We can imagine visiting these worlds, landing on their surfaces, and hiking across their valleys and mountains. Above us, the neutron star would glitter in the sky, a tiny point of light.

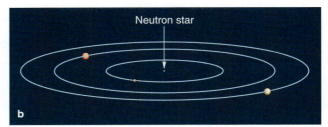

FIGURE 14-12

(a) The dots in this graph are observations showing that the period of pulsar PSR 1257+12 varies from its average value by a fraction of a billionth of a second. The blue line shows the variation that would be produced by planets orbiting the pulsar. (b) As the planets orbit the pulsar, they cause it to wobble by less than 800 km, a distance that is invisibly small in this diagram. (*Adapted from data by Alexander Wolszczan*)

| | REVIEW CRITICAL INQUIRY | |

How can a neutron star be found at X-ray wavelengths?

First, we should remember that a neutron star is very hot because of the heat released when it contracts to a radius of 10 km. It could easily have a surface temperature of 1,000,000 K, and Wien's law (Chapter 7) tells us that such an object will radiate most intensely at a very short wavelength typical of X rays. However, we know that the total luminosity of a star depends on its surface temperature and its surface area, and a neutron star is so small it can't radiate much energy. X-ray telescopes have found such neutron stars, but they are not easy to locate.

There is, however, a second way a neutron star can radiate X rays. If a normal star in a binary system loses mass to a neutron star companion, the inflowing matter will hit with so much energy that it will be heated to very high temperatures. It may form a very hot accretion disk that can radiate intense X rays easily detectable by X-ray telescopes orbiting above Earth's atmosphere.

If you discovered a pulsar, what observations would you make to determine whether it was young or old, single or a member of a binary system, alone or accompanied by planets?

Perhaps the strangest planets in the universe are those orbiting pulsars. But however strange a pulsar planet may be, we can imagine going to one. Our next topic, in contrast, seems beyond the reach of even our imaginations.

14-2 BLACK HOLES

We have now studied white dwarfs and neutron stars, two of the three end states of dying stars. Now we turn to the third—black holes.

Although the physics of black holes is difficult to discuss without sophisticated mathematics, simple logic is sufficient to predict that they should exist. Our problem is to consider their predicted properties and try to confirm that they exist. What objects observed in the heavens could be real black holes? More difficult than the search for neutron stars, the quest for black holes has nevertheless met with success.

To begin our discussion of black holes, we must consider a simple question. How fast must an object travel to escape from the surface of a celestial body? The answer will lead us to black holes.

ESCAPE VELOCITY

Suppose we threw a baseball straight up. How fast must we throw it if it is not to come down? Of course, gravity would pull back on the ball, slowing it, but if the ball were traveling fast enough to start with, it would never come to a stop and fall back. Such a ball would escape from Earth. The escape velocity is the initial velocity an object needs to escape from a celestial body (❚ Figure 14-13). (See Chapter 5.)

Whether we are discussing a baseball leaving Earth or a photon leaving a collapsing star, the escape velocity depends on two things: the mass of the celestial body and the distance from the center of mass to the escaping object. If the celestial body has a large mass, its gravity is strong, and we need a high velocity to escape; but if we begin our journey farther from the center of mass, the velocity needed is less. For example, to escape from Earth, a spaceship would have to leave Earth's surface at 11 km/s (25,000 mph), but if we could launch spaceships from the top of a tower 1000 miles high, the escape velocity would be only 8.8 km/s (20,000 mph). If we could make an object massive enough or small enough, its escape velocity could be greater than the speed of light. Relativity tells us that nothing can travel faster than the speed of light, so even photons, which have no mass, would be unable to escape. Such a small, massive object could never be seen because light could not leave it.

Long before Einstein and relativity, the Rev. John Mitchell, a British gentleman astronomer, realized that Newton's laws of gravity and motion had peculiar consequences. In 1783, Mitchell pointed out that an object 500 times the radius of the sun but of the same density would have an escape velocity greater than the speed of light. Then, "all light emitted from such a body would be made to return towards it." Mitchell didn't know it, but he was talking about a black hole.

FIGURE 14-13
Escape velocity, the velocity needed to escape from a celestial body, depends on mass. The escape velocity at the surface of a very small, low-mass body would be so low we could jump into space. Earth's escape velocity is much larger, about 11 km/s (25,000 mph).

SCHWARZSCHILD BLACK HOLES

If the core of a star collapses and contains more than about three solar masses, no known force can stop it. The object cannot stop collapsing when it reaches the size of a white dwarf because degenerate electrons cannot support the weight. It cannot stop when it reaches the size of a neutron star because degenerate neutrons cannot support the weight. No force remains to stop the object from collapsing to zero radius.

As an object collapses, its density and the strength of its surface gravity increase; and if an object collapses to zero radius, its density and gravity become infinite. Mathematicians call such a point a **singularity;** but in physical terms it is difficult to imagine an object of zero radius. Some theorists believe that a singularity is impossible and that when we better understand the laws of physics, we will discover that the collapse halts before the radius zero. Astronomically, it seems to make little difference.

If the contracting core of a star becomes small enough, the escape velocity in the region around it is so large that no light can escape. We can receive no information about the object or about the region of space near it, and we refer to this region as a **black hole.** If the core of an exploding star collapsed into a black hole, the core would vanish without a trace. Thus a supernova remnant that lacks a central pulsar may harbor a black hole (❙ Figure 14-14).

The boundary of the black hole is called the **event horizon,** because any event that takes place inside the event horizon is invisible to an outside observer. To see how a black hole can exist, we must consider general relativity.

In 1916, Albert Einstein published a mathematical theory of space and time that became known as the general theory of relativity. Einstein treated space and time as a single entity—space-time. His equations showed that gravity could be described as a curvature of space-time, and almost immediately the astronomer Karl Schwarzschild found a way to solve the equations to describe the gravitational field around a single, nonrotating, electrically neutral lump of matter. That solution contained the first general relativistic description of a black hole; nonrotating, electrically neutral black holes are now known as Schwarzschild black holes. In most cases in astronomy, we can use the Schwarzschild

ACTIVE FIGURE 14-15

A black hole forms when an object collapses to a small size (perhaps to a singularity) and the escape velocity in its neighborhood is so great that light cannot escape. The boundary of this region is called the event horizon because any event that occurs inside is invisible to outside observers. The radius of the region is R_s, the Schwarzschild radius.

Ace❀Astronomy™ Go to AceAstronomy and click Active Figures to see "Schwarzschild Radius." Take control of this diagram.

solution to think about black holes. We will see later in this chapter what difference rotation makes.

Schwarzschild's solution shows that if matter is packed into a small enough volume, then space-time curves back on itself. Objects can still follow paths that lead into the black hole, but no path leads out, so nothing can escape, not even light. Thus, the inside of the black hole is totally beyond the view of an outside observer. The event horizon is the boundary between the isolated volume of space-time and the rest of the universe, and the radius of the event horizon is called the **Schwarzschild radius, R_S**—the radius within which an object must shrink to become a black hole (❙ Figure 14-15).

Although Schwarzschild's work was highly mathematical, his conclusion is quite simple. The Schwarzschild radius (in meters) depends only on the mass of the object (in kilograms):

$$R_S = \frac{2GM}{c^2}$$

In this simple formula, G is the gravitational constant, M is the mass, and c is the speed of light. A bit of arithmetic shows that a 1-solar-mass black hole will have a Schwarzschild radius of 3 km, a 10-solar-mass black hole will have a Schwarzschild radius of 30 km, and so on (❙ Table 14-1). Even a very massive black hole would not be very large.

Every object with mass has a Schwarzschild radius, but not every object is a black hole. For example, Earth

X-ray image

FIGURE 14-14

Some supernova remnants contain no neutron star perhaps because the supernova formed a black hole instead. However, this supernova remnant, formed by Tycho Brahe's supernova of 1572, has the chemical composition typical of the remains of a type Ia supernova. Such supernovae are believed to destroy the white dwarf entirely and do not leave behind a neutron star or a black hole. (*John P. Hughes, Rutgers University*)

TABLE 14-1
The Schwarzschild Radius

Object	Mass ($M_\odot$)	Radius
Star	10	30 km
Star	3	9 km
Star	2	6 km
Sun	1	3 km
Earth	0.000003	0.9 cm

has a Schwarzschild radius of about 1 cm, but it could become a black hole only if we squeezed it inside that radius. Fortunately, Earth will not collapse spontaneously into a black hole because its mass is less than the critical mass of about 3 solar masses. Only exhausted stellar cores more massive than this can form black holes under the sole influence of their own gravity. In this chapter, we are interested in black holes that might originate from the deaths of stars. These black holes would have masses larger than 3 solar masses. In later chapters, we will encounter black holes whose masses might exceed a million solar masses.

Do not think of black holes as giant vacuum cleaners that will pull in everything in the universe. A black hole is just a gravitational field, and at a reasonably large distance its gravity is no greater than that of a normal object of similar mass. If the sun were replaced by a 1-solar-mass black hole, the orbits of the planets would not change at all. The gravity of a black hole becomes extreme only when we approach close to it. The universe contains many black holes. So long as we and other objects stay at a safe distance from the black holes, they have no catastrophic effects.

BLACK HOLES HAVE NO HAIR

Theorists who study black holes are fond of saying, "Black holes have no hair." By that they mean that once matter forms a black hole, it loses almost all of its normal properties. A black hole made of a collapsed star will be indistinguishable from a black hole made from peanut butter and fake-fur mittens. Once the matter is inside the event horizon, it retains only three properties—mass, angular momentum, and electrical charge.

The Schwarzschild black hole is represented by a solution to Einstein's equations for the special case where the object has only mass. Schwarzschild black holes do not rotate or have charge. The solutions for rotating or charged black holes (or for rotating, charged black holes) are more difficult and have been found in only the last few decades. Generally, rotating, charged black holes are similar to Schwarzschild black holes.

It seems that astronomers need not worry about charged black holes because stars, whose collapse presumably forms black holes, cannot have large electrostatic charges. Suppose that we could give the sun a large positive charge. It would begin to repel protons in its corona and attract electrons and would soon return to neutral charge. Thus, we should expect stars and black holes to be electrically neutral.

But everything in the universe seems to rotate, and collapsing stars spin rapidly as they conserve angular momentum. Thus, we should probably expect black holes to have angular momentum. In 1963, New Zealand mathematician Roy P. Kerr found a solution to Einstein's equations that describes a rotating black hole. This is now known as the **Kerr black hole.**

The mass of a black hole curves neighboring space-time, and the Kerr solution shows that the rotation of a black hole drags space-time around with it. The **ergosphere** is a region outside the event horizon in which space-time rotates with the rotating black hole so powerfully that nothing could avoid being dragged along. No one has ever approached a rotating black hole, so we have no idea what it might feel like to enter a region where space-time was moving past us.

The word *ergosphere* comes from the Greek word *ergo,* meaning "work," because the rotating space-time in the ergosphere can do work on a particle; that is, the particle can gain energy. In particular, the Kerr solution shows that a particle that enters the ergosphere can break into two pieces, one falling into the black hole and the other escaping with more energy than it had when it entered. Thus, energy can be extracted from a rotating black hole, and, as a result, the black hole slows its rotation very slightly.

The Kerr solution is a fascinating bit of theoretical physics, but it has an important application in astronomy. Almost certainly black holes rotate, and matter falling into black holes must pass through the ergosphere. Thus, we may eventually find situations where energy is extracted from rotating black holes.

A LEAP INTO A BLACK HOLE

Before we can search for real black holes, we must understand what theory predicts about the appearance of a black hole. To explore that idea, we can imagine that we leap, feet first, into a Schwarzschild black hole.

If we were to leap into a black hole of a few solar masses from a distance of an astronomical unit, the gravitational pull would not be very large, and we would fall slowly at first. Of course, the longer we fell and the closer we came to the center, the faster we would travel. Our wristwatches would tell us that we fell for about 65 days before we reached the event horizon.

Our friends who stayed behind would see something different. They would see us falling more slowly as we came closer to the event horizon because, as explained by general relativity, clocks slow down in curved space-time. This is known as **time dilation.** In fact, our friends would never actually see us cross the event horizon. To them we would fall more and more slowly until we seemed hardly to move. Generations later, our descendants could focus their telescopes on us and see us still inching closer to the event horizon. We, however, would have sensed no slowdown and would conclude that we had crossed the event horizon after only about 65 days.

Other relativistic effects would make it difficult to see us. As light travels out of a gravitational field, it loses energy, and its wavelength grows longer. This is known as a **gravitational red shift** (▌Table 14-2). Light leaving us and traveling away from the black hole would suffer a larger and larger gravitational red shift. In addition, as our inward velocity grew higher and higher, a relativistic effect would cause more and more of the light leaving us to be emitted in the forward direction. This would reduce the amount of light leaving us and traveling outward, making us even more difficult to see. Although we would notice none of these effects as we fell toward the black hole, our friends would need to observe at longer wavelengths and with larger telescopes to detect us.

While these relativistic effects seem merely peculiar, other effects would be quite unpleasant. Imagine again that we are falling feet first toward the event horizon of a black hole. We would feel our feet, which would be closer to the black hole, being pulled in more strongly than our heads. This is a tidal force, and at first it would be minor. But as we fell closer, the tidal force would become very large. Another tidal force would compress us as our left side and our right side both fell toward the center of the black hole. For any black hole with a mass like that of a star, the tidal forces would crush us laterally and stretch us longitudinally long before we reached the event horizon (▌Figure 14-16). The friction from such severe distortions of our bodies would heat us to millions of degrees, and we would emit X rays and gamma rays. (Needless to say, this would render us inoperative as thoughtful observers.)

Some years ago a popular book suggested that we could travel through the universe by jumping into a black hole in one place and popping out of another somewhere far across space. That might make for good science fiction, but tidal forces would make it an unpopular form of transportation even if it worked.

Our imaginary leap into a black hole is not frivolous. We now know how to find a black hole: Look for a strong source of X rays. It may be a black hole into which matter is falling.

THE SEARCH FOR BLACK HOLES

Do black holes really exist? Beginning in the 1970s, astronomers searched for observational evidence that their theories were correct. They tried to find one or more objects that were obviously black holes. That very

TABLE 14-2 The Gravitational Red Shift	
Object	Red Shift (percent)
Sun	0.0002
White dwarf	0.01
Neutron star	20
Black hole event horizon	Infinite

difficult search is a good illustration of how the unwritten rules of science help us understand nature (Window on Science 14-2).

A black hole alone is totally invisible because nothing can escape from the event horizon. But a black hole into which matter is flowing would be a source of X rays. Of course, X rays can't escape from inside the

FIGURE 14-16
Leaping feet first into a black hole, a person of normal proportions (left) would be distorted by tidal forces (right) long before reaching the event horizon around a typical black hole of stellar mass. Tidal forces would stretch the body lengthwise while compressing it laterally. Friction from this distortion would heat the body to high temperatures.

WINDOW ON SCIENCE 14-2

Natural Checks on Fraud in Science

Fraud is actually quite rare in science. The nature of science makes fraud difficult, and the way scientists publish their research makes it almost impossible. In fact, we can think of science as a set of unwritten rules of behavior that have evolved to prevent scientists from lying to one another or to themselves, even by accident.

Suppose for a moment that we wanted to commit scientific fraud. We would have to invent data supposedly obtained from experiment or observation. We might invent X-ray data supposedly obtained by observing an X-ray binary star. Or if we were interested in theory, we might invent a fraudulent mathematical calculation of the physics going on in an X-ray binary. We might get away with it for a short time, but one of the most important rules in science is that good results must be reproducible. Other people must be able to repeat our observations, experiments, and calcula-

tions. In fact, most scientists routinely repeat other scientists' work as a way of getting started on a research topic. As soon as someone tries to repeat our fraudulent research, we will be caught. The more important a scientific result is, the sooner other scientists will repeat it, so we don't have much of a chance of getting away with scientific fraud. In this way, science is self-correcting.

Even if we could invent some convincing scientific research, we would probably have difficulty publishing it. When a scientist submits an article to a scientific journal, it is subject to peer review. That is, the editor of the journal sends the article to one or two other experts in the field for comment and suggestions. These reviewers often make helpful suggestions, but they may also point out errors that have to be fixed before the journal can publish the article. In some cases, an article may be so flawed the editor will refuse to publish it

at all. If we submitted our fraudulent research on X-ray binaries, the reviewers would almost certainly notice things wrong with it, and it would never get into print.

Scientists know the rules, and they use them. If someone makes a big discovery and is interviewed by the press, scientists will begin asking, "Has this work been published in a peer-reviewed journal yet?" That is, they want to know if other experts have checked the work. Until research is published, it isn't really official, and most scientists would treat the results with care.

Fraud isn't impossible in science. Some cases have even been in the national news. Big grant money is a terrible temptation. But in science, "the truth will out." Because of the way scientists reproduce research and because of the way research is published, scientific fraud is quite rare.

---

event horizon, but X rays emitted by the heated matter flowing into the black hole could escape if the X rays were emitted before the matter crossed the event horizon. An isolated black hole will not have much matter flowing into it, but a black hole in a binary system might receive a steady flow of matter transferred from the companion star. Thus, we can search for black holes by searching among X-ray binaries.

Some X-ray binaries, such as Hercules X-1, contain a neutron star, and they will emit X rays much as would a binary containing a black hole. We can tell the difference between a neutron star and a black hole in an X-ray binary in two ways. If the compact object emits pulses, we know it is a neutron star. Otherwise, we must depend on the mass of the object. If the compact object has a mass greater than 3 solar masses, the object can't be a neutron star, and we can conclude that it must be a black hole.

The first X-ray binary suspected of harboring a black hole was Cygnus X-1, the first X-ray object discovered in Cygnus. It contains a supergiant B0 star and a compact object orbiting each other with a period of 5.6 days. Matter flows from the B0 star as a strong stellar wind, and some of that matter enters a hot accretion disk around the compact object (▌ Figure 14-17). The accretion disk is about 5 times larger in diameter than the orbit of Earth's moon, and the inner few hundred kilometers of the disk have a temperature of about 2 million Kelvin—hot enough to radiate X rays. The com-

pact object is invisible, but Doppler shifts in the spectrum reveal the motion of the B0 star around the center of mass of the binary. By making assumptions about the mass of the B0 star and the geometry of the orbit,

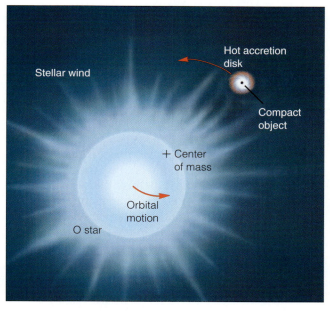

FIGURE 14-17
The X-ray source Cygnus X-1 is a supergiant O star and a compact object orbiting each other. Gas from the O star's stellar wind flows into the hot accretion disk, and the X rays we detect come from the disk.

TABLE 14-3
Nine Black-Hole Candidates

Object	Location	Companion Star	Orbital Period	Mass of Compact Object
Cygnus X-1	Cygnus	B0 Supergiant	5.6 days	$>3.8\ M_\odot$
LMC X-3	Dorado	B3 main-sequence	1.7 days	$6.7 \pm 1.1\ M_\odot$
A0620-00	Monocerotis	K main-sequence	7.75 hours	$10 \pm 5\ M_\odot$
V404 Cygni	Cygnus	K main-sequence	6.47 days	$12 \pm 2\ M_\odot$
J1655-40	Scorpius	F main-sequence	2.61 days	$6.9 \pm 1\ M_\odot$
QZ Vul	Vulpecula	K main-sequence	8 hours	$10 \pm 4\ M_\odot$
4U 1543-47	Lupus	A main-sequence	1.123 days	$2.7–7.5\ M_\odot$
V4641 Sgr	Sagittarius	B supergiant	2.81678 days	$8.7–11.7\ M_\odot$
XTE J1118+480	Ursa Major	K main-sequence	0.170113 days	$>6\ M_\odot$

astronomers can calculate the mass of the compact object—roughly 5 solar masses, well above the maximum for a neutron star.

To confirm that black holes exist, astronomers needed to find a conclusive example, an object that couldn't be anything else. Cygnus X-1 didn't quite succeed. Astronomers testing the hypothesis that it contained a black hole were quick to point out that the B0 star might not be a normal star. If it were of lower mass, the compact object might not be as massive as it seemed. Also, there could be a third star in the system confusing the analysis. The compact object in the Cygnus X-1 binary system cannot be shown conclusively to be more massive than 3 solar masses. Most astronomers believe it is a black hole, but the loophole remains, and more conclusive examples have been found.

As X-ray telescopes have found more X-ray objects, the list of black hole candidates has grown to a few dozen. A few of these objects, such as the first two in ▮ Table 14-3, contain massive stellar companions, either giants, supergiants, or massive main-sequence stars. This makes such systems difficult to analyze, because such massive companions dominate the system. A good example of this sort is LMC X-3 (LMC refers to the Large Magellanic Cloud, a small galaxy near our own). The compact object in LMC X-3 has a mass of about 10 solar masses. One reason astronomers think the compact object is so massive is that it distorts the shape of the B main-sequence star into an egg shape; and, as the system rotates, the light from the B star varies because we see the side of the egg and then the end. By analyzing the light change, astronomers can determine the shape of the B star and from that find the mass of the compact object.

Many of the black hole candidates are binary systems in which the normal star is a lower-mass main-sequence star. Such systems do not remain X-ray sources continuously, but suffer X-ray nova outbursts as matter flows rapidly into the accretion disk. A year or so after an outburst, the flow has stopped, the accretion disk has dimmed, and astronomers can detect the spectrum of the main-sequence star. The spectral type and Doppler motions of the main-sequence companion reveal the mass of the compact object with only limited uncertainty. The lower-mass stellar companion makes these X-ray nova systems easier to analyze than are systems with massive companions.

A number of examples of these X-ray nova systems are shown in Table 14-3. A0620-00 is an old nova that erupted again in 1975. It contains an ordinary main-sequence K star and a compact object that orbit each other with a period of 7.75 hours. From the orbital motion and the distortion of the K star, astronomers conclude the compact object must have a mass between 5 and 15 solar masses. Three of the best understood examples are V404 Cygni; J1655-40, also known as Nova Scorpii 1994; and QZ Vul. The compact objects in these systems seem much too massive to be anything but black holes.

The growing list of X-ray binaries with compact objects exceeding 3 solar masses has convinced astronomers that black holes really do exist. The problem now is to understand how these objects interact with matter flowing into them through accretion disks to produce X rays, gamma rays, and jets of matter.

Ace⟲Astronomy™ Go to AceAstronomy and click Active Figures to see "End States of Stars." Compare the deaths of high-, medium-, and low-mass stars.

REVIEW CRITICAL INQUIRY

How can a black hole emit X rays?

Once a bit of matter falling into a black hole crosses the event horizon, no light or other electromagnetic radiation it emits can escape from the black hole. The matter becomes lost to our view. But if it emitted radiation before it crossed the event horizon, that radiation could escape, and we could detect it. Furthermore, the powerful gravitational field near

Astronomers are confident that black holes really exist. Now the problem is to understand how compact objects can explain the exotic objects we see in the sky.

14-3 COMPACT OBJECTS WITH DISKS AND JETS

Neutron stars and black holes seem to be exotic objects, but they generate equally exotic phenomena. By studying those phenomena, we can learn more about the strange objects.

X-RAY BURSTERS

Beginning in the 1970s, X-ray telescopes revealed that some objects emit irregularly spaced bursts of X rays. Typically, bursts that follow a long quiet period are especially large (▌ Figure 14-18), and this suggests that some mechanism is accumulating energy that is released by the bursts. The longer the quiet phase, the more energy accumulates.

These **X-ray bursters** are thought to be binaries containing neutron stars. The X-ray burster 4U 1820−30 in the globular cluster NGC6624, for example, appears to be a neutron star pulling mass away from a white dwarf (▌ Figure 14-19). Matter from the white dwarf first flows into an accretion disk and then falls to the surface of the neutron star where the impact heats the gas. Hydrogen fuses on the surface of the star, leaving helium ash to accumulate. When the helium reaches a depth of about a meter, it fuses explosively into carbon and produces an X-ray burst. The rapid increase in brightness and the total amount of energy produced fit well with theoretical models of an explosion on the surface of an object as small as a neutron star. Of course, mass continues to accumulate, producing burst after burst. Notice the similarity with the mechanism that produces nova explosions on the surfaces of white dwarfs.

Dozens of X-ray bursters are known, and astronomers suspect that they occur in binary systems where matter flows through accretion disks and falls into neutron stars.

ACCRETION DISK OBSERVATIONS

When material falls into a neutron star or black hole it forms an accretion disk, and high-speed observations have been able to reveal some of the processes that go on in such disks.

When a blob of matter gets caught in an accretion disk, it can orbit so fast its orbital period is measured in thousandths of a second. Because the blob of matter looses energy by friction, it spirals inward, and its orbital period grows shorter. This process can be detected as a quick flicker as the disk emits a short series of pulses of electromagnetic radiation. Disks around some neutron stars produce pulses with a period as short as 0.00075 seconds, as blobs of material orbit only a dozen kilometers from the center of the neutron star. Because the pulse trains last only a few cycles, and because the period of the pulses grows rapidly shorter, they are called **quasi-periodic oscillations (QPOs)**. Do not confuse these rapid flickering pulses with the regular pulses emitted by a pulsar.

a

b

FIGURE 14-18

(a) X-ray bursters emit bursts of X rays that rise to full intensity in a few seconds and then fade in about 20 seconds. (b) The longer the interval since the preceding burst, the brighter the next burst will be.

FIGURE 14-19

(a) At visible wavelengths, the center of star cluster NGC6624 is crowded with stars. (b) In the ultraviolet, one object stands out, the X-ray burster 4U 1820−30 consisting of a neutron star orbiting a white dwarf. (c) An artist's conception shows matter flowing from the white dwarf into an accretion disk around the neutron star. *(a and b, Ivan King and NASA/ESA)*

Visual-wavelength image

a

X-ray source 4U 1820−30

UV image

b

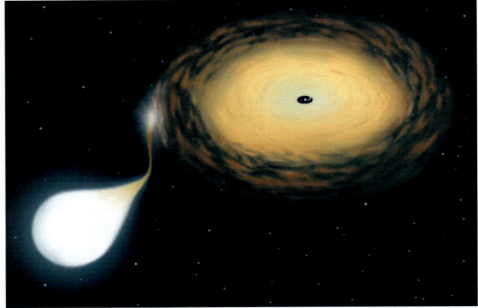

c

QPOs can be observed in J1550-564, a star shedding mass into an accretion disk around a black hole. The flow of mass is irregular and causes X-ray flares. Short flickering strings of pulses are seen with periods as short as 0.003 second, and the period decreases rapidly. This suggests we are seeing material at some boundary in the accretion disk moving rapidly inward toward orbits of shorter and shorter period.

Cyg X-1 appears to be a star shedding mass into a black hole of about 10 solar masses. Pulse trains have been detected with periods of 0.018 seconds. The pulses grow dimmer and faster, which seems to mean the material is spiraling into the black hole. But no impact is seen, as we would expect of material falling into a neutron star. Evidently, the pulses fade away because the material approaches the event horizon and a powerful gravitational red shift stretches the photons to such long wavelengths they cannot be detected.

Further evidence of an event horizon was found in a detailed study of 12 X-ray novae. Six of the novae consisted of binary systems containing neutron stars, and bursts of energy were detected, produced by in-falling material striking the surfaces of the neutron stars. In contrast, six of the novae were systems containing black holes, and no bursts were seen from impacts on a surface. Apparently, in these systems, the material spiraled inward and disappeared across event horizons (▌Figure 14-20).

Because space-time is so strongly curved near a compact object, very small orbits are unstable. Once material spirals inward to that smallest possible orbit, it must quickly fall into the object. Yet the Rossi X-ray Timing Explorer satellite found an X-ray binary containing a black hole in which a QPO had a period of 0.002 second. The smallest possible orbit around the black hole in that system has a period of a little over

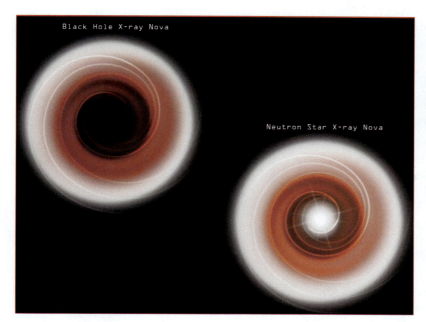

Black Hole X-ray Nova

Neutron Star X-ray Nova

FIGURE 14-20

Gas spiraling into an accretion disk grows hot, and as it nears the central object, a strong gravitational red shift makes it appear redder and dimmer. Systems containing a neutron star emit bursts of energy when the gas hits the surface of the neutron star, but such bursts are not seen for systems containing black holes. In those systems, the matter vanishes across the event horizon. *(NASA/CXC/SAO)*

0.003 second. This appears to be evidence that the spinning black hole is dragging space-time with it as theory predicts and making smaller, shorter period orbits stable. Astronomers are searching for more of these very-short-period QPOs, because they confirm the Kerr solution's prediction that spinning black holes drag space-time.

High-speed observations at many wavelengths are allowing astronomers to follow the rapid processes that occur inside accretion disks. Those same accretion disks can eject high-energy jets extending out into space.

JETS OF ENERGY FROM COMPACT OBJECTS

Our first impression of a black hole might suggest that it is impossible to get any energy out of such an object. In later chapters, we will meet galaxies that extract vast amounts of energy from massive black holes, so we should pause here to see how a compact object can produce energy.

Whether a compact object is a black hole or a neutron star, it has a strong gravitational field. Any matter flowing into that field is accelerated inward; and, because it must conserve angular momentum, it flows into an accretion disk made so hot by friction that the inner regions can emit X rays and gamma rays. Thermal and magnetic processes can cause the inner parts of the accretion disk to eject high-speed jets of matter in opposite directions along the axis of rotation (∎ Figure 14-21). This process is similar to the bipolar outflows ejected by protostars, but it is much more powerful. We have seen in the X-ray image on page 285 that the Crab Nebula pulsar is ejecting jets of highly excited gas. The Vela pulsar does the same (Figure 14-6). Systems containing black holes can also eject jets. The black hole can-

didate J1655-40 is observed at radio wavelengths to be sporadically ejecting oppositely directed jets at 92 percent the speed of light.

One of the most powerful examples of this process is an X-ray binary called SS433. Its optical spectrum shows sets of spectral lines that are Doppler-shifted by about one-fourth the speed of light, with one set shifted to the red and one set shifted to the blue. Furthermore, the two sets of lines shift back and forth across each other with a period of 164 days. News media reported that astronomers had discovered an object that was both approaching and receding at a fantastic speed, but astronomers recognized the Doppler shifts as evidence of oppositely directed jets.

Apparently, SS433 is a binary system in which a compact object thought to be a neutron star or possibly a black hole pulls matter from its companion star and forms an extremely hot accretion disk. Jets of high-temperature gas blast away in beams aimed in opposite directions. As the disk precesses, it sweeps these beams around the sky once every 164 days, and we see light from gas trapped in both beams. One beam produces a red shift, and the other produces a blue shift (∎ Figure 14-22). SS433 is a prototype that illustrates how the gravitational field around a compact object can produce powerful beams of radiation and matter. We will meet this phenomenon again when we study peculiar galaxies.

GAMMA-RAY BURSTS

The cold war has an odd connection to neutron stars and black holes. In 1963, a nuclear test ban treaty was signed; and, by 1968, the United States was able to put a series of Vela satellites in orbit to watch for nuclear tests that were violations of the treaty. A nuclear detonation emits gamma rays, so the Vela satellites were

FIGURE 14-21

In this artist's impression, matter from a normal star at left flows into an accretion disk around a compact object at right. Processes in the spinning disk eject gas and radiation in jets perpendicular to the disk. The jet inclined slightly toward us has a blue shift, and the jet inclined slightly away from us has a red shift. (NASA)

designed to watch for bursts of gamma rays coming from Earth. The experts were startled when the satellites began detecting about one gamma-ray burst a day coming from space. When those data were finally declassified, astronomers realized that the bursts might be coming from neutron stars or black holes. These objects are now known as **gamma-ray bursters.**

The Compton Gamma Ray Observatory reached orbit in 1991 and immediately began reporting gamma-ray bursts at the rate of a few a day. The intensity of the gamma rays rises to a maximum in seconds and then fades away quickly; a burst is usually over in seconds or minutes. Furthermore, the Compton Observatory discovered that the gamma-ray bursts were coming from all over the sky and not from any particular region. This helped astronomers sort out the different theories.

Some theories proposed that the gamma-ray bursts were being produced among the stars in our galaxy, but the Compton Observatory observation that the bursts were coming from all over the sky eliminated that possibility. If the gamma-ray bursts were produced among stars in our galaxy, we would expect to see them most often from along the Milky Way, where many stars are located. The observation of bursts occurring all over the sky meant that bursts were probably coming from distant galaxies.

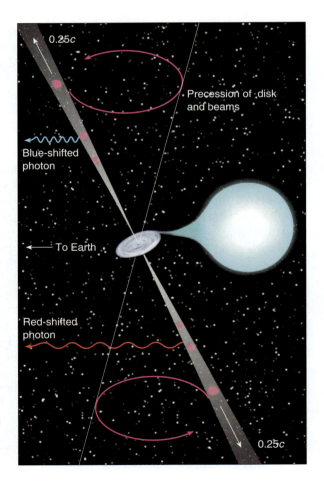

FIGURE 14-22

The generally accepted model of SS433 includes a compact object with a very hot accretion disk producing beams of radiation with embedded blobs of gas traveling at one-fourth the velocity of light. The precession of the disk swings the beams around a conical path every 164 days.

Gamma-ray bursts are hard to study because they occur without warning and fade so quickly; but, in 1997, the Italian-Dutch satellite BeppoSAX began giving Earth's astronomers immediate notice when a gamma-ray burst occurred, along with an accurate position in the sky. When astronomers quickly looked at those positions at visible wavelengths, they could see the fading glow of the explosion (▌ Figure 14-23). The spectra of these glowing clouds of gas reveal that most gamma-ray bursts are in very distant galaxies.

Given the intensity of a gamma-ray burst and its distance, astronomers can calculate its luminosity. If the gamma-ray bursts occur at very great distances, out among the galaxies, then they must be very powerful—at least as powerful as the most violent supernova explosions. Or perhaps the eruption blasts gamma rays out in beams. In that case, the explosion doesn't have to be quite as violent and could still be detected at very great distance if a beam were pointed at Earth. But that would mean that we don't see most of these events because their beams do not happen to be pointed toward us. In that case, the eruptions would have to be much more common than they appear.

Some bursts of lower-energy (soft) gamma rays repeat and are therefore known as **soft gamma-ray repeaters (SGRs).** Only a few are known, and they lie along the Milky Way. Astronomers suspect that they occur on young neutron stars that have magnetic fields as much as 1000 times stronger than the average neutron stars. We have discussed these magnetars earlier. Evidently, such a highly magnetic neutron star can produce bursts of gamma rays when shifts in the magnetic field break the crust of the neutron stars. One of these objects produced a burst of gamma rays that reached Earth on August 27, 1998, and temporarily ionized Earth's upper atmosphere, disrupting radio communication worldwide.

Bursts of higher-energy gamma rays are spread uniformly across the sky and do not appear to repeat, so they can't be explained by magnetars in our galaxy. These gamma-ray bursts must be produced by violent explosions occurring in distant galaxies. Two theories have been proposed. One suggestion is that gamma-ray bursts occur when two neutron stars orbiting each other lose energy by gravitational radiation and fall together. Theoretical models show that the merger of two neutron stars would produce a tremendous eruption lasting only seconds as the neutron stars ripped each other apart with tides and vanished into a newborn black hole. Such an event could produce bursts of gamma rays possibly in beams.

Host galaxy

Visual-wavelength image

FIGURE 14-23

Alerted by gamma-ray detectors on satellites, observers used one of the VLT 8.2-meter telescopes on a mountaintop in Chile to image the location of a gamma-ray burst only hours after the burst. The image at top left shows that fading glow of the eruption. The image at top right, recorded 13 years before, reveals no trace of an object at the location of the gamma-ray burst. The Hubble Space Telescope image at bottom was recorded a year later and reveals a very faint galaxy at the location of the gamma-ray burst. *(ESO and NASA)*

The second theory proposes that gamma-ray bursts are produced when an extremely massive star exhausts its nuclear fuels. For the most massive stars, the resulting supernova explosion can be smothered by the massive outer layers of the star, and the star may collapse directly into a black hole. These have been called **hypernovae** or **collapsars.** Although the hypernova may produce no supernova explosion, it can emit jets of gamma rays because its rotation slows the collapse of its equatorial regions. The rotating star would collapse first along its poles, and moments later the equatorial regions would fall in. Models suggest this could produce powerful bursts of gamma rays shooting out of the poles of rotation. These hypernovae are thought to be much more rare than normal supernovae because they are produced only by the most massive stars.

The growing evidence seems to favor the hypernovae theory over the merging neutron star theory, but this may be a case where similar phenomena are produced by different processes. Binary neutron stars are common, and mergers should occur occasionally. We may eventually find that many gamma-ray bursts are produced by hypernovae but that some are produced by neutron star mergers. The study of gamma-ray bursts is still new and may hold further surprises.

Incidentally, if a gamma-ray burst occurred only 1600 ly from Earth, the distance to a nearby binary pulsar, the gamma-ray burst would shower the Earth with radiation equivalent to a 10,000-megaton nuclear blast. (The largest bombs ever made were a few megatons.) The gamma rays could create enough nitric oxide in the atmosphere to produce intense acid rain and could destroy the ozone layer and expose life on Earth to deadly levels of solar ultraviolet radiation. Gamma-ray bursts may occur relatively near the Earth as often as every few 100 million years and could be one of the causes of the mass extinctions that show up in the fossil record.

It may seem odd that such rare events as merging neutron stars and hypernovae are blamed for something so common that gamma-ray telescopes observe one or more every day. But we must remember that these events are so powerful they can be detected over very great distances. There may be 30,000 neutron star binaries in each galaxy, so mergers must occur now and then in any galaxy. Massive stars explode as hypernovae only rarely in any one galaxy, but we can detect these explosions among a vast number of galaxies. Thus the 1000 or so gamma-ray bursts that the Compton Observatory detected each year appear to be coming from the final deaths of stars far across space.

REVIEW **CRITICAL INQUIRY**

Why do fluctuations from accretion disks have such short periods?

When we see an X-ray flicker coming from an accretion disk, we are seeing a blob of hot gas orbiting the neutron star or black hole at a very small radius. Earth takes a year to orbit the sun once, but it is very far from the sun. A blob of gas in an accretion disk around a neutron star is orbiting an object with a mass of about a solar mass, but it may have an orbit only a few dozen kilometers in radius. The orbital period must be very small, as we would expect from Kepler's third law.

The flickers we see coming from accretion disks tell us about fantastic processes going on there. Why do these trains of pulses fade away?

Neutron stars and black holes give rise to violent processes and emit high-energy radiation, but they are only single stars. When we turn our attention to galaxies, we study communities containing billions of stars. The processes there are much more powerful.

SUMMARY

If the remains of a star collapse with a mass greater than the Chandrasekhar limit of 1.4 solar masses, the object cannot reach stability as a white dwarf. It must collapse to the neutron star stage with a radius of about 10 km and a density equal to that of an atomic nucleus. Such a neutron star can be supported by the pressure of its degenerate neutrons. But if the mass is greater than 2 to 3 solar masses, the degenerate neutrons cannot stop the collapse, and the object must become a black hole.

Theory predicts that a neutron star should rotate very fast, be very hot, and have a strong magnetic field. Such objects have been identified as pulsars, sources of pulsed radio energy. Pulsars are evidently spinning neutron stars that emit beams of radiation from their magnetic poles. As they spin, they sweep the beams around the sky; and, if the beams sweep over Earth, we detect pulses. This is known as the lighthouse model. The spinning neutron star slows as it radiates energy into space, mostly as a pulsar wind, and glitches in its pulsation appear to be caused by starquakes in the rigid crust of the neutron star or by changes in the circulation of the fluid interior.

Dozens of pulsars have been found in binary systems. One such binary pulsar is losing orbital energy at the rate predicted by the theory of gravitational waves. In some binary systems, such as Hercules X-1, mass flows into a hot accretion disk around the neutron star and causes the emission of X rays. In other systems, called X-ray bursters, the accumulation of fuel on the neutron star causes periodic outbursts of X rays.

If a collapsing star has a mass greater than 2 to 3 solar masses, it must contract to a very small size—perhaps to a singularity, an object of zero radius. Near such an object, gravity is so strong not even light can escape, and we term the region a black hole. The surface of this region, called the event horizon, marks the boundary of the black hole. The Schwarzschild radius is the radius of this event horizon, amounting to only a few kilometers for black holes of stellar mass.

If we were to leap into a black hole, our friends who stayed behind would see two relativistic effects. They would see our clock slow relative to their own clock because of time dilation. Also, they would see our light red-shifted to longer wavelengths. We would not notice these effects, but we would feel powerful tidal forces that would deform and heat our mass until we grew hot enough to emit X rays. Any X rays we emitted before crossing the event horizon could escape.

To search for black holes, we must look for binary star systems in which mass flows into a compact object and emits X rays. If the mass of the compact object is greater than about 3 solar masses, then the object is presumably a black hole. A number of such objects have been located.

Accretion disks produce rapid, violent phenomena. X-ray bursters are caused by matter flowing through an accretion disk onto a neutron star. Flickering in the light from accretion disks is caused by the rapid orbital motion of matter in the disk. In many cases, the rapid rotation of the disk and the inflow of matter produce high-energy jets perpendicular to the disk. Soft gamma-ray repeaters appear to be produced by magnetars, highly magnetized neutron stars. True gamma-ray bursters appear to be produced by the merger of two neutron stars or by hypernovae.

NEW TERMS

neutron star

pulsar

lighthouse model

pulsar wind

glitch

magnetar

gravitational radiation

millisecond pulsar

singularity

black hole

event horizon

Schwarzschild radius (R_S)

Kerr black hole

ergosphere

time dilation

gravitational red shift

X-ray burster

quasi-periodic oscillations (QPOs)

gamma-ray burster

soft gamma-ray repeater (SGR)

hypernova

collapsar

REVIEW QUESTIONS

Ace ◐ Astronomy™ Assess your understanding of this chapter's topics with additional quizzing and animations at **http:// astronomy.brookscole.com/seeds8e**

1. Why is there an upper limit to the mass of neutron stars? Why is that upper limit not well known?

2. Explain in detail why we expect neutron stars to be hot, spin fast, and have strong magnetic fields.

3. Why can't we use visual-wavelength telescopes to locate neutron stars?

4. Why does the short length of pulsar pulses eliminate normal stars as possible pulsars?

5. What do we mean when we say "Every pulsar is a neutron star, but not every neutron star is a pulsar"?

6. According to our model of a pulsar, if a neutron star formed with no magnetic field at all, could it be a pulsar? Why or why not?

7. Why did astronomers first assume that the millisecond pulsar was very young?

8. Why do we suspect that only very fast pulsars can emit visible pulses?

9. If the sun were replaced by a 1-solar-mass black hole, how would Earth's orbit change?

10. If the sun has a Schwarzschild radius, why isn't it a black hole?

11. How can a black hole emit X rays?

12. What do we mean when we say "Black holes have no hair"?

13. What evidence do we have that black holes really exist?

14. How can mass transfer into a compact object produce jets of high-speed gas?

15. How does an X-ray burster resemble a nova?

16. Discuss the possible causes of gamma-ray bursters.

DISCUSSION QUESTIONS

1. Has the existence of neutron stars been sufficiently tested to be called a theory, or should it be called a hypothesis? What about the existence of black holes?

2. Why would you expect an accretion disk around a star the size of the sun to be cooler than an accretion disk around a compact object?

3. In this chapter, we imagined what would happen if we jumped into a Schwarzschild black hole. From what you have read, what do you think would happen to you if you jumped into a Kerr black hole?

PROBLEMS

1. If a neutron star has a radius of 10 km and rotates 642 times a second, what is the speed of the surface at the neutron star's equator in terms of the speed of light?

2. Suppose that a neutron star has a radius of 10 km and a temperature of 1,000,000 K. How luminous is it? (*Hint:* See Chapter 9.)

3. A neutron star and a white dwarf have been found orbiting each other with a period of 11 minutes. If their masses are typical, what is their average separation? Compare the separation with the radius of the sun, 7×10^5 km. (*Hint:* See Chapter 9.)

4. If the accretion disk around a neutron star has a radius of 2×10^5 km, what is the orbital velocity of a particle at its outer edge? (*Hint:* Use circular velocity, Chapter 5.)

5. What is the escape velocity from the surface of a typical neutron star? How does that compare with the speed of light? (*Hint:* See Chapter 5.)

6. If Earth's moon were replaced by a typical neutron star, what would the angular diameter of the neutron star be as seen from Earth? (*Hint:* Use the small-angle formula, Chapter 3.)

7. If the inner accretion disk around a black hole has a temperature of 1,000,000 K, at what wavelength will it radiate the most energy? What part of the spectrum is this in? (*Hint:* Use Wien's law, Chapter 7.)

8. What is the orbital period of a bit of matter in an accretion disk 2×10^5 km from a 10-solar-mass black hole? (*Hint:* Use circular velocity, Chapter 5.)

9. If SS433 consists of a 20-solar-mass star and a neutron star orbiting each other every 13.1 days, what is their average separation? (*Hint:* See Chapter 9.)

CRITICAL INQUIRIES FOR THE WEB

1. Imagine that you are on a mission to explore one of the pulsar planets noted in this chapter. What would you find there? Look on the Web for information about pulsars and the known pulsar planets, and describe what you might encounter on such a mission.

2. What would you experience if you were to pilot a spacecraft near a black hole? Visit related Internet sites to determine what the gravitational effects and general environment would be like. Also, use the Internet to find the limits of human tolerance to strong gravitational forces. (*Hint:* Look for information about astronaut training and find out how many Gs a human can withstand.) Using these sources, give a brief account of what your voyage would be like.

3. Search for information about gamma-ray bursters. What is the latest news in this developing story?

EXPLORING *THESKY*

1. The Crab Nebula pulsar is located in the Crab Nebula, also known as M1. Locate the nebula, zoom in, and compare its shape and size with Figure 14-4. (*TheSky* does not show the pulsar.)

 Visit the Seeds *Foundations of Astronomy* companion Web site for critical thinking exercises, articles, and additional readings from InfoTrac College Edition, Brooks/Cole's online student library.

THE MILKY WAY GALAXY

Jane was watching Mrs. Corry splashing the glue on the sky and Mary Poppins sticking on the stars. . . .
"What *I* want to know," said Jane, "is this: Are the stars gold paper or is the gold paper stars?"
There was no reply to her question and she did not expect one. She knew that only someone
very much wiser than Michael could give her the right answer.

P. L. Travers, Mary Poppins

2Mass

Chapter 15

GUIDEPOST

This chapter plays three parts in our cosmic drama. First, it introduces the concept of a galaxy. Second, it discusses our home, the Milky Way Galaxy, a natural object of our curiosity. Third, it elaborates our story of stars by introducing us to galaxies, the communities in which stars exist.

Science is based on the interaction of theory and evidence, and this chapter will show a number of examples of astronomers using evidence to test theo-

ries. If the theories seem incomplete and the evidence contradictory, we should not be disappointed. Rather, we must conclude that the adventure of discovery is not yet over.

We struggle to understand our own galaxy as an example. We will extend the concept of the galaxy in Chapters 16 and 17 on normal and peculiar galaxies. We will then apply our understanding of galaxies in Chapter 18 to the study of the universe as a whole.

P. L. Travers wrote her story about Mary Poppins, the supercalifragilisticexpialidocious nanny, in 1934, and at that time no one could answer the question that Jane asked in the quotation given on the previous page. But by the 1960s, astronomers could assure her that the gold paper was stars, or rather that the atomic elements of which gold paper is made were created inside generation after generation of stars. How the stars made the elements is one of the keys to understand our Milky Way Galaxy.

It isn't obvious that we live in a galaxy. We are inside, and we see nearby stars scattered all over the sky. From a dark site far from city lights, we can see great clouds of stars in our galaxy making a faint band of light circling the sky (▌Figure 15-1a). The ancient Greeks named that band *galaxies kuklos,* the "milky circle." The Romans changed the name to *via lactia,* "milky road" or "milky way." We will see in this chapter how the evidence tells us that we live inside a great wheel of stars, a galaxy.

Almost every object you see in the sky is part of our Milky Way Galaxy. Exceptions include the **Magellanic Clouds,** small irregular galaxies located near our galaxy in space. They are visible as faint patches of haze in the southern sky. Another exception is the Andromeda Galaxy, just visible to our unaided eyes as a faint patch of light in the constellation Andromeda. (Consult the star charts at the end of this book to locate the Andromeda

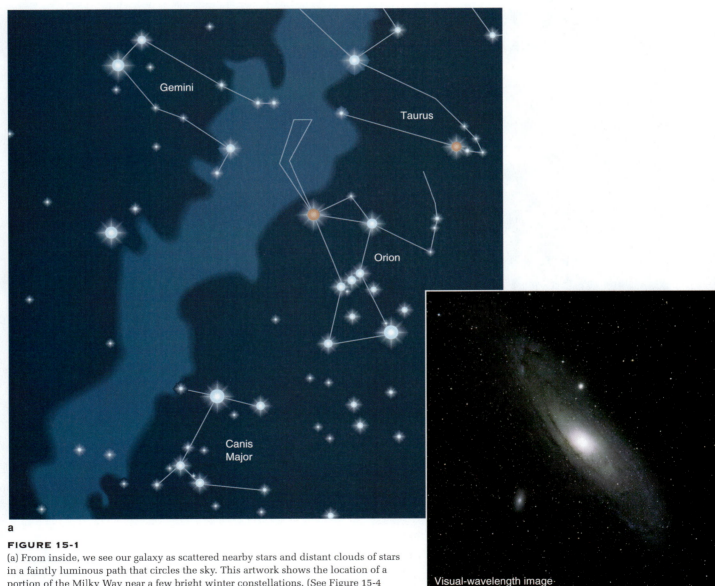

a

FIGURE 15-1

(a) From inside, we see our galaxy as scattered nearby stars and distant clouds of stars in a faintly luminous path that circles the sky. This artwork shows the location of a portion of the Milky Way near a few bright winter constellations. (See Figure 15-4 and the star charts at the end of this book to further locate the Milky Way in the sky.) (b) This photograph of the Andromeda Galaxy, a spiral galaxy about 2.3 million ly from us, shows us what our Milky Way Galaxy would look like if we could view it from such a great distance. *(AURA/NOAO/NSF)*

Visual-wavelength image

b

Galaxy.) We will see evidence that our galaxy looks much like the disk-shaped Andromeda Galaxy (Figure 15-1b).

The largest telescopes can detect an estimated 100 billion extremely faint, hazy objects which astronomers, drawing on the Greek word for milk, call galaxies. They appear to be other systems of stars similar to our own. Our goal in this chapter is to learn about our Milky Way Galaxy—how it formed and evolved and why it takes the shape it does. As always, we begin with observations, the evidence upon which all astronomy is based.

15-1 THE NATURE OF THE MILKY WAY GALAXY

The secret to studying any science lies in the simple question, "How do we know?" Because all scientific knowledge is based on evidence and logical arguments about natural processes, we need to know facts. But we also need to understand how we know those facts. Thus, we set the stage in the next section with a few facts, but our real interest is in how we know what our galaxy is like.

THE STRUCTURE OF OUR GALAXY

We live in a disk-shaped galaxy with beautiful spiral arms reaching from its center to its edge. Astronomers commonly give the diameter of the galaxy as 25,000 pc or 75,000 ly, with the sun located about two thirds of the way from the center to the edge (Figure 15-2). The diameter of our galaxy is not known very accurately, as you can see from the two numbers given above. Convert 25,000 pc into light-years, and you get significantly more than 75,000 ly. That these two numbers don't quite match is a warning that we don't know the size of our galaxy to better than about 10 percent.

The disk of the galaxy contains most of the stars and nearly all of the gas and dust. Although most of the stars in the disk are middle- to lower-main-sequence stars like the sun, a few are giants, and fewer still are brilliant O and B stars. These hot, luminous stars light up the disk and make it bright. It is difficult to judge just how big the disk is, because it is filled with gas and dust that dim starlight and make it difficult to see distant stars in the disk. Also, large dusty clouds block our view in many directions within the disk. In most directions in the plane of the disk, we can't see farther than a few **kiloparsecs.** (A kiloparsec is 1000 pc.)

If we look up or down out of the disk, we look away from the dust and gas, and we can see out into the **halo** of our galaxy, a spherical cloud of stars and star clusters that contains almost no gas and dust. The halo completely encloses the disk of the galaxy. The size of the halo is uncertain, but it is at least a few times larger than the disk.

ACTIVE FIGURE 15-2

An artist's conception of our Milky Way Galaxy, seen face-on and edge-on. Note the position of the sun and the distribution of globular clusters in the halo. Only the inner halo is shown here.

Ace Astronomy™ Go to AceAstronomy and click Active Figures to see "Explorable Milky Way" and go deeper into this diagram.

Around the center of our galaxy lies the **nuclear bulge,** a flattened cloud of stars about 3 kpc in radius. It contains little gas and dust but billions of stars similar to those found in the halo.

This quick portrait of our galaxy is just a collection of bare facts. It raises two important questions that will guide us through the rest of this chapter. How do we know what our galaxy is like? How did our galaxy get this way?

FIRST STUDIES OF THE GALAXY

How do astronomers know what our galaxy is like? Humanity has known of the Milky Way, the hazy path of light around the sky, since antiquity, but not until Galileo looked at the sky with his telescope in 1610 did anyone know the Milky Way was made of stars. Little more was known for two centuries after Galileo lived.

By the mid-18th century, astronomers generally understood that the stars were other suns, but they had little understanding of how the stars were distributed in space. One of the first people to study this problem was the English astronomer Sir William Herschel (1738–1822). He and his sister Caroline (1750–1848) mapped the stars in three dimensions by assuming that the sun was located inside a great cloud of stars and that they could see to the edges of the cloud in any direction. They believed that by counting the number of stars visible in different directions, they could gauge the relative distance to the edge of the cloud. If their telescope revealed few stars in one direction, they assumed that the edge was not far away, and if they saw many stars in another direction, they assumed that the edge was very distant. Calling their method "star gauges," they counted stars in 683 directions in the sky and outlined a model of the star cloud. Their data showed that the cloud was a great disk with the sun near the center, and using the technology of their day as an analogy, they called it the "grindstone model" (❙ Figure 15-3).

This model of the star system explains the most obvious feature of the Milky Way—it is a glowing path that circles the sky. The grindstone model reproduces this as a disk of stars with the sun near the center. Unfortunately, there was no way for the Herschels to find the diameter of the disk.

Generations of astronomers studied the size of our star system, culminating with Jacobus C. Kapteyn (1851–1922). In the early 20th century, he analyzed the brightness, number, and motion of stars and concluded that our star system is a disk about 10 kpc in diameter and about 2 kpc thick, with the sun near the center.

As we saw in the previous section, modern astronomers believe our galaxy is much larger and that the sun is not at the center. Why did these first studies of the galaxy underestimate its size and mistakenly place the sun near the center? The studies were misled because astronomers assumed that space was empty and that they could see to great distances. Today, we know that the disk of our galaxy is filled with gas and dust that dim stars and block our view. We can see only a small region near us in the disk of our galaxy. How we discovered the true size of our galaxy and came to understand the nature of a galaxy is one of the great adventures of human knowledge. It began about a century ago when a man studying star clusters and a woman studying variable stars unlocked one of nature's biggest secrets—that our star system is a galaxy.

DISCOVERING THE GALAXY

It seems odd to say that astronomers discovered something that is all around us, but until early in the 20th century, no one knew that the star system we live in is a galaxy. That began to change in the second decade of the 20th century, when a young astronomer named Harlow Shapley (1885–1972) discovered how big our star system really is. Besides being one of the turning points of modern astronomy, Shapley's study illustrates one of the most common techniques in astronomy. If we want to know how we know things in astronomy, then Shapley's story is well worth tracing in detail.

Shapley began by noticing that open star clusters lie along the Milky Way but that more than half of all globular clusters lie in or near the constellation Sagittarius (❙ Figure 15-4). The globular clusters seemed to be distributed in a great cloud whose center was located in the direction of Sagittarius. Shapley assumed that the orbital motion of these clusters was controlled by the gravitation of the entire star system, and thus the

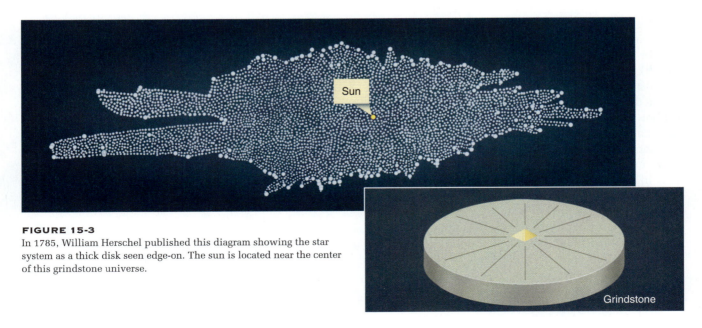

FIGURE 15-3
In 1785, William Herschel published this diagram showing the star system as a thick disk seen edge-on. The sun is located near the center of this grindstone universe.

Open cluster M52
in Cassiopeia

Open clusters lie
along the Milky Way
all around the sky.

Visual-wavelength image

FIGURE 15-4

Nearly half of cataloged globular clusters (red dots) are located in or near Sagittarius and Scorpius. A few of the brighter globular clusters, labeled with their catalog designations, are visible in binoculars or small telescopes. Constellations are shown as they appear above the southern horizon on a summer night as seen from latitude 40° N, typical for most of the United States. *(M52: NOAO/AURA/NSF; M19: Doug Williams, N.A. Sharp/ NOAO/AURA/NSF)*

Globular cluster M19

Visual-wavelength image

Globular clusters are scattered over the entire sky but are strongly concentrated toward Sagittarius.

center of the star system could not be near the sun but must lie somewhere toward Sagittarius.

To find the distance to the center of the star system and thus the size of the star system as a whole, Shapley needed to find the distance to the star clusters, but that was difficult to do. The clusters are much too far away to have measurable parallaxes. They do, however, contain variable stars (Chapter 12), and those were the lampposts Shapley needed to find the distances to the clusters.

Shapley knew of the work of Henrietta S. Leavitt (1868–1921), who in 1912 had shown that a Cepheid variable star's period of pulsation was related to its brightness. Leavitt worked on stars whose distances were unknown, so she was unable to find their absolute magnitudes or luminosities, but astronomers realized that the Cepheids could be a powerful tool in astronomy if their true luminosities could be discovered.

Cepheids are giant stars and so are relatively rare. None lies close enough to Earth to have measurable par-

allax. Their proper motions—slow movement across the sky—can be measured, and the more distant a star is, the smaller its proper motion tends to be. Thus proper motions contain clues to distance. Ejnar Hertzsprung (codiscoverer of the H–R diagram) made an early attempt to determine the absolute magnitudes of the Cepheids, but Shapley went further. He found 11 Cepheids with measured proper motions and used a statistical process to find their average distance and thus their average absolute magnitude. That meant he could erase Leavitt's apparent magnitudes from the period–luminosity diagram (see Figure 12-13) and write in absolute magnitudes. All of the stars in the diagram were thus calibrated as lampposts that astronomers could use to find distances (Window on Science 15-1), including the RR Lyrae stars that Shapley saw in the globular clusters.

Finding distance using Cepheid variable stars is so important in astronomy we should pause to illustrate the process. Suppose we studied a star cluster and discovered that it contained a type I Cepheid with a period

Astronomers often say that Shapley "calibrated" the Cepheids for the determination of distance, meaning that he did all the detailed background work so that the Cepheids could be used to find distances. Other astronomers could use the Cepheids without repeating the detailed calibration.

Calibration is actually very common in science. Chemists, for instance, have carefully calibrated the colors of certain compounds against acidity. They can

quickly measure the acidity of a solution by dipping into it a slip of paper containing the indicator compound and looking at the color. They don't have to repeat the careful calibration every time they measure acidity.

Engineers in steel mills have calibrated the color of molten steel against its temperature. They can use a hand-held device that measures the color of a ladle of molten steel and then look up the temperature in a table. They don't have to

repeat the calibration every time. Astronomers have made the same kind of color–temperature calibration for stars.

As you read about any science, notice how calibrations are used to simplify common measurements. But notice, too, how important it is to get the calibration right. An error in calibration can throw off every measurement made with that calibration. Some of the biggest errors in science have been errors of calibration.

---

of 10 days and an average apparent magnitude of 10.5. How far away is the cluster? From the period–luminosity diagram (Figure 12-13), we see that the absolute magnitude of the star must be about -3. Then the distance modulus is:

$$m - M_v = 10.5 - (-3) = 13.5$$

We could use this distance modulus and a table such as Table 9-1 to estimate the distance to the cluster as somewhere between 1000 and 16,000 pc. We can be more accurate if we solve the magnitude–distance formula (Chapter 9) for distance:

$$d = 10^{(m_v - M_v + 5)/5}$$

Now we can substitute the apparent magnitude and absolute magnitude, and we discover that the distance is equal to $10^{3.7}$. A pocket calculator tells us that the distance is about 5000 pc. This is one of the most common calculations in astronomy.

Once Shapley had calibrated the period–luminosity relation, he could use it to find the distance to any cluster in which he could identify variable stars. He could measure the apparent magnitudes of the variable stars in a cluster from a photographic plate, and he could find the periods of the variable stars from a series of photographic plates. Because the periods were very short, he knew the variable stars he saw in the clusters were the lowest-luminosity variable stars (known today as RR Lyrae stars). That allowed him to read their absolute magnitude off of the period–luminosity diagram. Once he knew the apparent magnitude and the absolute magnitude, he could calculate the distance to the star cluster.

This worked well for the nearer globular clusters, but the variable stars in the more distant clusters are too faint to detect. Shapley estimated the distances to these more distant clusters by calibrating the diameters of the clusters. For the clusters whose distance he knew, he could use their angular diameters and the small-angle formula to calculate linear diameters in parsecs. He found that the nearby clusters are about 25 pc in diameter, which he assumed is the average diameter of all globular clusters. He then used the angular diame-

ters of the more distant clusters to find their distances. This is another illustration of calibration in astronomy.

Shapley later wrote that it was late at night when he finally plotted the direction and distance to the globular clusters and found that, just as he had supposed, they formed a great swarm whose center lay many thousands of light-years in the direction of Sagittarius. Evidently the center of the star system was not near the sun but rather was far away in Sagittarius (❚ Figure 15-5). He called the only other person in the building, a cleaning lady, and the two stood looking at his graph as he explained that they were the only two people on Earth who understood that humanity lives, not at the center of a small star system, but in the suburbs of a vast wheel of stars.

Why did astronomers before Shapley think we lived near the center of a small star system? Space is filled with gas and dust that dim our view of distant stars. When we look toward the band of the Milky Way, we can see only the neighborhood near the sun. Most of the star system is invisible, and, like travelers in a fog, we seem to be at the center of a small region. Shapley was able to see the globular clusters at greater distances because they lay outside the plane of the star system and are not dimmed very much by the gas and dust.

Building on Shapley's work, other astronomers began to suspect that some of the faint patches of light visible through telescopes were other star systems like our own. We will continue this story in the next chapter when we discuss galaxies in general. It is sufficient here to note that within a few years astronomers had found evidence that the faint patches of light were other galaxies much like our own Milky Way Galaxy.

Shapley's study of star clusters led to the discovery that we live in a galaxy and that the universe is filled with similar galaxies. Like Copernicus, Shapley moved us from the center to the suburbs. Nevertheless, our home galaxy is dramatically beautiful. To get a true impression of the great galaxy in which we live, we must conduct a careful analysis.

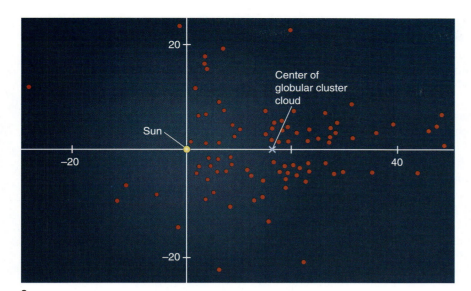

a

FIGURE 15-5

(a) Shapley's study of globular clusters showed that they were not centered on the sun, at the origin of this graph, but rather formed a great cloud centered far away in the direction of Sagittarius. Distances on this graph are given in thousands of parsecs. (b) Looking toward Sagittarius, we see nothing to suggest this is the center of the galaxy. Gas and dust block our view. Only the distribution of globular clusters told Shapely the sun lay far from the center of the star system. *(Daniel Good)*

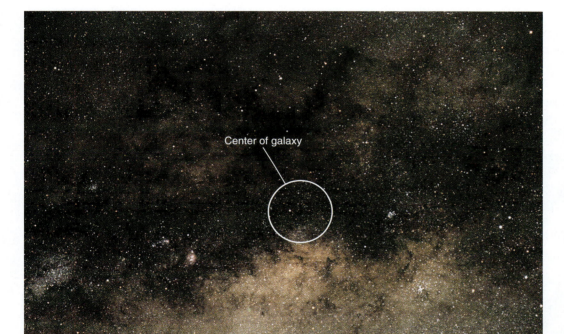

b

AN ANALYSIS OF THE GALAXY

We want to know how we know, and we are now ready to explore our modern knowledge of the galaxy. We can survey the galaxy's components, understand how we have learned about them, and begin to understand what they tell us about the nature of our home galaxy.

We can divide our galaxy into two main components whose characteristics will help us understand how our galaxy formed. The **disk component** of the galaxy contains most of the stars, gas, and dust that orbit within the disk. In contrast, the **spherical component** includes the nuclear bulge and the halo.

In addition to stars, the disk of the galaxy contains vast amounts of gas and dust—the interstellar medium. We saw in Chapter 10 how we can learn about this material. It contains mostly hydrogen gas and helium gas and is irregularly distributed as dense clouds separated by less dense regions. The densest of these clouds are the giant molecular clouds within which stars form (Chapter 11), and thus nearly all star formation in our galaxy takes place in the disk. The most massive of these newborn stars are the most luminous, the O and B stars. They light up the disk and make it glow brightly. Some of these stars are hot enough to ionize the nearby gas to produce HII regions, and thus the disk contains all of the emission nebulae in our galaxy.

Although we can survey the location and distance of the stars (Chapter 9), we cannot cite a single number for the thickness of the disk because it lacks sharp boundaries. Stars become less crowded farther from the central plane of the galaxy. Also, the thickness of the disk depends on the kind of object we study. Stars like the sun with ages of a few billion years lie within about 500 pc above and below the central plane. But the youngest stars, including the O and B stars, and the gas and dust from which these young stars are forming are confined to a disk only about 100 pc thick (Figure 15-6). With a diameter of 25 kpc, the disk is thinner than a thin pizza crust.

From the shape and size of our galaxy, astronomers conclude that we live in a spiral galaxy (Figure 15-7). The most dramatic features of these disk galaxies are the spiral arms—long spiral patterns of bright stars, HII regions, star clusters, and clouds of gas and dust. The sun is located on the inside edge of one of these spiral arms. We will see later in this chapter how we can observe the spiral arms in our own galaxy.

We see two kinds of star groupings in the disk of our galaxy. Associations are groups of 10 to a few hundred stars so widely scattered in space that their mutual gravity cannot hold the association together. From the turnoff points in their H–R diagrams (Chapter 12) we can tell they are very young groups of stars. We see them moving together through space (Figure 15-8) because they formed from a single gas cloud and haven't had time to wander apart. Two kinds of association—O and B associations and T Tauri associations (Chapter 11)— are located along the spiral arms.

The second kind of cluster in the disk is the open cluster (Figure 15-4), a group of 100 to a few thousand stars in a region about 25 pc in diameter. Because they have more stars in less space than associations, open clusters are more firmly bound by gravity. Although they lose stars occasionally, they can survive for a long time, and the turnoff points in their H–R diagrams give ages from a few million to a few billion years.

In contrast to the disk, the halo of our galaxy contains almost no gas and dust and only a thin scattering of stars and globular clusters. Because the halo contains no dense gas clouds, it cannot make new stars. Halo stars are old, cool, lower-main-sequence stars, red giants, and white dwarfs. It is difficult to judge the extent of the halo, but it could extend far beyond the edge of the visible disk and may be slightly flattened like a thick hamburger bun.

The halo contains roughly 200 globular clusters, each of which contains 50,000 to a million stars in a sphere about 25 pc in diameter (Figure 15-4). Because they contain so many stars in such a small region, the clusters are very stable and have survived for billions of years. From the turnoff points in their H–R diagrams, we know they are about 11 billion years old.

The nuclear bulge at the center of the galaxy is the most crowded part of the spherical component. Visible above and below the plane of the galaxy and through gaps in the obscuring dust, the nuclear bulge contains stars that are old and cool like the stars in the halo.

FIGURE 15-6 .

In these infrared images, the entire sky has been projected onto ovals with the center of the galaxy at the center. The Milky Way extends from left to right. In the near infrared, the nuclear bulge is prominent, and dust clouds block our view. At longer wavelengths, the dust emits significant energy and glows brightly. (*Near-IR: 2Mass; Far-IR: DIRBE Image courtesy Henry Freudenreich*)

Our analysis of the components of our galaxy leave us wondering about one critical question. How much matter does our galaxy contain? To answer that question, we must watch our galaxy rotate.

THE MASS OF THE GALAXY

To find the mass of an object, astronomers must observe it in orbital motion as, for example, in a binary star system. We don't live long enough to see our galaxy rotate significantly, but astronomers can observe the radial velocities, proper motions, and distances of stars and then calculate their orbits. The results can tell us the mass of the galaxy and give us hints about its origin.

Stars in the disk of the galaxy follow nearly circular orbits in the plane of the disk (❚ Figure 15-9a). By observing the radial velocities of other galaxies in various directions, astronomers can conclude that the sun is moving at about 220 km/s in the direction of Cygnus as it orbits the center of our galaxy in a nearly circular path. Given the distance to the center of our galaxy, 8.5 kpc, we can conclude that the sun has an orbital period of about 240 million years.

Halo stars and globular clusters, however, follow highly elongated orbits tipped steeply to the plane of the disk (Figure 15-9b). Although we show these orbits as elliptical, they are more like rosettes because of the influence of the thick bulge. Where halo stars pass through the disk, they are cutting across the regular traffic of disk stars and seem to have very high velocities. Thus, they are called **high-velocity stars.** The dramatic difference between the motion of halo stars and disk stars will be important evidence when we discuss the formation of the galaxy in the next section.

The orbital motion of the sun can tell us the mass of the galaxy. When we studied binary stars in Chapter 9, we saw that the total mass (in solar masses) of a binary star system equals the cube of the separation of the stars a (in AU) divided by the square of the period P (in years):

$$M = \frac{a^3}{P^2}$$

The Sombrero Galaxy

Visual-wavelength image

a

NGC 2997

Visual-wavelength image

b

FIGURE 15-7
Our disk galaxy resembles the spiral galaxies visible through large telescopes. (a) Seen edge-on, spiral galaxies have nuclear bulges and dust-filled disks. *(Todd Boroson/NOAO/AURA/NSF)* (b) Seen face-on, the spiral arms are dramatically outlined by hot O and B stars, emission nebulae, and dust. *(ESO)*

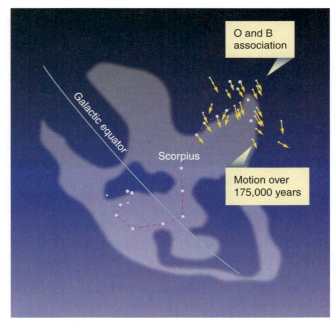

O and B association

Galactic equator

Scorpius

Motion over 175,000 years

FIGURE 15-8
Many of the stars in the constellation Scorpius are members of an O and B association that has formed recently from a single cloud of gas. As the stars orbit the center of our galaxy, they are moving together southwest along the Milky Way.

a

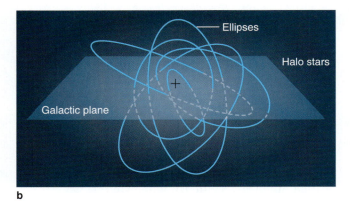

b

FIGURE 15-9

(a) Stars in the galactic disk have nearly circular orbits that lie in the plane of the galaxy. (b) Stars in the halo have randomly oriented, highly elongated orbits.

The radius of the sun's orbit around the galaxy is about 8500 pc, and each parsec contains 206,265 AU. Multiplying, we find that the radius of the sun's orbit is 1.75×10^9 AU. The orbital period is 240 million years, so the mass is:

$$M = \frac{(1.75 \times 10^9)^3}{(240 \times 10^6)^2} = 0.93 \times 10^{11} \, M_\odot$$

This is only a rough estimate because it does not include any mass outside the orbit of the sun. Adding an estimate for the mass outside the sun's orbit gives us a total mass for our galaxy of about 4×10^{11} solar masses.

The rotation of our galaxy is actually much more interesting than the description above. Stars at different distances from the center revolve around the center of the galaxy with different periods, so three stars near each other will draw apart as time passes (❚ Figure 15-10). This motion is called differential rotation. (Recall that we defined differential rotation in Chapter 8.) To fully describe the rotation of our galaxy, we must graph orbital velocity versus radius—a **rotation curve** (❚ Figure 15-11).

If all of the mass of the galaxy were concentrated near its center, then we would expect to see orbital velocities fall as we moved away from the center. This is what we see in our solar system, where nearly all of the mass is concentrated in the sun, and it is called **Keplerian motion** (a reference to Kepler's laws). The best observations of the rotation curve of the Milky Way

FIGURE 15-10

The differential rotation of the galaxy means that stars at different distances from the center have different orbital periods. In this example, the star just inside the sun's orbit has a shorter period and gains on the sun, while the star outside falls behind.

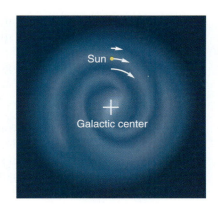

show orbital velocities constant or rising in the outer disk rather than falling. These higher orbital velocities indicate that the larger orbits enclose more mass and seem to tell us that our galaxy is more massive than it appears to be.

The evidence tells astronomers that the extra mass lies in an extended halo sometimes called a **galactic corona.** It may extend up to 10 times farther than the edge of the visible disk and could contain a trillion solar masses. Much of this mass is invisible, so astronomers conclude it is not emitting or absorbing light and refer to it as **dark matter.** At least some of the mass in the galactic corona is made up of low-luminosity stars and white dwarfs, but much of the mass must be some other form of matter. We will return to the problem of the dark matter in the following two chapters. It is one of the fundamental problems of modern astronomy.

REVIEW CRITICAL INQUIRY

If gas and dust block our view, how do we know how big our galaxy is?

The gas and dust in our galaxy block our view only in the plane of the galaxy. When we look away from the plane of the galaxy, we look out of the gas and dust and can see to great distances. The globular clusters are scattered through the halo with a strong concentration in the direction of Sagittarius. When we look above or below the plane of the galaxy, we can see those clusters, and careful observations reveal variable stars in the clusters. By using a modern calibration of the period–luminosity diagram, we can find the distance to those clusters. If we assume that the distribution of the clusters is controlled by the gravitation of the galaxy as a whole, then we can find the distance to the center of the galaxy by finding the center of the distribution of globular clusters.

In the critical analysis of ideas, it is important to ask "How do we know?", and that is especially true in science. For example, what measurements and assumptions tell us the total mass of our galaxy?

FIGURE 15-11
The rotation curve of our galaxy is plotted here as orbital velocity versus radius. Data points show measurements made by radio telescopes with error bars showing the uncertainty in the measurements. The fitted rotation curve (solid line) does not decline outside the sun's orbit as we would expect for Keplerian motion (dashed line). Rather, the curve is approximately flat at great distances, suggesting that the galaxy contains significant mass outside the orbit of the sun. *(Adapted from a diagram by Françoise Combes)*

The differences between the disk and spherical components extend even to their chemical composition. As we will see in the next section, those differences hold clues to the origin of our galaxy.

15-2 THE ORIGIN OF THE MILKY WAY

Just as dinosaurs left behind fossilized footprints, our galaxy has left behind a fossil footprint of its youth. The stars of the spherical component are old and must have formed long ago when the galaxy was very young.

STELLAR POPULATIONS

In the 1940s, astronomers realized that there are two families of stars in the galaxy. They form and evolve in similar ways, but they differ in the abundance of atoms that are heavier than helium—atoms that astronomers call **metals.** (Note that this is not the way the word "metal" is used by nonastronomers.) **Population I stars** are metal rich, containing 2 to 3 percent metals, whereas **population II stars** are metal poor, containing only about 0.1 percent metals or less. The difference is dramatically evident in spectra (▌ Figure 15-12).

Population I stars belong to the disk component of the galaxy and are sometimes called disk population stars. They have nearly circular orbits in the plane of the galaxy and are relatively young stars that formed within the last few billion years. The sun is a population I star, as are the type I Cepheid variables discussed in Chapter 12.

Population II stars belong to the spherical component of the galaxy and are sometimes called the halo population stars. These stars have randomly tipped, elliptical orbits and are old stars. The metal-poor globular clusters are part of the halo population, as are the RR Lyrae and type II Cepheids.

Since the discovery of stellar populations, astronomers have realized that there is a gradation between populations. Extreme population I stars are found only in the spiral arms. Slightly less metal-rich population I stars, called intermediate population I stars, are located throughout the disk. The sun is such a star. Stars even less metal rich, such as stars in the nuclear bulge, belong to the intermediate population II. The most metal-poor stars are those in the halo, including those in globular clusters. These are extreme population II stars.

Why do the disk and halo stars have different metal abundances? The answer to this question is the key to our understanding of the history of our galaxy.

The different populations of stars differ in other ways such as type of orbit and age (▌ Table 15-1). These are clues to the formation of our galaxy. To follow these clues we must consider the origin of the chemical elements.

THE ELEMENT-BUILDING PROCESS

We can use our understanding of stellar structure and stellar evolution (Chapters 11 to 13) to understand the origin of the chemical elements heavier than helium—elements that astronomers call metals. When the universe began, it contained only hydrogen and helium atoms. (We will review the evidence that supports this statement in Chapter 18.) All of the remaining chemical elements have been produced by **nucleosynthesis,** the process that fuses hydrogen and helium inside stars to make the heavier elements (Window on Science 15-2).

When our galaxy formed sometime after the beginning of the universe, the gas contained about 90 percent hydrogen atoms, 10 percent helium atoms, and almost no heavier atoms. Thus, the first stars to form were metal poor. Succeeding generations of stars manufactured heavier atoms, and the metal abundance gradually increased.

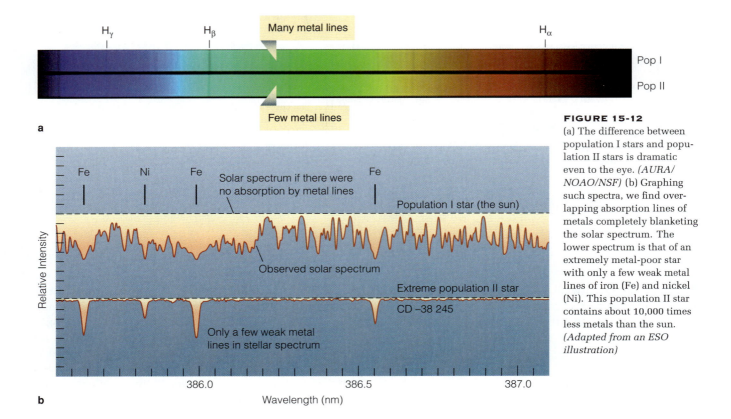

FIGURE 15-12
(a) The difference between population I stars and population II stars is dramatic even to the eye. *(AURA/NOAO/NSF)* (b) Graphing such spectra, we find overlapping absorption lines of metals completely blanketing the solar spectrum. The lower spectrum is that of an extremely metal-poor star with only a few weak metal lines of iron (Fe) and nickel (Ni). This population II star contains about 10,000 times less metals than the sun. *(Adapted from an ESO illustration)*

The most massive stars fuse helium into carbon, nitrogen, oxygen, and heavier elements up to iron (Chapter 11), and when those stars die in supernova explosions, traces of those elements are spread back into the interstellar medium. Furthermore, the supernova explosion itself can fuse nuclei into atoms heavier than iron. These are rare atoms, such as gold, silver, uranium, and platinum, but they, too, get spread back into space during supernova explosions.

Elements heavier than helium are quite rare in the universe. Traditionally, astronomers graph the abundance of the elements using an exponential scale, but if we replot the data using a linear scale we see how rare these atoms are (❚ Figure 15-13).

When we look at population II stars, such as those in the halo, we are looking at the survivors of the early generations of stars in our galaxy. Those stars formed from gas that was metal poor, and thus their spectra show only weak metal lines. The survivors of these early generations are low-mass, long-lived stars that don't make heavy elements. Furthermore, any elements they might have made are locked in their cores. Thus, the spectrum we see tells us the composition of the gas from which they formed and that the gas was metal poor. Population I stars, such as the sun, formed more recently, after the interstellar medium had been enriched in metals, and their spectra show stronger metal lines. Stars forming now show the strongest metal lines.

TABLE 15-1
Stellar Populations

| | Population I | | Population II | |
	Extreme	Intermediate	Intermediate	Extreme
Location	Spiral arms	Disk	Nuclear bulge	Halo
Metals (%)	3	1.6	0.8	Less than 0.8
Shape of orbit	Circular	Slightly elliptical	Moderately elliptical	Highly elliptical
Average age (yr)	100 million and younger	0.2–10 billion	2–10 billion	10–13 billion

Understanding Nature as Processes

One of the most common organizing themes in science is the concept of a process—a sequence of events that leads to some result or condition. Scientists try to discover the natural processes that shape the universe. Telling the stories of these processes explains why the universe is the way it is. At the same time, recognizing these processes and learning to tell their stories simplify much of science.

For example, astronomers try to assemble the sequence of events that led to the formation of the chemical elements. We know that the first stars formed from nearly pure hydrogen and helium and that they made certain heavier elements in their cores. Supernova

explosions made other elements and mixed the elements back into the interstellar medium. New stars formed from this gas, and generation after generation of stars added to the heavier elements that we are made of. Telling that story explains much of the science of the stars.

As you study any science, be alert for processes as organizing themes. Science, at first glance, seems to be nothing but facts, but one of the ways to organize the facts is to fit them into the story of a process. If you study biology and learn the process by which plant cells know whether to become roots, stems, or leaves, you will have organized a great deal of factual information into a story

that is not only easy to remember but also makes sense.

When you see a process in science, ask yourself a few basic questions. What conditions prevailed at the beginning of the process? What sequence of steps occurred? Can some steps occur simultaneously, or must one step occur before another can occur? What is the final state that this process produces? Most important, what evidence do we have that nature really works this way?

Not everything in science is a process, but identifying a process will help you remember and understand a great many details and principles.

GALACTIC FOUNTAINS

Supernova remnants are rich in metals and eventually mix back into the interstellar medium. In fact, supernovae continuously stir the interstellar medium. But a larger-scale process may be even more efficient at spreading newly created metals throughout the disk of the galaxy where they can be incorporated into the newly forming stars.

In Chapter 10, we saw how neighboring supernova remnants can merge to form a superbubble of hot gas (Figure 10-7b). A supernova remnant may be only a few tens of parsecs in diameter, but a superbubble can be more than ten times bigger. The denser gas of the galac-

tic disk tends to confine an expanding superbubble, but the disk is only a few hundred parsecs thick. If a superbubble breaks out of the denser gas of the galactic disk, it could spew hot gas high above the plane of the galaxy in a **galactic fountain** (❙ Figure 15-14). As this gas cools and falls back into the disk, it could spread metals through the galaxy.

Galactic fountains have not yet been observed directly, but isolated clouds of hot gas have been found high in the halo, and cool clouds have been found falling toward the disk. It seems very likely that this process contributes to spreading metals through the disk. Even without direct observational confirmation

FIGURE 15-13
The abundance of the elements in the universe. (a) When the elements are plotted on an exponential scale, we see that elements heavier than iron are about a million times less common than iron and that all elements heavier than helium (the metals) are quite rare. (b) The same data plotted on a linear scale provide a more realistic impression of how rare the metals are. Carbon, nitrogen, and oxygen make small peaks near atomic mass 15, and iron is just visible in the graph.

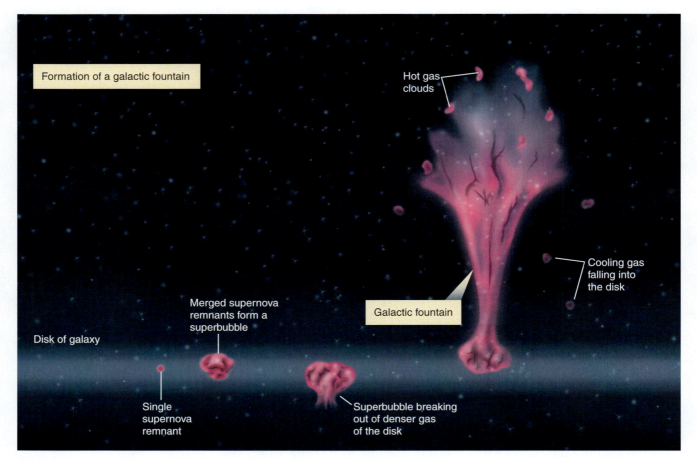

FIGURE 15-14

In this edge-on view of the galactic disk, we see that superbubbles, formed of multiple supernova remnants, can be roughly as large as the thickness of the disk. Once a superbubble breaks out of the denser gas of the disk, it is believed to produce a galactic fountain. The metal-enriched gas could flow out into the halo where it would cool and fall back to enrich the interstellar medium in metals.

of galactic fountains, we can be sure that supernovae create metals and mix them back into the interstellar medium where they can be incorporated into newly forming stars.

Because the metal abundance of newborn stars increases as the galaxy ages, we can use metal abundance as an index to the ages of the components of our galaxy. The halo is old, and the disk is younger. Such ages can help us understand the formation of the galaxy.

THE AGE OF THE MILKY WAY

Because we know how to find the age of star clusters, we can estimate the age of our galaxy. Nevertheless, uncertainties make our easy answer hard to interpret.

The oldest open clusters are 9 to 10 billion years old. These ages come from the turnoff point in the H–R diagram (see Chapter 12), but finding the age of an old cluster is difficult because old clusters change so slowly. Also, the exact location of the turnoff point depends on

chemical composition, which differs slightly among clusters. Finally, open clusters are not strongly bound by their gravity, so even older open clusters may have dissipated as their stars wandered away. Thus, open clusters tell us the galactic disk is at least 10 billion years old.

Globular clusters have faint turnoff points in their H–R diagrams and are clearly old, but finding these ages is difficult. Clusters differ slightly in chemical composition, which must be accounted for in calculating the stellar models from which ages are determined. Also, to find the age of a cluster, astronomers must know the distance to the cluster. Precise parallaxes from the Hipparcos satellite have allowed astronomers to more precisely calibrate the Cepheid variable stars, and careful studies with the newest large telescopes have refined the chemical compositions and better defined the H–R diagrams of globular clusters. The best analysis of all of the data suggests that the average globular clusters are about 11 billion years old. Some globular clusters are younger than that, and some are older. Studies of the oldest globular clusters suggest that the halo of our galaxy is at least 13 billion years old.

Both populations and clusters tell us that the disk is younger than the halo. We can combine these ages with the process of nucleosynthesis to tell the story of our galaxy.

THE HISTORY OF THE MILKY WAY GALAXY

In the 1950s, astronomers began to develop a hypothesis to explain the formation of our galaxy. Recent observations, however, are forcing a reevaluation of that traditional hypothesis.

The traditional hypothesis says that the galaxy formed from a single large cloud of gas roughly 14 billion years ago. As gravity pulled the gas together, the cloud began to fragment into smaller clouds, and, because the gas was turbulent, the smaller clouds had random velocities. Stars and star clusters that formed from these fragments went into orbits that had a wide range of shapes; a few were circular, but most were elliptical, and some were extremely elliptical. The orbits were also inclined at different angles, so the orbits produced a spherical cloud of stars. Of course, these first stars were metal poor because no stars had existed earlier to enrich the gas with metals. This is the way the traditional hypothesis explains the formation of the spherical component of the galaxy.

The second stage of this hypothesis accounts for the disk component. The contracting gas cloud was roughly spherical at first, but the turbulent motions canceled out, as do eddies in recently stirred coffee, leaving the cloud with uniform rotation. A rotating, low-density cloud of gas cannot remain spherical. A star is spherical because its high internal pressure balances its gravity; but, in a low-density cloud, the pressure cannot support the weight. Like a blob of pizza dough spun in the air, the cloud must flatten into a disk (■ Figure 15-15).

This contraction into a disk took billions of years, with the metal abundance gradually increasing as generations of stars were born from the gradually flattening gas cloud. The stars and globular clusters of the halo were left behind by the cloud when it was spherical, and subsequent generations of stars formed in flatter distributions. The gas distribution in the galaxy now is so flat that the youngest stars are confined to a disk only about 100 parsecs thick. These stars are metal rich and have nearly circular orbits.

This traditional hypothesis accounts for many of the Milky Way's properties. Advances in technology, however, have improved astronomical observation, and, beginning in the 1980s, contradictions arose. For example, many halo stars have metal abundances similar to those of globular clusters and may have escaped

Origin of the Halo and Disk

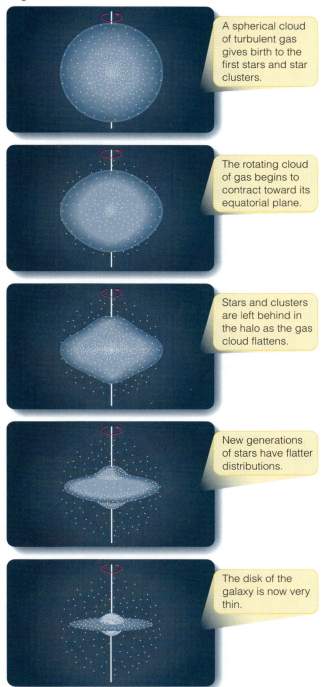

A spherical cloud of turbulent gas gives birth to the first stars and star clusters.

The rotating cloud of gas begins to contract toward its equatorial plane.

Stars and clusters are left behind in the halo as the gas cloud flattens.

New generations of stars have flatter distributions.

The disk of the galaxy is now very thin.

FIGURE 15-15
The traditional theory for the origin of our galaxy begins with a spherical gas cloud that flattens into a disk.

from such clusters, but some stars are even more metal poor. One star contains 200,000 times less metals than the sun. These stars may be older than the globular clusters. Also, some of the oldest stars in our galaxy are in the central bulge, not in the halo. Furthermore, not all globular clusters have the same age, and the younger clusters seem to be in the outer halo. The traditional hypothesis says that the halo formed first and that the clusters within it should have a uniform age.

Ages are a key problem for the traditional hypothesis. The globular clusters are about 10 to 13 billion years old, but the disk seems much younger. The oldest open star clusters in the disk are only 9 to 10 billion years old. Also, white dwarfs tend to accumulate over the eons, and there aren't as many white dwarfs in the disk as there should be if the disk were 10 to 13 billion years old. A successful hypothesis should account for these ages.

Another problem is that the oldest stars are metal poor but not metal free. There must have been at least a few massive stars to create these metals before the formation of the oldest stars we see in the halo.

Can we modify the traditional hypothesis to explain these observations? Perhaps the galaxy began with the contraction of a gas cloud to form the central bulge and the later accumulation of the halo from gas clouds that had been slightly enriched in metals by an early generation of massive stars. This would explain the age of the central bulge and the metals in the oldest stars.

The disk could have formed later as the gas that was already in the galaxy flattened and as more gas fell into the galaxy and settled into the disk. Perhaps entire galaxies were captured by the growing Milky Way Galaxy. In 2003, astronomers announced the discovery of a ring containing millions of stars circling our galaxy about twice as far from the center as the sun (❚ Figure 15-16). The ring may be the remains of a small galaxy that was absorbed by the Milky Way roughly 10 billion years ago. (We will see dramatic evidence in the next chapter that such galaxy mergers do occur.)

If the young Milky Way Galaxy absorbed a few small but partially evolved galaxies, then some of the globular clusters we see in the halo may be hitchhikers. This would explain the range of globular cluster ages. In fact, some theories hold that our galaxy formed by a rapid merger of a number of smaller pregalaxy clouds of gas and stars with later additions of in-falling gas and small merging galaxies.

The problem of the formation of our galaxy is frustrating because the theories are incomplete. We prefer certainty, but astronomers are still gathering observations and testing hypotheses. We see an older theory that has proven to be inadequate to explain the observations, and we see astronomers attempting to refine the observations and devise new theories. The metal abundances and ages of the stars in our galaxy seem to be important clues, but metal abundance and age do not tell the whole story. Astronomer Bernard Pagel was thinking of this when he said, "Cats and dogs may have the same age and metallicity, but they are still cats and dogs."

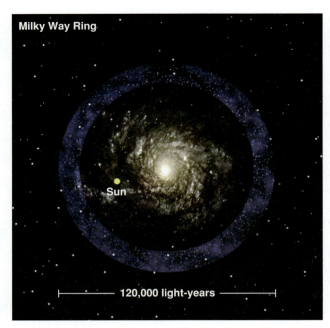

Milky Way Ring

Sun

120,000 light-years

FIGURE 15-16

Surveys of many faint stars have uncovered a ring of stars that encircles the Milky Way Galaxy. One astronomer described it as the "smoking gun" to show that our galaxy has absorbed at least one smaller galaxy in its past. Similar but less dramatic streams of stars have been found throughout the Milky Way Galaxy's halo, suggesting that such mergers have occurred repeatedly. *(Rensselaer Polytechnic Institute and the Sloan Digital Sky Survey [www.sdss.org])*

REVIEW CRITICAL INQUIRY

Why do metal-poor stars have the most elongated orbits?
Of course, the metal abundance of a star cannot affect its orbit, so our analysis must not confuse cause and effect with the relationship between these two factors. Both chemical composition and orbital shape depend on a third factor—age. The oldest stars are metal poor because they formed before there had been many supernova explosions to create and scatter metals into the interstellar medium. Those stars formed long ago when the galaxy was young and motions were not organized into a disk, and they tended to take up randomly shaped orbits, many of which are quite elongated. Thus, today, we see the most metal-poor stars following the most elongated orbits.

Nevertheless, even the oldest stars we can find in our galaxy contain some metals. They are metal poor, not metal free. How does that constrain our theories?

The story of our galaxy is an astronomical work in progress, but that is not the only mystery hidden in the Milky Way. The next two sections discuss two special problems that astronomers are trying to understand—the nature of spiral arms and the secret of the galactic nucleus.

UV + visual-wavelength image

a

FIGURE 15-17

(a) Images of other galaxies, such as NGC1232 shown here, show the spiral arms illuminated by clouds of young stars. The hot O and B stars in these clouds give the arms a blue tint. *(European Southern Observatory)* (b) Segments of spiral arms near the sun appear when we plot the location of O and B associations and the ionized nebulae that often surround such associations. These groups of massive stars have very short lives, so we find them only in regions of recent star formation. Dust blocks our view, and we can see only a few kiloparsecs in the plane of the Milky Way, but in the region we can see near the sun, O and B associations trace out segments of three spiral arms as diagrammed here.

Perseus arm

1 kpc

Orion-Cygnus arm

Sun

To center

Sagittarius arm

b

15-3 SPIRAL ARMS

The most striking feature of galaxies like the Milky Way is the system of spiral arms that wind outward through the disk. These arms contain swarms of hot, blue stars, clouds of dust and gas, and young star clusters. These young objects hint that the spiral arms involve star formation. As we try to understand the spiral arms, we face two problems. First, how can we be sure our galaxy has spiral arms when our view is obscured by gas and dust? Second, why doesn't the differential rotation of the galaxy destroy the arms? The solution to both problems involves star formation.

TRACING THE SPIRAL ARMS

Studies of other galaxies show us that spiral arms contain hot, blue stars (❙ Figure 15-17a). Thus, one way to study the spiral arms of our own galaxy is to locate these stars. Fortunately, this is not difficult, since O and B stars are often found in associations and, being very bright, are easy to detect across great distances. Unfortunately, at these great distances their parallax is too small to measure, so their distances must be found by other means, usually by spectroscopic parallax (see Chapter 9).

O and B associations in the sky are not located randomly but reveal parts of three spiral arms near the sun, which have been named for the prominent constellations through which they pass (Figure 15-17b). If we could penetrate the gas and dust, we could locate other O and B associations and trace the spiral arms farther; but, like a traveler in a fog, we see only the region near us.

Objects used to map spiral arms are called **spiral tracers.** O and B associations are good spiral tracers because they are bright and easy to see at great distances. Other tracers include young open clusters, clouds of hydrogen ionized by hot stars (emission nebulae), and certain higher-mass variable stars.

Notice that all spiral tracers are young objects. O stars, for example, live for only a few million years. If their orbital velocity is about 250 km/s, they cannot have moved more than about 500 pc since they formed. This is less than the width of a spiral arm. Because they don't live long enough to move away from the spiral arms, they must have formed there.

The youth of spiral tracers gives us an important clue about spiral arms. Somehow they are associated with star formation. Before we can follow this clue, however, we must extend our map of spiral arms to show the entire galaxy.

RADIO MAPS OF SPIRAL ARMS

The dust that blocks our view at visual wavelengths is transparent at radio wavelengths because radio waves are much longer than the diameter of the dust particles. When we point a radio telescope at a section of the Milky Way, we receive 21-cm radio signals coming from cool hydrogen in spiral arms at various distances across the galaxy. Fortunately, the signals can be unscrambled by measuring the Doppler shifts of the 21-cm radiation, except in the direction toward the nucleus. There, the orbital motions of gas clouds are perpendicular to the line of sight, and all of the radial velocities are zero. Thus, the radio map shown in ❚ Figure 15-18 reveals spiral arms throughout the disk of the galaxy, except in the wedge-shaped region toward the center.

The analysis of the 21-cm radial velocity data requires that astronomers estimate the orbital velocity of the clouds at different distances from the center of the galaxy. Because these velocities are not precisely known and because turbulent motions in the gas distort the radial velocities, we can't trace a perfect spiral pattern in Figure 15-18. We can be confident our galaxy has spiral arms, but we can't see the overall pattern.

Radio astronomers can also use the strong radio emission from carbon monoxide (CO) to map the location of giant molecular clouds in the plane of the galaxy. Recall from Chapter 11 that these clouds are sites of active star formation. Maps constructed from such observations reveal that the giant molecular clouds are located in large structures that resemble segments of spiral arms (❚ Figure 15-19).

Radio maps combined with studies at infrared and visible wavelengths paint a picture of our galaxy. Clearly we live in a spiral galaxy, but the spiral pattern appears slightly irregular with branches and spurs and gaps. The stars we see in Orion, for example, appear to be in a detached segment of a spiral arm, a spur. Some astronomers argue that our galaxy has a four-armed spiral pattern, but most believe it has a classic two-armed pattern with branches and spurs. The nuclear bulge of our galaxy does not seem to be a sphere but rather is an elongated bar pointing partially away from us. The spiral arms may spring from a ring that encloses this bar. In the next chapter we will see that such structures are common in other galaxies. Thus we can construct a best guess at what our galaxy might look like for comparison with other galaxies (❚ Figure 15-20).

The most important feature in the radio maps is easy to overlook—spiral arms are regions of higher gas density. Spiral tracers told us that the arms contain young objects, and we suspected active star formation. Now radio maps confirm our suspicion by telling us that the material needed to make stars is abundant in spiral arms.

THE DENSITY WAVE THEORY

Just what are spiral arms? We can be sure they are not physically connected structures like bands of magnetic field holding the gas in place. If they were, the strong

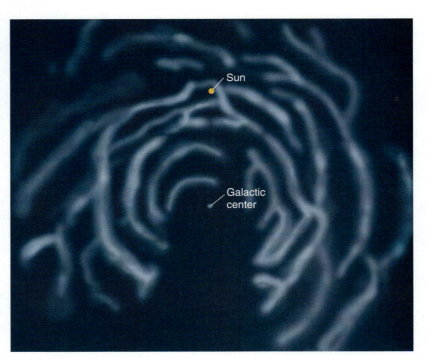

FIGURE 15-18

This map of our galaxy was made using 21-cm radio observations, and it confirms that our galaxy has spiral arms. Branches and spurs in the spiral arms are visible. The conical region around and beyond the center of the galaxy is the region where 21-cm radiation from gas clouds at different distances cannot be unscrambled. *(Adapted from a radio map by Gart Westerhout)*

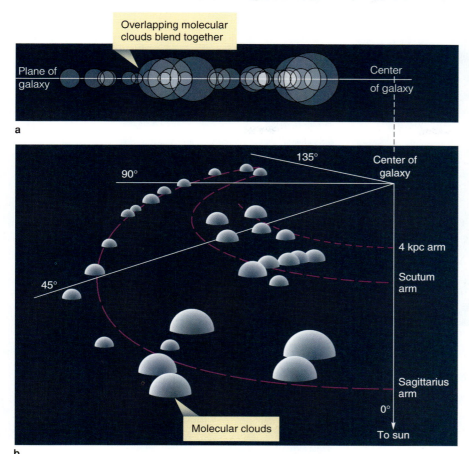

a

b

FIGURE 15-19

(a) At the wavelength emitted by the CO molecule we find many molecular clouds along the Milky Way, but they overlap in a confusing jumble. By using a model of the rotation of the galaxy and the radial velocities of the clouds, radio astronomers can find the distances to each cloud and use them to map spiral arms. (b) This map represents the view from a point 2 kpc directly above the sun. The molecular clouds, shown as hemispheres extending above the plane of our galaxy, are located along spiral arms. Angles in this diagram are galactic longitudes. *(Adapted from a diagram by T. M. Dame, B. G. Elmegreen, R. S. Cohen, and P. Thaddeus)*

differential rotation of the galaxy would destroy them within a billion years. They would get wound up and torn apart like paper streamers caught on the wheel of a speeding car. Yet spiral arms are common in galaxies (❚ Figure 15-21) and must last billions of years.

Astronomers believe that spiral arms are dynamically stable—they retain the same appearance even though the gas, dust, and stars in them are constantly changing. To see how this works, think of the traffic jam behind a slow-moving truck. Seen from an airplane, the

a

b

FIGURE 15-20

(a) This reconstruction of our galaxy is based on far-infrared observations of dust. The model includes a central bar surrounded by a ring from which spring four spiral arms. *(Courtesy Henry Freudenreich)* (b) This painting, based on radio and optical studies of the galaxy, includes a bar-shaped nucleus and a number of irregular arms with branches and spurs. *(Painting by M. A. Seeds, based on a study by G. De Vaucouleurs and W. D. Pence)*

Visual-wavelength image

FIGURE 15-21
Typical of spiral galaxies, M83 has a two-armed spiral pattern with prominent dust lanes and bright O and B stars along the arms. The galaxy also contains branches and spurs. *(European Southern Observatory)*

The Spiral Density Wave

Orbiting gas clouds overtake the spiral arm from behind.

Spiral arm

Gas cloud

Center of galaxy

Massive stars

Lower mass stars

The compression of a gas cloud triggers star formation.

Massive stars are highly luminous and light up the spiral arm.

The most massive stars die quickly.

Low-mass stars live long lives but are not highly luminous.

FIGURE 15-22
According to the density wave theory, star formation occurs as gas clouds pass through spiral arms.

traffic jam would be stable, moving slowly down the highway. But any given car would approach from behind, slow down, work its way to the front of the jam, pass the truck, and resume speed. The individual cars in the jam are constantly changing, but the traffic jam itself is dynamically stable.

In the **density wave theory,** the spiral arms are dynamically stable regions of compression that move slowly around the galaxy, just as the truck moves slowly down the highway. Gas clouds, moving at orbital velocity around the galaxy, overtake the slow-moving arms from behind and slam into the gas already in the arms. The sudden compression of the gas can trigger the collapse of the gas clouds and the formation of new stars (see Chapter 11). The newly formed stars and the remaining gas eventually move on through the arm and emerge from the front of the slow-moving arm to resume their travels around the galaxy (❚ Figure 15-22). Thus, whenever we look we see a spiral arm, although the stars, gas, and dust are constantly changing.

Science depends on evidence, so we should ask what evidence we have that spiral arms are caused by spiral density waves. For one thing, the stars in spiral arms fit the theory well. We find stars of all masses forming in spiral arms, but the O and B stars are the brightest. They live such short lifetimes that they die before they can move out of the spiral arm. Thus, the presence of these stars along with gas and dust assures us that spiral arms are sites of star formation. Lower-mass stars, such as the sun, also form in spiral arms. But because they live so long, they can leave the arms and circle the galaxy many times. The sun probably formed

as part of a cluster in a spiral arm roughly 5 billion years ago, left that cluster, and has circled the galaxy about 20 times, passing through spiral arms many times.

Giant molecular clouds and thick dust in the arms (❚ Figure 15-23) are evidence that material is compressed in the arms and spawns active star formation. These clouds can convert up to 30 percent of their mass into

FIGURE 15-23

This chance alignment of a small galaxy in front of a larger, more distant galaxy illustrates the nature of spiral arms. The small galaxy's spiral arms are bright with O and B stars where they are silhouetted against the darkness of space; but, seen against the glare of the distant galaxy, the arms are revealed to contain large amounts of dusty gas. *(NASA and The Hubble Heritage Team)*

stars. The Orion complex and the Ophiuchus dark cloud are just two of these regions filled with infrared protostars. Such a cloud can form stars for 10 to 100 million years before it is disrupted by heat and light from the most massive new stars. The cloud needs about this long to pass through a spiral arm.

The evidence seems to fit the spiral density wave theory well, but the theory has two problems. First, how is the complicated spiral disturbance started and sustained? Spiral density waves should slowly die out in a billion years or so. Something must regenerate the spiral wave. Mathematical models show that the galaxy is naturally unstable to certain disturbances, just as a guitar string is unstable to certain vibrations. Any sudden disturbance—the rumble of a passing truck, for example—can set the string vibrating at its natural frequencies. Similarly, minor fluctuations in the galaxy's disk, such as supernova explosions, might regenerate the waves.

Another suggestion is that gravitational disturbances can generate a spiral density wave. Observations show that the center of our galaxy is not a sphere but a bar. The rotation of such a bar could disturb the disk of the galaxy and stimulate the formation of a spiral density wave. A close encounter or collision with a nearby galaxy could also produce a spiral pattern.

The second problem for the density wave theory involves the spurs and branches in the arms of our own and other galaxies. Computer models of density waves produce regular, two-armed spiral patterns. Some galaxies, called grand-design galaxies, do indeed have symmetric two-armed patterns, but others do not. Some

galaxies have a great many short spiral segments, giving them a fluffy appearance. These galaxies have been termed **flocculent**, meaning "woolly" (Figure 15-24). Our galaxy is probably an intermediate galaxy. How can we explain these variations in the spiral pattern? Perhaps the answer lies in a process that sustains star formation once it begins.

STAR FORMATION IN SPIRAL ARMS

Star formation is a critical process in the creation of the spiral patterns that we see. It makes spiral arms visible and may shape the spiral pattern itself.

The spiral density wave creates spiral arms by the gravitational attraction of the stars and gas flowing through the arms. Even if there were no star formation at all, rotating disk galaxies could form spiral arms. But without star formation to make young, hot, luminous stars, the spiral arms would be difficult to see. It is the star formation that lights up the spiral arms and makes them so prominent. Thus, star formation helps determine what we see when we look at a spiral galaxy.

But star formation can also control the shape of spiral patterns if the birth of stars in a cloud of gas can renew itself and continue making more new stars. Consider a newly formed star cluster with a single massive star. The intense radiation from that hot star can compress nearby parts of the gas cloud and trigger further star formation (Figure 15-25). Also, massive stars evolve so quickly that their lifetimes are only an instant in the history of a galaxy. When they explode as supernovae, the expanding gasses can compress neighboring clouds of gas and trigger more star formation. Examples of such a **self-sustaining star formation** have been found. The Orion complex consisting of the Great Nebula in Orion and the star formation buried deep in the dark interstellar clouds behind the nebula is such a region.

Self-sustaining star formation can produce growing clumps of new stars, and the differential rotation of the galaxy can drag the inner edge ahead and let the outer edge lag behind to produce a cloud of star formation shaped like a segment of a spiral arm, a spur (Figure 15-26). Mathematical models of galaxies filled with such segments of spiral arms do have a spiral appearance, but they lack the bold, two-armed spiral that astronomers refer to as the grand-design spiral pattern (Figure 15-27). Rather, astronomers suspect that self-sustaining star formation can produce the branches and spurs so prominent in flocculent galaxies, but only the spiral density wave can generate the beautiful two-armed spiral patterns.

Our discussion of star formation in spiral structure illustrates the importance of natural processes. The spiral density wave creates graceful arms, but it is the star formation in the arms that makes them stand out so prominently. Self-sustaining star formation can act in some galaxies to modify the spiral arms and produce branches and spurs. In some galaxies, it can make the

a

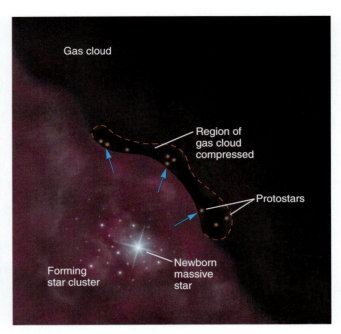

Gas cloud

Region of gas cloud compressed

Protostars

Newborn massive star

Forming star cluster

FIGURE 15-25

The energetic birth of a massive star compresses nearby gas clouds, triggering the gravitational contraction of the gas to form new stars.

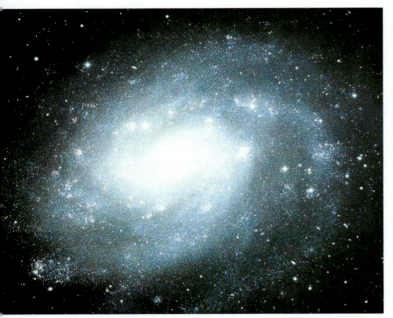

b

Visual-wavelength images

FIGURE 15-24

(a) Grand-design spiral galaxies, such as M100, have two dominant spiral arms even if they have minor spurs and branches. (b) Flocculent or woolly galaxies have a spiral appearance but lack two distinct spiral arms, as in the case of NGC300 shown here. *(Anglo-Australian Telescope Board)*

REVIEW CRITICAL INQUIRY

Why can't we use solar-type stars as spiral tracers?

Sometimes the timing of events is the critical factor in an analysis. In this case, we must think about the evolution of stars and their orbital periods around the galaxy. Stars like the sun live about 10 billion years, but the sun's orbital period around the galaxy is 240 million years. The sun almost certainly formed when a gas cloud passed through a spiral arm, but since then the sun has circled the galaxy many times and has passed through spiral arms often. Thus, the sun's present location has nothing to do with the location of any spiral arms. An O star, however, lives only a few million years. It is born in a spiral arm and lives out its entire lifetime before it can leave the spiral arm. We find short-lived stars such as O stars only in spiral arms, but G stars are found all around the galaxy.

The spiral arms of our galaxy must make it beautiful in photographs taken from a distance, but we are trapped inside it. How do we know that the spiral arms we can trace out near us actually extend across the disk of our galaxy?

Spiral arms are a puzzle of great beauty, but the center of our galaxy hides a puzzle of great power.

15-4 THE NUCLEUS

The most mysterious region of our galaxy is its very center, the nucleus. At visual wavelengths, this region is totally hidden from us by gas and dust that dim the

spiral pattern flocculent. By searching out and understanding the details of such natural processes, we can begin to understand the overall structure and evolution of the universe around us.

Self-Sustaining Star Formation

Differential rotation drags the inner edge of a gas cloud ahead of its outer edge.

Center of galaxy

A cloud can become elongated by continuing differential rotation.

Star formation in a gas cloud can produce massive stars whose high luminosity...

and supernova explosions can compress surrounding gas and trigger more star formation.

If star formation continues long enough, a cloud can be elongated into a spiral segment.

FIGURE 15-26
Continuing star formation may be able to produce long clouds of young stars that look like segments of spiral arms.

light by 30 magnitudes. If a trillion (10^{12}) photons of light left the center of the galaxy on a journey to Earth, only one would make it through the gas and dust. Thus, visual-wavelength photos tell us nothing about the nucleus. Observations at radio and infrared wavelengths paint a picture of tremendously crowded stars orbiting the center at high velocity. To understand what is hap-

Enhanced visual image

FIGURE 15-27
The beautiful symmetry of the grand-design spiral pattern is clear in this image of the Whirlpool Galaxy, M51. Young star clusters containing hot, bright stars and ionized hydrogen (pink regions) are located along the arms, while dark lanes of dust mark the inner edges of the arms. *(NASA and The Hubble Heritage Team. STScI/AURA)*

pening at the center of our galaxy, we must carefully compare observations and theories.

OBSERVATIONS

If we observe the Milky Way on a dark night, we might notice a slight thickening in the direction of the constellation Sagittarius, but nothing specifically identifies this as the direction of the heart of the galaxy. Even Shapley's study of globular clusters identified the center only approximately. When radio astronomers turned their telescopes toward Sagittarius, they found a complex collection of radio sources, with one, **Sagittarius A*** (abbreviated Sgr A* and usually pronounced "sadge A-star"), lying at the expected location of the galactic core.

Observations show that Sgr A* is less than a few astronomical units in diameter but is a powerful source of radio energy. The tremendous amount of infrared radiation coming from the central area appears to be produced by crowded stars and by dust warmed by those stars. But what could be as small as Sgr A* and produce so much energy?

Study ▮ "Sagittarius A*" on pages 328 and 329 and notice three important points. First, observations at visual wavelengths are useless in the study of Sgr A*. Only

Sagittarius A*

The constellation of Sagittarius is so filled with stars and with gas and dust we can see nothing at visual wavelengths of the center of our galaxy. At infrared wavelengths, as shown at bottom left, we see the glow of the cool stars, gas, and dust at the center of our galaxy.

The image below is a wide-field radio image of the center of our galaxy. Many of the features are supernova remnants (SNR), and a few are clouds of star formation. Peculiar features such as threads, the Arc, and the Snake may be gas trapped in magnetic fields. At the center lies Sagittarius A, the center of our galaxy.

Arc

Radio image

NRL

The radio map above shows Sgr A and the Arc filaments, 50 parsecs long. The image was made with the VLA radio telescope. The contents of the white box are shown on the opposite page.

Sgr D HII

Sgr D SNR

SNR 0.9 + 0.1

Apparent angular size of the moon for comparison

Sgr B2

Sgr B1

Arc

Sgr A

New SNR 0.3 + 0.0

Threads

The Cane

Background galaxy

Threads

NRAO

Infrared photons with wavelengths longer than 4 microns (4000 nm) come almost entirely from warm interstellar dust. The radiation at these wavelengths coming from Sagittarius is intense, and that tells us that the region contains lots of dust and is crowded with stars that warm the dust.

Sgr C

Coherent structure?

The Pelican

Snake

Sgr E

Radio image

SNR 359.1 − 00.5

Infrared image

2MASS

This high-resolution radio image of Sgr A (the white boxed area on the opposite page) reveals a spiral swirl of gas around an intense radio source known as Sgr A*, the presumed central object in our galaxy. About 3 pc across, this spiral lies in a low-density cavity inside a larger disk of neutral gas. The arms of the spiral are thought to be streams of matter flowing into Sgr A* from the inner edge of the larger disk (drawing at right).

Sgr A*

Radio image

N. Killeen and Kwok-Yung Lo

Evidence of a Black Hole at the Center of Our Galaxy

Star S2 near Sgr A*

ESO

Only one in a trillion photons of visible light makes it through the gas and dust from the center of our galaxy to our eyes. But one in ten infrared photons survives. The infrared image at left shows individual stars in the central light-year around Sgr A*.

March 1992

Orbit of star S2

Sept. 2002

Sgr A*

1 light-day

The star S2 has been observed since early 1992, and, using active optics and large telescopes, astronomers have been able to observe the star orbiting Sgr A*. Crosses show the uncertainty in the observations.

The orbital period of S2 is 15.2 years, and the semimajor axis of its orbit is 950 AU. The orbital motion of S2 combined with the motion of other stars in the region suggest a mass for Sgr A* of 2.6 million solar masses.

Our solar system is half a light-day in diameter.

At its closest, S2 comes within 17 light-hours of Sgr A*. Alternative theories that Sgr A* is a cluster of stars, neutron stars, or black holes are eliminated. Only a black hole could contain so much mass in so small a region.

A black hole with a mass of 2.6 million solar masses would have an event horizon no bigger than the white dot shown in this orbital diagram.

The Chandra X-ray telescope has imaged Sgr A* as shown here.

NASA/CXC/SAO

The evidence of a massive black hole at the center of our galaxy seems conclusive. It is much too massive to be the remains of a dead star, however, and astronomers conclude that it probably formed as the galaxy first took shape.

A slow dribble of only 0.0002 solar masses of gas flowing into the black hole could produce a hot accretion disk and release enough energy to power Sgr A*. A sudden increase as when a star falls in could produce an eruption.

at the longer wavelengths of infrared or radio waves, or at the very short wavelengths of X rays, can we image the features near the center of our galaxy. A second thing to notice is the evidence of star formation in the gas clouds near the center. Not only do we see clouds of gas forming new stars, but we also see supernova remnants produced by the deaths of massive stars, and those massive stars must have been born recently. Third, notice the evidence that Sgr A* is a supermassive black hole into which gas is flowing.

A supermassive black hole is an exciting idea, but we must notice the difference between adequacy and necessity. A supermassive black hole is adequate to explain the observations, but could there be other explanations? For example, astronomers have suggested that gas flowing inward could trigger tremendous bursts of star formation. Such theories have been considered and tested against the evidence, but none appears to be adequate to explain the observations. So far, the only theory that seems adequate is that our galaxy is home to a supermassive black hole. That is, a central black hole appears to be not only adequate but necessary to explain the observations. In later chapters we will see that such supermassive black holes are found at the centers of many galaxies.

REVIEW CRITICAL INQUIRY

Why do we think the center of our galaxy contains a large mass?

The best way to measure the mass of an astronomical object is to watch something orbit around it. We can then use Kepler's third law to find the mass inside the orbit. Because of gas and dust, we can't see to the center of our galaxy at visual wavelengths, but infrared observations can detect individual stars orbiting Sgr A*. The star S2 has been particularly well observed, but a number of stars can be followed as they orbit the center. The sizes and periods of these orbits, interpreted using Kepler's laws, tell us that Sgr A* contains roughly 2.6 million solar masses.

In addition to these orbital motions, the large amount of infrared radiation at wavelengths longer than 4 microns (4000 nm) also implies that vast numbers of stars are crowded into the center. Why?

Our observations of the Milky Way Galaxy suggest it is a place of great beauty and great power. Is it normal? Are all galaxies like ours? To answer that question we must survey other galaxies, which we begin in the next chapter.

SUMMARY

Our galaxy is a typical spiral galaxy with a disk about 25,000 pc in diameter. The sun is located about two-thirds of the way from the center to the edge. The disk contains nearly all of the gas and dust in the galaxy and most of the star formation. It is illuminated by the brilliant O and B stars. The nuclear bulge at the center of the galaxy is about 3000 pc in radius and contains stars similar to the large spherical halo of stars and globular clusters.

Gas and dust in the disk block our view, so until the 20th century, astronomers assumed that we were located at the center of a grindstone-shaped star cloud. Harlow Shapley was able to calibrate variable stars to find the distance to globular clusters, and he concluded that our galaxy was much bigger and that we were not located in the center. Soon after, other astronomers concluded that ours was only one of many galaxies in the universe.

The disk component of our galaxy contains metal-rich population I stars moving in nearly circular orbits in the plane of the disk. The spherical component, made up of the nuclear bulge and the halo, contains metal-poor population II stars moving in randomly tipped and highly elongated orbits. Observations of the rotation curve of our galaxy reveal that orbital velocities do not fall in the outer disk, suggesting that the galaxy contains large amounts of unseen mass commonly thought to be located in an extended halo.

The distribution of populations through the galaxy suggests a hypothesis for the formation of our galaxy from a single cloud of gas and dust that gradually contracted into a disk shape. Evidence suggests that the story is more complicated, however, and many astronomers now suspect that the nuclear bulge and halo formed first and the disk formed later as more gas fell into the galaxy. The galaxy has evidently absorbed smaller galaxies as well.

The very youngest stars lie along spiral arms within the disk. The most massive live such short lives that they don't have time to move from their place of birth in the spiral arms. Maps of the spiral tracers and cold hydrogen clouds reveal the spiral pattern of our galaxy. The spiral arms are also outlined by giant molecular cloud complexes. These molecular clouds are sites of star formation.

The spiral density wave theory suggests that spiral arms are regions of compression that move through the disk. When an orbiting gas cloud smashes into the compression wave, the gas cloud forms stars. Self-sustaining star formation is an important process that may modify the arms as the birth of massive stars triggers the formation of more stars by compressing neighboring clouds.

Dust hides the nucleus of our galaxy at visible wavelengths, but radio and infrared observations can penetrate the dust. Such observations reveal that the central region is crowded with stars orbiting at high velocity. This tells us that the central few light years must contain millions of solar masses. At the very center lies the radio source Sagittarius A*, an object no larger than a few AU in diameter. Theorists suspect that the object is a supermassive black hole that is drawing in matter through an accretion disk.

NEW TERMS

Magellanic Clouds

kiloparsec (kpc)

halo

nuclear bulge

disk component

spherical component

high-velocity star

rotation curve

Keplerian motion

galactic corona

dark matter

metals

population I star

population II star

nucleosynthesis

galactic fountain

spiral tracers

density wave theory

flocculent

self-sustaining star
  formation

Sagittarius A*

REVIEW QUESTIONS

Ace◐Astronomy™ Assess your understanding of this chapter's topics with additional quizzing and animations at **http:// astronomy.brookscole.com/seeds8e**

1. Why is it difficult to specify the dimensions of the disk and halo?

2. Why didn't astronomers before Shapley realize how large the galaxy is?

3. What evidence do we have that our galaxy has a galactic corona?

4. Explain why some star clusters lose stars more slowly than others.

5. Contrast the motion of the disk stars and that of the halo stars. Why do their orbits differ?

6. Why do high-velocity stars have lower metal abundance than the sun?

7. Why are metals less abundant in older stars than in younger stars?

8. Why are all spiral tracers young?

9. Why couldn't spiral arms be physically connected structures? What would happen to them?

10. Why does self-sustaining star formation produce clouds of stars that look like segments of spiral arms?

11. Describe the kinds of observations we can make to study the galactic nucleus.

12. What evidence do we have that the nucleus of the galaxy contains a supermassive energy source that is very small in size?

DISCUSSION QUESTIONS

1. How would this chapter be different if interstellar dust did not scatter light?

2. Why doesn't the Milky Way circle the sky along the celestial equator or the ecliptic?

PROBLEMS

1. Make a scale sketch of our galaxy in cross section. Include the disk, sun, nucleus, halo, and some globular clusters. Try to draw the globular clusters to scale.

2. Because of dust, we can see only about 5 kpc into the disk of the galaxy. What percentage of the galactic disk can we see? (*Hint:* Consider the area of the entire disk and the area we can see.)

3. If the fastest passenger aircraft can fly 0.45 km/s (1000 mph), how long would it take to reach the sun? the galactic center? (*Hint:* 1 pc = 3×10^{13} km.)

4. If a typical halo star has an orbital velocity of 250 km/s, how long does it take to pass through the disk of the galaxy? Assume that the disk is 1000 pc thick.

5. If the RR Lyrae stars in a globular cluster have apparent magnitudes of 14, how far away is the cluster? (*Hint:* See Figure 12-13.)

6. If intersellar dust makes an RR Lyrae variable star look 1 magnitude fainter than it should, by how much will we overestimate its distance? (*Hint:* Use the magnitude–distance formula or Table 9-1.)

7. If a globular cluster is 10 minutes of arc in diameter and 8.5 kpc away, what is its diameter? (*Hint:* Use the small-angle formula.)

8. If we assume that a globular cluster 4 minutes of arc in diameter is actually 25 pc in diameter, how far away is it? (*Hint:* Use the small-angle formula.)

9. If the sun is 5 billion years old, how many times has it orbited the galaxy?

10. If the true distance to the center of the galaxy is found to be 7 kpc and the orbital velocity of the sun is 220 km/s, what is the minimum mass of the galaxy? (*Hint:* Use Kepler's third law.)

11. What temperature would interstellar dust have to have to radiate most strongly at 100 μm? (*Hints:* 1 μm = 1000 nm. Use Wien's law, Chapter 7.)

12. Infrared radiation from the center of our galaxy with a wavelength of about 2 μm (2×10^{-6} m) comes mainly from cool stars. Use this wavelength as λ_{max} and find the temperature of the stars.

13. If an object at the center of our galaxy has a linear diameter of 10 AU, what will its angular diameter be as seen from Earth? (*Hint:* Use the small-angle formula, Chapter 3.)

CRITICAL INQUIRIES FOR THE WEB

1. Henrietta Leavitt discovered the period–luminosity relation for Cepheids while working on the staff at Harvard College Observatory under Edward Pickering. She was one of several women "computers" on staff there a century ago. Search the Web for information on Leavitt, her colleagues, and their work at Harvard. List three of the women employed by Pickering and note their contributions to astronomy. What was life like for a woman in astronomy at the beginning of the 20th century?

2. What if we lived near the center of the galaxy? Search the Web for research and information on the distribution of material near the center of our galaxy. Based on what you find, speculate as to how the sky would appear from a planet associated with a star near the galactic center.

3. Who was Wilhelm Heinrich Walter Baade, and why did he win the Bruce Medal? What other winners of the Bruce Medal do you recognize?

EXPLORING *THESKY*

1. Locate Sagittarius and examine the shape of the Milky Way there and the profusion of globular clusters. (*Hint:* To turn on Messier object labels, use **Labels** and **Setup** under the **View** menu.)

2. Locate the following globular clusters: M3, M4, M5, M10, M12, M13, M15, M22, M55, M92. Where are they located in the sky? (*Hint:* Use **Find** under the **Edit** menu.)

3. Compare the distribution of globular clusters with that of open clusters. (*Hint:* Use **Filters** under the **View** menu to turn off everything but globular clusters, the Milky Way, the Galactic Equator, and Constellation Boundaries. Use the thumbwheel at the bottom of the sky window to rotate the sky. Now repeat with globular clusters off and open clusters on.)

 Visit the Seeds *Foundations of Astronomy* companion Web site for critical thinking exercises, articles, and additional readings from InfoTrac College Edition, Brooks/Cole's online student library.

GALAXIES

A hypothesis or theory is clear, decisive, and positive, but it is believed
by no one but the man who created it. Experimental findings, on the other hand,
are messy, inexact things which are believed by everyone
except the man who did that work.

Harlow Shapley, Through Rugged Ways to the Stars

Bill Schoening/AURA/NOAO/NSF

Chapter 16

GUIDEPOST

The preceding chapter was about our Milky Way Galaxy, an important object to us but only one of the many billions of galaxies visible in the sky. We can no more understand galaxies by understanding a single example, the Milky Way, than we could understand humanity by understanding a single person. This chapter expands our horizon to discuss the different kinds of galaxies and their complex histories.

We take two lessons from this chapter. First, galaxies are not solitary beasts; they collide and interact with each other. Second, most of the matter in the universe is invisible. The galaxies we see are only the tip of a cosmic iceberg.

We will carry the lessons of this chapter into the next, where we will discuss violently active galaxies, and on into Chapter 18, where we discuss the universe as a whole.

VIRTUAL LABORATORIES

ASTRONOMICAL DISTANCE SCALES

THE HUBBLE LAW

Science fiction heroes flit effortlessly between the stars, but almost none voyages between the galaxies. As we leave our home galaxy, the Milky Way, behind, we voyage out into the vast depths of the universe, out among the galaxies, space so deep it is unexplored even in fiction.

Less than a century ago, astronomers did not understand that there were galaxies. Nineteenth-century telescopes revealed faint nebulae scattered among the stars, and some were spiral. Astronomers argued about the nature of these nebulae, but it was not until the 1920s that astronomers understood that some were other galaxies much like our own; and it was not until recent decades that astronomical telescopes could reveal the tremendous beauty and intricacy of the galaxies (▌Figure 16-1).

In this chapter, we will try to understand how galaxies form and evolve. We will discover that their complex shapes reflect long periods of peaceful star formation and episodes of cosmic violence.

Before we can build theories, however, we must gather some basic data concerning galaxies. How many different kinds of galaxies are there? How big are they?

How massive are they? That is, we must characterize the family of galaxies just as we characterized the family of stars in Chapter 9.

16-1 THE FAMILY OF GALAXIES

Like leaves on the forest floor, galaxies carpet the sky. Pick any spot on the sky away from the dust and gas of the Milky Way, and you are looking deep into space. Photons that have traveled for billions of years enter your eye, but they are too few to register on your retina. Only the largest telescopes can gather enough light to detect the most distant galaxies.

When astronomers picked a seemingly empty spot on the sky near the Big Dipper and used the Hubble Space Telescope to record an amazing time exposure of 10 days' duration, they found thousands of galaxies crowded into the image. This image, now known as a Hubble Deep Field, contains a few relatively nearby galaxies and others that are over 10 billion light-years away (▌Figure 16-2). The entire sky appears to be as thickly carpeted with galaxies as is the Hubble Deep Field.

Less than a century ago, humanity did not know that it lived in a galaxy or that the universe was filled

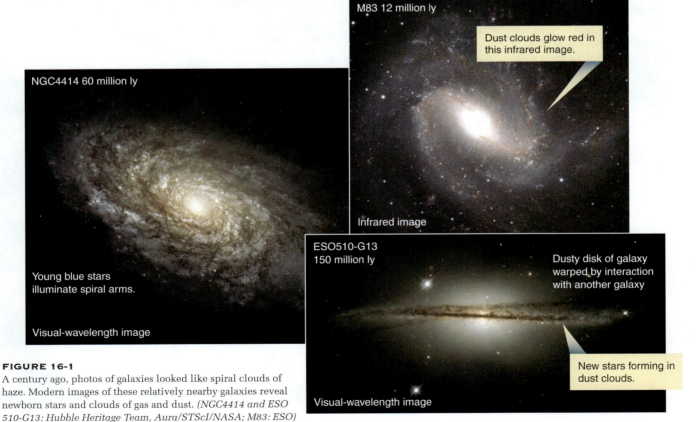

FIGURE 16-1
A century ago, photos of galaxies looked like spiral clouds of haze. Modern images of these relatively nearby galaxies reveal newborn stars and clouds of gas and dust. *(NGC4414 and ESO 510-G13: Hubble Heritage Team, Aura/STScI/NASA; M83: ESO)*

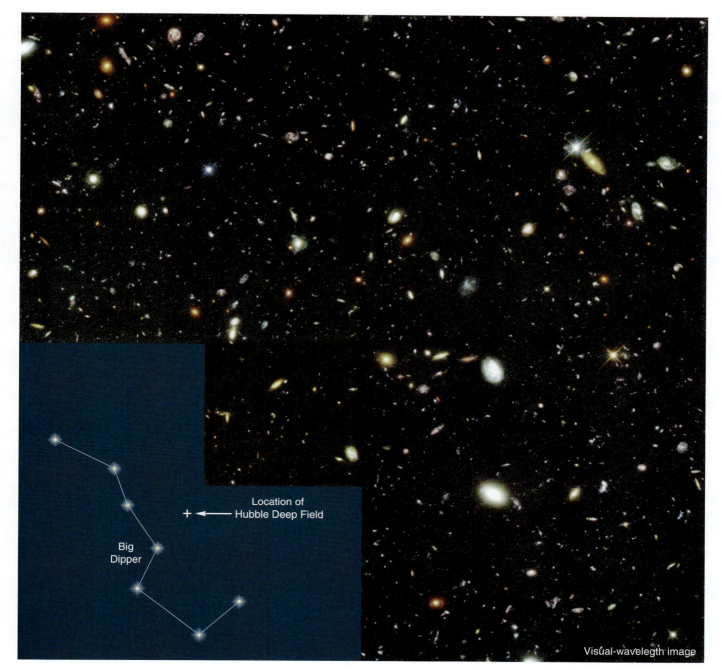

Location of
Hubble Deep Field

Big
Dipper

Visual-wavelegth image

FIGURE 16-2

An apparently empty spot on the sky only 1/30 the diameter of the full moon contains over 1500 galax-ies in this extremely long time exposure known as the Hubble Deep Field. Presumably the entire sky is similarly filled with galaxies. *(R. Williams and the Hubble Deep Field Team, STScI, NASA)*

with other galaxies. The discovery of galaxies is one of the great stories of astronomy.

THE DISCOVERY OF GALAXIES

Galaxies are faint objects, so they were not noticed until telescopes had grown large enough to gather sig-nificant amounts of light. In 1845, William Parsons, third Earl of Rosse in Ireland, built a telescope 72 in. in

diameter. It was, for some years, the largest telescope in the world. Before the invention of astronomical pho-tography (in the late 1800s), Parsons had to view the faint nebulae directly at the eyepiece and sketch their shapes. He noticed that some have a spiral shape, and they became known as **spiral nebulae.** Parsons immedi-ately concluded that the spiral nebulae were great spi-rals of stars. The German philosopher Immanuel Kant in 1755 had proposed that the universe was filled with

Searching for Clues Through Classification in Science

Classification is one of the most basic and most powerful of scientific tools. It is often a way to begin studying a topic, and it often leads to dramatic insights. Charles Darwin, for example, sailed around the world with a scientific expedition aboard the ship HMS *Beagle*. Everywhere he went, he studied the living things he saw and tried to classify them. For example, he studied the tortoises, mockingbirds, and finches he saw on the Galápagos Islands. His classifications of these and other animals eventually led him to think about evolution and natural selection. Scientists in many fields depend on carefully designed systems of classification.

Classification is an everyday mode of thought. We all use classifications when we order lunch, buy shoes, catch a bus, and so on. In science, classifications reveal the relationships between different kinds of objects and at the same time save the scientist valuable time by bringing order out of seeming disorder. An economist can classify different kinds of businesses and does not have to analyze each of the millions of businesses as if it were unique. Classifications of minerals, plants, psychological learning styles, modes of transportation, or sandwiches bring order to the world and help us deal with information.

Astronomers use classifications of galaxies, stars, moons, telescopes, and more. Whenever you encounter a scientific discussion, look for the classifications on which it is based. Classifications are the orderly framework on which much of science is built.

---

great wheels of stars that he called **island universes.** Parsons adopted that term for the spiral nebulae.

Not everyone agreed that the spiral nebulae were island universes. An alternative point of view was that we lived in a great star system in an otherwise empty universe. Sir William Herschel had counted stars in different directions and had shown that the star system was shaped approximately like a grindstone (Chapter 15). Beyond the edge of this grindstone universe, space was supposed to be a limitless void. The spiral nebulae, in this view, were vortexes of gas or faint stars within the star system.

The nature of the spiral nebulae could not be resolved in the 19th century because the telescopes were not large enough and the photographic plates were not sensitive enough. The spiral nebulae remained foggy swirls on the best photographs, and astronomers continued to disagree on their true nature. In April 1920, two astronomers debated the issue at the National Academy of Science in Washington, DC. Harlow Shapley of Mt. Wilson Observatory had recently shown that the Milky Way was a much larger star system than had been thought. He argued that the spiral nebulae were nearby nebulae within the star system. Heber D. Curtis of Lick Observatory argued that the spiral nebulae were island universes. Historians of science mark this **Shapley–Curtis Debate** as a turning point in modern astronomy, but it was inconclusive. Shapley and Curtis cited the right evidence but drew incorrect conclusions. The disagreement was finally resolved, as is often the case in astronomy, with a bigger telescope.

On December 30, 1924, Mt. Wilson astronomer Edwin Hubble (namesake of the Hubble Space Telescope) announced that he had taken photographic plates of a few bright galaxies with the new 100-in. telescope. Not only could he detect individual stars, but he could identify some of them as Cepheid variables with apparent magnitudes of about 18. For the Cepheids, which are supergiants, to look that faint, they had to be very distant, and thus the spiral nebulae were external to the Milky Way star system. They were galaxies.

Although it was spiral galaxies that first attracted the attention of astronomers, Hubble quickly realized that there were other kinds of galaxies. Not all galaxies had spiral shapes.

THE SHAPES OF GALAXIES

Look at the galaxies in the Hubble Deep Field, and you will see galaxies of different shapes. Some are spiral, some are elliptical, and some are irregular. Astronomers classify galaxies according to their shape using a system developed by Edwin Hubble in the 1920s. Such systems of classification are a fundamental technique in science (Window on Science 16-1).

Study ❚ "Galaxy Classification" on pages 338 and 339 and notice two important points. First, the amount of gas and dust in a galaxy strongly influences its appearance. Galaxies rich in gas and dust have active star formation and contain hot, bright stars. Such galaxies tend to be bluer and contain emission nebulae. Galaxies that are poor in gas and dust contain few or none of these highly luminous stars, so those galaxies look redder and have a much more uniform appearance.

Second, notice the wide range in the shapes of galaxies. Some look like spherical- or football-shaped clouds of stars, but others are thin disks. Still others are so irregular they look like scrambled galaxy.

Spiral galaxies are clearly disk shaped, but the true, three-dimensional shape of elliptical galaxies isn't obvious from images. Some are spherical, but the more elongated elliptical galaxies could be shaped like flattened spheres (bun shaped) or like elongated spheres (football shaped). Some may even have three different diameters; they are longer than they are thick and thicker than they are wide, like a flattened loaf of bread. Statistical studies suggest that all of these shapes are represented among elliptical galaxies.

Traditionally astronomers have classified galaxies according to their appearance in visual-wavelength images, but infrared images reveal the location of the vast majority of stars—stars that are not as luminous as the hot stars that produce most of the light. These studies suggest that the structure of many galaxies is dominated by the gravitational fields of large masses of cool stars. For example, some spiral galaxies that appear to lack bars have massive bars evident in infrared images. As we think about galaxies we must remember that the classification of galaxies tells us where the light comes from in a galaxy and not necessarily where most of the mass is located.

Why do galaxies look so different? Later in this chapter we will try to understand what determines the stellar content and shape of a galaxy. Our basic classification system will be an important tool.

HOW MANY GALAXIES?

Looking at the Hubble Deep Field forces even the most sophisticated astronomer to ask what seems to be a silly question. Just how many galaxies are there? The answer is impressive.

Two Hubble Deep Fields have been imaged, one in the northern hemisphere and one in the southern hemisphere, and they are both filled with galaxies. A new effort called GOODS (Great Observatories Origins Deep Survey) is continuing such studies using an improved visual-wavelength camera on the Hubble Space Telescope, X-ray observations by the Chandra spacecraft, and infrared observations by the Spitzer spacecraft. All of these studies are revealing tremendous numbers of distant galaxies in the small regions of the sky being studied. If the entire sky is similarly covered with galaxies, there must be over 100 billion galaxies visible with existing telescopes. Surely there are other galaxies too distant or too faint to see.

We might also wonder what proportion of the galaxies is elliptical, spiral, and irregular, but that is a difficult question to answer. In catalogs of galaxies, about 70 percent are spiral, but that is because spiral galaxies are gaudy and easy to notice. Spiral galaxies contain hot, bright stars and clouds of ionized gas. Most ellipticals are fainter and harder to notice. Small galaxies such as dwarf ellipticals and dwarf irregulars are very common, but they are hard to detect. From careful studies we can conclude that ellipticals are more common than spirals, and that irregulars make up only about 25 percent of all galaxies.

REVIEW CRITICAL INQUIRY

What color are galaxies?

Different kinds of galaxies have different colors, depending mostly on how much gas and dust they contain. If a galaxy contains large amounts of gas and dust, it probably contains lots of young stars, and a few of those young stars will be massive, hot, luminous O and B stars. They will give the galaxy a distinct blue tint. In contrast, a galaxy that contains little gas and dust will probably contain few young stars. It will lack O and B stars, and the most luminous stars in such a galaxy will be red giants. They will give the galaxy a red tint. Because the light from a galaxy is a blend of the light from billions of stars, the colors are only tints. Nevertheless, the most luminous stars in a galaxy determine the color. From this we can conclude that elliptical galaxies tend to be red and the disks of spiral galaxies tend to be blue.

Of course, the halo of a spiral galaxy would be both dimmer and redder than the disk. Why?

If we want to create a theory to explain the shapes of galaxies, we need answers to some basic questions. How big are galaxies? How luminous are they? Are they all the same mass, or are some more massive than others? To answer those questions, we must discover the basic properties of galaxies. We begin that task in the next section.

16-2 MEASURING THE PROPERTIES OF GALAXIES

Looking beyond the edge of our Milky Way Galaxy, astronomers find many billions of galaxies. Great clusters of galaxies, some containing thousands of galaxies, fill space as far as we can see. We can recognize the different types of galaxies, but what are the properties of these star systems? What are the diameters, luminosities, and masses of galaxies? Just as in our study of stellar characteristics (Chapter 9), the first step in our study of galaxies is to find out how far away they are. Once we know a galaxy's distance, its size and luminosity are relatively easy to find. Later in this section, we will see that finding mass is more difficult.

DISTANCE

The distances to galaxies are so large that it is not convenient to measure them in light-years, parsecs, or even kiloparsecs. Instead, we will use the unit **megaparsec (Mpc),** or 1 million pc. One Mpc equals 3.26 million ly, or approximately 2×10^{19} miles.

To find the distance to a galaxy, we must search among its stars, nebulae, and star clusters for familiar objects whose luminosity or diameter we know. Such objects are called **distance indicators** because we can use them to find the distance to a galaxy. Most distance indicators are objects whose brightness is known, and astronomers often refer to them as **standard candles.** If we can find a standard candle in a galaxy we can judge its distance.

Galaxy Classification

AURA/NOAO/NSF

The most commonly used system of galaxy classification is based on the appearance of galaxies in photographs made at visual wavelengths.

Elliptical galaxies are round or elliptical, contain no visible gas and dust, and lack hot, bright stars. They are classified with a numerical index ranging from 1 to 7; E0s are round, and E7s are highly elliptical. The index is calculated from the largest and smallest diameter of the galaxy used in the following formula and rounded to the nearest integer.

$$\frac{10(a - b)}{a}$$

Outline of an E6 galaxy

Visual-wavelength image

M87 is a giant elliptical galaxy classified E1. It is a number of times larger in diameter than our own galaxy and is surrounded by a swarm of over 500 globular clusters.

The Leo 1 dwarf elliptical galaxy is not many times bigger than a globular cluster.

Anglo-Australian Telescope Board

Visual

Spiral galaxies contain a disk and spiral arms. We do not see their halo stars, but presumably all spiral galaxies have halos. Spirals contain gas and dust and hot, bright O and B stars. The presence of short-lived O and B stars alerts us that star formation is occurring in these galaxies. Sa galaxies have larger nuclei, less gas and dust, and fewer hot, bright stars. Sc galaxies have small nuclei, lots of gas and dust, and many hot, bright stars. Sb galaxies are intermediate.

Anglo-Australian Telescope Board

Sa

Visual   **NGC 3623**

Ace Astronomy™

Go to AceAstronomy and click Active Figures to see "Galaxy Types" and review the classification of galaxies.

Anglo-Australian Telescope Board

BAR

Sb

NGC 3627   **Visual**

Sc

Roughly 2/3 of all spiral galaxies are **barred spiral galaxies** classified SBa, SBb, and SBc. They have an elongated nucleus with spiral arms springing from the ends of the bar. Our own galaxy is a barred spiral.

NGC 1365   **Visual**

NGC 2997   **Visual**

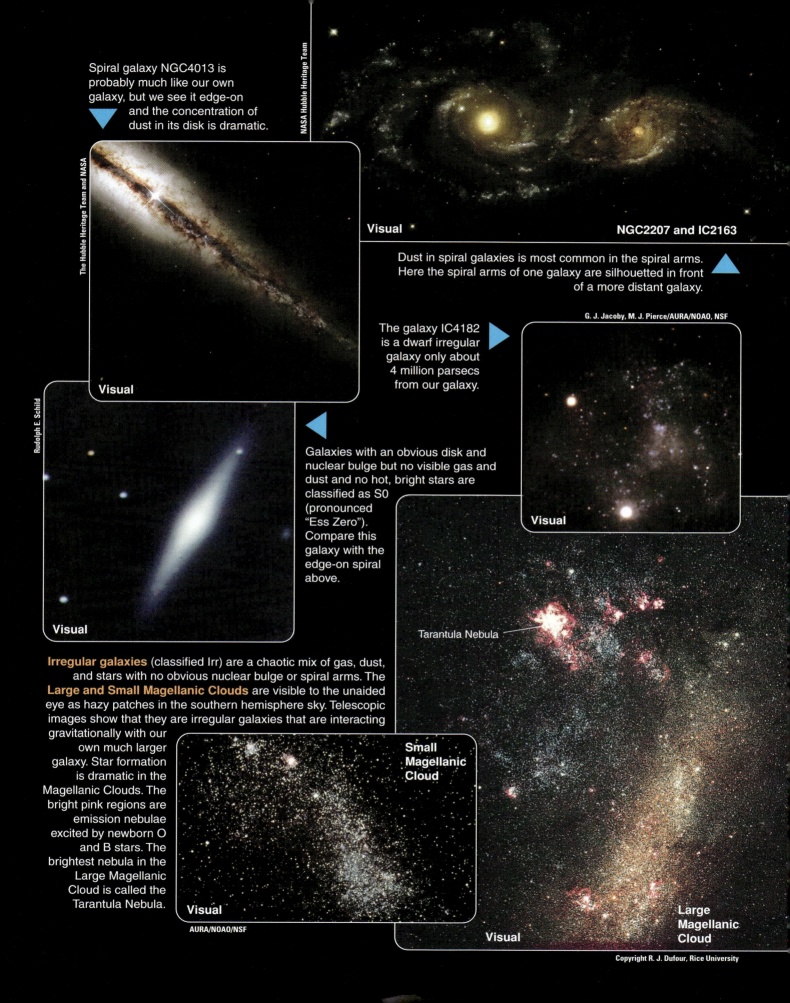

Spiral galaxy NGC4013 is probably much like our own galaxy, but we see it edge-on and the concentration of dust in its disk is dramatic.

NASA Hubble Heritage Team

The Hubble Heritage Team and NASA

Visual

Visual

NGC2207 and IC2163

Dust in spiral galaxies is most common in the spiral arms. Here the spiral arms of one galaxy are silhouetted in front of a more distant galaxy.

G. J. Jacoby, M. J. Pierce/AURA/NOAO, NSF

The galaxy IC4182 is a dwarf irregular galaxy only about 4 million parsecs from our galaxy.

Rudolph E. Schild

Galaxies with an obvious disk and nuclear bulge but no visible gas and dust and no hot, bright stars are classified as S0 (pronounced "Ess Zero"). Compare this galaxy with the edge-on spiral above.

Visual

Visual

Irregular galaxies (classified Irr) are a chaotic mix of gas, dust, and stars with no obvious nuclear bulge or spiral arms. The **Large and Small Magellanic Clouds** are visible to the unaided eye as hazy patches in the southern hemisphere sky. Telescopic images show that they are irregular galaxies that are interacting gravitationally with our own much larger galaxy. Star formation is dramatic in the Magellanic Clouds. The bright pink regions are emission nebulae excited by newborn O and B stars. The brightest nebula in the Large Magellanic Cloud is called the Tarantula Nebula.

Tarantula Nebula

Small Magellanic Cloud

Large Magellanic Cloud

Visual

Visual

AURA/NOAO/NSF

Copyright R. J. Dufour, Rice University

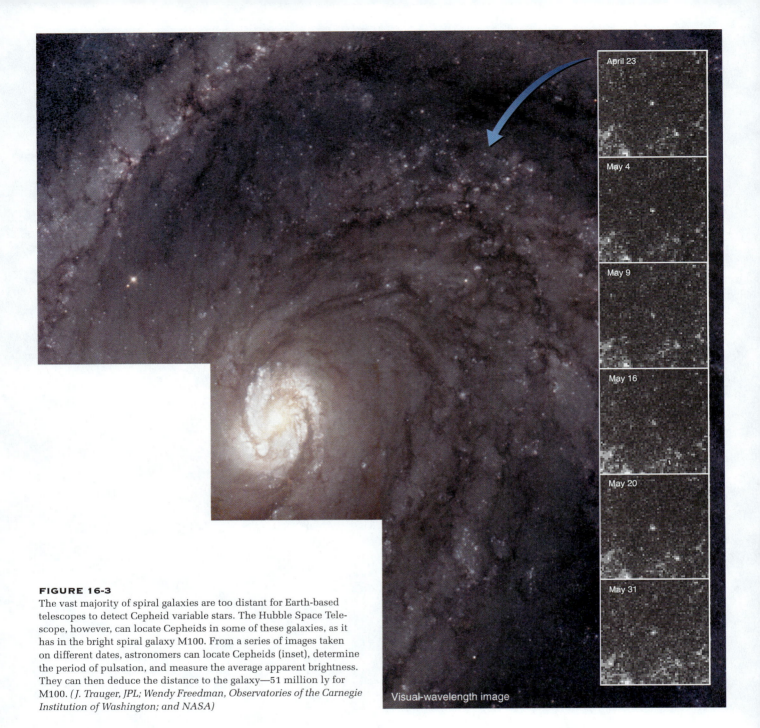

The inset images are labeled: April 23, May 4, May 9, May 16, May 20, May 31

Visual-wavelength image

FIGURE 16-3

The vast majority of spiral galaxies are too distant for Earth-based telescopes to detect Cepheid variable stars. The Hubble Space Telescope, however, can locate Cepheids in some of these galaxies, as it has in the bright spiral galaxy M100. From a series of images taken on different dates, astronomers can locate Cepheids (inset), determine the period of pulsation, and measure the average apparent brightness. They can then deduce the distance to the galaxy—51 million ly for M100. (*J. Trauger, JPL; Wendy Freedman, Observatories of the Carnegie Institution of Washington; and NASA*)

Because their period is related to their luminosity (see Figure 12-13), Cepheid variable stars are reliable distance indicators. If we know the period of the star's variation, we can use the period–luminosity diagram to learn its absolute magnitude. By comparing its absolute and apparent magnitudes, we can find its distance. ▍Figure 16-3 shows a galaxy in which the Hubble Space Telescope detected Cepheids.

Even with the Hubble Space Telescope, Cepheids are not visible in galaxies much beyond 80 million ly (25 Mpc), so astronomers must search for less common but brighter distance indicators and calibrate them using nearby galaxies containing visible Cepheids. For example, by studying nearby galaxies with distances known from Cepheids, astronomers have found that the brightest globular clusters have absolute magnitudes of about −10. If we find globular clusters in a more distant galaxy, we can assume that the brightest have similar absolute magnitudes and thus calculate the distance.

Another distance indicator is the cloud of ionized hydrogen called an HII region (Chapter 10) that forms around a very hot star. Astronomers have attempted to calibrate these, but they have not been as reliable as Cepheids because some are larger and brighter than others. Also, the size of HII regions seems to depend on the kind of galaxy they occupy, and so astronomers have not been able to calibrate them as accurately as other distance indicators.

Planetary nebulae have proven to be very useful distance indicators, which is surprising considering how faint their central star is. These central stars are small and hot. Their small surface area limits the amount of energy they can radiate, and their high temperature means they radiate most of their energy in the ultraviolet. To our eyes, these stars look faint, but the gas of the surrounding planetary nebula absorbs the ultraviolet radiation and reradiates it as visible emission lines, such as the oxygen emission line at a wavelength of 500.7 nm. Images of galaxies recorded at such wavelengths reveal planetary nebulae glowing like brightly colored paper lanterns illuminated from within. Astronomers have been able to calibrate the planetary nebulae for distance measurements by observing the absolute magnitude of planetary nebulae in nearby galaxies such as the Andromeda Galaxy, galaxies whose distances are known from other distance indicators. When planetary nebulae are detected in a more distant galaxy, a quick comparison of their apparent and absolute magnitudes reveals the distance to the galaxy. Notice once again the importance of calibration (Window on Science 15-1).

When a supernova explodes in a distant galaxy, astronomers rush to observe it. Type Ia supernovae, those produced by the collapse of a white dwarf, have been calibrated. By locating Cepheid variable stars in nearby galaxies that have been home to type Ia supernovae, astronomers have learned that these type of supernovae all reach the same absolute magnitude at maximum. Look at the image of the small galaxy IC4182 on page 339. It is an important link in this calibration because it is nearby and its Cepheids are visible and because astronomers saw a type Ia supernova there in 1937. When astronomers see a type Ia supernova in a more distant galaxy, they observe the apparent magnitude of the supernova at maximum and compare that with absolute magnitude these supernovae reach at maximum to find the distance to the galaxy. As we will see in Chapter 18, this is a critical calibration in modern astronomy.

At the greatest distances, astronomers must calibrate the total luminosity of the galaxies themselves. For example, studies of nearby galaxies show that an average galaxy like our Milky Way Galaxy has a luminosity about 16 billion times the sun's. If we see a similar galaxy far away, we can measure its apparent magnitude and calculate its distance. Of course, it is important to

Visual-wavelength image

FIGURE 16-4

Only the general characteristics of the galaxies are recognizable in this cluster of galaxies 4 to 6 billion light-years away. The bright blue galaxy at upper left is probably a foreground galaxy much closer than the cluster. The larger galaxies in this cluster are about the size of our Milky Way Galaxy. The diagonal line at lower right was produced by an artificial Earth satellite. *(K. Ratnatunga, R. Griffiths, Carnegie Mellon University, and NASA)*

recognize the different types of galaxies, and that is difficult to do at great distances (Figure 16-4). Averaging the distances to the brightest galaxies in a cluster can reduce the uncertainty in this method.

Notice how astronomers use calibration (Window on Science 15-1) to build a **distance scale** reaching from the nearest galaxies to the most distant visible galaxies. Often astronomers refer to this as the distance pyramid or the distance ladder because each step depends on the steps below it. Of course, the foundation of the distance scale rests on the Cepheid variable stars and our understanding of the luminosities of the stars in the H–R diagram, which ultimately rest on measurements of the parallax of stars.

The most distant visible galaxies are roughly 10 billion ly (3000 Mpc) away, and at such distances we see an effect akin to time travel. When we look at a galaxy millions of light-years away, we do not see it as it is now but as it was millions of years ago when its light began the journey toward Earth. Thus, when we look at a distant galaxy, we look back into the past by an amount called the **look-back time,** a time in years equal to the distance to the galaxy in light-years.

You may have experienced look-back time if you have ever made a long-distance phone call carried by satellite or watched a TV newscaster interview someone on the other side of the world via satellite. A half-second delay occurs as a radio signal carries a question 23,000 miles out to a satellite then back to Earth and then carries the answer out to the satellite and again back to Earth. That half-second look-back delay can make people hesitate on long-distance phone calls and produces a seemingly awkward delay in intercontinental TV interviews.

The look-back time to nearby objects is usually not significant. The look-back time across a football field is a tiny fraction of a second. The look-back time to the moon is 1.3 seconds, to the sun only 8 minutes, and to the nearest star about 4 years. The Andromeda Galaxy has a look-back time of about 2 million years, a mere eye blink in the lifetime of a galaxy. But when we look at more distant galaxies, the look-back time becomes an appreciable part of the age of the universe. We will see evidence in Chapter 18 that the universe began 14 billion years ago. Thus, when we look at the most distant visible galaxies, we are looking back 10 billion years to a time when the universe may have been significantly different. This effect will be important in our discussions in this and the next chapter.

THE HUBBLE LAW

Although astronomers find it difficult to measure the distance to a galaxy, they often estimate such distances using a simple relationship that was first noticed at about the same time astronomers were beginning to understand the nature of galaxies.

In 1913, V. M. Slipher at Lowell Observatory reported on the spectra of faint, nebulous objects in the sky. Their spectra seemed to be composed of a mixture of stellar spectra: Some had Doppler shifts that suggested rotation; most had red shifts as if they were receding; and the faintest had the largest red shifts. Within two decades, astronomers concluded that the faint objects were galaxies similar to our own Milky Way and that the galaxies are indeed receding from us in a general expansion.

In 1929, Edwin Hubble and Milton Humason published a graph that plotted the velocity of recession versus distance for a number of galaxies. The points in the graph fell along a straight line (▌Figure 16-5). The straight-line relation from Hubble's diagram can be written as the simple equation

$$V = Hd$$

That is, the velocity of recession V equals the distance d in millions of parsecs times the constant H. This relation between red shift and distance is known as the **Hubble law,** and the constant H is known as the **Hubble constant.**

Modern attempts to measure H have been difficult because of the uncertainty in the distances to galaxies.

FIGURE 16-5
Edwin Hubble's first diagram of the velocities and distances of galaxies did not probe very deeply into space. It did show, however, that the galaxies are receding from one another.

During the last decades of the 20th century, astronomers were caught up in a great controversy over the value of H. Different teams measuring distance differently got answers that ranged from 50 to 100 km/s/Mpc.* The controversy was settled by better observations made with the Hubble Space Telescope and with new-generation giant telescopes on Earth. The value of H appears to be about 70 km/s/Mpc.

The Hubble law is important because it is commonly interpreted to show that the universe is expanding. In Chapter 18, we will discuss this expansion, but here we will use the Hubble law as a practical way to estimate the distance to a galaxy.

Simply stated, a galaxy's distance equals its velocity of recession divided by H. For example, if a galaxy has a radial velocity of 700 km/s, and H is 70 km/s/Mpc, then the distance to the galaxy is 700 divided by 70, or about 10 Mpc. This makes it relatively easy to estimate the distances to galaxies, because a large telescope can photograph the spectrum of a galaxy and determine its velocity of recession even when it is too distant to have visible distance indicators.

DIAMETER AND LUMINOSITY

The distance to a galaxy is the key to finding its diameter and its luminosity. We can easily photograph a galaxy and measure its angular diameter in seconds of arc. If we know the distance, we can use the small-angle formula (Chapter 3) to find its linear diameter. Also, if we measure the apparent magnitude of a galaxy, we can use the distance to find its absolute magnitude and thus its luminosity (Chapter 9).

*H has the units of a velocity divided by a distance. These are usually written as km/s/Mpc, meaning km/s per Mpc.

Visual-wavelength image

FIGURE 16-6

The small galaxy cluster known as Hickson Compact Group 87 contains three spiral galaxies and one elliptical. That means the four galaxies lie at the same distance from Earth, and the differences in size among the galaxies are real. Some are much bigger than others. The edge-on spiral at bottom is very large compared to the small spiral at the center. *(Hubble Heritage Team, STScI/AURA/NASA)*

The results of such observations show that galaxies differ dramatically in size and luminosity. Irregular galaxies tend to be small, 1 to 25 percent the size of our galaxy, and of low luminosity. Although they are common, they are easy to overlook. Our Milky Way Galaxy is large and luminous compared with most spiral galaxies, though we know of a few spiral galaxies that are even larger and more luminous—nearly four times larger in diameter and about 10 times more luminous. Elliptical galaxies cover a wide range of diameters and luminosities. The largest, called giant ellipticals, are five times the size of our Milky Way. But many elliptical galaxies are very small, dwarf ellipticals, only 1 percent the diameter of our galaxy.

To put galaxies in perspective, we can use an analogy. If our galaxy were an 18-wheeler, the smallest dwarf galaxies would be the size of pocket-size toy cars, and the largest giant ellipticals would be the size of jumbo jets. Even among spiral galaxies that looked similar, the sizes could range from that of a go-cart to that of a small yacht (❙ Figure 16-6).

Clearly, the diameter and luminosity of a galaxy do not determine its type. Some small galaxies are irregular, and some are elliptical. Some large galaxies are spiral, and some are elliptical. We need more data before we can build a theory for the origin and evolution of galaxies.

Of the three basic parameters that describe a galaxy, we have found two—diameter and luminosity. The third, as was the case for stars, is more difficult to discover.

MASS

Although the mass of a galaxy is difficult to determine, it is an important quantity. It tells us how much matter the galaxy contains, which gives us clues to the galaxy's origin and evolution. In this section, we will examine two fundamental ways to find the masses of galaxies.

One way to find the mass of a galaxy is to watch it rotate. We know the stars in the outer parts of the galaxy are in orbit, so we can use Newton's laws to find the mass. All we need to know is the radius of the galaxy in AU and the orbital period of the stars in years at the galaxy's outer edge. Then we can use Newton's version of Kepler's third law to find the mass just as we found the mass of binary stars in Chapter 9.

Recall from the previous section that it is easy to find the size of a galaxy if we know the distance. We just measure its angular size and use the small-angle formula (Chapter 3).

We can find the orbital velocity of the stars from the Doppler effect. If we focused the image of the galaxy on the slit of a spectrograph, we would see a bright spectrum formed by the bright nucleus of the galaxy, but we would also see fainter emission lines produced by ionized gas in the disk of the galaxy. Because the galaxy rotates, one side moves away from us and one side moves toward us, so the emission lines would be red shifted on one side of the galaxy and blue shifted on the other side. We could measure those changes in wavelength, use the Doppler formula to find the velocities, and plot a diagram showing the velocity of rotation at different distances from the center of the galaxy—a diagram called a **rotation curve.** The artwork in ❙ Figure 16-7a shows the process of creating a rotation curve, and Figure 16-7b shows a real galaxy, its spectrum, and its rotation curve.

Of course, if we know the orbital velocity and the radius of a star's orbit around a galaxy, we can calculate its orbital period. The circumference of the orbit (2π times the radius) divided by the orbital velocity equals the orbital period. So knowing the velocity and size of an orbit easily tells us the orbital period, and from that we can find the mass.

The rotation curve of a galaxy contains all the information we need to find the mass of the galaxy, and the method is known as the **rotation curve method.** This is the most accurate way to find the mass of a galaxy, but it works only for the nearer galaxies, whose rotation curves can be observed. More distant galaxies appear so small we cannot measure the radial velocity at different points along the galaxy. Also, recent studies of our own galaxy and others show that the outer parts of the rotation curve do not decline to lower

FIGURE 16-7

(a) In the artwork in the upper half of this diagram, the astronomer has placed the image of the galaxy over a narrow slit so that light from the galaxy can enter the spectograph and produce a spectrum. A very short segment of the spectrum shows an emission line red shifted on the receding side of the rotating galaxy and blue shifted on the approaching side. Converting these Doppler shifts into velocities, the astronomer can plot the galaxy's rotation curve (right). (b) Real data are shown in the bottom half of this diagram. Galaxy NGC2998 is shown over the spectrograph slit, and the segment of the spectrum includes three emission lines. *(Courtesy Vera Rubin)*

velocities (Figure 15-11). This shows that the outermost visible parts of some galaxies do not travel more slowly and tells us that the galaxies contain large amounts of mass outside this radius, perhaps in extended galactic coronas like the one that seems to surround our own galaxy (see Chapter 15). Because the rotation curve method can be applied only to nearby galaxies and because it cannot determine the masses of galactic coronas, we must look at the second way to find the masses of galaxies.

The **cluster method** of finding galactic mass depends on the motions of galaxies within a cluster. If we measure the radial velocities of many galaxies in a cluster, we find that some velocities are larger than others because of the orbital motions of the individual galaxies in the cluster. Given the range of velocities and the size of the cluster, we can ask how massive a cluster of this size must be to hold itself together with this range of velocities. Dividing the total mass of the cluster by the number of galaxies in the cluster yields the average mass of the galaxies. This method contains the built-in assumption that the cluster is not flying apart. If it is, our result is too large. Because it seems likely that most clusters are held together by their own gravity, the method is probably valid.

A related way of measuring a galaxy's mass is called the **velocity dispersion method.** It is really a ver-

sion of the cluster method. Instead of observing the motions of galaxies in a cluster, we observe the motions of matter within a galaxy. In the spectra of some galaxies, broad spectral lines indicate that stars and gas are moving at high velocities. If we assume the galaxy is bound by its own gravity, we can ask how massive it must be to hold this moving matter within the galaxy. This method, like the one before, assumes that the system is not coming apart.

Measuring the masses of galaxies tells us two things. First the range of masses is wide—from 10^{-6} as much as that of the Milky Way Galaxy to 50 times more (❚ Table 16-1). And second, there is more to a galaxy than meets the eye.

TABLE 16-1
The Properties of Galaxies*

	Elliptical	Spiral	Irregular
Mass	0.0001–50	0.005–2	<0.0005–0.15
Diameter	0.01–5	0.2–1.5	0.05–0.25
Luminosity	0.00005–5	0.005–10	0.00005–0.1

*In units of the mass, diameter, and luminosity of the Milky Way.

SUPERMASSIVE BLACK HOLES IN GALAXIES

Rotation curves show the motions of the outer parts of a galaxy, but it is possible to detect the Doppler shifts of stars orbiting close to the centers of galaxies. Although these motions are not usually shown on rotation curves, they reveal something astonishing.

Measurements show that the stars near the centers of most galaxies are orbiting very rapidly. To hold stars in such small, short-period orbits, the centers of galaxies must contain masses of a million to a few billion solar masses. Yet no object is visible, so most astronomers believe that the nuclei of galaxies contain supermassive black holes. We saw in Chapter 15 that the Milky Way contains a supermassive black hole at its center. Evidently that is typical of galaxies.

Such supermassive black holes cannot be the remains of a dead star. Rather, these black holes must have formed as the galaxy formed or they must have accumulated over billions of years as matter sank into the centers of the galaxies. Measurements show that the masses of these supermassive black holes are related to the mass of nuclear bulges. A galaxy with a large nuclear bulge has a supermassive black hole whose mass is greater than that in a galaxy with a small nuclear bulge. Rare galaxies lacking nuclear bulges also lack supermassive black holes (⏸ Figure 16-8). This suggests that the supermassive black holes formed long ago as the galaxies formed. Of course, matter has continued to drain into the black holes, but they do not appear to have grown dramatically since they formed.

A billion-solar-mass black hole sounds like a lot of mass, but note that it is roughly 1 percent of the mass of a galaxy. The 2.6 million-solar-mass black hole at the center of the Milky Way Galaxy contains only a thousandth of a percent of the mass of the galaxy. In the next chapter we will discover that these supermassive black holes can produce titanic eruptions, but they represent only a small fraction of the mass of a galaxy.

DARK MATTER IN GALAXIES

Given the size and luminosity of a galaxy, we can make a rough guess as to the amount of matter it should contain. We know how much light stars produce, and we know about how much matter there is between the stars, so it is quite possible to estimate very roughly the mass

Visual-wavelength image

FIGURE 16-8

The Pinwheel Galaxy is near our Milky Way Galaxy in space, and it can be studied in detail. The velocities at the center of the Pinwheel Galaxy are low, showing that it does not contain a supermassive black hole at its center. It also lacks a nuclear bulge, and that confirms the observation that the mass of a supermassive black hole is related to the size of a galaxy's nuclear bulge. *(Bill Schoening/AURA/NOAO/NSF)*

of a galaxy from its luminosity. But when astronomers measure the masses of galaxies, they often find that the measured masses are much too large. We discovered this effect in Chapter 15 when we studied the rotation curve of our own galaxy. This seems to be true of most galaxies. Measured masses of galaxies amount to 10 to 100 times more mass than we can see. Because this mass is not visible, it is called dark matter.

X-ray observations reveal more evidence of dark matter. X-ray images of galaxy clusters show that many of them are filled with very hot, low-density gas. The amount of gas present is much too small to account for the dark matter. Rather, the gas is important because it is very hot and its rapidly moving atoms have not leaked away. Evidently the gas is held in the cluster by a strong gravitational field. To provide enough gravity to hold the hot gas, the cluster must contain much more matter than what we see. The detectable galaxies in the Coma cluster, for instance, amount to only a small fraction of the total mass of the cluster (⏸ Figure 16-9).

Dark matter is not an insignificant issue. Observations of galaxies and clusters of galaxies tell us that approximately 95 percent of the matter in the universe is

Visual-wavelength image

X-ray image

a

b

FIGURE 16-9

(a) The Coma cluster of galaxies contains at least 1000 galaxies and is especially rich in E and S0 galaxies. Two giant galaxies lie near its center. Only the central area of the cluster is shown in this image. If the cluster were visible in the sky, it would span 8 times the diameter of the full moon. *(L. Thompson © NOAO)* (b) In false colors, this X-ray image of the Coma cluster shows it filled and surrounded by hot gas. The box outlines the area of the cluster shown in part a. *(Simon D. M. White, Ulrich G. Briel, and J. Patrick Henry)*

dark matter; the universe we see—the kind of matter that we and the stars are made of—has been compared to the foam on an invisible ocean. We will cite further evidence of dark matter when we discuss cosmology in Chapter 18.

Dark matter is difficult to detect, and it is even harder to explain. Some astronomers have suggested that dark matter consists of low-luminosity white dwarfs and brown dwarfs scattered through the halos of galaxies. Both observation and theory support the idea that galaxies have massive extended halos, and searches for white dwarfs and brown dwarfs in the halo of our galaxy have been successful. Nevertheless, the searches have not turned up enough of these low-luminosity objects to make up all of the dark matter.

The dark matter can't be hidden in vast numbers of black holes and neutron stars, because we don't see the X rays these objects would emit. Remember, we need roughly 10 times more dark matter than visible matter, and that many black holes and neutron stars should produce X rays that are easy to detect.

Because observations seem to tell us the dark matter can't be hidden in normal objects, some theorists have suggested that the dark matter is made up of unexpected forms of matter. Until recently neutrinos were thought to be massless, but studies now suggest they have a very small mass. Thus they may represent part of the dark matter.

Dark matter remains one of the fundamental unresolved problems of modern astronomy. We will return to this problem in Chapter 18 when we try to understand how dark matter affects the nature of the universe, its past, and its future.

REVIEW CRITICAL INQUIRY

Why do we have to know the distance to a galaxy to find its mass?

Often the critical analysis of a problem is complicated by interrelationships between factors. Usually a bit of care is enough to reveal those connections. To find the mass of a galaxy, we find the size of the orbits and the orbital periods of stars at the galaxy's outer edge, and then use Kepler's third law to find the mass inside the orbits. Measuring the orbital velocity of the stars as the galaxy rotates is easy if we can obtain a rotation curve from a spectrum. But we must also know the radii of the stars' orbits in meters or astronomical units. That is where the distance comes in. To find the radii of the orbits, we must know the distance to the galaxy. Then we can use the small-angle formula to convert the radius in seconds of arc into a radius in parsecs or AU. If our measurement of the distance to the galaxy isn't accurate, we will get inaccurate radii for the orbits, and we will compute an inaccurate mass for the galaxy.

Many different measurements in astronomy depend on the calibration of the distance scale. For example, what would happen to our measurements of the diameters and luminosities of the galaxies if astronomers discovered that the Cepheid variable stars were slightly more luminous than had been believed?

We have gathered the basic data to help us characterize galaxies. That is, we can describe their general characteristics. Now we turn to the more difficult and more interesting question. How did galaxies get to be the way they are?

16-3 THE EVOLUTION OF GALAXIES

Our goal in this chapter has been to build a theory to explain the evolution of galaxies. In Chapter 15 we discussed the origin of our own Milky Way Galaxy; presumably, other galaxies formed similarly. But why did some galaxies become spiral, some elliptical, and some irregular? An important clue to that mystery lies in the clustering of galaxies.

CLUSTERS OF GALAXIES

Single galaxies are rare. Most occur in clusters containing a few to a few thousand galaxies in a volume 1 to 10 Mpc across (❚ Figure 16-10). Our Milky Way Galaxy is a member of a cluster containing slightly over three dozen galaxies, and surveys have cataloged over 2700 other clusters within 4 billion ly.

Visual-wavelength image

FIGURE 16-10

The Hercules galaxy cluster is named after the constellation in which it is found. It is a small cluster containing roughly a hundred galaxies both elliptical and spiral. It lies over 360 million light-years from our galaxy. *(NOAO/AURA/NSF)*

For purposes of our study, we can sort clusters of galaxies into rich clusters and poor clusters. **Rich clusters** contain a thousand or more galaxies, many elliptical, scattered through a volume roughly 3 Mpc (10^7 ly) in diameter. Such a cluster is nearly always condensed; that is, the galaxies are concentrated toward the cluster center. And, at the center, such clusters often contain one or more giant elliptical galaxies.

The Virgo cluster is an example of a rich cluster. It contains over 2500 galaxies located about 17 Mpc (55 million ly) away. The Virgo cluster, like most rich clusters, is centrally condensed and contains the giant elliptical galaxy M87 at its center (page 338).

X-ray observations have found that many of these rich clusters are filled with a hot gas—an intracluster medium. Some of this gas has presumably been driven out of galaxies by supernovae explosions, but much of it appears to be left over from the earliest generations of galaxy formation.

Poor clusters contain fewer than a thousand (and often only a few) galaxies spread through a region that can be as large as a rich cluster. That means the galaxies are more widely separated.

Our Milky Way Galaxy is a member of a poor cluster known as the Local Group (❚ Figure 16-11a). The total number of galaxies in the Local Group is uncertain, but it probably contains more than three dozen galaxies scattered irregularly through a volume roughly 1 Mpc in diameter. Of the brighter galaxies, 15 are elliptical, 4 are spiral, and 13 are irregular.

The total number of galaxies in the Local Group is uncertain because some lie in the plane of the Milky Way Galaxy and are difficult to detect. For example, a small dwarf galaxy, known as the Sagittarius Dwarf, has been found on the far side of our own galaxy where it is almost totally hidden behind the stars clouds of Sagittarius (Figure 16-11b). Astronomers in Dwingeloo, the Netherlands, located another spiral galaxy, dubbed Dwingeloo 1, hidden behind the Milky Way (Figure 16-11c). Dwingeloo 1 and its two neighbor galaxies, Maffie I and Maffie II, are a bit too far away to be part of our Local Group, but the large spiral galaxy known as the Andromeda Galaxy shown in Figure 15-1b is always included. Our Local Group surely contains other small galaxies hidden behind the stars, gas, and dust of the Milky Way.

The Local Group illustrates the subclustering found in poor galaxy clusters. The two largest galaxies, the Milky Way and the Andromeda Galaxy, are the centers of two subclusters. The Milky Way is accompanied

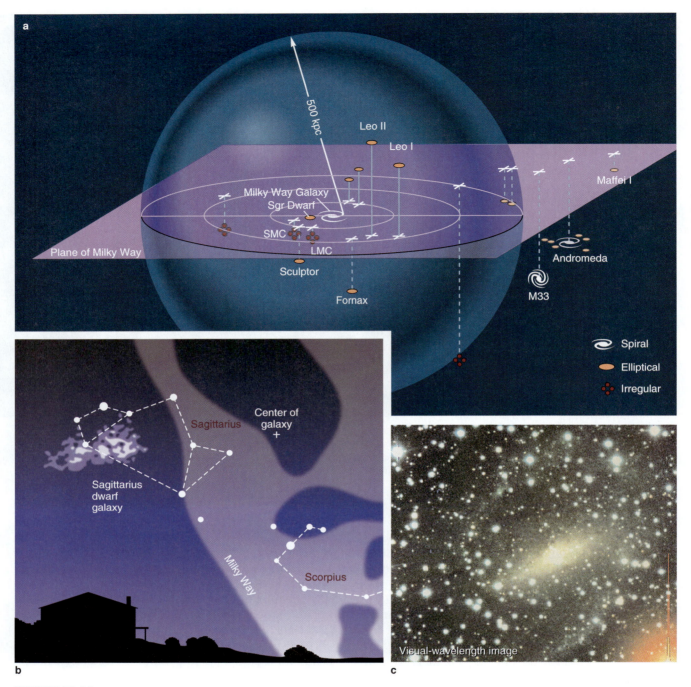

FIGURE 16-11

(a) The Local Group. Our galaxy is located at the center of this diagram. The vertical lines giving distances from the plane of the Milky Way are solid above the plane and dashed below. (b) The Sagittarius Dwarf Galaxy (Sgr Dwarf) lies on the other side of our galaxy. If we could see it in the sky, it would be 17 times larger than the full moon. (c) Spiral galaxy Dwingeloo 1, as large as our own galaxy, was not discovered until 1994 because it is hidden behind the Milky Way. Note the faint spiral pattern. Nearly all individual stars here are in our galaxy. *(Courtesy Shaun Hughes and Steve Maddox, Royal Greenwich Observatory; and Ofer Lahav and Andy Loan, University of Cambridge)*

by the Magellanic Clouds and seven other dwarf galaxies. The Andromeda Galaxy is attended by more dwarf elliptical galaxies (two of which are visible in Figure 15-1b) and a small spiral, M33 (Figure 16-8). The Andromeda Galaxy and our Milky Way Galaxy, with their retinue of smaller galaxies, are moving toward each other and will almost certainly collide in a few billion years.

The critical clue hidden in the clusters of galaxies is the relative abundance of the different types of galax-

Visual-wavelength images

FIGURE 16-12

The colliding galaxies NGC4038 and NGC4039 are known as the Antennae because their long curving tails resemble the antennae of an insect. Earth-based photos (left) show little detail, but a Hubble Space Telescope image (right) reveals the collision of two galaxies producing thick clouds of dust and raging star formation creating roughly a thousand massive star clusters such as the one at the top (arrow). Such collisions between galaxies are common. (*Brad Whitmore, STScI, and NASA*)

ies. In general, rich clusters tend to contain 80 to 90 percent E and S0 galaxies and a few spirals. Poor clusters contain a larger percentage of spirals; and, among isolated galaxies, those that are not in clusters, 80 to 90 percent are spirals. Galaxies that are crowded together tend to be E or S0 rather than spiral. Somehow the environment around a galaxy helps determine its type.

Astronomers suspect that collisions between galaxies are an important process. In a rich cluster, the galaxies are crowded much closer together, and they must collide with each other more often than galaxies in a poor cluster. Could such collisions explain the excess of elliptical galaxies in the crowded rich clusters? Astronomers have discovered that galaxy smashups are both entertaining and important.

COLLIDING GALAXIES

Galaxies should collide fairly often. The average separation between galaxies is only about 20 times their diameter. Like two blindfolded elephants blundering about at random under a circus tent, galaxies should bump into each other once in a while. Stars, on the other hand, almost never collide. In the region of the galaxy near the sun, the average separation between stars is about 10^7 times their diameter. Thus, collision between two stars is about as likely as collision between two blindfolded gnats flitting about at random in a football stadium.

Large telescopes reveal hundreds of galaxies that appear to be colliding with other galaxies. One of the most famous pairs of colliding galaxies, NGC4038 and

NGC4039, are called the Antennae because the long tails resemble the antennae of an insect (❚ Figure 16-12). In addition to having tails, some interacting galaxies are connected by bridges of gas, dust, and stars.

Galaxies do not need to collide directly to affect each other. If two galaxies pass near each other, their gravitational fields distort their shapes, drawing out long tails of gas, dust, and stars. Sometimes bridges of material are created that connect the interacting galaxies. These are tidal effects. The gravitation of one galaxy causes tides that distort the other galaxy and vice versa. Computer models show that even when galaxies rush past each other rapidly without interpenetrating, the distortions can be severe. Astronomers have begun to refer to this as "harassment" to distinguish it from direct collisions. One of the most famous pairs of interacting galaxies is called the Mice because of the tail-like deformities flung out as the galaxies whip around their common center of mass and tidally harass each other (❚ Figure 16-13).

A collision between galaxies can last hundreds of millions of years, but we can watch it happen in computer models. The Whirlpool Galaxy, M51, provides a good example of such a model (❚ Figure 16-14). Models show that the galaxies passed near each other but did not actually penetrate. The smaller galaxy passed behind the larger galaxy.

Unusual **ring galaxies** consist of a bright nucleus surrounded by a ring (❚ Figure 16-15). Models show that they are produced by a galaxy passing roughly perpendicularly through the disk of the larger galaxy. Indeed, many ring galaxies have nearby companions.

Visual-wavelength image

a

b

FIGURE 16-13

(a) The Mice are a pair of colliding galaxies with peculiar tails (shown here in a false-color image). *(NOAO)* (b) A computer model of a close encounter between normal galaxies shows how the centers of mass whip around each other, producing tails similar to the Mice. Colors of dots show the original membership of the stars and reveal that the interacting galaxies have exchanged mass. *(Illustration by Allen Beechel)*

As colliding galaxies approach each other, they carry tremendous orbital momentum. Models show that tidal forces can convert some of the orbital motion into random motion among the stars, thus robbing the galaxies of orbital momentum. Such galaxies may lose enough orbital momentum that they fall back together and merge.

A small galaxy merging with a larger galaxy will be pulled apart gradually, and its stars will spread through the larger galaxy in a process called **galactic cannibalism.** Computer simulations of mergers show that galaxies throw off shells of stars as the nuclei of the two galaxies spiral into each other, and such shells have been observed around real galaxies (❙ Figure 16-16).

Evidence of collisions often appears in the motions inside a galaxy. The tails on NGC7252 (❙ Figure 16-17a) resemble the tails on mice, and when the Hubble Space Telescope imaged the center of the galaxy it found a small, backward-spinning spiral. Evidently, NGC7252 was created about a billion years ago when two oppositely spinning galaxies collided and merged. The apparently normal galaxy M64 reveals its secrets in radio maps (Figure 16-17b). The center is rotating backward compared to the outer galaxy—the product of a merger between counter-rotating galaxies.

Evidence of past mergers also appears in the multiple nuclei in some giant elliptical galaxies (Figure 16-17c). These galaxies are often located in crowded rich clusters of galaxies, and the extra nuclei appear to be the dense centers of smaller galaxies that have been cannibalized and only partly digested.

At infrared wavelengths, astronomers are able to image galaxies that are a hundred times more luminous than our Milky Way Galaxy, but they are deeply shrouded in dust and thus are dim at visible wavelengths. They are evidently undergoing tremendous bursts of star formation that are producing dust and warming it with starlight until the warm dust emits vast amounts of infrared radiation. The best images of these **ultra-luminous infrared galaxies** show distorted bridges and tails. Many of these galaxies are undergoing multiple collisions as three or more galaxies are merging and presumably triggering firestorms of star formation.

Our own Milky Way Galaxy is an unreformed cannibal. The ring of stars around its outer edge (Figure 15-16) and streams of misplaced population I stars in its halo are traces of smaller galaxies it has digested. It is currently pulling the Sagittarius Dwarf galaxy (Figure 16-11b) apart, and astronomers find elongated clouds of gas and stars leading from the two Magellanic Clouds back to the disk of our galaxy. They must have passed through the disk recently and will eventually fall back and be cannibalized. Perhaps the Magellanic Clouds are irregular galaxies because they have been severely harassed by the Milky Way Galaxy.

We can find further evidence of harassment among galaxies by looking at the Virgo cluster. Studies with the Hubble Space Telescope and with very large telescopes on Earth have identified free-floating stars and planetary nebulae between the galaxies in the Virgo cluster. Evidently a significant number of the stars do not lie inside the galaxies but have been freed to wander alone through the cluster as galaxies have harassed each other.

The new evidence for collisions, harassment, and cannibalism among galaxies is very strong. These are the processes we need to understand in order to tell the story of the galaxies.

Galaxy Interactions and Spiral Arms

In this computer model, two uniform disk galaxies pass near each other.

The small galaxy passes behind the larger galaxy so they do not actually collide.

Tidal forces deform the galaxies and trigger the formation of spiral arms.

The upper arm of the large galaxy passes in front of the small galaxy.

A photo of the well-known Whirlpool Galaxy resembles the computer model.

Visual

FIGURE 16-14

An interaction between two galaxies that takes hundreds of millions of years can be followed step by step in a computer model. *(Illustration by Allen Beechel; photo: NOAO)*

THE ORIGIN AND EVOLUTION OF GALAXIES

The test of any scientific understanding is whether we can put all the evidence and theory together to tell the history of the objects we study. Can we describe the ori-

Visual-wavelength image

FIGURE 16-15

The Cartwheel Galaxy was once a normal spiral galaxy, but a few hundred thousand years ago one or the other of the two smaller galaxies at the right passed through the center of the larger galaxy. Tidal forces generated an expanding wave of star formation and a central core of newborn stars heavily reddened by dust. Such ring galaxies appear to be the fleeting consequences of bulls-eye collisions. Soon after the collision, the violent star formation exhausts itself, and normal spiral arms may re-form. This is suggested in the Cartwheel Galaxy by the faint spiral structure inside the outer ring. *(Kirk Borne, STScI and NASA)*

gin and evolution of the galaxies? Just a few decades ago, it would have been impossible, but the evidence from space telescopes and new-generation telescopes on Earth combined with advances in computer modeling and theory allow astronomers to outline the story of the galaxies.

As we begin, we should eliminate a few older ideas immediately. It is easy to imagine that galaxies evolve from one type to another. But an elliptical galaxy cannot become a spiral galaxy or an irregular galaxy, because ellipticals contain almost no gas and dust from which to make new stars. Elliptical galaxies can't be young galaxies. We can also argue that spiral and irregular galaxies cannot evolve into elliptical galaxies because spiral and irregular galaxies contain both young and old stars. The old stars mean that spiral and irregular galaxies can't be young. The galaxy classes tell us something important, but a single galaxy does not change from one class to another any more than a cat can change into a dog.

Another old idea held that galaxies that form from rapidly rotating clouds of gas would have lots of angular momentum and would contract slowly to form disk-shaped spiral galaxies. Clouds of gas that rotated less rapidly would contract faster, form stars quickly, use up all the gas and dust, and become elliptical galaxies. The evidence clearly shows that galaxies didn't form

In this computer model of two galaxies interacting, only one galaxy is shown.

The galaxy spirals around the center of mass shown by the cross.

The motion of the galaxy throws off shells of stars as the galaxies merge.

Models predict that galaxies formed of mergers could lie inside shells of stars.

Faint shells of stars shown in this negative image surround elliptical galaxy NGC3923.

▶ Visual

FIGURE 16-16

As two galaxies merge, they spiral around their common center of mass, which could produce shells of stars. Such faint shells are seen around some galaxies. *(Model courtesy François Schweizer and Alan Toomre; photo courtesy David Malin, Anglo-Australian Telescope Board)*

that are very luminous in the infrared because a collision has triggered a burst of star formation that is heating the dust (▌ Figure 16-18). The warm dust reradiates the energy in the infrared. Supernovae in such a galaxy can blow away any remaining gas and dust that don't get used up making stars. A few collisions and mergers could leave a galaxy with no gas and dust from which to make new stars. Astronomers now suspect that most ellipticals are formed by the merger of at least two or three galaxies.

In contrast, spirals seem never to have suffered major collisions. Their thin disks are delicate and would be destroyed by tidal forces in a collision with a massive galaxy. Also, they retain plenty of gas and dust and continue making stars. Of course, spiral galaxies can cannibalize smaller galaxies. We see evidence in our own galaxy, and some astronomers suspect that much of the halo consists of cannibalized galaxies. But our Milky Way Galaxy has never collided with a similarly large galaxy. Such a collision would trigger star formation using up gas and dust, scramble the orbits of stars, and destroy the thin disk. In a few billion years, our galaxy may merge with the approaching Andromeda Galaxy, and the final result may be an elliptical galaxy.

Barred spiral galaxies may be the product of tidal interactions. Mathematical models show that bars are not stable and eventually dissipate. It may take tidal interactions with other galaxies to regenerate the bars. Because roughly half of all spiral galaxies have bars, we can suspect that these tidal interactions are common.

Other processes can alter galaxies. The S0 galaxies retain a disk shape but lack gas and dust. They may have lost gas and dust during a starburst, but another process involves their motion through the gas in a cluster of galaxies. A galaxy orbiting through such a gas would encounter a tremendous wind blowing its gas and dust away. A disk-shaped spiral galaxy could be reduced to an S0 galaxy. This may also explain the small dwarf ellipticals. They are too small to be the product of galaxy mergers, but they have somehow lost their gas and dust. In contrast, the irregular galaxies may be small fragments ripped from larger galaxies.

Other factors must influence the evolution of galaxies. The hot gas in galaxy clusters warns us that galaxies do not form in isolation. Cooler gas clouds may fall into galaxies and add material for star formation. Also, we need to consider differences in the initial clouds of gas from which galaxies form. Astronomers are just beginning to understand which factors are important in the birth of galaxies.

THE FARTHEST GALAXIES

Observations with the largest and most sophisticated telescopes are taking us back to the age of galaxy formation. At great distances the look-back time is so great that we see the galaxies as they were long ago, and we

and evolve in isolation. Collisions and mergers dominate the history of the galaxies.

The ellipticals appear to be the product of galaxy mergers, which triggered star formation that used up the gas and dust. In fact, we see many **starburst galaxies**

FIGURE 16-17

Evidence of mergers. (a) The twisted shape of NGC7252 suggests a collision, and a Hubble Space Telescope image reveals a small spiral of young stars spinning backward in the heart of the larger galaxy. This is thought to be the remains of two oppositely rotating galaxies that merged about a billion years ago. *(François Schweizer, Carnegie Institution of Washington, and Brad Whitmore, STScI)* (b) The apparently normal galaxy M64 is shown at left in a radio map that shows radial velocity as color. Red regions are receding from us and blue regions approaching. At the center of the galaxy, the rotation is reversed, evidently the consequence of a merger between counter-rotating galaxies. *(Robert Braun, NRAO)* (c) Giant elliptical galaxies in rich clusters sometimes have multiple nuclei, thought to be the densest parts of smaller galaxies that have been absorbed and partly digested. *(Michael J. West)*

discover that there were more spirals then and fewer ellipticals. On the whole, galaxies long ago were more compact and more irregular than they are now. We can even see that galaxies were closer together long ago; about a third of all distant galaxies are in close pairs, but only 7 percent of nearby galaxies are in pairs. The observational evidence clearly supports the hypothesis that galaxies have evolved by merger.

At great distances, astronomers see tremendous numbers of small, blue, irregularly shaped galaxies that have been called blue dwarfs. The blue color tells us they are rapidly forming stars, but the role of these blue dwarfs in the formation of galaxies is not clear. They may be clouds of gas and stars that were eventually absorbed in the formation of larger galaxies, or they may have used up their gas and dust and have faded to obscurity.

At the limits of the largest telescopes, astronomers see faint red galaxies (❚ Figure 16-19). They look red to us because of their great red shifts. The floods of ultraviolet light emitted by these star-forming galaxies have been shifted into the far red part of the spectrum. We must be seeing them at a time when the universe was only a billion years old. These galaxies appear to be among the first to begin shining after the beginning of the universe, a story we will discuss in Chapter 18.

FIGURE 16-18

Starburst galaxy NGC3310 is forming stars at a prodigious rate. Several hundred star clusters, some containing as many as a million stars, outline the spiral arms in a blue glow. From the blue color of the clusters, astronomers conclude that the burst of star formation began about 100 million years ago, probably when a companion galaxy collided with NGC3310. *(Hubble Heritage Team, STScI/ AURA/NASA)*

Before we can consider the universe as a whole, we must examine the galaxies from a different perspective. Some galaxies are suffering tremendous eruptions as their cores blast radiation and matter outward. In the next chapter, we will discover that these peculiar galaxies are closely related to collisions and mergers between galaxies.

SUMMARY

Astronomers did not realize that the spiral nebulae visible in larger telescopes were other galaxies until Edwin Hubble detected Cepheid variable stars in a few nearby galaxies. That announcement in 1924 showed that the universe was filled with galaxies much like our own Milky Way.

We can divide galaxies into three classes—elliptical, spiral, and irregular—with subclasses giving the galaxy's shape or the amount of gas and dust present. The galaxy types appear to reflect different histories of star formation. The elliptical galaxies contain little gas and dust and cannot make many new stars. The spiral and irregular galaxies contain large amounts of gas and dust and are still forming stars.

To measure the properties of galaxies, we must first find their distances. For the nearer galaxies, we can judge distances using distance indicators, objects whose luminosities are known. The best of these standard candles are the Cepheid

Visual-wavelength image

FIGURE 16-19

Arrows point to three very distant red galaxies. The look-back time to these galaxies is so large we see them when the universe was only a billion years old. Such highly red-shifted galaxies are believed to be among the first to form stars and begin shining after the beginning of the universe about 14 billion years ago. *(NASA, H.-J. Yan, R. Windhorst and S. Cohen, Arizona State University)*

variable stars. Once the distances to some galaxies are known, astronomers can calibrate other distance indicators to piece together a distance scale. Some of the most-used distance indicators are globular clusters, planetary nebulae, and supernovae. For the most distant galaxies, astronomers must depend on a calibration of the luminosity of galaxies themselves.

The Hubble law shows that the radial velocity of a galaxy is proportional to its distance. Thus, we can use the Hubble law to estimate distances. The galaxy's radial velocity divided by the Hubble constant is its distance in megaparsecs.

The masses of galaxies can be measured in two ways—the rotation curve method and the cluster method. The first method is the most accurate but it is applicable only to nearby galaxies.

Many measurements of the mass of galaxies show that galaxies contain many times more mass than what we can see. This dark matter amounts to roughly 90 percent of the matter in the universe. Although some mass has been found as brown dwarfs, white dwarfs, neutrinos with mass, and so on, most of the dark matter seems to be made up of an as yet unrecognized form of matter.

The origin and evolution of galaxies is not fully understood, but it is clear that interaction between galaxies is important. Close encounters, collisions, and mergers of two galaxies can cause dramatic tidal distortions and bursts of star formation. Stripping can remove gas and dust from a galaxy. These processes appear to be common among galaxies. The Magellanic Clouds are being cannibalized by our own Milky Way Galaxy.

Normal elliptical galaxies are apparently formed by the merger of spiral galaxies. The average, bright elliptical galaxy may contain the merged remains of two or three spiral galaxies. Dwarf ellipticals are different systems and appear to be small galaxies stripped of gas and dust by their motion through the intracluster gas and by interactions with larger galaxies. According to this theory, the spiral galaxies have not experienced collisions with large galaxies very often since they were formed.

NEW TERMS

spiral nebula	Hubble constant (H)
island universe	rotation curve
Shapley–Curtis Debate	rotation curve method
elliptical galaxy	cluster method
spiral galaxy	velocity dispersion method
barred spiral galaxy	rich cluster
irregular galaxy	poor cluster
megaparsec (Mpc)	ring galaxy
distance indicator	galactic cannibalism
standard candle	ultraluminous infrared
distance scale	galaxy
look-back time	starburst galaxy
Hubble law	

REVIEW QUESTIONS

1. If a civilization lived on a planet in an E0 galaxy, do you think it would have a Milky Way? Why or why not?

2. Why can't the evolution of galaxies go from elliptical to spiral? from spiral to elliptical?

3. If all elliptical galaxies had three different diameters, we would never see an elliptical galaxy with a circular outline on a photograph. True or false? Explain your answer. (*Hint:* Can a football ever cast a circular shadow?)

4. What is the difference between an Sa and an Sb galaxy? between an S0 and an Sa galaxy? between an Sb and an SBb galaxy? between an E7 and an S0 galaxy?

5. Why wouldn't white dwarfs make good distance indicators?

6. Why isn't the look-back time important among nearby galaxies?

7. Explain how the rotation curve method of finding a galaxy's mass is similar to the method used to find the masses of binary stars.

8. Explain how the Hubble law permits us to estimate the distances to galaxies.

9. How can collisions affect the shape of galaxies?

10. What evidence do we have that galactic cannibalism really happens?

11. Describe the future evolution of a galaxy that we now see as a starburst galaxy. What will happen to its interstellar medium?

12. Why does the intracluster medium help determine the nature of the galaxies in a cluster?

DISCUSSION QUESTIONS

1. From what you know about star formation and the evolution of galaxies, do you think the Infrared Astronomy Satellite should have found irregular galaxies to be bright or faint in the infrared? Why or why not? What about starburst galaxies? What about elliptical galaxies?

2. Imagine that we could observe a gas cloud at such a high look-back time that it is just beginning to form one of the first galaxies. Further, suppose we discovered that the gas was metal rich. Would that support or contradict our understanding of galaxy formation?

PROBLEMS

1. If a galaxy contains a type I (classical) Cepheid with a period of 30 days and an apparent magnitude of 20, what is the distance to the galaxy?

2. If you find a galaxy that contains globular clusters that are 2 seconds of arc in diameter, how far away is the galaxy? (*Hints:* Assume that a globular cluster is 25 pc in diameter and use the small-angle formula.)

3. If a galaxy contains a supernova that at its brightest has an apparent magnitude of 17, how far away is the galaxy? (*Hint:* Assume that the absolute magnitude of the supernova is -19.)

4. If we find a galaxy that is the same size and mass as our Milky Way Galaxy, what orbital velocity would a small satellite galaxy have if it orbited 50 kpc from the center of the larger galaxy?

5. Find the orbital period of the satellite galaxy described in Problem 4.

6. If a galaxy has a radial velocity of 2000 km/s and the Hubble constant is 70 km/s/Mpc, how far away is the galaxy? (*Hint:* Use the Hubble law.)

7. If you find a galaxy that is 20 minutes of arc in diameter, and you measure its distance to be 1 Mpc, what is its diameter?

8. We have found a galaxy in which the outer stars have orbital velocities of 150 km/s. If the radius of the galaxy is 4 kpc, what is the orbital period of the outer stars? (*Hints:* 1 pc = 3.08×10^{13} km, and 1 yr = 3.15×10^7 seconds.)

9. A galaxy has been found that is 5 kpc in radius and whose outer stars orbit the center with a period of 200 million years. What is the mass of the galaxy? On what assumptions does this result depend?

10. Among the globular clusters orbiting a distant galaxy, the fastest is traveling 420 km/s and is located 11 kpc from the center of the galaxy. Assuming the globular cluster is gravitationally bound to the galaxy, what is the mass of the galaxy? (*Hint:* The galaxy had a slightly faster globular cluster, but it escaped some time ago. What is the escape velocity?)

CRITICAL INQUIRIES FOR THE WEB

1. How far out into the universe can we see Cepheid variables? Research sources on the Internet to find other galaxies whose distances have been found through observation of Cepheids. List the galaxies in which Cepheids have been identified and the distances determined from these data.

2. Who was Milton Humason, and how did he get started in astronomy?

3. In the early 1900s the nature of the "spiral nebulae" was not well understood. In 1920 a "great debate" was held between Harlow Shapley, who held that these objects were relatively nearby swirling clouds of gas, and Heber Curtis, who saw them as distant "island universes." Use the Internet to find information about the debate, outline the lines of evidence used by the two participants to present their views, and explain who was right and who was wrong.

4. Locate a Web page dedicated to the Messier Catalogue—a list of galaxies, clusters, and nebulae that is often used as a list of targets for small telescopes. Be sure that your destination includes images of the objects. For each of the galaxies in the Messier list, determine its Hubble classification. You may be given this information at the site, but examine the images to see if the features of these galaxies conform to a particular Hubble type.

EXPLORING THESKY

1. Locate the Andromeda Galaxy, also known as M31, and its companion galaxies. Zoom in on it and estimate its angular size compared to the full moon. (*Hint:* Use **Find** under the **Edit** menu.)

2. Take a survey of galaxies and see how many are spiral and how many are elliptical. Does your method produce a fair count or are you biased to notice only the brighter galaxies? (*Hint:* Use **Filters** under the **View** menu to turn off everything but **Galaxies** and **Mixed Deep Sky** objects. Make sure the Messier labels are switched on using the **Labels** and **Setup** under the **View** menu.)

3. Locate the Sombrero Galaxy (M104). Study the photographs and discuss this galaxy's special properties. Zoom in on it and estimate its angular size compared to the moon.

4. Study the distribution of galaxies and notice how they cluster together. Can you find the Virgo cluster? Zoom in until more galaxies appear and then scroll north to find the Coma Cluster. Zoom in on Leo to find the cluster there. What other clusters can you find?

5. Describe the galaxy located near the south celestial pole.

 Visit the Seeds *Foundations of Astronomy* companion Web site for critical thinking exercises, articles, and additional readings from InfoTrac College Edition, Brooks/Cole's online student library.

GALAXIES WITH ACTIVE NUCLEI

Virtually all scientists have the bad habit of displaying feats of virtuosity
in problems in which they can make some progress and
leave until the end the really difficult central problems.

M. S. Longair, High Energy Astrophysics

NASA

GUIDEPOST

This chapter is important for two reasons. First, it draws together ideas from many previous chapters to show how nature uses the same basic rules on widely different scales. Matter flowing into a protostar, into a white dwarf, into a neutron star, or into the heart of a galaxy must obey the same laws of physics, so we see the same geometry and the same phenomena. The only difference is the level of violence.

Second, this chapter is important because the most distant objects we can see in the universe are the most luminous galaxies, and many of those are erupting in outbursts and are thus peculiar. By studying these galaxies, our attention is drawn out in space to the edge of the visible universe and back in time to the earliest stages of galaxy formation. In other words, we are led to think of the origin and evolution of the universe, the subject of the next chapter.

VIRTUAL LABORATORY

ACTIVE GALACTIC NUCLEI

Explosion? Eruption? Detonation? Colossal? Titanic? Words fail us when we try to describe the violence that astronomers find at the center of some galaxies. If we say a massive star explodes as a supernova, what word shall we use for the eruption of an entire galaxy? Astonishing it may be, but violence in the centers of galaxies is easily detected across billions of light-years, and our double goal in this chapter is to explain the origin of these eruptions and relate them to the formation and evolution of galaxies.

The adventure of studying active galaxies began when radio astronomers first noticed such objects in the 1950s. In the decades that followed, astronomers used all parts of the electromagnetic spectrum to study these galaxies and piece together the puzzle.

Finding the solution was a long and difficult journey, but modern astronomers generally agree on the basic facts. Eruptions in the hearts of active galaxies are caused by supermassive black holes. We have see evidence in the previous two chapters that our own galaxy and most other galaxies contain supermassive black holes at their centers. In this chapter, we study not only the behavior of these strange objects, but also the process by which astronomers came to understand them.

A few astronomers have objected to the quotation that opens this chapter, perhaps because it seems to belittle scientists. But it is a realistic description of how scientists attack a problem. Big problems are made of little problems, and scientists begin by solving the little problems, learning from those solutions, and moving ever closer to the center of the big problem. Perhaps no better example can be found than modern astronomy's attack on "the really difficult central problem" in the nuclei of active galaxies.

17-1 ACTIVE GALAXIES

All galaxies, including our Milky Way, emit radio energy from neutral hydrogen, from molecules, from pulsars, and so on. In the 1950s, radio astronomers discovered that some galaxies emit as much as 10 million times more radio energy than a normal galaxy. These galaxies were called **radio galaxies.** Beginning in the 1960s, astronomers could send spacecraft above Earth's atmosphere, and they discovered that many of the radio galaxies also emit powerfully at infrared, ultraviolet, and X-ray wavelengths. As a result, these galaxies are now called **active galaxies** in recognition of the many wavelengths at which they radiate.

As we will see in this section, strong evidence suggests that the energy that powers active galaxies arises

Ace⟲Astronomy™ The AceAstronomy icon throughout the text indicates an opportunity for you to test yourself on key concepts and to explore animations and interactions on the AceAstronomy Web site at **http://astronomy.brookscole .com/seeds8e**

in their central nuclei. One of the great adventures of modern astronomy has been the quest to understand these **active galactic nuclei (AGN).** What are the mysterious and powerful AGN?

SEYFERT GALAXIES

In 1943, Mount Wilson astronomer Carl K. Seyfert published his study of spiral galaxies. Observing at visual wavelengths, Seyfert found that some spiral galaxies have small, highly luminous nuclei (❙ Figure 17-1b). Today, these **Seyfert galaxies** are recognized by the peculiar spectra of these bright nuclei.

The light from a normal galaxy comes from stars, and its spectrum is the combined spectra of many millions of stars. Thus, we see the principal absorption lines found in stellar spectra, mainly hydrogen Balmer lines and lines of calcium II. But the spectra of the Seyfert galaxy nuclei contain broad emission lines of highly ionized atoms. Emission lines suggest a hot, low-density gas, and ionized atoms suggest that the gas is very excited. The widths of the spectral lines suggest large Doppler shifts produced by high velocities in the nuclei; gas approaching us would produce blue-shifted spectral lines, and gas going away from us would produce red-shifted lines. The combined light would produce broad spectral lines. The velocities at the centers of Seyfert galaxies are roughly 10,000 km/s, about 30 times greater than velocities at the centers of normal galaxies.

About 2 percent of spiral galaxies appear to be Seyfert galaxies, and they are classified into two categories. Type 1 Seyfert galaxies are very luminous at X-ray and ultraviolet wavelengths and have the typical broad emission lines in their spectra. Type 2 Seyfert galaxies lack strong X-ray emission and have emission lines that are narrower than those of type 1 Seyferts but still broader than spectral lines produced by a normal galaxy.

In the 1960s, astronomers discovered that the brilliant nuclei of Seyfert galaxies fluctuate rapidly. A Seyfert nucleus can change its brightness by 50 percent in a few months. As we saw in Chapter 14, an astronomical body cannot change its brightness in a time shorter than the time it takes light to cross its diameter. If the Seyfert nucleus can change in a few months, then it cannot be larger in diameter than a few light-months. In spite of their small size, the cores of Seyfert galaxies produce tremendous amounts of energy. The brightest emit a hundred times more energy than the entire Milky Way. Something in the centers of these galaxies not much bigger than our solar system produces a galaxy's worth of energy.

The shapes of Seyferts can give us clues to their energy source. They are three times more common in interacting pairs of galaxies than in isolated galaxies. Also, about 25 percent have peculiar shapes suggesting tidal interactions with other galaxies (❙ Figure 17-2a). This statistical evidence (Window on Science 17-1) hints that Seyfert galaxies may have been triggered into activity by collisions or interactions with companions.

a

b

FIGURE 17-1

(a) Seyfert galaxy NGC1566 looks like a normal spiral galaxy, but short exposure images reveal that its nucleus is very small and very luminous. *(AATB)* (b) The small, bright nucleus of this Seyfert galaxy in the southern constellation Circinus is ejecting gas at high speeds in a bright, V-shaped cone located just above the nucleus (arrow). Higher above the disk, the ejected gas is visible as the bright pink regions extending to the top of the image. Presumably a similar cone of ejected gas is hidden below the disk of the galaxy. *(Andrew S. Wilson, University of Maryland; Patrick L. Shopbell, Caltech; Chris Simpson, Subaru Telescope; Thaisa Storchi-Bergmann and F. K. B. Barbosa, UFRGS, Brazil; Martin J. Ward, University of Leicester, U.K.; and NASA)*

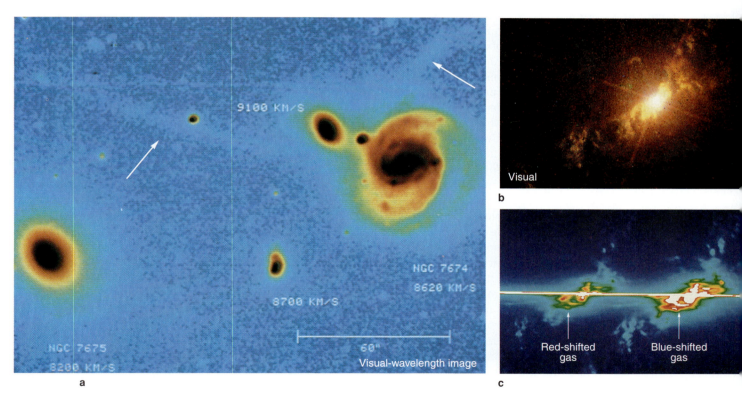

a

b

c

FIGURE 17-2

(a) Seyfert galaxy NGC7674 is distorted with tails to upper right and upper left (arrows) in this false-color image. Note the companion galaxies. *(John W. Mackenty, Institute for Astronomy, University of Hawaii)* (b) Seyfert galaxy NGC4151 has a brilliant nucleus in visible light images (top). The Space Telescope Imaging Spectrograph reveals Doppler shifts showing that part of the expelled gas has a blue shift (to the right) and is approaching us and part has a red shift (to the left) and is receding. *(John Hutchings, Dominion Astrophysical Observatory; Bruce Woodgate, GSFC/NASA; Mary Beth Kaiser, Johns Hopkins University; and the STIS Team)*

It Wouldn't Stand Up in Court: Statistical Evidence

Notice that some scientific evidence is statistical. For example, we might note that Seyfert galaxies are three times more likely to have a nearby companion than a normal galaxy is. This is statistical evidence because we can't be certain that any single Seyfert galaxy will prove to have a companion. Yet the probability is higher than if it were a normal galaxy, and that leads us to suspect that interactions between galaxies are involved. Such statistical evidence can tell us something in general about the cause of Seyfert eruptions.

Statistical evidence is common in science, but it is inadmissible in most courts of law. The American legal system is based on the principle of reasonable doubt, so most judges would not allow statistical evidence of guilt. For example, plaintiffs have had great trouble suing

tobacco companies for causing their cases of lung cancer, even though there is a clear statistical link between smoking and lung cancer. The statistics tell us something in general about smoking but cannot be used to prove that any specific case of lung cancer was caused by smoking. There are other causes of lung cancer, so there is always a reasonable doubt as to the cause of any specific case.

We can use statistical evidence in science because we do not demand that the statistics demonstrate anything conclusively about any single example. To continue our example of Seyfert galaxies, we don't demand that the statistics predict that any specific galaxy has a companion. Rather, we use the statistical evidence to gain a general insight into the cause of active galactic nuclei. That is, we are trying to under-

stand galaxies as a whole, not to convict any single galaxy.

Of course, if we surveyed only a few galaxies, our statistics might not be very good, and a critic might be justified in making the common complaint, "Oh, that's only statistics." For example, if we surveyed only four galaxies and found that three had companions, our statistics would not be very reliable. But if we surveyed 1000 galaxies, our statistics could be very good indeed, and our conclusions could be highly significant.

Thus, scientists can use statistical evidence if it passes two tests. It cannot be used to draw conclusions about specific cases, and it must be based on large enough samples so the statistics are significant. With these restrictions, statistical evidence can be a powerful scientific tool.

Some Seyferts are expelling matter in oppositely directed flows (Figure 17-2b), a geometry we have seen on smaller scales when matter flows into accretion disks around neutron stars or black holes.

Examine Figure 17-2c. This image of the core of Seyfert galaxy NGC4151 is a spectrum obtained with a special spectrograph that records an image at whatever wavelengths are bright in an object's emission spectrum. This image shows that some of the gas near the center of the galaxy has a high velocity toward us and some has a high velocity away from us. These high velocities are typical of gas in the centers of Seyfert galaxies.

All of this evidence leads modern astronomers to suspect that the cores of Seyfert galaxies contain supermassive black holes—black holes with masses as high as a billion solar masses. The gas in the centers of Seyfert galaxies is traveling so fast it would escape from a normal galaxy. Only very large central masses could exert enough gravity to hold the gas inside the nuclei, and that suggests supermassive black holes. Encounters with other galaxies could throw matter into a black hole; and, as we have seen in Chapter 14, large amounts of energy can be liberated by matter flowing into a black hole. We will expand this theory later, but first we must search for more evidence of supermassive black holes by turning our attention to galaxies that emit powerful radio signals.

DOUBLE-LOBED RADIO SOURCES

Beginning in the 1950s, radio astronomers found that some sources of radio energy in the sky consisted of

pairs of radio-bright regions. When optical telescopes studied the locations of these **double-lobed radio sources,** the photographs revealed galaxies located between the two radio lobes. Apparently, the central galaxies were producing the radio lobes.

Study ▌ "Cosmic Jets and Radio Lobes" on pages 362 and 363 and notice four important points. First, the geometry suggests that radio lobes are inflated by jets of excited gas emerging from the central galaxy. This has been called the **double-exhaust model.** Second, notice how observations such as the presence of synchrotron radiation and hot spots give us evidence in support of the model. The third thing to notice is how the complex shapes of some jets and radio lobes can be explained by the motions of the active galactic nuclei. A good example of this is 3C449 (the 449th source in the *Third Cambridge Catalog of Radio Sources*) with its twisting radio lobes. Finally, notice how these jets seem to be related to matter falling into a central black hole. We have seen similar jets produced by rotating disks in the case of protostars and neutron stars; although the details are not understood, it is presumably the same process that produces all of these jets.

Study the image of NGC5128 on page 362 closely for a further clue. In visual-wavelength images, the galaxy looks like a round elliptical galaxy surrounded by a belt of dust and gas. Doppler shifts reveal that the spherical cloud of stars is rotating about an axis that lies in the plane of the dust disk. That is, the galaxy appears to have two different axes of rotation. Furthermore, bright blue stars have been identified in the dust ring, and those stars must have formed recently. All of

Scientific Arguments

An argument can be a shouting match, but another definition of the word is "a discourse intended to persuade." Scientists construct arguments as part of the business of science not because they want to persuade others they are right, but because they want to test their ideas. For example, we can construct a scientific argument to show that the radio lobes that flank some active galaxies have their origin in jets coming from the cores of the galaxies. If we do our best and our argument is not persuasive, we begin to suspect we are on the wrong track. But if our argument seems convincing, we gain confidence that we are beginning to understand radio lobes.

A scientific argument is a logical presentation of evidence and theory with interpretations and explanations that help us understand some aspect of nature. Geologists might construct an argument to explain volcanism on mid-ocean islands, and the argument could include almost anything—maps, for example, or mineral samples compared with seismic evidence and mathematical models of subsurface magma motions. Such an argument might involve the physics of radioactive decay or observations of the shapes of bird beaks on the islands. The scientists would be free to include any evidence or theory that helps persuade, but they must observe one fundamental rule of scientific argument: They must be totally honest. The purpose of a scientific argument is to test our understanding, not to win votes or sell soap. Dishonesty in a scientific argument is self-deluding, and scientists consider such dishonesty the worst possible behavior.

On the other hand, scientists are human, and they often behave no better than other humans. They sometimes defend their own theories and attack opposing theories more vigorously than the evidence warrants. They may unintentionally overlook contradictory evidence or unconsciously misinterpret data. Of course, if a scientist did these things intentionally, we would be shocked because he or she would be violating the most serious rules of honesty in science. In the struggle to understand how nature works, however, it is not surprising to see that human passions rise and that scientific debate becomes heated. That is why science is so exciting—it is a struggle to understand how nature works. Nonetheless, no matter how scientists behave in the excitement of a controversy, they all aspire to that ideal of the scientific argument—the cool, logical presentation of evidence that leads to a new understanding of nature.

Much of this book or any other scientific book consists of scientific arguments—logical presentations of evidence. As you read about any science, look for the arguments and practice organizing and presenting them to test your own understanding. The "Critical Inquiry" that ends each chapter section in this book is intended to illustrate how arguments are constructed. Use the same plan when you develop your own arguments.

---

this is evidence of a recent collision, and we will see later that collisions between galaxies can trigger eruptions in their cores.

Drawing on all of the evidence available and guided by their understanding of the behavior of light and matter, astronomers have constructed a convincing scientific argument (Window on Science 17-2) that active galactic nuclei and radio lobes are powered by violent events in the core. Now we must consider the evidence that the central body is really a black hole.

TESTING THE BLACK HOLE HYPOTHESIS

So far we have skirted the "really difficult central problem." Are the central energy sources in active galactic nuclei really supermassive black holes? The evidence takes two forms—observations of size and observations of motion.

Earlier in this chapter we saw evidence that the cores of some active galaxies are very small. The cores of Seyfert galaxies fluctuate in brightness over very short time periods, and that must mean they are very small, so small that astronomers know of nothing except a supermassive black hole that could produce so much energy in such a small region.

The motion of stars and gas near the centers of galaxies is important because it can tell us the amount of mass located there. Consider the giant elliptical galaxy M87 (page 338). It has a very small, bright nucleus and a visible jet of matter 1800 pc long racing out of its core (❙ Figure 17-3). Radio observations show that the nucleus must be no more than a light-week in diameter. This evidence is suggestive, but observations of motions around the center are decisive. A high-resolution image shows that the core lies at the center of a spinning disk and the jet lies along the axis of the disk. Only 60 ly from the center, the gas in the disk is orbiting at 750 km/s. Substituting radius and velocity into the equation for circular velocity (Chapter 5) tells us that the central mass must be roughly 2.4 billion solar masses. Only a supermassive black hole could pack so much mass into such a small region.

The evidence of rapidly rotating disks at the centers of galaxies is dramatic. Although we think of elliptical galaxies as being nearly free of dust and gas, some have visible dust and gas around their centers. Presumably this dust and gas is left over from recent mergers with smaller galaxies. By placing the slit of the spectrograph across the core of such galaxies, astronomers can sometimes detect Doppler shifts produced by rapidly rotating disks of gas; one side of the disk comes toward us and produces a blue shift, and one side goes away from us and produces a red shift (❙ Figure 17-4).

One such galaxy, NGC7052, contains a dust disk about 570 pc in radius. Spectra show that a smaller disk only 57 pc from the center rotates with an orbital

Cosmic Jets and Radio Lobes

Many radio sources consist of two bright lobes — double-lobed radio sources — with a galaxy, often a peculiar or distorted galaxy, located between them. Evidence suggests these active galaxies are emitting jets of high-speed gas that inflate the lobes as cavities in the intergalactic medium. Where the jets impact the far side of the cavities, they create hot spots.

Size of Milky Way Galaxy

Hot spot

Radio image

Jet

Jet

Hot spots lie on the leading edge of a lobe where the jet pushes into the surrounding gas.

Hot spot

NRAO

Cygnus A, the brightest radio source in Cygnus, is a pair of lobes with jets leading from the nucleus of a highly disturbed galaxy. In this false-color image, the areas of strongest radio signals are shown in red and the weakest in blue. Because the radio energy detected is synchrotron radiation, astronomers conclude that the jets and lobes contain very-high-speed electrons, usually called relativistic electrons, spiraling through magnetic fields about 1000 times weaker than Earth's field. The total energy in a radio lobe is about 10^{53} J — what we would get if we turned the mass of a million suns directly into energy.

Visible galaxy

Used with permission, Fosbury R. A. E., Vernet, J., Villar-Martin, M., Cohen, M. H., Ogle, P. M., Tran, H. D. & Hook, R. N. 1998, Optical continuum structure of Cygnus A. "KNAW colloquium on: The most distant radio galaxies," Amsterdam, 15–17 October 1997, Roettgering H, Best P and Lehnert M eds, Reidel, astro-ph/9803310

Radio galaxy NGC5128 lies between two radio lobes, and the combined radio and X-ray image at the right shows a high-energy jet at the very center of the galaxy pointing to the upper left into the northern radio lobe.

NASA/CXC/NRAO/VLA

Infrared

E. Schreier, STScI, NASA

Radio = Red
X-ray = Blue

An infrared image of the center of the galaxy NGC5128 reveals a small, bright disk of hot gas surrounding the nucleus.

If the outer radio lobes of Centaurus A were visible to our eyes, they would look 10 times larger than the full moon.

X-ray: NASA/CXC/M. Karovska et al., Radio: NRAO/VLA/Schiminovich, et al., Radio: NRAO/VLA/J. Condon et al., Optical: Digitized Sky Survey U.K. Schmidt Image/STScI.

Radio (green)

X-ray (blue)

Visual + radio + X-ray

NGC5128 appears to be a giant elliptical galaxy experiencing a collision with a dusty spiral galaxy. It is known to radio astronomers as Centaurus A, an inner pair of radio lobes and a larger outer pair.

Richard A. Perley

3C449

Size of visible galaxy

Radio image

The source of jets in the giant elliptical galaxy 3C449 appears to be in orbital motion about the nucleus and may be the result of a merger.

Radio galaxy 3C129 appears to be moving through the intergalactic medium as its source of jets precesses to produce kinks.

3C129

Radio image

Laurence Rudnick, University of Minnesota, and Jack Burns, University of New Mexico

Direction of motion

Computer model

Vincent Icke, University of Minnesota and *The Astrophysical Journal*, published by the University of Chicago Press, © 1981 American Astronomical Society

Ace Astronomy ™

Go to AceAstronomy and click Active Figures to see "Jet Deflection" and take control of your own model of a moving radio galaxy.

The radio source 3C75 is produced by the interaction of two galaxies with active nuclei. The visible galaxies in these images would be many times the size of their active cores.

Radio image

3C75

NRAO/AUI, F. N. Owen, C. P. O'Dea, and M. Inoue

High-energy jets appear be caused by matter flowing into a supermassive black hole in the core of an active galaxy. Conservation of angular momentum forces the matter to form a whirling accretion disk around the black hole. How that produces a jet is not entirely understood, but it appears to involve magnetic fields that are drawn into the accretion disk and tightly wrapped to eject high-temperature gas. The twisted magnetic field confines the jets in a narrow beam and causes synchrotron radiation.

Jets from active galaxies may have velocities from thousands of kilometers per second up to a large fraction of the speed of light. Compare this with the jets in bipolar flows, where the velocities are only a few hundred kilometers per second. Active-galaxy jets can be millions of light-years long. Bipolar-flow jets are typically a few light-years long. The energy is different, but the geometry is the same.

Black hole

Magnetic field lines

Accretion disk

Excited matter traveling at very high speeds tends to emit photons in the direction of travel. Consequently, a jet pointed roughly toward us will look brighter than a jet pointed more or less away. This may explain why some radio galaxies have one jet brighter than the other, as in Cygnus A shown at the top of the opposite page. It may also explain why some radio galaxies appear to have only one jet. The other jet may point generally away from us and be too faint to detect, as in the case of NGC5128, also shown on the opposite page.

Adapted from a diagram by Ann Field, NASA, STScI.

The jet at the center of giant elliptical galaxy M87 is only a few percent the diameter of the entire galaxy. The jet is emerging at nearly half the speed of light. Hubble Space Telescope images reveal a small, rapidly spinning disk at the center of the galaxy with the jet emerging along the axis of the disk. Very-high-resolution radio observations show that the jet originates in a tiny active region at the center of the disk. *(M87: AURA/NOAO/NSF; jet: NASA/STScI; radio image: NRAO and J. Biretta)*

Ace⟲Astronomy™ Go to AceAstronomy and click Active Figures to see "M31 at Many Wavelengths." Different orbiting telescopes give us dramatically different views of a nearby galaxy.

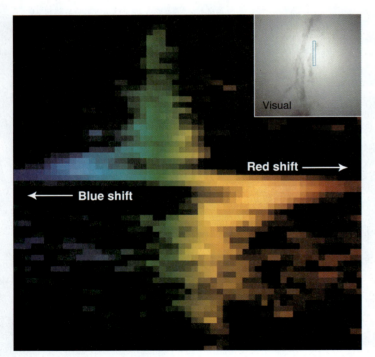

FIGURE 17-4

The elliptical galaxy M84 contains dust rings and an active core. Placing the slit of a spectrograph across the nucleus, astronomers found that spectral lines have a blue shift on one side of the nucleus and a red shift on the other side—evidence of rotation. Colors have been added to this figure to illustrate the blue shift and red shift. The actual width of the line is only about 2 nm. *(Gary Bower, Richard Green, NOAO; The STIS Instrument Definition Team; and NASA)*

velocity of 155 km/s (❚ Figure 17-5). The equation for circular velocity (Chapter 5) tells us the disk must enclose a mass of about 300 million solar masses. This galaxy ejects high-speed jets of ionized gas, but the jets are not perpendicular to the dust disk. Apparently, the central black hole and its accretion disk are tipped at an angle to the dust disk, which was evidently produced by a recent merger.

While the observational evidence for supermassive black holes continues to grow, two problems remain. Astronomers are not sure how these huge black holes formed. They appear to have formed when the galaxies first took shape, but the process is unknown. Also, astronomers do not know how a spinning accretion disk produces jets of mass and electromagnetic radiation. The mechanism is believed to involve magnetic fields, but the details are not understood. Modern astronomers now work to model the complex processes that occur around these supermassive black holes.

THE SEARCH FOR A UNIFIED MODEL

When a field of research is young, scientists often find many seemingly different phenomena such as double-lobed radio galaxies, Seyfert galaxies, cosmic jets, and so on. As the research matures, the scientists begin seeing similarities and eventually are able to unify the different phenomena as different aspects of a single process. This is the real goal of science, to organize evi-

Visual-wavelength images

FIGURE 17-5

Near the center of active elliptical galaxy NGC7052, a dust disk circles a massive nucleus. Deep in a "haze" of billions of stars, the near side of the dust disk obscures more stars than the far side and looks darker. Scattering of shorter-wavelength photons in the dust gives it a red tint. Velocities near the center suggest a central black hole containing 300 million solar masses. *(Roeland P. van der Marel, STScI, Frank C. van den Bosch, University of Washington, and NASA)*

dence and theory in logical arguments that explain how nature works (Window on Science 17-2). Astronomers studying active galaxies are now struggling to find a **unified model** of active galaxy cores.

It seems clear that the cores of active galaxies contain supermassive black holes surrounded by accretion disks that are extremely hot near the black hole but cooler further out. Theory predicts that the central cavity in the disk where the black hole lurks is very narrow but that the disk there is "puffed up" and thick. This means the black hole may be hidden deep inside this narrow central well. The hot inner disk seems to be the source of the jets often seen coming out of active galaxy cores, but the process by which jets are generated is not understood. The outer part of the disk, according to calculations, is a fat, dense doughnut (a torus) of dusty gas (▌ Figure 17-6).

What we see when we view the core of an active galaxy must depend on how this disk is tipped with respect to our line of sight. We should note, at this point, that the accretion disk may be tipped at a steep angle to the plane of a galaxy, so just because we see a galaxy face-on doesn't mean we are looking at the accretion disk face-on. Thus the important factor is not the inclination of the galaxy but the inclination of the accretion disk.

For example, in 1968 astronomers realized that an object they thought was an irregular variable star in the constellation Lacerta was actually the core of an active galaxy. The visible spectrum is featureless, but the spectrum of surrounding nebulosity is that of a giant elliptical galaxy. The object, and others like it known as **BL Lac objects** or **blazars,** are 10,000 times more luminous than the Milky Way Galaxy and fluctuate in only hours. They are believed to be the cores of active galaxies in which we happen to be looking directly into

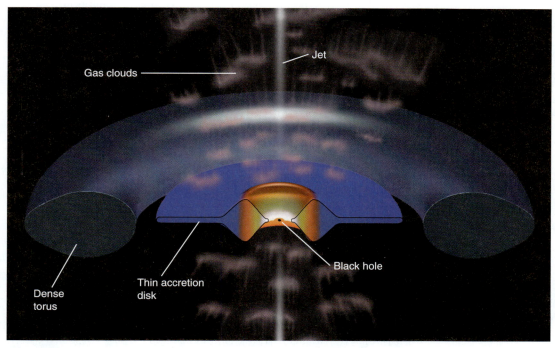

FIGURE 17-6

An accretion disk around a massive black hole. Matter flowing inward passes first through a large, opaque torus; then into a thinner, hotter disk; and finally into a hot, narrow cavity around the black hole. Gas clouds above and below the disk are excited by radiation from the hot disk. This diagram is not to scale. The central cavity may be only 0.01 pc in radius, while the outer torus may be 1000 pc in radius.

the brilliant jets. In this case, the accretion disk is apparently face-on, and we are looking directly into the central cavity around the black hole—down the dragon's throat.

If the accretion disk is tipped slightly, we may not be looking directly into the jet, but we may be able to see some of the intensely hot gas in the central cavity. This region emits broad spectral lines because the gas is orbiting at high velocities and the high Doppler shifts smear out the lines. Seyfert 1 galaxies may be explained by this phenomenon.

If we view the accretion disk from the edge, we cannot see the central area at all because the thick dusty torus blocks our view. Instead we see radiation emitted by gas lying above and below the central disk. Because this gas is farther from the center, it orbits more slowly and has smaller Doppler shifts. Thus we see narrower spectral lines. This might account for the Seyfert 2 galaxies.

This unified model is far from complete. The actual structure of accretion disks is poorly understood, as is the process by which the disks produce jets. Furthermore, the Seyfert galaxies are different from the giant elliptical galaxies that have double radio lobes. Also, our discussion of unification has ignored important factors such as differences in the rate at which mass flows inward and the influence of magnetic fields. A unified model does not explain all of the differences among active galaxies. Rather, it gives us some clues to what is happening in the cores of active galaxies.

BLACK HOLES AND GALAXY FORMATION

For decades, astronomers have confronted the central knotty problem of active galaxies: What makes a galaxy erupt? The solution to that problem seems at hand, and astronomers are going beyond that problem to explore the formation of galaxies.

The cores of active galaxies erupt when matter flows into their central supermassive black holes. A growing mass of evidence shows that active galaxies contain supermassive black holes. NGC4261, for example, is an active galaxy with a rapidly rotating core containing a 1.2 billion-solar-mass black hole (Figure 17-7).

Furthermore, studies show that nearly all galaxies contain supermassive black holes, but most are dormant (Figure 17-8). Evidently the cores of galaxies become active when interactions with other galaxies throw matter into their central black holes, and mathematical models of interacting galaxies show that such encounters are capable of throwing matter into the core of a galaxy (Figure 17-9). The importance of collisions is shown by the prevalence of companions to Seyfert galaxies and the distorted shapes of active galaxies.

Astronomers have been able to measure the masses of a few dozen supermassive black holes, and, as discussed in the previous chapter, they have discovered that the mass of the black hole is correlated with the mass of the nuclear bulge. Galaxies with massive nuclear bulges have massive black holes at their centers. This includes both elliptical galaxies and spiral galaxies with nuclear bulges. But there is no relationship between the masses of the central black holes and the disks of galaxies. This must mean that the supermassive black holes were built by the same process that formed the bulges. Galaxy formation may have begun with matter falling together to form bulges in which a small fraction of the mass fell all the way to the centers of the bulges to form supermassive black holes. Such a process would release a tremendous amount of energy. In contrast, the disks of galaxies are very delicate and are not correlated with the mass of the supermassive black holes. Disks may have formed later as additional material fell into orbit around the central bulges.

Perhaps the history of most galaxies includes a violent youth when matter fell together rapidly and cre-

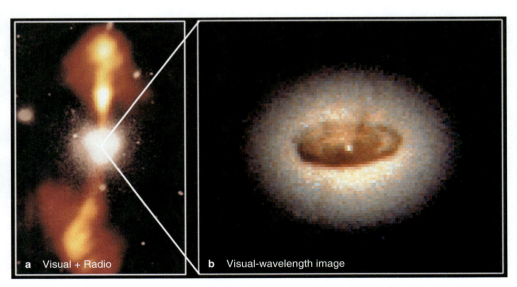

FIGURE 17-7
(a) Galaxy NGC4261 looks like a distorted fuzzy blob in an Earth-based photograph at visible wavelengths (white), but radio telescopes reveal jets leading into radio lobes (orange). (b) A Hubble Space Telescope image of the core of the galaxy reveals a bright central spot surrounded by a disk with its axis pointing at the radio lobes. *(L. Ferrarese, Johns Hopkins University, and NASA)*

a   Visual + Radio

b   Visual-wavelength image

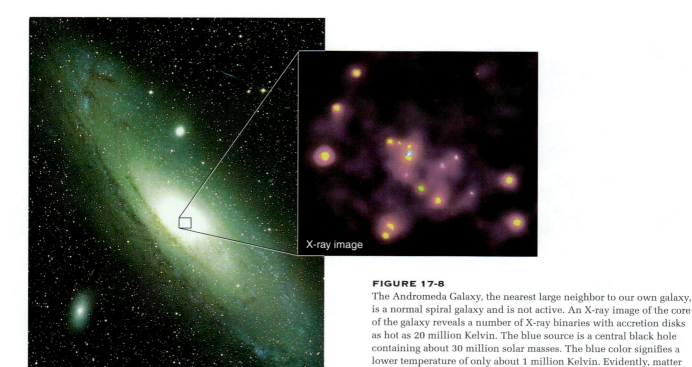

Visual image

X-ray image

FIGURE 17-8
The Andromeda Galaxy, the nearest large neighbor to our own galaxy, is a normal spiral galaxy and is not active. An X-ray image of the core of the galaxy reveals a number of X-ray binaries with accretion disks as hot as 20 million Kelvin. The blue source is a central black hole containing about 30 million solar masses. The blue color signifies a lower temperature of only about 1 million Kelvin. Evidently, matter is not flowing rapidly into the black hole, and it is dormant. *(AURA/ NOAO/NSFa; NASA/CXC/SAO)*

ated short-lived outbursts and giant black holes. If so, the supermassive black holes we see now are scars left from that ancient age of violence.

REVIEW **CRITICAL INQUIRY**

Construct an argument to show that double-lobed radio galaxies and Seyfert galaxies have similar energy sources. Observations show that double-lobed radio sources are produced by two jets flowing out of a galaxy's core and inflating the radio lobes. The evidence of small size and very-high-speed motions in the core show that these galaxies must contain large amounts of mass in small regions. Seyfert galaxies also have small nuclei, as shown by their rapid fluctuations, and the high velocities in their cores imply that the cores must contain very large amounts of mass. Consequently both kinds of objects are suspected to contain supermassive black holes at their centers.

The preceding paragraph is a quick summary of a logical argument citing evidence and theory to reach a conclusion. Construct an argument to propose that galaxy interactions are necessary to trigger active galactic nuclei.

Observational evidence and basic theory have led us to the unified model of active galactic nuclei. Although the model seems successful, there are still difficulties, and astronomers continue to struggle to understand the details. Nevertheless, the unified model may help us understand the mysterious quasars.

17-2 QUASARS

The largest telescopes detect multitudes of faint, starlike points of light with peculiar emission spectra, objects called **quasars** (also known as *quasi-stellar objects,* or *QSOs*). Although astronomers now recognize quasars as some of the most distant visible objects in the universe, they were a mystery when they were first identified. The discovery of these objects in the 1960s and the struggle to understand them comprise a classic example of how scientists explore hypotheses, gather evidence, and build confidence in new ways to understand nature. We begin this section by telling that story, a story that will lead us to think about the depths of the universe in both space and time.

THE DISCOVERY OF QUASARS

In the early 1960s, radio interferometers (see Chapter 6) showed that a number of radio sources are much smaller than normal radio galaxies. Photographs of the location of these radio sources did not reveal a galaxy, not even a faint wisp, but rather a single starlike point of light. The first of these objects so identified was 3C48, and later the source 3C273 was added. Though these objects emitted radio signals like those from radio galaxies, they were obviously not normal radio galaxies. Even the most distant photographable galaxies look fuzzy, but these objects looked like stars. Their spectra, however, were totally unlike stellar spectra, so the objects were called quasi-stellar objects (❚ Figure 17-10).

In this computer model, one galaxy is a disk of points, and the mass of the other galaxy is represented as a single body.

The large galaxy is kept centered in the field as the two galaxies whip past each other.

The outer parts of the large galaxy are distorted and take on the shape of spiral arms.

Mass is also thrown inward in the large galaxy in large enough quantities to cause…

a violent eruption as it flows into a supermassive black hole at the center of the large galaxy.

FIGURE 17-9
A computer model of the gravitational interaction between two passing galaxies shows that matter is not only thrown outward but is also thrown inward. *(G. Byrd, M. Valtonen, B. Sundelius, and L. Valtaoja)*

For a few years, the spectra of quasars were a mystery. A few unidentifiable emission lines were superimposed on a continuous spectrum. But, in 1963, Maarten Schmidt at Hale Observatories tried redshifting the hydrogen Balmer lines to see if they could

be made to agree with the lines in 3C273's spectrum. At a red shift of 15.8 percent, three lines clicked into place (❚ Figure 17-11). Other quasar spectra quickly yielded to this approach, revealing even larger red shifts.

The red shift z is the change in wavelength $\Delta\lambda$ divided by the unshifted wavelength λ_0:

$$\text{red shift} = z = \frac{\Delta\lambda}{\lambda_0}$$

Astronomers commonly use the Doppler formula to convert these red shifts into radial velocities. Thus, the large red shifts of the quasars imply large apparent velocities of recession.

Quasar spectra contain bright emission lines. The brightest are the Balmer lines of hydrogen, but other spectral lines also appear. The red shifts of quasars can be so large that spectral lines can be shifted completely out of the visible spectrum, and lines in the ultraviolet can be shifted into the visible spectrum. ❚ Figure 17-12 shows the spectra of four quasars with relatively low red shifts in which the H_α Balmer line normally seen in the red part of the spectrum is shifted into the infrared. Quasars have been found with even larger red shifts.

To understand the significance of these large red shifts and the large velocities of recession they imply, we must draw on the Hubble law. Recall that the Hubble law states that galaxies have apparent velocities of recession proportional to their distances, and thus the distance to a quasar is equal to its apparent velocity of recession divided by the Hubble constant. The large red shifts of the quasars imply that they must be at great distances.

The red shift of 3C273 is 0.158, and the red shift of 3C48 is 0.37. These are large red shifts, but not as large as the largest then known for galaxies, about 1.0. Soon, however, quasars were found with red shifts much larger than that of any known galaxy. Some quasars are evidently so far away that galaxies at those distances are very difficult to detect. Yet the quasars are easily photographed. This leads us to the conclusion that quasars must have 10 to 1000 times the luminosity of a large galaxy. Quasars must be superluminous.

Soon after quasars were discovered, astronomers detected fluctuations in brightness over times as short as a few months. Recall from our discussion of pulsars (Chapter 14) that an object cannot change its brightness appreciably in less time than it takes light to cross its diameter. The rapid fluctuations in quasars showed that they are small objects, not more than a few light-months in diameter.

Thus, by the late 1960s astronomers faced a problem: How could quasars be superluminous but also be very small? What could make 10 to 1000 times more energy than a galaxy in a region less than a light-year in diameter? Since that time, evidence has accumu-

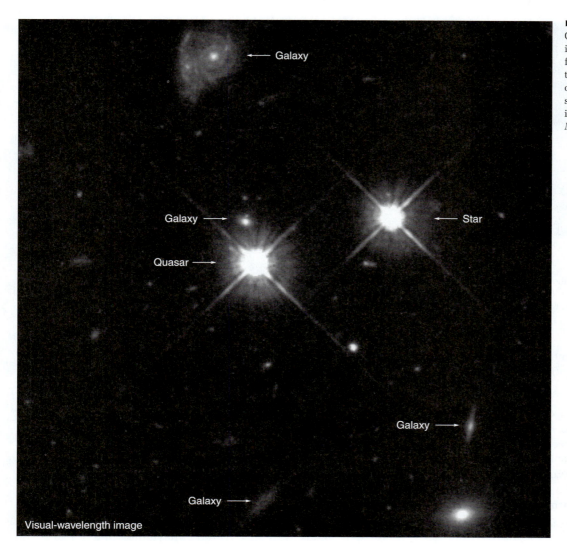

FIGURE 17-10
Quasars have starlike images clearly different from the images of distant galaxies. The spectra of quasars are unlike the spectra of stars or galaxies. *(C. Steidel, Caltech, NASA)*

Galaxy ←

Galaxy →

Quasar →

← Star

Galaxy →

Galaxy →

Visual-wavelength image

lated that quasars are the active cores of very distant galaxies, and the rest of this chapter will discuss that evidence. The key, as in most astronomical arguments, is distance. The distance to quasars is so important to our discussion and so much larger than the distances we have considered before that we need to pay special attention to how astronomers estimate the distances to these objects.

QUASAR DISTANCES

Astronomers estimate the distance to a quasar by using the Hubble law and dividing the quasar's velocity of recession by the Hubble constant (Chapter 16). Thus, knowing the distance to a quasar depends critically on knowing its red shift and from that its velocity. Many quasars have red shifts greater than 1, and if we put such large red shifts into the Doppler formula we discussed in Chapter 7, we get radial velocities greater than the velocity of light, which is unreasonable. Quasar red shifts greater than 1 do make

sense, however, if we consider the nature of very large red shifts.

First we must note that the Doppler formula given in Chapter 7 is only an approximation good for low velocities. It is correctly called the classical Doppler formula, and it works fine if the velocity is significantly less than the velocity of light. For very high velocities, however, we need to use the **relativistic Doppler formula** derived from Einstein's theory of relativity. It relates the radial velocity V_r to the speed of light c and the red shift z:

$$\frac{V_r}{c} = \frac{(z+1)^2 - 1}{(z+1)^2 + 1}$$

The relativistic Doppler formula cannot produce a radial velocity greater than the speed of light, no matter how large z becomes. For example, suppose a quasar had a red shift of 2. The classical Doppler formula would give the unreasonable result that the radial velocity was twice the speed of light, but when we put 2 into the relativistic Doppler formula we discover that

FIGURE 17-11

(a) This image of 3C273 shows the bright quasar at the center surrounded by faint fuzz. Note the jet protruding to lower right. (b) The spectrum of 3C273 (top) contains three hydrogen Balmer lines red-shifted by 15.8 percent. The drawing shows the unshifted positions of the lines. *(Courtesy Maarten Schmidt)*

$z + 1$ is 3, and therefore $(z + 1)^2$ is 9. Then the radial velocity divided by the speed of light is:

$$\frac{V_r}{c} = \frac{(9) - 1}{(9) + 1} = \frac{8}{10} = 0.8$$

Thus, the quasar has an apparent velocity of recession of 80 percent the speed of light.

■ Figure 17-13 compares the classical Doppler formula and the relativistic Doppler formula. At low velocities they give the same result, but at higher velocities the classical formula gives velocities that are too high. The important point for us is that no matter how large the red shift becomes, the velocity never quite equals the speed of light.

Now that we feel some comfort with quasar red shifts greater than 1, we should note that the red shifts associated with the expansion of the universe are not really Doppler shifts. The exact relation between the red shifts of distant galaxies and their expansion velocities is similar to the relativistic Doppler formula in that red shifts greater than 1 imply velocities near but not exceeding the velocity of light. But the exact relation depends critically on parameters such as the average density of the universe. Consequently, we can't use a Doppler formula to convert quasar red shifts into velocities. But we can easily conclude that the large red shifts imply very large distances, and that red shifts greater than 1 are not unreasonable. (We will discuss these red shifts produced by the expansion of the universe in the next chapter.)

If quasars are as far away as their large red shifts indicate, their look-back times are very large, and we see them as they were when the universe was significantly younger. In the next chapter, we will see that the universe is believed to be about 14 billion years old.

Thus, we see the most distant quasars as they were when the universe was only a few billion years old.

What evidence do we have that quasars really are located at these great distances? A discovery made in 1979 provides a dramatic illustration of the tremendous distances to quasars. The object cataloged as 0957+561 lies just a few degrees west of the bowl of the Big Dipper and consists of two quasars separated by only 6 seconds of arc. The spectra of these quasars are identical and even have the same red shift of 1.4136. Quasar spectra do resemble each other, but in detail they are as different as fingerprints. When two quasars so close together were discovered to have the same red shift and identical spectra, astronomers concluded that they were separate images of the same quasar.

The two images are formed by a **gravitational lens,** an effect first predicted by Einstein in 1936. The gravitational field of a galaxy between us and a distant quasar acts as a lens, bending the light from the quasar and focusing it into multiple images. A few dozen gravitational lenses are known (■ Figure 17-14); and, in most cases, the lensing galaxy is so far away it is difficult to detect. This provides powerful evidence that quasars really are distant objects as their large red shifts imply.

For more evidence that quasars are very distant, look again at Figure 17-10. Just above the image of the quasar lies the small, oval image of an elliptical galaxy, which, evidenced by its red shift, is about 7 billion light-years away. The spectrum of the quasar contains absorption lines with the same red shift as the elliptical galaxy. Evidently the lines are produced by the thin gas in the elliptical galaxy. From this we conclude that the quasar, though very bright, is actually farther away than the distant elliptical galaxy.

About 10 percent of quasars have absorption lines as well as emission lines in their spectra. These absorp-

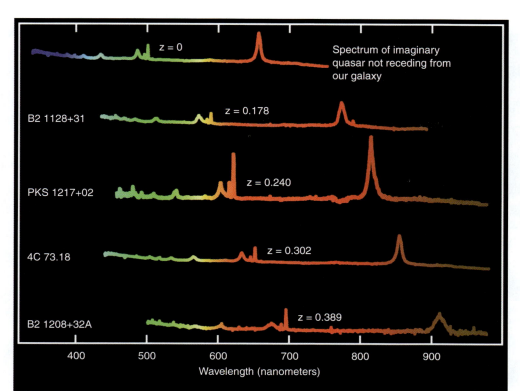

FIGURE 17-12
Spectra of four quasars are compared here with an idealized spectrum for an imaginary quasar that has no velocity of recession. The first three Balmer lines of hydrogen are visible, plus lines of other atoms, but the red shifts of the quasars move these lines to longer wavelengths. Nevertheless, the relative spacing of the spectral lines is unchanged, and astronomers can recognize the lines even with a large red shift. (*C. Pilachowski, M. Corbin, AURA/NOAO/NSF*)

tion lines have a smaller red shift than the corresponding emission lines, and, in some cases, multiple sets of absorption lines are observed, each with its individual red shift. Some of the lines are hydrogen lines, but some are lines of metals, heavier elements such as carbon, silicon, and magnesium. Apparently, these absorption lines are formed when the light from the quasar passes through objects on its way to Earth. The hydrogen lines must be formed as the light from the quasar passes through low-density clouds of hydrogen between the galaxies. The metal lines must be formed when the light passes through galaxies on its way to Earth. But the galaxies themselves are so distant that they are not visible, further evidence that quasars are very distant objects.

The quasars are very distant and must be superluminous and quite small. To understand quasars, we must examine their observed characteristics and compare that evidence with the unified model of active galaxies.

EVIDENCE OF QUASARS IN DISTANT GALAXIES

How do astronomers know that quasars are located in distant galaxies? Thousands of quasars have been found, and the largest known red shifts are a bit over 6, meaning that the objects are exceedingly far away. The lookback time to these objects is so large that we see them as they were when the universe was only 10 percent its present age. At such great distances it is difficult to detect the host galaxy because of the glare of the quasar.

Quasars with lower red shifts are not quite so far away, however, and fuzzy wisps are visible around such objects. By blocking the glare of the quasar, astronomers were able to photograph the spectrum of this quasar fuzz and found that it is the same as the spectrum of a normal galaxy with the same red shift as the quasar. For example, 3C273 appears to lie in a giant elliptical galaxy.

Moreover, some quasars have faint objects near them in the sky. The spectra of these faint objects are the same as the spectra of galaxies, and they have the same red shift as that of the quasar they accompany. This is evidence that these quasars are located in galaxies that are part of larger clusters of galaxies (■ Figure 17-15).

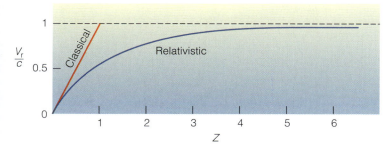

FIGURE 17-13
At high velocities, the relativistic Doppler formula must be used in place of the classical approximation. Note that no matter how large Z gets, the velocity can never equal that of light.

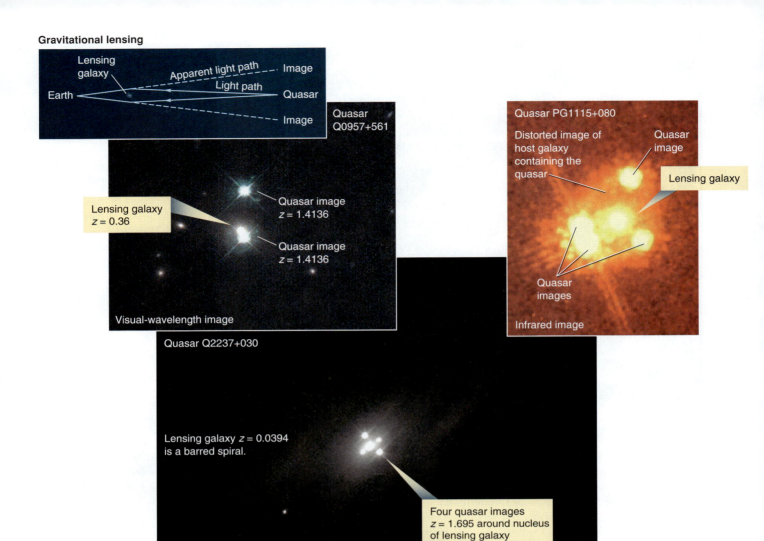

Gravitational lensing

Lensing galaxy
Apparent light path —— Image
Light path
Earth —— Quasar
—— Image

Quasar Q0957+561

Lensing galaxy
z = 0.36

Quasar image
z = 1.4136

Quasar image
z = 1.4136

Visual-wavelength image

Quasar PG1115+080

Distorted image of host galaxy containing the quasar

Quasar image

Lensing galaxy

Quasar images

Infrared image

Quasar Q2237+030

Lensing galaxy z = 0.0394 is a barred spiral.

Four quasar images
z = 1.695 around nucleus of lensing galaxy

Visual-wavelength image

FIGURE 17-14

Gravitational lensing can occur when a relatively nearby lensing galaxy is aligned with a distant quasar. This can produce multiple images of the quasar. In the case of the quasar at upper right, the host galaxy is visible as a faint, distorted ring around the lensing galaxy. Gravitational lenses give us direct evidence that quasars are at the great distances suggested by their red shifts. *(Q0957+561: George Rhee, NASA, STScI; color composite courtesy Bill Keel; PG1115+080: Chris Impy, Univ. of Arizona and NASA; Q2237+030: Image by W. Keel from data in the NASA/ESA Hubble Space Telescope archive, originally obtained with J. Westphal as Principal Investigator)*

Some quasars have characteristics related to active galaxies. For example, quasar 3C273 seems to be ejecting a jet much like the jets seen in active galaxies. By blocking the brilliance of the quasar, astronomers have been able to image other details that suggest active galaxies (❙ Figure 17-16). Some quasars even have radio lobes, which further suggests that they are related to active galaxies.

Photographs made with the largest telescopes on Earth and with the Hubble Space Telescope give us dramatic evidence. The galaxies that host quasars are in many cases clearly visible with these large telescopes. Furthermore, the galaxies are distorted or have nearby companions (❙ Figure 17-17). The host galaxies appear to be involved in collisions and interactions with other galaxies, or they have the distorted shapes that we

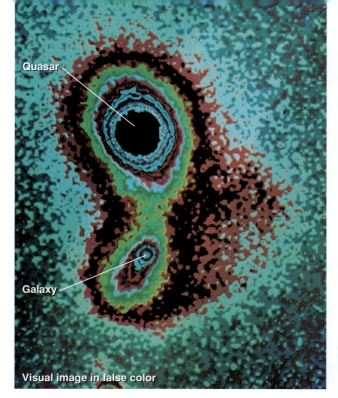

FIGURE 17-15

Quasar 0351+026 is located near a faint galaxy that has the same red shift and thus, presumably, the same distance from Earth as the quasar. The extended red region around the quasar (fuzz) is typical of lower-red-shift quasars and reveals the spectrum of a normal galaxy. Thus, this image shows two interacting galaxies, one of which contains a quasar. *(AURA/NOAO/NSF)*

recognize as the result of such collisions. Because we suspect that galaxy collisions can cause the core of a galaxy to erupt, we can also suspect that quasars may have been triggered into existence in the cores of galaxies by interactions between galaxies.

SUPERLUMINAL EXPANSION

As astronomers gathered more and more evidence and grew ever more confident that quasars were located in distant galaxies, they hit a snag. A few quasars were found that seem to be expanding faster than the velocity of light, a behavior called **superluminal expansion** that is forbidden by the theory of relativity. Thus we must consider this peculiar phenomenon before we can go further with our discussion of quasars.

Radio astronomers discovered superluminal expansion by using very long baseline interferometry (VLBI)—the interconnection of radio telescopes in different parts of the world. Such a network of antennae can resolve details about 0.002 second of arc in diameter. At the distance of quasar 3C273, for example, this corresponds to a few parsecs.

For a few quasars these radio maps show small blobs near the quasars, and maps made over a few years show these blobs moving away from the quasar. Quasar 3C273 has a blob that is moving about 0.0008 second of arc farther from the quasar each year (❙ Figure 17-18). If we believe that the quasar's red shift of 0.16 arises from the expansion of the universe, the quasar is about 960 Mpc distant, and the small-angle formula tells us that the blob is moving away from the quasar at a speed of 3.8 pc per year. This is the same as 12 ly per year, which is impossible. The theory of relativity clearly shows that nothing can travel faster than the velocity of light.

This superluminal expansion has been found in a number of quasars, and it is also seen in blazars. For quasars the apparent velocities of expansion are usually greater than five times the velocity of light, and for blazars the velocity is usually less than five times the velocity of light. As we have seen earlier in this chapter, we suspect that blazars are active galaxy cores in which one of the jets is pointed almost directly at us. This is an important clue.

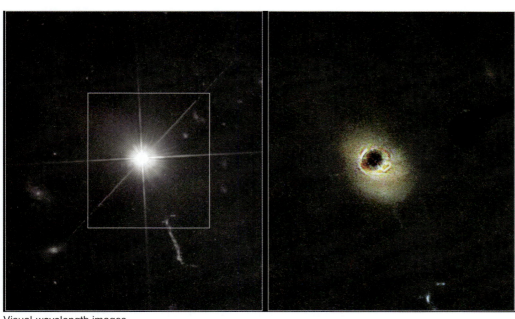

Visual-wavelength images

FIGURE 17-16

Quasar 3C273 is so bright it drowns out details around it (left). By blocking the brilliant light of the quasar (right), the Hubble Space Telescope can detect curving arms of stars in the galaxy that contains the quasar and can detect the inner parts of the jet extending to lower right. *(J. Bahcall, IAS; A. Marte, H. Ford, JHU; M. Clampin, G. Hartig [STScI]; G. Illingworth, UCO/Lick Observatory; ACS Science Team NASA and ESA)*

FIGURE 17-17

The bright object at the center of each of these images is a quasar. Fainter objects near the quasar are galaxies distorted by collisions. Compare the upper left ring-shaped galaxy with Figure 16-15 and compare the tail in the lower left image with the Antennae galaxies in Figure 16-12. (*J. Bahcall, Institute for Advanced Study, Mike Disney, University of Wales, NASA*)

but it clearly shows that a blob of gas in a relativistic jet can appear to travel faster than light if the jet points nearly toward Earth.

In fact, this interesting bit of physics has been observed closer to home. The jet in M87 has been observed to have an apparent velocity along a portion of its length of 2.5 times the speed of light. Evidently, the jet twists and turns as it leaves the nucleus of the galaxy, and one of those turns points the jet almost directly at Earth. Although the gas is moving at slightly over 0.9 times the speed of light, along that segment pointing toward us the gas appears to move 2.5 times faster than light. Even closer to home, within our own galaxy, a binary star containing a neutron star or a black hole has been observed ejecting material at speeds of about 1.25 times the speed of light. The gas is actually traveling at only 0.92 times the speed of light, but the material was ejected within 19° of our line of sight, and that produces the apparent superluminal motion.

Once the nature of superluminal expansion was understood, it was no longer an objection to quasars being located in very distant galaxies. Today the evidence seems very strong, and astronomers are able to combine evidence and theory in a logical argument that describes how a quasar works.

One way to solve the problem is to assume that the quasar is local, by which astronomers mean nearby. If 3C273 were only 10 Mpc away, then the blob is moving at a velocity of less than 15 percent the velocity of light. But, of course, all the other evidence points to quasars being very far away.

The clue is the jets in blazars that appear to be pointed almost straight at us. If the quasar is ejecting a jet of matter at nearly the velocity of light and that jet is pointed nearly toward Earth, a blob of matter in the jet could appear to move away from the quasar faster than the velocity of light. This is called the **relativistic jet model.**

To see how this can happen, consider ▌Figure 17-19. A blob of matter in the jet emits radiation that will reach Earth in the year 2005. In the next 35 years, that radiation travels 35 ly toward Earth, but the blob, traveling at 98 percent the velocity of light, travels 34 ly closer to Earth. Radiation that it emits from this second position is only 1 ly behind the radiation it emitted 35 years before. The newly emitted radiation will arrive on Earth in 2006, and we will see the blob move in only 1 year a distance that actually took 35 years. This is a simplified explanation that omits certain relativistic effects,

A MODEL QUASAR

The most satisfying part of science is combining evidence with theory to understand some aspect of nature. We have learned to interpret quasar red shifts, and we have reviewed the basic characteristics of quasars. Can we now assemble the evidence and theories in a logical argument that describes quasars? That is, can we create a model of a quasar?

Our model quasar is based on the geometry of a rotating accretion disk ejecting jets along its axis of rotation (page 363). Disks are very common in the universe. We have discovered rotating disks around protostars, neutron stars, and black holes. Our own galaxy has a disk shape, and in a later chapter we will see that our solar system has a disk shape. We have also seen that some disks can produce powerful jets of matter and radiation directed along the axis of rotation. Bipolar flows from protostars are low-energy examples, while the jets expelled by the X-ray binary SS433 are much more powerful. Jets expelled by active galactic nuclei are even more powerful. To create our model quasar, we imagine a galaxy containing this disk-jet structure at its core. Our model is, in fact, just an extension of the model of active galactic nuclei discussed earlier in this chapter.

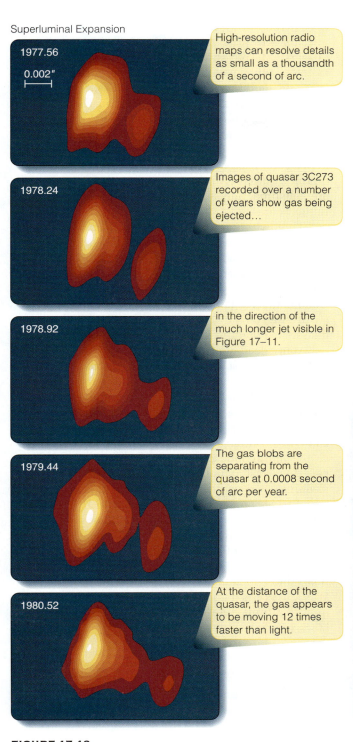

Superluminal Expansion

1977.56

0.002"

> High-resolution radio maps can resolve details as small as a thousandth of a second of arc.

1978.24

> Images of quasar 3C273 recorded over a number of years show gas being ejected…

1978.92

> in the direction of the much longer jet visible in Figure 17–11.

1979.44

> The gas blobs are separating from the quasar at 0.0008 second of arc per year.

1980.52

> At the distance of the quasar, the gas appears to be moving 12 times faster than light.

FIGURE 17-18
Very-high-resolution radio images of a few quasars show gas that seems to be moving faster than the speed of light.

To make the energy typical of a quasar we need a black hole of 10 million to a billion solar masses surrounded by a large, opaque disk. The innermost part of the accretion disk, much less than a light-year in diameter, would be intensely hot. If matter amounting to

about 1 solar mass per year flows through the accretion disk into the black hole, the energy released could power a quasar. Exactly how some of this matter and energy are focused into two oppositely directed high-speed jets isn't well understood, but magnetic fields are probably involved. In any case, as matter flows into the black hole, some is ejected in jets and a great deal of energy is radiated away. One astronomer described the situation by commenting, "Black holes are messy eaters."

This model could explain the different kinds of radiation we receive from quasars (see Figure 17-6). A small percentage of quasars are strong radio sources, and the radio radiation may come from synchrotron radiation produced in the high-energy gas and magnetic fields in the jets. Much of the light we see from a quasar is spread in a continuous spectrum, and that is the light that fluctuates quickly. Thus, it must come from a small region. Our model could produce this light in the innermost region surrounding the black hole. Because this is a very small region, roughly the size of our solar system, rapid fluctuations can occur, probably because of random fluctuations in the flow of matter into the

FIGURE 17-19
The relativistic jet model explains superluminal expansion by supposing that the quasar emits a jet that points nearly at Earth. If a blob of gas in the jet, traveling at nearly the velocity of light, emits radio energy toward Earth at two different times separated by 35 years, the first bundle of energy will have only a 1 ly lead. As the energy arrives at Earth, the blob will appear to move in 1 year a distance that actually took 35 years.

accretion disk. The emission lines in quasar spectra don't fluctuate rapidly, and that suggests they are produced in a larger region. In our model, emission lines are emitted by clouds of gas surrounding the core in a region many light-years in diameter. These gas clouds are excited by the intense synchrotron radiation streaming out of the central cavity.

Just as in the unified model of active galactic nuclei, the orientation of the accretion disk is important (see Figure 17-6). If the disk faces us so that one of its jets points directly toward us, we may see one kind of quasar. But if the disk is tipped slightly so the jet is not pointing straight at us, we may see a slightly different kind of quasar. Astronomers are now using this unified model to sort out the different kinds of quasars and active galaxies so we can understand how they are related. For example, using infrared radiation to penetrate dust, astronomers observed the core of the double-lobed radio galaxy Cygnus A (see page 362) and found an object much like a quasar. Astronomers have begun to refer to such objects as buried quasars.

As in the case of active galaxies, many quasars are found in distorted galaxies that are interacting with nearby companions. Such a collision could throw matter into a central black hole and trigger a quasar outburst.

QUASARS THROUGH TIME

When we look at a distant quasar, the look-back time is large, and we see the universe as it was long ago. That means we might be able to tell the story of how quasars formed, evolved, and died out. Perhaps we can trace the history of quasars through time.

In the next chapter, we will see evidence that the universe began about 14 billion years ago. The first clouds of gas began forming stars and falling together to form galaxies, and we suspect that some of that matter formed supermassive black holes at the centers of star clouds that became the nuclear bulges of galaxies. The abundance of matter flooding into these black holes could have triggered outbursts that we see as quasars.

We should also note that galaxies were closer together when the universe was young and had not expanded very much. Because they were closer together, the forming galaxies collided more often, and we have seen how collisions between galaxies could throw matter into supermassive black holes and trigger eruptions. We observe that quasars are often located in host galaxies that are distorted as if they were interacting with other galaxies.

The evidence suggests that when the universe was young, forming galaxies created supermassive black holes and, aided by interactions with other galaxies, large amounts of matter fell into those black holes, triggering quasar eruptions.

Quasars are most common with red shifts of about 2 and less common with red shifts above 2.7. The largest quasar red shifts are over 6, and such high-red-shift quasars are quite rare. Evidently, when we look at quasars with red shifts of 2 or so, we are looking back to an age when galaxies were actively forming, colliding, and merging. In that age of quasars, they were about 1000 times more common than they are now. When we look back to higher red shifts, we see fewer quasars because we are looking back to an age when the universe was so young it had not yet begun to form many galaxies. Nevertheless, even during the age of quasars, quasar eruptions must have been unusual. At any one time, only a few galaxies had quasars erupting in their cores.

Then where are all the dead quasars? An astronomer commented recently, "There is no way to get rid of supermassive black holes, so all of the galaxies that had short-lived quasars still have those black holes." Why don't we see those dead quasars today? Actually, astronomers have found quite a few. The quasar 3C273 is surprisingly near us in space. Evidently, the age of quasars is not quite over.

Many dead quasars are not truly dead—only sleeping. We have discovered that most galaxies contain supermassive black holes, and those black holes may have suffered quasar eruptions when the universe was younger, when galaxies were closer together, and when dust and gas were more plentiful. Quasar eruptions became less common as galaxies became more stable and as the abundance of gas and dust in the centers of galaxies was exhausted. The dormant black holes sleeping at the centers of the galaxies today can be triggered to become active galactic nuclei by collisions. Such AGN are much less energetic than quasars, evidently because there is now much less gas and dust available to feed the supermassive black holes.

Most galaxies contain sleeping supermassive black holes into which matter slowly trickles. Our own Milky Way Galaxy is a good example. It could have been a quasar a long time ago, but today its supermassive black hole is resting.

REVIEW CRITICAL INQUIRY

Why are most quasars so far away?

To analyze this question we must combine two factors. First, quasars are the active cores of galaxies, and only a small percentage of galaxies contain active cores. In order to sample a large number of galaxies in our search, we must extend our search to great distances. Most of the galaxies lie far away from us because the amount of space we search increases rapidly with distance. Just as most seats in a baseball stadium are far from home plate, most galaxies are far from our Milky Way Galaxy. Thus, most of the galaxies that might contain quasars lie at great distances.

But a second factor is much more important. The farther we look into space, the farther back in time we look. It seems that there was a time in the distant past when quasars were more common, and thus we see most of those quasars at large look-back times, meaning at large distances. For these two reasons, most quasars lie at great distances.

If our model of quasars is correct, then quasars must be triggered into eruption. What observational evidence supports that argument?

Our study of active galaxies has led us far out in space and back in time to quasars. The light now arriving from the most distant quasars left them when the universe was only a fraction of its present age. The quasars stand at the very threshold of the study of the history of the universe itself, the subject of the next chapter.

SUMMARY

Some galaxies have peculiar properties. Seyfert galaxies have small, luminous nuclei whose spectra contain emission lines broadened by violently moving gas. Double-lobed radio galaxies emit radio energy from the lobes on either side of the galaxies. Some have jets extending from the core of the galaxy into the radio lobes. Some giant elliptical galaxies have small, energetic cores, some of which have jets of matter rushing outward.

The proposed explanation for double-lobed radio galaxies is called the double-exhaust model. This model assumes that the galaxy ejects jets of relativistic particles in two beams that push into the intergalactic medium and blow up the radio lobes like balloons. The impact of the beams on the intergalactic medium produces the hot spots detected on the leading edges of many radio lobes. We can see instances where galaxies moving through the intergalactic medium leave behind trails of hot gas from their jets. Spiral jets suggest that the source of the jets is precessing.

The source of energy in the active galaxies is believed to be supermassive black holes in their centers. M87, a giant elliptical galaxy with a small, active core and a high-velocity jet, contains very high orbital velocities close to its center. This implies that its small core contains billions of solar masses, probably in the form of a supermassive black hole.

Active galactic nuclei seem to occur in galaxies involved in collisions or mergers. Seyfert galaxies, for example, are three times more common in interacting galaxies than in isolated galaxies. This suggests that an encounter between galaxies can throw matter into the central regions of the galaxies, where it can feed a black hole and release energy. Galaxies not recently involved in collisions will not have matter flowing into their central black hole and will not have active galactic nuclei.

The unified model of active galaxies supposes that what we see depends on the orientation of the black hole and its accretion disk. If we can see into the core, we see broad spectral lines and rapid fluctuations. If we see the disk edge-on, we see narrower spectral lines. If the jet points directly at us, we see a BL Lac object, also known as a blazar.

The quasars appear to be related objects. Their spectra show emission lines with very large red shifts. Their large red shifts imply that they are very distant, and the fact that they are visible at all implies that they are superluminous. Because they fluctuate rapidly, we conclude that they must be very small. A typical quasar can be 100 times more luminous than the entire Milky Way Galaxy but only a few times larger than our solar system.

Because quasars lie at great distances, we see them as they were when the universe was young. The highest-resolution images show that many quasars are the active cores of very distant galaxies. Furthermore, these galaxies are often distorted and have close companions. That suggests that the quasars contain supermassive black holes and are triggered into activity by interactions between galaxies. Quasars may also be linked to the formation of supermassive black holes as material fell together to form the first galaxies. In any case, heavy mass flow into a supermassive black hole could heat the gas and produce the properties of quasars we observe.

Superluminal expansion appears as blobs of material that appear to be rushing away from some quasars at speeds as high as 12 times the speed of light. Of course, it is impossible for the matter to exceed the speed of light, but we can understand it as matter traveling at slightly less than the speed of light in a jet directed almost exactly toward Earth. Then we would see the illusion of superluminal expansion. Thus, superluminal expansion does not contradict our understanding of quasars.

NEW TERMS

radio galaxy	unified model
active galaxy	BL Lac object
active galactic nucleus (AGN)	blazar
	quasar
Seyfert galaxy	relativistic Doppler formula
double-lobed radio source	gravitational lens
double-exhaust model	superluminal expansion
hot spot	relativistic jet model

REVIEW QUESTIONS

Ace Astronomy™ Assess your understanding of this chapter's topics with additional quizzing and animations at **http://astronomy.brookscole.com/seeds8e**

1. What is the difference between a type 1 Seyfert galaxy and a type 2 Seyfert galaxy? How might we explain the differences?

2. What evidence do we have that radio lobes are inflated by jets from active galactic nuclei?

3. What is the significance of spiral jets from galaxies?

4. What evidence do we have that AGNs contain black holes?

5. What properties of SS433 resemble the energy source in an AGN?

6. If you located a galaxy that had not interacted with another galaxy recently, would you expect it to be active or inactive? Why?

7. Seyfert galaxies are three times more common in close pairs of galaxies than in isolated galaxies. What does that suggest?

8. What evidence do we have that quasars are small?

9. What evidence do we have that quasars are very far away?

10. How do quasars resemble the AGN in Seyfert galaxies?

11. How does our model quasar account for the different components of a quasar's spectrum?

12. What does it mean that quasars are most common at a red shift of about 2?

13. Where are all the dead quasars?

5. If the active core of a galaxy contains a black hole of 10^6 solar masses, what will the orbital period be for matter orbiting the black hole at a distance of 0.33 AU? (*Hint:* See circular velocity, Chapter 5.)

6. If a quasar is 1000 times more luminous than an entire galaxy, what is the absolute magnitude of such a quasar? (*Hint:* The absolute magnitude of a bright galaxy is about -21.)

7. If a quasar in Problem 6 were located at the center of our galaxy, what would its apparent magnitude be? (*Hint:* Use the magnitude–distance formula.)

8. What is the apparent velocity of recession of 3C48 if its red shift is 0.37?

9. If the Hubble constant is 70 km/s/Mpc, how far away is the quasar in Problem 8? (*Hint:* Use the Hubble law.)

10. The hydrogen Balmer line Hβ has a wavelength of 486.1 nm. It is shifted to 563.9 nm in the spectrum of 3C273. What is the red shift of this quasar? (*Hint:* What is $\Delta\lambda$?)

DISCUSSION QUESTIONS

1. Do you think that our galaxy has ever been an active galaxy? Could it have hosted a quasar when it was young?

2. If a quasar is triggered in a galaxy's core, what would it look like to people living in the outer disk of the galaxy? Could life continue in that galaxy? (Begin by deciding how bright a quasar would look seen from the outer disk, considering both distance and dust.)

PROBLEMS

1. The total energy stored in a radio lobe is about 10^{53} J. How many solar masses would have to be converted into energy to produce this energy? (*Hints:* Use $E = m_0 c^2$. One solar mass equals 2×10^{30} kg.)

2. If the jet in NGC5128 is traveling 5000 km/s and is 40 kpc long, how long will it take for gas to flow from the core of the galaxy out to the end of the jet?

3. Cygnus A is 225 Mpc away, and its jet is about 50 seconds of arc long. What is the length of the jet in parsecs? (*Hint:* Use the small-angle formula.)

4. Use the small-angle formula to find the linear diameter of a radio source with an angular diameter of 0.0015 second of arc and a distance of 3.25 Mpc.

CRITICAL INQUIRIES FOR THE WEB

1. What object currently holds the distinction as the farthest known galaxy? Search the Web for information on this distant object and find out its red shift and distance. What is the look-back time for this object?

2. The text discusses the superluminal expansion of material from 3C273, but how common is this faster-than-light effect? Search the Internet for information on superluminal motion related to quasars and blazars and list several examples—with the expansion speeds implied by their motion. What is the highest superluminal expansion rate you can find?

3. Gravitational lenses were first predicted by Einstein in 1936 but were not observed until recently. Search the Web for instances of gravitational lensing of galaxies and quasars. For a particular case, discuss how the lens effect allows astronomers to determine information about the lensing and/or lensed objects that might not have been available without the alignment.

 Visit the Seeds *Foundations of Astronomy* companion Web site for critical thinking exercises, articles, and additional readings from InfoTrac College Edition, Brooks/Cole's online student library.

COSMOLOGY IN THE 21ST CENTURY

The Universe, as has been observed before, is an unsettlingly big place,
a fact which for the sake of a quiet life most people tend to ignore.

Douglas Adams, The Restaurant at the End of the Universe

K. Lanzetta, SUNY, R. Schaller for STScI, and NASA

Chapter 18

GUIDEPOST

This chapter marks a watershed in our study of astronomy. Since Chapter 1, our discussion has focused on learning to understand the universe. Our outward journey has discussed the appearance of the night sky, the birth and death of stars, and the interactions of the galaxies. Now we reach the limit of our journey in space and time, the origin and evolution of the universe as a whole.

The ideas in this chapter are the biggest and the most difficult in all of science. Our imaginations can hardly grasp such ideas as the edge of the universe and the first instant of time. Perhaps it is fitting that the biggest questions are the most challenging.

But this chapter is not an end to our story. Once we complete it, we will have a grasp of the nature of the universe, and we will be ready to focus on our place in that universe—the subject of the rest of this book.

VIRTUAL LABORATORIES

EVIDENCE FOR DARK MATTER

FATE OF THE UNIVERSE

What is the biggest number? A billion? How about a billion billion? That is only 10^{18}. How about 10^{100}? That very large number is called a googol, a number that is at least a billion billion times larger than the total number of atoms in all of the galaxies we can observe. Can you name a number bigger than a googol? Try a billion googols, or, better yet, a googol of googols. You can go even bigger than that—10 raised to a googol. (That's a googolplex.) No matter how large a number you name, we can immediately name a bigger number. That is what infinity means—big without limit.

If the universe is infinite in size, then you can name any distance you want, and the universe is bigger than that. Try a googol to the googol light-years. If the universe is infinite, there is more universe beyond that distance.

Is the universe infinite or finite? In this chapter, we try to answer that question and others, not by playing games with big numbers, but through **cosmology,** the study of the universe as a whole.

Studying the entire universe is difficult because we have expectations about the universe that are in some cases wrong. This chapter is carefully organized to begin with our common expectations, test them against observations, compare them with theories, and build a modern understanding of cosmology. Like wilderness explorers, we must proceed carefully step by step to reach our goal.

18-1 INTRODUCTION TO THE UNIVERSE

Everyone knows what a gold mine looks like, but few people have actually visited a gold mine. If you explored a real gold mine, you might be surprised at what you find. Similarly, we all have an impression of the universe as a vast depth filled with galaxies (■ Figure 18-1), but as we begin exploring the universe, we must sort out our expectations so they do not mislead us.

THE EDGE–CENTER PROBLEM

In our daily lives, we are accustomed to boundaries. Rooms have walls, athletic fields have boundary lines, countries have borders, oceans have shores. It is natural for us to think of the universe as having an edge, but that idea leads us astray.

If the universe had an edge, imagine going to that edge. What would you find there: a wall of cardboard? A great empty space? Nothing? We are not talking about

Ace✩Astronomy™ The AceAstronomy icon throughout the text indicates an opportunity for you to test yourself on key concepts and to explore animations and interactions on the AceAstronomy Web site at **http://astronomy.brookscole.com/seeds8e**

Visual + infrared image

FIGURE 18-1

The entire sky is filled with galaxies. Some lie in clusters of thousands, and others are isolated in nearly empty voids between the clusters. In this image of a typical spot on the sky, bright objects with spikes caused by diffraction in the telescope are nearby stars. All other objects are galaxies ranging from the nearby face-on spiral at upper right to the most distant galaxies visible only in the infrared, shown as red in this composite image. *(R. Williams, STScI HDF-South Team, NASA)*

an edge to the distribution of matter, but an edge to space itself, so presumably we could not reach beyond the edge and feel around.

An edge to the universe seems to violate common sense, and modern cosmologists assume that the universe cannot have an edge. That is, the universe must be unbounded.

If the universe has no edge, then it cannot have a center. We find the centers of things—galaxies, globular clusters, oceans, pizzas—by referring to their edges. If we agree that the universe cannot have an edge, then it cannot have a center.

We must think carefully about the universe and avoid using the ideas of an edge or a center. Those ideas are misleading. Of course, if the universe is infinite, then it has no edge or center. We have no edge–center problem if we think of the universe as infinite. Could the universe be finite—could it contain a limited volume? We will see later in this chapter how the universe could be finite but have no edge or center. Before we go further with that idea, we must deal with another of our expectations, the beginning of the universe.

THE NECESSITY OF A BEGINNING

We have all noticed that the night sky is dark. However, reasonable assumptions about the geometry of the universe can lead us to the conclusion that the night sky should glow as brightly as a star's surface. This conflict between observation and theory is called **Olbers's para-**

Thinking About an Abstract Idea: Reasoning by Analogy

"The economy is overheating and it may seize up," an economist might say. Economists like to talk in analogies because economics is often abstract, and one of the best ways to think about abstract problems is to find a more approachable analogy. Rather than discuss the details of the national economy, we might make conclusions about how the economy works by thinking about how a gasoline engine works. Of course, if our analogy is not a good one, it can mislead us.

Much of astronomy is abstract, and cosmology is the most abstract subject in astronomy. Furthermore, cosmology is highly mathematical, and unless we are prepared to learn some of the most difficult mathematics known, we must use analogies, such as lines of sight in a forest.

Reasoning by analogy is a powerful technique. An analogy can reveal unexpected insights and lead us to further discoveries. But carrying an analogy too far can be misleading. We might compare the human brain to a computer and that would help us understand how data flow in and are processed and how new data flow out. But our analogy is flawed. Data in computers are stored in specific locations, but memories are stored in the brain in a distributed form. No single brain cell holds a specific memory. If we carry the analogy too far, it can mislead us. Whenever we reason by analogy, we should be alert for potential problems.

As you study any science, be alert for analogies. They are tremendously helpful, but we must always take care not to carry them too far.

---

dox after Heinrich Olbers, a Viennese physician and astronomer, who discussed the problem in 1826.

However, Olbers's paradox is not Olbers's, and it isn't a paradox. The problem of the dark night sky was first discussed by Thomas Digges in 1576 and was further analyzed by astronomers such as Johannes Kepler in 1610 and Edmund Halley in 1721. Olbers gets the credit through an accident of scholarship on the part of modern cosmologists who did not know of previous discussions. What's more, Olbers's paradox is not a paradox. We will

be able to understand why the night sky is dark by revising our assumptions about the nature of the universe.

To begin, let's state the so-called paradox. Suppose we assume that the universe is infinite and uniformly filled with stars. (The aggregation of stars into galaxies makes no difference to our argument.) If we look in any direction, our line of sight must eventually reach the surface of a star. Look at ▌ Figure 18-2, where we use the analogy of lines of sight in a forest (Window on Science 18-1). When we are deep in a forest, every line of sight

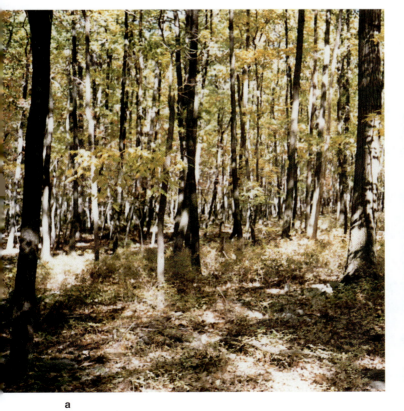

a

FIGURE 18-2

(a) Every direction we look in a forest eventually reaches a tree trunk, and we cannot see out of the forest. *(Photo courtesy Janet Seeds)* (b) If the universe is infinite and uniformly filled with stars, then any line from Earth should eventually reach the surface of a star. This assumption predicts that the night sky should glow as brightly as the surface of the average star, a puzzle commonly referred to as Olbers's paradox.

b

terminates on a tree trunk. In our example, every line of sight out into space should terminate on the surface of a star, and the entire sky should be as bright as the surface of an average star. It should not get dark at night.

Of course, the more distant stars would be fainter than nearby stars because of the inverse square law. However, the farther we look into space the larger the volume we see, and the two effects cancel out. Then given our assumptions, every spot on the sky must be occupied by the surface of a star, and the night sky should not be dark.

Imagine the entire sky glowing with the brightness of the surface of the sun. The glare would be overpowering. In fact, the radiation would rapidly heat Earth and all other celestial objects to the average temperature of the surface of the stars, a few thousand degrees. Thus, we can pose Olbers's paradox in another way: Why is the universe so cold?

Olbers assumed that the sky was dark because clouds of matter in space absorb the radiation from distant stars. But this interstellar medium would gradually heat up to the average surface temperature of the stars, and the gas and dust clouds would glow as brightly as the stars.

It is interesting to note that Copernicus and other astronomers of his time saw no paradox in the dark night sky because they believed in a finite universe. Once they looked beyond the sphere of stars, there was nothing to see except, perhaps, the dark floor of heaven. Soon after the time of Galileo and Newton, western astronomers began to think of an infinite universe, and that led to the paradox.

Today, cosmologists believe they understand why the sky is dark, and the universe is cold. Olbers's paradox makes the incorrect prediction that the night sky should be bright because it is based on an incorrect assumption. The universe is not eternal. That is, it is not infinitely old.

This solution to Olbers's paradox was first suggested by Edgar Allan Poe in 1848. He proposed that the night sky was dark because the universe was not infinitely old but had been created at some time in the past. The more distant stars are so far away that light from them has not reached us yet. That is, if we look far enough, the look-back time is greater than the age of the universe, and we look back to a time before stars began to shine. Thus, the night sky is dark because the universe is not infinitely old.

This is a powerful idea because it clearly illustrates the difference between the universe and the observable universe. The universe is everything that exists, and it could be infinite. But the **observable universe** is the part that we can see. We will learn later that the universe is about 14 billion years old. In that case, the observable universe has a radius of about 14 billion light-years. Do not confuse the observable universe, which is finite, with the universe as a whole, which could be infinite.

When we described Olbers's paradox, we assumed without noticing that the universe was infinitely old. This illustrates the importance of assumptions in cosmology and serves as a warning that our commonsense expectations are not dependable. All of astronomy is reasonably unreasonable—that is, reasonable assumptions often lead to unreasonable results. That is especially true in cosmology, so we must proceed with care.

Our expectation that the universe be eternal and unchanging is clearly wrong. Olbers's paradox makes a beginning a necessity. In the next section we will discover that we should not expect the universe to be static; observations show that it is expanding.

COSMIC EXPANSION

In the early 20th century, astronomers began photographing the spectra of galaxies, and they noticed that the spectral lines in galaxy spectra had red shifts. In 1929, Edwin P. Hubble published his discovery that the size of the red shift was proportional to the distance to the galaxy. (We discussed this as the Hubble law in Chapter 16, where we used it to estimate the distances to galaxies.) When interpreted using the Doppler effect, these red shifts imply that the galaxies are receding from each other and that the universe is expanding.

The red shifts of the galaxies are dramatic. Nearby galaxies have small red shifts, but when we look at more distant galaxies, the red shifts are quite large. ▌ Figure 18-3 shows spectra of galaxies in galaxy clusters at various distances. The Virgo cluster is relatively nearby, and its red shift is small. The Hydra cluster is very distant, and its red shift is so large that the two dark lines formed by once-ionized calcium are shifted from the near ultraviolet well into the visible part of the spectrum.

The important point about Hubble's discovery is that the red shifts are proportional to distance. That shows that the galaxies are receding from each other and the universe is expanding uniformly. But notice that the expansion does not imply that we are at the center of the universe. To see why, look at ▌ Figure 18-4, which shows an analogy in which we imagine baking raisin bread. As the dough rises, the raisins are pushed away from each other uniformly at velocities that are proportional to distance. Two raisins that were originally close to each other are pushed apart slowly, but two raisins that were far apart, having more dough between them, are pushed apart faster. If bacterial astronomers lived on a raisin in our raisin bread, they could observe the red shifts of the other raisins and derive a bacterial Hubble law. They could conclude that their universe was expanding uniformly. But it does not matter which raisin the bacterial astronomers lived on. There is no center to the expansion, and they would get the same Hubble law no matter what raisin they lived on. Similarly, there is no center to the expansion

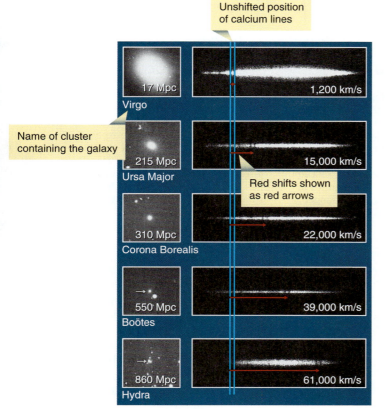

Unshifted position of calcium lines

Name of cluster containing the galaxy

Red shifts shown as red arrows

17 Mpc	1,200 km/s
Virgo	
215 Mpc	15,000 km/s
Ursa Major	
310 Mpc	22,000 km/s
Corona Borealis	
550 Mpc	39,000 km/s
Boötes	
860 Mpc	61,000 km/s
Hydra	

ACTIVE FIGURE 18-3

These galaxy spectra extend from the near ultraviolet at left to the blue part of the visible spectrum at right. The two dark absorption lines of once-ionized calcium are prominent in the near ultraviolet. The red shifts in galaxy spectra are expressed here as velocities of recession. Note that the apparent velocity of recession is proportional to distance. *(Caltech)*

Ace⊙Astronomy™ Go to AceAstronomy and click Active Figures to see "Cosmic Red Shift." Graph your own observations of red shift and distance.

of the universe and astronomers in any galaxy will see the same law of expansion.

When we look at Figure 18-4, we see the edge of the loaf of raisin bread and we can identify a center. This happens because our raisin bread analogy breaks down when we look at the crust of the bread. Remember that our universe cannot have an edge or a center, so there can be no center to the expansion. We must think and speak very carefully as we discuss cosmology (Window on Science 18-2).

THE NECESSITY OF A BIG BANG

The discovery of the expansion of the universe led astronomers to an astonishing conclusion. The universe must have begun with an event of cosmic fury.

Imagine that we have a videotape of the expanding universe, and we run it backward. We see the galaxies moving toward each other. There is no center to the expansion of the universe, so we do not see galaxies approaching a single spot; but, rather, we see the distances between all galaxies decreasing, and eventually the galaxies begin to merge. If we run our videotape far enough back, we see the matter and energy of the universe compressed into a high-density, high-temperature state. The expanding universe must have begun from this moment of extreme conditions, which modern astronomers call the **big bang.**

We must not allow ourselves to think of the big bang as an explosion. We have concluded that the universe cannot have a center, so the big bang cannot have occurred at a single point. The entire universe had been filled with the high-density, high-temperature state we call the big bang. Our challenge in this chapter is to understand how that could have happened without a center or an edge. We will resolve that problem when we refine our understanding of space and time.

ACTIVE FIGURE 18-4

The raisin bread analogy for the expansion of the universe. As the dough rises, raisins are pushed apart with velocities proportional to distance. The expansion has no center. A colony of bacteria living on any raisin will find that the red shifts of the other raisins are proportional to their distances.

Ace⊙Astronomy™ Go to AceAstronomy and click Active Figures to see "Raisin Bread." You can watch raisin bread rise.

Why Scientists Speak Carefully: Words Lead Thoughts

There are certain words we should never say, not even as a joke. We should never call our friend "fool," for example, because we might begin to think of our friend as a fool. Words lead thoughts. It works in advertising and politics, and it can work in science, too. Using words carelessly can lead us to think carelessly.

In science, there are certain ways to say things, not because scientists are sticklers for good grammar, but because if we say things wrongly we begin to think things wrongly. For example, a biologist would never let us say a beehive knows it must store food. "No, don't say it that way," the biologist would object. "The hive is just a collection of individual creatures and it can't 'know' anything. Even the individual bees don't really 'know' things. The instinctive behavior of individual bees causes food to accumulate in the hive. That's the way to say it."

All scientists are careful of language because careless words can mislead us.

We would never refer to the center of the universe, for example, but we must also be careful not to say things like "galaxies flying away from the big bang." Those words imply a center to the expansion of the universe, and we know the expansion must be centerless. Rather, we should say "galaxies flying away from each other."

Whatever science you study, notice the customary ways of using words. It is not just a matter of convention. It is a matter of careful thought.

For now, we can solve an easier problem. How long ago did the universe begin? We can estimate the age of the universe with a simple calculation. If we must drive to a city 100 miles away and we can travel 50 miles per hour, we divide distance by rate of travel and learn the travel time—in this example, 2 hours. To find the age of the universe, we can divide the distance between two galaxies by the velocity with which they are separating and find out how long they have taken to reach their present separation.

Dividing distance by velocity tells us the age of the universe, and the Hubble constant (Chapter 16) simplifies our task further. The Hubble constant H has the units km/s per Mpc, which is a velocity divided by a distance. If we calculate $\frac{1}{H}$ we have a distance divided by velocity. To make the division give us an age, we must convert megaparsecs to kilometers, and then the distances will cancel out and we will have an age in seconds. To get years, we must divide by the number of seconds in a year. If we make these simple changes in units, the age of the universe in years is approximately 10^{12} divided by H in its normal units, km/s/Mpc.

$$T \approx \frac{1}{H} \times 10^{12} \text{ years}$$

This is known as the **Hubble time.** For example, if H is 70 km/s/Mpc, then the age of the universe can be no older than $10^{12}/70$, which equals 14 billion years.

The Hubble time gives us an estimate of the age of the universe. We will fine tune our estimate later in this chapter, but for the moment we can conclude that basic observations of the recession of the galaxies require that the universe began with a big bang about 14 billion years ago.

Our instinct is to think of the big bang as a historical event, like the Gettysburg Address—something that happened long ago and happened in a particular place. But the big bang isn't really over. The look-back time makes it possible for us to observe the big bang directly. The look-back time to nearby galaxies is only a few million years, and the look-back to more distant galaxies is a large fraction of the age of the universe (Figure 18-5). Suppose we look between the distant galaxies, back to the time of the big bang. We should be able to detect the hot gas that filled the universe long ago.

Although our imaginations try to visualize the big bang as a localized event, we must keep firmly in mind that the big bang did not occur at a single place but filled the entire volume of the universe. The matter of which we are made was part of that big bang, so we are inside the remains of that event, and the universe continues to expand around us. We cannot point to any

Visual + infrared image

FIGURE 18-5

This faint galaxy (arrow) is believed to be so distant that its light has been traveling for most of the history of the universe. We see it as it was about 1 billion years after the big bang. The galaxy is bright in the infrared but it is not detectable at visual wavelengths. *(Ken Lanzetta and Amos Yahil, Stony Brook and NASA)* particular place and say, "The big bang occurred over there." The big bang occurred everywhere, and, in whatever direction we look, at great distance we look back to the age when the universe was filled with hot gas (▌ Figure 18-6).

The radiation that comes to us from this great distance has a tremendous red shift. The most distant visible objects are faint galaxies and quasars, with red shifts a bit over 6. In contrast, the radiation from the hot gas of the big bang has a red shift of about 1100. Thus the light emitted by the big bang gases arrives at Earth as infrared radiation and short radio waves. We can't see it with our eyes, but we should be able to detect it with infrared and radio telescopes. Unlike the Gettysburg Address, the big bang isn't over, and we should be able to see it happening if we can detect the radiation it emitted. That amazing discovery is the subject of the next section.

THE COSMIC BACKGROUND RADIATION

If radiation is now arriving from the big bang, then it should be detectable. The story of that discovery begins in the mid-1960s when two Bell Laboratories physicists, Arno Penzias and Robert Wilson, were using a horn antenna to measure the radio brightness of the sky (▌ Figure 18-7a). Their measurements showed a peculiar noise in the system, which they at first attributed to droppings from pigeons living inside the antenna. Perhaps they would have enjoyed cleaning the antenna more if they had known they would win the 1978 Nobel Prize for physics for the discovery they were about to make.

When the antenna was cleaned, they measured the brightness of the sky at radio wavelengths and again found the low-level radio noise. The pigeons were innocent, but what was causing the signal?

The explanation for the noise goes back to 1948, when George Gamow predicted that the early big bang would be very hot and would emit copious black body radiation. A year later, physicists Ralph Alpher and Robert Herman pointed out that the large red shift of the big bang would lengthen the wavelengths of the radiation into the far infrared and radio part of the spectrum. There was no way to detect this radiation until the mid-1960s, when Robert Dicke at Princeton concluded the radiation should be just strong enough to detect with newly developed techniques. Dicke and his team began building a receiver.

When Penzias and Wilson heard of Dicke's work, they recognized the noise they had detected as radiation from the big bang, the **cosmic microwave background radiation.**

The background radiation was measured at many wavelengths as astronomers tried to confirm that it did indeed follow a black body curve. Some measurements could be made from the ground, but balloon and rocket observations were critical and difficult. A few disturbing

a   A region of the universe during the big bang

b   A region of the universe now

c
The present universe as it appears from our galaxy

FIGURE 18-6
Three views of a small region of the universe centered on our galaxy. (a) During the big bang, the region is filled with hot gas and radiation. (b) Later, the gas has formed galaxies, but we can't see the universe this way because the look-back time distorts what we see. (c) Near us we see galaxies, but farther way we see young galaxies (dots), and at a great distance we see radiation (arrows) coming from the hot gas of the big bang.

observations suggested that the radiation departed inexplicably from a black body curve and thus might not come from the big bang.

In November 1989, the orbiting Cosmic Background Explorer (COBE) began taking the most accurate measurements of the background radiation ever made. Within two months, the verdict was in (❚ Figure 18-7b). To an accuracy of better than 1 percent, the radiation follows a black body curve with a temperature of 2.725 ± 0.002 K—in good agreement with accepted theory.

It may seem strange that the hot gas of the big bang seems to have a temperature of only 2.7 K, but recall the tremendous red shift. We see light that has a red shift of about 1100—that is, the wavelengths of the photons are about 1100 times longer than when they were emitted. The gas clouds that emitted the photons had a temperature of about 3000 K, and they emitted black body radiation with a λ_{max} of about 1000 nm (Chapter 7). Although this is in the near infrared, the gas would also have emitted enough visible light to glow orange-red. But the red shift has made the wavelengths about 1100 times longer, so λ_{max} is about 1 million nm (equivalent to 1 mm). Thus, the hot gas of the big bang seems to be 1100 times cooler, about 2.7 K.

The cosmic microwave background radiation is evidence that a big bang really did occur. In fact, the evidence is so strong that an alternative theory of the universe was abandoned. The **steady state theory,** popular with some astronomers during the 1950s and 1960s, held that the universe was eternal and unchanging. If the universe had no beginning, then it could never have had a big bang. According to the theory, new matter continuously appeared from nothing to maintain the density of the expanding, steady-state universe. The discovery of the cosmic microwave background radiation in 1965 led astronomers to abandon the steady-state theory within a few years. We know there was a big bang because, using radio and infrared telescopes, we can see it happening.

THE STORY OF THE BIG BANG

Simple observations of the darkness of the night sky and the red shifts of the galaxies tell us that the universe must have begun with a big bang. In fact, we have direct observational evidence of the big bang in the form of the cosmic microwave background radiation. Theorists can combine these observations with modern physics to tell the story of how the big bang occurred. As we review this story, remember that the big bang did not occur at a specific place. It filled the entire volume of the universe from that first moment.

We cannot begin our history at time zero, because we do not understand the physics of matter and energy under such extreme conditions, but we can come close. If we could visit the universe when it was very young, only 10 millionths of a second old, for instance, we

a

$T = 2.725 \pm 0.002$ K

b

FIGURE 18-7
(a) Robert Wilson (left) and Arno Penzias pose before the horn antenna with which they discovered the cosmic microwave background radiation. *(AT&T Archives)* (b) The background radiation from the big bang is observed in the infrared and radio part of the spectrum. At first, the critical measurements near the peak of the curve had to be made from balloons and rockets, but the COBE satellite observed from orbit. These COBE data show that the radiation fits a black body curve with a temperature of 2.725 K.

would find it filled with high-energy photons having a temperature well over 1 trillion (10^{12}) K and a density greater than 5×10^{13} g/cm³, nearly the density of an atomic nucleus. When we say the photons had a given temperature, we mean that the photons were the same as black body radiation emitted by an object of that temperature. Thus, the photons in the early universe were gamma rays of very short wavelength and therefore very high energy. When we say that the radiation had a certain density, we refer to Einstein's equation $E = m_0 c^2$. We can express a given amount of energy in the form of radiation as if it were matter of a given density.

If photons have enough energy, two photons can combine and convert their energy into a pair of particles—a particle of normal matter and a particle of **antimatter.** When an antimatter particle meets its matching particle of normal matter—when an antiproton meets a normal proton, for example—the two particles annihilate each other and convert their mass into energy in the form of two gamma rays. In the early universe, the photons had enough energy to produce proton–antiproton pairs; but when the particles collided, they converted their mass back into photons. Thus, the early universe was a dynamic soup of energy flickering from photons into particles and back again.

While all this went on, the expansion of the universe cooled the gamma rays to lower energy. By the time the universe was 0.0001 second old, its temperature had fallen to 10^{12} K. By this time, the average energy of the gamma rays had fallen below the energy equivalent to the mass of a proton or a neutron, so the gamma rays could no longer produce such heavy particles. The particles combined with their antiparticles and quickly converted most of the mass into photons.

It would seem from this that all of the protons and neutrons should have been annihilated with their antiparticles, but for reasons that are poorly understood a small excess of normal particles existed. For every billion protons annihilated by antiprotons, one survived with no antiparticles to destroy it. Consequently, we live in a world of normal matter, and antimatter is very rare.

Although the gamma rays did not have enough energy to produce protons and neutrons, they could produce electron–positron pairs, which are about 1800 times less massive than protons and neutrons. This continued until the universe was about 4 seconds old, at which time the expansion had stretched and cooled the gamma rays to the point where they could no longer create electron–positron pairs. The protons, neutrons, and electrons of which our universe is made were produced during the first 4 seconds of its history.

This soup of hot gas and radiation continued to cool and eventually began to form atomic nuclei. High-energy gamma rays can break up a nucleus, so the formation of such nuclei could not occur until the universe had cooled somewhat. By the time the universe was about 2 minutes old, protons and neutrons could link to form deuterium, the nucleus of a heavy hydrogen atom; by the end of the next minute, further reactions began converting deuterium into helium. But no heavier atoms could be built because no stable nuclei exist with atomic weights of 5 or 8 (in units of the hydrogen atom). Cosmic element building during the big bang had to proceed rapidly, step by step, like someone hopping up a flight of stairs (❚ Figure 18-8). The lack of stable nuclei at atomic weights of 5 and 8 meant there were missing steps in the stairway, and the step-by-step reactions had great difficulty jumping over these gaps. A tiny amount of lithium was produced but nothing heavier.

By the time the universe was 30 minutes old, it had cooled sufficiently that nuclear reactions had stopped. About 25 percent of the mass was helium nuclei, and the rest was in the form of protons—hydrogen nuclei. This is the cosmic abundance we see today in the oldest stars. (The heavier elements, remember, were built by nucleosynthesis inside many generations of massive stars.) The cosmic abundance of helium was fixed during the first minutes of the universe.

At first, the universe was dominated by radiation. The gamma rays interacted continuously with the matter, and they cooled together as the universe expanded. The gas was ionized because it was too hot for the nuclei to capture electrons to form neutral atoms, and the free electrons made the gas very opaque. A photon could not travel very far before it collided with an electron and was deflected. Thus, radiation and matter were locked together.

As the young universe expanded, it went through three important changes. First, when the universe reached an age of roughly 50,000 years, the density of the energy present as photons became less than the density of the gas. Before this, matter could not clump together because the intense sea of photons smoothed the gas out. Once the density of the radiation fell below that of matter, the matter could begin to draw together and form the clouds that eventually became stars and galaxies.

As the universe continued to expand, the ionized gas became less dense, which meant that the free electrons were spread farther apart. As the universe reached the age of a few hundred thousand years, the second important change began. The free electrons were spread so far apart, the photons could travel for thousands of parsecs before getting scattered off an electron. That is, the universe began to grow more transparent. At about the same time, the third change happened. As the

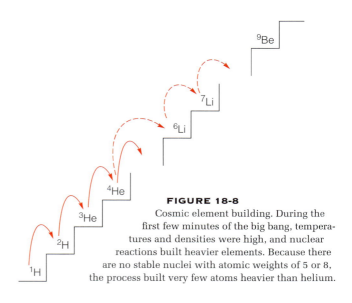

FIGURE 18-8
Cosmic element building. During the first few minutes of the big bang, temperatures and densities were high, and nuclear reactions built heavier elements. Because there are no stable nuclei with atomic weights of 5 or 8, the process built very few atoms heavier than helium.

Dense ionized gas · Low-density ionized gas · Neutral gas after recombination

a · b · c

FIGURE 18-9

Photons scatter from electrons (blue) easily but hardly at all from the much more massive protons (red). (a) When the universe was very dense and ionized, photons could not travel very far before they scattered off an electron. This made the gas opaque. (b) As the universe expanded, the electrons were spread further apart, and the photons could travel farther; this made the gas more transparent. (c) After recombination, most electrons were locked to protons to form neutral atoms, and the gas was highly transparent.

falling temperature of the universe reached 3000 K, protons were able to capture and hold free electrons to form neutral hydrogen, a process called **recombination.** As the free electrons were gobbled up, the gas finally became transparent and the photons could travel through the gas without being absorbed (Figure 18-9). Although the gas continued to cool, the photons no longer interacted with the gas, and consequently the photons retained the black body temperature that the gas had at recombination. Those photons, with a black body temperature of 3000 K, are what we observe today as the cosmic microwave background radiation. Remember that the large red shift makes that gas appear to have a temperature of about 2.7 K.

Recombination left the gas of the big bang neutral and transparent. At first the universe was filled with the glow of the hot gas; but, as the universe expanded and cooled, the glow faded into the infrared, and the universe entered what cosmologists call the **dark age,** a period lasting hundreds of millions of years during which the universe expanded in darkness.

The dark age ended as the first stars began to form. The gas from which these first stars formed contained almost no metals and was consequently highly transparent. Models show that, because the first stars formed from this metal-poor gas, they were very massive, very luminous, and very short lived. That first violent burst of star formation produced enough ultraviolet light to begin ionizing the gas, and today's astronomers, looking back to the most distant visible quasars and galaxies, can see traces of that **reionization** of the universe. (Figure 18-10)

 Figure 18-11 summarizes our story of the big bang, from the formation of helium in the first three minutes, through energy–matter equality, recombination, and finally reionization. Reionization marks the end of the dark ages and the beginning of the age of stars and galaxies in which we live today.

FIGURE 18-10

In this artist's conception of reionization, the first stars produce floods of ultraviolet photons that ionize the gas in expanding bubbles. Such a storm of star formation ended an age when the universe had expanded in darkness. Spectra of the most distant quasars reveal that those first galaxies were surrounded by neutral gas that had not yet been fully ionized. Thus the look-back time allows modern astronomers to observe the age of reionization. *(K. Lanzetta, SUNY, A. Schaller for STScI, and NASA)*

So far in this chapter we have compared our simple expectations with basic observations, and we have discovered some fundamental properties of the universe. Now it is time to stretch our imaginations and refine our expectations. The universe is much more interesting than we have imagined to this point.

18-2 THE SHAPE OF SPACE AND TIME

Our simple expectations have led us to a puzzle. How can the universe expand if it does not have an edge or a center? Solving that problem is the key to understanding modern cosmology, but our everyday expectations about how the world works are not a help. To solve the puzzle, we must put our reasonable expectations on hold and look carefully at how space and time behave on cosmic scales.

LOOKING AT THE UNIVERSE

Let's begin by carefully examining the general appearance of the universe. Not only do we see no edge to the universe and no center, there is no preferred direction.

The gas and dust in our own galaxy block our view of the distant universe along the Milky Way, but once we correct for that, the universe looks the same over the entire sky. Astronomers refer to this as **isotropy,** the property of being the same in all directions. No matter what direction we look we see the same kinds of galaxies and galaxy clusters. The universe seems isotropic.

Furthermore, the cosmic microwave background radiation is isotropic. Astronomers must correct for Doppler shifts caused by the motion of the Earth around the sun, the sun around our galaxy, and our galaxy around the center of mass of the local group, but those are only local variations. Once those local variations are taken into account, the background radiation is almost perfectly uniform over the entire sky.

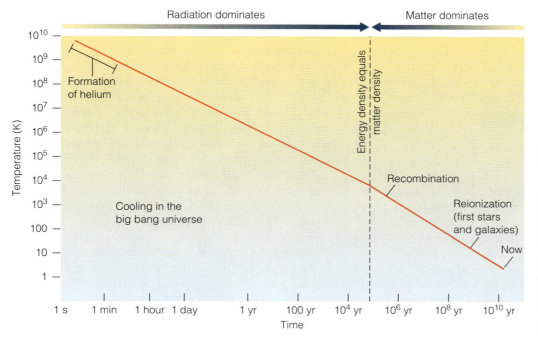

FIGURE 18-11

During the first few minutes of the big bang, some hydrogen was fused to produce helium, but the universe quickly became too cool for such fusion reactions to occur. The rate of cooling increased as matter began to dominate over radiation. Recombination freed the radiation from the influence of the gas, and reionization was caused by the birth of the first stars.

It certainly appears that the universe is the same everywhere. Of course, there are local variations; some points in space are inside galaxies, and some are isolated far from any galaxy, but overall, in its general properties, one place in the universe is the same as any other place. That is, there are no special places in the universe. We cannot point at any particular location in the universe and say, "That location is different from any other place in the universe." Astronomers refer to this as **homogeneity,** the property of being uniform. On the largest scales, the universe appears to be homogeneous.

Astronomers observe that the universe is isotropic and homogeneous on the largest scales, and that leads to an assumption that is critically important in cosmology. The **cosmological principle** says that any observer in any galaxy sees the same general features of the universe. Of course some observers live in spiral galaxies and some live in elliptical galaxies, but those are only local variations. According to the cosmological principle, when we look at the general features of the universe, we should all see the same kind of universe no matter where our planets are located.

Evolutionary changes are not included in the cosmological principle. If the universe is expanding and the galaxies are evolving, observers living at different times may see galaxies at different stages. The cosmological principle says that once observers correct for evolutionary changes, they should see the same general features.

The cosmological principle is actually an extension of the Copernican principle. Copernicus said that Earth is not in a special place; it is just one of a number of planets orbiting the sun. The cosmological principle says that there are no special places in the universe. Local irregularities aside, one place is just like another. Our location in the universe is typical of all other locations.

If we accept the cosmological principle, then we may not imagine that the universe has an edge or a center. Such locations would not be the same as all other locations. Observations confirm this conclusion. No matter what direction we look in the sky, no matter where we look in space, the universe is generally the same. We see no edge and no center.

THE SHAPE OF THE EXPANDING UNIVERSE

How can the universe expand if it does not have an edge or a center? We are now ready to resolve that central puzzle of cosmology. To do so we must understand the nature of space and time. Remember that we are small creatures, and we live on a small world. The distances and times we experience are quite small. It should not surprise us that space and time are actually more complicated than our expectations suggest.

Study ▌"The Nature of Space-Time" on pages 392 and 393 and notice three important ideas. First, the red

shifts of the galaxies are not produced by the Doppler effect. The galaxies do not move at high velocities through space. Space is expanding and carrying the galaxies away from each other and stretching protons to longer and redder wavelengths. Second, notice that the curvature of space-time allows the universe to have no edge or center even if it is finite. Finally, notice that we could make measurements in our three-dimensional universe to determine the curvature of space-time.

It seems to violate our common sense that the universe can expand without having a center or an edge, but we must not depend on our limited imagination as three-dimensional creatures. We must speak carefully and thoughtfully when we discuss cosmology so as not to mislead ourselves (Window on Science 18-2). We must not try to visualize the expansion of the universe as an outer edge moving into previously unoccupied space. Open, flat, or closed, the universe has no edge, so it does not need additional room to expand. The universe contains all of the volume that exists, and the expansion is a gradual change in the nature of space-time that causes that volume to increase.

MODEL UNIVERSES

Modern cosmologists have been able to use general relativity to construct mathematical models of the universe under different assumptions For example, they can describe the nature of the universe if it is open, closed, or flat. These models have had a strong influence on the development of cosmology.

The general curvature of the entire universe is determined by its density. If the average density of the universe is equal to the **critical density,** or 9×10^{-30} g/cm^3, space-time will be flat. If the average density of the universe is less than the critical density, the universe is negatively curved and open. If the average density is greater than the critical density, the universe is positively curved and closed.

We decided earlier that an open universe and a flat universe are both infinite. If their behavior is ruled entirely by their matter density, both kinds of universes will expand forever. The gravitation of the matter in the universe, present as a curvature of space-time, will cause the expansion to slow, but, if the universe is open, it will never come to a stop (▌Figure 18-12). If the universe is flat, it will barely slow to a stop after an infinite time. Thus the models predict that an open or flat universe will expand forever, and the galaxies will eventually become black, cold, solitary islands in a universe of darkness.

These models predict a different fate for a closed universe. In a closed universe, the gravitational field, present as curved space-time, is sufficient to slow the expansion to a stop and make the universe contract. Eventually, the contraction will compress all matter and energy back into the high-energy, high-density

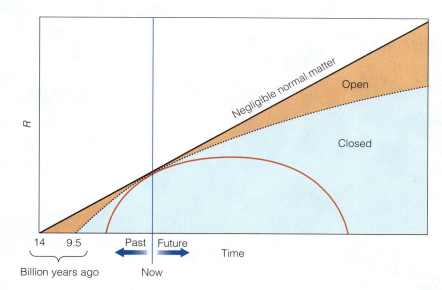

FIGURE 18-12

If the behavior of the universe is determined by the density of matter and energy, then its fate is linked to its geometry. In these models, R is a measure of the extent to which the universe has expanded. Open-universe models expand without end, and the corresponding curves fall in the region shaded orange. Closed models expand and then contract back to a high-density state (red curve). Curves representing closed models fall in the region shaded blue. The dotted line represents a flat universe, the dividing line between open and closed models. Note that the estimated age of the universe depends on the rate at which the expansion is slowing down. (This figure assumes $H = 70$ km/s/Mpc.)

state from which the universe began. This end to the universe, a big bang in reverse, has been called the big crunch. Nothing in the universe could avoid being destroyed by this "crunch."

Some theorists have suggested that the big crunch will spring back to produce a new big bang and a new expanding universe. This theory, called the **oscillating universe theory,** predicts that the universe undergoes alternate stages of expansion and contraction. Theoretical work, however, suggests that successive bounces of an oscillating universe would be smaller until the oscillation ran down, and the theory is no longer taken seriously.

Notice in Figure 18-12 that we must know the geometry of the universe before we can determine the age. Earlier we calculated the Hubble time $\frac{1}{H}$ as an estimate of the age of the universe. Figure 18-12 shows that the Hubble time is the age of a model universe that contains so little matter its expansion is hardly slowed at all. If the density of a model universe is higher, the expansion will be slowed, and that means the true age will be less that $\frac{1}{H}$. If the universe is flat, its true age will be $\frac{2}{3}$ of $\frac{1}{H}$. If H equals 70 km/s/Mpc, then $\frac{1}{H}$ equals 14 billion years. But if the universe is flat and its evolution is controlled by its curvature, then its true age is $\frac{2}{3}$ of 14 billion years, which equals a bit over 9 billion years. This is less than the age of the globular clusters, so the value of H combined with theoretical models of the universe suggest that the density of normal matter should be quite low.

Whether the universe is open, closed, or flat depends on its density, but it is quite difficult to measure the density of the universe. We could count galaxies in a given volume, multiply by the average mass of a galaxy, and divide by the volume; but we are not sure of the average mass of a galaxy, and many small galaxies are too faint to see even if they are nearby. Studies of light coming from distant galaxies suggest that large

amounts of normal matter lie in great, nonluminous clouds between the galaxies, so our estimates must include that matter. Also, we must include the mass equivalent to all the energy in the universe because mass and energy are related. When all of the mass in the universe is added up, it provides a density of less than 5 percent of the critical density needed to close the universe.

Many astronomers will admit that, for theoretical reasons we will discuss later, they prefer to believe the universe is flat. Science, however, is not ruled by preferences or theory, but rather by evidence. Observations of normal matter do not reveal enough density to make the universe flat, but we have not included an important factor. We have clear evidence that there is a large amount of dark matter in the universe. Could there be enough dark matter to make the universe flat?

DARK MATTER IN COSMOLOGY

There is more to the universe than meets the eye. In Chapter 16, we discovered that galaxies have much stronger gravitational fields than we would expect based on the amount of matter we see. Even when we add in the nonluminous gas and dust that we expect to find in galaxies, their gravitational fields are much stronger than we would expect. We concluded that galaxies and clusters of galaxies must contain dark matter.

This is not a small issue. Judging by their gravitational fields, galaxies and clusters of galaxies contain as much as 10 times more matter than we expect from what we see. To correct for this dark matter, we would not just add a small percentage. We have to multiply by a *factor* as large as 10.

If we have any doubt how insubstantial the visible universe is, we have only to look at gravitational lensing of distant galaxies. In Chapter 17 we saw how light from distant quasars could be bent by the gravitational

The Nature of Space-Time

In 1929, Edwin Hubble discovered that the red shifts of the galaxies are proportional to distance — a relationship now known as the Hubble law. It was taken as dramatic evidence that the universe is expanding — that the galaxies are moving away from each other. That leads us to consider not only the nature of the expansion but the nature of space and time.

Distance is the separation between two points in space, and time is the separation between two events. Albert Einstein showed how to relate space to time and treat the whole as space-time. We can think of space-time as a canvas on which the universe is painted, a canvas that can stretch.

Astronomers often express the red shifts of the galaxies as velocities, but these red shifts aren't really Doppler shifts. They are caused by the expansion of space-time.

That explains why the relativistic Doppler formula in Chapter 16 isn't quite correct. It describes the motion of an object through space and not the behavior of space itself.

Notice that our description of the expansion of the universe means the galaxies are not really moving any more than the raisins are swimming through the bread dough. Except for orbital motion among neighboring galaxies, the galaxies are motionless in space-time, and it is space-time that is stretching like a rubber sheet and carrying the galaxies away from each other.

Could the universe be finite and still have no edge?

How can it expand if it has no edge?

How can the universe expand if there is no space outside for it to expand into?

To answer these questions we must use Einstein's theory of general relativity, which predicts that the presence of mass can curve space-time, a curvature we experience as gravity. Furthermore, the theory predicts that even empty space-time can have a curvature, so what we see in the universe depends on the overall curvature of space-time. Finally, the theory shows that space-time could have the property of expansion. That is, the expansion of space-time is not caused by external influences, but is part of space-time itself.

What are the red shifts?

A distant galaxy emits a short-wavelength photon toward our galaxy.

Grid shows expansion of space-time.

The expansion of space-time stretches the photon to longer wavelength as it travels.

The farther the photon has to travel, the more it is stretched.

When the photon arrives at our galaxy, we see it with a longer wavelength — a red shift that is proportional to distance.

The stretching of space-time not only moves the galaxies away from each other, but it lengthens the wavelength of photons traveling through space-time.

To think about space-time curvature, we can use an analogy of a two-dimensional ant on an orange. If he is truly two-dimensional, he will not be able to understand up and down. He can only travel forward and back, right and left. Then as he walks over the surface of his spherical universe, he will eventually realize he has been everywhere because his universe is covered by his footprints. He will conclude that he lives in a finite universe that has no edge and no center.

Similarly, our universe could be curved back on itself so that it could be finite but we would never find an edge. To find out how our universe is really curved, astronomers must study very distant objects.

Notice that the center of the orange cannot be the center of the ant's two-dimensional universe because the center of the orange does not lie on the surface. That is, the center of the orange is not part of the ant's universe.

Aha!

Analogy Ahead

Don't forget: These analogies are only two-dimensional. When we think about our universe, we must think of a three-dimensional universe. It is as difficult for us to imagine curvature in our three-dimensional universe as it is for ants to imagine curvature in their two-dimensional universes.

There are three possible ways our universe might be curved. It could have positive curvature, analogous to the surface of a sphere. Models of such universes are called **closed universes** because they are finite. Of course, ours could be a **flat universe** with zero curvature. Another possibility is that our universe has negative curvature, analogous to a two-dimensional universe shaped like a saddle or a potato chip. Models of these universes are called **open universes.**

A positively curved universe, also called a closed universe, is finite and has no edge. In our two-dimensional analogy, ants on an orange would notice that the areas of large circles were less than πr^2.

Hey! C'mere!

A flat universe must be infinite or it will have an edge. In a two-dimensional analogy, ants on a flat sheet of paper would find that all circles have areas of πr^2.

Although our ant might be unable to sense a third dimension, it could still measure the curvature of its universe by drawing circles. On a flat universe, the area of a circle would always be πr^2 no matter how big the circle was. But on a spherical surface, large circles would contain less than πr^2. On a saddle-shaped surface, large circles would contain more than πr^2. In the same way, we can detect the curvature of our three-dimensional universe, but we must make measurements over very large distances.

Whether our universe is closed, flat, or open, it cannot have an edge or a center.

What?

A negatively curved universe, also called an open universe, must be infinite or it will have an edge. In a two-dimensional analogy, ants on a saddle shape will discover that large circles have areas greater than πr^2.

Visual-wavelength image

FIGURE 18-13

The mass of galaxy cluster 0024+1654 bends the light of a much more distant galaxy to produce arcs that are actually distorted images of the distant galaxy. This gravitational lens effect reveals that many galaxy clusters contain much more mass than is visible as galaxies. Thus galaxy clusters must contain large amounts of dark matter. *(W. N. Colley and E. Turner, Princeton, and J. A. Tyson, Bell Labs, and NASA)*

field of an intervening galaxy to create false images of the quasar. Using the largest telescopes in the world, astronomers are finding that light from the most distant galaxies can be bent by the mass of a nearby cluster of galaxies. The distant galaxies are extremely faint and blue, perhaps because of rapid star formation. As the light from these faint galaxies passes through a nearby cluster of galaxies, the images of the distant galaxies can be distorted into short arcs like rainbows (❚ Figure 18-13). The amount of gravitational lensing can tell astronomers the mass of the nearby galaxy cluster, and, in some cases, the mass is over 10 times greater than the mass visible as galaxies. Evidently such clusters contain over 90 percent dark matter.

Looking at the universe of visible galaxies is like looking at a ham sandwich and seeing only the mayonnaise. Most of the universe is invisible, and most astronomers now believe that the invisible matter is not the normal matter of which you and the stars are made. To follow that line of evidence, we must look for the atoms made during the big bang.

During the first few minutes of the big bang, nuclear reactions converted some protons into helium and a small amount into other elements. How much of these elements was created depends critically on the density of the material. Deuterium, for example, is an isotope of hydrogen in which the nucleus contains a proton and a neutron. The amount of deuterium produced in the big bang depends strongly on the density of normal matter. If there were a lot of normal particles such as protons and neutrons, then they would have collided with the deuterium nuclei and converted them into helium. If the density of normal particles was less, more deuterium would survive.

Lithium is another nucleus that could have been made in small amounts during the big bang. Figure 18-8 shows that there is a gap between helium and lithium; there is no stable nucleus with atomic mass 5, so regular nuclear reactions during the big bang could not convert helium into lithium. If, however, the density of normal matter such as protons and neutrons was high enough, a few nuclear reactions could have leaped the gap and produced a few lithium atoms.

Deuterium is so easily converted into helium that none can be made in stars. In fact, stars destroy what deuterium they have by converting it into helium. Lithium too is destroyed in stars. Using the largest telescopes, astronomers have been able to measure the chemical abundance of gas clouds near quasars. The look-back times to these gas clouds is so great that we see them as they were before stars could have altered the abundance of the elements. The observed abundances of deuterium and lithium tell us that the normal matter in the big bang, the protons and neutrons, could not have amounted to more than 5 percent of the mass needed to make the universe flat (▌ Figure 18-14). Yet observations show that galaxies and galaxy clusters contain large amounts of dark matter. The protons and neutrons that make up normal matter belong to a family of subatomic particles called *baryons*, so cosmologists believe that the dark matter must be **nonbaryonic matter**. No more than about 5 percent of the mass in the universe can be baryonic; the dark matter must be nonbaryonic.

This nonbaryonic matter must be made up of particles that don't often interact with normal matter. If the nonbaryonic particles don't interact very often, we would not notice them except for their gravitational influences. Particles called axions and neutralinos, for example, are predicted by some theories but have never been detected in the laboratory. If they exist, their mass could exert gravitational forces, and they could be part of the dark matter. Data from a large particle detector show that related particles called WIMPs, weakly interacting massive particles, may indeed exist, but the result is still being tested. To what extent WIMPs contribute to the dark matter is unknown.

For some years, astronomers thought that neutrinos could be an important part of the dark matter. We saw in Chapter 8 how observations made by underground detectors suggested that neutrinos oscillate, and that is important because, according to quantum mechanics, if neutrinos oscillate, they cannot have zero mass. Although the mass of neutrinos has not been measured, it is clearly very small. Even a small mass might be important because there are so many neutrinos in the universe—10^8 neutrinos for every normal particle.

Theoretical models of galaxy formation in the early universe tell us that the dark matter can't be mostly neutrinos. Dark matter composed of neutrinos and similar particles that travel at or near the speed of light is called **hot dark matter.** Such fast-moving particles do not clump together easily and could not have stimulated the formation of objects as small as galaxies and clusters of galaxies. The most successful models of galaxy formation require that the dark matter be made up of **cold dark matter,** meaning the particles move slowly and can clump into smaller structures. WIMPs, for example, are massive and slow moving and could be part of the cold dark matter.

Nonbaryonic dark matter does not interact significantly with normal matter or with photons, which is why we can't detect it easily. But that means that dark matter was not affected by the intense radiation that prevented normal matter from contracting into denser clouds right after the big bang. The gravitation of dark matter could pull it into clouds; and, once the density of radiation fell low enough, normal matter could begin falling into the clouds of dark matter to form the first galaxies. Thus, cold, nonbaryonic dark matter may have given the universe a head start in forming galaxies.

Now that we understand more about dark matter we can evaluate the possibility that the halos of galaxies contain large numbers of very-low-mass stars, brown dwarfs, or planets too faint to see. These **MACHOs,** for Massive Compact Halo Objects, can be detected if they pass between our telescopes on Earth and a distant star. The gravitational field of the MACHO acts as a gravitational lens, focusing the light of the star and making it grow brighter for a period of a few weeks as the MACHO passes between the star and Earth. Extensive searches have detected such events (▌ Figure 18-15) but not in large numbers. Furthermore, searches have turned up large numbers of white dwarfs in the halo of our galaxy. However common these halo objects are, they are made of baryonic matter, and the abundance of deuterium and lithium tells us that the dark matter is nonbaryonic. It does not appear that MACHOs and white dwarfs can make up a significant part of the dark matter.

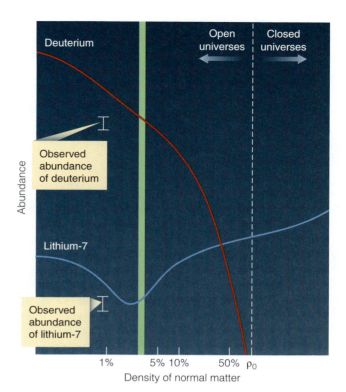

FIGURE 18-14

This graph plots the abundance of deuterium and lithium-7 versus the present density of the universe. Observations of the abundance of deuterium and lithium-7 are uncertain, but they limit the density of normal matter in the universe to a narrow range (green bar). The density of normal matter cannot be more than 5 percent of the critical density ρ_0.

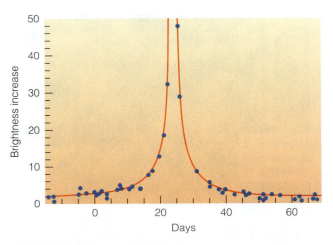

FIGURE 18-15
Measurements of the brightness of a distant star (dots) increased and then decreased exactly as predicted by general relativity (curve) for a star's passage behind a gravitational lens caused by a MACHO. In this example, the star brightened by a factor of over 40.

Although there is good evidence that dark matter exists, it has proven difficult to identify. No form of dark matter has been found that is abundant enough to provide the critical density needed to make the universe flat. In fact, all forms of dark matter appear to add up to less than 30 percent of the critical density. As we will see later in this chapter, there is more to the universe than meets the eye, and more even than the dark matter.

REVIEW CRITICAL INQUIRY

Why do astronomers think that dark matter can't be baryonic?

Isotopes like deuterium and lithium-7 were produced in the first minutes of the big bang, and the abundance of those elements depends strongly on the density of protons and neutrons. Because these particles belong to the family of particles called baryons, astronomers refer to normal matter as baryonic. Measurements of the abundance of deuterium and lithium-7 show that the universe cannot contain more baryons than 5 percent of the critical density. Yet observations of galaxies and galaxy clusters show that dark matter must make up almost 30 percent of the critical density. Consequently, astronomers conclude that the dark matter must be made up of nonbaryonic particles.

Finding the dark matter is important because the density of matter in the universe determines its curvature. How does curvature allow us to avoid an edge or center in a finite model of the universe?

By adding the curvature of space-time and effects of dark matter to our cosmology, we have made our theories much more sophisticated. We are now ready to explore the most recent and most exciting advances in cosmology.

18-3 21ST-CENTURY COSMOLOGY

Cosmology made dramatic advances around the beginning of the 21st century. It may be that astronomers are finally beginning to understand the overall nature of the universe.

INFLATION

In 1980, astronomers faced a problem. The big bang models of the universe could not explain two important features of the universe. To solve those two problems, a new theory of the big bang was created, and that theory has been startlingly successful. To introduce the new theory, we begin with the two problems.

One of the problems is called the **flatness problem.** The universe seems to be balanced near the boundary between an open and a closed universe. That is, it seems nearly flat. Given the vast range of possibilities, from zero to infinite, it seems peculiar that the density of the universe is within a factor of 10 of the critical density that would make it flat. If dark matter is as common as it seems, the density may be even closer than a factor of three to being perfectly flat.

Even a small departure from critical density when the universe was young would be magnified by subsequent expansion. To be so near critical density now, the density of the universe during its first moments must have been within 1 part in 10^{49} of the critical density. So the flatness problem is: Why is the universe so nearly flat?

The second problem with the big bang theory is the isotropy of the primordial microwave background radiation. When we correct for the motion of our galaxy, we see the same background radiation in all directions to at least 1 part in 1000. Yet when we look at background radiation coming from two points in the sky separated by more than a degree, we look at two parts of the big bang that were not causally connected when the radiation was emitted. That is, when recombination occurred and the gas of the big bang became transparent to the radiation, the universe was not old enough for any signal to have traveled from one of these regions to the other. Thus, the two spots we look at did not have time to exchange heat and even out their temperatures. Then how did every part of the entire big bang universe get to be so nearly the same temperature by the time of recombination? This is called the **horizon problem** because the two spots are said to lie beyond their respective light-travel horizons.

The key to these two problems and to others involving subatomic physics may lie with the theory called the **inflationary universe** because it predicts a sudden expansion when the universe was very young, an expansion even more extreme than that predicted by the big bang theory.

To understand the inflationary universe, we must recall that physicists know of only four forces—gravity, the electromagnetic force, the strong force, and the weak force (Chapter 8). We are familiar with gravity, and the electromagnetic force is responsible for making magnets stick to refrigerator doors and cat hair stick to wool sweaters charged with static electricity. The strong force holds atomic nuclei together, and the weak force is involved in certain kinds of radioactive decay.

For many years, theorists have tried to unify these forces; that is, they have tried to describe the forces with a single mathematical law. A century ago, James Clerk Maxwell showed that the electric force and the magnetic force were really the same effect, and we now count them as a single electromagnetic force. In the 1960s, theorists succeeded in unifying the electromagnetic force and the weak force in what they called the electroweak force, effective only for processes at very high energy. At lower energies, the electromagnetic force and the weak force behave differently. Now theorists have found ways of unifying the electroweak force and the strong force at even higher energies. These new theories are called **grand unified theories**, or **GUTs**.

According to the inflationary universe, the universe expanded and cooled until about 10^{-35} second after the big bang, when it became so cool that the forces of nature began to disconnect from each other and behave in different ways. This released tremendous energy, which suddenly inflated the universe by a factor between 10^{20} and 10^{30} (❚ Figure 18-16). At that time the part of the universe that we can see now, the entire observable universe, was no larger than the volume of an atom, but it suddenly inflated to the volume of a cherry pit and then continued its slower expansion to its present extent.

That sudden inflation can solve the flatness problem and the horizon problem. The sudden inflation of the universe would have forced whatever curvature it had toward zero, just as inflating a balloon makes a small spot on its surface flatter. Thus, we now see the universe nearly flat because of that sudden inflation long ago. In addition, because the part of the universe we can see was once no larger in volume than an atom, it had plenty of time to equalize its temperature before the inflation occurred. Now we see the same temperature for the background radiation in all directions.

The inflationary universe is based, in part, on quantum mechanics, and a slightly different aspect of quantum mechanics may explain why there was a big bang at all. Theorists believe that a universe totally empty of matter could be unstable and decay sponta-

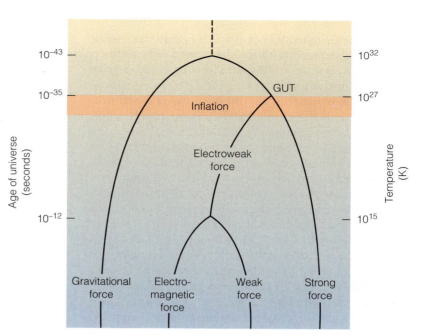

FIGURE 18-16
When the universe was very young and hot (top), the four forces of nature were indistinguishable. As the universe began to expand and cool, the forces separated and triggered a sudden inflation in the size of the universe.

neously by creating pairs of particles until it was filled with the hot, dense state we call the big bang. This theoretical discovery has led some cosmologists to believe that the universe could have been created by a chance fluctuation in space-time. In the words of physicist Frank Wilczyk, "The reason there is something instead of nothing, is that 'nothing' is unstable."

The inflationary theory predicts that the universe is almost perfectly flat. That is, the true density must equal the critical density. A theory can never be used as evidence, but the beauty of the inflationary theory has given many cosmologists confidence that the universe must be flat. Observations, however, seem to tell us the universe does not contain enough matter (baryonic plus dark) to be flat, so we must search further to understand the nature of the universe.

THE ACCELERATION OF THE UNIVERSE

Ever since Hubble discovered the expansion of the universe, astronomers have known what to expect. They have also known that expectations are not reliable; and, as the 21st century approached, astronomers made a mind-boggling discovery that contradicted everyone's expectations.

Both common sense and mathematical models suggest that as the galaxies recede from each other, the expansion should be slowed by gravity trying to pull the galaxies toward each other. How much the expansion is slowed should depend on the amount of matter in the universe. If the density of matter is less than the critical density, the expansion should be slowed only slightly, and the universe should expand forever. If the density of matter in the universe is greater than the critical density, the expansion should be slowing down dramatically, and the universe should eventually stop expanding and begin contracting. Notice that this is the same as saying a low-density universe should be open and a high enough density universe should be closed.

For decades, astronomers struggled to measure the Hubble constant and detect the slowing of the expansion. A direct measurement of the rate of slowing would reveal the true curvature of the universe. This was one of the key projects for the Hubble Space Telescope, and two teams of astronomers spent years making the measurements.

Detecting a change in the rate of expansion is a difficult project because it requires accurate measurements of the distances to very remote galaxies, and both teams used the same technique. They calibrated type Ia supernovae as distance indicators. A type Ia supernova occurs when a white dwarf gains matter from a companion star, exceeds the Chandrasekhar limit, and collapses in a supernova explosion. Because all such white dwarfs should collapse at the same mass limit, they should all produce explosions of the same luminosity, and that makes them good distance indicators. The teams calibrated type Ia supernovae by locating such supernovae occurring in nearby galaxies whose distance was known from Cepheid variables and other reliable distance indicators. Once the peak luminosity of type Ia supernovae had been determined, they could be used to find the distance to much more distant galaxies.

Both teams announced their results in 1998. The expansion of the universe is not slowing down. It is speeding up! That is, the expansion of the universe is accelerating (❙ Figure 18-17).

The announcement that the expansion of the universe is accelerating was exciting, but astronomers immediately began testing it. It depends critically on the calibration of type Ia supernovae as distance indicators, and some astronomers suggested that the calibration might be wrong (Window on Science 15-1). Some unknown kind of dust might make very distant supernovae too faint without making them too red. No reddening had been detected, so normal dust seemed to be ruled out. Or perhaps the very distant supernovae were too faint because we see them at great look-back times, and some unknown process made those early supernovae a bit less luminous than more recent supernovae. Those problems with the calibration seem to be ruled out by a supernova that occurred in 1997. Known as SN1997ff, it was recognized in 2001 as a type Ia supernova with the very large red shift of 1.7, implying a distance of 10 billion light-years (❙ Figure 18-18). The ob-servations of the maximum brightness are uncertain, but they seem to rule out problems with the calibration of type Ia supernovae. The universe really does seem to be expanding faster and faster.

If the expansion of the universe is accelerating, then there must be a force of repulsion in the universe, and astronomers are struggling to understand what it could be. One possibility leads us back to 1916.

When Albert Einstein published his theory of general relativity in 1916, he recognized that his equations describing space and time implied that space had to contract or expand. The galaxies could not float unmoving in space because their gravity would pull them together. The only solutions seemed to be a universe that was contracting under the influence of gravity or a universe in which the galaxies were rushing away from each other so rapidly that gravity could not pull them together. In 1916, astronomers did not yet know that the universe was expanding, so Einstein made what he later said was a mistake.

To balance the attractive force of gravity, Einstein added a constant to his equations called the **cosmological constant,** represented by an upper-case lambda (Λ). The constant plays the mathematical role of a force of repulsion that balances the gravitation in the universe so it does not have to contract or expand. Thirteen years later, in 1929, Edwin Hubble announced that the universe was expanding, and Einstein said introducing the cosmological constant was his biggest blunder. Modern astronomers aren't so sure.

One explanation for the acceleration of the universe is that the cosmological constant is not zero but rather represents a force that drives a continuing acceleration in the expansion of the universe. Because the cosmological constant remains constant with time, the universe would have to have experienced this acceleration throughout its history.

Another solution is to suppose that totally empty space, the vacuum, contains energy, which drives the acceleration. This is an interesting possibility because for years theoretical physicists have discussed an energy inherent in empty space. Astronomers have begun referring to this energy of the vacuum as **quintessence.** Unlike the cosmological constant, quintessence need not remain constant over time, so it may be more successful in explaining the observed distribution of galaxies and galaxy clusters.

Whichever explanation is right, the cosmological constant or quintessence, the acceleration alerts us to a form of energy spread throughout space. From Einstein's most famous quotation, $E = mc^2$, we know that energy and mass are equivalent, so the tremendous energy that drives the repulsion behaves like a mass spread through space, and cosmologists think that the invisible mass–energy can provide the density needed to make the universe flat. Astronomers refer to this en-

11 March 1997

Host galaxy

Supernova

13 March 1997

16 March 1997

29 March 1997

5 April 1997

Visual-wavelength images

a

FIGURE 18-17
(a) A type Ia supernova erupts in a very distant galaxy and begins to fade. By calibrating these supernovae, astronomers were able to find the distances to some of the farthest visible galaxies. *(ESO)* (b) Type Ia supernovae in distant galaxies are about 25 percent too faint, which must mean the galaxies are farther away from us than they would be in a universe expanding at a constant rate. This diagram shows observations of supernovae compared with a decelerating, flat universe dominated by dark matter (blue line). The red line shows the relationship for an accelerating, flat universe.

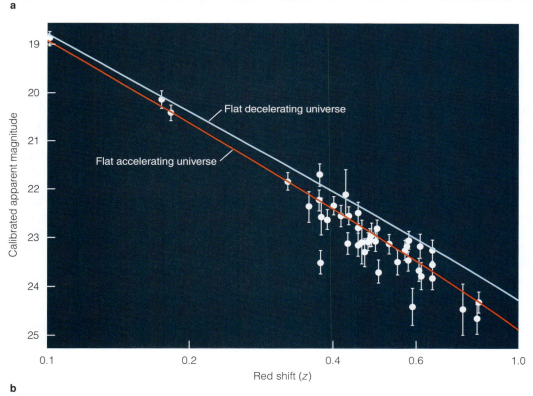

Flat decelerating universe

Flat accelerating universe

Calibrated apparent magnitude

Red shift (z)

b

ergy as **dark energy,** the energy inherent in empty space that drives the acceleration of the universe but does not contribute to the formation of starlight or the cosmic microwave background radiation.

The inflationary theory of the universe was proposed to solve two problems, the horizon problem and the flatness problem, but it makes the specific predic-tion that the universe is flat. The discovery of dark energy seems to confirm that discovery. Baryonic matter plus dark matter makes up about a third of the critical density, and dark energy appears to make up two-thirds. That is, the total density of the universe seems to equal the critical density, which means that the geometry of the universe is flat.

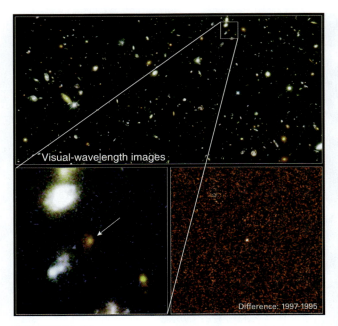

FIGURE 18-18

In a 1997 follow-up image of the Hubble Deep Field made in 1995, astronomers noticed a faint galaxy that was brighter in the second image (lower left). Subtracting the earlier image from the later image revealed a supernova cataloged as SN1997ff (lower right). The unexpected faintness of the supernova confirms that the universe is accelerating. *(Adam Riess, STScI/NASA)*

Acceleration helps with another problem. If the Hubble constant equals 70 km/ s/Mpc, then the Hubble time is about 14 billion years. If the universe is flat, then the age of the universe is two-thirds of the Hubble time, which equals about 9 billion years. That is younger than the globular clusters, and that is a problem that has been hard to explain. However, if the expansion of the universe has been accelerating, then it must have been expanding slower in the past, and that means it can be older than two-thirds of the Hubble time. The latest estimates suggest that the true age of the universe is about 14 billion years, and that solves the age problem.

Acceleration opens many possibilities that astronomers had not considered before. For example, the fate of the universe may not be determined by its geometry. A closed universe that is accelerating, for example, does not have to fall back. It could expand forever, depending on the way the acceleration changes with time.

The discovery of an acceleration of the universe is one of the most important discoveries in modern astronomy, but it leaves us wanting confirmation for this amazing result. Is there some other way to show that the universe is flat? One way to measure the curvature of space-time is to draw circles on it and measure the areas of the circles. But that is only practical in a simple cartoon analogy. In reality, astronomers must measure the size of things at great look-back times, and that measurement is yielding yet more exciting results, which are described in the next section.

THE ORIGIN OF STRUCTURE AND THE CURVATURE OF THE UNIVERSE

For a number of years, astronomers have studied the distribution of galaxies in space, what they refer to as structure. Those studies have led to three astonishing discoveries that are changing our understanding of the universe.

When we look at galaxies in the sky we see them in clusters ranging from a few to thousands, and those clusters appear to be grouped into **superclusters** (❚ Figure 18-19). The Local Supercluster, in which we live, is a roughly disk-shaped swarm of galaxy clusters 50 to 75 Mpc in diameter. By measuring the red shifts and positions of over 100,000 galaxies in great slices across the sky, astronomers have been able to create maps revealing that the superclusters are distributed in long, narrow filaments and thin walls that outline great voids nearly empty of galaxies (❚ Figure 18-20). These filaments and walls of superclusters stretching half a billion light-years are the largest structures in the universe. No larger objects exist.

These vast structures are a problem for cosmologists because the cosmic microwave background radiation is very uniform, and that means the gas of the big bang must have been extremely uniform at the time of recombination. Yet the look-back time to the furthest galaxies is about 93 percent of the way back to the big bang. How did the uniform gas at the time of recombination coagulate so quickly to form galaxies? Worse, how did that highly uniform gas form such large structures as the filaments, voids, and walls?

Baryonic matter is so rare in the universe it does not have enough gravity to pull itself together rapidly. Astronomers believe the nonbaryonic dark matter must provide the gravity, and models have attempted to describe how dark matter can form structure. As we have seen earlier in this chapter, hot dark matter does not clump into small enough structures, so the dark matter must be cold dark matter.

But what started the clumping of matter? Theorists say that subatomic quantum fluctuations in space would have been stretched at the moment of inflation to very large but very subtle variations in gravitational fields that could have stimulated the formation of clusters, filaments, and walls. The structure we see in Figure 18-20 may be the ghostly traces of quantum fluctuations in the infant universe.

The inflationary theory of the universe makes very specific predictions about the sizes of the fluctuations we should see in the cosmic microwave background radiation. Observations made by the COBE satellite in 1992 detected the largest variations, but detecting the smaller variations is critical for testing the theory. A half-dozen teams of astronomers have built specialized telescopes to make these observations. Some have flown under balloons high in the atmosphere, and others have observed from the ice of Antarctica. The most ex-

FIGURE 18-19
The distribution of brighter galaxies in the sky reveals the great Virgo cluster (center), containing over 1000 galaxies only about 17 Mpc away. Other clusters fill the sky, such as the more distant Coma cluster just above the Virgo cluster in this diagram. The Virgo cluster is linked with others to form the Local Supercluster.

tensive observations have been made by the Wilkinson Microwave Anisotropy Probe (WMAP), a robotic infrared telescope that observed from space.

The background radiation is very isotropic—it looks almost exactly the same in all directions. However, when the average intensity is subtracted from each spot on the sky, small irregularities are evident. That is, some spots on the sky look a tiny bit hotter and brighter than other spots. (❙ Figure 18-21). Those irregularities contain lots of information.

If the irregularities in the cosmic microwave background radiation were produced by inflation, then their size must depend on the velocity of the pressure waves (sound) that could travel through the gas at the time of recombination. Theory predicts most of the irregularities should be about one degree in diameter if the universe is flat. If the universe were open, the most common irregularities would be smaller. Careful measurements of the size of the irregularities in the cosmic background radiation show that the observations fit the

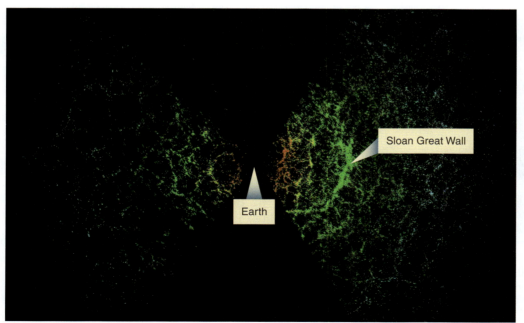

FIGURE 18-20
Nearly 70,000 galaxies are plotted in this double slice of the universe extending outward in the plane of Earth's equator. The nearest galaxies are shown in red and the more distant in green and blue. The galaxies form filaments and walls enclosing empty voids. The Sloan Great Wall is almost 1.4 billion light-years long and is the largest known structure in the universe. The most distant galaxies in this diagram are roughly 3 billion light-years from Earth. (*Sloan Digital Sky Survey*)

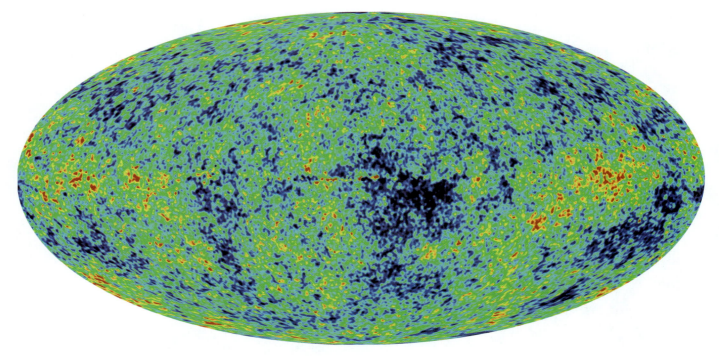

FIGURE 18-21
Data from the WMAP spacecraft were used to make this far-infrared map of the entire sky. The infrared glow from dust in our solar system and in the Milky Way Galaxy has been removed to reveal tiny irregularities in the cosmic microwave background radiation. Red and yellow spots are slightly warmer than the green and blue spots. *(NASA/WMAP Science Team)*

theory very well for a flat universe (▌ Figure 18-22). Not only is the theory of inflation confirmed, an exciting result itself, but these data tell us that the universe is flat, which indirectly confirms the existence of dark energy and the acceleration of the universe.

Cosmologists can analyze the irregularities using sophisticated mathematics to find out how commonly the spots of different sizes recur. The mathematics con-

firm that spots about 1 degree in diameter are the most common, but spots of other sizes recur as well, and it is possible to plot a graph such as ▌ Figure 18-23 to show how common different size irregularities are.

The results from the WMAP observations tell us a great deal about the universe. The universe is flat, accelerating, and will expand forever. The age of the universe derived from the data is 13.7 billion years. Fur-

Simulated data
Closed model universe
Larger spots

Simulated data
Flat model universe
Spots about 1 degree

Simulated data
Open model universe
Smaller spots

Observational data

FIGURE 18-22
The size of the irregularities in the background radiation can be seen in this diagram. Compare the observations at right with the three simulations at left. The observed size of the irregularities fits best with cosmological models having flat geometry. Detailed mathematical analysis confirms our visual impressions. The universe is flat. *(Courtesy of the BOOMERANG Collaboration)*

thermore, the smaller peaks in the curve reveal that the universe contains 4 percent baryonic (normal) matter, 23 percent dark matter, and 73 percent dark energy. The Hubble constant is confirmed to be 71 km/s/Mpc. The inflationary theory is confirmed, and the data support the cosmological constant version of dark energy, although quintessence is not ruled out. Hot dark matter is ruled out. The dark matter needs to be cold dark matter to clump together so rapidly after the big bang. In fact, the first stars began to produce light when the universe was only about 200 million years old. This is much earlier than most astronomers had expected.

WMAP and other studies of the cosmic microwave background radiation have revolutionized cosmology. At last, astronomers have accurate observations against which to test theories. The basic constants are known to a percent or so. On reviewing these results, one cosmologist announced that "Cosmology is solved!" but that may be premature. We don't understand dark matter or the dark energy that drives the acceleration, so over 90 percent of the universe is not understood. Hearing this, another astronomer suggested a better phrase was "cosmology in crisis." Certainly there are further mysteries to be explored, but cosmologists are growing more confident that they can describe the overall properties of the universe.

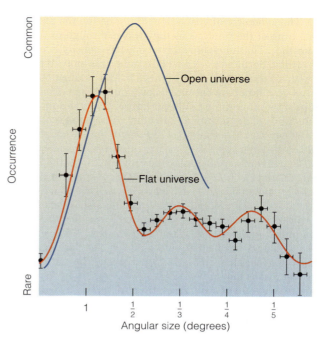

FIGURE 18-23

This graph shows how commonly irregularities of different sizes occur in the cosmic microwave background radiation. Irregularities of about 1 degree in diameter are most common. Models of the universe that are open or closed are ruled out. The data fit a flat model of the universe very well. Crosses on data points show the uncertainty in the measurements

REVIEW CRITICAL INQUIRY

How does inflation theory solve the flatness problem?

The flatness problem can be stated as a question: Why is the universe so nearly flat? After all, the density of matter in the universe could be anything from zero to infinite, but the observed density of baryonic and dark matter adds up to nearly 30 percent of the critical density that would produce a flat universe. Furthermore, the density must have been astonishingly close to the critical density when the universe was very young. Otherwise it would not be so close now. The inflationary theory solves this problem by proposing that the universe underwent a moment of rapid inflation when it was a tiny fraction of a second old. That inflation drove the universe toward flatness just as the inflation of a balloon makes a spot on the balloon nearly flat.

Understanding theory in cosmology is critically important, but science depends on evidence. What evidence do we have that the expansion of the universe is accelerating?

Although we have traced the origin of the universe, the origin of the elements, and the birth and death of stars, we have left out one important class of objects—planets. How do Earth and other planets fit into this grand scheme of origins? That is the subject of the following chapters.

SUMMARY

We conclude that it is impossible for the universe to have an edge, and because we find the centers of things by reference to their edges, we also conclude that the universe cannot have a center.

The fact that the night sky is dark leads us to conclude that the universe is not infinitely old. If it were infinite in extent and age, then every spot on the sky would glow as brightly as the surface of a star. This problem, known incorrectly as Olbers's paradox, forces us to conclude that the universe had a beginning.

Edwin Hubble's 1929 discovery that the red shift of a galaxy is proportional to its distance is known as the Hubble law. Tracing this expansion backward in time, we come to an initial high-density, high-temperature state commonly called the big bang. Although the expanding universe began from this big bang, it has no center. The galaxies recede from each other.

The cosmic microwave background radiation is black body radiation with a temperature of 2.73 K uniformly spread over the entire sky. It is the light from the big bang freed from the gas at the moment of recombination and red shifted by a factor of 1100. The background radiation is strong evidence that the universe began with a big bang.

During the first three minutes of the big bang, nuclear fusion converted some of the hydrogen into helium but was unable to make many other heavy atoms because no stable nuclei exist with weights of 5 or 8.

The universe appears to be isotropic and homogeneous, leading to the cosmological principle, the assumption that the universe looks the same from every location.

The geometry of space-time explains the red shift as a stretching of the photons as they travel through expanding space-time. The universe could be open if its density is less than the critical density and closed if its density is more. If its density equals the critical density, it is flat. The dark matter in the universe is important because it contributes to the density and helps determine the overall curvature. Observations of the abundance of deuterium and lithium-7 tell us that baryonic matter can make up roughly 4 percent of the critical density. Cold dark matter appears to make up less than 30 percent.

The inflationary theory of the universe solves the flatness problem and the horizon problem by proposing that the universe expanded dramatically when it was a tiny fraction of a second old. Inflation predicts that the universe is flat.

Observations of type Ia supernovae reveal that the expansion of the universe is accelerating because of the cosmological constant or because of quintessence. The dark energy of empty space makes up about 73 percent of the critical density and makes the universe flat.

Observation of structure in the universe—irregularities in the cosmic microwave background radiation and the clusters of galaxies—provide strong evidence that the universe is flat and accelerating. These irregularities appear to be subatomic quantum fluctuations in space-time magnified when inflation occurred. The observations show that the age of the universe is 13.7 billion years.

NEW TERMS

cosmology	open universe
Olbers's paradox	critical density
observable universe	oscillating universe theory
big bang	nonbaryonic matter
Hubble time	hot dark matter
cosmic microwave background radiation	cold dark matter
	MACHOs
steady-state theory	flatness problem
antimatter	horizon problem
recombination	inflationary universe
dark age	grand unified theories (GUTs)
reionization	
isotropy	cosmological constant (Λ)
homogeneity	quintessence
cosmological principle	dark energy
closed universe	supercluster
flat universe	

REVIEW QUESTIONS

Ace Astronomy™ Assess your understanding of this chapter's topics with additional quizzing and animations at **http://astronomy.brookscole.com/seeds8e**

1. How does the darkness of the night sky tell us something important about the universe?

2. How can we be located at the center of the observable universe if we accept the Copernican principle?

3. Why can't an open universe have a center? Why can't a closed universe have a center?

4. What evidence do we have that the universe is expanding? that it began with a big bang?

5. Why couldn't atomic nuclei exist when the universe was younger than 2 minutes?

6. Why is it difficult to determine the present density of the universe?

7. How does the inflationary universe theory resolve the flatness problem? the horizon problem?

8. If the Hubble constant were really 100 km/s/Mpc, much of what we understand about the evolution of stars and star clusters must be wrong. Explain why. (*Hint:* What would the age of the universe be?)

9. Why do we conclude that the universe must have been very uniform during its first million years?

10. What is the difference between hot dark matter and cold dark matter? What difference does it make to cosmology?

11. What evidence do we have that the expansion of the universe is accelerating?

12. What evidence do we have that the universe is flat?

DISCUSSION QUESTIONS

1. Do you think Copernicus would have accepted the cosmological principle? Why or why not?

2. If we reject any model of the universe that has an edge in space because we can't comprehend such a thing, shouldn't we also reject any model of the universe that has a beginning or an ending? Are those just edges in time, or is there a difference?

PROBLEMS

1. Use the data in Figure 18-3 to plot a velocity–distance diagram, find *H*, and estimate the Hubble time.

2. If a galaxy is 8 Mpc away from us and recedes at 456 km/s, what is *H*? What is the Hubble time? How old would the universe be if it were flat and there were no acceleration?

3. At what wavelength of maximum intensity did the gas of the big bang radiate at the time of recombination? By what factor is that different from the wavelength of maximum of the cosmic microwave background radiation?

4. If the average distance between galaxies is 2 Mpc, and the average mass of a galaxy is 10^{11} solar masses, what is the average density of matter in the universe? (*Hints:* The volume of a sphere is $\frac{4}{3}\pi r^3$, and the mass of the sun is 2×10^{33} g.)

5. Figure 18-12 is based on an assumed Hubble constant of 70 km/s/Mpc. How would you change the diagram to fit a Hubble constant of 50 km/s/Mpc?

6. Hubble's first estimate of the Hubble constant was 530 km/s/Mpc. If his distances were too small by a factor of 7, what answer should he have obtained?

7. What was the maximum age of the universe predicted by Hubble's first estimate of the Hubble constant?

8. If the value of the Hubble constant were found to be 60 km/s/Mpc, what would the Hubble time be? How old would the universe be if it were flat and there were no acceleration? How would acceleration change your answer?

CRITICAL INQUIRIES FOR THE WEB

1. What is the latest news concerning acceleration and the flatness of the universe? Search the Web for the most recent observations.

2. The steady-state theory was once a rival cosmology of the big bang. Search for Web sites that provide information on steady-state cosmology. (Be careful to locate legitimate sites that discuss the theory, rather than sites where individuals use steady-state ideas as part of nonscientific arguments on cosmology.) What were the key predictions of steady-state cosmology? How has recent evidence led to its decline?

3. Search for the Web pages of large surveys of galaxies, such as the 2dF Deep Field Survey and the Sloan Digital Sky Survey. What are they discovering about the largest and most distant objects in the universe?

4. What is the Microwave Anistropy Probe? Where is it? What are the latest results?

EXPLORING *THESKY*

1. Search for galaxy clusters. (*Hint:* Use the **View** menu to turn off everything but galaxies, deep sky objects, constellations, and labels. Zoom in until the field of view is 40° or smaller and the fainter galaxies appear. Search for clusters of galaxies in Ursa Major, Canes Venatici, Lynx, Virgo, and Coma Bereneces.)

2. Locate small clusters of galaxies and compare the brightness of the galaxies with those in large clusters. Can you tell that the smaller clusters tend to be further away, or is the range of galaxy luminosities too great? (*Hint:* Click on a galaxy to find its magnitude.)

3. Can you find galaxies and clusters of galaxies along the Milky Way? Turn on the Milky Way and search for galaxy clusters within its outline.

 Visit the Seeds *Foundations of Astronomy* companion Web site for critical thinking exercises, articles, and additional readings from InfoTrac College Edition, Brooks/Cole's online student library.

THE ORIGIN OF THE SOLAR SYSTEM

What place is this?
Where are we now?

Carl Sandburg, Grass

Chapter 19

Moon, UCO/Lick Observatory; all planets, NASA

GUIDEPOST

The preceding 18 chapters have described the origin, structure, and evolution of the physical universe, but they have neglected one important class of objects—planets. In this chapter, we can look back on what we have learned and find our place in the universe. We live on a planet. What does that mean? Where do we fit?

Each time we have studied a new object, we have asked how it formed and how it evolved to its present state. We have done that with stars and galaxies and the universe, so it is appropriate to begin our discussion of the solar system by considering its origin.

Another reason for discussing the origin of the solar system here is to give ourselves a framework into which we can fit the planets as we discuss them in the chapters that follow. Without a theoretical framework, science is nothing but a jumble of facts. With a good framework in hand, we will be ready to make sense of the solar system through the next six chapters.

VIRTUAL LABORATORY

EXTRASOLAR PLANETS

Microscopic creatures live in the roots of your eye-lashes. Don't worry. Everyone has them, and they are harmless.* They hatch, fight for survival, mate, lay eggs, and die in the tiny spaces around the roots of our eyelashes without doing us any harm. Some live in renowned places—the eyelashes of a glamorous movie star—but the tiny beasts are not self-aware; they never stop to say, "Where are we?" Humans are more intelligent; we have the ability and the responsibility to wonder where we are in the universe and how we came here.

In this chapter, we begin exploring our solar system: the sun and its family of planets. We must avoid, however, the great pitfall of science. We must avoid memorizing facts instead of understanding nature. Jupiter is 317.83 times more massive than Earth. That is a fact, and it is boring. The excitement of science comes from understanding nature. We want to understand how the sun and planets formed, and that will tell us *why* Jupiter is so massive.

Facts fit together to build understanding, so we begin our study of the solar system by developing a unifying hypothesis for the origin of the sun and planets. As we learn facts about the solar system we can use them to test and improve our hypothesis, and that will increase our understanding of nature. A unifying hypothesis is like a basket into which we can put facts. The facts are easier to remember because they make sense, and the combination of hypothesis and supporting facts represents real knowledge—real understanding—and not just memory.

We have no choice; we must understand our solar system out of self-preservation. We live on a planet with 6 billion other humans, and we are altering our world's environment in unplanned and unknown ways. We could make Earth uninhabitable. The deadly deserts of Mars and sulfuric acid fogs of Venus may help us understand our own more comfortable world.

We must also study planets as astronomers. It seems likely that roughly half of all stars have planetary systems, so there are more planets in the universe than stars. Also, the chemical reactions that give rise to life seem to need the moderate conditions of the surface of planets. To search for life in the universe, we must understand planets.

Above all, we study the solar system because it is our home in the universe. Because we are an intelligent species, we have the right and the responsibility to wonder what we are. Our kind have inhabited this solar system for at least a million years, but only within the last hundred years have we begun to understand what a solar system is. Like sleeping passengers on a train, we waken, look out at the passing scenery, and mutter: "What place is this? Where are we now?"

Demodex folliculorum has been found in 97 percent of individuals and is a characteristic of healthy skin.

19-1 THE GREAT CHAIN OF ORIGINS

We are linked through a great chain of beginnings that leads backward through time to the first instant when the universe began about 14 billion years ago. The gradual discovery of the links in that chain is one of the most exciting adventures of the human intellect. In earlier chapters, we discussed some of that story: the origin of the universe in the big bang, the formation of galaxies, the origin of stars, and the growth of the chemical elements. Here we will explore further to consider the formation of planets.

EARLY HYPOTHESES

The earliest theories for Earth's origin are myths and folktales that go back beyond the beginning of recorded history. In addition, almost all religions contain an account of the origin of the world. It was not until about the time of Galileo that philosophers began searching for rational explanations for natural phenomena. While people like Copernicus, Kepler, and Galileo tried to find logical explanations for the motions of Earth and the other planets, other philosophers began thinking about the origin of the planets.

The first rational theory for Earth's origin was proposed by the French philosopher and mathematician René Descartes (1596–1650). Because he lived and wrote before the time of Newton, Descartes did not have the concept of gravitation as the dominant force in the universe. Rather, he believed that force was communicated by contact between bodies and that the entire universe was filled with vortices of whirling particles. In 1644, he proposed that the sun and planets formed when a large vortex contracted and condensed. Thus, his hypothesis explained the general properties of the solar system known at the time.

A century later, in 1745, the French naturalist Georges-Louis de Buffon (1707–1788) proposed an alternative hypothesis that the planets were formed when a passing star collided with or passed close to the sun and pulled matter out of the sun and the star. The matter condensed to form the planets, and they fell into orbit around the sun (❚ Figure 19-1a). This **passing star hypothesis** was popular off and on for two centuries, but it contains serious flaws. First, stars are very small compared to the distances between them, and thus they collide very infrequently. In the entire history of our galaxy, stars have probably collided only a few times.

a

b

FIGURE 19-1

(a) The passing star hypothesis was catastrophic. It proposed that the sun was hit by or had a very close encounter with a passing star and that matter torn from the sun and the star formed planets orbiting the sun. The theory is no longer accepted. (b) Laplace's nebular hypothesis was evolutionary. It suggested that a contracting disk of matter conserved angular momentum, spun faster, and shed rings of matter that formed planets. The hypothesis could not explain the sun's low angular momentum and was never fully accepted, even though the modern theory shares some of its features.

More important, the gas pulled from the sun and the star would be much too hot to condense to make planets. Furthermore, even if planets formed, they would not go into stable orbits.

The hypotheses of Descartes and Buffon fall into two broad categories. Descartes's hypothesis is **evolutionary** in that it calls upon common, gradual events to produce the sun and planets. If it is correct, stars with planets are very common. Buffon's hypothesis, on the other hand, is **catastrophic.** It calls upon unlikely, sudden events to produce the solar system, and thus it implies that solar systems are very rare. While our imaginations may be tempted by colliding stars, modern hypotheses for the origins of the planets are evolutionary, with, as we will see, a few astonishing catastrophes thrown in (Window on Science 19-1).

Our modern theory of the origin of the solar system had its true beginning with Pierre-Simon de Laplace

Two Kinds of Theories: Evolution and Catastrophe

Many theories in science can be classified as either evolutionary, in that they involve gradual processes, or catastrophic, in that they depend on specific, unlikely events. Scientists have generally preferred evolutionary theories. Nevertheless, catastrophic events do occur.

Something in us prefers catastrophic theories, perhaps because we like to see spectacular violence from a safe distance, which may explain the success of movies that include lots of car crashes and explosions. Also, cataclysmic theories resonate with Old Testament accounts of catastrophic events and special acts of creation. Thus, we have an understandable interest in catastrophic theories.

Nevertheless, most scientific theories are evolutionary. Such theories do not depend on unlikely events or special acts of biblical creation, and thus they are open to scientific investigation. For example, geologists much prefer theories of mountain building that are evolutionary, with the mountains being pushed up slowly as centuries pass. All the evidence of erosion and the folding of rock layers show that the process is gradual. Because most such natural processes are evolutionary, scientists sometimes find it difficult to accept any theory that depends on catastrophic events.

We will see in this and later chapters that catastrophes do occur. Earth, for example, is bombarded by debris from space, and some of those impacts are very large. As you study astronomy or any other natural science, notice that most theories are evolutionary but that we must allow for the possibility of unpredictable catastrophic events.

(1749–1827), the brilliant French astronomer and mathematician. In 1796, he combined Descartes's vortex with Newton's gravity to produce a model of a rotating cloud of matter contracting under its own gravitation and flattening into a disk—the **nebular hypothesis.** As the disk grew smaller, it had to conserve angular momentum and spin faster and faster. Laplace reasoned that, when it was spinning as fast as it could, the disk would shed its outer edge to leave behind a ring of matter. Then the disk could contract further, speed up again, and leave another ring. Thus, the contracting disk would leave behind a series of rings, each to become a planet circling the newborn sun at the center of the disk (Figure 19-1b).

According to the nebular hypothesis, the sun should be spinning very rapidly, or, to put it another way, the sun should have most of the angular momentum of the solar system. (Recall from Chapter 5 that angular momentum is the tendency of a rotating object to continue rotating.) As astronomers studied the planets and the sun, however, they found that the sun rotated slowly and that the planets moving in their orbits had most of the angular momentum in the solar system. In fact, the rotation of the sun contains only about 0.3 percent of the angular momentum of the solar system. Because the nebular hypothesis could not explain this **angular momentum problem,** it was never fully successful, and astronomers toyed with various versions of the passing star hypothesis for over a century.

In contrast to the astronomy of earlier centuries, 20th-century astronomy applied modern physics to the stars and galaxies and so illuminated our origins. We can now trace that great chain of origins that began with the big bang and led to the matter of which we are made. We will review the story of the origin of matter, and then we will add to it the story of how that matter formed our world.

A REVIEW OF THE ORIGIN OF MATTER

Look at your thumb. The matter in your thumb came into existence within minutes of the beginning of the universe. Astronomers have strong evidence that the universe began in an event called the big bang (Chapter 18), and by the time the universe was three minutes old, the protons, neutrons, and electrons in your thumb had come into existence. You are made of very old matter.

Although those particles formed quickly, they were not linked together as they are today. Most of the matter was hydrogen, and about 25 percent was helium. Very few heavier atoms were made in the big bang. Although helium is very rare in our bodies, your thumb contains many of those ancient hydrogen atoms unchanged since the universe began.

Within a few hundred million years after the big bang, matter began to collect to form galaxies containing billions of stars. Astronomers understand that nuclear reactions inside stars combine low-mass atoms such as hydrogen to make heavier atoms (Chapter 8). Generation after generation of stars cooked the original particles, linking them together to build atoms such as carbon, nitrogen, and oxygen (Chapter 15). Look at your thumb. Even the calcium in the bone and the iron in the blood were assembled inside stars.

The atoms heavier than iron were created by rapid nuclear reactions that can only occur when a massive star explodes (Chapter 13). Gold and silver are rare in our bodies, but iodine is critical in your thyroid gland, and there are no doubt a few of those iodine atoms circulating through your thumb at this very moment, thanks to the violent deaths of massive stars.

Our galaxy contains at least 100 billion stars, of which our sun is one. It formed from a cloud of gas and dust about 5 billion years ago, and the atoms in your thumb were part of that cloud. How the sun took shape, how the cloud gave birth to the planets, how those atoms in your thumb found their way onto Earth and into you is the story of this chapter. As we explore the origin of our solar system, keep in mind the great chain of origins that created the atoms. As the geologist Preston Cloud remarked, "Stars have died that we might live."

THE SOLAR NEBULA HYPOTHESIS

The stars gave birth to the heavy atoms of which our world is made, and the modern theory of the origin of the planets is based on star formation. The **solar nebula hypothesis** proposes that the planets were formed from the disk of gas and dust that surrounded the sun as it formed (❚ Figure 19-2).

As stars form in contracting clouds, they remain surrounded by cocoons of dust and gas, and the rotation of the cloud causes that dust and gas to form a spinning disk around the protostar. When the center of the star grows hot enough to ignite nuclear reactions, its surface quickly heats up, becomes more luminous, and blows away the gas and dust cocoon.

Modern theories suppose that planets form in the rotating disks of gas and dust around young stars. Infrared observations of T Tauri stars, for instance, show that some are surrounded by gas clouds rich in dust, and spectra show that these stars are blowing away their nebulae at speeds up to 200 km/s. Astronomers find clear evidence that stars are surrounded by disks of gas. Bipolar flows from protostars (Chapter 11) were the first evidence of such disks, but modern techniques can image the disks directly (Figure 11-6). Our own planetary system probably formed in such a disk-shaped cloud around the sun. When the sun became luminous enough, the remaining gas and dust were blown away into space, leaving the planets orbiting the sun. Notice that the solar nebula hypothesis also suffers from the angular momentum problem. If we are to adopt the theory, we must understand why the sun now rotates so slowly, a puzzle we will consider later in this chapter.

If the solar nebula hypothesis is correct, then our Earth and the other planets of the solar system formed billions of years ago as the sun condensed from the interstellar medium. If that is true, then planets form as a by-product of star formation, and most stars should have planets. Before we go further, we should ask if there is any way to check this prediction against the evidence. Can we see planets orbiting other stars?

PLANET-FORMING DISKS

If the solar nebula theory is correct, then planets should be common, and we could test the theory by looking for planets orbiting other stars, objects astronomers call **extrasolar planets.**

The Solar Nebula Hypothesis

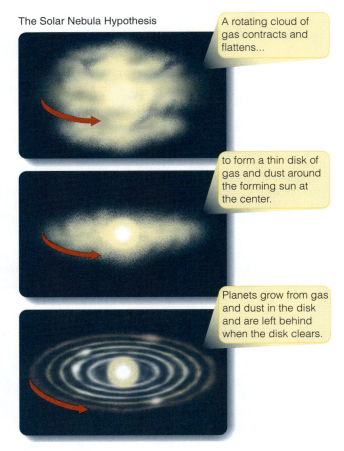

A rotating cloud of gas contracts and flattens...

to form a thin disk of gas and dust around the forming sun at the center.

Planets grow from gas and dust in the disk and are left behind when the disk clears.

FIGURE 19-2
The solar nebula hypothesis proposes that the planets formed along with the sun.

Unfortunately, imaging an extrasolar planet is a bit like photographing a bug crawling on the bulb in a distant searchlight. Planets are small and shine by reflected light. Even a planet as big as Jupiter, 11 times Earth's diameter, reflects little light. If it orbited the nearest star to the Earth, it would be 4 ly away and would be only 22nd magnitude, a very faint image. Furthermore, it would be located only 4 seconds of arc from its star. Current telescopes can't image extrasolar planets directly.

Nevertheless, astronomers have found evidence of extrasolar planets by looking for the dust that accompanies planets and by looking for the motions that planets cause as they orbit their stars.

Infrared astronomers have found very cold, low-density dust disks around stars such as Beta Pictoris. These stars are believed to have completed their formation, so they are clearly older than the newborn stars in Orion. The dusk disk around Beta Pictoris is about 20 times the diameter of our solar system and, like the other known low-density disks, has an inner zone with even lower density. These inner regions may be places where planets have formed (❚ Figure 19-3). Such tenuous dust disks are sometimes called "debris disks" because they are understood to be debris released in col-

Glare of star Beta Pictoris hidden behind central mask

Warp and rings in disk may be caused by one or more planets.

Visual-wavelength image

Size of Pluto's orbit

Dust grains contain ices and silicates.

Star

Narrow ring of dust may be confined by one or more planets.

Glare of star HR4796A hidden behind mask

Diameter of Neptune's orbit

Visual

The K2V star Epsilon Eridani is not very bright in the far infrared.

Clumps in ring of dust may be related to planets.

Far-infrared image

Size of Pluto's orbit

ACTIVE FIGURE 19-3

Dust disks have been detected orbiting a number of stars, but, in the visible part of the spectrum, the dust is at least 1000 times fainter than the stars. In the far infrared, the stars are not bright, and the dust can be detected. Warps, rings, and clumps in these disks suggest the gravitational influence of planets. *(Beta Pic: C. Burrows and J. Krist, STScI and NASA; HR4769A: A. Weinberger, E. Becklin, UCLA, G. Schneider, U. of Arizona, and NASA; Eps Eri: Joint Astronomy Center.)*

Ace⟲Astronomy™ Go to AceAstronomy and click Active Figures to see "Circumstellar Disk" and observe a model disk at different wavelengths.

lisions among small bodies such as comets and asteroids. Our own solar system contains such dust in its inner regions, and astronomers believe it has an extensive debris disk of cold dust extending far beyond the orbits of the planets.

Both visible and radio wavelength observations detect dense disks of gas orbiting young stars. For example, at least 50 percent of the stars in the Orion nebula are encircled by dense disks of gas and dust (▌Figure 19-4 and page 231). A young star is visible at the center of each disk, and astronomers estimate that the disks contain at least a few times Earth's mass in a region a few times larger in diameter than our solar system. The Orion star-forming region is only a few million years old, so it does not seem likely that planets could have formed in these disks yet. Furthermore, the intense radiation from the hot stars in the area is evaporating the disks so fast that planets may never have a chance to grow large. The important point for astronomers is that so many of these young stars have disks. Evi-

dently, disks of gas and dust are a common feature of star formation.

The Hubble Space Telescope can detect dense disks of gas and dust around young stars in a slightly different way. The disks show up by the shadows they cast in the nebulae that surround the newborn stars (▌Figure 19-5). These disks are related to the formation of bipolar flows (Figure 11-6).

Notice the difference between the two kinds of disks that astronomers have found. The low-density dust disks such as the one around Beta Pictoris are produced by dust from collisions among comets and asteroids. Such disks are evidence that planetary systems have already formed. The dense disks of gas and dust such as those seen round the stars in Orion are sites where planets could be forming right now.

A number of infrared objects have been found in Orion that have masses too low to be stars, but many astronomers hesitate to call them planets because they do not orbit stars; they float freely through space. Are they brown dwarfs, failed stars, planets? We discussed

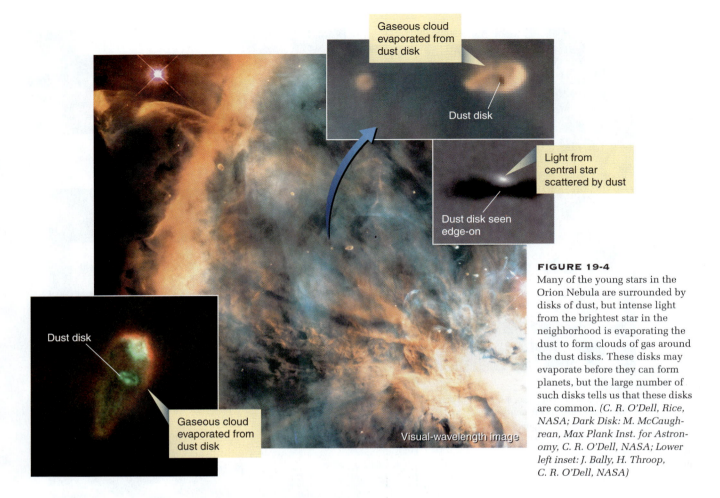

Gaseous cloud evaporated from dust disk

Dust disk

Gaseous cloud evaporated from dust disk

Light from central star scattered by dust

Dust disk seen edge-on

Dust disk

Visual-wavelength image

FIGURE 19-4
Many of the young stars in the Orion Nebula are surrounded by disks of dust, but intense light from the brightest star in the neighborhood is evaporating the dust to form clouds of gas around the dust disks. These disks may evaporate before they can form planets, but the large number of such disks tells us that these disks are common. *(C. R. O'Dell, Rice, NASA; Dark Disk: M. McCaughrean, Max Plank Inst. for Astronomy, C. R. O'Dell, NASA; Lower left inset: J. Bally, H. Throop, C. R. O'Dell, NASA)*

this puzzle in Chapter 12. The controversy is really an argument over the definition of the word *planet,* a word that astronomers will have to define more carefully now that these lower-mass objects are being found.

PLANETS ORBITING OTHER STARS

Detecting the dust and gas of planet formation or low-mass brown dwarfs suggests that planets form easily, but the direct detection of planets orbiting another star is difficult. Nevertheless, such objects have been found. Although such a planet is too faint and too close to its star to be imaged directly, it can be detected from the motion of the star it orbits. As the planet and the star orbit their common center of mass, the star wobbles a bit, just as a person walking an excited dog gets tugged about.

The first planet detected this way orbits the star 51 Pegasi. As the planet circles the star, the star wobbles slightly, and this very small motion of the star is detectable as Doppler shifts in the star's spectrum (❙ Figure 19-6a). From the motion of the star and estimates of the star's mass, astronomers can deduce that the planet has half the mass of Jupiter and orbits only 0.05 AU from the star. Half the mass of Jupiter amounts to 159 Earth

masses, so this is a large planet. Note also that it orbits very close to its star.

Astronomers were not surprised by the announcement that a planet orbited 51 Pegasi; for years astronomers had assumed that many stars had planets. Nevertheless, they greeted the discovery with typical skepticism (Window on Science 19-2). That skepticism led to careful tests of the data and further observations that confirmed the discovery. In fact, over 100 planets have been discovered in this way, including at least three planets orbiting the star Upsilon Andromedae—a true planetary system (Figure 19-6b).

Another way to search for planets is to look for changes in the brightness of the star when the orbiting planet crosses in front of the star. The decrease in light is very small, but it is detectable, and astronomers have used this technique to detect three different planets as they crossed in front of their stars. All three planets are roughly the size of Jupiter and even more massive. These observations further confirm the existence of extrasolar planets.

Notice how the techniques used to detect these planets resemble techniques used to study binary stars. Most of the planets were discovered using the same observational methods used to study spectroscopic

Visual-wavelength images

FIGURE 19-5

Disks around young stars are evident in these Hubble Space Telescope infrared images. The stars are so young that material is still falling inward and being illuminated by the light from the star. Dark bands across the nebulae (arrows) are caused by dense disks of gas and dust that orbit the stars. Note that the star at lower left is double. *(D. Padgett, IPAC/Caltech, W. Brandner, IPAC, K. Stapelfeldt, JPL, and NASA)*

binaries, but a few were found by observing the stars as if they were eclipsing binaries (Chapter 9).

The planets discovered so far tend to be massive and have short periods because lower-mass planets or longer-period planets are harder to detect. Low-mass planets don't tug on their stars very much, and present-day spectrographs can't detect the very small velocity changes that these gentle tugs produce. Planets with longer periods are harder to detect because Earth's astronomers have not been making high-precision observations for many years. Jupiter takes 11 years to circle the sun once, so it will take years for astronomers to see the longer-period wobbles produced by planets lying farther from their stars. Consequently we should

not be surprised that most of the first planets discovered are massive and have short periods.

The new planets may seem odd for another reason. Our experience in our own solar system is that the large planets lie farther from the sun. How could big planets form so near their stars? Theorists find that planets that form in an especially dense disk of matter could spiral inward. Thus it is possible for a few planets to become the massive, short-period planets that we can detect most easily.

A few of the newly discovered extrasolar planets have elliptical orbits, and that seems odd compared with our solar system in which the planetary orbits are nearly circular. Theorists point out, however, that planets

FIGURE 19-6

(a) Small Doppler shifts in the spectrum of the star 51 Pegasi over a period of a month show it wobbling as it is orbited by a planet with a period of 4.2 days. (b) With the rapid wobble caused by the innermost planet removed, the data for Upsilon Andromedae show the influence of two outer planets. The outermost planet orbits the star in 2.6 years, and the middle planet orbits in 0.66 years. The innermost planet has a period of only 4.6 days. (*Adapted from data by Geoffrey Marcy and Paul Butler*)

a

b

may interact in some young planetary systems and can be thrown into elliptical orbits. This is probably rare among planetary systems, but we find these extreme systems more easily because they tend to produce big wobbles.

The preceding paragraphs should reassure us that massive planets in small orbits or in elliptical orbits are not outrageous. At present we know only one planetary system well, our own. As we find more extrasolar planets, we will get a better understanding of what is normal and what is rare.

The discovery of extrasolar planets gives astronomers added confidence in the solar nebula theory. The theory predicts that planets are common, and astronomers are finding them orbiting many stars.

REVIEW CRITICAL INQUIRY

Why does the solar nebula theory imply planets are common?
Often, the implications of a theory are more important in its analysis than its own conjecture about nature. If the solar nebula theory is correct, the planets of our solar system formed from the disk of gas and dust that surrounded the sun as it condensed from the interstellar medium. If that is

WINDOW ON SCIENCE 19-2

Scientists: Courteous Skeptics

"Scientists are just a bunch of skeptics who don't believe in anything." That is a common complaint about scientists, but it misinterprets the fundamental characteristic of the scientist. Yes, scientists are skeptical about new ideas and discoveries, but scientists do hold strong beliefs about how nature works. Ultimately, scientists try to tell stories about how nature works, and skepticism is the scientists' guide.

Some people think that "telling a story" is telling a fib. The stories that scientists tell are exactly the opposite; perhaps we can call them antifibs, because they are as true as scientists can make them. The stories are explanations that have been tested over and over to make them accurate descriptions of nature. Skepticism is the tool scientists use to test every aspect of an explanation.

When a planet was discovered orbiting 51 Pegasi, astronomers were skeptical—not because they thought the observations were wrong, but because that is how science works. Every observation is tested, and every discovery is confirmed. Only an idea that survives many tests begins to be accepted as a scientific truth.

To nonscientists, this makes scientists seem like irritable skeptics, but among scientists it is not bad manners to say, "Really, how do you know that?" or "Why do you think that?" or "Show me the evidence!"

Scientists try to tell stories about how nature works, stories that are sometimes called theories. A story by John Steinbeck can be a brilliant work of art invented to help us understand some aspect of our lives, and another story,

such as a TV script, can be little more than chewing gum for the mind. But scientists' stories are different in a critical way. They are true. To create such stories, scientists cannot create facts to fit their story. Rather, every link in a scientific story must be based on evidence and logic. Of course, evidence can be misunderstood, and logical errors happen. To test every aspect of their story, scientists must be continuously skeptical. That is the only way to discover those scientific truths that help us understand how nature works.

Skepticism is not a refusal to hold beliefs. Scientists often believe sincerely in their theories once those theories have been tested over and over. Rather, skepticism is the tool scientists use to find those natural principles worthy of belief.

---

true, then it should be a common process. Most stars should form with disks of gas and dust around them, and so we would expect them to have planets.

Without knowing any more about the theory, the implication that planets are common suggests a way we could test the theory. If we look at a large number of stars and find lots of planets, the solar nebula theory is confirmed. But if we search and find few planets, it is contradicted. What observational evidence do we have so far on this issue?

Astronomers are finding extrasolar planets orbiting many stars, and that suggests that planetary systems are common. Evidently, planets form easily in the disks of gas and dust around newborn stars, so it is time to construct a detailed theory to describe planet formation. We can begin by surveying our own solar system to discover the distinguishing characteristics of planetary systems. Those are the characteristics that our theory must eventually explain.

19-2 A SURVEY OF THE SOLAR SYSTEM

To test our theories, we must search the present solar system for evidence of its past (Window on Science 19-3). In this section, we survey the solar system and compile a list of its most significant characteristics, potential clues to how it formed.

We begin with the most general view of the solar system. It is, in fact, almost entirely empty space (Figure 1-7). Imagine that we reduce the solar system until Earth is the size of a grain of table salt, about 0.3 mm (0.01 in.) in diameter. The moon is a speck of pepper about 1 cm (0.4 in.) away, and the sun is the size of a small plum 4 m (13 ft) from Earth. Mercury, Venus, and Mars are grains of salt. Jupiter is an apple seed 20 m (66 ft) from the sun, and Saturn is a smaller seed over 36 m (120 ft) away. Uranus and Neptune are slightly larger than average salt grains, and Pluto, the farthest planet (on average), is a speck of pepper over 150 m (500 ft) from the central plum. We would need a powerful microscope to detect the asteroids. These tiny specks of matter scattered around the sun are the remains of the solar nebula.

REVOLUTION AND ROTATION

The planets revolve* around the sun in orbits that lie close to a common plane. The orbit of Mercury, the closest planet to the sun, is tipped 7° to Earth's orbit, and Pluto's orbit is tipped 17.2°. The rest of the planets' orbital planes are inclined by no more than 3.4°. Thus, the solar system is basically disk-shaped.

The rotation of the sun and planets on their axes also seems related to this disk shape. The sun rotates

---

*Recall from Chapter 2 that English distinguishes between the words *revolve* and *rotate*. A planet revolves around the sun but rotates on its axis. Cowboys in the old west didn't carry revolvers. They carried rotators.

Reconstructing the Past from Evidence and Hypothesis

Scientists often face problems in which they must reconstruct the past. Some of these reconstructions are obvious, such as an archaeologist excavating the ruins of a burial tomb, but others are less obvious. In each case, success requires the interplay of hypotheses and evidence to re-create a past that no longer exists.

The reconstruction of the past is obvious when we use the chemical abundance of stars to reconstruct the story of the formation of our galaxy, but a biologist studying a centipede is also reconstructing the past. How did this creature come to have a segmented body with so many legs? How did it develop the metabolism that allows it to move quickly and hunt prey? Although the problem might at first seem to be one of mere anatomy, the scientist must reconstruct an environment that no longer exists.

Geologists are fond of saying "the present is the key to the past," and that is true for any scientific problem that requires us to reconstruct the past. Because the past is over, we cannot observe it directly, so we must use the evidence of the present to test hypotheses about the nature of the past.

The grand interplay of evidence and theory is the distinguishing characteristic of science, and nowhere do we see it more clearly than when scientists reconstruct the past.

with its equator inclined only 7.25° to Earth's orbit, and most of the other planets' equators are tipped less than 30°. The rotations of Venus, Uranus, and Pluto are peculiar, however. Venus rotates backward compared with the other planets, and both Uranus and Pluto rotate on their sides (with their equators almost perpendicular to their orbits). We will discuss these planets in detail in Chapters 22 and 24, but later in this chapter we will try to understand how they could have acquired their peculiar rotations.

Apparently, the preferred direction of motion in the solar system—counterclockwise as seen from the north—is also related to its disk shape. All the planets revolve counterclockwise around the sun, and with the exception of Venus, Uranus, and Pluto they rotate counterclockwise on their axes. Furthermore, nearly all of the moons in the solar system, including Earth's moon, orbit around their planets counterclockwise. With only a few exceptions, most of which are understood, revolution and rotation in the solar system follow a disk theme.

TWO KINDS OF PLANETS

Perhaps the most striking clue we have to the origin of the solar system comes from the division of the planets into two categories: the small Earthlike worlds and the giant Jupiterlike worlds. The difference is so dramatic we can hardly keep from shouting, "Aha, this must mean something."

Study ▌ "Terrestrial and Jovian Planets" on pages 418 and 419 and notice three important points. First, the two kinds of planets are distinguished by their location. The four inner planets are quite different from the next four outward. Also, notice that the planets differ dramatically in mass and density. Jupiter's family of planets is much more massive but much less dense than Earth's family. Finally, notice that craters are common all through the solar system. The only reason we don't see craters on Jupiter and its family is that those planets do not have solid surfaces.

The division of the planets into two families is a clue to how our solar system formed, but we can learn little from the individual planets because they have evolved since they formed. For further clues we must look at smaller objects that have remained largely unchanged since the birth of the solar system.

SPACE DEBRIS

The sun and planets are not the only remains of the solar nebula. The solar system is littered with three kinds of space debris: asteroids, comets, and meteoroids. Although these objects represent a tiny fraction of the mass of the system, they are a rich source of information about the origin of the planets.

The **asteroids,** sometimes called minor planets, are small rocky worlds, most of which orbit the sun in a belt between the orbits of Mars and Jupiter. Roughly 25,000 asteroids have been identified, of which about 2000 follow orbits that bring them into the inner solar system, where they can occasionally collide with a planet. Earth has been struck many times in its history. Some asteroids are located in Jupiter's orbit, some have been found beyond the orbit of Saturn, and a growing number are being discovered among and even beyond the outermost planets.

About 200 asteroids are more than 100 km (60 mi) in diameter, and over 27,000 are estimated to be more than 10 km (6 mi) in diameter. There are probably 500,000 that are larger than 1 km (0.6 mi) and billions that are smaller. Because even the largest are only a few hundred kilometers in diameter, Earth-based telescopes can detect no details on their surfaces, and the Hubble Space Telescope can image only the largest features.

We do however have clear evidence that asteroids are irregularly shaped cratered worlds. A number of spacecraft have visited asteroids and sent back photos. For instance, the NEAR spacecraft rendezvoused with the asteroid Eros, went into orbit, and studied the asteroid in detail. Like most asteroids, Eros is an irregular, rocky body pocked by craters (▌ Figure 19-7). We

Visual-wavelength images

FIGURE 19-7

(a) Over a period of three weeks, the NEAR spacecraft approached the asteroid Eros and recorded a series of images arranged here in an entertaining pattern showing the irregular shape and 5-hour rotation of the asteroid. Eros is 34 km (21 mi) long. (b) This close-up of the surface of Eros shows an area about 11 km (7 mi) from top to bottom. *(Johns Hopkins University, Applied Physics Laboratory, NASA)*

will discuss these observations in detail in Chapter 25, but in our quick survey of the solar system we can note that all evidence suggests the asteroids have suffered many impacts from collisions with other asteroids.

Older theories proposed that the asteroids are the remains of a planet that broke up, but modern astronomers recognize the asteroids as the debris left over by a planet that failed to form at a distance of 2.8 AU from the sun. When we discuss the formation of planets later in this chapter, we must be prepared to explain why material in the solar nebula failed to form a planet at a distance of 2.8 AU.

In contrast to asteroids, the brightest **comets** are impressively beautiful objects (❚ Figure 19-8a). Most comets are faint, however, and are difficult to locate even at their brightest. A comet may take months to sweep through the inner solar system, during which time it appears as a glowing head with an extended tail of gas and dust.

The beautiful tail of a comet can be longer than an astronomical unit, but it is produced by an icy nucleus only a few tens of kilometers in diameter. In its long,

elliptical orbit, the nucleus remains frozen and inactive while it is far from the sun. As its orbit carries the nucleus into the inner solar system, the sun's heat begins to vaporize the ices, releasing gas and dust. The pressure of sunlight and the solar wind pushes the gas and dust away, forming a long tail. The motion of the nucleus along its orbit, the pressure of sunlight, and the outward flow of the solar wind can create comet tails that are long and straight or gently curved, but in either case the tails of comets always point generally away from the sun (Figure 19-8b).

For decades astronomers described comet nuclei as dirty snowballs, meaning that they were icy bodies with a little bit of embedded rock and dust. Starting with the passage of Comet Halley in 1986, astronomers have found growing evidence that comet nuclei are not made of dirty ice but rather of icy dirt. That is, the nuclei are at least 50 percent rock and dust. One astronomer has suggested we replace the dirty snowball model with the icy mud-ball model.

The nuclei of comets are icy bodies left over from the origin of the planets. Thus, we can conclude that at

Terrestrial and Jovian Planets

The distinction between the terrestrial planets and the Jovian planets is dramatic. The inner four planets, Mercury, Venus, Earth, and Mars, are terrestrial planets, *meaning they are small, dense, rocky worlds with little or no atmosphere. The outer four planets, Jupiter, Saturn, Uranus, and Neptune, are* Jovian planets, *meaning they are large, low-density worlds with thick atmospheres and liquid interiors. Pluto does not fit this scheme, being small but low density; we will see in a later chapter that it is a very special world and might not be considered a planet at all.*

Planetary orbits to scale. The terrestrial planets lie quite close to the sun, whereas the Jovian planets are spread far from the sun outside the asteroid belt. The elliptical shape of Pluto's orbit is visible here.

The planets and the sun to scale. Saturn's rings would just reach from Earth to the moon.

Mercury

Venus

Sun

Earth

Moon

Mars

Jupiter

Saturn

Uranus

Neptune

Pluto

Jupiter

Mercury

Venus

Mars

Earth

Asteroids

Saturn

Uranus

Neptune

Pluto

Of the terrestrial planets, Earth is most massive, but the Jovian planets are much more massive. Jupiter is over 300 Earth masses, and Saturn is nearly 100 Earth masses. Uranus and Neptune are 15 and 17 Earth masses.

Mercury is only 40 percent larger than Earth's moon, and its weak gravity cannot retain a permanent atmosphere. Like the moon, it is covered with craters from meteorite impacts.

Mercury

Earth's moon

Craters are common on all of the surfaces in the solar system that are strong enough to retain them. Earth has about 150 impact craters, but many more have been erased by erosion. Besides the planets, the asteroids and nearly all of the moons in the solar system are scarred by craters. Ranging from microscopic to hundreds of kilometers in diameter, these craters have been produced over the ages by meteorite impacts. When astronomers see a rocky or icy surface that contains few craters, they know that the surface is young.

NASA

UCO/Lick Observatory

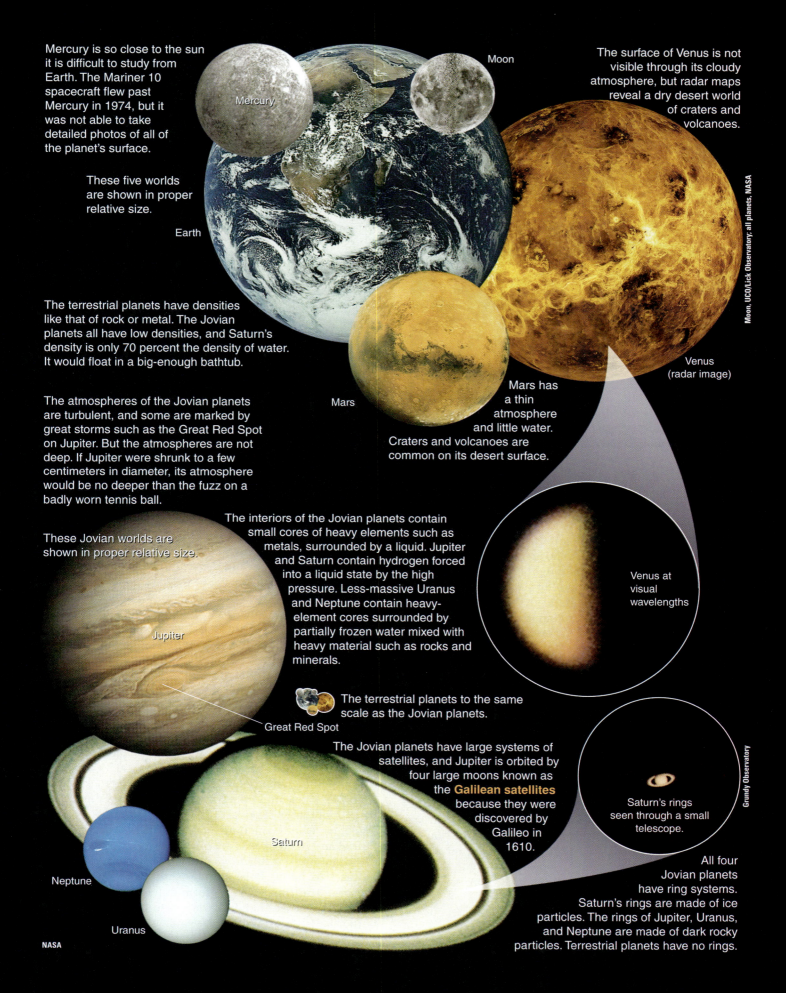

Mercury is so close to the sun it is difficult to study from Earth. The Mariner 10 spacecraft flew past Mercury in 1974, but it was not able to take detailed photos of all of the planet's surface.

These five worlds are shown in proper relative size.

Mercury

Moon

Earth

The surface of Venus is not visible through its cloudy atmosphere, but radar maps reveal a dry desert world of craters and volcanoes.

Moon, UCO/Lick Observatory; all planets, NASA

The terrestrial planets have densities like that of rock or metal. The Jovian planets all have low densities, and Saturn's density is only 70 percent the density of water. It would float in a big-enough bathtub.

The atmospheres of the Jovian planets are turbulent, and some are marked by great storms such as the Great Red Spot on Jupiter. But the atmospheres are not deep. If Jupiter were shrunk to a few centimeters in diameter, its atmosphere would be no deeper than the fuzz on a badly worn tennis ball.

Venus
(radar image)

Mars has a thin atmosphere and little water. Craters and volcanoes are common on its desert surface.

Mars

The interiors of the Jovian planets contain small cores of heavy elements such as metals, surrounded by a liquid. Jupiter and Saturn contain hydrogen forced into a liquid state by the high pressure. Less-massive Uranus and Neptune contain heavy-element cores surrounded by partially frozen water mixed with heavy material such as rocks and minerals.

These Jovian worlds are shown in proper relative size.

Venus at visual wavelengths

Jupiter

The terrestrial planets to the same scale as the Jovian planets.

Great Red Spot

The Jovian planets have large systems of satellites, and Jupiter is orbited by four large moons known as the **Galilean satellites** because they were discovered by Galileo in 1610.

Saturn's rings seen through a small telescope.

Grundy Observatory

Neptune

Saturn

Uranus

All four Jovian planets have ring systems. Saturn's rings are made of ice particles. The rings of Jupiter, Uranus, and Neptune are made of dark rocky particles. Terrestrial planets have no rings.

NASA

Visual-wavelength image.

a

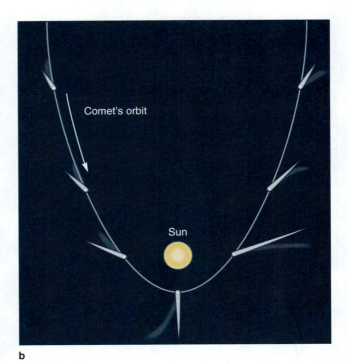

Comet's orbit

Sun

b

FIGURE 19-8

(a) A comet may remain visible in the evening or morning sky for weeks as it moves through the inner solar system. Comet West was in the sky during March 1976. (b) A comet in a long, elliptical orbit becomes visible when the sun's heat vaporizes its ices and pushes the gas and dust away in a tail. *(Celestron International)*

least some parts of the solar nebula were rich in ices. We will see later in this chapter how important ices were in the formation of the Jovian planets, and we will discuss comets in more detail in Chapter 25.

Unlike the stately comets, **meteors** flash across the sky in momentary streaks of light (▌ Figure 19-9). They are commonly called "shooting stars." Of course, they are not stars but small bits of rock and metal falling into Earth's atmosphere and bursting into incandescent vapor about 80 km (50 mi) above the ground because of friction with the air. This vapor condenses to form dust, which settles slowly to Earth, adding about 40,000 tons per year to the planet's mass.

Technically, the word *meteor* refers to the streak of light in the sky. In space, before its fiery plunge, the object is called a **meteoroid,** and any part of it that survives its fiery passage to Earth's surface is called a **meteorite.** Most meteoroids are specks of dust, grains of sand, or tiny pebbles. Almost all the meteors we see in the sky are produced by meteoroids that weigh less than 1 g. Only rarely is one massive enough and strong enough to survive its plunge and reach Earth's surface.

Thousands of meteorites have been found, and we will discuss their particular forms in Chapter 25. We mention meteorites here for one specific clue they can give us concerning the solar nebula: Meteorites can tell us the age of the solar system.

THE AGE OF THE SOLAR SYSTEM

If the solar nebula theory is correct, the planets should be about the same age as the sun. The most accurate way to find the age of a celestial body is to bring a sample into the laboratory and determine its age by analyzing the radioactive elements it contains.

When a rock solidifies, it incorporates known percentages of the chemical elements. A few of these elements are radioactive and can decay into another element, called the daughter element. For example, the isotope of uranium ^{238}U can decay into an isotope of lead, ^{206}Pb. The **half-life** of a radioactive element is the time it takes for half of the atoms to decay. The half-life of ^{238}U is 4.5 billion years. Thus, the abundance of a radioactive element gradually decreases as it decays, and the abundance of the daughter element gradually increases (▌ Figure 19-10). If we know the abundance of the elements in the original rock, we can measure the present abundance and find the age of the rock. For example, if we studied a rock and found that only 50 percent of the ^{238}U remained and the rest had become ^{206}Pb, we would conclude that one half-life must have passed and the rock was 4.5 billion years old.

Uranium isn't the only radioactive element used in radioactive dating. Potassium (^{40}K) decays with a half-life of 1.3 billion years to form calcium (^{40}Ca) and argon (^{40}Ar). Rubidium (^{87}R) decays to strontium (^{87}Sr) with a half-life of 47 billion years. Any of these ele-

FIGURE 19-9

A meteor is the streak of glowing gases produced by a bit of material falling into Earth's atmosphere. Friction with the air vaporizes the material about 80 km (50 mi) above Earth's surface. *(Daniel Good)*

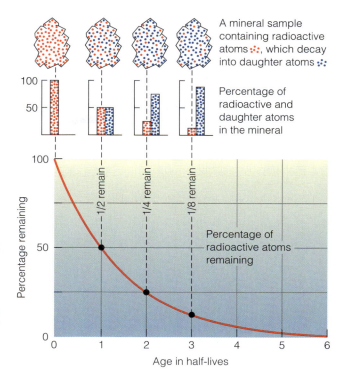

A mineral sample containing radioactive atoms ⋮, which decay into daughter atoms ⋮

Percentage of radioactive and daughter atoms in the mineral

Percentage of radioactive atoms remaining

ACTIVE FIGURE 19-10

The radioactive atoms in a mineral sample (red) decay into daughter atoms (blue). Half the radioactive atoms are left after one half-life, a fourth after two half-lives, an eighth after three half-lives, and so on.

Ace⊙Astronomy™ Go to AceAstronomy and click Active Figures to see "Radioactive Decay" and graph the decay of different radioactive isotopes.

ments can be used as a radioactive clock to find the age of mineral samples.

Of course, to find a radioactive age, we need a sample in the laboratory, and the only celestial bodies from which we have samples are Earth, the moon, Mars, and meteorites.

The oldest Earth rocks so far discovered and dated are tiny zircon crystals from Australia 4.3 billion years old. That does not mean that Earth formed 4.3 billion years ago. The surface of Earth is active, and the crust is continually destroyed and reformed from material welling up from beneath the crust (see Chapter 20). Thus, the age of these oldest rocks tells us only that Earth is *at least* 4.3 billion years old.

One of the most exciting goals of the Apollo lunar landings was bringing lunar rocks back to Earth's laboratories, where they could be dated. Because the moon's surface is not being recycled like Earth's, some parts of it might have survived unaltered since early in the history of the solar system. Dating the rocks showed the oldest to be 4.48 billion years old. Thus, the solar system must be *at least* 4.48 billion years old.

Although no one has yet been to Mars, a few meteorites found on Earth have been identified by their

chemical composition as having come from Mars. Most of these have ages of only 1.3 billion years, but one has an age of about 4.6 billion years. Thus, Mars must be at least that old.

Another important source for determining the age of the solar system is meteorites. Radioactive dating of meteorites yields a range of ages, with the oldest about 4.56 billion years old. This figure is widely accepted as the age of the solar system.

Meteorites also contain tiny grains containing mostly silicon and carbon, and the composition of those grains shows that they are from the interstellar medium. They cannot be dated radioactively, but they must have formed long before our solar system and were incorporated in the nebula that contracted to form our sun and our planetary system. These tiny particles are messengers from beyond the beginning of our solar system.

One last celestial body deserves mention: the sun. Astronomers estimate the age of the sun to be about 5 billion years, but this is not a radioactive date because they cannot obtain a sample of solar material. Instead, they estimate the age of the sun from mathematical models of the sun's interior. This yields an age of about 5 billion years plus or minus 1.5 billion years, a number that is in agreement with the age of the solar system derived from the age of meteorites.

TABLE 19-1
Characteristic Properties of the Solar System

1. Disk shape of the solar system
   Orbits in nearly the same plane
   Common direction of rotation and revolution
2. Two planetary types
   Terrestrial—inner planets; high density
   Jovian—outer planets; low density
3. Planetary ring systems for Jupiter, Saturn, Uranus, and Neptune
4. Space debris—asteroids, comets, and meteors
   Composition
   Orbits
5. Common ages of about 4.6 billion years for Earth, the moon, Mars, meteorites, and the sun

Apparently, all the bodies of the solar system formed at about the same time some 4.6 billion years ago. We can add this as the final item to our list of characteristic properties of the solar system (▌Table 19-1).

REVIEW CRITICAL INQUIRY

In what ways is the solar system a disk?

First, the general shape of the solar system is that of a disk. The planets follow orbits that lie in nearly the same plane. The orbit of Mercury is inclined 7° to the plane of Earth's orbit, and the orbit of Pluto is inclined a bit more than 17°. Thus, the planets follow orbits confined to a thin disk with the sun at its center.

Second, the motions of the sun and planets also follow this disk theme. The sun and most of the planets rotate in the same direction, counterclockwise as seen from the north, with their equators near the plane of the solar system. Also, all of the planets revolve around the sun in that same direction. Thus, our solar system seems to prefer motion in the same direction, which further reflects a disk theme.

One of the basic characteristics of our solar system is its disk shape, but another dramatic characteristic is the division of the planets into two groups. What are the distinguishing differences between the terrestrial and Jovian planets?

We have completed our survey of the solar system and gathered the preliminary evidence we can use to evaluate the solar nebula hypothesis. Now it is time to ask how the planets formed from the solar nebula and took on the characteristics we see today.

19-3 THE STORY OF PLANET BUILDING

The challenge for modern planetary astronomers is to compare the characteristics of the solar system with the solar nebula theory and tell the story of how the planets formed.

THE CHEMICAL COMPOSITION OF THE SOLAR NEBULA

Everything we know about the solar system and star formation suggests that the solar nebula was a fragment of an interstellar gas cloud. Such a cloud would have been mostly hydrogen with some helium and tiny traces of the heavier elements.

This is precisely what we see in the composition of the sun (see Table 7-2). Analysis of the solar spectrum shows that the sun is mostly hydrogen, with a quarter of its mass being helium and only about 2 percent being heavier elements. Of course, nuclear reactions have fused some hydrogen into helium, but this happens in the sun's core and has not affected its surface composition. Thus, the composition revealed in its spectrum is essentially the composition of the gases from which it formed.

This must have been the composition of the solar nebula, and we can see that composition reflected in the chemical compositions of the planets. One factor, however, can dramatically alter the composition of a planet. We must recognize that planets grow by two processes. Planets begin growing by the sticking together of solid bits of matter. Only when a planet has grown to a mass of about 15 Earth masses does it have enough gravitation to begin capturing gas directly from the solar nebula in a process called **gravitational collapse.** The Jovian planets began forming by the aggregation of bits of rock and ice. Once they accumulated enough mass, they began growing by gravitational collapse. That is, they could capture large amounts of gas, mostly hydrogen and helium, directly from the solar nebula. Jupiter and Saturn grew rapidly and captured the most hydrogen and helium, whereas Uranus and Neptune grew more slowly and didn't capture gas as rapidly as Jupiter and Saturn. Thus, the Jovian worlds grew to be low-density, hydrogen-rich planets, with Jupiter and Saturn the largest and least dense.

The terrestrial planets, in contrast, contain very little hydrogen and helium. These small planets have such low masses they were unable to keep the hydrogen and helium from leaking into space. Indeed, because of their small masses, they were probably unable to capture very much of these lightweight gases from the solar nebula. The terrestrial planets are dense worlds because they are composed of the heavier elements from the solar nebula.

Although we can see how the chemical composition of the solar nebula is reflected in the present composition of the sun and planets, a very important question remains: How did gas and dust in the solar nebula come together to form the solid matter of the planets? We must answer that question in two stages. First, we must understand how the gas formed billions of small solid particles, and then we must explain how these particles built the planets.

THE CONDENSATION OF SOLIDS

The key to understanding the process that converted the nebular gas into solid matter is the variation in density among solar system objects. We have already noted that the four inner planets are high-density, terrestrial bodies, whereas the outermost planets are low-density, giant planets (except for Pluto, which is low in density but not a giant). This division is due to the different ways gases condensed into solids in the inner and outer regions of the solar nebula.

Even among the four terrestrial planets, we find a pattern of subtle differences in density. Merely listing the observed densities of the terrestrial planets does not reveal the pattern, because Earth and Venus, being more massive, have stronger gravity and have squeezed their interiors to higher densities. We must look at the **uncompressed densities**—the densities the planets would have if their gravity did not compress them. These densities (■ Table 19-2) show that in general the closer a planet is to the sun, the higher its uncompressed density.

This density variation originated when the solar system first formed solid grains. The kind of matter that condensed in a particular region would depend on the temperature of the gas there. In the inner regions, the temperature may have been 1500 K or so. The only materials that can form grains at this temperature are compounds with high melting points, such as metal oxides and pure metals, which are very dense. Farther out in the nebula it was cooler, and silicates (rocky material) could condense. These are less dense than metal ox-

TABLE 19-3
The Condensation Sequence

Temperature (K)	Condensate	Planet (Estimated Temperature of Formation; K)
1500	Metal oxides	Mercury (1400)
1300	Metallic iron and nickel	
1200	Silicates	
1000	Feldspars	Venus (900)
680	Troilite (FeS)	Earth (600)
		Mars (450)
175	H₂O ice	Jovian (175)
150	Ammonia–water ice	
120	Methane–water ice	
65	Argon–neon ice	Pluto (65)

ides and metals. In the cold outer regions, ices of water, methane, and ammonia could condense. These are low-density materials.

The sequence in which the different materials condense from the gas as we move away from the sun is called the **condensation sequence** (■ Table 19-3). It suggests that the planets, forming at different distances from the sun, accumulated from different kinds of materials. Thus, the inner planets formed from high-density metal oxides and metals, and the outer planets formed from low-density ices.

We must also remember that the solar nebula did not remain the same temperature throughout the formation of the planets but may have grown progressively cooler. Thus, a particular region of the nebula may have begun by producing solid particles of metals and metal oxides but, after cooling, began producing particles of silicates. Allowing for the cooling of the nebula makes our theory much more complex, but it also makes the processes by which the planets formed much more understandable.

THE FORMATION OF PLANETESIMALS

In the development of a planet, three groups of processes operate. First, grains of solid matter grow larger, eventually reaching diameters ranging from a few centimeters to kilometers. The larger of these objects, called **planetesimals,** are believed to be the bodies that the second group of processes collects into planets. Finally, a third set of processes clears away the solar nebula. The study of planet building is the study of these three groups of processes.

According to the solar nebula theory, planetary development in the solar nebula began with the growth of dust grains. These specks of matter, whatever their

TABLE 19-2
Observed and Uncompressed Densities

Planet	Observed Density (g/cm³)	Uncompressed Density (g/cm³)
Mercury	5.44	5.30
Venus	5.24	3.96
Earth	5.50	4.07
Mars	3.94	3.73
(Moon)	3.36	3.40

composition, grew from microscopic size by two processes: condensation and accretion.

A particle grows by **condensation** when it adds matter one atom at a time from a surrounding gas. Thus, snowflakes grow by condensation in Earth's atmosphere. In the solar nebula, dust grains were continuously bombarded by atoms of gas, and some of these stuck to the grains. A microscopic grain capturing a gas atom increases its mass by a much larger fraction than a gigantic boulder capturing a single atom. Thus, condensation can increase the mass of a small grain rapidly, but as the grain grows larger, condensation becomes less effective.

The second process is **accretion,** the sticking together of solid particles. You may have seen accretion in action if you have walked through a snowstorm with big, fluffy flakes. If you caught one of those "flakes" on your mitten and looked closely, you saw that it was made up of many tiny, individual flakes that had collided as they fell and accreted to form larger particles. In the solar nebula the dust grains were, on the average, no more than a few centimeters apart, so they collided frequently. Their mutual gravitation was too small to hold them to each other, but other effects may have helped. Static electricity generated by their passage through the gas could have held them together, as could compounds of carbon that might have formed a sticky surface on the grains. Ice grains might have stuck together better than some other types. Of course, some collisions might have broken up clumps of grains; on the whole, however, accretion must have increased grain size. If it had not, the planets would not have formed.

There is no clear distinction between a very large grain and a very small planetesimal, but we can consider an object a planetesimal when its diameter becomes a kilometer or so. Objects larger than a centimeter were subject to new processes that tended to concentrate them. One important effect may have been that the growing planetesimals collapsed into the plane of the solar nebula. Dust grains could not fall into the plane because the turbulent motions of the gas kept them stirred up, but the larger objects had more mass, and the gas motions could not have prevented them from settling into the plane of the spinning nebula. This would have concentrated the solid particles into a thin plane about 0.01 AU thick and would have made further planetary growth more rapid.

This collapse of the planetesimals into the plane is analogous to the flattening of a forming galaxy. However, an entirely new process may have become important once the plane of planetesimals formed. Computer models show that the rotating disk of particles should have been gravitationally unstable and would have broken up into small clouds (▌ Figure 19-11). This would further concentrate the planetesimals and help them coalesce into objects up to 100 km (60 mi) in diameter.

FIGURE 19-11
Gravitational instabilities in the rotating disk of planetesimals may have forced them to collect in clumps, accelerating their growth.

Thus, the theory proposes that the nebula became filled with trillions of solid particles ranging in size from pebbles to tiny planets. As the largest began to exceed 100 km in diameter, new processes began to alter them, and a new stage in planet building began, the growth of protoplanets.

THE GROWTH OF PROTOPLANETS

The coalescing of planetesimals eventually formed **protoplanets,** massive objects destined to become planets. As these larger bodies grew, new processes began making them grow faster and altered their physical structure.

If planetesimals collided at orbital velocities, it is unlikely that they would have stuck together. The average orbital velocity in the solar system is about 10 km/s (22,000 mph). Head-on collisions at this velocity would have vaporized the material. However, the planetesimals were moving in the same direction in the nebular plane and thus avoided head-on collisions. Instead, they merely rubbed shoulders at low relative velocities. Such collisions would have been more likely to fuse them than to shatter them.

In addition, some adhesive effects probably helped. Sticky coatings and electrostatic charges on the surfaces of the smaller planetesimals probably aided formation of larger bodies. Collisions would have fragmented some of the surface rock, but if the planetesimals were large enough, their gravity would have held on to some fragments, forming a layer of soil composed entirely of crushed rock. Such a layer on the larger planetesimals may have been effective in trapping smaller bodies.

The largest planetesimals would grow the fastest because they had the strongest gravitational field. Not only could they hold on to a cushioning layer to trap fragments, but their stronger gravity could also attract additional material. These planetesimals probably grew quickly to protoplanetary dimensions, sweeping up more and more material. When massive enough, they trapped some of the original nebular gas to form primitive atmospheres. At some point, they crossed the boundary between planetesimals and protoplanets.

To trace the growth of a protoplanet, we can think of the formation of Earth. In its simplest form, our theory of protoplanet growth supposes that all the planetesimals had about the same chemical composition. The planetesimals accumulated gradually to form a planet-size ball of material that was of homogeneous composition throughout. Once the planet formed, heat began to accumulate in its interior from the decay of short-lived radioactive elements, and this heat eventually melted the planet and allowed it to differentiate. **Differentiation** is the separation of material according to density. When the planet melted, the heavy metals such as iron and nickel settled to the core, while the lighter silicates floated to the surface to form a low-density crust. The story of planet formation from planetesimals of similar composition is shown in the left half of ▌Figure 19-12.

In this simple form, the theory predicts that Earth's present atmosphere was not its first. The first atmosphere consisted of small amounts of gases trapped from the solar nebula—mostly hydrogen and helium. They were later driven off by the heat, aided perhaps by outbursts from the infant sun, and new gases that baked from the rocks formed a secondary atmosphere. This creation of a planetary atmosphere from the planet's interior is called **outgassing.**

We can improve this simple theory of planet formation in two ways. First, it seems likely that the solar nebula cooled during the formation of the planets, so they did not accumulate from planetesimals of common composition. As planet building began, the first particles to condense in the inner solar system were metals and metal oxides, so the protoplanets may have

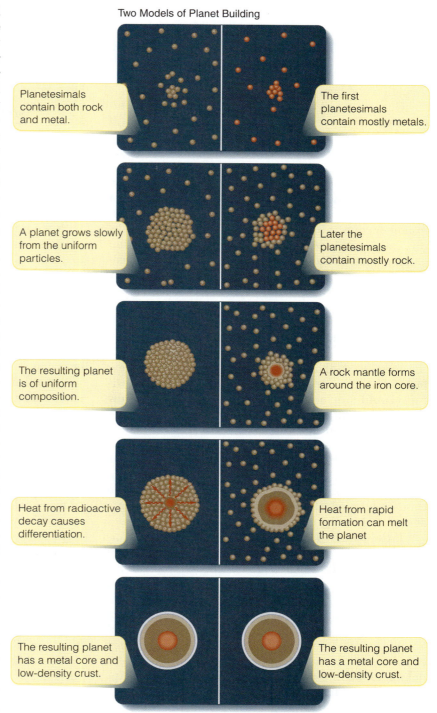

Two Models of Planet Building

Planetesimals contain both rock and metal.

The first planetesimals contain mostly metals.

A planet grows slowly from the uniform particles.

Later the planetesimals contain mostly rock.

The resulting planet is of uniform composition.

A rock mantle forms around the iron core.

Heat from radioactive decay causes differentiation.

Heat from rapid formation can melt the planet

The resulting planet has a metal core and low-density crust.

The resulting planet has a metal core and low-density crust.

FIGURE 19-12

If the temperature of the solar nebula changed during planet building, the composition of the planetesimals may have changed. The simple model at left assumes no change occurred, but the model at the right incorporates a change from metallic to rocky planetesimals.

begun by accreting metallic cores. Later, as the nebula cooled, more silicates could form, and the protoplanets added silicate mantles. A second improvement in the theory proposes that the planets grew so rapidly that the heat released by the in-falling particles, the **heat of formation,** did not have time to escape. This heat rapidly accumulated and melted the protoplanets as they formed. If this is true, then the planets would have differentiated as they formed. This improved story of planet formation is shown in the right half of Figure 19-12.

Our improved theory suggests that Earth's first atmosphere was not captured from the solar nebula. If Earth formed in a molten state, then there was never a time when it had a primitive atmosphere of hydrogen and helium accumulated from the solar nebula. Rather, the gases of the atmosphere were released by the molten rock as the protoplanet grew. Those gases would not have included much water, however, so some astronomers now think that Earth's water and much of its present atmosphere accumulated late in the formation of the planets as Earth swept up volatile-rich planetesimals forming in the cooling solar nebula. Such icy planetesimals may have formed in the outer parts of the solar nebula and been scattered by encounters with the Jovian planets in a bombardment of comets.

According to the solar nebula theory, the Jovian planets began growing by the same processes that built the terrestrial planets. But the Jovian planets grew faster. In the inner solar nebula only metals and silicates could form solids, so the terrestrial planets grew slowly. The outer solar nebula not only contained solid bits of metals and silicates, but it was a blizzard of ices. The Jovian planets grew rapidly and quickly became massive enough to grow by gravitational collapse as they drew in large amounts of gas from the solar nebula. The terrestrial planets could not grow from ices, so they developed slowly and never became massive enough to grow by gravitational collapse.

If the planets formed in this way, the Jovian planets must have grown to their present size in about 10 million years. How do astronomers know this? Almost all T Tauri stars that have gas disks are younger than 10 million years; older T Tauri stars have blown their gas disks away. So the Jovian planets must have been complete within 10 million years or so. The terrestrial planets grew from solids and not from the gas, so they could have continued to grow by accretion from solid debris left behind when the gas was blown away. Apparently, the terrestrial planets were almost fully formed within 30 million years. Models of the solar nebula suggest that all planet growth ended and the last of the solar nebula was blown away by 100 million years.

The solar nebula theory has been very successful in explaining the formation of the solar system. But there are some problems, and the Jovian planets are the troublemakers.

THE JOVIAN PROBLEM

Two observations are making it hard to explain the formation of the Jovian parents, and that is forcing astronomers to expand and revise their theories of planet formation.

The first of the two observations is the discovery of lots of extrasolar planets. Those planets are all massive enough to be Jovian worlds. Also, the three planets that have been observed crossing in front of their stars blocked enough light to show that they are as big as Jovian planets. All of this tells us that Jovian planets are common.

The second observation is the discovery that the gas and dust disks around newborn stars don't last long. Images show that the disks around the young stars in the Orion nebula (Figure 19-4) are being evaporated by the intense ultraviolet radiation from hot stars with the nebula. It seems that nearly all stars form in clusters containing massive stars, so this must happen to most disks. Even if a disk did not evaporate quickly, the gravitational influence of the crowded stars should quickly strip away the outer parts of the disk. These observations seem to mean that disks don't last longer than about 7 million years at most, and many must evaporate within 100,000 years or so. That's not long enough to grow a Jovian planet. Yet Jovian worlds are common.

Mathematical models of the solar nebula have been computed using specially built computers running programs that take weeks to finish a calculation. The results show that the rotating solar nebula may have become unstable and formed outer planets by direct gravitational collapse. That is, massive planets may have been able to form from the gas without first forming a dense core by accretion. Jupiters and Saturns form in these calculated models with a few hundred years.

If the Jovian planets formed in this way, they could have formed quickly before the solar nebula disappeared. Terrestrial planets grow from solid bits of material in the inner nebula, so they could have continued to grow by accretion even after the disk was cleared of gas and smaller particles of dust.

This new insight into the formation of the outer planets may help explain the formation of Uranus and Neptune. They are so far from the sun that accretion could not have built them rapidly. The gas and dust of the solar nebula must have been sparse out there, and Uranus and Neptune orbit so slowly they would not have swept up material very rapidly. The conventional view is that they grew by accretion so slowly that they never became quite massive enough to begin growing by gravitational collapse. In fact, it is hard to understand how they could have reached their present mass if they grew by accretion so far from the sun. Theoretical calculations show that they might have formed closer to the sun in the region of Jupiter and Saturn and then could have been shifted outward by gravitational interactions. In any case, the formation of Uranus and Neptune is part of the Jovian problem.

The solar nebula theory proposes that the planets formed by accreting a core and then, if they became massive enough, by gravitational collapse. The new models suggest that the outer planets may have skipped the core accretion phase. Whatever the details of the story are, we can now understand the distinguishing characteristics of the solar system.

EXPLAINING THE CHARACTERISTICS OF THE SOLAR SYSTEM

Table 19-1 contains the list of distinguishing characteristics of the solar system. Any theory of the origin of the solar system should explain these characteristics.

The disk shape of the solar system is inherited from the solar nebula. The sun and planets revolve and rotate in the same direction because they formed from the same rotating gas cloud. The orbits of the planets lie in the same plane because the rotating solar nebula collapsed into a disk, and the planets formed in that disk.

The solar nebula hypothesis is evolutionary in that it calls on continuing processes to gradually build the planets. To explain the rotation of Venus, Uranus, and Pluto, however, we may need to consider catastrophic events. Uranus rotates on its side. This might have been caused by an off-center collision with a massive planetesimal. In Chapter 24 we will see evidence that Pluto's highly inclined rotation may be the result of a similar collision. Two theories have been proposed to explain the backwards rotation of Venus. Theoretical models suggest that the sun can produce tides in the thick atmosphere of Venus and eventually reverse its rotation—an evolutionary theory. It is also possible that the rotation of Venus was altered by an off-center impact late in the planet's formation, and that is a catastrophic theory.

The second item in Table 19-1, the division of the planets into terrestrial and Jovian worlds, can be understood through the condensation sequence. The terrestrial planets formed in the inner part of the solar nebula, where the temperature was high and most compounds remained gaseous. Only compounds such as the metals and silicates could condense to form solid particles. Planets must begin forming by the sticking together of solid particles because a small blob of matter does not have enough gravitation to capture gas. Thus, the planets that began growing in the inner solar system had to form mainly from metals and silicates. Consequently, they became the small, dense terrestrial planets.

In contrast, the Jovian planets formed in the outer solar nebula, where the lower temperature allowed the gas to form large amounts of ices, perhaps three times more ices than silicates. Thus, the Jovian planets grew rapidly and became massive.

The heat of formation (the energy released by infalling matter) was tremendous for these massive planets. Jupiter must have grown hot enough to glow with a luminosity of about 1 percent that of the present sun. However, because it never got hot enough to generate nuclear energy as a star would, it never generated its own energy. Jupiter is still hot inside. In fact, both Jupiter and Saturn radiate more heat than they absorb from the sun.

The Jovian planets are low-density worlds because they grew in the outer solar nebula where there were large amounts of low-density ices in addition to bits of metal and silicates. Also, Jupiter and Saturn are so massive they have been able to grow by drawing in the cool gas directly from the solar nebula. The terrestrial planets could not do this because they never became massive enough and because the gas in the inner nebula was hotter and more difficult to capture.

When planets were first discovered orbiting other stars, some astronomers wondered if the solar nebula theory was quite right. The newly discovered planets are massive, so they are presumably Jovian, but they are close to their stars where a terrestrial planet should be. Also, some of the new planets follow elliptical orbits. Models of planet formation resolved the problem. If a Jovian planet forms in a disk rich in planetesimals, it may sweep up mass so fast it migrates inward toward its star. Also, interactions among massive planets could throw a planet into an elliptical orbit. Thus the newly discovered planets probably do not pose a problem for the solar nebula explanation of terrestrial and Jovian planets.

A glance at the solar system suggests that we should expect to find a planet between Mars and Jupiter at the present location of the asteroid belt. Mathematical models suggest that we have asteroids there, and not a planet, because Jupiter grew into such a massive planet it was able to gravitationally disturb the motion of nearby planetesimals. The bodies that should have formed a planet just inward in the solar system from Jupiter were broken up, thrown into the sun, or ejected from the solar system. The asteroids we see today are the last remains of the rocky planetesimals.

The comets, in contrast, are evidently the last of the icy planetesimals. Some may have formed in the outer solar nebula beyond Neptune and Pluto, but many probably formed among the Jovian planets where ices could condense easily. Mathematical models show that the massive Jovian planets could have ejected some of these icy planetesimals into the far outer solar system. In a later chapter, we will see evidence that a comet forms when one of these icy bodies falls into the inner solar system.

The large satellite systems of the Jovian worlds may contain two kinds of moons. Some moons may have formed in orbit around the forming planet in a miniature of the solar nebula. But some of the smaller moons may be captured planetesimals and asteroids. The large masses of the Jovian planets would have made it easier for them to capture satellites.

In Table 19-1, we noted that all four Jovian worlds have ring systems, and we can understand this by considering the large mass of these worlds and their remote location in the solar system. A large mass makes it easier for a planet to hold onto orbiting ring particles; and, being farther from the sun, the ring particles are not as easily swept away by the pressure of sunlight and the solar wind. It is hardly surprising, then, that the terrestrial planets, low-mass worlds located near the sun, have no planetary rings.

The last entry in Table 19-1 is the common ages of solar system bodies, and the solar nebula hypothesis has no difficulty explaining that characteristic. If the hypothesis is correct, then the planets formed at the same time as the sun and should have roughly the same age. Although we have mineral samples for radioactive dating only from Earth, the moon, Mars, and meteorites, so far the ages seem to agree.

The solar nebula hypothesis can account for the distinguishing characteristics of the solar system, but there is yet another test we should apply to the hypothesis. What about the problem that troubled Laplace and his nebular hypothesis? What about the angular momentum problem? If the sun and planets formed from a contracting nebula, the sun should have been left spinning very rapidly. That is, it should have most of the angular momentum in the solar system, and instead it has very little. The solar nebula hypothesis allows us to explain the angular momentum problem by thinking about the relationship between the origin of planetary systems and star formation. What we know about star formation tells us that the sun must have had a strong stellar wind when it was young. Also, a rapidly spinning star is likely to generate a strong magnetic field. Like a great paddle wheel, the sun's magnetic field must have stuck out into space, and gas flowing away from the sun would have been dragged around with the rotating magnetic field. This would have slowed the sun and left it rotating slowly. Astronomers now believe that the angular momentum problem is no longer an objection to the solar nebula hypothesis.

Our general understanding of the origin of the solar system also explains the origin of asteroids, meteors, and comets. They appear to be the last of the debris left behind by the solar nebula. These objects are such important sources of information about the history of our solar system that we will discuss them in detail in Chapter 25. But for now we ask: What happened to the solar nebula?

CLEARING THE NEBULA

The sun probably formed along with many other stars in a swirling nebula. Observations of young stars in Orion suggest that radiation from nearby hot stars would have evaporated the disk of gas and dust around the sun and that the gravitational influence of nearby stars would have pulled gas away. Even without the external effects, four internal process would have gradually destroyed the solar nebula.

The most important of these internal processes was **radiation pressure.** When the sun became a lumi-

nous object, light streaming from its surface pushed against the particles of the solar nebula. Large bits of matter like planetesimals and planets were not affected, but low-mass specks of dust and individual gas atoms were pushed outward and eventually driven from the system. This is not a sudden process, and it may not have occurred at the same time everywhere in the nebula. Before sunlight could begin clearing the outer nebula, it first had to push its way through the inner nebula.

The second effect that helped clear the nebula was the solar wind, the flow of ionized hydrogen and other atoms away from the sun's upper atmosphere. This flow is a steady breeze that rushes past Earth at about 400 km/s (250 mi/s). When the sun was young, it may have had an even stronger solar wind, and irregular fluctuations in its luminosity, like those observed in young stars such as T Tauri stars, may have produced surges in the wind that helped push dust and gas out of the nebula.

The third effect for clearing the nebula was the sweeping up of space debris by the planets. All of the old, solid surfaces in the solar system are heavily cratered by meteorite impacts (Figure 19-13). Earth's moon, Mercury, Venus, Mars, and most of the moons in the solar system are covered with craters. A few of these craters have been formed recently by the steady rain of meteorites that falls on all the planets in the solar system, but most of the craters we see appear to have been formed roughly 4 billion years ago in what is called the **heavy bombardment,** as the last of the debris in the solar nebula was swept up by the planets. We will find more evidence of this bombardment in later chapters. Thus, many of the last remaining solid objects in the solar nebula were swept up by the planets soon after they formed.

The fourth effect was the ejection of material from the solar system by close encounters with planets. If a small object such as a planetesimal passes close to a planet, it can gain energy from the planet's gravitational field and be thrown out of the solar system. Ejection is most probable in encounters with massive planets, so the Jovian planets were probably very efficient at ejecting the icy planetesimals that formed in their region of the nebula.

Attacked by the radiation and gravity of nearby stars and racked by internal processes, the solar nebula could not survive very long. Once the gas and dust were gone, the planets could no longer gain mass, and planet building ended.

REVIEW CRITICAL INQUIRY

Why are there two kinds of planets in our solar system?

Planets begin forming from solid bits of matter, not from gas. Consequently, the kind of planet that forms at a given a   b   Visual-wavelength images

FIGURE 19-13
Every old, solid surface in the solar system is scarred by craters. (a) Earth's moon is scarred by craters ranging from basins hundreds of kilometers in diameter down to microscopic pits. (b) The surface of Mercury as photographed by a passing spacecraft shows vast numbers of overlapping craters. *(NASA)*

distance from the sun depends on the kind of compounds that can condense out of the gas to form solid particles. In the inner parts of the solar nebula, the temperature was so high that most of the gas could not condense to form solids. Only metals and silicates could form solid grains, and thus the innermost planets grew from this dense material. Much of the mass of the solar nebula consisted of hydrogen, helium, water vapor, and other gases, and they were present in the inner solar nebula but couldn't form solid grains. The small terrestrial planets couldn't grow from these gases, so the terrestrial planets are small and dense.

In the outer solar nebula, the composition of the gas was the same, but it was cold enough for water vapor to condense to form ice grains. Because hydrogen and oxygen are so abundant, there was lots of ice available. Thus, the outer planets grew from solid bits of metal and silicate combined with large amounts of ice. The outer planets grew so rapidly that they became massive enough to capture gas directly, and thus they became the hydrogen- and helium-rich Jovian worlds.

The condensation sequence combined with the solar nebula hypothesis gives us a way to understand the difference between the terrestrial and Jovian planets. How do we understand the difference between asteroids and comets?

The great chain of origins leads from the first instant of the big bang through the birth of galaxies, the formation of stars, the origin of our solar system, and finally to us. One of the most elegant and complex links in that chain is planet building. As we explore the solar system in detail in the following chapters, we must stay alert for further clues to the birth of the planets.

SUMMARY

René Descartes proposed that the solar system formed from a contracting vortex of matter, and Buffon later suggested that a passing star pulled matter out of the sun to form the planets. While Buffon's hypothesis was catastrophic, Laplace's nebular hypothesis was evolutionary. The nebular hypothesis required a contracting nebula to leave behind rings that formed each planet, but it could not explain the low angular momentum of the sun. The modern solar nebula hypothesis suggests that the sun was slowed by its magnetic field and mass loss. Disks of gas and dust have been found around young stars, so astronomers suspect that planetary systems are common. Over a hundred planets have been found orbiting other stars.

The solar nebula hypothesis proposes that the solar system began as a contracting cloud of gas and dust that flattened into a rotating disk. The center of this cloud eventually became the sun, and the planets formed in the disk of the nebula. The properties of this nebula and the processes that built the planets explain many of the characteristic properties of the solar system.

One of the most striking properties of the solar system is its disk shape. The orbits of the planets lie in nearly the same plane, and they all revolve around the sun in the same direction, counterclockwise as seen from the north. With only three exceptions, the planets rotate counterclockwise around axes roughly perpendicular to the plane of the solar system. This disk shape and the motion of the planets appear to have originated in the solar nebula.

Another striking feature of the solar system is the division of the planets into two families. The terrestrial planets, which are small and dense, lie in the inner part of the system. The Jovian planets are large, low-density worlds in the outer part of the system. In general, the closer a planet lies to the sun, the higher its uncompressed density.

The solar system is now filled with smaller bodies such as asteroids, comets, and meteors. The asteroids are small, rocky worlds, most of which orbit the sun between Jupiter and Mars. They appear to be material left over from the formation of the solar system.

Another important characteristic of the solar system bodies is their similar ages. Radioactive dating tells us that Earth, the moon, Mars, and meteorites are no older than about 4.6 billion years. Thus, it seems our solar system took shape about 4.6 billion years ago.

According to the condensation sequence, the inner part of the nebula was so hot that only high-density minerals could form solid grains. The outer regions, being cooler, condensed to form icy material of lower density. The planets grew from these solid materials, with the denser planets forming in the inner part of the nebula and the lower-density Jovian planets forming farther from the sun.

Planet building began as dust grains growing by condensation and accretion into planetesimals ranging from a kilometer to hundreds of kilometers in diameter. These planetesimals settled into a thin plane around the sun and accumulated into larger bodies, the largest of which grew the fastest and eventually became protoplanets.

Once a planet had formed from a large number of planetesimals, heat from radioactive decay could have melted it and allowed it to differentiate into a dense metallic core and a lower-density silicate crust. In fact, it is possible that the solar nebula cooled as the protoplanets grew so that the first planetesimals were metallic and later additions were silicate. It is also likely that the planets grew rapidly enough that the heat of formation released by the in-falling material melted the planets and allowed them to differentiate as they formed.

Jupiter and Saturn grew rapidly in the outer solar nebula. Some models suggest they were able to form quickly without first accreting a core from solid materials. In any case, they grew into massive, low-density worlds by attracting gas from the solar nebula. Heat of formation raised their temperatures very high when they were young, and Jupiter and Saturn still radiate more heat then they absorb from the sun. In contrast, Uranus and Neptune grew more slowly in the outer solar nebula and never became massive enough to capture large amounts of gas.

Once the sun became a luminous object, it cleared the nebula as its light and solar wind pushed material out of the system. The planets helped by absorbing some planetesimals and ejecting others from the system. Once the solar system was clear of debris, planet building ended.

NEW TERMS

passing star hypothesis	meteorite
evolutionary hypothesis	half-life
catastrophic hypothesis	gravitational collapse
nebular hypothesis	uncompressed density
angular momentum problem	condensation sequence
solar nebula hypothesis	planetesimal
extrasolar planets	condensation
terrestrial planet	accretion
Jovian planet	protoplanet
Galilean satellites	differentiation
asteroid	outgassing
comet	heat of formation
meteor	radiation pressure
meteoroid	heavy bombardment

REVIEW QUESTIONS

Ace Astronomy™ Assess your understanding of this chapter's topics with additional quizzing and animations at http://astronomy.brookscole.com/seeds8e

1. What were the objections to the passing star hypothesis? to the nebular hypothesis?

2. What produced the helium now present in the sun's atmosphere? in Jupiter's atmosphere? in the sun's core?

3. What produced the iron in Earth's core and the heavier elements like gold and silver in Earth's crust?

4. What evidence do we have that disks of gas and dust are common around young stars?

5. According to the solar nebula theory, why is the sun's equator nearly in the plane of Earth's orbit?

6. Why does the solar nebula theory predict that planetary systems are common?

7. Why do we think the solar system formed about 4.6 billion years ago?

8. If you visited another planetary system, would you be surprised to find planets older than Earth? Why or why not?

9. Why is almost every solid surface in our solar system scarred by craters?

10. What is the difference between condensation and accretion?

11. Why don't terrestrial planets have rings like the Jovian planets?

12. How does the solar nebula theory help us understand the location of asteroids?

13. How does the solar nebula theory explain the dramatic density difference between the terrestrial and Jovian planets?

14. If you visited some other planetary system in the act of building planets, would you expect to see the condensation sequence at work, or was it unique to our solar system?

15. Why do we expect to find that planets are differentiated?

16. What processes cleared the nebula away and ended planet building?

DISCUSSION QUESTIONS

1. In your opinion, should all solar systems have asteroid belts? Should all solar systems show evidence of an age of heavy bombardment?

2. If the solar nebula hypothesis is correct, then there are probably more planets in the universe than stars. Do you agree? Why or why not?

PROBLEMS

1. If you observed the solar system from the nearest star (1.3 pc), what would the maximum angular separation be between Jupiter and the sun? (*Hint:* Use the small-angle formula.)

2. The brightest planet in our sky is Venus, which is sometimes as bright as apparent magnitude −4 when it is at a distance of about 1 AU. How many times fainter would it look from a distance of 1 parsec (206,265 AU)? What would its apparent magnitude be? (*Hints:* Remember the inverse square law, Chapter 5; see Chapter 2.)

3. What is the smallest-diameter crater you can identify in the photo of Mercury on page 418? (*Hint:* See Appendix A, Properties of the Planets, to find the diameter of Mercury in kilometers.)

4. A sample of a meteorite has been analyzed, and the result shows that out of every 1000 nuclei of ^{40}K originally in the meteorite, only 100 have not decayed. How old is the meteorite? (*Hint:* See Figure 19-10.)

5. In Table 19-2, which object's observed density differs least from its uncompressed density? Why?

6. What composition might we expect for a planet that formed in a region of the solar nebula where the temperature was about 100 K?

7. Suppose that Earth grew to its present size in 1 million years through the accretion of particles averaging 100 g each. On the average, how many particles did Earth capture per second? (*Hint:* See Appendix A to find Earth's mass.)

8. If you stood on Earth during its formation as described in Problem 7 and watched a region covering 100 m², how many impacts would you expect to see in an hour?

(*Hints:* Assume that Earth had its present radius. The surface area of a sphere is $4\pi r^2$.)

9. The velocity of the solar wind is roughly 400 km/s. How long does it take to travel from the sun to Pluto?

CRITICAL INQUIRIES FOR THE WEB

1. How does our solar system compare with the others that have been found? Search the Internet for sites that give information about planetary systems around other stars. What kinds of planets have been detected by these searches so far? Discuss the selection effects (see Window on Science 25-2) that must be considered when interpreting these data.

2. The process of protoplanetary accretion is still not well understood. Search the Web for current research in this field. From the results of your search, outline the basic steps in the formation of a protoplanet through accretion. What specific factors are important in these models of planet building? Do these models produce planetary systems similar to the ones we know to exist?

3. How is radioactive dating carried out on meteorites and rocks from surfaces of various bodies in the solar system? Look for Web sites on the details of radioactive dating, and summarize the methods used to uncover the abundances of radioactive elements in a particular sample. (*Hint:* Try looking for information on how a particular meteorite—for example, the Martian meteorite ALH84001—was studied, what age range was determined, and what radioactive elements were used to arrive at the age.)

EXPLORING *THESKY*

1. Look at the solar system from space. Notice how thin the disk of the solar system is and how inclined the orbits of Jupiter and Mercury are. (*Hint:* Under the **View** menu, choose **3D Solar System Mode** and then zoom in or out. Tip the solar system up and down to see it edge-on.)

2. Look at the solar system from space and notice how small the orbits of the inner planets are compared to the orbits of the outer planets. They make two distinct groups. (*Hint:* Use **3D Solar System Mode.**)

3. Watch the comets orbiting around the sun. Can you locate the comet C/198 M5 (Linear)? What is its orbit like? (*Hint:* Use **3D Solar System Mode** and set the time step to 30 days (**30d**).)

 Visit the Seeds *Foundations of Astronomy* companion Web site for critical thinking exercises, articles, and additional readings from InfoTrac College Edition, Brooks/Cole's online student library.

PLANET EARTH

Nature evolves. The world was different yesterday.

Preston Cloud, Cosmos, Earth and Man

Chapter 20

GUIDEPOST

Astronomy has been described as the science of everything above the clouds. Planetary astronomers, however, must also think about what lies below the clouds because Earth is the basis for comparison with all other Earthlike planets. We know Earth well, and we can apply what we know about Earth to other worlds.

There is another reason for studying Earth in an astronomy course. Astronomy is really about us. Astronomy is exciting and fascinating because it helps us understand what we are and where we are in the universe. Thus, we cannot omit Earth from our discussion—it is where we are.

The next two chapters will discuss the Earthlike planets, but that will not end our thoughts of Earth. The moons of the giant outer worlds will seem Earthlike in strange ways, and our discussion of the smaller bodies of our solar system will alert us to the dangers Earth faces. Throughout the rest of this book, we will remain painfully aware of the fragile beauty of our planet.

VIRTUAL LABORATORIES

 PLANETARY GEOLOGY

 PLANETARY ATMOSPHERES AND THEIR RETENTION

Stars die, but planets don't. That should give us a feeling of security, but we know that the survival of our planet is limited by the life expectancy of our star. Furthermore, although planets don't die, they do evolve, and our planet has changed dramatically since it formed. Mountains rise and weather away, the climate warms and then cools, and the continents themselves drift and grind against each other. To fully understand how our own planet works, we need to know how planets in general evolve. Why is Venus hot and Mars cold? Why does Mercury have many craters and Earth only a few?

To answer such questions, we must use **comparative planetology,** the study of planets through comparison and contrast, and Earth is our basis of comparison. One reason why we begin with Earth is that it is the planet we know best. We can study it carefully and then draw comparisons to other worlds. But another reason is that our home is a planet of extremes. The land we see above the oceans is quite different from the land under the oceans (▌ Data File Two). Earth's interior is molten and generates a magnetic field. Its crust is active, with moving sections that push against each other and trigger earthquakes, volcanoes, and mountain building. Even Earth's atmosphere is extreme. Processes have altered Earth's air from its original composition to a highly unusual, oxygen-rich sea of gas. Once we understand Earth's complex properties, the remaining planets in our system should be easier to comprehend.

In our study of Earth, we will find a four-stage history of planetary development. The moon and all the terrestrial planets have passed through these stages, although differences in the way the planets were altered by these stages have produced dramatically different worlds. The moon, for example, is much like Earth, but its evolution has been dramatically altered by its smaller size. As we explore the solar system, we will discover not entirely new processes but rather familiar effects working in slightly different ways.

20-1 THE EARLY HISTORY OF EARTH

Like all the terrestrial planets, Earth formed from the inner solar nebula about 4.6 billion years ago. As it took form, it began to change, passing through four developmental stages (▌ Figure 20-1). Processes that occurred during these four stages determined the present condition of Earth's interior, crust, and atmosphere.

Ace Astronomy™ The AceAstronomy icon throughout the text indicates an opportunity for you to test yourself on key concepts and to explore animations and interactions on the AceAstronomy Web site at **http://astronomy.brookscole .com/seeds8e**

EARTH
Data File Two

A shaded relief map of Earth with the oceans removed reveals the geological features on the continents and under the oceans. *(National Geophysical Data Center)*

Average distance from the sun	1.00 AU (1.495979×10^8 km)
Eccentricity of orbit	0.0167
Maximum distance from the sun	1.0167 AU (1.5210×10^8 km)
Minimum distance from the sun	0.9833 AU (1.4710×10^8 km)
Inclination of orbit to ecliptic	0°
Average orbital velocity	29.79 km/s
Orbital period	1.00 y (365.25 days)
Period of rotation (with respect to the sun)	$24^h00^m00^s$
Inclination of equator to orbit	23°27′
Equatorial diameter	12,756 km
Mass	5.976×10^{24} kg
Average density	5.497 g/cm^3 (4.07 g/cm^3 uncompressed)
Surface gravity	1.0 Earth gravity
Escape velocity	11.2 km/s
Surface temperature	−50° to 50°C (−60° to 120°F)
Average albedo*	0.39
Oblateness	0.0034

*Albedo is the fraction of light striking an object that the object reflects. The albedo of a perfect reflector is 1.

Four Stages of Planetary Development

Differentiation produces a dense core, thick mantle, and low-density crust.

The young Earth was heavily bombarded in the debris-filled early solar system.

Flooding by molten rock and later by water can fill lowlands.

Slow surface evolution continues due to geological processes and erosion.

FIGURE 20-1
The four stages of planetary development are illustrated for Earth.

FOUR STAGES OF PLANETARY DEVELOPMENT

The first stage of planetary evolution is *differentiation,* the separation of material according to density. Earth now has a dense core and a lower-density crust, and that structure must have originated very early.

Differentiation would have occurred easily if Earth was molten when it was young. Two sources of heat could have heated Earth. First, heat of formation would be created by in-falling material. A meteorite hitting Earth at high velocity converts most of its energy of motion into heat, and the in-falling of a large number of meteorites could release tremendous heat. If Earth formed rapidly, this heat would have accumulated much more rapidly than it could leak away, and Earth may have been molten when it formed. A second source of heat

requires more time to develop. The decay of radioactive elements trapped in the Earth releases heat gradually; but, as soon as Earth formed, that heat would have begun to accumulate and could have helped melt Earth to facilitate differentiation. Most of Earth's radioactive elements are now concentrated in the crust, where they continue to warm and soften the rock layers.

Earth formed by material falling together, but meteorites could have left no trace until a crust solidified. Once Earth had a hard surface, the meteorites could form craters. This second stage in planetary evolution, *cratering,* was violent. The heavy bombardment was intense because the solar nebula was filled with rocky and icy debris, and the young Earth was battered by meteorites that pulverized the newly forming crust. The largest meteorites blasted out crater basins hundreds of kilometers in diameter. As the solar nebula cleared, the amount of debris decreased, and the level of cratering fell to its present low level. Although meteorites still occasionally strike Earth and dig craters, cratering is no longer the dominant influence on Earth's geology. As we compare other worlds with Earth, we will discover traces of this intense period of cratering, the heavy bombardment, on every old surface in the solar system.

The third stage, *flooding,* no doubt began while cratering was still intense. The fracturing of the crust and the heating caused by radioactive decay allowed molten rock just below the crust to well up through fissures and flood the deeper basins. We will discuss such flooded basins on other worlds, such as the moon, but all traces of this early lava flooding have been destroyed by later geological activity in Earth's crust. On Earth, flooding continued as the atmosphere cooled and water fell as rain, filling the deepest basins to produce the first oceans. Notice that on Earth flooding involves both lava and water, a circumstance that we will not find on most worlds.

The fourth stage, *slow surface evolution,* has continued for the last 3.5 billion years or more. Earth's surface is constantly changing as sections of crust slide over each other, push up mountains, and shift continents. Also, moving air and water erode the surface and wear away geological features. Almost all traces of the first billion years of Earth's geology have been destroyed by the active crust and erosion.

EARTH AS A PLANET

We are studying Earth in this chapter in order to use it as a basis for comparison with other worlds. Of course, four of the worlds in our solar system, the Jovian planets, are dramatically different from Earth, so we will not try to compare them with Earth. But the four terrestrial planets are rocky worlds much like Earth, and many of the moons in the solar system have geology that we can compare with Earth and its four stages of development.

All terrestrial planets pass through these four stages, so in that respect Earth is a good basic reference planet for comparative planetology. Some planets have emphasized one stage over another, and some planets have failed to progress fully through the four stages. Nevertheless, Earth is a good standard of comparison. Every major process on any rocky world in our solar system is represented in some form on Earth. Clearly, Earth is a highly complex world; for that reason, it makes a good standard of comparison.

On the other hand, Earth is peculiar in two ways. First, it has large amounts of liquid water on its surface. Fully 75 percent of its surface is covered by this liquid, and no other planet in our solar system is known to have such extensive liquid water on its surface. Not only does water fill the oceans, but it evaporates into the atmosphere, forms clouds, and then falls as rain. Water falling on the continents flows downhill to form rivers that flow back to the sea, and, in so doing, the water produces extensive erosion. Entire mountain ranges can literally dissolve and wash away in only a few tens of millions of years, less than 1 percent of Earth's total age. We will not see such intense erosion on most worlds. Liquid water is, in fact, a rare material on most planets.

Our planet is special in a second way. Some of the matter on the surface of this world is alive, and a small part of that living matter is aware. We do not know how the presence of living matter has affected the evolution of Earth, but this process seems to be totally missing from other worlds in our solar system. Furthermore, the thinking part of the life on Earth, humankind, is actively altering our planet. It does not seem likely that we have made major changes in Earth yet, but our study of Earth as a standard of comparison with other worlds will also give us new insight into our own planet and how we may be altering it.

REVIEW CRITICAL INQUIRY

Why do we think Earth went through a stage of cratering?
Recall from the previous chapter that the planets formed by the accretion of planetesimals from the solar nebula. The protoearth may have been molten as it formed, but as soon as it grew cool enough to form a solid crust, the remaining planetesimal impacts would have formed craters. Thus, we can reason from the solar nebula hypothesis that Earth should have been cratered. But we can't use a theory as evidence to support some other theory. To find real observational evidence, we need only look at the moon. The moon has craters, and so does every other old surface in our solar system. There must have been a time, when the solar system was young, when there were large numbers of objects striking all the planets and moons and blasting out craters. If it happened to other worlds in our solar system, it must have happened to Earth, too.

Of course, we don't see those craters on Earth today. Why not?

As we explore our solar system in the chapters that follow, Earth will be our standard of comparison. Our goal in this chapter is to study Earth's interior, crust, and atmosphere in order to establish a basis for comparison with other planets. Only by understanding Earth in detail can we understand planets in general.

20-2 THE SOLID EARTH

Although we might think of Earth as made of solid rock, it is in fact neither entirely solid nor entirely rock. The thin crust seems solid, but it floats and shifts on a semiliquid layer of molten rock just below the crust. Below that lies a deep, rocky mantle surrounding a core of liquid metal. Much of what we see on the surface of Earth is determined by its interior.

EARTH'S INTERIOR

The theory of the origin of planets from the solar nebula predicts that Earth should have melted and differentiated into a dense metallic core with a low-density silicate crust. But did it differentiate? Clearly, Earth's surface is made of silicates, but what of the interior?

High temperature and tremendous pressure in Earth's interior make any direct exploration impossible. Even the deepest oil wells extend only a few kilometers down and don't reach through the crust. It is quite impossible to drill far enough to sample Earth's core. Yet Earth scientists have studied the interior and found clear evidence that Earth did differentiate.

This analysis of Earth's interior is possible because earthquakes produce vibrations called **seismic waves,** which travel through the interior and eventually register on sensitive detectors, known as **seismographs,** all over the world (❚ Figure 20-2). Two kinds of seismic waves are important to us here. The **P,** or **pressure, waves** are much like sound waves in that they travel as a region of compression. As a P wave passes, particles of matter vibrate back and forth parallel to the direction of wave travel (❚ Figure 20-3a). In contrast, the **S,** or **shear, waves** move as displacements of particles perpendicular to the waves' direction of travel (Figure 20-3b). Thus, S waves distort the material but do not compress it. Normal sound waves are pressure waves, whereas the vibrations in a bowl of jelly are shear waves. Because P waves are compression waves, they can move through a liquid, but S waves cannot. A glass of water can't shimmy like jelly because a liquid does not have the rigidity required to transmit S waves.

The P and S waves caused by an earthquake do not travel in straight lines or at constant speed within Earth. The waves may reflect off boundaries between layers of different density, or they may be refracted as they pass through such a boundary. In addition, the gradual increase in density toward Earth's center means

FIGURE 20-2

A seismograph in northern Canada made this record of seismic waves from an earthquake in Mexico. The first vibrations, *P* waves, arrived 11 minutes after the quake, but the slower *S* waves took 20 minutes to make the journey.

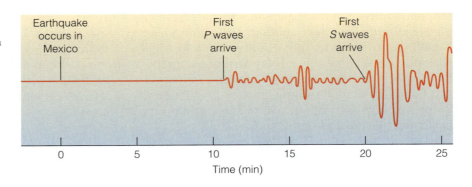

the speed of sound is greater toward the center. Sound waves approaching such a region tend to curve away (refract), so seismic waves do not travel in straight lines through Earth's interior but are refracted to follow curved paths. Earth scientists can use the arrival times of reflected and refracted seismic waves from distant earthquakes to construct a picture of Earth's interior.

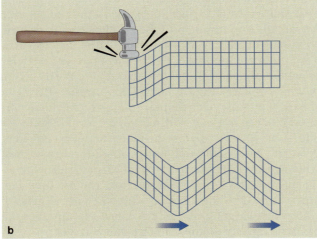

FIGURE 20-3

(a) *P*, or pressure, waves, like sound waves in air, travel as a region of compression. (b) *S*, or shear, waves, like vibrations in a bowl of jelly, travel as displacements perpendicular to the direction of travel. *S* waves tend to travel more slowly than *P* waves and cannot travel through liquids.

Such studies show that the interior consists of three parts: a central core, a thick mantle, and a thin crust. *S* waves give us an important clue to the nature of the core, because no such waves are observed to travel through the core. When an earthquake occurs, no direct *S* waves register on seismographs on the opposite side of Earth, as if the core were casting a shadow (❙ Figure 20-4). This shows that the core is mostly liquid, and the size of the shadow fixes the size of the core at about 55 percent of Earth's radius. Mathematical models predict that the core is hot (about 6000 K), dense (about 14 g/cm³), and composed of iron and nickel.

Earth's core is as hot as the surface of the sun, but the tremendous pressure inside Earth keeps the material from vaporizing. The core is instead a liquid. Nearer the center the material is under higher pressure, which in turn raises the melting point so high that the material cannot melt at the existing temperature (❙ Figure 20-5). Thus, there is an inner core of solid iron and nickel. Estimates suggest the inner core's radius is about 22 percent that of Earth.

The **mantle** is the layer of dense rock that lies between the molten core and the crust. The paths of seismic waves in the mantle show that it is not molten, but it is not precisely solid either. Mantle material behaves like a **plastic,** a material with the properties of a solid but capable of flowing under pressure. The asphalt used in paving roads is a common example of a plastic. It shatters if struck with a sledgehammer, but it bends under the weight of a heavy truck. Just below Earth's crust, where the pressure is less than at greater depths, the mantle is most plastic.

Earth's rocky crust is made up of low-density rocks and floats on the denser mantle. The crust is thickest under the continents, up to 60 km thick, and thinnest under the oceans, where it is only about 10 km thick. Unlike the mantle, the crust is brittle and breaks much more easily than the plastic mantle.

Perhaps no region is more immediate yet more inaccessible than Earth's core. Only a few thousand kilometers from where we sit, Earth's interior is beyond our immediate reach. Yet we can send our imagination where our bodies can never go (Window on Science

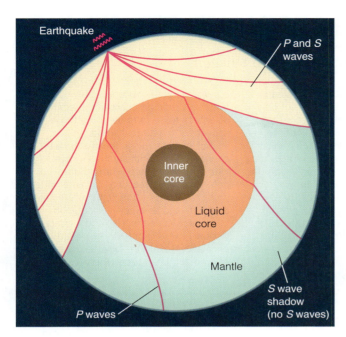

ACTIVE FIGURE 20-4

P and *S* waves give us clues to Earth's interior. That no direct *S* waves reach the far side shows that Earth's core is liquid. The size of the *S* wave shadow tells us the size of the outer core.

Ace⬣Astronomy™ Go to AceAstronomy and click Active Figures to see "Seismic Waves" and follow *P* and *S* waves as they travel through Earth's interior.

20-1). Earth's seismic activity gives us evidence of Earth's innermost secrets. But there is another source of evidence about Earth's interior—its magnetic field.

THE MAGNETIC FIELD

Apparently, Earth's magnetic field is a direct result of its rapid rotation and its molten metallic core. The internal heat forces the liquid core to churn with convection while Earth's rotation turns the core about an axis.

This core is a highly conductive iron–nickel alloy, a better electrical conductor than copper, the material commonly used for electrical wiring. The rotation of this convecting, conducting liquid generates Earth's magnetic field in a process called the dynamo effect (⬣ Figure 20-6). This is believed to be the same process that generates the solar magnetic field in the convective layers of the sun, and we will meet it again when we explore other planets.

Earth's magnetic field protects it from the solar wind. Blowing outward from the sun at about 400 km/s, the solar wind consists of ionized gases carrying a small part of the sun's magnetic field. When the solar wind encounters Earth's magnetic field, it is deflected like water flowing around a boulder in a stream. The surface where the solar wind is first deflected is called the **bow shock,** and the cavity dominated by Earth's magnetic field is called the **magnetosphere** (⬣ Figure 20-7a). High-energy particles from the solar wind leak into the magnetosphere and become trapped within Earth's magnetic field to produce the **Van Allen belts** of radiation. We will see in later chapters that all planets that have magnetic fields have bow shocks, magnetospheres, and radiation belts.

One dramatic result of Earth's magnetic field is the aurora, glowing rays and curtains of light in the upper atmosphere (Figure 20-7b). The solar wind carries charged particles past Earth's extended magnetic field, and this generates tremendous electrical currents that flow into Earth's atmosphere near the north and south magnetic poles. The currents ionize gas atoms in Earth's atmosphere, and when the ionized atoms capture electrons and recombine, they emit light as if they were part of a vast "neon" sign. Thus, the spectrum of an aurora is an emission spectrum.

Although we are confident that Earth's magnetic field is generated within its molten core, many mysteries remain. For example, rocks retain traces of the magnetic field in which they solidify, and some contain

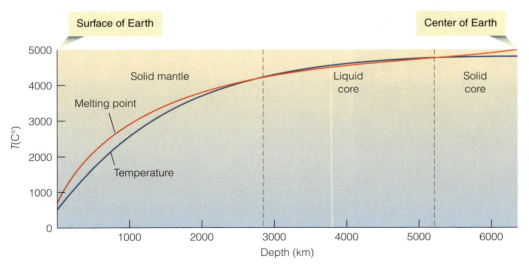

FIGURE 20-5

Theoretical models combined with observations of the velocity of seismic waves reveal the temperature inside Earth (blue line). The melting point of the material (red line) is determined by its composition and by the pressure. In the mantle and in the inner core, the melting point is higher than the existing temperature, and the material is not molten.

fields that point backward. That is, they imply that Earth's magnetic field was reversed at the time they solidified. Careful analysis of such rocks indicates that Earth's field has reversed itself every million years or so, with the north magnetic pole becoming the south magnetic pole and vice versa. These reversals are poorly understood, but they may be related to changes in the core convection.

Convection in Earth's core is important because it generates the magnetic field. As we will see in the next section, convection in the mantle constantly remakes Earth's surface.

EARTH'S ACTIVE CRUST

Earth's crust is composed of lower-density rock that floats on the mantle. The image of a rock floating may seem odd, but recall that the rock of the mantle is very dense. Also, just below the crust, the mantle rock tends to be highly plastic, so great sections of low-density crust do indeed float on the mantle like great lily pads floating on a pond.

The motion of the crust and the erosive action of water make Earth's crust highly active. Look at ▌ "The Active Earth" on pages 440 and 441 and notice three important points. First, the motion of crustal plates produces much of the geological activity we see on Earth. Earthquakes, volcanism, and mountain building are all linked to motions in the crust. Second, notice how features on Earth's surface give us clear evidence that the crustal plates are moving. Third, notice that most of the geological features we know—mountain ranges, the Grand Canyon, and even the outline of the continents—are recent products of Earth's active surface. Earth's surface is constantly renewed. The oldest rocks on Earth, small crystals called zircons from western Australia, are 4.3 billion years old. Most of the crust is much younger than that. Most of the mountains and valleys you see around you are no more than a few tens of millions of years old.

Plate tectonics explains why Earth contains so few impact craters. The moon is richly cratered, but Earth contains only about 150 impact craters. Plate tectonics and erosion have destroyed all but the most recent craters on Earth.

As astronomers, we see Earth's geology dominated by two dramatic forces. Just below the thin crust of

FIGURE 20-6
The dynamo effect couples convection in the liquid core with Earth's rotation to produce electric currents that are believed to be responsible for Earth's magnetic field.

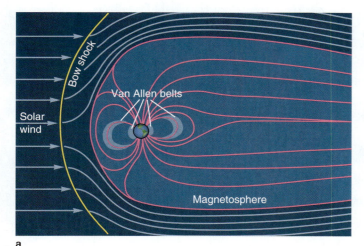

a

FIGURE 20-7

(a) Earth's magnetic field dominates space around Earth by deflecting the solar wind and trapping high-energy particles in radiation belts. (b) Electrical currents in Earth's magnetic field flow down into Earth's atmosphere near the north and south magnetic poles and excite atoms to emit photons, which we see as aurorae. The pink and green glows in this photo as well as the vertical "curtains of light" are aurorae. A meteor (shooting star) is visible below wisps of dark clouds silhouetted against the aurora. *(Jimmy Westlake)*

b

solid rock lies a churning molten interior that rips the crust to fragments and pushes the pieces about like bits of algae on a pond. The second force modifying the crust is water. It falls as rain and snow and tears down the mountains, erodes the river valleys, and washes any raised ground into the sea. Like angry children, tectonics and erosion build mountains and continents and then rip them to nothing.

REVIEW CRITICAL INQUIRY

What evidence do we have that Earth has a molten metallic core?

In science, the critical analysis of ideas must eventually return to evidence. In this case, the evidence is indirect because we can never visit Earth's core. Seismic waves from distant earthquakes pass through Earth, but a certain kind of wave, the S waves, cannot pass through the core. Because the S waves cannot propagate through a liquid, we conclude that Earth's core is a liquid. Earth's magnetic field gives us further evidence of a metallic core. The theory for the generation of magnetic fields, the dynamo effect, requires a rotating liquid, convective, conducting core. If the core were not a liquid, it would not be able to generate a magnetic field. Thus, we have two different kinds of evidence to tell us that our planet has a liquid core.

Every theory must be supported by evidence. What evidence do we have to support the theory of plate tectonics?

For the astronomer, accustomed to the longest view of cosmic time, Earth's geology seems not so much the

The Active Earth

Our world is an astonishingly active planet. Not only is it rich in water and therefore subject to rapid erosion, but its crust is divided into moving sections called plates. Where plates spread apart, lava wells up to form new crust; where plates push against each other, they crumple the crust to form mountains. Where one plate slides over another, we see volcanism. This process is called **plate tectonics,** *referring to the Greek word for "builder." (An architect is literally an arch builder.)*

Midocean rise

Red Sea

Midocean rise

A typical view of planet Earth

Mountains are common on Earth, but they erode away rapidly because of the abundant water.

William K. Hartmann

Janet Seeds

Evidence of plate tectonics was first found in ocean floors, where plates spread apart and magma rises to form **midocean rises** made of rock called **basalt,** a rock typical of solidified lava. Radioactive dating shows that the basalt is younger near the midocean rise. Also, the ocean floor carries less sediment near the midocean rise. As Earth's magnetic field reverses back and forth, it is recorded in the magnetic fields frozen into the basalt. This produces a magnetic pattern in the basalt that shows that the sea floor is spreading away from the midocean rise.

A **subduction zone** is a deep trench where one plate slides under another. Melting releases low-density magma that rises to form volcanoes such as those along the northwest coast of North America, including Mt. St. Helens.

Subduction zone

Pacific Ocean

S. America

Plate motion

Melting

Midocean rise

Atlantic Ocean

Africa

Plate motion

Plate motion

Ocean floor

Mantle

Subduction zone

Hawaiian-Emperor chain

Subduction zone

Hawaii

Midocean rise

Appalachian Mountains

Midocean rise

Andes mountains

Red Sea

Ural Mountains

Himalaya Mountains

Midocean rise

Subduction zone

Hot spots caused by rising magma in the mantle can poke through a plate and cause volcanism such as that in Hawaii. As the Pacific plate has moved northwestward, the hot spot has punched through to form a chain of volcanic islands, now mostly worn below sea level.

Folded mountain ranges can form where plates push against each other. For example, the Ural Mountains lie between Europe and Asia, and the Himalaya Mountains are formed by India pushing north into Asia. The Appalachian Mountains are the remains of a mountain range pushed up when North America was pushed against Africa.

The floor of the Atlantic Ocean is not being subducted. It is locked to the continents and is pushing North and South America away from Europe and Africa at about 3 cm per year, a motion called *continental drift.* Radio astronomers can measure this motion by timing pulsars from European and from American radio telescopes.

Roughly 200 million years ago, North and South America were joined to Europe and Africa. Evidence of that lies in similar fossils and similar rocks and minerals found in the matching parts of the continents. Notice how North and South America fit against Europe and Africa like a puzzle.

National Geophysical Data Center

Hawaii

The floor of the Pacific Ocean is sliding into subduction zones in many places around its perimeter. This pushes up mountains such as the Andes and triggers earthquakes and active volcanism all around the Pacific in what is called the Ring of Fire. In places such as southern California, the plates slide past each other, causing frequent earthquakes.

Not long ago, Earth's continents came together to form one continent.

Pangaea broke into a northern and a southern continent.

Notice India moving north toward Asia.

The continents are still drifting on the highly plastic upper mantle.

Yellow lines on this globe mark plate boundaries. Red dots mark earthquakes since 1980. Earthquakes within the plate, such as those at Hawaii, are related to volcanism over hot spots in the mantle.

Continental Drift

Pangaea

200 million years ago

Laurasia

Gondwanaland

135 million years ago

65 million years ago

Today

A **rift valley** forms where continental plates begin to pull apart. The Red Sea has formed where Africa has begun to pull away from the Arabian peninsula.

Plate tectonics pushes up mountain ranges and causes bulges in the crust, and water erosion wears the rock away. The Colorado River began cutting the Grand Canyon only about 10 million years ago when the Colorado plateau warped upward under the pressure of moving plates. That sounds like a long time ago, but it is only 0.01 billion years. A mile down, at the bottom of the canyon, lie rocks 0.57 billion years old, the roots of an earlier mountain range that stood as high as the Himalayas. It was pushed up, worn away to nothing, and covered with sediment long ago. Many of the geological features we know on Earth have been produced by very recent events.

Mike Seeds

Formation of Grand Canyon
Age of dinosaurs
Formation of Earth
Breakup of Pangaea
Heavy bombardment
First animals emerge on land
?
Oldest fossil life

| 4.6 | 4 | 3 | 2 | 1 | Now |

Billions of years ago

history of slow-moving rock as it does the churning surface of a pot of boiling soup. And just as steam rises from boiling soup, vapors rise from Earth's crust. Our atmosphere not only acts to erode the rocky surface, but some of it came from the crust as a by-product of geological activity.

20-3 THE ATMOSPHERE

We cannot complete our four-stage history of Earth without mentioning the atmosphere. Not only is it necessary for life, but it is also intimately related to the crust. It affects the surface by eroding geological features through wind and water, and in turn the chemistry of Earth's surface affects the composition of the atmosphere.

ORIGIN OF THE ATMOSPHERE

Until recent years, Earth scientists believed that Earth's atmosphere had developed slowly by outgassing, the release of gases from a planet's interior. According to this picture, Earth formed by the slow accumulation of planetesimals into a large, cold ball of matter whose composition was uniform from surface to center (❚ Figure 19-12). Early Earth might have attracted small amounts of gases such as hydrogen, helium, and methane from the solar nebula to form a **primeval atmosphere.** The slow decay of uranium, thorium, and potassium heated Earth's interior, melted it, caused it to differentiate into core, mantle, and crust, and triggered widespread volcanism. When a volcano erupts, 50 to 80 percent of the gas released is water vapor. The rest is mostly carbon dioxide, nitrogen, and smaller amounts of sulfur gases such as hydrogen sulfide—the rotten-egg gas that we smell when we visit geothermal pools and geysers such as those at Yellowstone National Park. These gases diluted the primeval atmosphere and produced a **secondary atmosphere** rich in carbon dioxide, nitrogen, and water vapor.

A modern understanding of how the solar system formed suggests that Earth formed so rapidly that it was melted by the heat released by in-falling material. This would have caused continuous outgassing, and the atmosphere would have grown rich in carbon dioxide, nitrogen, and water vapor from the beginning.

For some years, astronomers have suspected that the water on Earth arrived late in the formation process as a bombardment of volatile-rich planetesimals. These icy bodies, the theory goes, were scattered by the growing mass of the outer planets, and the inner solar system was bombarded by a storm of comets. This theory has faced a serious objection. Spectroscopic studies of a few comets revealed that the ratio of deuterium to hydrogen in comets does not match the ratio in the water on Earth. Some astronomers think this means that the

water now on Earth could not have arrived in comet-like planetesimals. However, studies of Comet Linear, which broke up in 1999 as it passed the sun, show that the water in that comet had a composition similar to water on Earth. There may be differences among comets. Those that formed far from the sun may be richer in deuterium, while comets that formed closer to the orbit of Jupiter than to the orbit of Neptune may have more Earthly water. This is an interesting controversy, and it reminds us that the origin of Earth's atmosphere and oceans is yet to be fully understood.

However the atmosphere originated, the mix of gases must have changed. The young atmosphere must have been rich in water vapor, carbon dioxide, and other gases. As the atmosphere cooled, the water condensed to form the first oceans, and as the oceans grew bigger, carbon dioxide began to dissolve in the water. Carbon dioxide is easily soluble in water—which is why carbonated beverages are so easy to manufacture—but the first oceans could not have absorbed all of the atmospheric carbon dioxide if the gas had not reacted with dissolved substances in the seawater to form silicon dioxide, limestone, and other mineral sediments. Thus, chemical reactions in the oceans transferred the carbon dioxide from the atmosphere to the seafloor.

Earth's lower atmosphere is now protected from ultraviolet radiation by an **ozone layer** about 25 km above the surface. However, the atmosphere of the young Earth did not contain free oxygen, so an ozone layer could not form, and the sun's ultraviolet radiation penetrated deep into the atmosphere. The energetic ultraviolet photons would have broken up weaker molecules such as water (H_2O). The hydrogen would have escaped to space, and the oxygen would have formed oxides in the crust. Earth's atmosphere could not reach its present composition (❚ Table 20-1) until it was protected by an ozone layer, and that required oxygen.

The origin of the atmospheric oxygen is linked to the origin of life, the subject of Chapter 26, but it is sufficient here to note that life must have originated within a billion years of Earth's formation. This life did not significantly alter the atmosphere, however, until nature invented photosynthesis about 3.3 billion years ago. Photosynthesis absorbs carbon dioxide from the air or water and utilizes it for plant growth, releasing oxygen. Because oxygen is a very reactive element, it combines easily with other compounds, and thus the oxygen abundance grew slowly at first. Apparently, the development of large, shallow seas along the continental margins half a billion years ago allowed ocean plants to manufacture oxygen faster than chemical reactions could consume it. Atmospheric oxygen then increased rapidly, and it is still increasing at the rate of about 1 percent every 36 million years.

Earth's climate is critically sensitive to many different factors. For example, the **albedo** of a world is the fraction of the sunlight that hits it that gets reflected

TABLE 20-1
Earth's Atmosphere

Gas	Percent by Weight
N_2	75.5
O_2	23.1
Ar	1.29
CO_2	0.05
Ne	0.0013
He	0.00007
CH_4	0.0001
Kr	0.0003
H_2O (vapor)	1.7–0.06

away. An albedo of 1 would be a perfectly white world, and an albedo of 0 would be perfectly black. Earth's overall albedo is 0.39, meaning it reflects back into space 39 percent of the sunlight that hits it. Much of this reflection is caused by clouds, and the formation of clouds depends critically on the presence of water vapor in the upper atmosphere, the temperature of the upper atmosphere, and the patterns of atmospheric circulation. Even a small change in any of these factors could change Earth's albedo and thus its climate. We are just beginning to understand all of the factors that have combined to assure our comfortable lives on the surface of our world.

HUMAN EFFECTS ON EARTH'S ATMOSPHERE

Earth is comfortable for human life because of our atmosphere. The temperature is moderate at sea level, but if we climb to the top of a high mountain, we find the temperature to be very low (▌Figure 20-8). Most clouds form at such altitudes. Higher still, we find the air much colder and so thin it cannot protect us from the intense ultraviolet radiation in sunlight. We can walk Earth's surface in safety because our atmosphere protects us, but we are altering our atmosphere in at least two serious ways: We are adding carbon dioxide (CO_2), and we are destroying ozone.

The concentration of CO_2 in our atmosphere is important to us because CO_2 can trap heat in a process called the **greenhouse effect** (▌Figure 20-9a). When sunlight shines through the glass roof of a greenhouse, it heats the benches and plants inside. The warmed interior radiates infrared radiation, but the infrared cannot get out through the glass. Heat is trapped within the greenhouse, and the temperature climbs until the glass itself grows warm enough to radiate heat away as fast as the sunlight enters. In Earth's atmosphere, the CO_2 admits sunlight, and the surface grows warm. However, the CO_2 is not transparent to infrared radiation, so some of the heat is trapped, and that helps keep our planet warm (Figure 20-9b). Without the greenhouse effect, our planet would be much colder, and large parts of it would be uninhabitable.

CO_2 is not the only greenhouse gas. Water vapor, methane, and other gases also help warm Earth, but CO_2 is the most important because the human race is increasing its concentration in our atmosphere. For billions of years, natural processes on Earth have removed CO_2 from the atmosphere and buried the carbon in the form of coal, oil, and natural gas. Since the beginning of the industrial revolution in the mid-19th century, our civilization has been digging up this carbon, burning it to get energy, and releasing CO_2 back into the atmosphere. At the same time, we are cutting down parts of the forests that absorb CO_2. Estimates are that the amount of CO_2 in our atmosphere will double during the next century.

The increased concentration of CO_2 will increase the greenhouse effect and warm Earth in what is known as **global warming.** The actual amount of warming is difficult to predict. The best mathematical models predict a warming between 1.5 and 4.5°C (3 to 8°F). Predictions are uncertain because Earth's climate depends on so many factors. A slight warming should increase water vapor in the atmosphere, and water vapor is another greenhouse gas that would enhance the warming. But increased water vapor could increase cloud cover, increase Earth's albedo, and partially reduce the greenhouse effect. High icy clouds, however, tend to enhance the greenhouse effect. Even small changes in

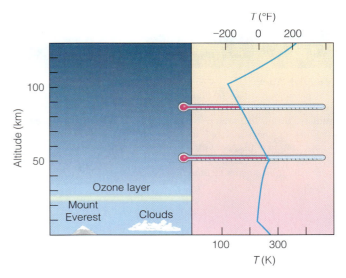

FIGURE 20-8
Thermometers placed in Earth's atmosphere at different levels would register the temperatures shown in the graph at the right. The lower few kilometers where we live are comfortable, but higher in the atmosphere the temperature is quite low. The ozone layer lies about 25 km above Earth's surface.

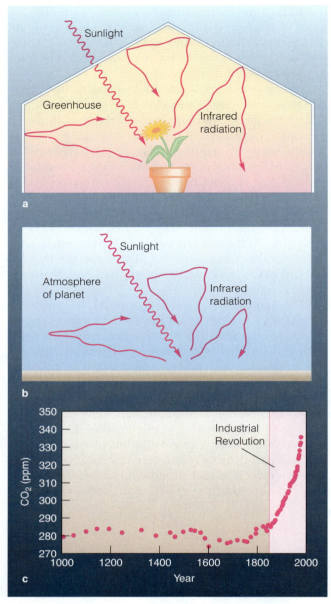

FIGURE 20-9

The greenhouse effect. (a) Short-wavelength light can enter a greenhouse and heat its contents, but the longer-wavelength infrared radiated by the warm interior cannot get out. (b) The same process can heat a planet's surface if its atmosphere contains CO_2. (c) The concentration of CO_2 in Earth's atmosphere as measured in Antarctic ice cores remained roughly constant until the Industrial Revolution. *(Adapted from a figure by Etheridge, Steele, Langenfelds, Francey, Barnola, and Morgan)*

$0.6°C$ (0.5 to 1.1°F). It is difficult to be sure how much of this change is due to the increased greenhouse effect and how much is natural variation in Earth's climate, but it is worrisome that general trends point toward warming. Mountain glaciers have melted back dramatically since the 19th century, and the warmest years of the last 100 years have all been recent.

Although the temperature increase is small now, it is a serious issue for the future. Even a small rise in temperatures can alter climate and dramatically affect agriculture, not only through rising temperatures but also through changes in rainfall. Also, a small increase in temperature will begin melting the polar caps, causing a rise in sea levels that will flood coastal regions and alter shore environments. A dramatic rise could cover huge low-lying areas such as the entire state of Florida.

There seems no doubt that our civilization is warming Earth through an enhanced greenhouse effect, but a remedy is difficult to imagine. We must reduce the amount of CO_2 and other greenhouse gases released to the atmosphere. But modern society depends on burning fossil fuels for energy. We must conserve our forests, but our growing population, especially in developing countries with large forest reserves, demands the wood and agricultural land produced when forests are cut.

Human influences on Earth's atmosphere go beyond the greenhouse effect. We are also reducing ozone in Earth's atmosphere. Ozone is an unstable molecule formed of three oxygen atoms (O_3). Ozone is produced in auto emissions, and it is a pollutant in city air. But ozone is produced naturally in our atmosphere at an altitude of about 25 km, as we have seen, and there it is beneficial. This ozone layer absorbs ultraviolet photons and prevents them from reaching the ground.

Certain chemicals called chlorofluorocarbons (CFCs), used in industrial processes and in refrigeration and air conditioning, can destroy ozone. As these CFCs escape into the atmosphere, they become mixed into the ozone layer and convert the ozone into normal oxygen molecules. Oxygen does not block ultraviolet radiation, so depleting the ozone layer will cause an increase in ultraviolet radiation at Earth's surface. In small doses, ultraviolet radiation can produce a suntan, but in larger doses it can cause skin cancers.

The ozone layer over the Antarctic may be especially sensitive to CFCs. Since the late 1970s, the ozone concentration has been falling over the Antarctic, and a hole in the ozone layer now develops over the continent each October—spring for Antarctica (❙ Figure 20-10). Satellite and ground-based measurements show that the same thing is happening at higher northern latitudes and that the amount of ultraviolet radiation reaching the ground is increasing. This is an early warning that human civilization is modifying Earth's atmosphere in a potentially dangerous way.

temperature may alter circulation patterns in the atmosphere and in the oceans, and the consequences of such changes are very difficult to predict.

Although the future is difficult to predict, the evidence does suggest Earth is growing warmer. Studies of the growth rings in very old trees show that Earth's climate has been cooling for the last 1000 years, but the 20th century reversed that trend with a rise of 0.3 to

a

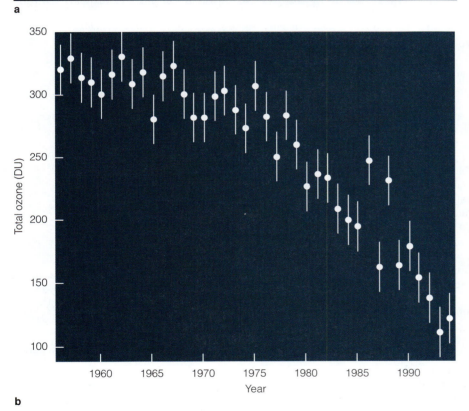

b

FIGURE 20-10

(a) Ozone concentrations in Dobson Units (DU) over Antarctica are shown here as red for highest concentration and violet for lowest. Since the 1970s, a hole in the ozone layer has developed over the South Pole. (No satellite data are available for 1995.) (b) Monthly average ozone each October above Halley Bay Station, Antarctica. Although ozone depletion is most dramatic above the South Pole, ozone concentrations have declined at all latitudes. The blue line shows the best statistical fit to the measurements. (*NASA/GSFC/TOMS and Glenn Carver*)

The CO_2 problem and the ozone problem in Earth's atmosphere are paralleled on Venus and Mars. When we study Venus in Chapter 22, we will discover a runaway greenhouse effect that has made the planet into a thermal Hades. On Mars we will discover an atmosphere without an ozone layer. A few minutes of sunbathing on Mars would kill you. Once again, we will learn more about our own planet by studying other planets.

REVIEW CRITICAL INQUIRY

Why is our atmosphere poor in carbon dioxide and rich in oxygen?

We might expect our atmosphere to be very rich in carbon dioxide, CO_2. Outgassing from Earth's crust releases mostly CO_2 and water vapor. Luckily for us, CO_2 is highly soluble in water, and Earth's surface temperature is just right for it to be covered with liquid water. The CO_2 dissolves in the oceans and combines with minerals in seawater to form deposits of silicon dioxide, limestone, and other mineral deposits. Thus, the CO_2 is removed from our atmosphere and buried in Earth's crust. Oxygen, in contrast, is highly reactive and forms oxides so easily it could not remain in our atmosphere very long unless it was continually replenished. We depend on life in the form of green plants to release oxygen into our atmosphere faster than chemical reactions can remove it. Were it not for liquid-water oceans, we would be choking in a thick CO_2 atmosphere with no free oxygen.

Why would an excess of CO_2 and a deficiency of free oxygen be harmful to all life on Earth in ways that go beyond mere respiration?

Lowland regions must have been flooded by molten lava when Earth was young, but on Earth, water can form a liquid, so the lowlands are now filled with oceans.

The human race is altering Earth's atmosphere in two dramatic ways: We are increasing the greenhouse effect, and we are reducing the ozone layer. The greenhouse effect is produced by carbon dioxide in the atmosphere, which traps heat and makes Earth warmer. As we burn fossil fuels and cut down forests, we add carbon dioxide to the atmosphere and reduce the plant life that cleans the air. Thus, we appear to be making Earth a warmer planet, and that could eventually alter our climate, melt back the polar ice caps, flood low regions, and reduce our ability to grow food.

Certain waste gases called chlorofluorocarbons are thinning the ozone layer 25 km above Earth's surface. Ozone absorbs ultraviolet radiation from the sun, and as we thin the ozone layer, we admit more ultraviolet radiation. Such high-energy photons can cause skin cancers.

Earth is a rich and beautiful world, and when we visit Venus and Mars, both of which have CO_2-rich atmospheres, we will appreciate Earth even more. But before we rush to compare Earth with planets that have atmospheres, we should examine two worlds that have no atmosphere at all—the moon and Mercury. As usual, the comparison of worlds will be revealing.

SUMMARY

Comparative planetology is the study of planets by comparison and contrast. We use Earth as the basis for our study of planets because we know it best and because it is a planet of extremes. It has an active crust and a water-rich atmosphere.

Like all terrestrial planets, Earth has passed through a four-stage history: (1) differentiation; (2) cratering; (3) flooding by lava, water, or both; and (4) slow surface evolution.

We know that Earth differentiated because seismic waves traveling through its interior reveal a core of nickel and iron, a thick mantle of dense rock, and a thin crust of low-density rock. Except for the very center, the core is molten, and circulation within the conducting liquid combined with Earth's rotation generates the planet's magnetic field through the dynamo effect.

Earth's low-density crust floats as great plates on the highly plastic upper mantle. The motions of these plates make Earth's surface active. The crust splits open along midocean rises where upwelling lava makes new crust, and old crust plunges into deep subduction zones along the coasts of some continents, where it is remelted. This plate tectonics causes volcanism and earthquakes and constantly recycles Earth's crust. Consequently, we see few old rocks and few impact craters.

Earth's atmosphere has been produced by gases released from the in-falling planetesimals that formed the planet. Carbon dioxide and water have been joined by other gases released by volcanism. Oxygen has been added by plant life.

NEW TERMS

comparative planetology	midocean rise
seismic waves	subduction zone
seismograph	basalt
pressure (P) waves	folded mountain range
shear (S) waves	rift valley
mantle	primeval atmosphere
plastic	secondary atmosphere
bow shock	ozone layer
magnetosphere	albedo
Van Allen belts	greenhouse effect
plate tectonics	global warming

REVIEW QUESTIONS

Ace⊙Astronomy™ Assess your understanding of this chapter's topics with additional quizzing and animations at **http://astronomy.brookscole.com/seeds8e**

1. How do we know that Earth differentiated?

2. What evidence do we have that Earth's metallic core is molten?

3. How are earthquakes in Hawaii different from those in southern California?

4. If Earth's core were not an electrical conductor, how would the magnetic field differ?

5. What evidence do we have that the Atlantic Ocean is growing wider?

6. How is our concept of Earth's first atmosphere related to the speed with which Earth formed from the solar nebula?

7. What has produced the oxygen in our atmosphere?

8. Why would an increase in CO_2 in Earth's atmosphere cause a rise in Earth's temperature?

9. Why would a decrease in the density of the ozone layer cause public health problems?

DISCUSSION QUESTIONS

1. If we visited a planet in another solar system and discovered oxygen in its atmosphere, what might we guess about its surface?

2. If liquid water is rare on the surface of planets, then most terrestrial planets must have CO_2-rich atmospheres. Why?

PROBLEMS

1. In Figure 20-2, the earthquake occurred 7440 km from the seismograph. How fast did the *P* waves travel in km/s? How fast did the *S* waves travel?

2. What percentage of Earth's volume is taken up by its metallic core?

3. If the Atlantic seafloor is spreading at 3 cm/year and is now 6400 km wide, how long ago were the continents in contact?

4. The Hawaiian-Emperor chain of undersea volcanoes is about 7500 km long. If the Pacific plate is moving 9.2 cm a year, how old is the oldest detectable volcano in the chain? What has happened to older volcanoes in the chain?

5. From Hawaii to the bend in the Hawaiian-Emperor chain is about 4000 km. Use the speed of plate motion given in Problem 4 to estimate how long ago the direction of plate motion changed. (The San Andreas Fault in southern California became active at about the same time!)

6. Calculate the age of the Grand Canyon as a percent of Earth's age.

CRITICAL INQUIRIES FOR THE WEB

1. Search the Web to find the location of the most recent earthquakes. The Earth quakes somewhere each day. Search for seismographs that will give you information about the most recent earthquakes.

2. Search for information about the arrangement of land masses on Earth before Pangaea. Were there previous supercontinents? What was Rodinia? What is the evidence?

3. Earth's magnetic poles are not coincident with its axis of rotation, nor are their positions fixed. Search for information about the location of the magnetic poles and their motions. Why do the positions change?

4. What are the oldest known rocks on Earth? Where do they come from, and how were their ages determined?

5. What evidence can you find about our changing climate? Is the rise in Earth's temperature related to the greenhouse effect? Scientific, industrial, and political leaders do not agree on this issue, so be prepared to analyze arguments with great care.

6. Can you locate a current image of the Antarctic ozone hole? Find data for the ozone concentration in the upper atmosphere over the place where you live. What is the latest news concerning ozone depletion?

 Visit the Seeds *Foundations of Astronomy* companion Web site for critical thinking exercises, articles, and additional readings from InfoTrac College Edition, Brooks/Cole's online student library.

THE MOON AND MERCURY: AIRLESS WORLDS

That's one small step for man . . . one giant leap for mankind.

Neil Armstrong, on the moon

Beautiful, beautiful. Magnificent desolation.

Edwin E. Aldrin, Jr., on the moon

NASA

Chapter 21

GUIDEPOST

The two preceding chapters have been preparation for the exploration of the planets. In this chapter, we begin that detailed study with two goals in mind. First, we search for evidence to test the solar nebula hypothesis for the formation of the solar system. Second, we search for an understanding of how planets evolve once they have formed.

The moon is a good place to begin because people have been there. This is an oddity in astronomy in that astronomers are accustomed to studying objects at a distance. In fact, many of the experts on the moon are not astronomers but geologists, and much of what we will study about the moon is an application of earthly geology.

While no one has visited Mercury, we will recognize it as familiar territory. It is much like the moon, so our experience with lunar science will help us understand Mercury as well as the other worlds we will visit in the chapters that follow.

VIRTUAL LABORATORIES

PLANETARY GEOLOGY

PLANETARY ATMOSPHERES AND THEIR RETENTION

When Neil Armstrong stepped out onto the moon on July 20, 1969, he spoke words for the history books. Edwin Aldrin, the second man to walk on the moon, spoke more directly about what he saw. The lunar landscape is indeed "magnificent desolation." Had Armstrong and Aldrin landed on Mercury, they would have had much the same view: an airless, rocky surface blasted to dust by over 4 billion years of impacts.

In this chapter, we begin our study of other worlds by applying comparative planetology to the study of the moon and Mercury. As we compare them with each other and with Earth, we will discover three important themes of planetary astronomy. First, we will see the importance of impact cratering. The surfaces of the moon and Mercury have been scarred by the impact of uncountable meteorites in a process that must have affected all the planets. Second, we will observe the importance of internal heat in worlds smaller than Earth. Third, we will discover the possibility of giant impacts so large they could shatter planets. By concentrating on these three principles, we will be able to organize the flood of details astronomers have learned about the moon and Mercury.

Our ultimate goal in the study of any world is to tell its story. How did the planet form, how has it evolved, and why is it the way it is? Each time we study a new planet, we will conclude our discussion by trying to combine what we have learned to tell the planet's story. For the moon and Mercury, we will use our three themes—cratering, internal heat, and giant impacts—to try to understand how small worlds can become such magnificently desolate places.

A purist might argue that the moon and Mercury can't be similar because one is a satellite and one is a planet. But both are worlds. They have had complex geological histories that we can compare with each other and with Earth's history. The moon and Mercury illustrate the fate of small, airless worlds.

21-1 THE MOON

Only 12 people have stood on the moon, but planetary astronomers know it well. The photographs, measurements, and samples brought back to Earth paint a picture of an airless, ancient, battered crust.

THE VIEW FROM EARTH

A few billion years ago, the moon must have rotated faster, but Earth is four times larger in diameter than the moon (▌Data File Three), and its tidal forces on the moon are strong. Earth's gravity raised tidal bulges on the moon, and friction in the bulges slowed the moon until it now rotates once each orbit to keep the same side facing Earth in what is called **tidal coupling.** This

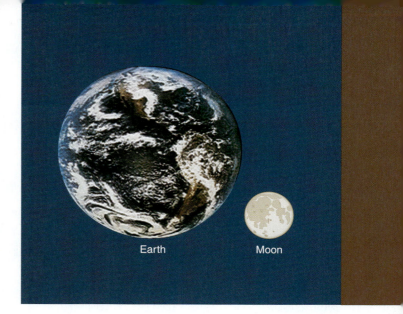

Earth   Moon

THE MOON
Data File Three

The moon's diameter is roughly 27 percent that of Earth.

Average distance from Earth	384,400 km (center to center)
Eccentricity of orbit	0.055
Maximum distance from Earth	405,500 km
Minimum distance from Earth	363,300 km
Inclination of orbit to ecliptic	5°9′
Average orbital velocity	1.022 km/s
Orbital period (sidereal)	27.321661 days
Orbital period (synodic)	29.5305882 days
Inclination of equator to orbit	6°41′
Equatorial diameter	3476 km
Mass	7.35×10^{22} kg (0.0123 $M_{\oplus}$)
Average density	3.36 g/cm^3 (3.35 g/cm^3 uncompressed)
Surface gravity	0.167 Earth gravity
Escape velocity	2.38 km/s (0.21 $V_{\oplus}$)
Surface temperature	−170° to 130°C (−274° to 266°F)
Average albedo	0.07

FIGURE 21-1

The side of the moon that faces Earth is a familiar sight. Craters have been named for famous scientists and philosophers, and the dark, smooth areas have been called seas (*mare* in Latin). Mare Imbrium is the Sea of Rains and Mare Tranquillitatis is the Sea of Tranquillity. There is, in fact, no water on the moon. *(Photo © UC Regents/Lick Observatory)*

Visual-wavelength image

means that we who live on Earth always see the same side of the moon; we never see the back. The moon's familiar face has shone down on Earth since long before there were humans (❚ Figure 21-1).

Even from Earth, we can tell that the moon is an airless world. How do we know? First, our understanding of gravity tells us that a world as small as the moon must have a low escape velocity (Data File Three), and gas atoms near its surface escape easily into space. Furthermore, we can see dramatic shadows near the **terminator,** the dividing line between daylight and darkness. There is no air on the moon to scatter light and soften shadows. We also see stars disappear behind the **limb** of the moon—the edge of its disk—without dimming. Clearly, the moon is an airless (and soundless) world.

The most dramatic features on the moon, regions that are visible even to our unaided eyes, are **mare,** Latin for "sea." The dark maria (plural of *mare*) that make up the face of the man in the moon were so named because early observers thought the great smooth expanses were seas. Even a small telescope will show that the maria are marked by ridges, faults, smudges, and craters and therefore can't be water. The maria are actually great lowland plains covering 17 percent of the moon's surface.

Not many decades ago, astronomers argued about the nature of the maria. Some said that they were great solidified lava flows, but others argued that they were vast oceans of dust blasted from the lunar mountains by billions of years of meteorites and drawn to the lowlands by the lunar gravity. Science fiction writers wrote fascinating stories about the dust oceans. For example, *A Fall of Moon Dust* by Arthur C. Clarke tells of tourists in a lunar sightseeing boat trapped when their boat sinks in the dust to the bottom of a mare. Although it is a wonderful rescue story, it can never happen. We know now that the maria are great lava flows with only a few centimeters of loose dust on their surfaces.

These lava flows suggest volcanism, but only small traces of past volcanic activity are visible from Earth. No major volcanic peaks are visible, and no active volcanism has been detected. The lava flows that created the maria happened long ago and were much too fluid to build peaks. We can see a few small domes pushed up by lava below the surface, and we can see long, winding channels called **sinuous rilles** (❚ Figure 21-2), often found near the edges of the maria. These channels were evidently cut by flowing lava. In some cases, such a channel may have had a roof of solid rock, forming a lava tube. When the lava drained away, meteorite impacts collapsed the roof to form a sinuous rille. Thus, the view from Earth gives us only hints of ancient volcanic activity associated with the maria.

Even a small telescope reveals craters on the moon. The lunar surface is scarred with million of craters formed by the impact of meteorites. Study ❚ "Impact Cratering" on pages 452 and 453 and notice three important points. First, impact craters have certain distinguishing characteristics such as their shape and the way they eject material across the lunar surface. Second, notice the range of sizes of impact craters from giant basins to microscopic pits. Finally, notice that most of the craters we see on the moon are old; they were formed long ago when the solar system was young.

Meteorites strike the moon all the time, but large impacts are rare. Meteorites a few tens of meters in diameter probably strike the moon every 50 years or so, but no one has ever seen such an impact with certainty. Small flashes of light have been seen on the dark side of the moon during showers of meteorites, but those impacts must have been made by very small objects. No newly formed craters have been found on the moon since the invention of the telescope. Large impacts on the moon and Earth are quite rare, and nearly all of the lunar craters we see through telescopes date from the solar system's youth.

Craters are a key element in planetary study. They will help us understand the moon, Mercury, and many other planets and moons throughout our solar system. We begin here by using craters to unravel the history of the lunar maria.

a

b   Visual-wavelength images

FIGURE 21-2

(a) The edge of Mare Imbrium is marked by high mountains and the sinuous rille called Hadley Rille (arrow). Such rilles were caused by lava flowing through a channel. Other rilles, straight faults, and many small craters are visible in the surface of the mare. (b) Apollo 15 landed near the 1200-foot-deep rille (circle in part a) and used a battery-powered rover to visit the rille and nearby mountains and craters. *(NASA)*

a

b   Visual-wavelength images

FIGURE 21-3

(a) The lunar highlands are saturated with craters. Note the traces of old craters nearly obliterated by more recent craters. (b) The lunar lowlands are filled with smooth maria such as Mare Humorum shown here. Note the generally circular shape of the mare, the cracks and wrinkles in the surface, and the partially flooded craters around its edge. *(NASA)*

HIGHLANDS AND LOWLANDS

The surface of the moon is divided into two dramatically different kinds of terrain. The highlands are heavily cratered and are about 3 km higher than the lowlands, which are filled by the mostly uncratered lava flows we see as maria. We can use impact cratering to understand these two types of terrain.

There are no folded mountain ranges on the moon. No Rockies or Andies or Urals wander across the moon. What appears to be a mountainous region is really the highly cratered lunar highlands. The surface there is saturated with craters, meaning that it would be impos-

sible to form a new crater without destroying the equivalent of one old crater (■ Figure 21-3a).

The maria fill the lunar lowlands, and they contain few craters (Figure 21-3b). Many of those that do scar the maria have bright rays, suggesting that they are younger craters. For examples, locate the craters Kepler and Copernicus in Figure 21-1.

We can conclude from this that the highlands are old regions where we see craters from early in the history of the solar system. Most of those craters were

Impact Cratering

The craters that cover the moon and many other bodies in the solar system were produced by the high-speed impact of meteorites of all sizes. Meteorites striking the moon travel 10 to 60 km/s and can hit with the energy of many nuclear bombs.

A meteorite striking the moon's surface can deliver tremendous energy and can produce an impact crater 10 or more times larger in diameter than the meteorite. The vertical scale is exaggerated here for clarity.

Lunar craters such as Euler, 27 km (17 mi) in diameter, look deep when we see them near the terminator where shadows are long, but a typical crater is only a fifth to a tenth as deep as its diameter, and large craters are even shallower.

Because craters are formed by shock waves rushing outward, by the rebound of the rock, and by the expansion of hot vapors, craters are almost always round, even when the meteorite strikes at a steep angle.

Debris blasted out of a crater is called **ejecta**, and it falls back to blanket the surface around the crater. Ejecta shot out along specific directions can form bright **rays**.

Euler

Visual-wavelength image

Bright ejecta blankets and rays gradually darken as sunlight darkens minerals and small meteorites stir the dusty surface. Bright rays are signs of youth. Rays from the crater Tycho, perhaps only 100 million years old, extend halfway around the moon.

Rays　**Tycho**

Visual

Impact Cratering

A meteorite approaches the lunar surface at high velocity.

On impact, the meteorite is deformed, heated, and vaporized.

The resulting explosion blasts out a round crater.

Slumping produces terraces in crater walls, and rebound can raise a central peak.

Ace Astronomy™

Go to AceAstronomy and click Active Figures to see "The Moon's Craters" and experiment with different energy impacts.

Rock ejected from distant impacts can fall back to the surface and form smaller craters called **secondary craters.** The chain of craters here is a 45-km-long chain of secondary craters produced by ejecta from the large crater Copernicus 200 km out of the frame to the lower right.

Visual

All photos: NASA

Plum Crater, 40 m (130 ft) in diameter, was visited by Apollo 16 astronauts. Note the many smaller craters visible. Lunar craters range from giant impact basins to tiny pits in rocks struck by **micrometeorites**, meteorites of microscopic size.

Lunar rover

Sun glare in camera lens

Mare Orientale

Visual-wavelength images

Solidified Lava

The energy of an impact can melt rock, some of which falls back into the crater and solidifies. When the moon was young, craters could also be flooded by lava welling up from below the crust.

In larger craters, the deformation of the rock can form one or more inner rings concentric with the outer rim. The largest of these craters are called **multiringed basins.** In Mare Orientale on the west edge of the visible moon, the outermost ring is almost 900 km (550 mi) in diameter.

Most of the craters on the moon were produced long ago when the solar system was filled with debris from planet building. As that debris was swept up, the cratering rate fell rapidly.

A few meteorites found on Earth have been identified chemically as fragments of the moon's surface blasted into space by cratering impacts. The fragmented nature of these meteorites tells us the moon's surface has been battered by impact craters.

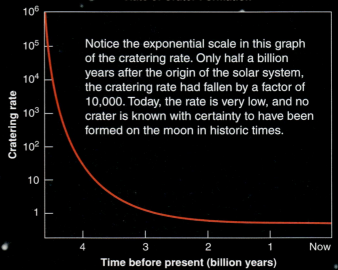

Rate of Crater Formation

Notice the exponential scale in this graph of the cratering rate. Only half a billion years after the origin of the solar system, the cratering rate had fallen by a factor of 10,000. Today, the rate is very low, and no crater is known with certainty to have been formed on the moon in historic times.

Cratering rate: 10^6, 10^5, 10^4, 10^3, 10^2, 10, 1

Time before present (billion years): 4, 3, 2, 1, Now

formed during the heavy bombardment over 4 billion years ago. The maria that fill the lowlands must have formed later, after the decline of the heavy bombardment. Thus the maria are younger than the highlands. These are called **relative ages** because we can tell which features formed first, but we can't give ages in years.

The cratering curve shown on page 453 can give us **absolute ages,** meaning ages in years. Studies of cratered surfaces in the solar system have allowed planetary scientists to calibrate the cratering curve. The lava flows that filled the lowlands and covered up the craters there could not have formed earlier than 3 to 4 billion years ago. Had they occurred earlier, the surfaces of the maria would show more craters.

Absolute ages derived from the cratering curve cannot be precise because planetary scientists suspect that the cratering rate may not have fallen as smoothly as the cratering curve suggests. Although the debris in the early solar system was gradually removed, bursts of cratering may have occurred. A major impact between large planetesimals might shower the solar system with fragments and produce a storm of crater impacts. Mathematical models suggest that the final formation of Uranus and Neptune might have flung icy planetesimals through the solar system like giant shrapnel. Thus the true cratering curve might have rises and dips superimposed on a gradual decline.

Whatever the exact timing of the heavy bombardment, we can be sure that the craters in the highlands are old and that the later lava flows that formed the maria covered over any craters that had been formed in the lowlands. In this way, we can begin telling the story of the moon's surface (Window on Science 21-1). To better understand the moon we must do more than just look from Earth. We must go there.

THE APOLLO MISSIONS

On May 25, 1961, President John Kennedy addressed Congress and committed the United States to landing a human being on the moon by 1970. The reasons for that decision are complicated and related more to economics and the stimulation of technology than to science, but the Apollo program became a fantastic adventure in science, a flight to the moon that changed how we think about Earth.

Flying to the moon is not particularly difficult; with powerful enough rockets and enough food, water, and air, it is a simple trip. Landing on the moon is more difficult but not impossible. The moon's gravity is only one-sixth that of Earth, and there is no atmosphere to disturb the trajectory of the spaceship. Getting to the moon is simple and landing is possible; the difficulty is doing both on one trip. The spaceship must carry food, water, and air for a number of days in space plus fuel and rockets for midcourse corrections and for a return to Earth. All of this adds up to a ship that is too

massive to make a safe landing on the lunar surface. The solution was to take two spaceships to the moon, one to ride in and one to land in (❚ Figure 21-4).

The command module was the long-term home and command center for the trip. Three astronauts had to live in it for a week, and it had to carry all the life-support equipment, navigation instruments, computers, power packs, and so on for a week's jaunt in space. The lunar landing module (LM for short) was tacked to the front of the command module like a bicycle strapped to the front of the family camper. It carried only enough fuel and supplies for the short trip to the lunar surface, and it was built to minimize weight and maximize maneuverability.

The weaker gravity of the moon made the design of the LM simpler. Landing on Earth requires reclining couches for the astronauts, but the trip to the lunar surface involved smaller accelerations. In an early version of the LM, the astronauts sat on what looked like bicycle seats, but these were later scrapped to save weight. The astronauts had no seats at all in the LM, and once they began their descent and acquired weight, they stood at the controls held by straps, riding the LM like two daredevils riding a rocket surfboard.

Lifting off from the lunar surface, the LM saved weight by leaving the larger descent rocket and support stage behind. Only the astronauts, their instruments,

Visual-wavelength image

FIGURE 21-4

The Apollo 12 lunar module is shown here about to descend to the lunar surface. The lunar module was designed to minimize weight. The astronauts stood without seats is a space hardly bigger than two telephone booths. The metal skin was so thin it was easily flexible, like metal foil, and the legs of the module, designed specifically for the moon and its weak gravity, would never have been able to support the lander's weight on Earth. *(NASA)* and the cargo of rocks returned to the command module orbiting above. Again, the astronauts in the LM blasted up from the lunar surface standing at the controls. The rocket engine that lifted them back into lunar orbit was not much bigger than a dishwasher.

The most complicated parts of the trip were the rendezvous and docking between the tiny remains of the LM and the command module, aided by radar systems and computers. The astronauts rejoined their colleague in the command module, transferred their moon rocks, and jettisoned the LM. Only the command module returned to Earth.

The first manned lunar landing was made on July 20, 1969. While Michael Collins waited in orbit, Neil Armstrong and Edwin Aldrin, Jr., took the LM down to the surface. Although much of the descent was controlled by computers, the astronauts had to override a number of computer alarms and take control of the LM to avoid a boulder-strewn crater bigger than a football field. Climbing down the ladder, Armstrong and then Aldrin stepped onto the lunar surface.

Between July 1969 and December 1972, 12 people reached the lunar surface and collected 380 kg (840 lb) of rocks and soil (▌ Table 21-1). The flights were carefully planned to visit different regions and thus develop a comprehensive history of the lunar surface.

The first flights went into relatively safe landing sites (▌ Figure 21-5a)—Mare Tranquillitatis for Apollo 11 and Oceanus Procellarum for Apollo 12. Apollo 13 was aimed at a more complicated site, but an explosion in an oxygen tank on the way to the moon ended all chances of a landing and nearly cost the astronauts

their lives. They succeeded in using the life support in the LM to survive.

The last four Apollo missions, 14 through 17, sampled geologically important places on the moon. Apollo 14 visited the Fra Mauro region, which is covered by ejecta from the impact that dug the basin now filled by Mare Imbrium. Apollo 15 visited the edge of Mare Imbrium at the foot of the Apennine Mountains and examined Hadley Rille (Figure 21-2b). Apollos 16 and 17 visited highland regions to sample the older parts of the lunar crust (▌ Figures 21-5b and 21-6). Almost all of the lunar samples from these six landings are now held at the Planetary Materials Laboratory at the Johnson Space Center in Houston. They are a national treasure containing clues to the beginnings of our solar system.

MOON ROCKS

Of all of the rock samples that the Apollo astronauts carried back to Earth, every one is igneous. That is, they formed by the cooling and solidification of molten rock. No sedimentary rocks were found, which is consistent with the moon never having had liquid water on its surface. In addition, the rocks were extremely dry. Almost all Earth rocks contain 1 to 2 percent water, either as free water trapped in the rock or as water molecules chemically bonded with certain minerals. But moon rocks contain no water at all.

Rocks from the lunar maria are dark-colored, dense basalts much like the solidified lava produced by the Hawaiian volcanoes (▌ Figure 21-7). These rocks are rich in heavy elements such as iron, manganese, and

TABLE 21-1
Apollo Lunar Landings

Apollo Mission*	Astronauts: Commander LM Pilot CM Pilot	Date	Mission Goals	Sample Weight (kg)	Typical Samples	Ages (10^9 y)
11	Armstrong Aldrin Collins	July 1969	First manned landing; Mare Tranquillitatis	21.7	Mare basalts	3.48–3.72
12	Conrad Bean Gordon	November 1969	Visit Surveyor 3; sample Oceanus Procellarum (mare)	34.4	Mare basalts	3.15–3.37
14	Shepard Mitchell Roosa	February 1971	Fra Mauro, Imbrium ejecta sheet	42.9	Breccia	3.85–3.96
15	Scott Irwin Worden	July 1971	Edge of Mare Imbrium and Apennine Mountains, Hadley Rille	76.8	Mare basalts; highland anorthosite	3.28–3.44 4.09
16	Young Duke Mattingley	April 1972	Sample highland crust; Cayley formation (ejecta); Descartes	94.7	Highland basalt; breccia	3.84 3.92
17	Cernan Schmitt Evans	December 1972	Sample highland crust; dark halo craters; Taurus–Littrow	110.5	Mare basalt; highland breccia; fractured dunite	3.77 3.86 4.48

*The Apollo 13 mission suffered an explosion on the way to the moon and did not land.

titanium, which give them their dark color. Some of the basalts are **vesicular**, meaning that they contain holes caused by bubbles of gas in the molten rock. Like bubbles in a carbonated beverage, these bubbles do not form while the magma is under pressure. Only when the rock flows out onto the surface, where the pressure is low, do bubbles appear. Thus, the presence of vesicular basalts assures us that these rocks formed in lava flows that reached the surface and did not solidify underground.

Absolute ages of the mare basalts can be found from the radioactive atoms they contain (Chapter 19),

a

b

FIGURE 21-5
A comparison of horizons. (a) Apollo 11, the first mission to the moon, landed on the dusty surface of Mare Tranquillitatis in the lunar lowlands. The flat lava plain is almost featureless, and the horizon is straight. (b) Apollo 17, the last lunar mission, landed at Taurus–Littrow in the lunar highlands. Note the rugged terrain and the irregular horizon. *(NASA)*

FIGURE 21-6
The landing sites of the six Apollo missions that reached the moon. For the names of lunar features, see Figure 21-1.

and these ages range from about 3.1 to 3.8 billion years. Thus, the lava flows that formed these rocks must have happened some time after the end of the heavy bombardment.

The highlands are composed of low-density rock containing calcium-, aluminum- and oxygen-rich minerals that would be among the first to solidify and float to the top of molten rock. The ages of the crustal rocks range from 4.0 to 4.5 billion years, significantly older than the mare basalts. The highlands contain **anorthosite,** a light-colored rock that contributes to the highlands' bright contrast with the dark, iron-rich basalts of the lowlands. The rocks of the highlands, although badly shattered by impacts, represent the moon's original low-density crust, whereas the mare basalts rose from the deep crust and upper mantle.

A large fraction of the lunar rocks are **breccias,** rocks that are made up of fragments of earlier rocks cemented together by heat and pressure. Evidently, meteorite impacts have broken up the rocks and fused them together time after time.

Both the highlands and the lowlands of the moon are covered by a layer of powdered rock and crushed fragments called the **regolith.** It is about 10 m deep on the maria but over 100 m deep in certain places in the highlands. About 1 percent of the regolith is meteoric

The Apollo astronauts found that all moon rocks are igneous, meaning they solidified from molten rock.

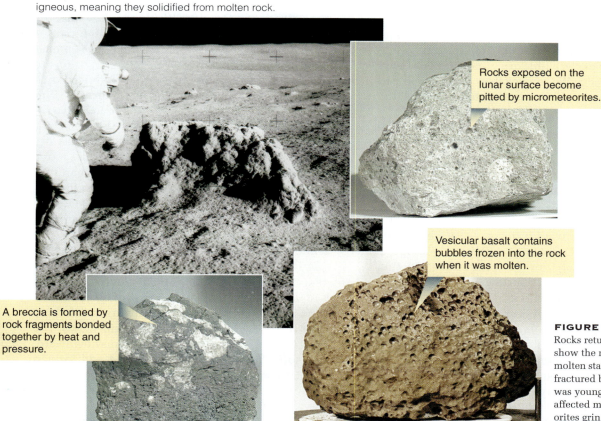

Rocks exposed on the lunar surface become pitted by micrometeorites.

Vesicular basalt contains bubbles frozen into the rock when it was molten.

A breccia is formed by rock fragments bonded together by heat and pressure.

FIGURE 21-7
Rocks returned from the moon show the moon formed in a molten state, that it was heavily fractured by cratering when it was young, and that it is now affected mainly by micrometeorites grinding away at surface rock. *(NASA)*

fragments; the rest is the smashed remains of moon rocks that have been ground down by the constant blasting of meteorites. The smallest meteorites, the micrometeorites, do the most damage by their constant sandblasting of the lunar surface. Thus, we see that impact cratering dominates the lunar surface and is even responsible for the lunar regolith.

The moon rocks are old, dry, igneous, and badly shattered by impacts. We can use these facts, combined with what we know of lunar features, to reconstruct the moon's past.

THE HISTORY OF THE MOON

The four-stage history of the moon is dominated by a single fact—it is small, only one-fourth the diameter of Earth. The escape velocity is low, so it has been unable to hold any atmosphere, and it cooled rapidly as its internal heat flowed outward into space. As we will see when we study other worlds, a planet's geology is largely driven by energy in the form of heat flowing outward from its interior (Window on Science 21-2). Small worlds have less heat and lose it more rapidly, so the moon's small size has been critical in its history.

The Apollo moon rocks show that the moon must have formed in a molten state. Planetary geologists now refer to the newborn moon as a sea of magma. Denser materials sank to form a small core; and, as the magma ocean cooled, low-density minerals floated to the top to form a low-density crust. In this way the moon must have differentiated into core, mantle, and crust. The moon has a low density and no internal magnetic field, so its dense core must be small and must contain little metallic iron. The radioactive ages of the moon rocks tell us that the surface solidified from 4.6 to 4.1 billion years ago.

The second stage, cratering, began as soon as the crust solidified, and the older highlands show that the cratering was intense during the first 0.5 billion years—during the heavy bombardment at the end of planet building.

The moon's crust was shattered to a depth of 10 kilometers or so, and the largest impacts formed giant multiringed crater basins hundreds of kilometers in diameter, such as Mare Orientale. This led to the third stage—flooding. Although Earth's moon cooled rapidly after its formation, some process such as radioactive decay heated material deep in the crust, and part of it melted. Molten rock followed the cracks up to the surface and flooded the giant basins with successive lava flows of dark basalts from 3.8 to 3.2 billion years ago. This formed the maria (▌Figure 21-8).

Some maria, such as Mare Imbrium, Mare Serenitatis, and Mare Crisium, retain their round shapes, but others are irregular because the lava overflowed the edges of the basin or because the shape of the basin was modified by further cratering. The moving lava formed channels that we see as sinuous rills (Figure 21-2). The weight of the maria pressed the crater basins downward, and the solidified lava was compressed and formed wrinkle ridges visible even in small telescopes (Figure 21-3b). The tension at the edges of the maria broke the hard lava to produce straight fractures and faults (Figure 21-2a). Later cratering and overlapping lava floods further modified the maria. Thus the maria are accumulations of features reflecting a complex history.

a

⊢————— 200 km —————⊣

Mare Imbrium

Mare Serenitatis

Mare Crisium

Aitken Basin

Near side of moon

Far side of moon

b

FIGURE 21-8

(a) Major impacts broke the lunar crust and produced large basins. Lava flooded up through the fractures and flooded the basins to produce maria. *(Adapted from a diagram by William Hartmann)* (b) In these images, color marks elevation, with red the highest regions and purple the lowest. Maria on the near side are generally circular, low, lava-filled basins. Note the circular outline of the labeled maria. The crust on the far side is thicker, and there is less flooding. Even the Aitken Basin contains little lava flooding. *(NASA/Clementine)*

Mare Imbrium is a dramatic example of how the great basins became the maria. Near the end of the heavy bombardment, roughly 4 billion years ago, a meteorite the size of Rhode Island struck the moon and blasted out a giant multiringed basin. The impact was so violent the ejecta blanketed 16 percent of the moon's surface. Apollo 14 landed at Fra Mauro to sample the ejecta (▌ Figure 21-9). After the cratering rate fell at the end of the heavy bombardment, lava flows welled up time after time and flooded the Imbrium Basin, burying all but the highest parts of the giant multiringed basin. The Imbrium Basin is now a large, generally round mare marked by only a few craters that have formed since the last of the lava flows (▌ Figure 21-10).

Our story of the moon might suggest that it was a violent place during this cratering phase, but large impacts were in fact rare; and, except for a continuous rain of smaller meteorites, the moon was, for the most part, a peaceful place even during the heavy bombardment. Centuries might pass between major impacts. Of course, a large impact might have threatened to bury us under ejecta or jolted us by seismic shocks. We could have felt the Imbrium impact anywhere on the moon, but had we been standing on the side of the moon opposite this impact, we would have been at the focus of seismic waves traveling around the moon from different directions. When the waves met under our feet, the surface would have jerked up and down by as much as 10 m. The place on the moon opposite the Imbrium Basin is a strangely disturbed landscape called **jumbled terrain.** We will find similar effects of large impacts on other worlds.

Studies of our moon show that its crust is thinner on the side toward Earth, perhaps due to tidal effects. Consequently, while lava flooded the basins on the Earthward side, it was unable to rise through the thicker crust to flood the lowlands on the far side. The largest impact basin in the solar system is the south-pole Aitkin Basin on the moon (Figure 21-8b). It is about 2500 kilometers (1500 miles) in diameter and as deep as 13 kilometers (8 miles) in places, but flooding has never filled it with smooth lava flows.

The fourth stage, slow surface evolution, is limited because our moon lacks water and has cooled rapidly. Flooding on Earth included water, but our moon has never had an atmosphere and thus has never had liquid water. With no air and no water, erosion is limited to the constant bombardment of micrometeorites and rare larger impacts.

The moon is small, and small worlds cool rapidly because they have a large ratio of surface area to volume, just as a small cupcake cools more rapidly than a large cake. The moon has lost much of its internal heat, and it is the outward flow of heat that produces geological activity. The crust of the moon rapidly grew thick and never divided into moving plates. There are no rift valleys or folded mountain chains on the moon. The last lava flows on the moon ended one or two billion years ago when the moon's temperature fell too low to maintain subsurface lava.

The overall terrain on the moon is almost fixed. On Earth a billion years from now, plate tectonics will have totally changed the shapes of the continents, and erosion will have long ago worn away the Rocky Mountains. On the moon, impacts will have formed only a few more large craters, and nearly all of the lunar scenery will be unchanged. Micrometeorites, however, will

FIGURE 21-9

Apollo 14 landed on rugged terrain suspected of being ejecta from the Imbrium impact. The large boulder here is ejecta that, at some time in the past, fell here from an impact beyond the horizon. *(NASA)*

have blasted the soil, erasing the footprints left by the Apollo astronauts and reducing the equipment they left behind to peculiar chemical contamination in the regolith at the six Apollo landing sites.

We have told the story of our moon's evolution in detail for later comparison with other planets and moons in our solar system, but we have skipped one important question. Where did Earth get such a large satellite?

THE ORIGIN OF EARTH'S MOON

Over the last two centuries, astronomers have developed three different hypotheses for the origin of Earth's moon, but these traditional ideas have failed to survive comparison with the evidence. A theory proposed in the 1980s may hold the answer. We begin by testing the three unsuccessful theories against the evidence.

The first of the three traditional theories, the **fission hypothesis,** supposes that the moon formed by the fission of Earth. If the young Earth spun fast enough, tides raised by the sun might make it break into two parts (❙ Figure 21-11a). If this separation occurred after Earth differentiated, the moon would have formed from crust material. Thus, the theory explains the moon's low density.

But the fission theory has problems. No one knows why the young Earth should have spun so fast, nearly ten times faster than today, nor where all that angular momentum went after the fission. In addition, the moon's orbit is not in the plane of Earth's equator, as it would be if it had formed by fission.

Origin of Mare Imbrium

Four billion years ago, an impact forms a multiringed basin over 1000 km (700 mi) in diameter.

Continuing impacts crater the surface but do not erase the high walls of the multiringed basin.

Archimedes

Impacts form a few large craters, and, starting about 3.8 billion years ago, lava floods low regions.

Repeated lava flows cover most of the inner rings and overflow the basin to merge with other flows.

Impacts continue, including those that formed the relatively young craters Copernicus and Kepler.

Kepler   Copernicus

FIGURE 21-10
Lava flooding after the end of the heavy bombardment filled a giant multiringed basin and formed Mare Imbrium. *(Don Davis)*

The second traditional theory is the **condensation** (or double-planet) **hypothesis.** It supposes that Earth and the moon condensed as a double planet from the same cloud of material (Figure 21-11b). However, if they formed from the same material, they should have the same chemical composition and density, which they don't. The moon is very poor in iron and related elements such as titanium and in volatiles such as water vapor and sodium. Yet the moon contains almost exactly the same ratios of oxygen isotopes as does Earth's mantle. Condensation cannot explain these compositional differences.

The third theory is the **capture hypothesis.** It supposes that the moon formed somewhere else and was later captured by Earth (Figure 21-11c). We might imagine the moon forming inside the orbit of Mercury, where the heat would prevent the condensation of solid metallic grains and only high-melting-point metal oxides could have solidified. A later encounter with Mercury could "kick" the moon out to Earth.

The capture theory is not popular because it requires highly unlikely events involving interactions with Mercury and Earth to move the moon from place to place. Scientists are always suspicious of explanations that require a chain of unlikely coincidences. Also, upon encountering Earth, the moon would have been moving so rapidly that Earth's gravity would have been unable to capture it without ripping the moon to fragments through tidal forces.

Until recently, astronomers were thus left with no acceptable theory to explain the origin of the moon, and they occasionally joked that the moon could not exist. But during the 1980s, planetary astronomers developed a new theory that combines the best aspects of the fission hypothesis and the capture hypothesis.

The **large-impact hypothesis** supposes that the moon formed from debris ejected into a disk around Earth by the impact of a large body. The impacting body may have been twice as large as Mars. Instead of saying that Earth was hit by a large body, we may be more nearly correct in saying that Earth and the moon resulted from the collision and merger of two very large planetesimals. The resulting large body became Earth, and the ejected debris formed the moon (Figure 21-12). Such an impact would have melted the proto-Earth, and the material falling together to form the moon would have been heated hot enough to melt. This fits our expectation that the moon formed as a sea of magma.

This hypothesis would explain a number of things. The collision had to occur at a steep angle to eject enough matter to make the moon. That is, the objects did not collide head-on. Such a glancing collision would have spun the resulting material rapidly enough to explain the observed angular momentum in the Earth–moon system. If the colliding planetesimals had already differentiated, the ejected material would be mostly iron-poor mantle and crust. Calculations show that the iron core of the impacting body would have fallen into the Earth. This would explain why the moon is so poor in iron and why the abundances of other elements are so similar to rocks from Earth's mantle. Also, the material that eventually became the moon would have remained in a disk long enough for volatile elements, which the moon lacks, to be lost to space.

FIGURE 21-11

Three traditional theories for the moon's origin. (a) Fission theories suppose that Earth and the moon were once one body and broke apart. (b) Condensation theories suppose that the moon formed from material near Earth. (c) Capture theories suggest that the moon formed elsewhere and was captured by Earth. None of these theories explains all the facts.

a

b

c

Thus, the moon may be the remains of a giant impact. Until recently, astronomers have been reluctant to consider such catastrophic events, but a number of lines of evidence suggest that planets have been affected by giant impacts. Consequently, the third theme we identified in the introduction to this chapter, giant impacts, has the potential to help us understand other worlds.

The moon's evolution has been restricted by its size. It is too small to have an atmosphere, and it cooled so rapidly that it never developed plate motion. Another world in our solar system has suffered from its diminutive size. Mercury, like the moon, is a small, rocky world, cratered by impacts and flooded by ancient lava flows driven by its internal heat.

21-2 MERCURY

Earth's moon and Mercury are similar in a number of ways. Their rotation has been altered by tides, their surfaces are heavily cratered, their lowlands are flooded in places by ancient lava flows, and both are small and airless and have ancient, inactive surfaces. Yet the differences between them are impressive and will help us understand the nature of these airless worlds.

Mercury is the innermost planet in the solar system (▌Data File Four), and thus its orbit keeps it near the sun in the sky. It is sometimes visible near the horizon in the evening sky after sunset or in the dawn sky just before sunrise. An Earth-based telescope shows Mercury passing through phases like the moon, but little detail is visible. Most of what we know about the surface of Mercury is based on photographs and measurements taken by the Mariner 10 spacecraft when it looped through the inner solar system in 1974 and 1975 (▌Figure 21-13). Until then, Mercury was a mystery world.

ROTATION AND REVOLUTION

During the 1880s, the Italian astronomer Giovanni Schiaparelli sketched the faint features he thought he saw on the disk of Mercury and concluded that the planet was tidally coupled to the sun just as the moon is tidally coupled to Earth. That is, he concluded that Mercury rotated on its axis to keep the same side facing the sun throughout its orbit. This was actually a very good guess because, as we will see, tidal coupling between rotation and revolution is common in the solar

The Large-Impact Hypothesis

A protoplanet nearly the size of Earth differentiates to form an iron core.

Core
Mantle
Crust

Mercury

Earth

Another body that has also formed an iron core strikes the larger body and merges, trapping most of the iron inside.

Iron-poor rock from the mantles of the two bodies forms a ring of debris.

Volatiles are lost to space as the particles in the ring begin to accrete larger bodies.

Eventually the moon forms from the iron-poor and volatile-poor matter in the disk.

FIGURE 21-12
Computer simulations of impacts between differentiated bodies suggest Earth's low-density, iron-poor moon might have formed from such an event.

system. But the rotation of Mercury is more complex than Schiaparelli thought.

In 1962, radio astronomers detected radio emissions from the planet and concluded that the dark side

MERCURY

Data File Four

In proportion to its size, Mercury has a very large metallic core.

Average distance from the sun	0.387 AU (5.79×10^7 km)
Eccentricity of orbit	0.2056
Maximum distance from the sun	0.467 AU (6.97×10^7 km)
Minimum distance from the sun	0.306 AU (4.59×10^7 km)
Inclination of orbit to ecliptic	7°00′16″
Average orbital velocity	47.9 km/s
Orbital period	87.969 days (0.24085 y)
Period of rotation	58.646 days (direct)
Inclination of equator to orbit	0°
Equatorial diameter	4878 km (0.382 $D_{\oplus}$)
Mass	3.31×10^{23} kg (0.0558 $M_{\oplus}$)
Average density	5.44 g/cm^3 (5.4 g/cm^3 uncompressed)
Surface gravity	0.38 Earth gravity
Escape velocity	4.3 km/s (0.38 $V_{\oplus}$)
Surface temperature	−173° to 330°C (−279° to 620°F)
Average albedo	0.1
Oblateness	0

Visual-wavelength images

FIGURE 21-13
Photomosaic of Mercury made by the Mariner 10 spacecraft. Caloris Basin is in shadow at left. Almost no surface detail is visible from Earth (inset). *(NASA, inset courtesy Lowell Observatory)*

was not as cold as it should have been if the planet kept one side perpetually in darkness. In 1965, radio astronomers made radar contact with Mercury. They used the 305-m Arecibo dish (Figure 6-21b) to transmit a pulse of radio energy at Mercury and then waited for the reflected signal to return. Doppler shifts in the reflected radio pulse showed that the planet was rotating with a period of only about 59 days, much shorter than the orbital period of 87.969 days.

Apparently, Mercury is tidally coupled to the sun such that it rotates not once per orbit but 1.5 times per orbit. Thus, its period of rotation is two-thirds its orbital period, or 58.65 days. This means that a mountain on Mercury directly below the sun at one place in its orbit will point away from the sun one orbit later and toward the sun after the next orbit (▌Figure 21-14).

If you flew to Mercury and landed your spaceship in the middle of the day side, the sun would be high overhead, and it would be noon. Your watch would show that almost 44 Earth days passed before the sun set in the west, and a total of 88 Earth days would have passed before the sun reached the midnight position. In those 88 Earth days, Mercury would have completed one orbit around the sun (Figure 21-14). It would require another entire orbit of Mercury for the sun to return to the noon position overhead. So a full day on Mercury is two Mercury years long!

The complex tidal coupling between the rotation and revolution of Mercury is an important illustration of the power of tides. Just as the tides in the Earth–

moon system have slowed the moon's rotation and locked it to Earth, so have the sun–Mercury tides slowed the rotation of Mercury and coupled its rotation to its revolution. We can refer to such a relationship as a **resonance.** We will see many such resonances as we explore the solar system.

Like its rotation, Mercury's orbital motion is also peculiar. In Chapter 5, we discussed the orbital motion of Mercury and saw that its elliptical orbit precesses faster than can be explained by Newton's laws but at just the rate predicted by Einstein's theory of general relativity. Thus, the orbital motion of Mercury is taken as a confirmation of the curvature of space-time as predicted by general relativity.

THE SURFACE OF MERCURY

Because the days are so long on Mercury and because it is close to the sun and too small to keep an atmosphere, the temperatures on Mercury are extreme. The highest daytime temperatures can exceed 600 K (330°C), although 500 K is a more usual high temperature. In the dark nights, with no atmosphere to retain heat, the surface can cool to 100 K (−173°C). Thus, the surface of Mercury cycles from very high temperatures to very low temperatures each Mercury day.

In appearance, Mercury looks much like Earth's moon. It is heavily battered, with craters of all sizes, including some large basins. Some craters are obviously old and degraded; others seem quite young and have bright rays of ejecta. However, a quick glance at photos of Mercury shows no large, dark maria like the moon's flooded basins.

The largest basin on Mercury is called Caloris Basin after the Latin word for "heat," recognition of its location at one of the two "hot poles," which face the sun at alternate perihelions. At the times of the Mariner encounters, the Caloris Basin was half in shadow (▌Figure 21-15a). Although half cannot be seen, the low angle of illumination is ideal for the study of the lighted half because it produces dramatic shadows.

The Caloris Basin is a gigantic multiringed impact basin 1300 km in diameter with concentric mountain rings up to 3 km high. The impact threw ejecta 600 to 800 km across the planet, and the focusing of seismic waves on the far side produced peculiar terrain (called lineated terrain) that looks much like the jumbled surface of the moon (Figure 21-15b and c). The Caloris Basin is partially filled with lava flows. Some of this lava may be material melted by the energy of the impact, but some may be lava from below the crust that leaked up through cracks. The weight of this lava and the response of the crust have produced deep cracks in the central lava plains. The geophysics of such large, multiringed crater basins is not well understood at present, but Caloris Basin seems to be the same kind of structure as the Imbrium Basin on the moon, but it has not been as deeply flooded.

The Rotation of Mercury

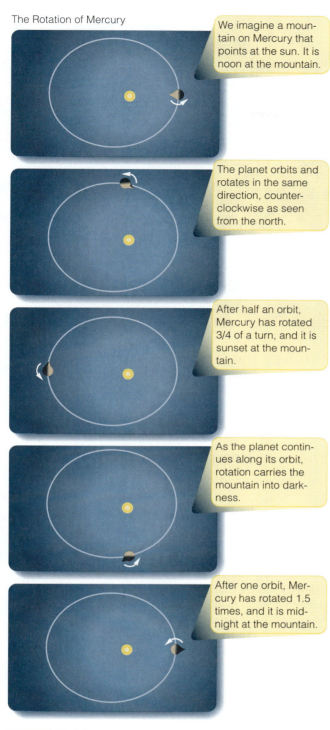

We imagine a mountain on Mercury that points at the sun. It is noon at the mountain.

The planet orbits and rotates in the same direction, counter-clockwise as seen from the north.

After half an orbit, Mercury has rotated 3/4 of a turn, and it is sunset at the mountain.

As the planet continues along its orbit, rotation carries the mountain into darkness.

After one orbit, Mercury has rotated 1.5 times, and it is midnight at the mountain.

FIGURE 21-14

Mercury's orbital motion and rotation are related. It orbits the sun in 88 days and rotates on its axis in two-thirds of that time. Yet one full day on Mercury from noon to noon takes two full orbits.

When planetary scientists began looking at the Mariner photographs in detail, they discovered something not seen on the moon. Mercury is marked by great curved cliffs called **lobate scarps** (❚ Figure 21-16). These seem to have formed when the planet cooled and shrank by a few kilometers, wrinkling its crust as a

FIGURE 21-15

(a) Only part of the giant Caloris Basin on Mercury is visible in this Mariner 10 photomosaic. The center of the basin is in darkness just beyond the lower left corner of the photo, and the outer ring is marked by a dashed line. This giant crater basin was produced by a large impact and subsequent lava flows. (b) Seismic waves from the impact were focused on the far side of Mercury and produced the jumbled terrain shown in part c. *(NASA)*

a

b

c

FIGURE 21-16

A lobate scarp (arrow) crosses craters, indicating that Mercury cooled and shrank, wrinkling its crust, after many of its craters had formed. (NASA)

Visual-wavelength image

drying apple wrinkles its skin. Some of these scarps are as high as 3 km and reach hundreds of kilometers across the surface. Other faults in Mercury's crust are straight and may have been produced by tidal stresses generated when the sun slowed Mercury's rotation.

THE PLAINS OF MERCURY

The most striking difference between Mercury and the moon is that Mercury lacks the great dark lava plains so obvious on the moon. Under careful examination, the Mariner 10 photographs show that Mercury has plains, two different kinds, in fact, but they are different from the moon's. Understanding the differences is the key to understanding the history of Mercury.

Much of Mercury's surface is old, cratered terrain (▌Figure 21-17), but other areas called **intercrater plains** are less heavily cratered. These plains are marked by meteorite craters less than 15 km in diameter and secondary craters produced by chunks of ejecta from larger impacts. Unlike the heavily cratered regions, the intercrater plains are not totally saturated with craters. This suggests that the intercrater plains were produced by lava flows that buried older terrain.

In contrast, smaller regions called **smooth plains** appear to be even younger than the intercrater plains. They have fewer craters and appear to be ancient lava flows. Much of the region around the Caloris Basin is composed of these smooth plains (Figure 21-17), and they appear to have formed soon after the Caloris impact.

Given the available evidence, most planetary astronomers believe that the plains of Mercury are solidified lava flows, in which case they are much like the maria on the moon. Unlike the maria, Mercury's lava plains are not significantly darker than the rest of the crust. Except for a few bright crater rays, Mercury's surface is a uniform gray with an albedo of only about 0.1. Thus, Mercury's lava plains are not as dramatically obvious on photographs as the much darker maria on our own moon.

THE INTERIOR OF MERCURY

One of the most striking differences between Mercury and the moon is the difference between their interiors. We have seen that the moon is a low-density world that contains no more than a small core of metals. Yet evidence suggests that Mercury has a large metallic core.

Mercury is over 60 percent denser than the moon. Yet Mercury's surface appears to be normal rock, so we must conclude that its interior contains a large core of dense metals, probably iron. In proportion to its size, Mercury must have a larger metallic core than Earth (see diagram in Data File Four).

If Mercury has a large metallic core that remains molten, then the dynamo effect should generate a magnetic field (Chapter 20). The Mariner 10 spacecraft, however, found a magnetic field only about 0.5 percent as strong as Earth's, and this weak field makes it difficult to understand the nature of the interior. Because Mercury is a relatively small world, about halfway between the size of Earth and the size of the moon, it should have lost most of its internal heat long ago. But if its metallic core is not molten, we would not expect to detect a magnetic field as strong as 0.5 percent of Earth's field. The verdict is still out on the nature of Mercury's large metallic core and weak magnetic field.

Similarly, it is difficult to explain the proportionally large size of the metallic core inside Mercury. In

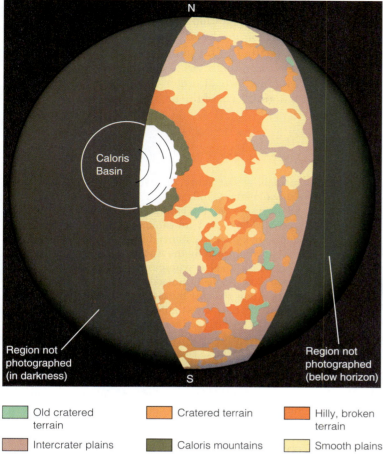

N

Caloris
Basin

Region not
photographed
(in darkness)

Region not
photographed
(below horizon)

S

| | Old cratered terrain | | Cratered terrain | | Hilly, broken terrain |
| | Intercrater plains | | Caloris mountains | | Smooth plains |

FIGURE 21-17

A geological map of part of Mercury shows the Caloris Basin at left and the hilly, broken terrain associated with that impact. The intercrater plains are interrupted by smooth plains and more heavily cratered terrain.

most part of the solar nebula, and, as we have seen, a giant impact may have robbed it of some of its lower-density rock and left it a small, dense world with a large metallic core.

Like the moon, Mercury suffered heavy cratering by debris in the young solar system. We don't know accurate ages of features on Mercury, but we can assume that cratering, the second stage of planetary evolution, occurred over the same period as did the cratering on the moon—during the heavy bombardment that lasted from formation until about 4 billion years ago. This intense cratering declined rapidly as the planets swept up the last of the debris left over from planet building.

The cratered surface of Mercury is not exactly like that of the moon. Because of Mercury's stronger gravity, the ejecta from a crater on Mercury would be thrown only about 65 percent as far as on the moon, and thus the ejecta from an impact does not blanket as much of the surface. Also, the intercrater plains appear to have formed when lava flows occurred during the early bombardment, burying the older surface, and then accumulated more craters. Sometime near the end of cratering, a planetesimal over 100 km in diameter smashed into the planet and blasted out the great multi-ringed Caloris Basin. Only parts of that basin have been flooded by lava flows. Finally, as cratering rapidly declined, the cooling interior contracted, and the crust broke to form the lobate scarps.

The third stage, flooding, began with lava flows filling some lowlands. The Caloris impact may have been so big it fractured the crust and triggered still more outpourings of molten lava. These later lava flows formed the smooth plains. Nevertheless, the age of lava flooding ended quickly, perhaps because the shrinkage of the planet squeezed off the lava channels to the surface. Mercury lacks a true atmosphere, so we would not expect flooding by water, but radar images show a bright spot at the planet's north pole (❙ Figure 21-18) that may be caused by ice trapped in perpetually shaded crater floors where the temperature never exceeds 60 K (−350°F).

The fourth stage in the story of Mercury, slow surface evolution, is now limited to micrometeorites, which grind the surface to dust; rare larger meteorites, which leave bright-rayed craters; and the slow but intense cycle of heat and cold, which weakens the rock at the surface. The planet's crust is now thick, and although its core may be partially molten, the heat flowing outward is unable to drive plate tectonics that would actively erase craters and build folded mountain ranges.

Chapter 19, we saw that planets that formed near the sun should incorporate more metals and thus be denser, but some models suggest that this cannot account for all of Mercury's large metallic core. One hypothesis that has been proposed involves a giant impact when Mercury was young, an impact much like the planet-shattering impact proposed to explain the origin of the moon. If the forming planet had differentiated and was then struck by a large planetesimal, the impact could have shattered the crust and mantle and blasted much of the lower-density material into space. The denser core could have survived, re-formed, and attracted some of the lower-density debris to form a thin mantle and so leave Mercury with a deficiency of low-density crust and mantle rock. Thus, both Mercury and the moon may be products of giant impacts.

A HISTORY OF MERCURY

Having reviewed the evidence, we must now try to tell the story of Mercury. Presumably, it formed in the inner-

FIGURE 21-18
A radar map of Mercury reveals a bright region at its north pole (top). Observers suspect that ice deposits may survive just under the surface or in regions that never receive direct sunlight. *(Martin Slade, JPL)*

Mercury, like the moon, is now an inactive world. Impact cratering, internal heat, and possible giant impacts have left their marks on these worlds, but they are now nearly dead. To find planets where the heat flowing outward from the interior still drives geological activity, we must look at larger worlds such as Venus and Mars, the subjects of the next chapter.

SUMMARY

The moon is, at first glance, dramatically different from Earth. We see two distinct kinds of terrain on the moon. The maria, named after seas, are lowland plains with few craters. They are great lava flows and are younger than the highlands. The highlands are lighter in color and heavily cratered.

Samples returned by the Apollo missions show that the maria are dark basalts, whereas the highlands are lighter-colored and lower-density minerals. Many samples are breccias, which show how severely the moon's crust has been fractured.

The ages of the mare samples range from about 3 to 3.8 billion years, but the highland samples are about 4.0 to 4.3 billion years old. Few samples are older than 4.4 billion years. This suggests that the moon's surface was molten until about 4.4 billion years ago.

As the moon's early crust solidified, cratering battered it and excavated great basins. Later, molten rock rose through fractures in the crust and flooded the basins to produce the maria. By the time the maria formed, cratering had declined, so the maria are marked by few craters.

The lunar surface now is evolving very slowly. Because the moon is small, it has cooled rapidly, its crust has grown thick, and it has never divided into moving plates. Now only meteorites alter the surface.

The large-impact hypothesis suggests that the Earth–moon system formed in the collision of two large planetesimals. The differentiated bodies formed a single large object that became the higher-density, iron-rich Earth. Ejected mantle material, poor in iron, formed a disk and eventually condensed to form the lower-density, iron-poor moon.

Mercury is a small world, only about 40 percent larger than the moon. Like the moon, it has lost any permanent atmosphere it might have had, and its interior has cooled rapidly. Debris from the formation of the solar system cratered Mercury's surface as it did the moon's, and lava flows buried parts of that terrain under the intercrater plains. The Caloris Basin, formed near the end of cratering, is the largest crater basin on Mercury. Soon after the Caloris impact, lava flows created the smooth plains. These lava plains have about the same color as the intercrater plains, so the contrast between plains and highlands is not as obvious on Mercury as it is on the moon.

The interior of Mercury must contain a large metallic core to account for the planet's high density. A slight shrinkage in the diameter of the planet at the time of the cooling of the core may have led to wrinkling of the crust. This would account for the lobate scarps that mark all of the photographed parts of Mercury.

REVIEW **CRITICAL INQUIRY**

Why don't Earth and the moon have lobate scarps?

At first glance, we might suppose that any world with a large metallic interior should have lobate scarps. When the metallic core cools and contracts, the world should shrink, and the contraction should wrinkle and fracture the brittle crust to form lobate scarps.

Of course, Earth has a fairly large metallic core; but, being a large world, it has not cooled very much, and the crust is not brittle. In fact, the heat flowing out of Earth's interior keeps the crust thin, flexible, and active. If any lobate scarps ever formed on Earth, they would have been quickly destroyed by plate tectonics.

The moon, unlike Mercury, does not have a large metal component in its interior. It may be that the rocky interior did contract slightly as the moon lost its internal heat, but that smaller contraction may not have produced major lobate scarps. Also, any surface features that formed early in the moon's history would have been destroyed by the heavy bombardment.

We can, in a general way, understand lobate scarps, but not all materials contract when they cool and solidify. Water, for instance, expands when it freezes. What kinds of features would you expect to find on small, cold, icy worlds that once had interiors of liquid water? Such icy worlds exist as the moons of the outer planets, and we will explore them in Chapter 23.

NEW TERMS

tidal coupling

terminator

limb

mare

sinuous rille

ejecta

ray

secondary crater

micrometeorite

multiringed basin

relative age

absolute age

vesicular basalt

anorthosite

breccia

regolith

jumbled terrain

fission hypothesis

condensation hypothesis

capture hypothesis

large-impact hypothesis

resonance

lobate scarp

intercrater plain

smooth plain

REVIEW QUESTIONS

Ace⊘Astronomy™ Assess your understanding of this chapter's topics with additional quizzing and animations at **http:// astronomy.brookscole.com/seeds8e**

1. How do we find the relative ages of the moon's maria and highlands?

2. How do we know that Copernicus is a young crater?

3. Why did the first Apollo missions land on the maria? Why were the other areas of more scientific interest?

4. Why do we suppose that the moon formed with a molten surface?

5. Why are so many lunar samples breccias?

6. What do the vesicular basalts tell us about the evolution of the lunar surface?

7. What evidence would we expect to find on the moon if it had been subjected to plate tectonics? Do we find such evidence?

8. Cite objections to the fission, condensation, and capture hypotheses.

9. How does the large-impact hypothesis explain the moon's lack of iron?

10. How does the tidal coupling between Mercury and the sun differ from that between the moon and Earth?

11. What is the difference between the intercrater plains and the smooth plains in terms of time of formation?

12. What evidence do we have that Mercury has a partially molten, metallic core?

DISCUSSION QUESTIONS

1. Old science-fiction paintings and drawings of colonies on the moon often showed very steep, jagged mountains. Why did the artists assume that the mountains would be more rugged than mountains on Earth? Why are lunar mountains actually less rugged than mountains on Earth?

2. From your knowledge of comparative planetology, propose a description of the view that astronauts would have if they landed on the surface of Mercury.

PROBLEMS

1. Calculate the escape velocity of the moon from its mass and diameter. (*Hint:* See Chapter 5.)

2. Why do small planets cool faster than large planets? Compare surface area to volume.

3. The smallest detail visible through Earth-based telescopes is about 1 second of arc in diameter. What size is this on the moon? (*Hint:* See Chapter 3.)

4. The trenches where Earth's seafloor slips downward are 1 km or less wide. Could Earth-based telescopes resolve such features on the moon? Why are we sure that such features are not present on the moon?

5. The Apollo command module orbited the moon about 200 km above the surface. What was its orbital period? (*Hint:* See Chapter 5.)

6. From a distance of 200 km above the surface of the moon, what is the angular diameter of an astronaut in a space suit? Could we have seen the astronauts from the command module? (*Hint:* See Chapter 3.)

7. If we transmitted radio signals to Mercury when it was closest to Earth and waited to hear the radar echo, how long would we wait?

8. Suppose we sent a spacecraft to land on Mercury and it transmitted radio signals to us at a wavelength of 10 cm. If we saw Mercury at its greatest angular distance west of the sun, to what wavelength would we have to tune our radio telescope to detect the signals? (*Hints:* See Data File Four, and the Doppler shift in Chapter 7.)

9. What would the wavelength of maximum be for infrared radiation from the surface of Mercury? How would that differ for the moon?

10. Calculate the escape velocity from Mercury. How does that compare with the escape velocity from the moon and from Earth?

CRITICAL INQUIRIES FOR THE WEB

1. What did the Apollo astronauts do while they were on the moon? Search for Web sites that describe the experiments performed during the astronauts' extravehicular activities (EVAs). What types of experiments/activities were performed? What did we learn from them?

2. Search for maps of the moon that show geological information such as chemical composition, elevation, age, and so on. Do the maps confirm your understanding of the lunar surface?

3. What would it be like to walk on the lunar surface? Apollo astronauts visited six different locations on the moon exploring the variations in lunar terrain. Describe the horizons and general relief of the landing locations of the different missions by exploring Web sites that provide lunar surface photography from the missions. What differences do you see between images from landings in highlands and in maria?

4. What plans are being made to send a spacecraft to Mercury? What kind of observations will be made, and what theories will be tested?

EXPLORING *THESKY*

1. Locate the moon in the sky. What constellation is it in? Find the moon's phase as a percentage of full, its angular diameter, and its distance from Earth. Use **Find** to locate the moon.

2. Follow the moon through one complete orbit measuring its distance and angular diameter. How do these change? Under **Tools,** use the **Moon Phase Calendar.** Change dates by clicking on **Data** and then **Site Information.**

3. Locate Mercury and zoom in until you can see its phase. Follow it as it orbits the sun and describe how its phase changes with its orbital position. Use **Find** to locate Mercury.

 Visit the Seeds *Foundations of Astronomy* companion Web site for critical thinking exercises, articles, and additional readings from InfoTrac College Edition, Brooks/Cole's online student library.

VENUS AND MARS

There wasn't a breath in that land of death . . .

Robert Service, The Cremation of Sam McGee

NASA

GUIDEPOST

The previous chapter grouped Earth's moon and Mercury together because they are similar worlds. This chapter groups Venus and Mars together because we might expect them to be similar. They are Earthlike in their size and location in the solar system, so it is astonishing to see how different they actually are. Much of this chapter is aimed at understanding how Venus and Mars evolved to their present states.

Neither Venus nor Mars can tell us much about the formation of the planets. Both planets have evolved since they formed. Nevertheless, we find further hints that the solar system was a dangerous place, with major impacts smashing the surfaces of the planets, a process we first suspected when we studied the moon and Mercury.

This chapter concludes our exploration of the Earthlike worlds. In the next two chapters, we will visit planets that give new meaning to the word "unearthly."

VIRTUAL LABORATORIES

PLANETARY GEOLOGY

PLANETARY ATMOSPHERES AND THEIR RETENTION

Without a space suit, you could live only about 30 seconds on the surface of Mars. The temperature might not be too bad. On a hot summer day at noon, the temperature might reach 20°C (68°F). But the air is 96 percent carbon dioxide, is deadly dry, and contains almost no oxygen. Also, the air pressure at the surface of Mars is only 0.01 that at the surface of Earth. Stepping out onto the surface of Mars without a space suit would be about the same as opening the door of an airplane flying at 30,000 m (100,000 ft) above Earth. (Commercial jets do not generally fly above 12,000 m, or 38,000 ft.)

No space suit would save you on Venus. The air pressure is 90 times Earth's, and the air is almost entirely carbon dioxide, with traces of various acids. Worse yet, the surface is hot enough to melt lead.

Why are Venus and Mars such rotten places for picnics? In many ways, they resemble Earth. They are made of silicates and metals, have atmospheres, and lie in the inner solar system. We might expect them to be similar. How did they get to be so different from each other and from Earth? Comparative planetology will give us some clues.

When we used comparative planetology to study the moon and Mercury in the previous chapter, we saw the importance of impact cratering and the outward flow of heat. As we study Venus and Mars, we will see the same effects, but they will be given new significance because of the larger size of Venus and Mars. In this chapter, we will discover the important principle that the size of a world determines how fast it loses its internal heat. Venus, roughly the size of Earth, has lost internal heat slowly, but Mars, intermediate in size between Earth and the moon, is nearly as cold and dead as the moon. Furthermore, these larger worlds have enough gravity to hold onto an atmosphere, and that alters the evolution of surface features.

Venus and Mars are more than just two more planets to add to our gallery of weird worlds. They give us an entirely new perspective on comparative planetology and thus a new perspective on Earth.

22-1 VENUS

An astronomer once became annoyed when someone referred to Venus as a planet gone wrong. "No," she argued, "Venus is probably a fairly normal planet. It is Earth that is peculiar. The universe probably contains more planets like Venus than like Earth." To under-

Ace⏾Astronomy™ The AceAstronomy icon throughout the text indicates an opportunity for you to test yourself on key concepts and to explore animations and interactions on the AceAstronomy Web site at **http://astronomy.brookscole .com/seeds8e**

stand how unusual, how special Earth is, we have only to compare it with Venus.

In many ways, Venus is a twin of Earth, and we might expect the two planets to be quite similar. Venus is nearly 95 percent the diameter of Earth (❙ Data File Five) and has a similar average density. It formed in the same general part of the solar nebula, so we might expect it to have a composition similar to Earth's. Planets of Earth's size cool slowly, so we might expect Venus to have a molten metallic interior and an active crust.

Unfortunately, we cannot see the surface of Venus because of perpetual thick, yellow-tinted white clouds that completely envelop the planet. From the time of Galileo to the early 1960s, astronomers could only speculate about Earth's twin, and science fiction writers supposed that Venus was a steamy swamp planet inhabited by strange creatures.

During the last few decades, astronomers have peered below the clouds using radar, 23 spacecraft have flown past or orbited Venus, and over a dozen have landed on its surface. The resulting picture of Venus is dramatically different from the murky swamps of fiction. In fact, the surface of Venus is drier than any desert on Earth and two times hotter than a hot kitchen oven.

If Venus is not a planet gone wrong, it is certainly a planet gone down a different evolutionary path. How did Earth's twin become so different?

THE ROTATION OF VENUS

Nearly all of the planets in our solar system rotate counterclockwise as seen from the north. Uranus and Pluto are exceptions, and so is Venus.

In 1962, radio astronomers were able to transmit a radio pulse of precise wavelength toward Venus and detect the echo returning some minutes later. That is, they detected Venus by radar. The echo was not a precise wavelength, however. Part of the reflected signal had a longer wavelength, and part had a shorter wavelength. Evidently the planet was rotating; radio energy reflected from the receding edge was red shifted, and radio energy reflected from the approaching edge was blue shifted. From this Doppler effect, the radio astronomers could tell that Venus was rotating once every 243.01 days. Because the west edge of Venus produced the blue shift, the planet had to be rotating in the backward direction.

This rotation period turns nearly the same face toward Earth each time the planets come closest together, and it is tempting to imagine that there is a tidal resonance locking the rotation of Venus to the orbital motion of Earth. However, a true resonance would require that Venus rotate 0.06 percent slower.

Why does Venus rotate backward? For decades, textbooks have suggested that proto-Venus was set spinning backward when it was struck off center by a large planetesimal. That is a reasonable possibility; we have suggested that a similar collision gave birth to Earth's moon. But there is an alternative. Sophisticated mathematical models suggest that a terrestrial planet with a molten core and a dense atmosphere can have its rotation gradually reversed by solar tides in its atmosphere. (Notice the contrast between the catastrophic theory of a giant impact and the evolutionary theory of atmospheric tides.)

Whichever theory explains the backward rotation of Venus, the near resonance with the motion of the Earth seems to be a coincidence. Earth is too small and too distant to raise tides on Venus and affect its rotation.

THE ATMOSPHERE OF VENUS

Although Venus is Earth's twin in size, the composition, temperature, and density of its atmosphere make it the most inhospitable of planets. About 96 percent of its atmosphere is carbon dioxide, and 3.5 percent is nitrogen. The remaining 0.5 percent is water vapor, sulfuric acid (H_2SO_4), hydrochloric acid (HCl), and hydrofluoric acid (HF). In fact, the thick clouds that hide the surface are believed to be composed of sulfuric acid droplets and microscopic sulfur crystals.

Soviet and American spacecraft have dropped probes into the atmosphere of Venus, and those probes have radioed data back to Earth as they fell toward the surface. These studies show that the cloud layers are much higher and much more stable than those on Earth. The highest layer of clouds, the layer we see from Earth, extends from 68 to 58 km above the surface (❙ Figure 22-1). For comparison, the highest clouds on Earth do not extend higher than about 16 km. Another cloud layer lies from 56 to 52 km, and a third, the densest, from 52 to 48 km. Thin haze extends down to about 33 km, below which the air is surprisingly clear.

These cloud layers are highly stable because the atmospheric circulation on Venus is much more regular than that on Earth. The heated atmosphere at the **subsolar point,** the point on the planet where the sun is directly overhead, rises and spreads out in the upper atmosphere. Convection circulates this gas toward the dark side of the planet and the poles, where it cools and sinks. This circulation produces 300-km/h jet streams in the upper atmosphere, which carry the atmosphere from east to west (the same direction the planet rotates) so rapidly that the entire atmosphere seems to rotate with a period of only four days. As evidence of these circulations, we can see a distinctive Y pattern in ultraviolet images of Venus such as that in Data File Five.

The high-speed winds in the upper atmosphere were studied in June 1985 when the Vega 1 and Vega 2 spacecraft, on their way to rendezvous with Comet Halley, flew past Venus and dropped instrumented balloons into the upper atmosphere. Floating at an altitude of 55 km, the 3.5-m-diameter balloons found winds as high as 240 km/h. In the 46 hours the balloons

UV image

VENUS
Data File Five

Venus is only 5 percent smaller than Earth, but its surface is a blazing hot desert and its atmosphere is perpetually cloudy. *(NASA)*

Average distance from the sun	0.7233 AU (1.082 × 10⁸ km)
Eccentricity of orbit	0.0068
Maximum distance from the sun	0.7282 AU (1.089 × 10⁸ km)
Minimum distance from the sun	0.7184 AU (1.075 × 10⁸ km)
Inclination of orbit to ecliptic	3°23′40″
Average orbital velocity	35.03 km/s
Orbital period	0.61515 y (224.68 days)
Period of rotation	243.01 days (retrograde)
Inclination of equator to orbit	177°
Equatorial diameter	12,104 km (0.95 $D_\oplus$)
Mass	4.870 × 10²⁴ kg (0.815 $M_\oplus$)
Average density	5.24 g/cm³ (4.2 g/cm³ uncompressed)
Surface gravity	0.903 Earth gravity
Escape velocity	10.3 km/s (0.92 $V_\oplus$)
Surface temperature	745 K (472°C, or 882°F)
Albedo (cloud tops)	0.76
Oblateness	0

a

UV image

b

ACTIVE FIGURE 22-1

(a) This ultraviolet image of Venus was made by the Galileo spacecraft. *(NASA)* (b) The four principal cloud layers in the atmosphere of Venus are shown at left. Thermometers inserted into the atmosphere at different levels would register the temperatures indicated by the red line in the graph at the right. The blue dashed line shows the temperatures in Earth's atmosphere for comparison. Note the difference in temperatures at the surfaces of the planets.

Ace🌀Astronomy™ Go to AceAstronomy and click Active Figures to see "Planetary Atmospheres" and explore this figure.

survived, they were carried one-third of the way around Venus by the high-altitude winds.

The details of this atmospheric circulation are not well understood, but it seems that the slow rotation of the planet is an important factor. On Earth, large-scale circulation patterns are broken into smaller cyclonic disturbances by Earth's rapid rotation. Because Venus rotates more slowly, its atmospheric circulation is not broken up into small cyclonic storms but instead is organized as a planetwide wind pattern.

Although the upper atmosphere is cool, the lower atmosphere is quite hot (Figure 22-1b). Instrumented probes that have reached the surface report that the temperature is 745 K (880°F), and the atmospheric pressure is 90 times that of Earth. Earth's atmosphere is 1000 times less dense than water, but on Venus the air is only 10 times less dense than water. If we could survive the unpleasant composition, intense heat, and high pressure, we could strap wings to our arms and fly.

The present atmosphere of Venus is extremely dry, but there is evidence that it once had significant amounts of water. As a Pioneer Venus spacecraft descended through the atmosphere of Venus in 1978, it detected about 150 times more deuterium per hydrogen atom than on Earth. This abundance of deuterium, the heavy isotope of hydrogen, could develop if ultraviolet radiation in sunlight broke water molecules into hydrogen and oxygen. The hydrogen would escape into space, but the heavier deuterium would escape more

slowly than normal hydrogen. Thus, the excess of deuterium is evidence that Venus once had more water. Calculations suggest that it may have once had enough water to make a planetwide ocean up to 25 meters deep. (For comparison, the water on Earth would make a uniform planetwide ocean 3000 meters deep.) Without an ozone layer to protect the atmosphere from ultraviolet radiation, the water was broken up and much of the hydrogen was lost to space. The oxygen presumably formed oxides in the soil. Thus, Venus is now a deadly dry world with only enough water vapor to make an ocean 0.3 meter deep.

Even if Venus once had lots of water, it was probably too warm for that water to fall from the atmosphere as rain and form oceans, and that lack of oceans is the biggest difference between Earth and Venus. Earth's oceans have made it a living paradise. The lack of oceans on Venus, through the greenhouse effect, has made it a hell.

THE VENUSIAN GREENHOUSE

We saw in Chapter 20 how the greenhouse effect warms Earth. Carbon dioxide (CO_2) is transparent to light but opaque to infrared (heat) radiation. Thus, energy can enter the atmosphere as light and warm the surface, but the surface cannot radiate an equal amount of energy back to space because of the CO_2. Like Earth, Venus has a greenhouse effect, but on Venus the effect is fearsome.

Whereas Earth's atmosphere contains only about 0.03 percent CO_2, the atmosphere of Venus contains 96 percent CO_2, and the surface of Venus is nearly twice as hot as a hot kitchen oven.

When Venus was young, it may have been cooler than it is now, but because it formed 30 percent closer to the sun than did Earth, it was warmer than Earth, and that unleashed processes that made it even hotter. As we saw in the previous section, even if Venus once had water, it probably never had large liquid-water oceans. CO_2 is highly soluble in water, but without large oceans to dissolve CO_2 and remove it from the atmosphere, Venus was trapped with a CO_2-rich atmosphere, and the greenhouse effect made the planet warmer. As the planet warmed, any small oceans that did exist evaporated, and Venus lost the ability to cleanse its atmosphere of CO_2. Volcanoes on Earth vent gases that are mostly CO_2 and water vapor, and presumably the volcanoes on Venus did the same. The water vapor was broken up by ultraviolet radiation, and the CO_2 accumulated in the atmosphere. The high temperature baked even more CO_2 out of the surface, and the atmosphere became even less transparent to infrared radiation. Consequently, the temperature rose further. This runaway greenhouse effect has made the surface so hot that even sulfur, chlorine, and fluorine have baked out of the rock and formed sulfuric, hydrochloric, and hydrofluoric acid vapors.

Earth avoided this runaway greenhouse effect because it was farther from the sun and cooler. Thus, it could form and preserve liquid-water oceans to absorb the carbon dioxide, which left a nitrogen atmosphere that was relatively transparent in some parts of the infrared. If all of the carbon dioxide in Earth's sediments was put back into the atmosphere, our air would be as dense as that of Venus, and we would suffer from a severe greenhouse effect. Recall from Chapter 20 how our use of fossil fuels and our destruction of forests is increasing the CO_2 concentration in our atmosphere and warming the planet. Venus warns us of what a greenhouse effect can do.

THE SURFACE OF VENUS

Given that the surface of Venus is hidden perpetually by clouds, is hot enough to melt lead, and suffers under crushing atmospheric pressure, it is surprising how much planetary scientists know about the geology of Venus. Early radar maps made from Earth penetrated the clouds and showed that it had mountains, plains, and some craters. The Soviet Union launched a number of spacecraft that landed on Venus, and although the spacecraft failed within an hour or so of landing, they did analyze the rock and snap a few photos. The rock seems to be basalt, typical of volcanism, and the photographs revealed dark-gray rocky plains bathed in a deep-orange glow caused by sunlight filtering down through the thick atmosphere (Figure 22-2).

Real geological studies of Venus became possible when spacecraft reached the planet. Both the United States and the Soviet Union sent spacecraft to Venus. Some probed the atmosphere, and others landed on the surface and sent back data before succumbing to the intense heat and pressure. Some of the most fruitful missions used radar to penetrate the clouds and map the surface. The NASA Pioneer Venus probe orbited Venus in 1978 and made radar maps showing features as small as 25 km in diameter. Later two Soviet Venera spacecraft mapped the north polar regions with a resolution of 2 km. From 1992 to 1994, the NASA Magellan spacecraft orbited Venus and created radar maps showing details as small as 100 m. These radar maps give us a unique look below the clouds.

a

b

Visual-wavelength images

FIGURE 22-2
The color of the surface of Venus appears orange in the original Venera 13 photograph (a), but analysis shows that the color is produced by the thick atmosphere. The corrected image (b) reveals slabs of gray rock. *(C. M. Pieters, through the Brown/Vernadsky Institute to Institute Agreement and the U.S.S.R. Academy of Sciences)*

FIGURE 22-3

Venus without its clouds: This mosaic of Magellan radar maps has been given an orange color to mimic the sunset coloration of daylight at the surface of the planet. The image shows scattered impact craters and volcanic regions such as Beta Regio and Atla Regio. *(NASA)*

The color of Venus radar maps is mostly arbitrary. Our eyes can't see radio waves, so the radar maps must be given some false colors to distinguish height or roughness or composition. Some maps use blues and greens for lowlands and yellows and reds for highlands. Coloring the lowlands blue like oceans on Earth maps might be deceptive, however, because there is no liquid water on Venus. Rather, Magellan scientists chose to use yellows and oranges for their radar maps in an effort to mimic the orange color of daylight caused by the thick atmosphere (❙ Figure 22-3). When we look at these orange images we need to remind ourselves that the true color of the rock is dark gray.

Radar maps of Venus can tell us a number of things about the surface. If we transmit a radio signal down through the clouds and measure the time until we hear the echo coming back up, we can measure the altitude of the surface. Part of the Magellan data is a detailed altitude map of the surface. We can also measure the amount of the signal that is reflected from each spot on the surface. Much of the surface of Venus is smooth lava flows that do not look bright in radar maps, but faults and uneven terrain look brighter. Some extremely rough lava flows contain billions of tiny crevices (❙ Figure 22-4) that bounce the radar signal around and shoot it back the way it came. These rough lava flows look bright in radar maps. Certain mineral deposits are also bright; consequently, radar maps do not tell us how the surface would look to our eyes, but rather give us information about altitude, roughness, and, in some cases, chemical composition.

❙ Figure 22-5 is a map of all of Venus except the polar regions. (Note that this map has been color coded by altitude.) By international agreement, the names of celestial bodies and features on celestial bodies are assigned by the International Astronomical Union, which has decided that all names on Venus should be female. There are only a few exceptions, such as the mountain Maxwell Montes, 50 percent higher than Mt. Everest. It was discovered in early Earth-based radar maps and named for James Clark Maxwell, the 19th-century physicist who first described electromagnetic radiation. Alpha Regio and Beta Regio were also discovered by radar before the naming convention was adopted. Other names on Venus are female, such as the highland regions Ishtar Terra and Aphrodite Terra, named for the Babylonian and Greek goddesses of love.

Radar maps show that the surface of Venus consists of low, rolling plains and highland regions. By comparison with Earth, we might be tempted to think of these regions as ocean floors and continents, but there is no liquid water on Venus. Rather, the rolling plains appear to be large-scale smooth lava flows, and the highlands are regions of deformed crust.

Craters on Venus tell us the surface is not old. With nearly 1000 impact craters on its surface, Venus has

FIGURE 22-4
Although it is nearly 1000 years old, this lava flow near Flagstaff, Arizona, is still such a rough jumble of sharp rock that it is dangerous to venture onto its surface. Rough surfaces are very good reflectors of radio waves and look bright in radar maps. Lava flows on Venus show up as bright regions in the radar maps because they are rough. *(M. A. Seeds)*

more craters than Earth but not nearly as many as the moon. The craters are uniformly scattered over the surface and look sharp and fresh (❙ Figure 22-6). With no water and a thick, sluggish atmosphere, there is little erosion on Venus, and the thick atmosphere protects the surface from small meteorites. Consequently there are no small craters. Putting all of this observational evidence together, planetary scientists conclude that the surface of Venus is older than Earth's surface but not nearly as old as the moon's. Unlike the moon, there are no old, cratered highlands. Lava flows have covered the entire surface within the last half-billion years.

VOLCANISM ON VENUS

Volcanism seems to dominate the surface of Venus. Much of Venus is covered by lava flows such as those photographed by the Venera landers (Figure 22-2). Also, volcanic peaks and other volcanic features are evident in the radar maps.

FIGURE 22-5
Magellan radar images have been combined and color coded to form this map of Venus. North and south polar regions are not shown. The highlands (yellow and red) consist mainly of Ishtar Terra, Aphrodite Terra, and Lada Terra. The dry lowlands are colored green and blue in this map. *(NASA)*

FIGURE 22-6

Impact crater Howe in the foreground of this Magellan radar image is 37 km in diameter. Craters in the background are 47 km and 63 km in diameter. This radar map has been digitally modified to represent the view from a spacecraft flying over the craters. *(NASA)*

As usual, we learn more about other worlds by comparing them with each other and with Earth. Study ▌ "Volcanoes" on pages 480 and 481 and notice three important ideas. First, there are two main types of volcanoes found on Earth. Second, notice that we can recognize the volcanoes on Venus and Mars by their shapes, even when the images are manipulated in computer

mapping (Window 22-1). The shield volcanoes found on Venus and Mars are the kind produced by hot-spot volcanism and not by plate tectonics. This is evidence that neither Venus nor Mars has been dominated by plate tectonics as on Earth. The third thing to notice is the large size of volcanoes on Venus and Mars. This too is evidence against moving tectonic plates on those worlds.

The radar image of Sapas Mons in ▌ Figure 22-7 shows a dramatic overhead view of this volcano, which is 400 km (250 mi) in diameter at its base and 1.5 km (0.9 mi) high. Many bright lava flows radiate outward covering older, less bright flows. Sapas Mons and other nearby volcanoes are located along a system of faults where rising magma breaks through the crust.

In addition to actual volcanos, radar images reveal other volcanic features on the surface of Venus. Lava channels are common on Venus, and they appear similar to the sinuous rills visible on Earth's moon. The longest channel on Venus is also the longest known lava channel in the solar system. It stretches 6800 km (4200 miles), roughly twice the distance from Chicago to Los Angeles. These channels are 1 to 2 km wide and can sometimes be traced back to collapsed areas where lava appears to have drained from beneath the crust.

Volcanism is evident on Venus in circular features called **coronae**, circular bulges up to 2100 km in diam-

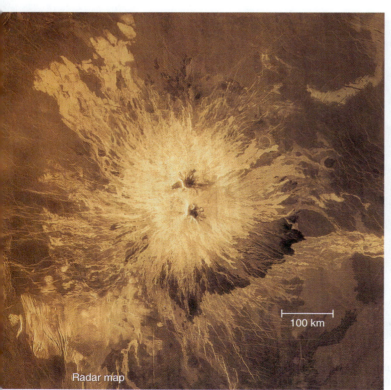

100 km

Radar map

a

b

FIGURE 22-7

(a) Volcano Sapas Mons, lying along a major fracture zone, is topped by two lava-filled calderas and flanked by rough lava flows. The orange color of this radar map mimics the orange light that filters through the thick atmosphere. *(NASA)* (b) Seen by light typical of Earth's surface, Sapas Mons might look more like this computer-generated landscape. Volcano Maat Mons rises in the background. The vertical scale has been exaggerated by a factor of 15 to reveal the shape of the volcanoes and lava flows. *(Copyright © 1992, David P. Anderson, Southern Methodist University)* eter containing volcanic peaks, lava flows, and bordered by fractures. The coronae appear to be caused by rising currents of molten magma below the crust that create an uplifted dome and then withdraw to allow the surface to subside. Coronae are sometimes accompanied by features called pancake domes, circular outpourings of viscous lava, and by domes and hills pushed up by molten rock below the surface. All of these features are shown in ❙ Figure 22-8.

Ishtar Terra may hold secrets to the history of Venus. Western Ishtar consists of a high volcanic plateau called Lakshmi Planum. It rises 4 km above the rolling plains and appears to have formed from lava flows from volcanic vents such as Collette and Sacajawea (❙ Figure 22-9). Although folded mountain ranges do not occur on Venus as they do on Earth, the areas north and west of Lakshmi Planum appear to be ranges of wrinkled mountains where horizontal motion of the crust has pushed up against Ishtar Terra. Faults and deep chasms are widespread over the surface of Venus. Thus there is reason to suspect that although we see no direct evidence of plate motion, there has been compression and wrinkling of the crust in some areas and stretching and faulting in other areas.

The history of Venus has been dominated by volcanism. While Earth's crust has broken into rigid moving plates, the surface of Venus seems more pliable. In any case, to understand Venus we will have to understand its volcanic history.

A HISTORY OF VENUS

Earth has passed through four major stages in its history (Chapter 20), and we have seen how the moon and Mercury were affected by the same stages. Venus, however, has had a peculiar passage through the four stages of planet formation, and its history is difficult to understand. We are not sure how the planet formed and differentiated, how it was cratered and flooded, or how its surface has evolved.

Venus formed only slightly closer to the sun than did Earth, so we might expect it to be a similar planet differentiated into a silicate mantle and a molten iron core. The density and size of Venus tell us that it must have a dense interior much like Earth's, but if the core is molten, then we might expect the dynamo effect to generate a magnetic field. No spacecraft has ever detected such a field around Venus. The magnetic field must be at least 25,000 times weaker than Earth's. Some theorists wonder if the core of the planet is solid. If it is solid, then we must wonder how Venus got rid of its internal heat so much faster than Earth did.

Because the planet lacks a magnetic field, the solar wind is deflected by the uppermost layers of the planet's atmosphere, thus forming a bow shock where the solar wind is slowed and deflected (❙ Figure 22-10). The magnetic field carried by the solar wind drapes over Venus like seaweed over a fishhook, forming a long tail within which ions flow away from the planet. We will see in Chapter 25 that comets, which also lack magnetic fields, interact with the solar wind in the same way.

Volcanoes

Molten rock (magma) is less dense than its surroundings and tends to rise. Where it bursts through Earth's crust, we see volcanism. The two main types of volcanoes on Earth give us good examples for comparison with those on Venus and Mars.

On Earth, **composite volcanoes** form above subduction zones where the descending crust melts and the magma rises to the surface. This forms chains of volcanoes along the subduction zone, such as the Andes along the west coast of South America.

Magma rising above subduction zones is not ▼ very fluid, and it produces explosive volcanoes with sides as steep as 30°.

Shield volcano   Lava flow

Oceanic crust   Magma chamber

A **shield volcano** is formed by highly fluid lava (basalt) that flows easily and creates low-profile volcanic peaks with slopes of 3° to 10°. The volcanoes of Hawaii are shield volcanoes that occur over a hot spot in the middle of the Pacific plate.

Magma collects in a chamber in the crust and finds its way to the surface through cracks.

Magma forces its way upward through cracks in the upper mantle and causes small, deep earthquakes.

A hot spot is formed by a rising convection current of magma moving upward through the hot, deformable (plastic) rock of the mantle.

Oceanic plate

Subduction zone

Composite volcanoes

Continental crust

Upper mantle

Chains of composite volcanoes are not found on Venus or Mars, which is evidence that subduction and plate motion does not occur on those worlds.

Upper mantle

Based on *Physical Geology*, 4th edition, James S. Monroe and Reed Wicander, Wadsworth Publishing Company. Used with permission.

Mount St. Helens exploded northward on May 18, 1980, killing 63 people and destroying 600 km² (230 mi²) of forest with a blast of winds and suspended rock fragments that moved as fast as 480 km/hr (300 mph) and had temperatures as hot as 350°C ▼ (660°F). Note the steep slope of this composite volcano.

The Cascade Range ▶ composite volcanoes are produced by an oceanic plate being subducted below North America and partially melting.

Pacific Ocean

Seattle

Washington

St Helens   Rainier

Portland   Hood

Oregon

Shasta

Nevada

California

Lassen

USGS

Volcano Gula Mons

Volcano Sif Mons

Radar map

NASA

Volcanoes on Venus are shield volcanoes. They appear to be steep sided in some images created from Magellan radar maps, but that is because the vertical scale has been exaggerated to enhance detail. The volcanoes of Venus are actually shallow-sloped shield volcanoes.

Volcanism over a hot spot results in repeated eruptions that build up a shield volcano of many layers. Such volcanoes can grow very large.

Vertical scale exaggerated

Hot spot

This computer model of a mountain with the vertical scale magnified 10 times appears to have steep slopes such as those of a composite volcano.

Mike Seeds

A true profile of the computer model shows the mountain has very shallow slopes typical of shield volcanoes.

If the crustal plate is moving, magma generated by the hot spot can repeatedly penetrate the crust to build a chain of volcanoes. Only the volcanoes over the hot spot are active. Older volcanoes slowly erode away. Such volcanoes cannot grow large because the moving plate carries them away from the hot spot.

Old volcanic island eroded below sea level

Plate motion

Hot spot

Ace Astronomy ™

Go to AceAstronomy and click Active Figures to see "Hotspot Volcanoes" and compare volcanism on Earth with that on Venus.

The volcanoes that make up the Hawaiian Islands have been produced by a hot spot poking upward through the middle of the moving Pacific plate.

NASA

Time since last eruption (million years)

5   3   1.5   1   0

Kauai   Oahu   Molokai   Maui   Hawaii   Active volcanoes

Plate motion

Lo'ihi—Newborn underwater volcano

The plate moves about 9 cm/yr and carries older volcanic islands northwest, away from the hot spot. The volcanoes cannot grow extremely large because they are carried away from the hot spot. New islands form to the southeast over the hot spot.

Olympus Mons is the largest volcano on Mars. It is a shield volcano 25 km (16 mi) high and 700 km (440 mi) in diameter at its base. Its vast size tells us that the crustal plate must have remained stationary over the hot spot. This is evidence that Mars has not had plate tectonics.

Olympus Mons contains 95 times more volume than the largest volcano on Earth, Mauna Loa in Hawaii.

Caldera from repeated eruptions

Digital elevation map

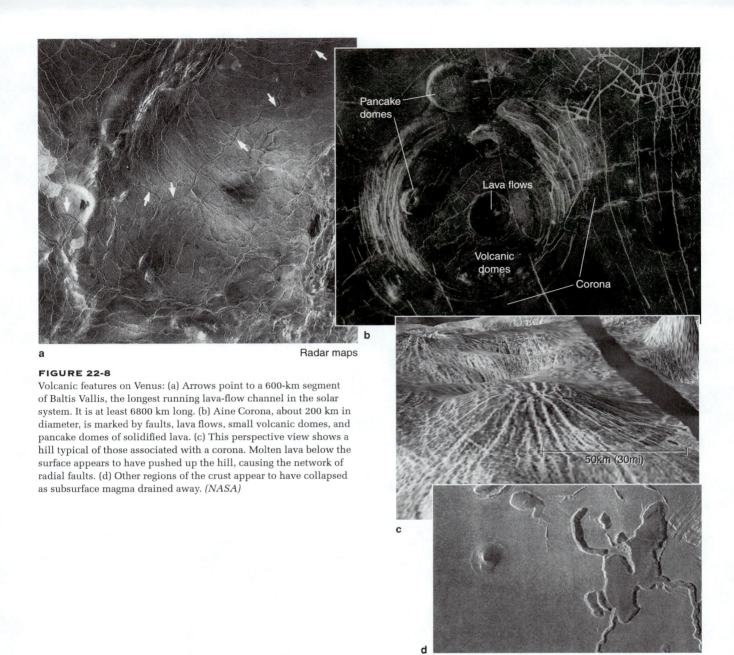

a

Radar maps

b

c

d

FIGURE 22-8

Volcanic features on Venus: (a) Arrows point to a 600-km segment of Baltis Vallis, the longest running lava-flow channel in the solar system. It is at least 6800 km long. (b) Aine Corona, about 200 km in diameter, is marked by faults, lava flows, small volcanic domes, and pancake domes of solidified lava. (c) This perspective view shows a hill typical of those associated with a corona. Molten lava below the surface appears to have pushed up the hill, causing the network of radial faults. (d) Other regions of the crust appear to have collapsed as subsurface magma drained away. *(NASA)*

The interaction of Venus with the solar wind is clear evidence that the planet has little or no magnetic field. Thus, we know little about the differentiation of the planet into core and mantle. The size of the core shown in Figure 22-10 is estimated by analogy with Earth's.

Calculations show that Venus and Earth have outgassed about the same amount of carbon dioxide, but Earth's oceans have dissolved it and converted it to sediments such as limestone. If all of Earth's carbon were dug up and converted back to carbon dioxide, our atmosphere would be about as dense as the air on Venus. This suggests that the main difference between Earth and Venus is the lack of water on Venus. Venus may have had small oceans when it was young; but, being closer to the sun, it was warmer, and the carbon dioxide in the atmosphere created a greenhouse effect

that made the planet even warmer. That process could have dried up any oceans that did exist and reduced the ability of the planet to purge its atmosphere of carbon dioxide. As more carbon dioxide was outgassed, the greenhouse effect grew even more severe. Thus, Venus seems to have been trapped in a runaway greenhouse effect.

Fully 70 percent of the heat from Earth's interior flows outward through midocean volcanism. But Venus lacks crustal rifts, and its numerous volcanoes cannot carry much of the heat out of the interior. Rather, Venus seems to get rid of its interior heat through large currents of hot magma that rise beneath the crust. We see coronae, lava flows, and volcanism above such currents. In fact, the surface rock is dark gray basalt typical of basalts found in ocean crust on Earth.

FIGURE 22-9

This radar map of Lakshmi Planum and Maxwell Montes is based on radar data from both Soviet and American spacecraft with ground-based radar data added (color). Violet and purple regions are smoothest, and orange is roughest. Note the volcanic calderas Collette and Sacajawea and the wrinkled mountains around Lakshmi Planum. *(USGS)*

True plate tectonics is not important on Venus. For one thing, the crust is very dry and is consequently about 12 percent less dense than Earth's crust. This low-density crust is more buoyant than Earth's crust and thus resists being pushed into the interior. Also, the crust is so hot it is halfway to the melting point of rock. Such hot rock is not very stiff, so it cannot form the rigid plates typical of plate tectonics on Earth.

There is no sign of plate tectonics on Venus, but rather evidence that convection currents below the crust are deforming the crust to create coronae and push up mountains such as Maxwell. Detailed measurements of the strength of gravity over the mountains on Venus show that some must be held up not by deep roots like mountains on Earth but by rising currents of magma. Other mountains, like those around Ishtar Terra, appear to be folded mountains caused by limited horizontal motions in the crust, driven perhaps by convection in the mantle.

The small number of craters on the surface of Venus hints that the entire crust has been replaced within the last half-billion years or so. This may have occurred in a planetwide overturning as the old crust broke up and sank and lava flows created a new crust. This could happen periodically on Venus, or the planet may have had a geology more like Earth's until a single resurfacing not too long ago. In either case, unearthly Venus may eventually tell us more about how our own world works.

REVIEW CRITICAL INQUIRY

What evidence do we have that Venus does not have plate tectonics?

On Earth, plate tectonics is identifiable by the worldwide network of faults, subduction zones, volcanism, and folded mountain chains that outline the plates. Although some of these features are visible on Venus, they do not occur in a planetwide network that outlines plates. Volcanism is widespread, but folded mountain ranges occur in only a few places, such as near Lakshmi Planum and Maxwell Montes. Rather than being dominated by the motion of rigid crustal plates, Venus may have a crust made more flexible by the heat, a crust dominated by rising plumes of molten rock that strain the crust to produce coronae and that break through to form volcanoes and lava flows.

At first glance, comparative planetology suggested that Earth and Venus should be sister worlds, but it seems they can be no more than distant cousins. We can blame the thick atmosphere of Venus for altering its geology, but that raises the question: Why isn't Earth's atmosphere similar?

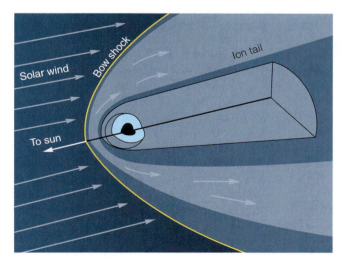

FIGURE 22-10

By analogy with Earth, the interior of Venus should contain a molten core (estimated here), but no spacecraft has detected a planetary magnetic field. Thus, Venus is unprotected from the solar wind, which strikes the planet's upper atmosphere and is deflected into an ion tail.

Venus is certainly an unearthly world, but the familiar principles of comparative planetology help us understand it. As we turn our attention to Mars, we will see the same principles in action, but we will discover that Mars is a world that lacks what Venus has in abundance—internal heat and air.

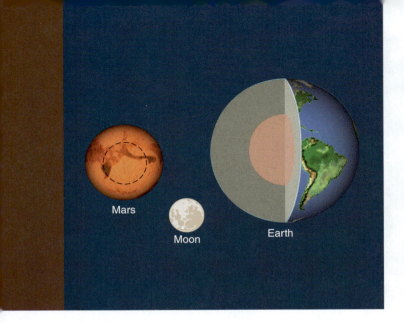

Mars

Moon

Earth

MARS

Data File Six

Mars is a medium-size world smaller than Earth but larger than the moon. Earth's core is hot and molten, while the moon is cold and may not have a molten core at all. The nature of the interior of Mars is unknown.

Average distance from the sun	1.5237 AU (2.279 × 10⁸ km)
Eccentricity of orbit	0.0934
Maximum distance from the sun	1.6660 AU (2.492 × 10⁸ km)
Minimum distance from the sun	1.3814 AU (2.066 × 10⁸ km)
Inclination of orbit to ecliptic	1°51'09"
Average orbital velocity	24.13 km/s
Orbital period	1.8808 y (686.95 days)
Period of rotation	24ʰ37ᵐ22.6ˢ
Inclination of equator to orbit	25°19'
Equatorial diameter	6792 km (0.53 $D_\oplus$)
Mass	0.6424 × 10²⁴ kg (0.1075 $M_\oplus$)
Average density	3.94 g/cm³ (3.3 g/cm³ uncompressed)
Surface gravity	0.379 Earth gravity
Escape velocity	5.0 km/s (0.45 $V_\oplus$)
Surface temperature	−140° to 20°C (−220° to 68°F)
Average albedo	0.16
Oblateness	0.009

22-2 MARS

Mercury and the moon are small. Venus and Earth are, for terrestrial planets, large. But Mars occupies an intermediate position (▌ Data File Six). It is twice the diameter of the moon but only 53 percent Earth's diameter. Its small size has allowed it to cool faster than Earth, and much of its atmosphere has leaked away. Its present carbon-dioxide atmosphere is only 1 percent as dense as Earth's.

In some ways, Mars is much like Earth. A day on Mars is nearly the same length as an Earth day—24 hours and 40 minutes—and its year lasts 1.88 Earth years. Also, just as Earth's axis is tipped 23.5°, that of Mars is tipped 25°. As the northern and southern hemispheres turn alternately toward the sun, seasonal changes are visible even through a small telescope. As spring comes to the southern hemisphere, the white polar cap shrinks and the grayish surface markings grow darker and, according to some observers, greener. At one time, these seasonal changes led some to believe that plant life flourished on Mars when spring thawed the polar cap. We will see later that this is not so.

In other ways, Mars is very different from Earth. Its surface has never broken into moving plates like those that create the topography on Earth. Instead, Mars is a one-plate planet, with its crust frozen into a solid layer but bearing signs of geological activity that include some of the largest volcanoes in the solar system.

THE CANALS OF MARS

Even before the space age brought us photographs of the Martian surface, the planet Mars was a mysterious landscape in the public mind. In the century following Galileo's first astronomical use of the telescope, astronomers discovered dark markings on Mars as well as bright polar caps. Timing the motions of the markings, they concluded that a Martian day was about 24 hours long, and the similarity with Earth's day further supported the belief that Mars was another Earth.

The so-called canals of Mars began life innocently enough in 1858 when the Jesuit astronomer Angelo Secchi referred to a region on Mars known later as Syrtis Major as *Atlantic Canale*. This is the first use of the Italian word *canale* (channel) to refer to a feature on Mars. Then, in the late summer of 1877 the Italian astronomer Giovanni Virginio Schiaparelli, using a telescope only 8.75 in. in diameter, thought he glimpsed fine, straight lines on Mars. He too used the Italian word *canali* (plural) for these lines, and the word was translated into English not as "channel," a narrow body of water, but as "canal," an artificially dug channel. The "canals of Mars" were born. Many astronomers could not see the canals at all, but others drew maps showing hundreds (▌ Figure 22-11).

In the decades that followed Schiaparelli's discovery, many people assumed that the canals were water-

FIGURE 22-11

(a) Early in the 20th century, Percival Lowell mapped canals over the face of Mars and concluded that intelligent life resided there. (b) Modern images recorded by spacecraft reveal a globe of Mars with no canals. Instead, the planet is marked by craters and, in some places, volcanoes. Both of these images are reproduced with south at the top, as they appear in telescopes. Lowell's globe is inclined more nearly vertically and is rotated slightly to the right compared with the modern globe. *(a, Lowell Observatory; b, U.S. Geological Survey)*

courses built by an intelligent race on Mars to carry water from the polar caps to the lower latitudes. Much of this excitement was generated by Percival Lowell, a wealthy Bostonian who founded Lowell Observatory in 1894, principally for the study of Mars. He not only mapped hundreds of canals but also popularized his results in books and lectures. Although some astronomers claimed the canals were merely illusions, by 1907 the general public was so sure that life existed on Mars that the *Wall Street Journal* could suggest that the most extraordinary event of the previous year had been "the proof by astronomical observations . . . that conscious, intelligent human life exists upon the planet Mars." Further sightings of bright clouds and flashes of light on Mars strengthened this belief, and some urged that gigantic geometrical diagrams be traced in the Sahara Desert to signal to the Martians that Earth, too, is inhabited. All seemed to agree that the Martians were older and wiser than humans.

This fascination with men from Mars was not a passing fancy. Beginning in 1912, Edgar Rice Burroughs wrote a series of 11 novels about the adventures of the Earthman John Carter lost on Mars. Burroughs made the geography of Mars, named by Schiaparelli after Mediterranean lands both real and mythical, into household words. He also gave his Martians green skin.

By Halloween night of 1938, people were so familiar with life on Mars that they were ready to believe that Earth had been invaded. When a radio announcer repeatedly interrupted dance music to report the landing of a spaceship in New Jersey, the emergence of monstrous beings, and their destruction of whole cities, thousands of otherwise sensible people rushed to their cars and fled in panic, not knowing that Orson Welles and other actors were dramatizing H. G. Wells's book *The War of the Worlds.*

The fascination with Mars, its canals, and its little green men lasted right up to July 15, 1965, when Mariner 4, the first spacecraft to fly past Mars, radioed back photos of the surface and proved that there are no canals and no Martians. Apparently, the canals are optical illusions produced by the human brain's astounding ability to assemble a field of disconnected marks into a coherent image. If your brain could not do this, the photos on these pages would be nothing but swarms of dots, and the images on the screen of a television set would never make sense. The brain of an astronomer looking for something at the edge of visibility is capable of connecting faint, random markings on Mars into the straight lines of canals.

Even today, Mars holds a fascination for the general public. The grocery store tabloids regularly run stories about a giant face carved on Mars by an ancient race. Although planetary scientists recognize it as nothing more than chance shadows in a photograph and dismiss the issue as a silly hoax, the stories persist. A

Visual-wavelength image

hundred years of speculation have given us high expectations for Mars. If there really was intelligent life on Mars and its representatives came to Earth, they would probably be a big disappointment to the readers of the tabloids.

THE ATMOSPHERE OF MARS

If we visited Mars, our first concern, even before we opened the door of our spaceship, would be the atmosphere. Is it breathable? Even for the astronomer observing safely from Earth, the atmosphere of Mars is a major concern. The gases that cloak Mars are critical to our understanding of the history of the red planet.

The air on Mars is 95 percent carbon dioxide, with a few percent each of nitrogen and argon. It contains almost no water vapor or oxygen, and its density at the surface of the planet—only 1 percent that of Earth's atmosphere—does not provide enough pressure to prevent liquid water from boiling into vapor. If you stepped outdoors on Mars without a space suit, your own body heat would make your blood boil.

Although the air is thin, it is dense enough to be visible in photographs (▌Figure 22-12). Haze and clouds come and go, and occasional weather patterns are visible. Winds on Mars can be strong enough to produce dust storms that envelope the entire planet. The reddish soil is evidently caused by iron oxides (rusts), and that tells us that the oxygen we humans would prefer to find in the atmosphere is locked in chemical compounds in the soil (▌Figure 22-13). The polar caps visible in photos are also related to the Martian atmosphere. The ices in the polar caps are frozen carbon dioxide (dry ice) with unknown amounts of frozen water underneath.

To understand Mars, we must ask why its atmosphere is so thin and dry and why the surface is rich in oxides. The answers to these questions lie in the origin and evolution of the Martian atmosphere.

Presumably, the gases in the Martian atmosphere were mostly outgassed from its interior. Volcanism on terrestrial planets typically releases carbon dioxide and water vapor plus other gases. Because Mars formed farther from the sun, we might expect that it incorporated more volatiles when it formed. But Mars is smaller than

Visual-wavelength image

FIGURE 22-12

The atmosphere of Mars is evident in this image made by the Hubble Space Telescope. The haze is made up of high, water-ice crystals in the thin CO_2 atmosphere. The spot at extreme left is the volcano Ascraeus Mons, 25 km (16 mi) high, poking up through the morning clouds. Note the north polar cap at the top. *(Philip James, University of Toledo; Steven Lee, University of Colorado, Boulder; and NASA)*

FIGURE 22-13

The red planet is truly red. Iron oxides in the soil and rock have given the surface of Mars a rusty red color. Note the dust on the airbags at the bottom of this 360-degree panorama recorded by the Mars rover Spirit, which landed on Mars on January 3, 2004. See Figure 22-20 for an image of the landing site, Gusev Crater, 160 km in diameter. Dust suspended in the atmosphere gives the sky a reddish tint in this true-color image. *(NASA)*

Earth, so it has had less internal heat to drive geological activity, and we might suspect that it has not outgassed as much as Earth. In any case, the outgassing occurred early in the planet's history, and Mars, being small, cooled rapidly and now releases little gas.

How much atmosphere a planet has depends on how rapidly it releases internal gas and how rapidly it loses gas from its atmosphere, and the rate at which a planet loses gas depends on its mass and temperature. The more massive the planet, the higher its escape velocity (Chapter 5) and the more difficult it is for gas atoms to leak into space. Mars has a mass less than 11 percent that of Earth, and its escape velocity is only 5 km/s, less than half Earth's. Thus, gas atoms can escape from it much more easily than they can escape from Earth.

The temperature of a planet's atmosphere is also important. If the gas is hot, its molecules have a higher average velocity and are more likely to exceed escape velocity. Thus, a planet that is near the sun and very hot is less likely to retain an atmosphere than a more distant, cooler planet. The velocity of a gas molecule, however, also depends on the mass of the molecule. On average, a low-mass molecule travels faster than a massive molecule. Thus, a planet loses its lowest-mass gases more easily because those molecules travel fastest.

We can see this principle of comparative planetology if we plot a diagram such as that in ▌ Figure 22-14. The points show the escape velocity versus temperature of the larger objects in our solar system. The temperature we use is the temperature of the gas that is in a position to escape. For the moon, which has essentially no atmosphere, this is the temperature of the sun-lit surface. For Mars, we use the temperature at the top of the atmosphere. The lines in Figure 22-14 show the typical velocity of the fastest-traveling examples of various molecules. At any given temperature, some water molecules, for example, travel faster than others, and it is the highest-velocity molecules that escape from a planet. The diagram shows that the Jovian planets have escape velocities so high that very few molecules can escape. Earth and Venus can't hold hydrogen, and Mars can hold only the more massive molecules. Earth's moon is too small to keep any of the common gases. We will refer to this diagram again as we study the atmospheres of other worlds.

Over the 4.5 billion years since Mars formed, it has lost some of the lower-mass gases. Water molecules are massive enough for Mars to keep, but ultraviolet radiation can break up the water molecules. On Earth, the ozone layer protects water vapor from ultraviolet radiation, but Mars never had an oxygen-rich atmosphere so it never had an ozone layer. Ultraviolet photons from the sun can penetrate deep into the atmosphere and break up molecules such as water. The hydrogen escapes, and the oxygen, a very reactive element, forms oxides in the soil—the oxides that make Mars the red planet. Thus, molecules too massive to leak into space can be lost if they break into lower-mass fragments.

The argon in the Martian atmosphere is evidence of a denser blanket of air in the past. Argon atoms are massive, almost as massive as a CO_2 molecule, and would not be lost easily. In addition, argon is inert and would not form compounds in the soil. The 1.6 percent argon in the atmosphere of Mars is evidently left over

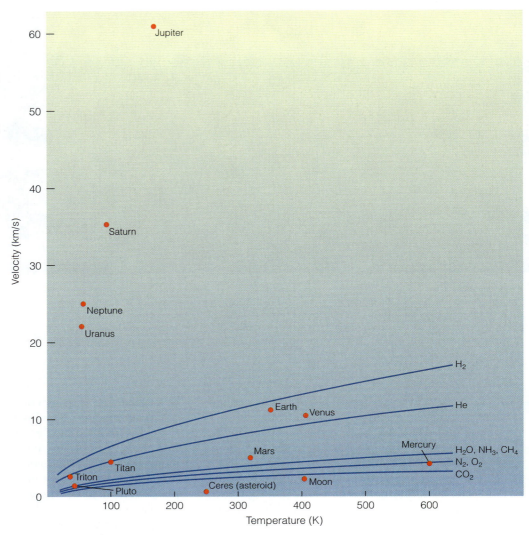

FIGURE 22-14

The loss of planetary gases. Dots represent the escape velocity and temperature of various solar-system bodies. The lines represent the typical highest velocities of molecules of various masses. The Jovian planets have high escape velocities and can hold on to even the lowest-mass molecules. Mars can hold only the more massive molecules, and the moon has such a low escape velocity that even the most massive molecules can escape.

from an ancient atmosphere that was 10 to 100 times denser than the present Martian air.

Finally, we must consider the interaction of the solar wind with the atmosphere of Mars. This is not an important process for Earth because Earth has a magnetic field that deflects the solar wind. Because Mars has no magnetic field, the solar wind interacts directly with the Martian atmosphere, and detailed calculations show that significant amounts of CO_2 could have been carried away by the solar wind over the history of the planet. This process would have been most efficient long ago, however, when the sun was more active and the solar wind was stronger. We must also wonder if Mars had a magnetic field when it was younger and still retained significant internal heat. If it had a molten, conducting core, it may have had a magnetic field to protect it from the solar wind.

Although we remain uncertain as to how much of an atmosphere Mars has had in its past and how much it has lost, it is a good example for our study of comparative planetology. When we look at Mars, we see what can happen to the atmosphere of a medium-size world.

Like its atmosphere, the geology of Mars is typical of intermediate-size worlds.

THE GEOLOGY OF MARS

Mars has been visited by a number of spacecraft, beginning with Mariner 4 in 1965. Some have orbited the planet, and a few have landed. The photographs and measurements transmitted back to Earth tell us that Mars is a world of contrasts, an ancient cratered world with deep canyons and giant volcanoes (❙ Figure 22-15).

A few generations ago, astronomers watched seasonal changes on Mars and wondered if they were seeing a cycle of growing plants. Evidently, these changes are caused by vast seasonal dust storms that cover dark rock and, as spring begins, strip the dust away. The markings grow darker and, because the entire planet has a reddish tint, the dark marks look greenish in contrast. Thus Mars is not a world of green fields, but a world of deadly dry deserts.

Photographs from the surface of Mars, such as Figure 22-13, show reddish deserts of broken rock. These

Visual-wavelength image

FIGURE 22-15

The red surface of Mars is marked by three giant volcanoes at left and a canyon, Valles Marineris, 4000 km long stretching across the bottom of this image made by the orbiting Viking spacecraft. Small round features are impact craters. *(NASA/USGS, courtesy Alfred S. McEwen)*

appear to be lava plains fractured by meteorite impacts, but they don't look much like the surface of Earth's moon. The atmosphere of Mars, thin though it is, protects the surface from the blast of micrometeorites that grinds moon rocks to dust. Also, the Martian dust storms may sweep fine dust away from some areas and leave larger rocks exposed.

Spacecraft orbiting Mars have imaged the surface and measured elevations to reveal that all of Mars is divided into two parts. The southern highlands are heavily cratered, which must mean that they are 2 to 3 billion years old. In contrast, northern lowlands are smooth (❙ Figure 22-16) and so remarkably free of craters that the lowlands must have been resurfaced roughly a billion years ago. Some astronomers have suggested that volcanic floods filled the northern lowlands and buried the craters there. Growing evidence, however, suggests that the northern lowlands were once filled by an ocean of liquid water. This is a controversial theory, which we will discuss later, but it serves to illustrate how planetary scientists must work from the features we see today to reconstruct the past history of a world (Window on Science 22-2).

FIGURE 22-16

These globes of Mars are color coded to show elevation. The northern lowlands lie about 4 km below the southern highlands. The largest volcanoes are very high (white), and the giant impact basins, Hellas and Argyre, are low. Note the depth of the canyon Valles Marineris. The two Viking spacecraft landed on Mars in 1976. Pathfinder landed in 1997. Rovers Spirit and Opportunity landed in 2004. *(NASA)*

WINDOW ON SCIENCE 22-2

The Present Is the Key to the Past

Geologists are fond of saying "The present is the key to the past." By that they mean that we can learn about Earth's history by looking at the present condition of Earth's surface. The position and composition of various rock layers in the Grand Canyon, for example, tell us that the western United States was once at the floor of an ocean. This principle of geology was astonishing when it was first formulated in the late 1700s, and it continues to be relevant today as we try to understand other worlds, such as Venus and Mars.

In the late 18th century, naturalists recognized that the present gave them clues to Earth's history. This was astonishing because most people assumed that Earth had no history. That is, they assumed either that Earth had been created in its present state as described in the Old Testament or that Earth was eternal. In either case, people commonly assumed that the hills and mountains they saw around them had always existed more or less as they were. The 18th-century naturalists began to see evidence that the hills and mountains were not eternal but were the result of past processes and were slowly changing. That gave birth to the idea that Earth had a history.

As the naturalists of the 18th century made the first attempts to thoughtfully and logically explain Earth's nature by looking at the evidence, they were inventing modern geology as a way of understanding Earth. What Copernicus, Kepler, and Newton did for the heavens in the 1600s, the first geologists did for Earth beginning in the late 1700s. Of course, the invention of geology as the study of Earth led directly to our modern attempts to understand the geology of other worlds.

Geologists and astronomers share a common goal: they are attempting to reconstruct the past (Window on Science 19-3). Whether we study Earth, Venus, or Mars, we are looking at the present evidence and trying to reconstruct the past history of the planet by drawing on observations and logic to test each step in the story. How did Venus get to be covered with lava, and how did Mars lose its atmosphere? The final goal of planetary astronomy is to draw together all of the available evidence (the present) to tell the story (the past) of how the planet got to be the way it is. Those first geologists of the late 1700s would be fascinated by the stories planetary astronomers tell today.

The cratering and volcanism on Mars confirm our understanding of Mars. It is larger than Earth's moon, so it cooled more slowly and its volcanism has continued longer than on the moon. But Mars is smaller than Earth and less geologically active, so some of its ancient cratered terrain has survived undamaged by volcanism.

The Martian volcanoes are shield volcanoes with shallow slopes showing that the lava flowed easily. Such volcanoes occur over hot spots of rising magma below the crust and are not related to plate tectonics. The largest volcano in the solar system is Olympus Mons on Mars. (See page 481.) The volcano Mauna Loa in Hawaii is so heavy it has sunk into Earth's crust to form an undersea moat. Olympus Mons is much larger than Mauna Loa but has not sunk into the crust of Mars, which is evidence that the crust of Mars is much thicker than the crust of Earth (▌ Figure 22-17).

Other evidence shows that the Martian crust has been thinner and more active than the moon's. Valles Marineris is a network of canyons that stretches 4000 km (2500 mi) long and up to 600 km (400 mi) wide (Figure 22-15). At its deepest, it is four times deeper than the Grand Canyon on Earth. It is long enough to stretch from New York to Los Angeles. The canyon has been produced by faults in the crust that allowed great blocks to sink. Landslides and some erosion have further modified the canyon. Although Valles Marineris is an old feature, it does show that the crust of Mars has been more active than the crusts of the moon or Mercury, worlds that lack such dramatic canyons.

The faults that created Valles Marineris seem linked at the west end to a great volcanic bulge in the crust of Mars called the Tharsis rise. Nearly as large as the United States, the Tharis rise extends 10 km (6 mi) above the mean radius of Mars. Tharsis is home to many smaller volcanoes, but on its summit lie three giants, and just off of its northwest edge lies Olympus Mons (▌ Figure 22-18). The origin of the Tharsis rise is not well understood, but it appears that magma rising from below the crust has broken through repeatedly to build a giant bulge of volcanic deposits. This bulge is large enough to have modified the climate and seasons on Mars and may be critical in understanding the history of the planet.

A similar uplifted volcanic bulge, the Elysium region visible in Figure 22-16, lies halfway around the planet. It appears to be similar to the Tharsis rise but it is more heavily cratered and so must be older—roughly 1.5 billion years old.

The vast sizes of the Tharsis rise and of Olympus Mons tell us that the crust of Mars has not been broken into moving plates. If a plate were moving over a hot spot, the rising magma would produce a long chain of shield volcanoes and not a single large peak (page 481). On Earth, the hot spot that creates the volcanic Hawaiian islands has punched through the moving Pacific plate repeatedly to produce the Hawaiian-Emperor island chain extending 7500 km (4700 mi) northwest across the Pacific seafloor. (See page 440.) No such

FIGURE 22-17

Olympus Mons, a shield volcano on Mars, is much larger than Mauna Loa on Earth but has not sunk into the Martian crust. This is evidence that Mars's crust is thicker than Earth's. In the overhead image, the edge of Olympus Mons is marked by a cliff a number of kilometers high rather than a moat. *(NASA)*

chain of volcanoes is evident on Mars, so we conclude that the crust is not divided into moving plates.

No spacecraft has ever photographed an erupting volcano on Mars, but it is possible that some of the volcanoes are still active. Craters in the youngest lava flows in the Elysium region show that the volcanoes may have been active as recently as a few million years ago. Mars may still retain enough heat to trigger an eruption, but the interval between eruptions may be very long.

HIDDEN WATER ON MARS

We would not expect to find liquid water on Mars. The atmospheric pressure is too low to keep it from boiling away to vapor. Nevertheless, some geological features seem to have been formed by moving water, and sophisticated measurements suggest that significant amounts of water are hidden in the crust.

When the two Viking spacecraft reached orbit around Mars in 1976, they found exciting hints that water once flowed over the surface. More recent spacecraft such as Mars Global Surveyor, which reached

Mars in 1997, have identified additional features that seem to be related to water. We must examine those features with care, however, because the water on Mars is not obvious at all.

❚ Figure 22-19 summarizes water-related features. **Outflow channels** appear to have been cut by massive floods carrying as much as 10,000 times the water flowing down the Mississippi River. In a matter of hours or days, such floods swept away geological figures and left outflow channels. One way to account for these features is to suppose that the water was frozen in the soil as permafrost. Melted by volcanic heat and pressurized under the weight of the overlying rock, water could have burst out to scour the surface in sudden floods. The number of craters formed on top of the outflow channels show that they are billions of years old.

In contrast, the **valley networks** look like meandering riverbeds that may have formed over long periods. Some streambeds appear to have flowed repeatedly over millions of years. The valley networks are located in the old, cratered, southern hemisphere, and they have about the same age. Evidently the atmosphere was denser 2 to 3 billion years ago.

As shown in Figure 22-19, some features suggest water in the crust of Mars, perhaps as ice. Jumbled valleys appear to be regions where water below the surface has drained away and allowed the terrain to collapse. Some impact craters appear to be surrounded by a "splash" of material, as if the crust was rich in water at the time of the impact. Slow seepage could gradually form a valley network over long periods. The low atmospheric pressure at the surface of Mars today could not prevent liquid water from boiling away to vapor, but water trapped in the crust might be under enough pressure to remain a liquid. Also, water frozen as ice could be trapped in the crust. This gives us reason to hope that water may yet be hidden in the crust of Mars.

More recent spacecraft have been able to photograph the surface of Mars at a much higher resolution than could the Viking orbiters. Those new images show evidence of water. In many places, the spacecraft found gullies leading down steep slopes, and the lack of craters formed on the gullies tells us they are young. Early interpretations suggested the gullies were cut by subsurface water released in a gush when an ice plug melted. Newer studies suggest the gullies could be de-

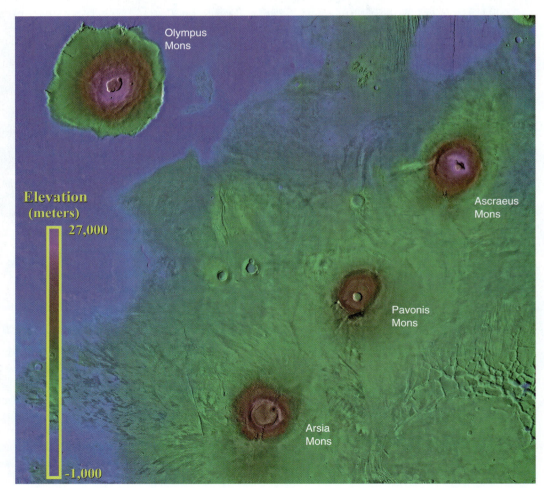

FIGURE 22-18
This color-coded image shows the Tharsis volcanic bulge as green, rising 6 km above the surrounding plains (blue). The three largest Tharsis volcanoes rise high above the surface. The giant volcano Olympus Mons at upper left rests on the lower plains but rises as high as Ascraeus Mons. Note the faults at lower right. Compare with Figures 22-16 and 22-17. (© Calvin J. Hamilton, Columbia, Maryland)

Outflow channels were produced by sudden massive floods.

Some regions appear to have collapsed because of the withdrawal of subsurface water.

Crater counts date outflow channels to a few billion years ago.

Gullies in steep slopes may be debris flows.

Crater counts show that these formations are very young.

Splash craters suggest water was present in the crust.

A valley network channel resembles a meandering riverbed.

A central channel suggests long-term flowing water.

Crater counts date valley networks to billions of years ago.

FIGURE 22-19

These visual-wavelength images made by the Viking orbiters and Mars Global Surveyor show some of the features that suggest liquid water on Mars. Outflow channels and valley networks are old, but some gullies may be quite recent. *(Malin Space Science Systems and NASA)*

bris flows caused by rock and soil sliding downhill after being released by evaporating ice. Another study suggests that the gullies formed beneath deep layers of melting snow. Choosing among these hypotheses is not easy, and we must keep in mind that more than one process may have formed gullies. In any case, the observation of young gullies are further evidence that there is still water on Mars.

The atmosphere of Mars contains a tantalizing clue. Recall that deuterium is an isotope of hydrogen that contains a proton and a neutron as a nucleus. Because deuterium is twice as heavy as normal hydrogen, it escapes from a planet's atmosphere more slowly. Earlier in this chapter we found that the deuterium in the

atmosphere of Venus was evidence that Venus once had more water. Observations of the Martian atmosphere show that deuterium is 5.5 times more abundant relative to hydrogen on Mars than on Earth. That suggests that Mars once had about 20 times more water than it does now.

The terrain at the edges of the northern lowlands has been compared to shorelines, and there is a growing suspicion that the northern lowlands were at least once filled with an ocean. Study Figure 22-16, where the lowlands have been color-coded blue, and notice the major outflow channels leading from the highlands into the lowlands northwest of the Viking 1 landing site

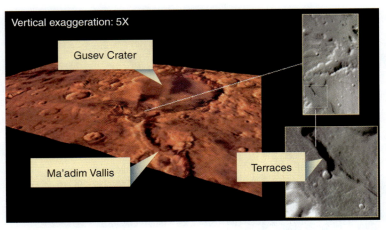

FIGURE 22-20
Lakes as big as Texas and New Mexico combined appear to have drained repeatedly through Ma'adim Vallis into the crater Gusev. Terraces in Ma'adim Vallis confirm water flow, and Gusev (160 km in diameter) shows signs of deltas and shorelines. The crater may have been filled by large lakes a number of times in its history. The rover Spirit landed in Gusev Crater in 2004 to search for traces of water. *(Kevin Hand, NASA/Ames)*

and southeast of the Pathfinder landing site. Growing evidence suggests that the northern lowlands were filled by a major ocean 2.5 to 3.5 billion years ago.

Other studies have located what appear to be major lakebeds. Ma'adim Vallis is a valley longer and deeper than the Grand Canyon on Earth. It appears to have been cut when the water overflowed an ancient lake and drained through the valley northward (❙ Figure 22-20). Such lakebeds and valleys occur in a number of places in the highlands and tell us that Mars was once a wetter world.

The Mars Odyssey Spacecraft, which reached orbit around Mars in February 2002, may have found the remains of that ancient water. Two instruments on Mars Odyssey are capable of detecting water within the top meter of the crust, and both agree that water ice makes up a significant fraction of the crust from latitude 60° north and from latitude −60° south. It isn't possible to determine how deep the ice goes, but the amount detected so far is enough to fill Lake Michigan twice.

If liquid water once flowed on Mars, then its climate must have been different in the past. The atmosphere of Mars must have been thicker long ago before much of the gas had time to leak away. Also, volcanism may have been more active in the past, and outpouring of volcanic gases could have added to the atmosphere, increased the greenhouse effect, and vaporized more of the polar caps to further thicken the atmosphere.

Further hints of climate change were found in the 1970s when the Viking orbiters photographed layered terrain near the polar caps. More modern Mars probes have imaged these layers in greater detail (❙ Figure 22-21). These layers are laid down year after year as dust accumulates on the polar caps and is then left behind when the polar caps vaporize in the spring. Over periods of thousands of years, deep layers can develop. What is significant is that newer layers are superimposed on older layers, showing that the climate on Mars has changed repeatedly.

Studies of geological formations near the south pole of Mars now span a number of years. Comparison of images made in 1999 with those made later show that the ice cap of dry ice seems to be vaporizing back a little further each spring. (❙ Figure 22-22).

If these layered regions and receding areas of polar cap show traces of climate cycles, it supports studies that show that the climate on Mars may change because of cyclic changes in the rotation and orbital revolution of the planet. Recall from Chapter 2 that Earth is affected by such Milankovitch cycles.

One astonishing bit of evidence of water on Mars is the analysis of rock samples from the planet. Of course, no astronaut has ever visited Mars and brought back a rock, but over the history of the solar system occasional impacts by asteroids have blasted giant craters on Mars and ejected bits of rock into space. A few of those bits of rock have fallen to Earth as meteorites, and nearly a dozen have been found and identified (❙ Figure 22-23). These meteorites include basalts, as we might expect from a planet so heavily covered by lava flows. They also contain small traces of water and minerals that are deposited by water. Chemical analysis shows that the magma from which the rocks solidified must have contained up to 1.8 percent water.

If all of the lava flows on Mars contained that much water and it was all outgassed, it could create a planetwide ocean 20 m deep. That isn't quite enough to explain all the flood features on Mars, but it does assure us that the planet once had more water.

Mars has water, but it is hidden. The deuterium in its atmosphere tells us it once had more water, but much of that has been lost. The climate has changed in a cyclic way, but the atmosphere has also gradually grown thinner. The oceans and lakes are gone. The last of the water on Mars is hidden in the crust.

When humans reach Mars, they will not need to dig far to find water, probably as ice. They can use solar power to break the water into hydrogen and oxygen. Hydrogen is fuel, and oxygen is the breath of life, so the water on Mars may prove to be buried treasure.

Even more exciting is the realization that Mars once had bodies of liquid water on its surface. It is a desert world now, but someday an astronaut may scramble down an ancient Martian streambed, turn over a rock, and find a fossil.

A HISTORY OF MARS

The four-stage history of Mars is a case of arrested development. Because of its small size, its interior cooled,

Layers of
ice and dust

Dark layers
containing sand

Sand
dunes

2 km

a

b

c

FIGURE 22-21

(a) The north polar cap of Mars is made of many regions of ice separated by narrow valleys free of ice. *(NASA)* (b) In an image near the edge of the polar cap, a set of light-colored layers of dust overlie a set of darker layers. Sand eroding from the lower layers is forming dunes. *(NASA/JPL/Malin Space Science Systems)* (c) In places, different sets of layers are superimposed, suggesting periodic changes in the Martian climate. *(Adapted from a diagram by J. A. Cutts, K. R. Balasius, G. A. Briggs, M. H. Carr, R. Greeley, and H. Masursky)*

During the second stage of planetary evolution, cratering, the crust of Mars was battered during the heavy bombardment as the last of the debris in the young solar system was swept up. The old southern hemisphere survives from this age about 4 billion years ago. The largest impacts blasted out the great basins like Hellas and Argyre.

The third stage of planet formation, flooding, included flooding by great lava flows that smoothed some regions. Volcanism in the Tharsis and Elysium regions was very active. Whether flooding included oceans of liquid water is controversial, because the present atmosphere is too thin to allow liquid water to exist on the surface. Nevertheless, the evidence that there was once water on Mars seems quite impressive.

The early atmosphere was evidently thicker, and liquid water carved the erosional features we see today. A liquid-water ocean may have once filled the northern-hemisphere lowlands, and water may have formed flood features and valley networks. As atmospheric gases leaked into space, the surface temperature fell, and what water remained became trapped in the soil as permafrost. Mars became a frozen world.

The history of Mars may hinge on climate variations. Recently calculated models suggest that Mars once may have rotated at a much steeper angle to its orbit, as much as 45°. This could have made the climate warmer and kept more of the CO_2 from freezing out at the poles. The rise of the Tharsis bulge could have tipped the axis to its present 25° and cooled the climate. That might account for the warmer, wetter climate implied by the flow features. Some calculations suggest that Mars goes through cycles as its rotational inclination fluctuates, and this may cause short-term variations in climate much like the ice ages on Earth. That might explain the layered terrain near the polar caps.

The fourth stage of planet formation is unremarkable on Mars. The crust of Mars is now too thick to be active. The planet has lost much of its internal heat, and now it lacks much of a molten core, as evidenced by the lack of a planetwide magnetic field. Although the crust was stressed by rising magma that produced the Tharsis and Elysium volcanic rises and the great canyon Valles Marineris, it is too thick for plate tectonics. Volcanism may still occur on Mars, but the crust

and geological activity did not last very long. Also because of its small size, it gradually lost some of its atmosphere, and its surface fell into a deep freeze.

In the first stage of planetary evolution, the planet seems to have differentiated into a core, mantle, and crust. There are no obvious traces of plate tectonics such as folded mountain ranges. Mars lacks a magnetic field, so its core cannot contain much molten iron. The Mars Global Surveyor spacecraft has detected traces of residual magnetic fields trapped in parts of the crust, and that suggests that Mars once had a magnetic field and presumably a molten iron core.

FIGURE 22-22
Each time spring comes to the southern hemisphere of Mars, the dry ice at the edge of the southern polar cap vaporizes, leaving pits and hills. Comparison of images made in 1999 (left) and in 2001 (right) shows that the polar cap is vaporizing further each spring. This suggests a gradual change in climate. Each of these images is 250 meters in diameter. *(NASA/JPL/Malin Space Science Systems)*

significant amounts of internal heat, and there is plenty of evidence that they have had rising convection currents of magma under their crusts. Of course, we wouldn't expect to see coronae on Earth, because its surface is rapidly modified by erosion and plate tectonics. Furthermore, the mantle convection on Earth seems to produce plate tectonics rather than coronae. Mars, however, is a smaller world and must have cooled faster. We see no evidence of plate tectonics, and we do see giant volcanoes that suggest rising plumes of magma erupting up through the crust at the same point over and over. Perhaps we see no coronae on Mars because the crust of Mars rapidly grew too thick to deform easily over a rising plume. On the other hand, perhaps we could think of the entire Tharsis bulge as a single giant corona.

Planetary astronomers haven't explored enough planets yet to see all the fascinating combinations nature has in store for us. Perhaps after we have visited a few hundred terrestrial planets, we will recognize the geology of Mars as typical of medium-size worlds. Of course, Mars is not medium in terms of its location. Of the terrestrial planets, Mars is the farthest from the sun. How has that affected the evolution of its atmosphere?

has grown too thick for much geological activity beyond slow erosion by wind-borne dust.

Our history of Mars is far from complete, but it does illustrate how the intermediate size of Mars has influenced both its atmosphere and its geology. Neither small nor large, Mars is a medium world.

REVIEW CRITICAL INQUIRY

Why doesn't Mars have coronae like those on Venus?
The coronae on Venus are believed to have been caused by rising currents of molten magma in the mantle, pushing upward under the crust and then withdrawing to leave the circular scars called coronae. Earth, Venus, and Mars have had

Our history of Mars must include a phenomenon that was not part of our histories of Mercury and Venus. Unlike those worlds, Mars has moons. These moons, and the moons we will study in the outer solar system, contain important clues to the origin of the planets.

22-3 THE MOONS OF MARS

If we could camp overnight on Mars, we might notice its two small moons, Phobos and Deimos. Phobos, shaped like a flattened loaf of bread 20 km × 23 km × 28 km, would appear less than half as large as Earth's full moon. Deimos, only 12 km in diameter and three times farther from Mars, would look only 1/15 the diameter of Earth's moon.

Both moons are tidally locked to Mars, keeping the same side facing the planet as they orbit. Also, both moons revolve around Mars in the same direction that Mars rotates, but Phobos follows such a small orbit that it revolves faster than Mars rotates. Thus, Phobos rises in the west and sets in the east.

ORIGIN AND EVOLUTION

Deimos and Phobos are typical of the small, rocky moons in our solar system (Figure 22-24). They are dark gray, with albedos of only about 0.06, and they have low densities, about 2 g/cm³.

a

b

FIGURE 22-23

(a) The impact crater Galle at right, known for obvious reasons as the Happy Face Crater, is 210 km in diameter. Such large impacts may occasionally eject fragments of the crust into space. A few fragments from Mars have fallen to Earth as meteorites. *(Malin Space Science Systems/NASA)* (b) Meteorite ALH84001, found in Antarctica, has been identified as an ancient rock from Mars. Minerals found in the meteorite were deposited in water and thus suggest that the Martian crust was once richer in water than it is now. *(NASA)*

Many of the properties of these moons hint that they are captured asteroids. In the outer parts of the asteroid belt, almost all asteroids are dark, low-density objects. Massive Jupiter, orbiting just outside the asteroid belt, can scatter such bodies throughout the solar system, so we should not be surprised if Mars, the next planet inside the asteroid belt, has captured a few of these as satellites.

However, capturing a passing asteroid into a closed orbit is not so easy that it happens often. The asteroid approaches the planet along a hyperbolic orbit and, if it is unimpeded, swings around the planet and disappears back into space. To convert the hyperbolic orbit into a closed orbit, the planet must slow the asteroid as it passes. Tidal forces might do this, but they would be rather weak. Interactions with other moons or grazing collisions with a thick atmosphere might also slow the asteroid.

Both satellites have been photographed by spacecraft, and those photos show that the satellites are heavily cratered. Such cratering could have occurred while the moons were either still in the asteroid belt or in orbit around Mars. In any case, the heavy battering has broken the satellites into irregular chunks of rock,

and they cannot pull themselves into smooth spheres because their gravity is too weak to overcome the structural strength of the rock. We will discover that low-mass moons are typically irregular in shape, whereas more massive moons are more spherical.

Photos of Phobos reveal a unique set of narrow, parallel grooves (Figure 22-24a). Averaging 150 m wide and 25 m deep, the grooves run from Stickney, the largest crater, to an oddly featureless region on the opposite side of the satellite. One theory suggests that the grooves are deep fractures produced by the impact that formed the crater. The featureless region opposite Stickney may be similar to the jumbled terrain on Earth's moon and on Mercury. All were produced by the focusing of seismic waves from a major impact on the far side of the body. High-resolution photographs show that the grooves are lines of pits, suggesting that the pulverized rock material on the surface has drained into the fractures or that gas, liberated by the heat of impact, escaped through the fractures and blew away the dusty soil.

The Mars Global Surveyor spacecraft revealed more about the dust. Observations made with the spacecraft's infrared spectrometer show that the moon's

a

b

c   Visual-wavelength images

FIGURE 22-24

The moons of Mars are too small to pull themselves into spherical shape. (a) Phobos has an impact crater 10 km in diameter at one end with grooves radiating away. *(Damon Simonelli and Joseph Ververka, Cornell University/NASA)* (b) A photo from the Mars Global Surveyor spacecraft shows features as small as 8 meters in diameter. *(Malin Space Science Systems/NASA)* (c) Deimos, smaller than Phobos, looks more uniform because of dusty soil covering the smaller features. *(NASA)*

surface cools quickly from −4°C to −112°C (from 25°F to −170°F) as it passes from sunlight into the shadow of Mars. Solid rock would retain heat and cool more slowly, so the dust must be at least a meter deep and very fine. In photos made by the spacecraft camera, the dust blankets the terrain, but the photos also show boulders a few meters in diameter thought to be ejected by impacts (Figure 22-24b).

Deimos not only has no grooves, it also looks smoother because of a thicker layer of dust on its surface (Figure 22-24c). This material partially fills craters and covers minor surface irregularities. It seems certain that Deimos experienced collisions in its past, so fractures may be hidden below the debris.

The debris on the surfaces of the moons raises an interesting question. How can the weak gravity of small bodies hold any fragments from meteorite impacts? Escape velocity on Phobos is only 12 m/s. An athletic astronaut could almost jump into space. Certainly, most fragments from impacts should escape, but some could fall back and accumulate on the surface.

Deimos, smaller than Phobos, has a smaller escape velocity. But it has more debris on its surface because it is farther from Mars. Phobos is close enough to Mars that most ejecta from impacts on Phobos will be drawn into Mars. Deimos, being farther from Mars, is able to keep a larger fraction of its ejecta. Phobos is so close to Mars that tides are making its orbit shrink, and it will fall into Mars or be ripped apart by tidal forces within about 100 million years.

Deimos and Phobos illustrate three principles. First, some satellites are probably captured asteroids. Second, small satellites tend to be irregular in shape and heavily cratered. And third, tidal forces can affect small moons and gradually change their orbits. We find even stronger tidal effects in Jupiter's satellite system (Chapter 23).

REVIEW CRITICAL INQUIRY

Why would we be surprised to find volcanism on Phobos or Deimos?

In discussing Earth's moon, Mercury, Venus, Earth, and Mars, we have seen illustrations of the principle that the larger a world is, the more slowly it loses its internal heat. It is the flow of that heat from the interior through the surface into space that drives geological activity such as volcanism and plate motion. A small world, like Earth's moon, cools quickly and remains geologically active for a shorter time than a larger world like Earth. Phobos and Deimos are not just small, they are tiny. However they formed, any interior heat would have leaked away very quickly; with no energy flowing outward, there can be no volcanism.

Some futurists suggest that our first human missions to Mars will be not to land on the surface but to build a colony on Phobos or Deimos. These plans speculate that there may be water deep inside the moons that colonists could use. What would happen to water released in the sunlight on the surface of such small worlds?

Venus is nearly the same size as Earth, but it is unlikely that humans will ever colonize its surface. The smaller and colder Mars, however, may someday be home to large colonies. In the next chapter, we turn our attention to Jupiter and Saturn, giant worlds so totally unlike Earth that we will need entirely new principles of comparative planetology to describe them.

SUMMARY

All terrestrial planets pass through a four-stage history: (1) differentiation; (2) cratering; (3) flooding of the crater basins by lava, water, or both; and (4) slow surface evolution. The importance of each of these processes in the evolution of a planet depends on the planet's mass and temperature.

Venus is Earth's twin in size and density, but the planet has evolved along divergent lines because it is slightly closer to the sun. The higher temperature evaporated any early oceans and prevented the absorption of carbon dioxide from the atmosphere. The accumulating carbon dioxide created a greenhouse effect that produced very high surface temperatures.

The crust of Venus is marked by low, rolling plains and higher uplands with some very high volcanoes. Coronae appear to be produced by rising currents of molten rock below the crust. Where the magma breaks through, it builds shield volcanoes and floods the surface with lava flows. The surface contains more impact craters than Earth but fewer than the moon, and astronomers suspect the entire planet has been resurfaced by volcanism within the last half billion years.

A number of spacecraft have landed on the surface of Venus, but none has survived for more than an hour or two. The surface temperature is about 745 K (880°F), and the atmosphere contains traces of sulfuric, hydrochloric, and hydrofluoric acids. Soviet spacecraft have taken photographs on the surface and transmitted them back to Earth. These photos show rocky terrain of broken and weathered rock. Analysis of samples at the landing sites suggests that the rocks are basalts.

No magnetic field has been detected around Venus, and that suggests that the core is not molten iron generating a magnetic field through the dynamo effect as in Earth's core. Another mystery is the retrograde rotation. This may have been produced by a major impact during formation or by tides caused by the sun.

Mars is smaller than Earth but larger than Mercury. Although early astronomers saw canals on Mars and thought it to be inhabited, we now know that it is a dry desert world. Its southern highlands are heavily cratered and old; the northern lowlands have been resurfaced and lack craters. Much of Mars has been altered by volcanism, and some of its shield volcanoes are very large. Some may still be active.

Mars may once have had liquid water flowing over its surface. Photographs returned by spacecraft show extensive traces of erosion by water. Most of these seem to have been cut by sudden flooding rather than by gradual runoff. Whatever water Mars still has is now frozen in the soil or in the polar caps.

The Martian atmosphere is now very thin because the low gravity has been unable to retain most gases. But features such as the streambeds and the layered terrain near the polar caps suggest that the Martian climate may have been warmer and wetter in the past and that it may change cyclically, perhaps due to orbital changes.

Deimos and Phobos, the two small moons of Mars, may be captured asteroids. Like most small moons, they are irregular in shape because their gravities are too weak to deform their rock into spheres.

NEW TERMS

subsolar point

composite volcano

shield volcano

corona

outflow channel

valley network

REVIEW QUESTIONS

Ace⊘Astronomy™ Assess your understanding of this chapter's topics with additional quizzing and animations at **http:// astronomy.brookscole.com/seeds8e**

1. Why might we expect Venus and Earth to be similar?

2. What evidence do we have that Venus and Mars once had more water than at present? Where did that water come from? Where did it go?

3. What features will you look for in future radar maps of Venus to search for plate tectonics?

4. Why doesn't Mars have mountain ranges like those on Earth? Why doesn't Earth have large volcanoes like those on Mars?

5. What were the canals on Mars? How do they differ from the dry streambeds on Mars?

6. Propose an explanation for the nearly pure carbon-dioxide atmospheres of Venus and Mars. How did Earth avoid such a fate?

7. What evidence do we have that the climate on Mars has changed?

8. How can we estimate the ages of the streambeds on Mars?

9. Why are Phobos and Deimos nonspherical? Why is Earth's moon much more spherical?

DISCUSSION QUESTIONS

1. From what you know of Earth, Venus, and Mars, do you expect the volcanoes on Venus and Mars to be active or extinct? Why?

2. If humans someday colonize Mars, the biggest problem may be finding water and oxygen. With plenty of solar energy beating down through the thin atmosphere, how might colonizers extract water and oxygen from the Martian environment?

PROBLEMS

1. How long would it take radio signals to travel from Earth to Venus and back if Venus were at its nearest point to Earth? at its farthest point from Earth?

2. The Pioneer Venus Orbiter circled Venus with a period of 24 hours. What was its average distance above the surface of Venus? (*Hint:* See Chapter 5.)

3. Calculate the velocity of Venus in its orbit around the sun.

4. What is the maximum angular diameter of Venus as seen from Earth? (*Hint:* Use the small-angle formula.)

5. If the Magellan spacecraft transmitted radio signals down through the clouds on Venus and heard an echo from a certain spot 0.000133 second before the main echo, how high is the spot above the average surface of Venus?

6. The smallest feature visible through an Earth-based telescope has an angular diameter of about 1 second of arc. If a canal on Mars was just visible when Mars was at its closest to Earth, how wide was the canal? (*Hint:* Use the small-angle formula.)

7. What is the maximum angular diameter of Phobos as seen from Earth? What surface features should we expect to see? (*Hint:* Use the small-angle formula.)

8. What is the maximum angular diameter of Phobos as seen from the surface of Mars?

9. Deimos is about 12 km in diameter and has a density of 2 g/cm^3. What is its mass? (*Hint:* The volume of a sphere is $\frac{4}{3}\pi r^3$.)

CRITICAL INQUIRIES FOR THE WEB

1. Have you ever wondered how the various features on the planets get their names? Surface features on Venus are (mostly) named for female figures from history and mythology. Who decides how planetary features are named? Look for information on planetary nomenclature and summarize the way different types of features on Venus are assigned names.

2. "Martians" have fascinated humans for the last century or more. Many online sources chronicle the representation of life on Mars by Earthlings throughout history and literature. Read about the Martians as represented by a particular literary work or nonfiction account and discuss to what extent they are (or are not) based on realistic views of the nature of Mars—in terms of our current understanding as well as the views of the period in which the work was written.

3. Spacecraft are studying Mars right now. Search for the latest discoveries and photographs.

EXPLORING *THESKY*

1. Venus and Mars exhibit phases similar to the moon's phases.

   a. Use *TheSky* to sketch the appearance of Venus and Mars at the present time. Also give the phase as a percentage of full.

   How to proceed: Use **Find** to give you a highly magnified view of a particular planet. Click on the planet to obtain **Object Information.**

   b. Explain the observed phase on the basis of geometrical relationship between the sun, Earth, and the planet as shown by the **3D Solar System Mode.**

 Visit the Seeds *Foundations of Astronomy* companion Web site for critical thinking exercises, articles, and additional readings from InfoTrac College Edition, Brooks/Cole's online student library.

JUPITER AND SATURN

There is something fascinating about science. One gets such wholesale
returns of conjecture out of such a trifling investment of fact.

Mark Twain, Life on the Mississippi

NASA

GUIDEPOST

As we begin this chapter, we leave behind the psychological security of planetary surfaces. We can imagine
standing on the moon, on Venus, or on Mars, but Jupiter
and Saturn have no surfaces. Thus, we face a new
challenge—to use comparative planetology to study
worlds so unearthly we cannot imagine being there.

One reason we find the moon and Mars of interest
is that we might go there someday. Humans may

become the first Martians. But the outer solar system
seems much less useful, and that gives us a chance
to think about the cultural value of science.

This chapter begins our journey into the outer
solar system. In the next chapter, we will visit worlds
out in the twilight at the edge of the sun's family.

VIRTUAL LABORATORIES

PLANETARY GEOLOGY

PLANETARY ATMOSPHERES AND THEIR RETENTION

a

Visual-wavelength image

Shadow of a moon

Great Red Spot

FIGURE 23-1

(a) The atmosphere of Jupiter is banded by turbulent belts of clouds that circulate parallel to the planet's equator. Oval spots are rotating storm systems, such as the Great Red Spot. The dark spot is the shadow of one of Jupiter's moons. (b) The Great Red Spot is roughly twice the diameter of Earth. This infrared image from the Galileo spacecraft shows lower clouds as blue, higher thin hazes as red, and the highest clouds as white. Note the turbulent circulation of the gases. *(NASA)*

Infrared false-color image

b

When Mark Twain wrote the sentences on the previous page he was poking gentle fun at science, but he was right. The exciting thing about science isn't the so-called facts, the observations in which scientists have great confidence. Rather, the excitement lies in the understanding that scientists get by rubbing a few facts together. Jupiter and Saturn are strange, distant worlds, but we can combine the available observations with known principles of comparative planetology and understand these giant planets.

Jupiter and Saturn have been studied from Earth for centuries, but much of what we know has been radioed back by space probes. Pioneer 10 and Pioneer 11 explored Jupiter and Saturn in the mid-1970s, but their instruments were not very sensitive. The Voyager 1 and Voyager 2 spacecraft were launched in 1977 on a mission to visit all four Jovian worlds, concluding when Voyager 2 flew past Neptune in 1989. In December 1995, the Galileo spacecraft reached Jupiter. While an instrumented probe dropped into Jupiter's atmosphere, the main Galileo spacecraft went into orbit to begin a long-term study of the planet and its moons. Throughout this discussion of the Jovian worlds, you will find images and data returned to Earth by these robotic explorers.

Ace✺**Astronomy**™ The AceAstronomy icon throughout the text indicates an opportunity for you to test yourself on key concepts and to explore animations and interactions on the AceAstronomy Web site at **http://astronomy.brookscole .com/seeds8e**

Jupiter and Saturn are unearthly, but because they obey the same laws of nature, we can understand their peculiar personalities. We begin with the closest and largest Jovian planet, named for the ruler of the Roman gods.

23-1 JUPITER

Jupiter is the most massive of the Jovian planets, containing nearly three-fourths of all the planetary matter in our solar system. This high mass accentuates some processes that are less obvious or nearly absent on the other Jovian worlds. Just as we used Earth as the basis of comparison for our study of the terrestrial planets, we will examine Jupiter in detail so we can use it as a standard in our comparative study of the other Jovian planets.

The most striking feature of Jupiter is its varicolored cloud belts, which make it look like a child's rubber ball (❚ Figure 23-1). We will examine these belts later and see that they may be related to liquid circulations deep inside the planet. However, first we must survey Jupiter in general.

Liquid hydrogen

Heavy-element core

Liquid metallic hydrogen

Earth   Jupiter

JUPITER

Data File Seven

A theoretical model of the interior of Jupiter. Note the oblateness of the planet.

Average distance from the sun	5.2028 AU (7.783×10^8 km)
Eccentricity of orbit	0.0484
Maximum distance from the sun	5.455 AU (8.160×10^8 km)
Minimum distance from the sun	4.951 AU (7.406×10^8 km)
Inclination of orbit to ecliptic	1°18′29″
Average orbital velocity	13.06 km/s
Orbital period	11.867 y (4334.3 days)
Period of rotation	$9^h55^m30^s$
Inclination of equator to orbit	3°5′
Equatorial diameter	142,900 km (11.20 $D_\oplus$)
Mass	1.899×10^{27} kg (317.83 $M_\oplus$)
Average density	1.34 g/cm³
Gravity at base of clouds	2.54 Earth gravities
Escape velocity	61 km/s (5.4 $V_\oplus$)
Temperature at cloud tops	−120°C (−180°F)
Albedo	0.51
Oblateness	0.0637

SURVEYING JUPITER

Jupiter is interesting because it is big, massive, mostly liquid hydrogen, and very hot inside. The preceding facts are common knowledge among astronomers, but we should demand an explanation of how they know these facts. Often the most interesting thing about a fact isn't the fact itself, but how we know it.

We know that Jupiter is a big planet because it looks big. At their closest points to Earth, Jupiter is about eight times farther away than Mars, but even a small telescope will reveal that the disk of Jupiter is more than two times bigger than the disk of Mars. If we use the small-angle formula, we can compute the diameter of Jupiter—a bit over 11 times Earth's diameter (❚ Data File Seven). That's a big planet!

We know that Jupiter is massive because its moons race around it at high speed. Io is the innermost of the four Galilean moons, and its orbit is just a bit larger than the orbit of our moon around Earth. Io streaks around its orbit in less than two days, whereas Earth's moon takes a month. Jupiter has to be a very massive world to hold on to such a rapidly moving moon (❚ Figure 23-2). In fact, we can put the radius of Io's orbit and its orbital period into Newton's version of Kepler's third law (Chapter 5) and calculate the mass of Jupiter—almost 318 times Earth's mass.

The size and mass of Jupiter are easy to find, but we might wonder how astronomers know that it is made of hydrogen. That fact is just a little bit harder to find, but the first step is to divide the mass by the volume and find Jupiter's average density, about 1.34 g/cm³. Of course, it is more dense at the center and less dense near the surface, but this average density tells us that Jupiter can't be made totally of rock. Rock has a density of 2.5 to 4 g/cm³, so Jupiter must contain less-dense material such as hydrogen.

Spectra recorded from Earth and from spacecraft visiting Jupiter tell us that the composition of Jupiter is much like that of the sun—it is mostly hydrogen and helium. This fact was confirmed when a probe from the Galileo spacecraft fell into the atmosphere in 1995 and radioed its results back to Earth. Jupiter is mostly hydrogen and helium with traces of heavier atoms forming molecules such as methane (CH_4), ammonia (NH_3), and water (❚ Table 23-1).

Just as astronomers can build mathematical models of the interiors of stars, they can use the equations that describe gravity, energy, and how matter responds to pressure to build mathematical models of the interior of Jupiter. The models tell us that a planet as big and massive as Jupiter, having a low density and composed mostly of hydrogen and helium, must contain only a small core of elements heavier than helium. Astronomers often refer to this as a "rocky" core, but the pressure is so high there is nothing rocklike about the material. We will call it a heavy-element core. Although the core is about the size of Earth, it must con-

FIGURE 23-2

It is obvious that Jupiter is a very massive planet when we compare Earth's moon with Jupiter's moon Io. Although Io is 10 percent farther from Jupiter, it travels 17 times faster in its orbit than does Earth's moon. Clearly, Jupiter's gravitational field is much stronger than Earth's.

tain roughly 15 Earth masses. The models also tell us that the pressure below the atmosphere is so great that hydrogen cannot exist as a gas; most of Jupiter is liquid hydrogen.

Even the shape of the planet is revealing. By watching the cloud belts, we can tell that Jupiter rotates rapidly, in about 10 hours. Simple measurements on a photograph show that it is slightly flattened; it is 6.37 percent larger in diameter, through its equator, than it is through its poles. Astronomers refer to this by saying that Jupiter's **oblateness,** its degree of flattening, is

0.0637. The amount of flattening depends on the speed of rotation and on the rigidity of the material. A rotating planet that is mostly liquid will be more oblate than a planet that is mostly rocky. Thus, astronomers can use the observed oblateness of Jupiter to check their models of the size of its heavy-element core.

It is easy to think of the heavy-element core as a ball of rock, but astronomers are quick to point out that it is very hot, about 30,000 K. That is five times hotter than the surface of the sun, but the pressure is so great that it keeps the material confined to the core of the planet. How do we know it is so hot? Because infrared observations show that Jupiter emits about 1.7 times as much energy as it receives from the sun. It must be very hot inside to have so much heat flowing outward into space.

The models of Jupiter's interior tell us that ordinary liquid hydrogen cannot support the pressure deep inside the planet. It is compressed into a state in which the electrons are free to move easily from atom to atom, and although the hydrogen can flow like a liquid, it can conduct electricity like a metal. This **liquid metallic hydrogen** does not occur naturally on Earth, but it has been created under tremendous pressure in laboratories.

TABLE 23-1		
Composition of Jupiter and Saturn (by mass)		
Molecule	Jupiter (%)	Saturn (%)
H_2	78	88
He	19	11
H_2O	0.0001	—
CH_4	0.2?	0.6
NH_3	0.5?	0.2

Basic observations and the known laws of physics can tell us a great deal about Jupiter. Its vast magnetic field can tell us even more.

JUPITER'S MAGNETIC FIELD

Astronomers detected a magnetic field around Jupiter as early as the 1950s, which is surprising considering that magnetic fields are invisible to the eye. In fact, Jupiter's giant magnetic field was first detected by radio.

In 1955, radio astronomers detected bursts of synchrotron radio waves coming from Jupiter. They called it **decameter radiation** because its wavelength was tens of meters. Synchrotron radiation is electromagnetic radiation from fast electrons spiraling in a magnetic field, so it was obvious that Jupiter had a magnetic field.

Because the radio signals fluctuated with a period of 9 hours 55.5 minutes, astronomers concluded that this period was the true period of rotation of Jupiter. This result is what we would expect from the dynamo theory for the generation of magnetic fields, which requires an electrically conducting fluid, convection driven by heat flowing outward, and rapid rotation. The liquid metallic hydrogen in Jupiter is a very good electrical conductor, and we have seen clear evidence that heat flows outward from inside Jupiter. That, coupled with Jupiter's rapid rotation, should generate a powerful dynamo effect. Thus, the fluctuation of the radio signals tells us the true rotation period of the planet's liquid interior.

By the early 1960s, radio astronomers had discovered short-wavelength **decimeter radiation,** with wavelengths a few tenths of a meter long. Although this type of radiation is also synchrotron radiation, radio telescopes revealed that it comes from a region at least three times larger than the disk of Jupiter. Detailed studies of the decimeter radiation showed that Jupiter's magnetic field was at least ten times stronger than Earth's, and, like Earth's field, was inclined about 10° to the axis of rotation.

Radio studies were helpful, but nothing is as good as a visit. In 1973 and 1974, two Pioneer spacecraft flew past Jupiter, followed in 1979 by two Voyager spacecraft. The Galileo spacecraft entered orbit around Jupiter in late 1995. These spacecraft found Jupiter surrounded by a magnetosphere over 100 times larger than Earth's.

The magnetic field deflects the solar wind and traps high-energy particles in radiation belts much more intense than Earth's. The radiation is more intense because Jupiter's magnetic field is stronger and can trap and hold higher-energy particles, mostly electrons and protons. The spacecraft passing through the radiation belts received radiation doses equivalent to a billion chest X rays—at least 100 times a lethal dose for a human. Some of the electronics on the spacecraft were damaged by the radiation.

Jupiter's rapid rotation has flattened its magnetosphere, and the charged particles are concentrated in a thin layer called the **current sheet** (❙ Figure 23-3). Because the magnetic field is inclined to the axis of rotation, the magnetosphere wobbles up and down as Jupiter rotates.

We have seen that Earth's magnetosphere produces aurorae, and the same is true on Jupiter. Charged

FIGURE 23-3

Jupiter's magnetic field traps particles from the solar wind to form powerful radiation belts. The rapid rotation of the planet forces the slightly inclined magnetic field to wobble as it rotates and confines most of the high-energy particles to a thin plane called the current sheet. Earth's magnetosphere and radiation belts are shown to scale (inset).

Visual-wavelength image

UV image

UV image

a

ACTIVE FIGURE 23-4

(a) Aurorae on Jupiter are strongest in rings around the magnetic poles. Aurora rings also occur on Earth, which explains why aurorae are seen at high latitudes but not lower latitudes. *(John Clarke, University of Michigan, NASA)* (b) The Io flux tube (blue curve) connects the moon Io to Jupiter. In the ultraviolet image, small spots of aurora mark the points where the flux tube enters Jupiter's atmosphere. Locate the same spots in part a of this figure. *(NASA)*

Ace⟳Astronomy™ Go to AceAstronomy and click Active Figures to see "Explorable Jupiter" and observe Jupiter at different wavelengths.

Visual   UV

b

particles in the magnetosphere leak downward along the magnetic field, and, where they enter the atmosphere, they produce aurorae 1000 times more powerful than those on Earth. The aurorae on Jupiter, like aurorae on Earth, occur in rings around the magnetic poles (▌Figure 23-4a).

Some of the heavier ions in the radiation belts appear to come from the inner Galilean moon, Io. As we will see later in this chapter, Io has active volcanoes that spew gas and ash. Because Io orbits with a period of 1.8 days, and Jupiter's magnetic field rotates in only 10 hours, the wobbling magnetic field rushes past Io at high speed, sweeping up stray particles, accelerating them to high energy, and spreading them around Io's orbit in a doughnut of ionized gas called the **Io plasma torus.**

Jupiter's magnetic field interacts with Io to produce a powerful electric current (about a million amperes) that flows through a curving path called the **Io flux tube** from Jupiter out to Io and back to Jupiter. Spots of bright aurora lie at the two points where the Io flux tube enters Jupiter's atmosphere (Figure 23-4b).

This Io flux tube explains a mystery. Back in 1955, when radio astronomers discovered the decametric radio signals, they noticed that the signals were stronger when Io was in certain parts of its orbit. Now we can understand that the electrical current flowing through the Io flux tube is at least partially responsible for generating the decametric radio signals.

Fluctuations in the aurorae tell us that the solar wind buffets Jupiter's magnetosphere, but some of the fluctuations seem to be caused by changes in the magnetic dynamo deep inside the planet. Thus, studies of the aurorae on Jupiter may help astronomers learn more about its liquid depths.

Much of the power and fascination of Jupiter is invisible to our eyes, but the swirling cloud belts are beautiful in their complexity.

JUPITER'S ATMOSPHERE

If you parachuted into Jupiter with a small rubber boat, expecting to go sailing on the liquid hydrogen ocean, you would be disappointed. Mathematical models of Jupiter tell us that there is no surface. Deep below the

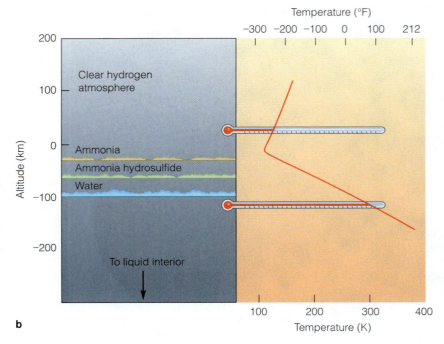

a

b

ACTIVE FIGURE 23-5

(a) In 1995, the Galileo probe dropped into Jupiter's atmosphere. After releasing its heat shield, it descended through the turbulent clouds transmitting measurements back to Earth. *(NASA)* (b) Thermometers inserted into the Jovian atmosphere would register temperatures as shown by the red line in the graph at right. Note the temperatures at which the three cloud layers form by extending them to the right until they touch the line on the graph. Read the temperature from the scale at the bottom and compare with Figure 20-8 and Figure 23-20b.

Ace◯Astronomy™ Go to AceAstronomy and click Active Figures to see "Planetary Atmospheres" and compare the atmospheres of the planets.

clouds, the temperature and pressure exceed the **critical point** for hydrogen, the temperature and pressure above which liquid hydrogen and gaseous hydrogen have the same density and are indistinguishable. As you parachuted down through Jupiter's atmosphere, the temperature, pressure, and density would rise, and gradually the gas would become a fluid. You would never splash down and your rubber boat would be useless. Below the clouds of Jupiter lies the largest ocean in the solar system—and it has no surface and no waves.

When we look at Jupiter, we don't see a surface. We see those parts of the atmosphere that contain clouds. These cloud layers lie deep inside a nearly transparent atmosphere of hydrogen and helium. We can detect this nearly transparent hydrogen atmosphere by noticing that Jupiter has limb darkening just as the sun does (Chapter 8). When we look near the limb of Jupiter (the edge of its disk), the clouds are much dimmer (Figure 23-1a) because it is nearly sunset or sunrise along the limb. If we were on Jupiter at this location, we would see the sun just above the horizon, and it would be dimmed by the atmosphere. In addition, light reflected from clouds must travel out at a steep angle through the

atmosphere to reach Earth, dimming the light further. Jupiter is brighter near the center of the disk because the sunlight shines nearly straight down and the cloud layers look brighter.

Jupiter's atmosphere is not deep. The Galileo probe fell through a transparent hydrogen atmosphere and then into the clouds (❚ Figure 23-5a); it was in total darkness by the time it reached 80 km below the highest clouds. It was destroyed by heat and pressure at a depth of 130 km, and models tell us that the gas of the atmosphere merges into the liquid interior about 1000 km below the clouds, which is not as deep as it seems when we recall that Jupiter's radius is 72,000 km. The entire atmosphere is only about 1 percent of the planet's radius. Earth's atmosphere is about 2.5 percent of Earth's radius. Furthermore, the actual clouds we see on Jupiter are confined to a layer less than 100 km deep.

Spectroscopic observations combined with models of the atmosphere suggest that Jupiter's clouds lie in three layers (Figure 23-5b). The highest layer is composed of ammonia crystals at a temperature of about 150 K (−180°F). Deeper in the atmosphere where it is warmer, ammonia hydrosulfide (NH_4SH) crystals form

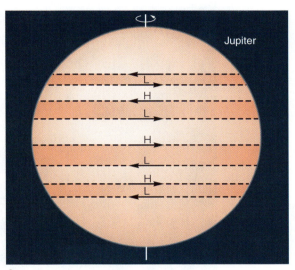

FIGURE 23-6

(a) In Jupiter's atmosphere belts are formed by low-pressure, sinking gas that is deeper in the atmosphere and not as bright as the higher-pressure rising gas in the zones. (b) In Earth's atmosphere, a wave circulation separates low- and high-pressure regions. (c) On Jupiter, low- and high-pressure regions take the form of belts of sinking gas and zones of rising gas.

a layer of clouds with a temperature of about 200 K. Deeper still lies a cloud layer of water droplets and ice crystals at a temperature of about 280 K. (Make special note of the temperatures of these cloud layers. We will find similar clouds at similar temperatures when we visit Saturn.)

The clouds of Jupiter are organized into dark **belts** and bright **zones.** Observations made by spacecraft show that the zones are cooler and higher than the belts. The zones appear to be high-pressure regions of rising gas that cools as it rises and forms clouds higher in the atmosphere where they receive more sunlight and look brighter (■ Figure 23-6a). The belts are low-pressure regions with sinking gas and lower clouds that are not as brightly lit.

Observations made by the Cassini spacecraft as it raced past Jupiter on its way to Saturn revealed that the dark belts contain small rising storm systems too small to see in previous images. Evidently, the general circulation usually attributed to the belts and zones is much more complex when we observe in more detail. Further understanding of the small-scale motions in Jupiter's atmosphere may have to await future planetary probes to Jupiter.

On Earth, high- and low-pressure regions are bounded by high-speed winds (Figure 23-6b), and the same is true on Jupiter. Winds blowing hundreds of kilometers per hour separate the belts and zones (Figure 23-6c).

The clouds on Jupiter are twisted and swirled by the winds, but the belts and zones are highly stable. Although their colors and brightness change sometimes, they have not changed their position since humans have been mapping them.

The colors of the clouds cannot be caused by ammonia, ammonia hydrosulfide, or water, which form colorless, white crystals. Rather, astronomers think that the colors are caused by traces of compounds containing sulfur or phosphorus modified by sunlight.

Mixed within the belt–zone circulation are light and dark spots, a few larger than Earth. Some appear and disappear in days; others last for years. Dark spots appear to be openings through which we see deeper, darker clouds, and white spots appear to be higher cloud circulations. The largest is the Great Red Spot shown in Figure 23-1, which has been in one of the southern zones for more than 330 years (■ Figure 23-7).

FIGURE 23-7

The Great Red Spot on Jupiter changes its appearance as shown in this eight-year sequence of images, but it has been visible for over 300 years. Rotating counter-clockwise with winds of 270 mi/hr, the spot is the largest storm in the solar system. *(Hubble Heritage Team STScI/AURA/ NASA and Amy Simon, Cornell U)*

Visual-wavelength images

In its longest dimension, it is twice the diameter of Earth. Pioneer, Voyager, and Galileo observations show that the Great Red Spot is higher than its surroundings. It is evidently formed by rising gas carrying heat upward from deep below the clouds and creating a vast, rotating storm.

On Earth, rising currents of hot gas can produce thunderheads that tower 15 kilometers and spawn powerful winds and tremendous lightning. The same is true on Jupiter. Voyager photos of the night side of Jupiter revealed lightning bolts among the clouds, and Galileo images show thunderheads up to 50 km high (white clouds in Figure 23-1b upper left).

Weather patterns on Earth are driven from above by heat from the sun, but weather patterns on Jupiter appear to be dominated by the interior. The Galileo probe falling into Jupiter's atmosphere found that the high-speed winds that bound the belts and zones extend deep into the atmosphere. Apparently, large-scale circulations in the liquid interior cause these winds and organize the overall belt–zone pattern (❙ Figure 23-8a). Heat rising from the interior drives convection and generates circulating storms. Evidently the highly complex weather we see on Jupiter is driven from below (Figure 23-8b).

The highly complex spacecraft that have visited Jupiter have given us a new understanding of the giant planet, but we must not confuse the technology of spacecraft with the science of comparative planetology (Window on Science 23-1). Science is about understanding nature, and Jupiter is an entirely new kind of planet in our study. In fact, Jupiter has a feature that we have not studied to this point. Jupiter has a ring.

WINDOW ON SCIENCE 23-1

Science, Technology, and Engineering

Sending spacecraft to another world is expensive, and it may seem pointless when that world seems totally hostile to human life. What practical value is there in sending a space probe to Jupiter? To resolve that question, we need to consider the distinction between science, technology, and engineering.

Science is nothing more than the logical study of nature, and the goal of science is understanding. Although much scientific knowledge proves to have tremendous practical value, the only goal of science is a better understanding of how nature works. Technology, in contrast, is the practical application of scientific knowledge to solve a specific problem. People trying to find a faster way to paint automobiles might use all the tools and techniques of science, but if their goal has some practical outcome, we should more properly call it technology rather than science. Engineering is the most practical form of technology. An engineer is likely to use well-understood technology to find a practical solution to a problem.

Of course, there are situations in which science and technology blur together. For example, humanity has a practical and tragic need to solve the AIDS problem; we need a cure and a preventative. Unfortunately, we don't understand the AIDS virus itself or viruses in general well enough to design a simple solution, and thus much of the work involves going back to basic science and trying to better understand how viruses interact with the human body. Is this technology or science? It is hard to decide.

We might describe science that has no known practical value as basic science or basic research. Our exploration of worlds such as Jupiter would be called basic science, and it is easy to argue that basic science is not worth the effort and expense because it has no known practical use. Of course, the problem is that we have no way of knowing what knowledge will be of use until we acquire that knowledge. In the middle of the 19th century, Queen Victoria is supposed to have asked physicist Michael Faraday

what good his experiments with electricity and magnetism were. He answered, "Madam, what good is a baby?" Of course, Faraday's experiments were the beginning of the electronic age. Many of the practical uses of scientific knowledge that fill our world—transistors, vaccines, plastics—began as basic research. Basic scientific research provides the raw materials that technology and engineering use to solve problems.

Basic scientific research has yet one more important use that is so valuable it seems an insult to refer to it as merely practical. Science is the study of nature, and as we learn more about how nature works, we learn more about what our existence in this universe means for us. The seemingly impractical knowledge we gain from space probes to other worlds tells us about our own planet and our own role in the scheme of nature. Science tells us where we are and what we are, and that knowledge is beyond value.

a

Visual false-color image

b

FIGURE 23-8

(a) The overall organization of Jupiter's belts and zones may be determined by large circulation patterns in the liquid interior. Such cylindrical circulations have been reproduced in laboratory experiments.
(b) The extreme complexity of Jupiter's clouds is evident in this color-enhanced image of the region near the Great Red Spot. Heat rising from the interior supplies the energy to generate these storms and circulations. *(NASA)*

JUPITER'S RING

Astronomers have known for centuries that Saturn has rings, but it was not until 1979, when the Voyager 1 spacecraft sent back photos, that Jupiter's ring was discovered. Less than 100 times as bright as Saturn's rings, the ghostly ring around Jupiter is a puzzle. What is it made of? Why is it there? A few simple observations help us understand.

Saturn's rings are made of bright ice chunks, but the particles in Jupiter's ring are very dark and reddish. We can conclude that the ring is rocky rather than icy.

We can also conclude that the ring particles are mostly microscopic. Photos of the ring show that it is very bright when illuminated from behind (Figure 23-9)—it is scattering light forward. Efficient **forward scattering** occurs when particles have diameters roughly the same as the wavelength of light, a few millionths of a meter. Large particles do not scatter light forward, so a ring filled with basketball-size particles would look dark when illuminated from behind. Forward scattering tells us that the ring is made of particles about the size of those in cigarette smoke.

The location of these particles near the planet is understandable. They orbit inside the **Roche limit,** the distance from a planet within which a moon cannot hold itself together by its own gravity. If a moon stays far from its planet, then the moon's gravity will be much greater than the tidal forces caused by the planet, and the moon will be able to hold itself together. If, however, the planet's moon comes inside the Roche limit, the tidal forces overcome its gravity and pull the moon apart. The International Space Station can orbit inside the Roche limit because it is held together by bolts and welds, but a moon held together by its own gravity cannot survive inside the Roche limit. If a planet and its moon have similar densities, the Roche limit is 2.44 planetary radii. Jupiter's main ring has an outer radius of 130,000 km (1.8 planetary radii) and lies inside the Roche limit, as do the rings of Saturn, Uranus, and Neptune.

Now we can understand the dust near Jupiter. If a dust speck gets knocked loose from a small moon inside the Roche limit, the moon's gravity cannot hold the dust speck. And the billions of dust specks in the ring can't pull themselves together to make a new moon because of tidal forces inside the Roche limit.

We can be sure that the ring particles are not old. The pressure of sunlight and the powerful magnetic

Visual-wavelength image

a

Visual false-color image

b

ACTIVE FIGURE 23-9

(a) The main ring of Jupiter, illuminated from behind, glows brightly in this photo made by the Galileo spacecraft from within Jupiter's shadow. (b) Digital enhancement and false color reveal the halo of ring particles that extends above and below the main ring. *(NASA)*

Ace Astronomy™ Go to AceAstronomy and click Active Figures to see "Roche Limit."

field alter the orbits of the particles, and they gradually spiral into the planet. Images show faint ring material extending down toward the cloud tops, and this is evidently dust specs spiraling into the planet. Dust is also lost from the ring as electromagnetic effects force it out of the plane of the ring to form a low-density halo above and below the ring (Figure 23-9b). Another reason the ring particles can't be old is that the intense radiation around Jupiter will grind the dust specks down to nothing in a century or so. Thus the rings we see today can't be material left over from the formation of Jupiter.

The rings of Jupiter must be continuously resupplied with new dust. Small moons that orbit near the outer edge of the rings may lose dust particles as they are hit by micrometeorite impacts. Observations made by the Galileo spacecraft show that the main ring is densest at its outer edge where the small moon Adrastea orbits and that another small moon, Metis, orbits inside the ring. Galileo images reveal much fainter rings called the **gossamer rings,** extending twice as far from the planet as the main ring. These gossamer rings are most dense at the orbits of two small moons, Amalthea and Thebe, again giving us evidence that the dust is being blasted into space by impacts on the inner moons.

Besides supplying the rings with particles, it is possible that the moons help confine the ring particles and keep them from spreading outward. This is an important process in planetary rings, and we will explore it in detail when we study the rings of Saturn later in this chapter.

Our exploration of Jupiter reveals that it is much more than just a big planet. It is the gravitational and magnetic center of an entire community of objects. Occasionally the community suffers an intruder.

COMET IMPACT ON JUPITER

Comets are very common in the solar system, and Jupiter, because of its strong gravity, probably gets hit by comets more often than most planets. But no one had ever seen it happen until 1994, when fragments from a disrupted comet smashed into Jupiter.

The comet, named Shoemaker–Levy 9 after its discoverers, Eugene and Carolyn Shoemaker and David H. Levy, was captured into orbit around Jupiter sometime in the past. In 1992, it passed inside Jupiter's Roche limit and was pulled into at least 21 pieces that looped out away from Jupiter in long elliptical orbits. At this point, the objects were discovered on a photograph. Drawn out into a long strand of small comets, the pieces fell back and slammed into Jupiter over a period of six days in July 1994 (❚ Figure 23-10).

The fragments were no bigger than a kilometer or so in diameter and contained a fluffy mixture of rock and ices. They hit Jupiter at speeds of about 60 km/s and released energy equivalent to a few million megatons of TNT. (At the height of the cold war, all the nu-clear weapons on Earth totaled about 80,000 megatons. The Hiroshima bomb was only 0.15 megaton.)

The impacts occurred just over Jupiter's horizon as seen from Earth, but they produced fireballs almost 3000 km high, and the rapid rotation of Jupiter brought the impact points within sight of Earth only 15 minutes later. Infrared telescopes easily spotted the glowing scars where the comets hit, and, as they cooled, the impact sites became dusty, dark spots at visible wavelengths as shown in Figure 23-10. Visible through even small telescopes, some of these dark smudges were larger than Earth.

The comet impact was an astonishing spectacle, but what can we learn from it? We can use the event in two different ways. First, astronomers used the impacts as probes of Jupiter's atmosphere. By making assumptions about the nature of Jupiter's atmosphere and by using the most powerful computers, astronomers created models of a high-velocity projectile penetrating into Jupiter's upper atmosphere. By comparing the observed impacts with the models, astronomers were able to fine-tune the models to better represent Jupiter's atmosphere. This method of comparing models with reality is a critical part of science; therefore, the impacts on Jupiter resulted in a better understanding of the nature of Jupiter's atmosphere.

Second, the spectacle helps us in a more general way. It reminds us that planets are hit by large objects such as asteroids and the heads of comets. Jupiter probably is hit every century or so. In 1690, the Italian astronomer Cassini observed a dark spot that appeared on Jupiter and changed in a pattern over a period of days. We now recognize that pattern as the scar of a comet impact. We have seen evidence for large impacts on Earth, the moon, Venus, Mercury, and Mars, and we will see further evidence as we continue our exploration of the solar system.

Comets hit planets all the time. The impact in 1994 was just the first such event to occur since the rise of modern astronomy. We will see in Chapter 25 that a comet or asteroid impact on Earth may have changed the climate and killed the dinosaurs. A solar system is a dangerous place to put an inhabited planet.

THE HISTORY OF JUPITER

Our goal in studying any planet is to be able to tell its story—to describe how it got to be the way it is. While we understand part of the story of Jupiter, we still have much to learn.

If the solar nebula theory for the origin of the solar system is correct, then Jupiter formed from the colder gases of the outer solar nebula, where ices were able to condense. Thus, Jupiter grew rapidly and eventually became massive enough to trap hydrogen and helium gas directly from the solar nebula. The denser material, such as rock and metal, sank to the center to form a

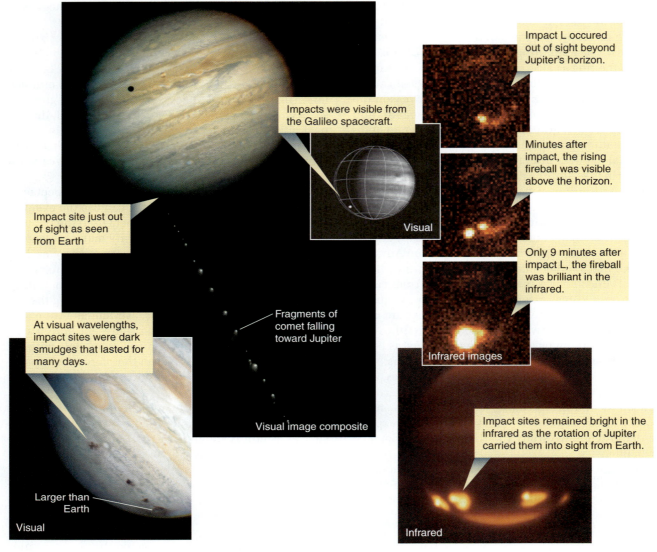

FIGURE 23-10

In 1994, fragments of Comet Shoemaker–Levy 9 struck Jupiter. Although these were the first comet impacts ever seen by Earth's inhabitants, such events are probably common occurrences in planetary systems. *(Composite and visual images: NASA; IR sequence: Mike Skrutskie; IR image: University of Hawaii)*

core, and we now see Jupiter as a low-density world of liquid hydrogen surrounding a hot core of heavier elements.

In the interior of Jupiter, the hydrogen takes the form of liquid metallic hydrogen, which is a very good electrical conductor. The planet's rapid rotation, coupled with the outward flow of heat from its hot interior, drives a dynamo effect that produces a powerful magnetic field. That vast magnetic field traps high-energy particles from the solar wind to form intense radiation belts.

The rapid rotation and large size of Jupiter cause the weather patterns to take the form of belt–zone circulation. Heat flowing upward from the interior causes rising currents in the bright zones, and cooler gas sinks in the dark belts. As on Earth, winds blow at the mar-

gins of these regions, and large spots appear to be cyclonic disturbances.

Although the age of planet building is long past, Jupiter continues to be hit by meteorites and occasional comets, as do all the planets. Any debris left over from the formation of Jupiter would have been blown away long ago by the solar wind, so the dust we find trapped in Jupiter's thin ring must be young. It probably comes from meteorites hitting the innermost moons.

Our study of Jupiter has been challenging because Jupiter lacks any surface—it is difficult to imagine being there. Most of the surface features and processes we found on the terrestrial planets are missing on Jupiter, but, as the prototype of the Jovian worlds, it earns its place as the ruler of the solar system.

The study of Jupiter is more meteorology than geology, but when we examine its moons, we find all the excitement of impact cratering and active volcanism.

23-2 JUPITER'S FAMILY OF MOONS

How many moons does Jupiter have? You will have to check the World Wide Web to get the latest figure. Astronomers are finding more and more small moons, and the count is now over two dozen. Most of these moons are small and rocky, and many are probably captured asteroids. Four of the moons, those discovered by Galileo and now called the Galilean moons, are large and have interesting geologies (❙ Figure 23-11).

Our study of the moons of Jupiter will illustrate three important principles in comparative planetology. First, a body's composition depends on the temperature of the material from which it formed. This is illustrated by the prevalence of ice as a building material in the outer solar system where sunlight is weak. The second principle is that cratering can tell us the age of a surface. Finally, we will see that internal heat has a powerful influence over the geology of these larger moons.

CALLISTO: THE ANCIENT FACE

The outermost of Jupiter's four large moons, Callisto is half again as large in diameter as Earth's moon. Like all of Jupiter's larger satellites, Callisto is tidally locked to its planet, keeping the same side forever facing Jupiter. From its gravitational influence on passing spacecraft, astronomers can calculate Callisto's mass, and dividing by its volume tells us that its density is 1.79 g/cm³. Ice has a density of about 1 and rock 2.5 to 4 g/cm³, so Callisto must be a mixture of rock and ice.

Photographs from the Voyager and Galileo spacecraft show that the surface of Callisto is dark, dirty ice heavily pocked with craters (❙ Figure 23-12). Old, icy surfaces in the solar system become dark because of dust added by meteorites and because meteorite impacts vaporize water, leaving the rock in the ice behind to form a dirty crust.

Spectra of the surface show that it is mostly a 50/50 mix of ice and rock, but some areas are ice free. Nevertheless, the slumped shape of craters suggests that the outer 10 km is mostly ice. This can be understood when we recall that the spectra tell us about the outer 1 mm of the surface, which can be quite dirty, while the shapes of craters tell us about the outermost 10 km, which could be rich in ice.

Visual-wavelength images

Size of Earth's moon

FIGURE 23-11
The Gallilean moons of Jupiter from left to right are Io, Europa, Ganymede, and Callisto. The circle shows the size of Earth's moon. *(NASA)*

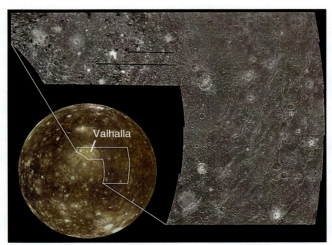

Visual-wavelength images

FIGURE 23-12

The dark surface of Callisto is dirty ice marked by craters. The youngest craters look bright, because they have dug down to cleaner ice. Valhalla is the 4000-km-diameter scar of a giant impact feature, one of the largest in the solar system. Valhalla is so large and old that the icy crust has flowed back to partially heal itself, and the outer rings of Valhalla are shallow troughs marking fractures in the crust. *(NASA)*

Delicate measurements of the shape of Callisto's gravitational field were made by the Galileo spacecraft, and they show that Callisto has never fully differentiated to form a dense core and a lower-density mantle. Its interior is a mixture of rock and ice. This is consistent with the observation that it has only a weak magnetic field of its own. A strong magnetic field would be

generated by the dynamo effect in a molten convecting core, and Callisto has no core. It does, however, interact with Jupiter's magnetic field in a way that suggests it has a layer of liquid water roughly 10 km thick about 100 km below its icy surface. Slow radioactive decay in its interior may provide enough heat to keep this layer of water liquid.

GANYMEDE: A HIDDEN PAST

The next Galilean moon inward is Ganymede, larger than Earth's moon, larger than Mercury, and over three-quarters the diameter of Mars. Its density is 1.9 g/cm³, and its influence on the Galileo spacecraft reveals that it has a rocky core, an ice-rich mantle, and a crust of ice 500 km thick. It may even have a small iron core. It is large enough for radioactive decay to have melted its interior when it formed, allowing iron to drain to its center.

The surface hints at an active past. Although a third of the surface is old, dark, and cratered, the rest is marked by bright parallel grooves. Because this bright **grooved terrain** (Figure 23-13a) contains fewer craters, it seems young.

Observations show that the bright terrain was produced when the icy crust broke and water flooded up from below. Some low-lying regions are smooth and appear to have been flooded by water. Spectra reveal concentrations of salts such as those that would be left behind by the evaporation of mineral-rich water. Also, some features in or near the bright terrain appear to be

a

b   Visual-wavelength images

FIGURE 23-13

(a) This color-enhanced image of Ganymede shows the frosty poles at top and bottom, the old dark terrain, and the brighter grooved terrain. (b) A band of bright terrain runs from lower left to upper right, and a collapsed area, a possible caldera, lies at the center in this image. Calderas form where subsurface liquid has drained away, and the bright areas do contain other features associated with flooding by water. *(NASA)* caldera formed when subsurface water drained away and the surface collapsed (Figure 23-12b).

The Galileo spacecraft found that Ganymede has a magnetic field about 10 percent as strong as the Earth's. It even has its own magnetosphere inside the larger magnetosphere of Jupiter. Theoretical calculations find it difficult to produce a magnetic field in a water-rich mantle, and there does not appear to be enough heat in Ganymede for it to have a molten metallic core. Astronomers wonder if its magnetic field is left over from a time when it was hotter and more active.

Ganymede's magnetic field fluctuates with the 10-hour period of Jupiter's rotation. The rotation of the planet sweeps its tilted magnetic field past the moon, and the two fields interact. That interaction reveals that the moon has a layer of liquid water about 170 km (100 mi) below its surface. The water layer may be about 5 km (3 mi) thick. It is possible that the water layer was thicker and closer to the surface long ago when the interior of the moon was warmer. That might explain the flooding that appears to have helped form the bright grooved terrain.

A possible heat source for Ganymede is **tidal heating,** the frictional heating of a body by changing tides (▌Figure 23-14a). In the past, Ganymede could have interacted with the other moons and entered a slightly elliptical orbit. Jupiter's gravity would have deformed the moon, and, as Ganymede followed its elliptical orbit, tides would have flexed it, and friction would have heated it. Such an episode of tidal heating might have driven a dynamo to produce a magnetic field and break the crust to make the grooved terrain.

Because massive planets like Jupiter draw debris inward, a moon orbiting near a massive planet should be struck by many meteorites (Figure 23-14b). Nevertheless, the bright terrain on Ganymede has few craters and appears to be only about 1 billion years old. It may date from an age when Ganymede was in an orbital resonance with other moons that forced it into a slightly elliptical orbit where it suffered strong tidal heating. In any case, the grooved face of Ganymede hides a fascinating past.

EUROPA: A HIDDEN OCEAN

The next Galilean moon inward is Europa, a bit smaller than Earth's moon. Europa has a density of about 3 g/cm³, so it must be mostly rock and metal. Yet its surface is ice.

Europa lies closer to Jupiter than Ganymede, so it should be exposed to more meteorite impacts (Figure 23-14b), but the icy crust of Europa is almost free of craters. A recent crater such as Pwyll is bright, but most are hardly more than blemishes in the ice (▌Figures 23-15a and b). Evidently the surface of Europa is active and erases craters almost as fast as they form. The number of impact scars on Europa suggests that the average

FIGURE 23-14

Two effects on planetary satellites. (a) Tidal heating occurs when changing tides cause friction within a moon. (b) The focusing of meteoroids exposes satellites in small orbits to more impacts than satellites in larger orbits receive.

age of its surface is only 10 million years. Other signs of activity include long cracks in the icy crust and regions where the crust has broken into sections that have moved apart as if they were icebergs floating on water (Figure 23-15c).

Even Europa's clean face tells us its surface is young. The albedo of the surface is 0.69, meaning that it reflects 69 percent of the light that hits it. This is a high albedo produced by the clean ice. We have discovered that old, icy surfaces tend to be very dark. Europa's high albedo must mean the surface is active.

Europa is too small to have retained much heat from its formation or from radioactive decay, and the Galileo spacecraft found that Europa has no magnetic field of its own. It cannot have a molten conducting core. Tidal heating, however, is important for Europa and apparently provides enough heat to keep the little moon active. In fact, the curving cracks in its crust reveal the shape of the tidal forces that flex it as it orbits Jupiter.

If you hiked on Europa with a compass in your hand, you would be in trouble. The moon has no magnetic field of its own, but Jupiter rotates rapidly and drags its field past the little moon. That induces a fluctuating magnetic field in Europa that would make your compass wander uselessly. That magnetic interaction reveals the presence of a liquid-water ocean lying about 15 km (10 mi) below the icy surface. The ocean may be as deep as 100 miles (▌Figure 23-16) and could contain twice as much water as all the oceans on Earth. It would be rich in dissolved minerals, which would make it a good electrical conductor and allow it to interact with Jupiter's magnetic field. No one knows what might be swimming through such an ocean.

FIGURE 23-15

(a) The icy surface of Europa is shown here in natural color. Many faults are visible on its surface, but very few craters. The bright crater is Pwyll, a young impact feature. (b) This circular bull's-eye is 140 km in diameter. It is the remains of an impact by an object about the size of a mountain. Notice the younger cracks and faults that cross the older impact feature. (c) Like icebergs on an arctic ocean, blocks of crust on Europa appear to have floated apart. Spectra show that the blue ice is stained by salts such as those that would be left behind by mineral-rich water welling up from below and evaporating. White areas are ejecta from the impact that formed Pwyll. *(NASA)*

a   b

c   Visual-wavelength images

FIGURE 23-16

The gravitational influence of Europa on the passing Galileo spacecraft shows that the little moon has differentiated into a dense core and rocky mantle. Magnetic interactions with Jupiter show that it has a liquid-water ocean below its icy crust. Heat produced by tidal heating could flow outward as convection in such an ocean and drive geological activity in the icy crust. *(NASA)*

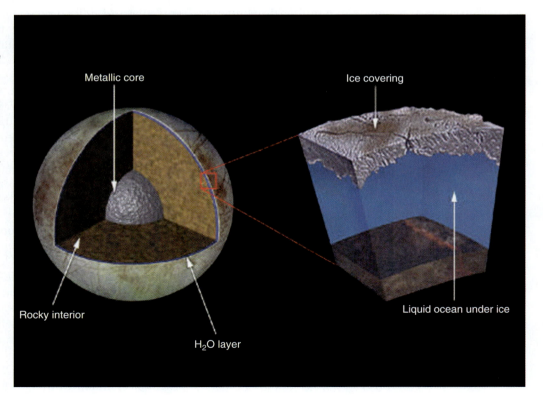

Metallic core   Ice covering

Rocky interior

H$_2$O layer   Liquid ocean under ice

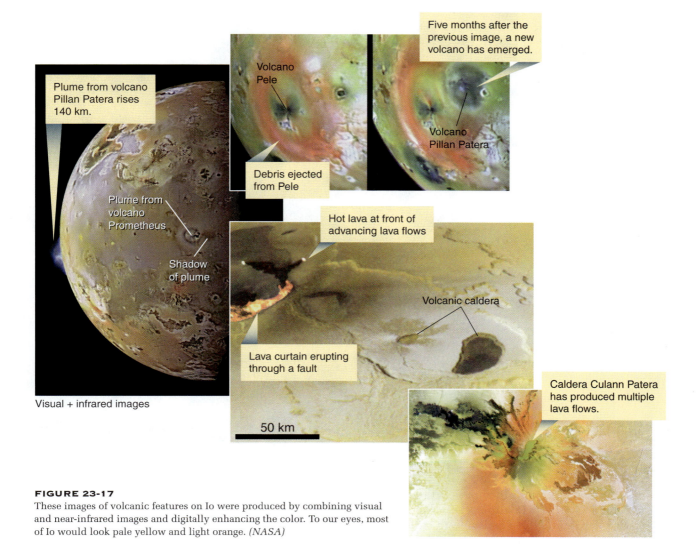

FIGURE 23-17

These images of volcanic features on Io were produced by combining visual and near-infrared images and digitally enhancing the color. To our eyes, most of Io would look pale yellow and light orange. *(NASA)*

The following labels appear in the figure:

Plume from volcano Pillan Patera rises 140 km.

Volcano Pele

Five months after the previous image, a new volcano has emerged.

Debris ejected from Pele

Volcano Pillan Patera

Plume from volcano Prometheus

Shadow of plume

Hot lava at front of advancing lava flows

Volcanic caldera

Lava curtain erupting through a fault

Caldera Culann Patera has produced multiple lava flows.

Visual + infrared images

50 km

Tidal heating makes Europa geologically active. Apparently, rising currents of water can break through the icy crust or melt patches. Many of the cracks show evidence they have spread apart and that fresh water has welled up and frozen between. In other regions, compression in the crust is revealed by networks of faults and low ridges. Compression on Earth pushes up mountain ranges, but no such ranges appear on Europa. The icy crust isn't strong enough to support ridges higher than a kilometer or so.

Orbiting deep inside Jupiter's radiation belts, Europa is bombarded by high-energy particles that damage the icy surface. Water molecules are freed and broken up to be dispersed into a doughnut-shaped cloud spread round Jupiter and enclosing Europa's orbit. Flying past Jupiter in 2002, the Cassini spacecraft was able to image this cloud of excited gas. Europa's gas cloud serves to warn us that moons orbiting deep inside a massive planet's radiation belts are exposed to a form of erosion that is entirely lacking on Earth's moon.

IO: BURSTING ENERGY

Geological activity is driven by heat flowing out of a planet's interior, and nothing could illustrate this principle better than Io, the innermost of Jupiter's Galilean moons. Photographs from the Voyager and Galileo spacecraft show no impact craters at all—surprising considering Jupiter's power to focus meteoroids inward (Figure 23-14b). No subtlety is needed to explain the missing craters. Over 150 active volcanoes are visible on the moon's surface, blasting sulfurous gas and ash out over the surface to bury any newly formed craters (❙ Figure 23-17). Unlike the dead or dormant outer Galilean moons, Io is bursting with energy.

Spectra reveal that Io has a tenuous atmosphere of gaseous sulfur and oxygen, but those gases can't be permanent. The erupting volcanoes pour out about one ton of gases per second, but, because of Io's low escape velocity, the gases leak into space easily. Also, any gas atoms that become ionized are swept away by Jupiter's

FIGURE 23-18

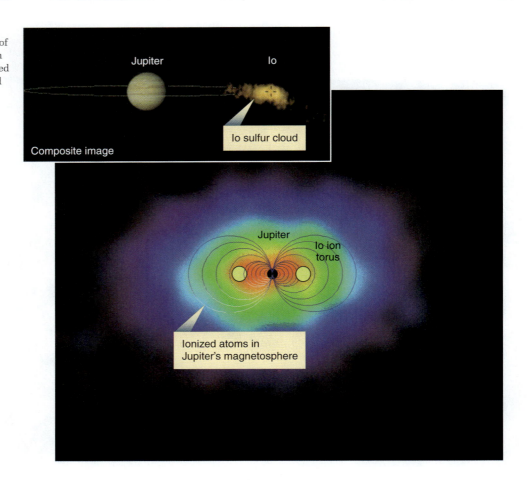

In the inset (right), a visual image of Jupiter has been combined with an image made at a wavelength emitted by sulfur atoms to reveal the cloud of gases around Jupiter's volcanic moon Io. *(NASA/JPL/Bruce A. Goldberg)* The larger map was created by the Cassini spacecraft and shows the size of Jupiter's magnetosphere and the location of the torus of ionized gases that enclose Io's orbit. A fainter torus of water atoms is located around the orbit of Europa. *(NASA/JPL/ Johns Hopkins Univ. Applied Physics Lab)*

rapidly rotating magnetic field. The ions produce a cloud of sulfur and sodium ions in a torus (a doughnut shape) enclosing Io's orbit (■ Figure 23-18). Io's neighborhood is dirty (and probably smelly) with sulfur. In fact, the reddish color of the small inner moon Amalthea may be caused by sulfur pollution from Io.

We can use basic observations to deduce the nature of Io's interior. From the density, 3.55 g/cm³, we can conclude that it is mostly rock. Spectra reveal no trace of water at all, so there is no ice on Io. It is the driest world in our solar system. The oblateness of Io caused by its rotation and by the slight distortion produced by Jupiter's gravity gives astronomers clues to its interior. Model calculations suggest it contains a modest core of metals and a deep rocky mantle.

The colors of Io have been compared to those of a badly made pizza. Certainly the reds, oranges, and browns of Io are caused by sulfur and sulfur compounds, but the crust is true rock. Infrared measurements show that volcanoes on Io erupt lava with a temperature over 1500°C (2700°F), about 300°C hotter than lavas on Earth. Sulfur on Io would boil at only 550°C. Also, a few isolated mountains exist that are as high as 18 km, twice the height of Mt. Everest. Sulfur is not strong enough to support such high mountains. The lava flows and most of the crust must be silicate rock and not sulfur.

Volcanism is continuous on Io. Plumes come and go over periods of months, but some, such as Pele, have been active since the Voyager spacecraft first visited Io in 1979 (Figure 23-17). Earth's volcanoes eject lava and ash because of water dissolved in the lava. As rising lava reaches Earth's surface, the sudden decrease in pressure allows the water to come out of solution in the lava. It is like popping the cork on a bottle of champagne. The water flashes into vapor and blasts material out of the volcano, a process that was responsible for the Mount St. Helens explosion in 1980. But Io is dry. Its volcanoes appear to be powered by sulfur dioxide dissolved in the magma. When the pressure is released, the sulfur dioxide boils out of solution and blasts gas and ash high above the surface in plumes up to 500 km high. Ash falling back to the surface produces debris layers around the volcanoes, such as that around Pele in Figure 23-17. Whitish areas on the surface are thought to be frosts of sulfur dioxide.

Great lava flows are visible carrying molten material downhill, where it buries the surface under layer after layer. Sometimes lava bursts upward through faults to form long lava curtains, a form of eruption seen in Hawaii. Both of these processes are shown in Figure 23-17.

What powers Io? It is bursting with energy, but it is only a few percent bigger than Earth's moon, which is cold and dead. Io is too small to have retained heat from its formation or to remain hot from radioactive decay. In fact, the energy blasting out of its volcanoes is about three times more energy than it could make by radioactive decay in its interior.

Io is heated by the same kind of tidal heating that is suspected in Ganymede's past and in Europa. Because Io is so close to Jupiter, tides are powerful and should have forced Io's orbit to become circular long ago. Io, however, has fallen in with a bad crowd. Io, Europa, and Ganymede are locked in an orbital resonance; in the time it takes Ganymede to orbit once, Europa orbits twice and Io four times. This gravitational interaction keeps the orbits slightly elliptical, and Io, being closest to Jupiter, suffers dramatic tides, with its surface rising and falling by about 100 m. The resulting friction is enough to melt the interior and drive volcanism. In fact, the energy flowing outward is continually recycling Io's crust. Deep layers melt, are spewed out through the volcanoes, fall back to cover the surface, and are later covered themselves until they are buried so deeply that they are again melted.

Io is the most active world in our solar system because it lives so close to Jupiter. The other Galilean moons are more distant and thus have had less dramatic histories of tidal heating and geological activity.

THE HISTORY OF THE GALILEAN MOONS

The goal of astronomy is to tell the story of the universe. Now we must test our understanding of the Galilean moons by trying to tell their story. To do that we not only draw on what we have learned about the moons, but we also draw on what we have learned about Jupiter and about the origin of the solar system (Chapter 19).

The minor moons of Jupiter are probably captured asteroids, but the Galilean moons seem to be primordial. That is, they formed with Jupiter. Also, they seem to be a set in that they share common features and their densities are related to their distance from Jupiter (▪ Table 23-2). This information suggests that the four moons formed in a disk-shaped nebula around Jupiter— a mini-solar nebula—in much the same way the planets formed from the solar nebula around the sun.

As Jupiter grew massive, it should have formed a hot, dense disk of matter around its equator. The moons could have condensed inside that disk with the innermost moons, Io and Europa forming from rocky material and the outer moons Ganymede and Callisto incorporating more ices. This theory follows the condensation sequence that led to rocky planets forming near the sun and ice-rich worlds forming farther away.

There are objections to this theory. The disk around Jupiter would have been dense and hot, and moons would have formed rapidly, perhaps in only a 1000 years. The heat of formation released as material fell

Name	Radius (km)	Density (g/cm³)	Orbital Period (days)
Io	1816	3.55	1.769
Europa	1563	3.04	3.551
Ganymede	2638	1.93	7.155
Callisto	2410	1.81	16.689

TABLE 23-2
The Galilean Satellites*

*For comparison, the radius of Earth's moon is 1738 km, and its density is 3.36 g/cm³.

into the moons would have heated them so hot they would have lost their water. Ganymede and Callisto are rich in water, and Callisto has never been hot enough to differentiate and form a core. Also, mathematical models show that moons orbiting in the dense disk would have lost orbital momentum and would have spiraled into Jupiter within a century.

A newer theory proposes that Jupiter's early disk was indeed dense and hot and may have created moons, but those moons spiraled into the planet and were lost. Only later as the disk grew thinner and cooler did the present Galilean moons begin to form. Additional material may have dribbled slowly into the disk, but the moons must have formed slowly enough to retain their water and avoid spiraling into Jupiter.

We can understand the interiors of the moons by thinking about tidal heating. The moons formed slowly, over perhaps 100,000 years, and were not heated severely by in-falling material. The inner moons, however, were cooked by tidal heating—possibly enhanced when an orbital resonance developed between Ganymede, Europa, and Io. The innermost moon, Io, was heated so much it lost all of its water, and Europa retained only a small amount. Ganymede was heated enough to differentiate but retained much of its water. Callisto, orbiting far from Jupiter and avoiding orbital resonances, was never heated enough to differentiate. The Galilean moons we see today appear to be the result of slow formation and tidal heating.

Io is bursting with energy and the Galilean satellite system is bursting with history. As we learn to understand that history, we will learn more about how worlds form and evolve.

REVIEW CRITICAL INQUIRY

What produces Io's internal heat?

Part of critical thinking is recognizing and using basic principles. In this case, we understand that small worlds lose their internal heat quickly and become geologically inactive.

Io is a bit smaller than Earth's moon, which is cold and dead, but Io is bursting with energy flowing outward.

Clearly, Io must have a powerful source of heat inside, and that heat source is tides. Io's orbit is slightly elliptical, so it is sometimes closer to Jupiter and sometimes farther away. That means that Jupiter's powerful gravity sometimes squeezes Io more than at other times, and the flexing of the little moon's interior produces heat through friction. Such tides would rapidly force Io's orbit to become circular—and then tidal heating would end and the planet would become inactive—except that the gravitation of the other moons keeps Io's orbit elliptical. Thus, it is the influence of its companions that keeps Io in such an active state.

Io has almost no impact craters, but Callisto has many. What does the distribution of craters on the Galilean satellites tell us about their history?

We have now applied the principles of comparative planetology not only to planets but also to moons. With the insight we have gained from Jupiter and its satellite system, we are ready to visit the most beautiful of the Jovian worlds.

23-3 SATURN

Saturn has played second fiddle to its own rings since Galileo first saw it in 1610. He didn't recognize the rings for what they are, but we know them now as one of the great sights in the solar system. Nevertheless, Saturn itself, only slightly smaller than Jupiter (▌ Data File Eight), is a fascinating planet with a few mysteries of its own. Our discussion of Saturn can make use of the principles we have learned from Jupiter.

PLANET SATURN

The basic characteristics of Saturn tell us about its composition and interior. Only about a third of the mass of Jupiter and 16 percent smaller in radius, Saturn has an average density of 0.69 g/cm³. It is less dense than water—it would float! Although Saturn rotates about as fast as Jupiter, it is almost twice

as oblate, so we can conclude that it must not contain a large core of heavy elements. Spectra show that its atmosphere is rich in hydrogen and helium (see Table 23-1), and models predict that it is mostly liquid hydrogen.

Infrared observations show that Saturn is radiating 1.8 times as much energy as it receives from the sun, telling us that heat is flowing out of its interior. It must be hot inside. In fact, it is hotter than we would expect. It should have lost more heat since it formed. Astronomers suspect that helium in the liquid hydrogen interior is condensing into droplets and falling inward. The falling droplets, releasing energy as they pick up speed, heat the planet. This heating is similar to the heating produced when a star contracts.

We can learn more about Saturn's interior from its magnetic field. The Voyager spacecraft found that Saturn's magnetic field is about 20 times weaker than Jupiter's. It also has correspondingly weak radiation belts. Models predict that the pressure inside Saturn, being weaker than that inside Jupiter, produces a smaller mass of liquid metallic hydrogen. Heat flowing outward must cause convection in this conducting core, and the rapid rotation drives a dynamo effect to produce the magnetic field. Unlike most magnetic fields, Saturn's field is not inclined to its axis of rotation, something we notice in ultraviolet images that show rings of aurorae around Saturn's poles (▌ Figure 23-19). This alignment between the magnetic axis and the axis of rotation is peculiar and isn't understood.

UV image

FIGURE 23-19
At ultraviolet wavelengths, the Hubble Space Telescope can photograph rings of aurorae surrounding Saturn's north and south magnetic poles. (Compare with Figure 23-4a.) *(J. T. Trauger, JPL, and NASA)*

Saturn's atmosphere is rich in hydrogen and contains belt–zone circulation, which appears to arise in the same way as that on Jupiter. Zones are higher clouds formed by rising gas, and belts are lower clouds formed by sinking gas.

But the clouds are not very distinct on Saturn (▌Figure 23-20a). Measurements from Voyager spacecraft tell us that Saturn's atmosphere is much colder than Jupiter's—something we would expect because Saturn is twice as far from the sun and receives only one-fourth as much solar energy per square meter. The clouds on Saturn form at about the same temperature as the clouds on Jupiter, but those temperatures are deeper in Saturn's cold atmosphere. Compare the cloud layers in Figure 23-20b with those in Figure 23-5b. The atmospheres of Jupiter and Saturn are very similar once we remember that Saturn is colder.

One dramatic difference between Jupiter and Saturn concerns the winds. On Jupiter, winds bound each of the belts and zones, but on Saturn the pattern is not the same. Saturn has fewer such winds, but they are much stronger. The eastward wind at the equator of Saturn, for example, blows at 500 m/s (1100 mph), roughly five times faster than the eastward wind at Jupiter's equator. The reason for this difference is not clear.

SATURN'S RINGS

Looking at the beauty and complexity of Saturn's rings, an astronomer once said, "The rings are made of beautiful physics." We might add that the physics is actually rather simple, but the result is one of the most astonishing sights in our solar system.

In 1609, Galileo was the first to see the rings of Saturn, but perhaps because of the poor optics in his telescopes, he did not recognize the rings as a disk. He drew Saturn as three objects—a central body and two smaller ones on either side. In 1659, Christian Huygens realized that the rings were a disk surrounding but not touching the planet.

Study "▌The Ice Rings of Saturn" on pages 524 and 525 and notice four things. First, the rings are made up of billions of ice particles. Second, notice that the rings can't be as old as Saturn. The rings must be replenished now and then by impacts on Saturn's icy moons. Third, where ring particles orbit with periods that are simply related to the orbital period of one of Saturn's inner moons, a resonance can create a gap in the rings. The fourth thing to notice is a related phenomenon: Small moons near the rings can shepherd the ring particles and confine them. Thus the rings of Saturn, and the rings of the other Jovian moons, are created by and controlled by the planet's moons. Without the moons, there would be no rings.

Saturn's rings are governed by simple physics, but understanding what the rings are has demanded over a century of human ingenuity. In 1859, James Clerk Maxwell

Visual

SATURN

Data File Eight

Saturn's rings of ice particles are bright and beautiful, but its belt–zone circulation is subdued. A moon and its shadow lie just above center. *(NASA)*

Average distance from the sun	9.5388 AU (14.27×10^8 km)
Eccentricity of orbit	0.0560
Maximum distance from the sun	10.07 AU (15.07×10^8 km)
Minimum distance from the sun	9.005 AU (13.47×10^8 km)
Inclination of orbit to ecliptic	2°29′17″
Average orbital velocity	9.64 km/s
Orbital period	29.461 y (10,760 days)
Period of rotation	$10^h39^m25^s$
Inclination of equator to orbit	26°24′
Equatorial diameter	120,660 km (9.42 $D_\oplus$)
Mass	5.69×10^{26} kg (95.147 $M_\oplus$)
Average density	0.69 g/cm^3
Gravity at base of clouds	1.16 Earth gravities
Escape velocity	35.6 km/s (3.2 $V_\oplus$)
Temperature at cloud tops	−180°C (−292°F)
Albedo	0.61
Oblateness	0.102

The Ice Rings of Saturn

The brilliant rings of Saturn are made up of billions of ice particles ranging from microscopic specks to chunks bigger than a house. Each particle orbits Saturn in its own circular orbit. Much of what we know about the rings has been learned from spacecraft visiting Saturn. From Earth, astronomers see three rings labeled A, B, and C. Voyager images reveal over a thousand ringlets within the rings.

Saturn's rings can't be leftover material from the formation of Saturn. The rings are made of ice particles, and the planet would have been so hot when it formed that it would have vaporized and driven away any icy material. Rather, the rings must be debris from collisions between passing comets and Saturn's icy moons. Such impacts should occur every 10 million years or so, and they would scatter ice throughout Saturn's system of moons. The ice would quickly settle into the equatorial plane, and some would become trapped in rings. Although the ice may waste away due to meteorite impacts and damage from radiation in Saturn's magnetosphere, new impacts could replenish the rings with fresh ice. The bright, beautiful rings we see today may be only a temporary enhancement caused by an impact that occurred since the extinction of the dinosaurs.

Earth to scale

NASA

Visual-wavelength image

NASA

Encke's division

Cassini's division

A ring

B ring

C ring
The Crepe ring

As in the case of Jupiter's ring, Saturn's rings lie inside the planet's Roché limit where the ring particles cannot pull themselves together to form a moon.

Because it is so dark, the C ring has been called the Crepe ring, referring to the black, semitransparent cloth associated with funerals.

An astronaut could swim through the rings. Although the particles orbit Saturn at high velocity, all particles at the same distance from the planet orbit at about the same speed, so they collide gently at low velocities. If we could visit the rings, we could push our way from one icy particle to the next. This artwork is based on a model of particle sizes in the A ring.

The C ring contains boulder-size chunks of ice, whereas most particles in the A and B ring are more like golf balls, down to dust-size ice crystals. Further, C ring particles are less than half as bright as particles in the A and B rings. The reason for this difference is unknown.

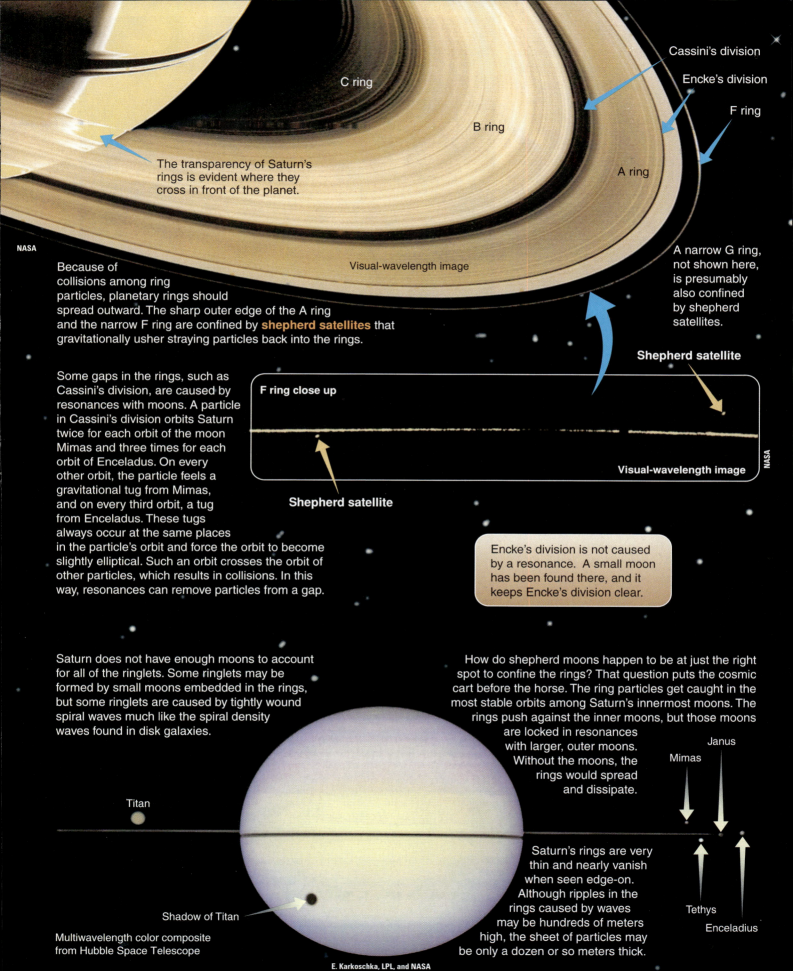

C ring

B ring

Cassini's division

Encke's division

F ring

A ring

The transparency of Saturn's rings is evident where they cross in front of the planet.

NASA

Visual-wavelength image

A narrow G ring, not shown here, is presumably also confined by shepherd satellites.

Because of collisions among ring particles, planetary rings should spread outward. The sharp outer edge of the A ring and the narrow F ring are confined by **shepherd satellites** that gravitationally usher straying particles back into the rings.

Some gaps in the rings, such as Cassini's division, are caused by resonances with moons. A particle in Cassini's division orbits Saturn twice for each orbit of the moon Mimas and three times for each orbit of Enceladus. On every other orbit, the particle feels a gravitational tug from Mimas, and on every third orbit, a tug from Enceladus. These tugs always occur at the same places in the particle's orbit and force the orbit to become slightly elliptical. Such an orbit crosses the orbit of other particles, which results in collisions. In this way, resonances can remove particles from a gap.

Shepherd satellite

F ring close up

Shepherd satellite

Visual-wavelength image

NASA

Encke's division is not caused by a resonance. A small moon has been found there, and it keeps Encke's division clear.

Saturn does not have enough moons to account for all of the ringlets. Some ringlets may be formed by small moons embedded in the rings, but some ringlets are caused by tightly wound spiral waves much like the spiral density waves found in disk galaxies.

How do shepherd moons happen to be at just the right spot to confine the rings? That question puts the cosmic cart before the horse. The ring particles get caught in the most stable orbits among Saturn's innermost moons. The rings push against the inner moons, but those moons are locked in resonances with larger, outer moons. Without the moons, the rings would spread and dissipate.

Mimas

Janus

Titan

Shadow of Titan

Multiwavelength color composite from Hubble Space Telescope

Saturn's rings are very thin and nearly vanish when seen edge-on. Although ripples in the rings caused by waves may be hundreds of meters high, the sheet of particles may be only a dozen or so meters thick.

Tethys

Enceladius

E. Karkoschka, LPL, and NASA

(for whom the large mountain on Venus is named) proved mathematically that solid rings would be unstable. Saturn's rings, he concluded, had to be made of particles. In 1867, Daniel Kirkwood demonstrated that gaps in the rings were caused by resonances with some of Saturn's moons. Modern astronomers find simple gravitational interactions that produce even more complex processes, such as waves that sweep through the rings creating the hundreds of ringlets visible in spacecraft images.

Electromagnetic phenomena are also present in the rings. Images recorded by spacecraft show radial features in the rings called **spokes** (∎ Figure 23-21). The differential rotation in the rings should erase such features rapidly, but the spokes survive and sweep around Saturn with the same period with which the planet's magnetic field rotates. Evidently, dust particles in the rings become electrostatically charged by interacting with sunlight and are then lifted out of the plane of the rings by Saturn's magnetic field.

Like a beautiful flower, the rings of Saturn are controlled by many different natural processes. When the Cassini spacecraft reaches orbit around Saturn in the summer of 2004, we will learn even more about the rings (Window on Science 23-2).

THE HISTORY OF SATURN

The farther we journey from the sun, the more difficult it is to understand the history of the planets. Any fully successful history of Saturn should explain its low density, its peculiar magnetic field, and its beautiful rings. We can't tell that fully successful story yet, but we can understand a few of the principles.

Certainly, we can understand that Saturn formed in the outer solar nebula, where ice particles were stable. It grew rapidly and became massive enough to capture hydrogen and helium gas directly from the nebula. The denser elements sank to the middle to

a

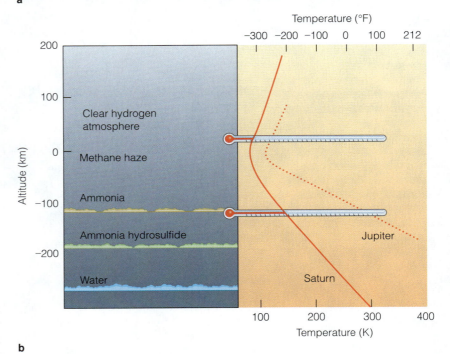

b

ACTIVE FIGURE 23-20

(a) Saturn's belt–zone circulation is not very distinct at visible wavelengths. These images were recorded when Saturn's southern hemisphere was tipped toward Earth. Compare these images with the visual-wavelength image in Data File 8. *(NASA and E. Karkoschka)* (b) Because Saturn is colder than Jupiter, the clouds form deeper in the hazy atmosphere. Notice that the three cloud layers on Saturn form at about the same temperature as do the three cloud layers on Jupiter. (Compare with Figure 23-5)

Ace◯Astronomy™ Go to AceAstronomy and click Active Figures to see "Planetary Atmospheres" and compare the atmospheres of the planets.

form a denser core, and the hydrogen formed a liquid mantle containing liquid metallic hydrogen. The outward flow of heat from the core is believed to drive convection currents in this mantle that, coupled with the rapid rotation of the planet, produce its magnetic field. Because Saturn is smaller than Jupiter, it has less liquid metallic hydrogen, and its magnetic field is weaker.

Visual-wavelength images

FIGURE 23-21

Radial spokes in the bright B ring are visible in these Voyager images. Note the motion of the spokes (arrow) between the two frames. Evidently, the spokes are produced by Saturn's magnetic field sweeping up dust in the rings. *(NASA)*

We might be tempted at first to suppose that the rings of Saturn are primordial, made of material left over from the formation of the planet. That seems unlikely, however. Saturn, like Jupiter, should have been very hot when it formed, and that heat would have vaporized and driven off any leftover material. Also, such a hot Saturn would have had a very distended atmosphere, which would have slowed ring particles by friction and caused the particles to fall into the planet.

Planetary rings do not seem to be stable over 4.6 billion years, so we must suppose that the ring material is more recent. Saturn's beautiful rings may have been produced within the last 100,000 years. One suggestion is that an icy planetesimal came within Saturn's Roche limit, and tides pulled it apart. At least some of the debris would settle into the ring plane. We will see in the next two chapters that these icy planetesimals are common in the outer solar system. Another possibility is that a comet struck one of Saturn's moons. Because both comets and moons in the outer solar system are icy, such a collision would produce icy debris. Bright planetary rings such as Saturn's may be temporary phenomena, produced when fresh ice is scattered into the rings and then wasting away as the ice is gradually lost.

TABLE 23-3

Organic Compounds Detected on Titan by a Voyager Infrared Spectrometer

C_2H_6	Ethane
C_2H_2	Acetylene
C_2H_4	Ethylene
C_3H_4	Methylacetylene
C_3H_8	Propane
C_4H_2	Diacetylene
HCN	Hydrogen cyanide
HC_3N	Cyanocetylene
C_2N_2	Cyanogen

In any contest, Saturn should win the ribbon for most beautiful planet. Its subtly marked disk and its delicate rings are both massive and graceful. Its extensive system of moons only adds to the mystery of this beautiful and curious world.

23-4 SATURN'S MOONS

Saturn has over two dozen known satellites—far too many for us to examine individually—but these moons share characteristics common to icy worlds. Most of them are small and dead, but one is big enough to have an atmosphere and perhaps even oceans.

TITAN

Saturn's largest satellite is a giant ice moon with a thick atmosphere and a mysterious surface. From Earth it is only a dot of light, and the cameras aboard the Voyager spacecraft were unable to penetrate its hazy atmosphere. Nevertheless, a few basic observations tell us a great deal about this strange world.

Titan is an icy world a bit larger than the planet Mercury and almost as large as Jupiter's moon Ganymede. Unlike these worlds, Titan has a thick atmosphere. Its escape velocity is low, but it is so far from the sun it is very cold. Thus, most gas atoms don't move fast enough to escape. (See Figure 22-14.) Its density is 1.9 g/cm^3, but its uncompressed density (Chap-

ter 19) is only 1.2 g/cm^3. Although it must have a rocky core, it must also contain a large amount of ices.

The surface of Titan is not visible in photographs because its atmosphere is filled with haze (▌ Figure 23-22a). Methane was detected spectroscopically in 1944, as were various hydrocarbons in 1970, but most of Titan's air is nitrogen with only a few percent methane. Sunlight converts methane (CH_4) into the gas ethane (C_2H_6) plus a collection of organic molecules* (▌ Table 23-3). Some of these molecules produce the smoglike haze; and, as the smog particles have gradually settled to the surface, they may have deposited a layer of smelly, organic goo 10 m to 1 km thick.

This goo is exciting because such organic molecules may have been the precursors of life on Earth. Thus, similar chemical reactions on Titan might have created life there. Although some scientists argue that the temperature on Titan is too low to create such molecules, future explorers will want to visit Titan to look for life. We will consider this further in Chapter 26.

Before you imagine floundering through this gooey landscape, we must add a bit of chemistry. Voyager measurements show that the atmospheric pressure at the surface is 50 percent greater than the air pressure on Earth, and the temperature is 94 K ($-290°F$). Under these conditions, methane and ethane form liquids, and in the 4.6 billion years since the solar system formed, sunlight could have produced enough ethane to make a layer 1 km deep—an ethane ocean flavored with liquid methane. (On Earth, methane is common natural gas. Both methane and ethane are flammable—but only on planets with free oxygen.)

Observations from Earth can tell us a little bit about the surface. Infrared observations can see through the hazy atmosphere and reveal that some parts of Titan are brighter than other parts (Figure 23-22b). That must mean that Titan isn't covered by a global ocean. The icy

*Organic molecules are common in living things on Earth but do not have to be derived from living things. One chemist defines an organic molecule as "any molecule with a carbon backbone."

Visual-wavelength image

a

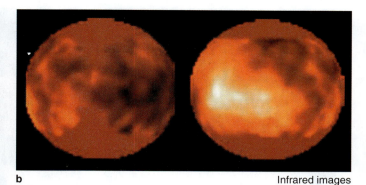

Infrared images

b

Infrared image

c

FIGURE 23-22

(a) A visual-wavelength photo of Titan taken by the Voyager space-craft shows a smoggy atmosphere with a circulation pattern near the north pole. (b) Infrared images taken by the Hubble Space Telescope show bright and dark areas on the surface. (c) An infrared image from the Keck I telescope shows a bright highland area at upper right and a darker area at lower left. *(a and b, NASA; c, Lawrence Livermore National Laboratories)*

crust must be exposed in at least some places. Radar observations have found that Titan often has a "shiny spot" like a highly polished sphere. This appears to be a reflection from the surface of a body of liquid. The radar observations hint that up to 75 percent of the surface may be covered by liquid oceans or lakes.

The methane in the atmosphere of Titan presents planetary scientists with a further puzzle. Sunlight should convert all of the methane into organic smog particles in only a million years. Obviously there is some source of fresh methane, and two theories have been proposed. Titan is big enough to have some internal heat that could drive geological activity, and that may release methane into the atmosphere. We can imagine methane volcanoes, but the process might be much less spectacular as methane seeps out of the icy crust. A second possibility is that Titan has a methane cycle much like Earth's water cycle. On Earth, water falls as rain, forms oceans, evaporates into the atmosphere, falls as rain, and so on. Titan may have a methane cycle, but it is not clear at present how that might operate.

Does Titan have vast oceans or small lakes? Is there organic goo on the surface or is it just ice? Is the sur-

face cratered or are there signs of geological activity? We will know soon. The Cassini spacecraft will reach Saturn in July 2004, and it will map Titan by radar and with infrared cameras. In January 2005, the spacecraft will drop a probe into Titan's atmosphere. Named Huygens after Christian Huygens, who discovered Titan in 1655, the probe will study the atmosphere and photograph the surface for more than two hours as it descends by parachute. It may survive for an hour or so on the surface and tell us by radio whether it landed on ice, on rock, on goo, or on waves.

Titan is an exciting world to think about because it might harbor life in its organic deposits, but it is also exciting because it is so totally unearthly.

THE SMALLER MOONS

In addition to Titan, the two Voyager spacecraft imaged eight smaller Saturnian moons. They have low densities and are evidently formed of rock and ice.

These smaller moons are heavily cratered and appear geologically dead, but a few show signs of past activity. Tethys is marked by a crack 3 km deep and 1500 km long; it stretches three-quarters of the way around

Visual-wavelength images

FIGURE 23-23

(a) Heavily cratered Tethys appears to be a dead world with a long fault in its icy crust (arrows). (b) The leading side of Iapetus (facing upper right) is much darker than the trailing side due to dark deposits on its surface. (c) Some portions of Enceladus have fewer craters and must have been resurfaced recently. This suggests that Enceladus has been active at some time since the end of the age of cratering. *(NASA)*

the moon (❙ Figure 23-23a). Similarly, the inner moon Mimas is marred by a crack 2 km deep. Such cracks may have formed when the interiors of the icy moons froze and expanded, breaking the crust—the opposite of what happened when the metallic core of Mercury solidified and contracted to form lobate scarps.

Because most of the moons are heavily cratered, we can conclude that they are geologically dead, but Enceladus may be an active moon (Figure 23-23c). Its surface contains areas of fewer craters, and some regions contain parallel grooves similar to those on Ganymede. These may be cracks filled by frozen water from the interior. The light yellow color of Enceladus also suggests that its surface is not old. It seems possible that eruptions of liquid water occasionally resurface parts of Enceladus with fresh ice. This would explain why the ice crystals of Saturn's extended E ring are more common near the orbit of Enceladus.

Whenever we see an active moon, we must wonder about the source of its internal energy. Enceladus is too small to have been heated significantly by radioactive decay, so we can suspect that tidal heating is responsible.

Iapetus is an asymmetric moon. Its trailing side, the side that always faces backward as it orbits Saturn, is about as bright as dingy snow. Its leading side, the side that always faces forward in its orbit, has an albedo of only 5 percent—about as dark as fresh asphalt on a highway (Figure 23-23b). The origin of this dark material is unknown, but theorists suggest that the little moon has swept up dark, silicon- and carbon-rich material on its leading side. This dust could be produced by meteorites striking the outermost moon, Phoebe.

Saturn's moon Hyperion is odd because it is not spherical. It is shaped like a battered hockey puck with

Visual-wavelength image

a

Titan

Enceladus Dione Iapetus Phoebe

Mimas Tethys Rhea Hyperion Earth's moon for comparison

b

FIGURE 23-24

(a) Hyperion is a small satellite that has not been able to pull itself into a spherical shape. (b) Larger moons have enough gravity to overcome the strength of the rock and ice and force themselves into spheres.

diameters of 250 km, 230 km, and 110 km (Figure 23-24). Mimas and Enceladus are only slightly larger and are quite spherical, so we must suppose that the gravitational field of a larger moon is sufficient to overcome the structural strength of ice and squeeze the moon into spherical shape. A small moon like Hyperion, located far from its parent planet where tidal heating is unimportant, may remain cold, rigid, irregular, and undifferentiated.

THE ORIGIN OF SATURN'S SATELLITES

Jupiter's four Galilean satellites seem clearly related to one another, and we concluded that they probably formed with Jupiter. No such systematic relationship links Saturn's satellites. We may be forced to conclude that they are captured icy planetesimals left over from the solar nebula or that the impacts of comets have so badly fractured them that they no longer show evidence of their common origin.

Understanding the origin of Saturn's moons is also difficult because the moons interact gravitationally in complicated ways, and thus the orbits they now occupy may differ from their earlier orbits. These interactions were revealed dramatically by the Voyager spacecraft.

We have previously discussed the small satellite that shepherds the outer edge of the A ring and the pair of moons that shepherds the F ring. An even more peculiar pair of moons is known as the coorbital satellites. These two irregularly shaped moonlets have orbits separated by only 100 km. Because one moon is about 200 km in diameter and the other about 100 km, they cannot pass in their orbits. Instead, the innermost moon gradually catches up with the outer moon. As the moons draw closer together, the gravity of the trailing

moon slows the leading moon and makes it fall into a lower orbit. Simultaneously, the gravity of the leading moon pulls the trailing moon ahead, and it rises into a higher orbit (Figure 23-25). The higher orbit has a longer period, so the trailing moon begins to fall behind the leading moon, which is now in a smaller, faster orbit. Thus, the moons exchange orbits and draw apart only to meet again and again. It seems very likely that these two moons are fragments of a larger moon destroyed by a major impact.

In addition to waltzing coorbital moons, the Voyager spacecraft discovered small moonlets trapped at the L_4 and L_5 Lagrangian points (see Figure 13-5) in the orbits of Dione and Tethys. These points of stability lie 60° ahead of and 60° behind the two moons, and small moonlets can become trapped in these regions. (We will see in Chapter 25 that asteroids are trapped in the Lagrangian points of Jupiter's orbit around the sun.) This gravitational curiosity is surely not unique to the Saturn system, and it warns us that interactions between moons can dramatically alter their orbits.

Some of Saturn's larger moons, especially Titan, may have formed with the planet, but the orbital relationships and intense cratering suggest that the moons have interacted and may have collided in the past. Many of the moons may be icy bodies that Saturn has captured since the formation of the solar system. As we continue our exploration of the outer solar system, we must be alert for the presence of such small, icy worlds.

REVIEW CRITICAL INQUIRY

What features on Enceladus suggest that it has been active?

The smaller moons of Saturn are icy worlds battered by impact craters, and thus we suspect that they are cold and old. Small worlds lose their heat quickly, and with no inter-

Coorbital Moons

The inner moon orbits faster and overtakes the outer moon.

The gravitational interaction pulls the outer moon backward and the inner moon forward.

As they approach, the inner moon moves to a higher orbit, and the outer moon sinks to a lower orbit.

The moons have changed orbits, and the inner moon begins gaining on the outer moon.

The inner moon will eventually overtake the outer moon from behind once again.

(Not to scale)

ACTIVE FIGURE 23-25

Saturn's coorbital moons follow nearly identical orbits. The moon in the lower orbit travels faster and always overtakes the other moon from behind. The moons interact and change orbits over and over again.

Ace✆**Astronomy**™ Go to AceAstronomy and click Active Figures to see "Coorbital Moons" and animate this diagram.

nal heat, there is no geological activity to erase impact craters. A small, icy world covered with craters is exactly what we would expect in the outer solar system. Enceladus, however, is peculiar. Although it is small and icy, its surface is not very dark, and some areas contain fewer craters than we would expect. In fact, some regions seem almost free of craters. Parallel grooves mark some regions of the little moon and suggest motion in the icy crust. These features should have been destroyed long ago by impact cratering, so we must suppose that the moon has been geologically active at some time since the end of heavy cratering at the conclusion of planet building.

Of course, if Enceladus has been active, then it must have had a source of heat in its interior. Only the outward flow of energy can drive geological activity. Because Enceladus is much too small to have kept itself warm from internal radioactive decay, we might suspect tidal heating.

If we look at Titan, Saturn's largest moon, we see quite a different world. Although Titan is only about 50 percent bigger than Earth's moon, it has a substantial atmosphere. How can such a small world keep an atmosphere?

Jupiter and Saturn reign over the solar system like their namesake ancient gods. Beyond them, out where the sunlight is dim and temperatures deadly cold, orbit three more planets that were unknown to ancient astronomers. In the next chapter, we will add these worlds to our study of comparative planetology.

SUMMARY

Jupiter and Saturn are so massive that they have been able to retain much of their hydrogen and helium. This excess of light elements gives them densities much lower than those of the terrestrial planets. In their interiors, the tremendous pressure forces the hydrogen into a liquid state, and the pressure is so high that hydrogen becomes a liquid metal, which is a good conductor of electricity. Electric currents within this material produce magnetic fields that trap particles from the solar wind to form radiation belts around the planets. Both Jupiter and Saturn are known to have radiation belts similar to Earth's Van Allen belts, and Jupiter's belts are very intense. Both planets have aurorae around their magnetic poles.

Jupiter is encircled by atmospheric dark belts, regions of sinking gas, and bright zones, regions of rising gas. Unknown trace chemicals give the clouds their colors. Although Saturn has the same cloud layers, its atmosphere is colder and the cloud layers form deeper. They are less distinct.

The Great Red Spot on Jupiter is a giant rotating storm produced by rising gas. Larger than Earth, it has lasted for at least three centuries. Jupiter's atmosphere contains similar smaller spots that are short lived, but such spots are rare on Saturn.

Jupiter's ring is dark in color but bright in forward scattering, so we know it is made up of dusty, rocky material. The dust falls into Jupiter or is destroyed by radiation in less than 100 years, so it must be constantly resupplied as meteorites chip material off of the innermost moons.

Saturn's rings are bright, icy, and highly complex. Much of the structure in the rings appears to be created by the gravitational influence of Saturn's moons, by waves that move through the rings, and by Saturn's magnetic field, which produces spokes. A few small moons have been located shepherding the narrow F ring and the outer edge of the A ring. It is most likely that the icy rings are composed of debris that is created by the breakup of icy moons or planetesimals.

Jupiter has a number of small moonlets that are probably captured asteroids. But the four Galilean satellites seem to have formed at about their present locations. The density of these moons decreases with increasing distance from Jupiter, suggesting that they formed in a small nebula around the condensing planet. Tidal heating, aided perhaps by resonances among the three inner moons, has heated the inner moons more than the outer moons, driving off water and making the inner moons more dense. This process does not seem to have occurred in the Saturn system, perhaps because Saturn is less massive and did not become as hot as Jupiter.

Jupiter's satellites have clearly been affected by tidal heating. Io is violently active with many volcanoes. Europa is not visibly active, but the lack of craters on its surface suggests that its icy crust is still active enough to erase craters. Ganymede and Callisto are both heavily cratered, inactive worlds, but the grooved terrain on Ganymede seems to date from a period when it was more active. How much tidal heating contributed to the formation of the Galilean satellites is not known.

Titan, Saturn's largest moon, is able to keep a thick atmosphere of nitrogen and methane because it is so cold. In fact, sunlight converts the methane into ethane and other organic materials that form a smog in the atmosphere and hide the moon's surface. Ethane falls to the surface and probably exists as liquid oceans or lakes, while falling smog particles create a deposit of organic goo.

Some of Saturn's moons are clearly inactive. Mimas, Tethys, and Dione are heavily cratered, icy bodies. Nevertheless, the presence of long, deep valleys formed by cracks in the icy crust shows that the worlds once had more internal heat.

Enceladus seems to have had an active history. It is cratered, but some regions are obviously much younger. Lava flows of water may have resurfaced portions of Enceladus. The E ring of Saturn is most dense near the orbit of Enceladus, and the ring particles may originate on the icy moon.

The debris in space craters the surfaces of moons most heavily on the side facing forward in their orbits. The trailing side is less cratered. Iapetus is bright on the trailing side and very dark on the leading side. Whether the black deposits are debris from some outside source or material from the interior is not known.

Many of Saturn's moons are small, irregular bodies that are either captured objects or the fragments of collisions. Hyperion is very irregular in shape.

The orbital dynamics of the smaller moons is complex. The coorbital moons share nearly identical orbits, whereas other moons occupy the Lagrangian points in the orbits of larger moons.

NEW TERMS

oblateness	zone
liquid metallic hydrogen	forward scattering
decameter radiation	Roche limit
decimeter radiation	gossamer rings
current sheet	grooved terrain
Io plasma torus	tidal heating
Io flux tube	shepherd satellite
critical point	spoke
belt	

REVIEW QUESTIONS

Ace⟳Astronomy™ Assess your understanding of this chapter's topics with additional quizzing and animations at **http:// astronomy.brookscole.com/seeds8e**

1. Why is Jupiter oblate? Do you expect all Jovian planets to be oblate? Why or why not?

2. How do the interiors of Jupiter and Saturn differ? How does this affect their magnetic fields?

3. What is the difference between a belt and a zone?

4. How can we be certain that Jupiter's ring does not date from the formation of the planet?

5. If Jupiter had a satellite the size of our own moon orbiting outside the orbit of Callisto, what would you predict for its density and surface features?

6. Why are there no craters on Io and few on Europa? Why should we expect Io to suffer more impacts per square kilometer than Callisto?

7. Why are the belts and zones on Saturn less distinct than those on Jupiter?

8. If Saturn had no moons, what do you suppose its rings would look like?

9. Suggest two theories for the origin of Saturn's rings.

10. How can Titan keep an atmosphere when it is smaller than airless Ganymede?

11. If you piloted a spacecraft to visit Saturn's moons and wanted to land on a geologically old surface, what features would you look for? What features would you avoid?

12. Why does the leading side of some satellites differ from the trailing side?

13. Why do we expect small satellites to be irregular in shape? Give examples and exceptions.

DISCUSSION QUESTIONS

1. Some astronomers argue that Jupiter and Saturn are unusual, while other astronomers argue that all solar systems should contain one or two such giant planets. What do you think? Support your argument with evidence.

2. Why don't the terrestrial planets have rings?

PROBLEMS

1. What is the maximum angular diameter of Jupiter as seen from Earth? What is the minimum? Repeat this calculation for Saturn and Titan. (*Hint:* Use the small-angle formula.)

2. The highest-speed winds on Jupiter are in the equatorial jet stream, which has a velocity of 150 m/s. How long does it take for these winds to circle Jupiter?

3. What are the orbital velocity and period of a ring particle at the outer edge of Jupiter's ring? at the outer edge of Saturn's A ring? (*Hints:* The radius of the edge of the A ring is 136,500 km. See Chapter 5.)

4. What is the angular diameter of Jupiter as seen from the surface of Callisto? (*Hint:* Use the small-angle formula.)

5. What is the escape velocity from the surface of Ganymede if its mass is 1.5×10^{26} g and its radius is 2628 km? (*Hint:* See Chapter 5.)

6. If we photograph the spectrum of Saturn and its rings, we should find light from one edge of the rings redshifted and light from the other edge blue-shifted. If we observe at a wavelength of 500 nm, what difference in wavelength should we expect between the two edges of the rings? (*Hints:* See Problem 3 and Chapter 7.)

7. What is the difference in orbital velocity between particles at the outer edge of Saturn's B ring and particles at the inner edge of the B ring? (*Hint:* The outer edge of the B ring has a radius of 117,500 km, and the inner edge has a radius of 92,000 km.)

8. What is the difference in orbital velocity between the two coorbital satellites if the semimajor axes of their orbits are 151,400 km and 151,500 km? The mass of Saturn is 5.7×10^{26} kg. (*Hint:* See Chapter 5.)

CRITICAL INQUIRIES FOR THE WEB

1. Saturn has a great many satellites. Locate information on these moons, including discovery dates (make sure you use an updated source that includes discoveries from the past few years). Excluding the satellite discoveries associated with the Voyager data, what observational circumstance do the discoveries of almost all of the moons of Saturn have in common?

2. The collision of Comet Shoemaker-Levy 9 with Jupiter was one of the first events to be publicized directly on the Web. Search for Web sites dedicated to this comet and its spectacular demise. How long were the effects of the impacts visible in the atmosphere of Jupiter? How might they have appeared differently if this event had occurred at Saturn?

3. What is the status of our understanding of the Jovian moon Europa? This moon was studied extensively by the Galileo mission. Visit Galileo-related Web sites to find the results of this exploration, and discuss our current ideas regarding the possibility of liquid water beneath Europa's crust. How might future missions to Europa probe such an underground ocean?

4. Cassini is scheduled to reach orbit around Saturn on July 1, 2004, and the Huygens probe will enter the atmosphere of Titan on January 14, 2005. Check the Cassini Web page to get the latest news from Saturn and Titan.

5. When will Saturn's rings next be edge-on to the Earth? When were they last edge-on? Find photographs of Saturn and its rings at different inclinations.

EXPLORING *THESKY*

1. Galileo recognized the significance of Jupiter's moons when he realized they were moving around Jupiter. Sketch the position of the four Galilean moons at intervals of 1 hour. Could you recognize their orbital motion in a long evening? (*Hint:* Use **Find** to **Center & Frame** Jupiter. Zoom in until you can see the moons. Use **Tracking Setup** to lock on Jupiter and then use the **Time Skip** buttons to make the moons move.)

2. Calculate the mass of Jupiter from your own observations of the orbital period and orbital radius of Jupiter's moons. (*Hint:* Use the equation for circular velocity in Chapter 5.)

3. Repeat Activity 2 for Saturn.

 Visit the Seeds *Foundations of Astronomy* companion Web site for critical thinking exercises, articles, and additional readings from InfoTrac College Edition, Brooks/Cole's online student library.

URANUS, NEPTUNE, AND PLUTO

A good many things go around in the dark besides Santa Claus.

Herbert Hoover

Floyd Herbert, LPL

GUIDEPOST

In the three previous chapters, we have used our tools of comparative planetology to study other worlds, and we continue that theme in this chapter. A second theme running through this chapter is the nature of astronomical discovery. Unlike the other planets in our solar system, Uranus, Neptune, and Pluto were discovered, and the story of their discovery helps us understand how science progresses.

As we probe the outer fringes of our planetary system in this chapter, we see strong evidence of smaller bodies that fall through the solar system and impact planets and satellites. The next chapter will allow us to study these small bodies in detail and will give us new evidence that our solar system formed from a solar nebula.

VIRTUAL LABORATORIES

PLANETARY GEOLOGY

PLANETARY ATMOSPHERES AND THEIR RETENTION

Out in the darkness beyond Saturn, out where sunlight is about 1000 times fainter than on Earth, there are things going around that Aristotle, Galileo, and Newton never imagined. They knew about Mercury, Venus, Mars, Jupiter, and Saturn, but our solar system includes three more planets, planets that were not discovered until after the invention of the telescope.

Uranus and Neptune are Jovian planets and are similar in size, about four times Earth's diameter. Thus, they are about half the size of Saturn. Through Earth-based telescopes, we see nothing but tiny disks, and until recently what little we knew of these planets was deduced from a few observations compared with Jupiter and Saturn. But in January 1986 the Voyager 2 spacecraft flew past Uranus and returned a wealth of photographs and measurements. In August 1989, Voyager 2 flew past Neptune, revealing a blue world with complex weather, rings, and moons. In just a few years, Uranus and Neptune went from obscure points of light to known worlds.

Pluto, however, is a more difficult subject. Through an Earth-based telescope, it looks like nothing more than a faint point of light, and even through the Hubble Space Telescope it is a tiny disk. Voyagers 1 and 2 left the solar system without passing near Pluto, so we have no good photos of the little world's surface. Nevertheless, we may be able to guess its nature by combining the limited observations with the principles of comparative planetology.

Uranus, Neptune, and Pluto differ from the other planets because they were not known from antiquity. They had to be discovered. The stories of these discoveries highlight the nature of scientific discovery.

THE DISCOVERY OF URANUS

William Herschel (❚ Figure 24-1) came from a musical family in Hanover, Germany, but immigrated to England as a young man and eventually obtained a prestigious job as the organist at the Octagon Chapel in Bath.

To compose exercises for his students and choral and organ works for the chapel, Herschel studied the mathematical principles of musical harmony from a book by Professor Robert Smith of Cambridge. The mathematics was so interesting that Herschel searched out other books by Smith, including a book on optics. Of course, an 18th-century book on optics written by a professor in England relied heavily on Isaac Newton's optics and described the principles of the reflecting telescope (see Chapter 6). Herschel and his brother Alexander began building telescopes, and William went on to study astronomy in his spare time.

Herschel's telescopes were similar to Newton's in that they had metal mirrors, but they were much larger. Newton's telescope had a mirror about 1 in. in diameter, but Herschel developed ways of making much larger mirrors. He soon had telescopes as long as 20 ft, although one of his favorite telescopes was 7 ft long with a mirror 6.2 in. in diameter. Using it, he began the research project that led to the discovery of Uranus.

Herschel did not set out to search for a planet; he was trying to detect stellar parallax produced by Earth's motion around the sun. No one had yet detected this effect, although by the 1700s all astronomers accepted Earth's motion. Galileo had pointed out that parallax might be detected if a nearby star and a very distant star lay so nearly in the same direction that they looked like

24-1 URANUS

In March 1781, Benjamin Franklin was in France raising money, troops, and arms for the American Revolution. George Washington and his colonial army were only six months away from the defeat of Cornwallis at Yorktown and the end of the war. In England, King George III was beginning to show signs of madness. And a German-speaking music teacher in the English resort city of Bath was about to discover the planet Uranus.

FIGURE 24-1
William Herschel is honored on this postage stamp issued to commemorate the passage of Comet Halley in 1986. The comet is shown in the distance beyond an artist's conception of the planet Uranus and its rings. Herschel never knew Uranus had rings. What he saw through his telescope was a tiny blue-green dot moving against the background of stars.

a very close double star through a telescope. In such a case, Earth's orbital motion would produce a parallactic shift in the position of the nearby star with respect to the more distant star (∎ Figure 24-2). Herschel began to examine all stars brighter than eighth magnitude to search for double stars, which might show parallax. That project alone took over two years.

On the night of March 13, 1781, Herschel set up his 7-ft telescope and continued his work. He later wrote, "In examining the small stars in the neighbourhood of H Geminorum, I perceived one that appeared visibly larger than the rest." As seen from Earth, Uranus is never larger in angular diameter than 3.7 seconds of arc, so Herschel's detection of the disk illustrates the quality of his telescope and his eye. At first he suspected that the object was a comet, but other astronomers quickly showed that it was a planet orbiting the sun beyond Saturn.

The discovery of Uranus made Herschel world famous. Since antiquity, astronomers had known the five planets—Mercury, Venus, Mars, Jupiter, and Saturn—and they supposed that the universe was complete. Herschel's discovery extended the classical universe. The English public accepted Herschel as their astronomer-hero, and, having named the new planet *Georgium Sidus* (George's Star) after King George III, Herschel received a royal pension. The former music teacher was welcomed into court society, where he later met and married a wealthy widow and took his place as one of the great English astronomers in history. With his new financial position, he built large telescopes on his estate and, with his sister Caroline, a talented astronomer herself, attempted to map the extent of the universe.

Continental astronomers were less than thrilled that an Englishman had made the great discovery, and even some professional English astronomers called Herschel's discovery a lucky accident. Herschel responded in his own defense by making three points. First, he had built some of the finest-quality telescopes then in existence. Second, he had been conducting a systematic research project and would have found Uranus eventually because he was inspecting all of the brighter stars visible with his telescope. And third, he had great experience seeing fine detail with his telescopes. As a musician, he knew the value of practice and applied it to the business of astronomical observing. In fact, astronomers had seen Uranus more than 17 times before Herschel, but each time they failed to notice that it was not a star. They plotted Uranus on their charts as if it was just a faint star.

This illustrates one of the ways in which scientific discoveries are made. Often, discoveries seem accidental, but on closer examination we find that the scientist has earned the right to the discovery through many years of study and preparation. To quote a common saying, "Luck is what happens to people who work hard."

Continental astronomers, especially the French, insisted that the new planet not be named after an English king. The French stubbornly called the planet Herschel, as did many other non-English astronomers. Some years later, German astronomer J. E. Bode suggested the name Uranus, the oldest of the Greek gods.

Over the half-century following the discovery of Uranus, astronomers noted that Newton's laws did not quite predict the observed position of the planet. Tiny variations in the orbital motion of Uranus eventually led to the discovery of Neptune, a story we will take up later in this chapter.

THE MOTION OF URANUS

Uranus follows a slightly elliptical orbit with an average distance from the sun of 19.18 AU and an orbital period of 84.013 years (∎ Data File Nine). The ancients thought of Saturn as the slowest of the planets, but Saturn orbits in a bit over 29 years. Uranus, being further from the sun, has a longer orbital period.

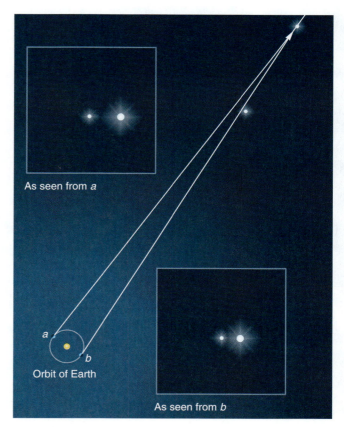

As seen from *a*

a

b

Orbit of Earth

As seen from *b*

FIGURE 24-2

A double star consisting of a nearby star and a more distant star could be used to detect stellar parallax. The separation between the stars would be different as seen from different places in Earth's orbit. The effect is actually much too small to detect by eye.

Heavy element core

Ices and rock

Hydrogen and helium

Earth

Uranus

URANUS

Data File Nine

About four times the diameter of Earth, Uranus is believed to contain a liquid interior and dense core.

Average distance from the sun	19.18 AU (28.69×10^8 km)
Eccentricity of orbit	0.0461
Maximum distance from the sun	20.1 AU (30.0×10^8 km)
Minimum distance from the sun	18.3 AU (27.4×10^8 km)
Inclination of orbit to ecliptic	0°46′23″
Average orbital velocity	6.81 km/s
Orbital period	84.013 y (30,685 days)
Period of rotation	17^h14^m
Inclination of equator to orbit	97°55′
Equatorial diameter	51,118 km (4.01 $D_\oplus$)
Mass	8.69×10^{25} kg (14.54 $M_\oplus$)
Average density	1.29 g/cm^3
Gravity	0.919 Earth gravity
Escape velocity	22 km/s (1.96 $V_\oplus$)
Temperature above cloud tops	−220°C (−364°F)
Albedo	0.35

The rotation of Uranus is peculiar. Earth rotates approximately upright in its orbit. That is, its axis of rotation is inclined only 23.5° from the perpendicular to its orbit. Uranus, in contrast, rotates on an axis that is inclined 97.9° from the perpendicular to its orbit. It rotates on its side (❙ Figure 24-3).

The seasons on Uranus are peculiar. In 1986, when Voyager 2 flew past Uranus, the planet was in the segment of its orbit in which its south pole faced the sun. Thus its southern hemisphere was bathed in continuous sunlight, and a creature living on Uranus (an unlikely possibility, as we will discover later) would have seen the sun circling near the planet's south celestial pole. This is shown at lower left in Figure 24-3. Once Uranus moves about a quarter of the way around its orbit, the sun will be above its equator, and our citizen of Uranus would see the sun rise and set with the rotation of the planet. This is shown at lower right in Figure 24-3. As Uranus continues along its orbit, the sun will approach the planet's north celestial pole, and the southern hemisphere of the planet will experience a lightless winter lasting 21 Earth years. In other words, the ecliptic on Uranus passes very near the planet's celestial poles, and the resulting seasons are extreme.

The high inclination of the axis of rotation has never been fully explained. Planetary astronomers generally propose that as Uranus formed from the icy planetesimals of the outer solar nebula, it had grown to nearly full size when it was struck at a sharp angle by a very large planetesimal, perhaps as large as Earth. That impact could have altered the axis of rotation. Such impacts may have been much more common in the outer solar system, where large amounts of ice could condense. We will see more evidence for large-impact events when we consider the satellites of Uranus. But first we must examine the atmosphere and interior of the planet.

THE ATMOSPHERE OF URANUS

Like Jupiter and Saturn, Uranus has no surface. The gases of its atmosphere—mostly hydrogen, 15 percent helium, and a few percent methane, ammonia, and water vapor—blend gradually with a fluid interior. Also like Jupiter and Saturn, the atmosphere of Uranus is rich in hydrogen and helium.

Seen through Earth-based telescopes, Uranus is a small, featureless blue disk. The blue color arises because the hydrogen-rich atmosphere contains methane, which is a good absorber of longer-wavelength photons. As sunlight penetrates into the atmosphere and is scattered back in our direction, the shorter-wavelength (blue) photons are more likely to avoid being absorbed. Consequently, the light entering our eye is richer in blue photons and poorer in red photons, giving the planet a blue color.

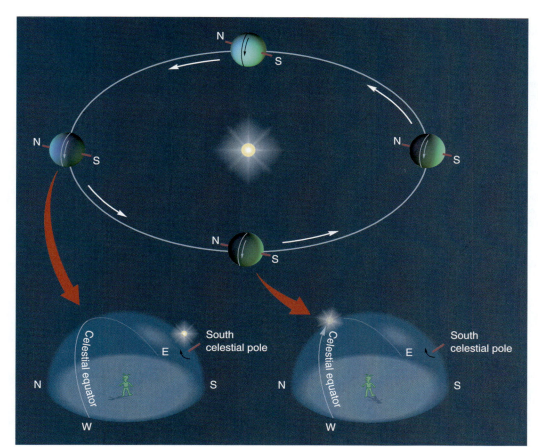

FIGURE 24-3

Uranus rotates on an axis that is tipped 97.9° from the perpendicular to its orbit, so its seasons are extreme. When one of its poles is pointed nearly at the sun (a solstice), a citizen of Uranus would see the sun near a celestial pole, and it would never rise or set. As it orbits the sun, the planet maintains the direction of its axis in space, and thus the sun moves from pole to pole. At the time of an equinox on Uranus, the sun would be on the celestial equator and would rise and set with each rotation of the planet.

As Voyager 2 drew closer to the planet in late 1985, astronomers studied the images radioed back to Earth. Uranus was a pale blue ball with no obvious clouds, and only when the images were carefully computer enhanced was any banded structure detected (❙ Figure 24-4a and b).

A few very high clouds of methane ice particles were detected (Figure 24-4c), and that allowed astronomers to measure the rotation period of the planet.

We can understand the appearance of the atmosphere by studying the temperature profile of Uranus

Visual-wavelength image

a

South Pole

Enhanced visual image

b

Enhanced visual image

c

FIGURE 24-4

(a) This Voyager 2 image of Uranus was made in 1986 and shows no cloud. Only when the image is computer enhanced, as in (b), is a banded structure visible. At the time, the axis of rotation was pointed nearly at the sun. (c) Under extreme computer enhancement, small methane clouds were visible. The geometry of the banding and the clouds suggests belt–zone circulation. *(NASA)*

The atmosphere of Uranus is much colder than that of Jupiter or Saturn, and the only visible cloud layer is one formed of methane ice crystals deep in the hydrogen atmosphere. Other cloud layers would be much deeper in the atmosphere and are not visible. The temperature profile of Neptune is similar to that of Uranus, and it has methane clouds at about the same place in its atmosphere.

Ace Astronomy™ Go to AceAstronomy and click Active Figures to see "Planetary Atmospheres" and compare the atmospheres of the planets.

shown in ❚ Figure 24-5. The atmosphere of Uranus is much colder than that of Saturn or Jupiter. Consequently, the three cloud layers of ammonia, of ammonia hydrosulfide, and of water that occur in the atmosphere of Jupiter and Saturn form very deep in the atmosphere of Uranus where it is warmer. We do not see these cloud layers because of the thick atmosphere of hydrogen through which we have to look. The clouds we do see on Uranus are clouds of methane ice crystals, which form at such a low temperature that they occur higher in the atmosphere of Uranus. Look again at Figure 24-5 and notice that there can be no methane clouds on Jupiter because that planet is too warm. The coldest part of the atmosphere of Saturn is just cold enough to form a thin methane haze high above its more visible cloud layers. (See Figure 23-20b.)

The clouds and atmospheric banding faintly visible on Uranus appear to be a form of belt–zone circulation, which is a bit surprising. Uranus rotates on its side, so solar energy strikes its surface in a geometry quite different than it does on Jupiter and Saturn. Evidently the belt–zone circulation is dominated by the rotation of the planet and not by the direction of sunlight.

Is Uranus always as featureless as it was then Voyager 2 flew past in early 1986? The Hubble Space Telescope has detected changing clouds on Uranus (❚ Figure 24-6), and those clouds appear to follow belt–zone patterns. Furthermore, since 1986, Uranus has moved on along its orbit, and spring has been coming to its northern hemisphere. As sunlight has reached these latitudes, clouds have appeared there and then dissipated.

We may be seeing part of a seasonal cycle on Uranus, but its year lasts 85 Earth years, so we must be patient.

THE INTERIOR OF URANUS

Astronomers cannot describe the interiors of Uranus and Neptune as accurately as they can the interiors of Jupiter and Saturn. Observational data are sparse, and the materials inside these planets are not as easy to model as simple liquid hydrogen.

The average density of Uranus, 1.29 g/cm³, tells us that the planet must contain a larger share of dense materials such as rock and ice than does Jupiter or Saturn. Nearly all models of the interior of Uranus contain three layers. The atmosphere is rich in hydrogen, helium, and methane. A deep mantle must contain large amounts of ices of water, methane, and ammonia, mixed with hydrogen and with silicate matter. This mantle is sometimes described as "liquid" or as "ice," but it is probably quite unlike the material these words suggest. The third layer in these three-layer models is a small core of rock. Because of the high pressure and high temperature, the material is probably not very rocklike. The term *rock* refers to its chemical composition and not its appearance. It would be better to call it a heavy-element core.

Planetary scientists make a subtle distinction among the Jovian planets. Jupiter and Saturn are sometimes called "gas giant planets," because they are rich in hydrogen and helium. A growing number of scientists are referring to Uranus and Neptune as "ice giant

Enhanced visual image

FIGURE 24-6

Images of Uranus recorded by the Hubble Space Telescope show cloud features in the atmosphere that were not seen when Voyager 2 flew past in 1986. This suggests that seasonal changes are taking place on Uranus as spring comes to the northern hemisphere. The rings of Uranus have been enhanced to make them visible in this image. *(NASA)*

planets," in recognition of the larger proportion of ices in their interiors.

Because Uranus has a much lower mass than Jupiter, its internal pressure is not high enough to produce liquid metallic hydrogen. Thus we might expect it to lack a strong magnetic field, but the Voyager 2 spacecraft found that Uranus has a magnetic field about 75 percent as strong as Earth's. Furthermore, the field is tipped 60° to the axis of rotation and is offset from the center of the planet about 30 percent of the planet's radius (❚ Figure 24-7). Theorists suggest that the magnetic field is produced by a dynamo effect operating, not at the center, but nearer the surface in a layer of liquid water with dissolved ammonia and methane. Such a material would be a good conductor of electricity, and the rotation of the planet coupled with circulation in the fluid could generate the magnetic field.

The high inclination of the magnetic field and its offset from the center of Uranus give us a good estimate of the true rotation period of the planet, otherwise difficult to estimate because of the lack of cloud features and high atmospheric winds. The magnetic field deflects the solar wind and traps some charged particles to create weak radiation belts trapped in a magnetosphere. High-speed electrons spiraling along the magnetic field produce synchrotron radio emission just as on Jupiter, and

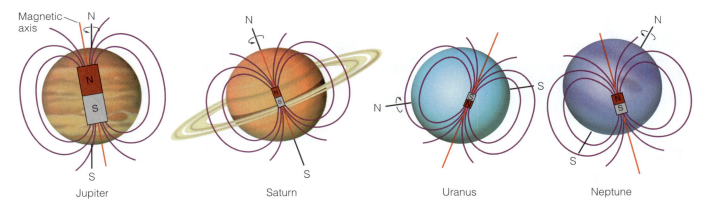

FIGURE 24-7

The magnetic fields of Uranus and Neptune are peculiar. While the magnetic axis of Jupiter is tipped only 10° from its axis of rotation and the magnetic axis of Saturn is not tipped at all, the magnetic axes of Uranus and Neptune are tipped at large angles. Furthermore, the magnetic fields of Uranus and Neptune are offset from the center of the planet. This suggests that the dynamo effect operates differently in Uranus and Neptune than in Jupiter and Saturn.

the Voyager 2 spacecraft detected this radiation fluctuating with a period of 17.24 hours—the period of rotation of the magnetic field and, presumably, the planetary interior.

The magnetic field and the high inclination of the planet produce some peculiar effects. As is the case for all planets, the solar wind deforms the magnetosphere and draws it out into a long tail extending away from the planet in the direction opposite the sun. The rapid rotation of Uranus and its high inclination make the magnetosphere and its long extension take on a corkscrew shape, which was partially mapped by Voyager 2. At the time Voyager 2 flew past in 1986, the south pole of Uranus was pointed nearly at the sun, and once each rotation the solar wind could pour directly down into the south magnetic pole. This interaction produced strong aurorae at both magnetic poles, which were detected by the spacecraft in the ultraviolet (■ Figure 24-8). In the years since, Uranus has moved farther around its orbit, so the geometry of its interaction with the solar wind is now different. Unfortunately, we have no spacecraft bound for Uranus, and thus we have no way to observe this interaction in detail.

The true rotation period of Uranus is an important clue to the nature of its interior when we combine it with the observed oblateness (0.024). Models of the interior of the planet cannot produce such a large oblateness at the observed rotation period if most of the heavy elements are concentrated at the center of the planet. Rather, the models predict that the planet has a small core of heavy elements surrounded by a deep layer of water mixed with heavier material, such as rock, in a slushy stew. The top of this dirty ocean merges with an overlying atmosphere made up mostly of hydrogen and helium.

Even the temperature of Uranus can tell us something about its interior. Jupiter and Saturn are warmer than we would expect given the amount of heat they receive from the sun, and we conclude that heat is leaking out from their hot interiors. Uranus, in contrast, is about the temperature we would expect for a world at its distance from the sun. Apparently, it has lost much of its interior heat. Yet it must have some internal heat to cause convection in the fluid mantle and thus drive the dynamo effect. The decay of natural radioactive elements would generate heat, but some astronomers have suggested that the slow settling of heavier elements through the fluid mantle could release sufficient energy to warm the interior.

Laboratory studies of methane show that it can break down under the temperature and pressure found inside Uranus and form various compounds plus pure carbon in the form of diamonds. If this happens in Uranus, the diamond crystals would fall inward, warming the interior through friction. Whatever diamonds might exist in Uranus are forever beyond our reach, but some of Uranus's internal heat may come from its diamond-making layers.

a   b

FIGURE 24-8

Aurorae on Uranus were detected in the ultraviolet by the Voyager 2 spacecraft when it flew by in 1986. These maps of opposite sides of Uranus show the location of aurorae near the magnetic poles. Recall that the magnetic field is highly inclined and offset from the planet's center, so the magnetic poles do not lie near the poles defined by rotation. In fact, the north magnetic pole (right) falls near the planet's equator. Also recall that Uranus is highly inclined and that in 1986 its south pole, located at the bottom of these diagrams, was pointed nearly at the sun. The white dashed line marks zero longitude. (*Courtesy Floyd Herbert, LPL*)

The interior of a world like Uranus seems at first glance to lie forever beyond our ken, but a few critical observations, combined with carefully constructed models, give us a glimpse of a fluid world. Unlike the interior, the rings of Uranus are exposed to view. The only problem is detecting the faint sunlight they reflect.

THE RINGS OF URANUS

Both Uranus and Neptune have rings that are more like those of Jupiter than those of Saturn. They are dark, faint, and confined by shepherd satellites. They are not easily visible from Earth, and the first hint that these planets had rings came from **occultations,** the passage of the planet in front of a star. Nevertheless, most of what we know about these ring systems comes from the observations of the Voyager 2 spacecraft.

Study ▌ "The Rings of Uranus and Neptune" on pages 544 and 545 and notice three important points. First, the particles that make up the rings are very dark and quite unlike the bright icy particles that orbit Saturn. Second, notice the importance of small moons imbedded among the rings. The moons shape and confine the rings into narrow hoops. Finally, notice how the ring particles must be occasionally resupplied by impacts on the moons.

As you read about planetary rings, notice their close relationship with moons. Because of collisions among ring particles, planetary rings tend to spread outward almost like an expanding gas. If a planet had no moons, its rings would spread out into a more and more tenuous sheet until they were gone. The spreading rings can be anchored by small moons, which get pushed outward very slowly. Through orbital resonances, those small moons can be anchored by larger, more distant moons that are so massive they do not get pushed outward. In this way, a system of moons can confine and preserve a system of planetary rings.

Planetary rings are beautiful blossoms created when debris falls into a system of small moons and becomes trapped in the most stable orbits among the moons. Without moons, there could be no rings.

THE MOONS OF URANUS

The five largest moons of Uranus (▌ Figure 24-9), from the outermost inward, are Oberon, Titania, Umbriel, Ariel, and Miranda. Umbriel and Ariel are names from Alexander Pope's *The Rape of the Lock,* and the rest are from Shakespeare's *A Midsummer Night's Dream* and *The Tempest* (an Ariel also appears in *The Tempest*). Spectra show that they contain frozen water, and their surfaces are very dark, so astronomers assumed they were made of dirty ices. Little more was known of the moons before Voyager 2 flew through the system.

The Voyager 2 cameras discovered ten moons too small to have been seen from Earth. Also, with the con-

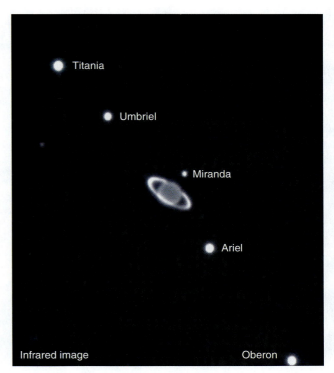

FIGURE 24-9
This infrared image shows Uranus, its disk, and its five major satellites. Because of photographic effects, the images of the satellites are much larger than the moons themselves. The largest, Titania, is nearly 32 times smaller in diameter than Uranus. Faint points of light to the left and right of the rings are the smaller moons Portia and Puck. Other faint points of light in this image are background stars. *(ESO)*

struction of new-generation telescopes and new imaging techniques, astronomers have found even more moons orbiting Uranus and there are almost certainly more still to be found.

The smaller moons are all as dark as coal. They are probably icy worlds with surfaces that have been darkened by impacts. Because they orbit near Uranus, they are inside the radiation belts, and the radiation may convert methane trapped in their ices into dark carbon deposits and further darken their surfaces.

The five large satellites are all tidally locked to Uranus, and thus their south poles were pointed toward the sun in 1986; it was these hemispheres that were photographed by Voyager 2. Their densities suggest that they contain relatively large rock cores surrounded by icy mantles (▌ Figure 24-10).

Oberon, the outermost of the large moons, has a dark, cratered surface, but it was once an active moon (▌ Figure 24-11a). A large fault crosses the sunlit hemisphere, and dark material, perhaps dirty water "lava," appears to have flooded the floors of some craters.

Titania is the largest of the five moons and has a heavily cratered surface, but it lacks the largest craters (Figure 24-11b). This suggests that soon after it formed, after most of the large debris had been swept up, Titania underwent an active phase in which internal melting

The Rings of Uranus and Neptune

Both Uranus and Neptune have rings made up of very dark particles orbiting in faint, narrow rings that are confined by small moons. Although they are similar, the two ring systems illustrate different processes that shape planetary rings.

The rings of Uranus were discovered in 1977, when Uranus crossed in front of a star. During this occultation, astronomers saw the star dim a number of times before and again after the planet crossed over the star. The dips in brightness were caused by rings circling Uranus.

Ace🌀Astronomy™

Go to AceAstronomy and cli[c]
Active Figures to see "Uranu[s]
Ring Detection" and watch ho[w]
the rings were discovere[d]

Notice the eccentricity of the ε ring. It lies at different distances on opposite sides of the planet.

More rings were discovered by Voyager 2. The rings are identified in different ways depending on when and how they were discovered.

Minutes from midoccultation

Intensity

δ γ β α α β γ δ

ε ε

Distance from center of Uranus (km)

The albedo of the ring particles is only about 0.015, darker than lumps of coal. If the rin[g] particles are made of methane-rich ices, radiation from the planet's radiation belts coul[d] break the methane down to release carbon and darken the ices. The same process ma[y] darken the icy surface of Uranian moons.

The narrowness of the rings suggests they are shepherded by small moons. Voyager 2 found Ophelia and Cordelia shepherding the ε ring. Other sma[ll] moons must be shepherding the other narrow rings. Such moons mu[st] be structurally strong to hold themselves together inside the planet's Roche limit.

Ophelia

Cordelia

The eccentricity of the ε ring is apparently caused by the eccentri[c] orbits of Ophelia and Cordelia.

Ring particles don't last forever as they collide with each other and are expose[d] to radiation. The rings of Uranus may need to be resupplied with fresh particles occasionally as impacts on icy moons scatter icy debris.

When the Voyager 2 spacecraft looked back at the rings illuminated from behind by the sun, the rings were not bright. That is, the rings are not bright in forward-scattered light. That means they must not contain much dust. The nine main rings contain particles no smaller than meter-sized boulders.

Uranus

Collisions among ring particles produce dust, which is thinly scattered inward from the λ ring, but the high, tenuous atmosphere of Uranus is slowing the dust particles and making them fall into the planet. The rings of Uranus actually contain very little dust.

Disk of Neptune

Visual-wavelength image

The brightness of Neptune is hidden behind the black bar in this Voyager 2 image. Two narrow rings are visible, and a wider, fainter ring lies closer to the planet. More ring material is visible between the two narrow rings.

The rings of Neptune are bright in forward-scattered light, as in the image above, and that tells us the rings contain significant amounts of dust. The ring particles are as dark as those that circle Uranus, however, so they probably contain methane-rich ice darkened by radiation.

Arc

When Neptune occulted stars, astronomers sometimes detected rings and sometimes did not. From that they concluded that Neptune might have ring arcs. Computer enhancement of this Voyager 2 visual-wavelength image shows arcs, regions of higher density, in the outer ring.

Arc

NASA image courtesy of Carolyn C. Porco

Enhanced visual image

The ring arcs visible in the outer ring appear to be generated by the gravitational influence of the moon Galatea, but other moons must also be present to confine the rings.

Neptune's rings lie in the plane of the planet's equator and inside the Roche limit. The narrowness of the rings suggests that shepherd moons must confine them, and a few such moons have been found among the rings. There must be more small moons to confine the rings.

Naiad

Galetea

Thalassa

Adams

LeVerrier

Despina

Galle

Neptune's rings have been given names associated with the planet's history. English astronomer Adams and French astronomer LeVerrier predicted the existence of Neptune from the motion of Uranus. The German astronomer Galle discovered the planet in 1846 based on LeVerrier's prediction.

Like the rings of Uranus, the ring particles that orbit Neptune cannot have survived since the formation of the planet. Occasional impacts on Neptune's moons must scatter debris and resupply the rings with fresh particles.

Neptune

FIGURE 24-10

The densities of the five major satellites of Uranus suggest that they contain relatively large rocky cores with mantles of dirty ice. The size of Earth's moon is shown for comparison.

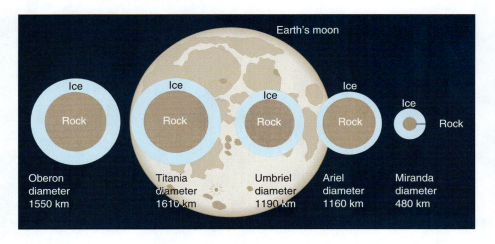

Earth's moon

Ice
Rock
Oberon
diameter
1550 km

Ice
Rock
Titania
diameter
1610 km

Ice
Rock
Umbriel
diameter
1190 km

Ice
Rock
Ariel
diameter
1160 km

Ice
Rock
Miranda
diameter
480 km

flooded the surface with water and erased the early craters. Since then, more craters have formed, but few are as large as those that were erased. Another clue to Titania's past activity is the network of faults that crosses its surface.

Umbriel, the third moon inward, is a very dark, cratered world with no sign of faults or surface activity (Figure 24-11c). It is the darkest of the moons, with an albedo of only 0.16 compared with 0.25 to 0.45 for the other moons. Planetary astronomers suspect that Umbriel's crust is a dark mixture of rock and ice. A bright crater floor in one region suggests that some clean ice may lie at shallow depths in some regions.

Ariel has the brightest surface of the five major satellites and shows clear signs of geological activity. It is crossed by faults over 10 km deep, and some regions appear to have been smoothed by resurfacing (Figure 24-11d). Crater counts show that the smoothed regions are younger than the other regions. Apparently, past melting in the interior has caused geological activity on the moon's surface. Ariel may have been subject to tidal heating caused by an orbital resonance with Miranda and Umbriel.

Miranda, the smallest of the five satellites discovered from Earth, is the most unusual. In fact, its surface is so peculiar that one astronomer referred to it as a "moon designed by a committee." Miranda is marked by oval patterns of grooves, which at first were dubbed *circi maximi* (singular, *circus maximus*) after the ancient Roman chariot-racing tracks (▮ Figure 24-12). Now known as **ovoids,** these features were originally thought to have been produced when Miranda was shattered by a major impact and the mutual gravity of the fragments pulled them back together to reassemble the little moon. More recent studies of the ovoids show that they are associated with faults, ice-lava flows, and rotated blocks of crust. These features suggest that a major impact was not involved. Rather, large convection currents in Miranda's icy mantle may have forced material upward to create the ovoids.

Certainly, Miranda has had a violent past. Near the equator, a huge cliff rises 20 km. An astronaut who slipped from the top of the cliff would fall for 10 minutes before hitting the surface. The cliff and the ovoids are old, however. Miranda is no longer geologically active, and we can read the history of its past on its disturbed surface.

The peculiar geology of Miranda illustrates the problem astronomers face in understanding the geology of the Jovian satellites. Has it been dominated by interior or exterior factors? If the ovoids were caused by convection in Miranda's icy mantle, then heat rising from its interior must be the dominant factor, and we must suppose that its interior was heated by tidal stress as in Jupiter's moon Io. Miranda is too small to have kept its heat of formation or to have been severely heated by radioactive decay. Tidal heating could have heated Miranda if its orbit was disturbed by a resonance with other satellites. Thus, we can understand how internal factors could have driven slow convection currents rising through Miranda's icy mantle to create its ovoids.

But external factors are also a possibility. Comets sweep through the solar system, and they must occasionally strike the planets and moons. The Jovian planets would retain no permanent scars of such impacts, as Earth's astronomers saw dramatically in the summer of 1994 when a comet struck Jupiter. The moons in the outer solar system, however, could be battered or even disrupted by such impacts. Icy debris from such impacts may occasionally resupply the rings of Uranus with fresh icy particles. In this way, the rings and moons appear to be related to each other.

Although planetary scientists now prefer internal heat to explain Miranda's geology, we must accept that external factors such as major impacts do occur. Thus, Miranda reminds us that modern astronomy recognizes the importance of catastrophic events in the history of the solar system. (See Window on Science 19-1.) We will see more suggestions of catastrophic events when we discuss the moons of Neptune.

c

d   Visual-wavelength images

FIGURE 24-11

(a) Oberon appears to have an old, inactive, heavily cratered surface. (b) Titania has a cratered surface crossed by deep valleys up to 100 km across—signs of an active past. (c) Umbriel is a dark, cratered world with no sign of faults or valleys. (d) This mosaic of images of Ariel shows it crossed by faults and broad valleys that appear to have been smoothed of earlier craters. *(a, b, and c, NASA; d, U.S. Geological Survey, Flagstaff, Arizona)*

A HISTORY OF URANUS

The challenge of comparative astronomy is to tell the story of a world, and Uranus may present the biggest challenge of all the objects in our solar system. Not only is it so distant that it is difficult to study, but it is also peculiar in many ways.

Presumably, Uranus formed from the solar nebula, as did the other Jovian planets, but it must have grown more slowly, perhaps because the solar nebula was less

Visual

a

b

FIGURE 24-12

Miranda, the smallest of the five major moons of Uranus, is only 480 km in diameter, but its surface shows signs of activity. This photomosaic of Voyager 2 images (a) reveals that it is marked by great oval systems of grooves. In the map (b), the smallest features detected are about 1.5 km in diameter. The cliff marked by an arrow near the bottom of the map is 20 km high. *(U.S. Geological Survey, Flagstaff, Arizona; map by P. Bridges)*

dense beyond Saturn. Also, objects orbit slowly in the outer solar system, so protoplanets must have grown slowly there. By the time the proto-Uranus had grown massive enough to begin trapping hydrogen, the sun had turned on, and the solar nebula was being blown away. As one astronomer said, "By the time Uranus got all dressed up, the party was over."

In fact, some theorists believe Uranus and Neptune could not have grown to their present size in the slow-moving orbits they now occupy. Highly complex mathematical models of planet formation in the solar nebula suggest that Uranus and Neptune might have formed closer to the sun, in the neighborhood of Jupiter and Saturn. Gravitational interactions among the growing planets could have moved Uranus and Neptune outward to their present locations. This is a controversial hypothesis, but it illustrates how uncertain the histories of the outer planets really are.

The highly inclined axis of Uranus may have originated late in its formation when it was struck by a planetesimal as large as Earth. That impact may have disturbed the interior of the world and caused it to lose much of its heat. Uranus now radiates less than 10 percent more heat than it receives from the sun. There is now just enough heat flowing outward to drive circulation in its slushy mantle and generate its magnetic field.

Impacts have been important in the history of the moons and rings of Uranus. All of the satellites are cratered, and some show signs of large impacts. Plan-

etesimals and the nuclei of comets striking the moons may create dust and debris that become trapped among the orbits of the smaller moons to produce the narrow rings. It seems very unlikely that the rings could have lasted since the formation of the planet. When the final history of Uranus is written, it will surely include some dramatic impact events.

REVIEW **CRITICAL INQUIRY**

How do we know what the interior of Uranus is like?

Obviously, we can't see inside the planet, so we are limited to a few basic observations that we must connect with models in a chain of inference to describe the interior. First, we can measure the size of Uranus in photographs, and we can find its mass by observing the orbital radius and period of its moons. Its mass divided by its volume gives us its density, which implies that, for a Jovian planet, it must contain a higher proportion of dense material such as ice and rock. To produce a planet of this size and density, models predict a core of heavy elements and a mantle of icy slush. Two other observations we can make are the oblateness of the planet and the rotation period, derived from the radio noise produced in the magnetic field. Models cannot make the planet as flattened as is observed if the heavy elements are highly concentrated toward the core, so we must assume that the mantle contains denser material, such as rock, mixed into the water and ice. We can also observe the highly inclined and offset magnetic field, which is believed to be generated in the mantle, and that tells us that the mantle must be a good electrical conductor, such as water with dissolved ammonia.

Going from observable properties to unobservable properties is the heart of astronomical research. For example, what observational properties of the rings of Uranus argue that small moons must orbit among the rings?

When Voyager 2 left Uranus, Earth's astronomers were left grappling with new planetary puzzles. Over the following years, some of the puzzles were explained, but in August 1989 Voyager 2 flew past Neptune, and the game began all over again.

24-2 NEPTUNE

Uranus and Neptune are often discussed together. They are about the same size and density and are probably very similar planets in some ways. They do differ, however, in certain respects. Neptune has a peculiar ring and a peculiar satellite system. Even the discovery of Neptune was different from the discovery of Uranus.

THE DISCOVERY OF NEPTUNE

In 1843, the young English astronomer John Couch Adams (1819–1892) completed his degree in astronomy and immediately began the analysis of one of the great problems of 19th-century astronomy. William Herschel had discovered Uranus in 1781, but earlier astronomers had seen the planet as early as 1690 and had plotted it on charts as a star. When 19th-century astronomers tried to combine this data, they found that no planet obeying Newton's laws of motion and gravity could follow such an orbit.

Some astronomers had suggested that the gravitational attraction of an undiscovered planet was causing the discrepancies. Adams began with the variation in the positions of Uranus, never more than 2 minutes of arc, and by October 1845 had calculated the orbit and position of the undiscovered planet. He sent his prediction to the Astronomer Royal, Sir George Airy, who passed it on to an observer. Neither Airy nor his observer took the prediction very seriously, and they were very dilatory in making the search.

Meanwhile, the French astronomer Urbain Jean Leverrier (1811–1877) made the same calculations and sent his prediction to Johann Galle at the Berlin Observatory. Galle received Leverrier's prediction on the afternoon of September 23, 1846, and, after searching for 30 minutes that evening, found Neptune. It was only 2° from Adams's predicted position.

When the English announced Adams's work, the French suspected that he had plagiarized the calculations, and the controversy was bitter. In fact, both men had performed the calculations correctly. But they could have been beaten to the discovery had Galileo paid less attention to Jupiter and more attention to what he saw in the background. Recent studies of Galileo's notebooks show that he saw Neptune on December 24, 1612, and again on January 28, 1613, but he plotted it as a star in the background of drawings of Jupiter. It is interesting to speculate about the response of the Inquisition had Galileo proposed that a planet existed beyond Saturn. Unfortunately for history (perhaps fortunately for Galileo), he did not recognize Neptune as a planet, and its discovery had to wait another 234 years.

The discovery of Neptune was a triumph for Newtonian physics: The three laws of motion and the law of gravity had proved sufficient to predict the position and orbit of an unseen planet. Thus, the discovery of Neptune was fundamentally different from the discovery of Uranus—the existence of Neptune was predicted from basic laws. Discovery by prediction is important in science because confirmation of the prediction serves as confirmation of the basic laws.

THE ATMOSPHERE AND INTERIOR OF NEPTUNE

Little was known about Neptune before the Voyager 2 spacecraft swept past it in August 1989. Seen from Earth, Neptune is a tiny, blue-green dot never more than 2.3 seconds of arc in diameter. Astronomers knew it was almost four times the diameter of Earth and about 4 percent smaller in diameter than Uranus (■ Data File Ten). Spectra revealed that its blue-green color was caused by methane in its hydrogen-rich atmosphere. Methane absorbs red light, making Neptune look blue-green. Neptune's density showed that it was a Jovian planet rich in hydrogen, but almost no detail is visible from Earth, so even its period of rotation was poorly known.

Voyager 2 passed only 4905 km above Neptune's cloud tops, closer than any spacecraft had ever come to one of the Jovian planets. The Voyager photos revealed that Neptune is marked by dramatic belt–zone circulation parallel to the planet's equator. Voyager 2 saw at least four cyclonic disturbances. The largest, dubbed the Great Dark Spot, looked similar to the Great Red Spot on Jupiter (■ Figure 24-13). The Great Dark Spot was located in the southern hemisphere and rotated counterclockwise, with a period of about 16 days. It appeared to be caused by gas rising from the planet's interior. When the Hubble Space Telescope began imaging Neptune in 1994, the Great Dark Spot was gone, and new cloud features were seen appearing and disappearing in Neptune's atmosphere (Figure 24-13). Thus, the cyclonic disturbances on Neptune are not nearly as long-lived as Jupiter's Great Red Spot.

Uranus Earth Neptune

Hydrogen and helium

Ices and rock

Heavy element core

NEPTUNE

Data File Ten

Neptune is a bit smaller than Uranus and four times larger in diameter than Earth. Its interior contains a rocky core with a deep mantle of slushy ices mixed with rock.

Average distance from the sun	30.0611 AU (44.971 × 10⁸ km)
Eccentricity of orbit	0.0100
Maximum distance from the sun	30.4 AU (45.4 × 10⁸ km)
Minimum distance from the sun	29.8 AU (44.52 × 10⁸ km)
Inclination of orbit to ecliptic	1°46′27″
Average orbital velocity	5.43 km/s
Orbital period	164.793 y (60,189 days)
Period of rotation	16^h3^m
Inclination of equator to orbit	28°48′
Equatorial diameter	49,500 km (3.93 $D_\oplus$)
Mass	1.030×10^{26} kg (17.23 $M_\oplus$)
Average density	1.66 g/cm³
Gravity	1.19 Earth gravities
Escape velocity	25 km/s (2.2 $V_\oplus$)
Temperature at cloud tops	−216°C (−357°F)
Albedo	0.35
Oblateness	0.027

The Voyager 2 images reveal other cloud features standing out against the deep blue of the methane-rich atmosphere. The white clouds are made up of crystals of frozen methane and range up to 50 km above the deeper layers, just where the temperature in Neptune's atmosphere is low enough for rising methane to freeze into ice crystals. (See Figure 24-5.) Presumably these features are related to rising convection currents that form clouds high in Neptune's atmosphere, where they catch sunlight and look bright. Observations made by the Hubble Space Telescope suggest that atmospheric activity on Neptune may be related in some way to solar activity, but further observations are needed to explore this connection.

As on the other Jovian worlds, high winds circle Neptune parallel to its equator, but the winds blow at very high velocities and tend to blow backward—against the rotation of the planet. Why Neptune should have such high winds is not understood, and it is part of the larger problem of the belt–zone circulation. Now that we have seen at least traces of belts and zones on all four of the Jovian planets (if we may take the faint clouds on Uranus as belt–zone features), we can ask what drives this circulation. Because the belts and zones remain parallel to the planet's equator even when the planet rotates at a high inclination, as in the case of Uranus, it seems reasonable to believe that the atmospheric circulation is dominated by the rotation of the planet and not by solar heating.

The same observations that helped define the interior of Uranus can be applied to Neptune. The interior is believed to contain a heavy-element core, smaller than that in Jupiter, surrounded by a deep mantle of slushy ice mixed with heavier material such as rock. Neptune's magnetic field is a bit less than half as strong as Earth's and is tipped 47° from the axis of rotation. It is offset 55 percent of the way to the surface (Figure 24-7). As in the case of Uranus, the field is probably generated by the dynamo effect acting in the conducting fluid mantle.

Neptune has more internal heat than Uranus, and part of that may be generated by radioactive decay in the minerals in its interior. Some of the energy may be released by denser material falling inward, including, as in the case of Uranus, diamond crystals formed by the disruption of methane.

Neptune is a tantalizing world just big enough to be imaged by the Hubble Space Telescope but far enough away to make it difficult to study. The data from Voyager 2 revealed one detail invisible from Earth—rings.

THE RINGS OF NEPTUNE

Astronomers on Earth saw hints of rings when Neptune occulted stars, but the rings were not firmly recognized until Voyager 2 flew past the planet in 1989.

Neptune in 1989

Great Dark Spot

Visual-wavelength image
from Voyager 2

1996

Visual-wavelength
images from
Hubble Space Telescope

1998

2002

FIGURE 24-13

Because Neptune is inclined almost 29 degrees in its orbit, it experiences seasons that each last about 40 years. Since 1989, spring has come to the southern hemisphere, and the weather has clearly changed, which is surprising because sunlight on Neptune is 900 times dimmer than on Earth. *(NASA, L. Sromovsky, and P. Fry, University of Wisconsin-Madison.)*

Study "The Rings of Uranus and Neptune" on pages 544 and 545 and notice how Neptune's rings compare to those of Uranus. Both sets of rings are made up of narrow rings confined by shepherd moons, but there are two important differences. First, notice how forward scattering tells us that Neptune's rings contain more dust than Uranus's rings. Second, notice a new way that rings can interact with moons; one of Neptune's moons is producing short arcs in the outermost ring.

The rings of Neptune resemble the rings of Uranus in one very important way. They can't be primordial. That is, they can't have lasted from the formation of Neptune. Evidently, impacts on moons occasionally scatter debris through the satellite system, and some of it falls into the rings in the most stable places among the orbits of the moons. Planetary rings are constantly being remade.

THE MOONS OF NEPTUNE

Neptune has only two moons that are visible from Earth, but Voyager 2 discovered six more small moons, and a few more small moons have been found using Earth-based telescopes. The two largest moons, Triton and Nereid, have been a puzzle for years because of their peculiar orbits.

Triton has a nearly circular orbit, but it travels backward—clockwise as seen from the north. This makes Triton the only large satellite in the solar system with a retrograde (backward) orbit. The other satellite visible from Earth, Nereid, goes in the right direction, but its orbit is highly elliptical and very large (Figure 24-14). Nereid takes 359.4 days to orbit Neptune once. Many astronomers have speculated that the orbits of the two moons are evidence of a violent event. An encounter with a massive planetesimal may have disturbed the moons, or Triton itself may have been captured into orbit during a close encounter with Neptune.

Voyager 2 found six more moons orbiting Neptune, but they all lie close to the planet among the rings. No new moons were found beyond the orbit of Triton. Some astronomers have suggested that Triton, in its retrograde orbit, would have consumed any moons near it.

The Voyager 2 photographs show that Triton is a highly complex moon. Although it is only 1360 km in radius (78 percent the radius of Earth's moon), it is so cold (34.5 K, or −398°F) that it can hold an atmosphere of nitrogen with some methane. Its air is 10^5 times less dense than Earth's atmosphere. Although a few wisps of haze are visible in the photographs, the atmosphere is transparent, and the surface is easily visible (Figure 24-15).

The surface of Triton is evidently composed of ices. Early analysis suggested that the ices were mostly water ice with methane and ammonia ice mixed in. However, more recent studies reveal that the surface

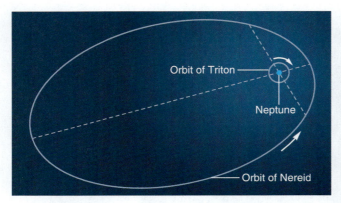

FIGURE 24-14

Nereid has a large, elliptical orbit around Neptune, but Triton follows a small, circular, retrograde orbit. At a distance of 30 AU from Earth, Triton is never more than 16 seconds of arc from the center of Neptune.

ice is dominated by frozen nitrogen with some methane, carbon monoxide, and carbon dioxide. This is consistent with the nitrogen-rich atmosphere. Some regions of the surface look dark and may be older terrain that has grown dark as sunlight has converted methane into dark organic compounds. The south pole of Triton had been turned toward the sun for 30 years when Voyager 2 flew past, and deposits of nitrogen frost in a large polar cap appeared to be vaporizing there and refreezing in the darkness of the north pole (Figure 24-15a). Thus, the cycle of nitrogen on Triton resembles in some ways the cycle of carbon dioxide on Mars.

The surface of Triton suggests that it has been active recently and may still be active. Few craters mark the surface of the icy world, and comparison with the rate at which we might expect meteorites to hit Triton suggests the average age of the surface is no more than 100 million years. Some process has erased older craters. Long linear features appear to be fractures in an icy crust, and some roughly round basins appear to have been flooded time after time by liquids from the interior (Figure 24-15b). Triton is much too cold for the liquid to be molten rock or even liquid water. Rather, the flooding may have been caused by water containing agents such as ammonia, which would lower the freezing point of the liquid. It isn't possible to say at present whether Triton is still active, but 100 million years isn't very long in the history of a world. It may still suffer periodic eruptions and flooding.

Another form of activity on Triton are the dark smudges visible in the bright nitrogen ices near the south pole. These may be caused by nitrogen ice beneath the crust. Warmed slightly by the sun, the nitrogen ice can change from one form of nitrogen ice to another and release heat that could vaporize some of the nitrogen. This nitrogen vapor could vent through the crust to form nitrogen gas geysers. Sunlight could convert some of the methane carried along with the nitro-

a

b   Visual-wavelength images

FIGURE 24-15

(a) Triton's south pole (bottom) had been in sunlight for 30 years when Voyager 2 flew past in 1989. The frozen nitrogen in the polar cap appears to be vaporizing, perhaps to refreeze in the darkness at the north pole. The dark smudges are produced when liquid nitrogen in the crust vaporizes and drives nitrogen geysers. Methane carried along in the nitrogen is turned to black compounds by sunlight and makes the black smudges. (b) Roughly round basins on Triton may be old impact basins flooded repeatedly by liquids from the interior. Notice the small number of craters on Triton, a clue that it is a partially active world. *(NASA)* gen into dark organic material, which would fall to the surface and account for the dark smudges on the ice.

Active worlds must have a source of energy, so we must ask where Triton gets its energy. Triton is big enough to retain some heat from low-level radioactivity in its interior. That may be enough to melt exotic ices and cause flooding on the surface. The nitrogen geysers appear to be powered not from the interior but from sunlight. In fact, Triton is very efficient at absorbing sunlight. Its thin atmosphere allows sunlight to fall on the crust, which is composed of ices that are partially transparent to light. As the light penetrates into the ice and is gradually absorbed, it is converted to heat. The ices, however, are not very transparent to infrared radiation, and the heat is trapped in the crust. Thus, Triton may be heated, in part, by an icy form of the greenhouse effect.

This low-level heating would not be enough to erase nearly all of Triton's craters, so some planetary astronomers wonder if Triton might have been captured by Neptune within the last billion years. Tidal forces from such a capture would cause enough tidal heating to melt Triton and totally resurface it. Triton may still be active with residual heat.

THE HISTORY OF NEPTUNE

Can we tell the story of Neptune? In some ways it seems to be a simple world, but its magnetic field is peculiar, and its moons and rings demand careful attention.

Neptune is much like Uranus, and we can assume that it formed from the solar nebula much as did Uranus. It is a mostly liquid world with a dense core. Its hazy atmosphere, marked by changing cloud patterns, hides a mantle of partially frozen water where astronomers suspect the dynamo effect generates its off-center magnetic field.

Neptune's satellite system suggests a peculiar history. Triton, the largest moon, revolves around Neptune in a retrograde orbit, while Nereid's long-period orbit is highly elliptical. These orbital oddities suggest that the satellite system may have been disturbed by an encounter with some other object, such as a large planetesimal or the capture of Triton. We have seen evidence in other satellite systems of impacts with large objects, so such an event is not unreasonable.

At least six smaller moons orbit Neptune, but they lie near its ring system. Because the rings are bright in forward scattering, we can conclude that they are rich in dust, and the shepherding of small satellites could confine their width and produce the observed arcs. Such delicate rings could not have survived since the origin of the planets, so we must suppose that the rings are occasionally supplied with fresh particles generated by collisions between comets and Neptune's moons. Once again, the evidence we see of major impacts in the solar system's history assures us that such impacts do occur.

REVIEW CRITICAL INQUIRY

Why is Neptune blue but its clouds white?

When we look *at* something, we really turn our eyes toward it and receive light *from* the object. The light we receive from Neptune is sunlight that is reflected from various layers of Neptune and journeys to our eyes. Because sunlight contains a distribution of photons of all visible wavelengths, it looks white to us, but sunlight entering Neptune's atmosphere must pass through hydrogen gas that contains a small amount of methane. While hydrogen is transparent, methane is a good absorber of longer wavelengths, so red photons are more likely to be absorbed than blue photons. Once the light is scattered from deeper layers, it must run this methane gauntlet again to emerge from the atmosphere, and again red photons are more likely to be absorbed. The light that finally emerges from Neptune and eventually reaches our telescopes is poor in longer wavelengths and thus looks blue.

The methane-ice-crystal clouds lie at high altitudes, so sunlight does not have to penetrate very far into Neptune's atmosphere to reflect off the clouds, and thus it loses many fewer of its red photons. The clouds look white.

This discussion shows how a careful, step-by-step analysis of a natural process can help us better understand how nature works. For example, explain where the energy might come from to keep Triton active.

Neptune was the last planet visited by Voyager 2, which is now journeying out of our solar system. It will not pass near any other worlds unless, millions of years from now, it enters some distant solar system. Nevertheless, we have one more planet to analyze.

24-3 PLUTO

Pluto, the farthest outpost of our solar system, is a small, icy world. Discovered through a combination of ingenuity, perseverance, and luck, it is now known to have a satellite.

THE DISCOVERY OF PLUTO

Percival Lowell (1855–1916) was fascinated with the idea that an intelligent race built the canals he thought he could see on Mars (Chapter 22). Lowell founded Lowell Observatory primarily for the study of Mars. Later, to improve the reputation of his observatory, he began to search for a planet beyond Neptune.

Lowell used the same method that Adams and Leverrier had used to predict the position of Neptune. Working from the observed irregularities in the motion of Neptune, Lowell predicted the location of an undiscovered planet of about 7 Earth masses, which, he concluded, would look like a 13th-magnitude stellar image in eastern Taurus. Lowell searched for the planet photographically until his death in 1916.

In the late 1920s, a 22-year-old amateur astronomer, Clyde Tombaugh, began using a homemade 9-in. telescope to sketch Jupiter and Mars from his family's wheat farm in western Kansas. He sent his drawings to Lowell Observatory, and the observatory director hired him (without an interview) to resume Lowell's search for the undiscovered planet. The young Tombaugh bought a one-way train ticket for Flagstaff not knowing what his new job would be like.

The observatory director set Tombaugh to work photographing the sky along the ecliptic around the predicted position of the planet. The search technique was a classic method in astronomy. Tombaugh obtained pairs of 14 × 17-inch glass plates exposed two to three days apart. To search a pair of plates, he mounted them in a blink comparator, a machine that allowed him to look through a microscope at a small spot on one plate and then at the flip of a lever see the same spot on the other plate. As he blinked back and forth, the star images did not move, but a planet would have moved along its orbit during the two to three days that elapsed before the second plate was exposed. So Tombaugh searched the giant plates, star image by star image, looking for an image that moved. A single pair of plates could contain 400,000 star images. He searched pair after pair and found nothing.

The observatory director turned to other projects, and Tombaugh, working alone, expanded his search to cover the entire ecliptic. For almost a year, Tombaugh exposed plates by night and blinked plates by day. Then on February 18, 1930, nearly a year after he had left Kansas, a quarter of the way through a pair of plates, he found a 15th magnitude image that moved (❙ Figure 24-16). He later remembered, "Oh, I thought, I had better look at my watch; this could be a historic moment. It was within about 2 minutes of 4 PM [MST]."

The discovery was announced on March 13, the 149th anniversary of the discovery of Uranus and the 75th anniversary of the birth of Percival Lowell. The planet was named Pluto after the god of the underworld and, in a way, after Lowell, because the first two letters in Pluto are the initials of Percival Lowell.

The discovery of Pluto seemed a triumph of discovery by prediction, but Tombaugh sensed something was wrong from the first moment he saw the image. It was moving in the right direction by the right amount, but it was 2.5 magnitudes too faint. Clearly, Pluto was not the 7-Earth-mass planet that Lowell had predicted. The faint image implied that Pluto was a small planet with a mass too low to seriously alter the motion of Neptune.

a   b   Visual-wavelength images

FIGURE 24-16
Pluto is small and far away, so its image is indistinguishable from that of a star on most photographs. Clyde Tombaugh discovered the planet in 1930 by looking for an object that moved relative to the stars on a pair of photographs taken a few days apart. *(Lowell Observatory photographs)*

Later analysis has shown that the variations in the motion of Neptune, which Lowell used to predict the location of Pluto, were random uncertainties of observation and could not have led to a trustworthy prediction. The discovery of the new planet only 6° from Lowell's predicted position was apparently an accident, which proves that if you search long enough and know what to expect, you are likely to find something (Window on Science 24-1).

PLUTO AS A PLANET

Pluto is very difficult to observe from Earth. Only a bit larger than 0.1 second of arc in diameter, the little planet is only 65 percent the diameter of Earth's moon and shows little surface detail. The best photos by the Hubble Space Telescope reveal large areas of light and dark terrain (▌ Data File Eleven), promising that Pluto will reveal itself to be an interesting world when spacecraft finally visit it.

Most planetary orbits in our solar system are nearly circular; but Pluto's is quite elliptical (see Figure 1-7). On the average, it is the most distant planet, but it can come closer to the sun than Neptune. In fact, from January 21, 1979, to March 14, 1999, Pluto was closer to the sun than Neptune. The planets will never collide, however, because Pluto's orbit is inclined 17°.

Orbiting so far from the sun, Pluto is cold enough to freeze most compounds we think of as gases, and

spectroscopic observations have found evidence of solid nitrogen ice on its surface with traces of frozen methane and carbon monoxide. The daytime temperature of about 50 K is enough to vaporize some of the nitrogen and carbon monoxide and a little of the methane to form a thin atmosphere around Pluto. This atmosphere was detected in 1988 when Pluto passed in front of (occulted) a distant star, and the starlight faded gradually rather than winking out suddenly.

Pluto's only known moon was discovered on photographs in 1978. It is very faint and about half the diameter of Pluto (▌ Figure 24-17). The moon was named Charon after the mythological ferryman who transports souls across the river Styx into the underworld.

The discovery of Charon was important for a number of reasons. Charon orbits Pluto in a nearly circular orbit in the plane of Pluto's equator. Observations show that the moon and Pluto are tidally locked to each other, and that Pluto's axis of rotation is highly inclined (▌ Figure 24-18). Furthermore, the orbital motion of Charon allows the calculation of the mass of Pluto. Charon orbits 19,640 km from Pluto with an orbital period of 6.387 days. Kepler's third law tells us that the mass of the system is 6.5×10^{-9} solar masses or about 0.2 Earth masses. Most of this mass is Pluto, which seems to be about 12 times more massive than Charon.

The mass of a body is important in astronomy because mass divided by volume is density. The density of Pluto is about 2 g/cm^3, and the density of Charon is

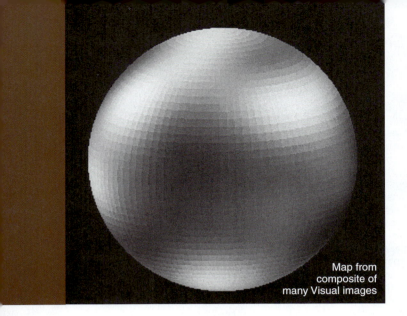

Map from
composite of
many Visual images

just a bit less. Pluto and Charon must contain about 35 percent ice and 65 percent rock.

Spectra show that Pluto's surface is mostly nitrogen ice with some methane, carbon monoxide, and possibly water ice. The nitrogen ice vaporizes most easily and nitrogen gas makes up most of the atmosphere. Spectra of Charon show that the small moon lacks most of the ices and has a surface that is mostly water ice. Perhaps it has lost its more volatile compounds because of its lower escape velocity. Water ice at Charon's surface temperature is no more volatile than is a piece of rock on Earth.

PLUTO

Data File Eleven

The Hubble Space Telescope is just able to reveal bright and dark regions on Pluto. Some of the bright regions may be deposits of bright methane ice. *(Alan Stern, Southwest Research Institute, Marc Buie, Lowell Observatory, NASA, ESA)*

Average distance from the sun	39.44 AU (59.00×10^8 km)
Eccentricity of orbit	0.2484
Maximum distance from the sun	49.24 AU (73.66×10^8 km)
Minimum distance from the sun	29.64 AU (44.34×10^8 km)
Inclination of orbit to ecliptic	17°9′3″
Average orbital velocity	4.73 km/s
Orbital period	247.7 y (90,465 days)
Period of rotation	$6^d 9^h 21^m$
Inclination of equator to orbit	119.6°
Equatorial diameter	2370 km (0.19 $D_\oplus$)
Mass	1.2×10^{22} kg (0.002 $M_\oplus$)
Average density	2.0 g/cm³
Surface gravity	0.06 Earth gravity
Escape velocity	1.2 km/s (0.11 $V_\oplus$)
Surface temperature	−230°C (−382°F)
Albedo	0.5

Visual-wavelength image

a

Visual-wavelength image

b

FIGURE 24-17

(a) A high-quality ground-based photo shows Pluto and its moon, Charon, badly blurred by seeing. *(NASA)* (b) The Hubble Space Telescope image clearly separates the planet and its moon and allows more accurate measurements of the position of the moon. *(R. Albrecht, ESA/ESO Space Telescope European Coordinating Facility, NASA)*

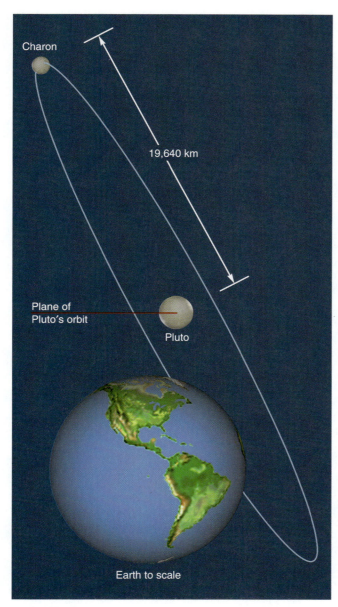

FIGURE 24-18
The nearly circular orbit of Charon is only a few times larger than the diameter of Earth. It is seen here nearly edge-on, and consequently it looks elliptical in this diagram. The orbit and the equator of Pluto are inclined 119.6° to the plane of Pluto's orbit around the sun.

Both Pluto and Charon go through dramatic seasons as they circle the sun in their highly inclined orbit. Seasons on Pluto and Charon are much like those on Uranus. This may cause changes in Pluto's atmosphere as it grows warmer when it is closest to the sun, as it was in the late 1980s, and then freezes as it draws away.

We strain our imaginations when we try to visualize the surface of Pluto. Old surfaces should be cratered, so we might imagine an icy, cratered world. The tidal lock between Pluto and Charon suggests an age of tidal heating that may have driven geological activity and

left traces on the surfaces of these worlds. However, the seasonal vaporization and refreezing of ices may have erased or covered much of the older surface with ices. Thus Pluto and Charon may be worlds unlike any we have seen before. Yet the biggest puzzle may be not their geology, but their origin.

THE ORIGIN OF PLUTO AND CHARON

Is Pluto really a planet? It is so dramatically different from its Jovian neighbors that it doesn't seem to fit into the outer solar system, and a few people have argued that it isn't a planet at all. In fact, this depends almost entirely on our arbitrary definition of the word *planet*. Although the news media find this debate interesting, astronomers are much more interested in the history of Pluto than in the definition of a word. A better question is: "How did Pluto form?"

An old theory proposed that Pluto was an escaped moon of Neptune. Pluto is much like Neptune's moon Triton. It is roughly the same density, and it is only 12 percent smaller; it could easily fit in with the icy moons of the Jovian planets. The disturbed orbits of Triton and Nereid could be the result of a complex interaction with a passing massive planetesimal in which Pluto might have been ejected from an orbit around Neptune. This theory has been mostly abandoned, because such an interaction seems unlikely. A newer hypothesis seems more probable.

Modern astronomers have begun thinking of Pluto as the largest of a family of small icy bodies that were once common in the outer solar system. Searches for other icy worlds beyond the orbit of Neptune have turned up hundreds, with hints that there are a vast number of more distant bodies. (We will discuss this belt of icy planetesimals in detail in Chapter 25.) Pluto is locked in a 3:2 orbital resonance with Neptune; that is, Neptune orbits the sun three times while Pluto orbits twice. This may give Pluto's orbital motion added stability. Indeed, many of the smaller icy bodies found beyond the orbit of Neptune are caught in resonances with Neptune. Thus, we can think of Pluto as the largest of the icy planetesimals that never became part of a larger planet.

The moon Charon is presumably another of these icy worlds. A collision between Charon and Pluto could have tilted the axis of rotation of Pluto and trapped Charon in an elliptical orbit around Pluto. Tidal forces between Pluto and Charon could then have forced the orbit of Charon to become nearly circular as it is today. If that happened, then we might hope to see dramatic evidence of tidal heating in Pluto and in Charon when spacecraft finally reach Pluto.

At present no spacecraft is heading for Pluto. Plans have been proposed, but no funds have been set aside. It may be a long time before we see the surface of these cold worlds, but our study of comparative planetology helps our imagination go where our cameras cannot.

REVIEW CRITICAL INQUIRY

What evidence do we have that cataclysmic impacts have occurred in our solar system?

We would expect objects that formed together from the solar nebula to rotate and revolve in the plane of the solar system, so the high inclination of Pluto and the orbit of Charon suggest a past collision or, at least, a close gravitational interaction with a passing body. And the Pluto–Charon system is not the only evidence for major impacts in the past. The peculiar orbits of Neptune's moons Triton and Nereid, the fractured and cratered state of the Jovian satellites, and the peculiar rotation of Uranus all hint that impacts and encounters with large planetesimals or the nuclei of comets have been important in the history of these worlds. Furthermore, the existence of planetary rings suggests that impacts have scattered small particles and replenished the ring systems. Even in the inner solar system, the backward rotation of Venus, the density of Mercury, and the formation of Earth's moon are seen as possible consequences of major impacts when the solar system was young.

Certainly, we do not imagine that planets bounce around like billiard balls, but we must recognize that the planets are not totally isolated now and may have been victims (or beneficiaries) of large impacts as they formed long ago. For example, how do the size and composition of Pluto and Charon, compared with the moons of the outer Jovian worlds, suggest an unexpected population of icy worlds in the outer solar system?

Like clues at the scene of a crime, the evidence we find in the outer solar system alerts us to the reality that all of the planets in our solar system have been and still are sitting ducks for impacts by the smaller objects that hurtle along their orbits around the sun. What are these objects? What can they tell us about the origin and evolution of our solar system? We explore these questions in the next chapter.

SUMMARY

Although Uranus was discovered unexpectedly in 1781, we should probably not refer to its discovery as accidental. In 1846, the position of Neptune was predicted from the orbital motion of Uranus, and Pluto was discovered in 1930 based on predictions derived from the motion of Neptune.

Uranus and Neptune are commonly classified as Jovian planets, but their interiors are different from the interiors of Jupiter and Saturn. The density, rotation periods, and oblateness of Uranus and Neptune tell us that they contain small cores of heavy materials, such as rock and metal, and deep mantles of partly frozen ices mixed with denser material, such as rock. Their atmospheres are rich in hydrogen and helium, and absorption of red light by small amounts of methane causes both planets to have a blue color.

The axis of rotation of Uranus is tipped 97.9° from the perpendicular to its orbit, and its south pole was facing the sun when Voyager 2 flew past. The spacecraft discovered that belts and zones lie deep in the hazy atmosphere. Thus, the rotation of the planet, rather than solar energy, dominates atmospheric circulation.

The rings of Uranus are very narrow and dark. Two shepherd satellites were discovered orbiting on either side of the ε ring, and other shepherd satellites are believed to bound the other rings. These satellites and the larger satellites appear to anchor the rings and prevent them from spreading and dissipating. The dark color of the ring particles appears to come from the methane-rich ices darkened by the radiation belts around the planet. The surfaces of the moons are also very dark.

The rings of Uranus are not very bright in forward scattering and thus must be very poor in small particles. Some dust was detected, but it seems to be falling into the planet due to the resistance of low-density hydrogen gas leaking out of the upper atmosphere.

Voyager 2 discovered ten small satellites orbiting Uranus, and more have been found from Earth-based telescopes. All of these moons are small and dark. The five largest moons show signs of impact cratering and some surface activity in the past. The innermost moon, Miranda, is marked by large ovals called ovoids. Planetary scientists believe that Miranda was heated by tides when it was young and that great convection currents stirred its partially frozen mantle and formed the ovoids.

Weather features on Uranus and Neptune do change. Voyager 2 found Uranus almost cloudless, and on Neptune it found the Great Dark Spot, a rotating storm much like the Great Red Spot on Jupiter. In the years since, the Hubble Space Telescope has observed the appearance of methane clouds on Uranus and the disappearance of the Great Dark Spot and the appearance of new weather patterns on Neptune.

Like Uranus, Neptune has a highly tilted off-center magnetic field presumably generated in a circulating mantle of slushy ices. Also like Uranus, Neptune has narrow, dim rings shepherded by small moons.

Triton, Neptune's largest moon, is an icy world with a thin atmosphere of nitrogen and methane. A polar cap of methane ice was visible as Voyager 2 flew past, but few impact craters mark the surface, hinting that the moon has been active. Nitrogen geysers detected in the south polar cap appear to be powered by heat from the sun.

Pluto and its moon Charon are rocky worlds with deep ice mantles. Pluto is large enough to keep a thin atmosphere of nitrogen, carbon monoxide, and methane. These two icy worlds may be among the largest of the icy planetesimals that filled the outer solar system during planet formation.

NEW TERMS

occultation

ovoid

REVIEW QUESTIONS

1. Describe the location of the equinoxes and solstices in the Uranian sky. What are seasons like on Uranus?

2. Why is belt–zone circulation difficult to detect on Uranus?

3. Discuss the origin of the rings of Uranus and Neptune. Cite evidence to support hypotheses.

4. How do the magnetic fields of Uranus and Neptune suggest that the mantles inside those planets are fluid?

5. If Neptune had no satellites at all, would you expect it to have rings? Why or why not?

6. Why might the surface brightness of ring particles and small moons orbiting Uranus and Neptune depend on whether those planets have magnetic fields?

7. Both Uranus and Neptune have a blue-green tint when observed through a telescope. What does that tell us about their composition?

8. How can small worlds like Triton and Pluto have atmospheres when a larger world such as Ganymede has none?

9. Why do we suspect that Triton has had an active past? What source of energy could power such activity?

10. If you visited Pluto and found Charon a full moon directly overhead, where would Charon be in the sky when it was new? when it was first quarter?

11. What evidence do we have that Pluto and Charon are made of mixtures of rock and ice?

DISCUSSION QUESTIONS

1. Why might it be unfair to describe William Herschel's discovery of Uranus as accidental? Why might it be unfair to describe the discovery of the rings of Uranus as accidental?

2. Suggest a single phenomenon that could explain the inclination of the rotation axis of Uranus, the orbits of Neptune's satellites, and the existence of Pluto's moon.

PROBLEMS

1. What is the maximum angular diameter of Uranus as seen from Earth? of Neptune? of Pluto? (*Hint:* Use the small-angle formula.)

2. One way to recognize a distant planet is by its motion along its orbit. If Uranus circles the sun in 84 years, how many seconds of arc will it move in 24 hours? (This does not include the motion of Earth. Assume a circular orbit for Uranus.)

3. What is the orbital velocity of Miranda around Uranus?

4. What is the escape velocity from the surface of Miranda? (*Hints:* Radius = 242 km. Assume that the density is 2 g/cm^3. See Chapter 5.)

5. The magnetosphere of Uranus rotates with the planetary interior in 17.24 hours. What is the velocity of the outer portion of the magnetic field just beyond the orbit of Oberon? (*Hint:* The circumference of a circle is $2\pi r$.)

6. If the ε ring is 60 km wide and the orbital velocity of Uranus is 6.81 km/s, how long a blink should we expect to see when the ring crosses in front of a star? Is this consistent with the data on page 544?

7. What is the escape velocity from the surface of an icy moon with a diameter of 20 km? (*Hints:* The density of ice is 1 g/cm^3. The volume of a sphere is $\frac{4}{3}\pi r^3$. See Chapter 5.)

8. What is the difference in the orbital velocities of the two shepherd satellites Cordelia and Ophelia? (*Hints:* Orbital radii = 49,800 km and 53,800 km. See Chapter 5.)

9. Repeat Problem 2 for Pluto. That is, ignoring the motion of Earth, how far across the sky would Pluto move in 24 hours? (Assume a circular orbit for Pluto.)

10. Given the size of Triton's orbit (R = 355,000 km) and its orbital period (P = 5.877 days), calculate the mass of Neptune. (*Hint:* See Chapter 5.)

CRITICAL INQUIRIES FOR THE WEB

1. Who was William Herschel? What can you find out about his life after he discovered Uranus?

2. What is weather on Uranus like now? Can you locate recent photos of cloud patterns there?

3. How many moons does Uranus have? More are discovered every year, it seems, so search for the latest discoveries.

4. What happened to John Couch Adams and Urbain Jean Leverrier after the discovery of Neptune? Did they make other important discoveries? What is the Adams Society? Can you find out what Leverrier did in the revolution of 1848?

5. Repeat Inquiries 2 and 3 for Neptune. Remember that finding new moons orbiting Neptune is difficult because it is farther away.

6. Who was Clyde Tombaugh? What can you find out about his life after he discovered Pluto?

 Visit the Seeds *Foundations of Astronomy* companion Web site for critical thinking exercises, articles, and additional readings from InfoTrac College Edition, Brooks/Cole's online student library.

METEORITES, ASTEROIDS, AND COMETS

When they shall
cry "PEACE, PEACE" then
cometh sudden destruction!
<u>COMET'S CHAOS?</u>—What
<u>Terrible events</u> will the
<u>Comet</u> bring?

*From a religious pamphlet predicting the end of the world
because of the appearance of Comet Kohoutek, 1973*

Chapter 25

GUIDEPOST

In Chapter 19, we began our study of planetary astronomy by asking how our solar system formed. In the five chapters that followed, we surveyed the planets, but we gained only limited insight into the origin of the solar system. The planets are big, and they have evolved as heat has flowed out of their interiors. In this chapter, we have our best look at unevolved matter left over from the solar nebula. These small bodies are, in fact, the last remains of the nebula that gave birth to the planets.

This chapter is unique in that it covers small bodies. In past chapters, we have used the principles of comparative planetology to study large objects—the planets. In this chapter, we see that the same principles apply to smaller bodies, but we also see that we need some new tools in order to think about the tiniest worlds in the solar system.

VIRTUAL LABORATORY

ASTEROIDS AND KUIPER BELT OBJECTS

FIGURE 25-1

Comet Hyakutake swept through the inner solar system in 1996 and was dramatic in the northern sky. Seen here from Kitt Peak National Observatory, the comet passed close to the north celestial pole (behind the observatory dome in this photo). Notice the Big Dipper below and to the left of the head of the comet. *(Courtesy Tod Lauer)*

In 1910, Comet Halley was spectacular. On the night of May 19, Earth actually passed through the tail of the comet—and millions of people panicked. The spectrographic discovery of cyanide gas in the tails of comets led many to believe that life on Earth would end. Householders in Chicago stuffed rags around doors and windows to keep out the gas, and bottled oxygen was sold out. Con artists in Texas sold comet pills and inhalers to ward off the noxious fumes. An Oklahoma newspaper reported (in what was apparently a hoax) that a religious sect tried to sacrifice a virgin to the comet.

Throughout history, bright comets have been seen as portents of doom. Even in our own time, the appearance of bright comets has generated predictions of the end of the world. Comet Kohoutek in 1973, Comet Halley in 1986, and Comet Hale–Bopp in 1997 caused deep concern among the superstitious. A bright comet moving slowly through the night sky is so out of the ordinary that we should not be surprised if it generates some instinctive alarm (Figure 25-1).

In fact, comets are graceful and beautiful visitors to our skies. Astronomers think of comets as messengers from the age of planet building. By studying comets, we learn about the conditions in the solar nebula from which planets formed. But comets are only the icy remains of the solar nebula; the asteroids are the rocky debris left over from planet building. Because we cannot easily visit comets and asteroids, we begin our discussion with the fragments of those bodies that fall to Earth—the meteorites.

Ace⟳Astronomy™ The AceAstronomy icon throughout the text indicates an opportunity for you to test yourself on key concepts and to explore animations and interactions on the AceAstronomy Web site at **http://astronomy.brookscole .com/seeds8e**

25-1 METEORITES

In the afternoon of November 30, 1954, Mrs. E. Hulitt Hodges of Sylacauga, Alabama, lay napping on her living room couch. An explosion and a sharp pain jolted her awake, and she found that a meteorite had smashed

TABLE 25-1
Meteor Showers

Shower	Dates	Hourly Rate	Radiant* R. A.	Radiant* Dec.	Associated Comet
Quadrantids	Jan. 2–4	30	15^h24^m	50°	
Lyrids	April 20–22	8	18^h4^m	33°	1861 I
η Aquarids	May 2–7	10	22^h24^m	0°	Halley?
δ Aquarids	July 26–31	15	22^h36^m	−10°	
Perseids	Aug. 10–14	40	3^h4^m	58°	1982 III
Orionids	Oct. 18–23	15	6^h20^m	15°	Halley?
Taurids	Nov. 1–7	8	3^h40^m	17°	Encke
Leonids	Nov. 14–19	6	10^h12^m	22°	1866 I Temp
Geminids	Dec. 10–13	50	7^h28^m	32°	

*R. A. and Dec. give the celestial coordinates (right ascension and declination) of the radiant of each shower.

through the ceiling and bruised her left leg. Mrs. Hodges is the only person known to have been injured by a meteorite. By the way, Mrs. Hodges lived right across the street from the Comet Drive-In Theater.

Meteorite impacts on homes are not common. Statistical calculations show that a meteorite should damage a building somewhere in the world about once every 16 months. About two meteorites large enough to produce visible impacts strike Earth each day, but most meteors are small particles ranging from a few centimeters down to microscopic dust. Earth gains about 40,000 tons of mass per year from meteorites of all sizes.

Recall from Chapter 19 that astronomers distinguish between the words *meteoroid, meteor,* and *meteorite.* A small body in space is a meteoroid, but once it begins to vaporize in Earth's atmosphere it is called a meteor. If it survives to strike the ground, it is called a meteorite.

We have two questions to consider about meteorites: Where in the solar system do these objects originate? What can meteorites tell us about the origin of the solar system? To answer these questions, we must consider the orbits of meteoroids and the minerals found in meteorites.

METEOROID ORBITS

Meteoroids are much too small to be visible through even the largest telescope. We see them only when they fall into Earth's atmosphere at 10 to 30 km/s, roughly 30 times faster than a rifle bullet, and are heated to glowing by friction with the air. The average meteoroid is about the mass of a paper clip and vaporizes at an altitude of about 80 km above Earth's surface, but it produces a bright streak of fire that we see as a meteor. The trail of a meteor points back along the path of the meteoroid, so

if we could study the direction and speed of meteors, we could get clues to the orbits of the meteoroids.

One way to backtrack meteor trails is to observe meteor showers. On any clear night, you can see 3 to 15 meteors an hour, but on some nights you can see a shower of dozens of meteors an hour that are obviously related to each other. To confirm this, try observing a meteor shower. Pick a shower from ▌Table 25-1, and on the appropriate night stretch out in a lawn chair and watch a large area of the sky. When you see a meteor, sketch its path on the appropriate sky chart from the back of this book. In just an hour or so you will discover that most of the meteors you see seem to come from a single area of the sky, the **radiant** of the shower (▌Figure 25-2a). In fact, meteor showers are named after the constellation from which they seem to radiate. Thus the Perseids are a meteor shower that radiates from the constellation Perseus in mid-August. The Leonids were spectacular in November of 2001, when observers saw hundreds per hour. The Leonids are not often that dramatic.

a

b

FIGURE 25-2
(a) Meteors in a meteor shower enter Earth's atmosphere along parallel paths, but perspective makes them appear to diverge from a radiant point. (b) Similarly, parallel railroad tracks appear to diverge from a point on the horizon.

Enjoying the Natural World

We can enjoy a meteor shower as Mother Nature's fireworks, but our enjoyment is much greater once we begin to understand what causes meteors and why meteors in a shower follow a pattern. When we know that meteor showers help us understand the origin of our world, the evening display of shooting stars is even more exciting. Science typically increases our enjoyment of the natural world by revealing the significance of things we might otherwise enjoy only in a casual way.

Everyone likes flowers, for example. We enjoy them in a casual way, admiring their colors and fragrances, their graceful petals and massed blooms. But botanists know that evolution has carefully designed flowers to attract insects and spread pollen. The bright colors attract insects, and the shapes of the flowers provide little runways so the insect will find it easy to land. Some color patterns even guide the insect in like landing lights at an airport, and many flowers, such as orchids and snapdragons, force the insect to crawl inside in just the right way to exchange pollen and fertilize the flower. The nectar is nothing more than bug bait. Once we begin to understand what flowers are for, a visit to a garden becomes not only an adventure of color and fragrance, but also an adventure in understanding the beauty of design and function, an adventure in meaning as well as an adventure of the senses.

And our understanding of a natural phenomenon helps us understand and enjoy related phenomena. For example, some flowers attract flies for pollination, and such blossoms smell like rotting meat. A few flowers depend on bats, and they open their blossoms at night. Many plants, such as pine trees, depend on the wind to spread their pollen, and those plants do not have colorful flowers at all; but flowers that depend on hummingbirds have long trumpet-shaped blossoms that just fit the hummingbirds' beaks.

The more science tells us about nature, the more enjoyable the natural world is. We can enjoy a meteor shower as nothing more than a visual treat, but the more we know about it, the more interesting and exciting it becomes. The natural world is filled with meaning, and science, as a way of discovering and understanding that meaning, gives us new opportunities to enjoy the world around us.

Observing a meteor shower is relaxing, and the show is natural fireworks, but it is even more exciting when we understand what a meteor shower tells us (Window on Science 25-1). The significant fact is that the meteors in a shower appear to come from a single point in the sky, the radiant. That must mean the meteoroids were traveling through space along parallel paths. When they encounter Earth and are vaporized in the upper atmosphere, we see their fiery tracks in perspective; they appear to come from a single radiant point, just as railroad tracks seem to come from a single point on the horizon (Figure 25-2b).

Studies of meteor-shower radiants reveal that these meteors are produced by bits of matter orbiting the sun along the paths of comets. The vaporizing head of the comet releases bits of debris that become spread along the entire orbit (❚ Figure 25-3). When Earth passes through this stream of material, we see a meteor shower. In some cases the comet has wasted away and is no longer visible, but in other cases the comet is still prominent though somewhere else along its orbit. For example, in May Earth comes near the orbit of Comet Halley, and we see the Eta Aquarids shower. Around October 20 Earth passes near the other side of the orbit of Comet Halley, and we see the Orionids shower.

Our pleasant observations of meteor showers tell us that at least some meteoroids come from comets. Even when there is no shower, we still see meteors, which are called **sporadic meteors** because they are not part of specific showers. Many of these are produced by stray bits of matter that were released long ago by comets. Such comet debris gets spread through the inner solar system, and bits fall into Earth's atmosphere continuously.

Another way to backtrack meteor trails is to photograph a meteor from two locations on Earth a few miles

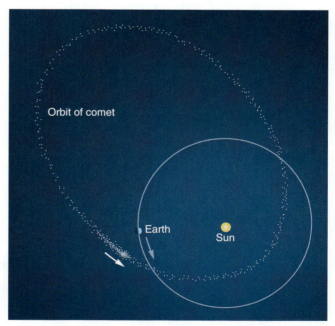

FIGURE 25-3

As a comet's ices evaporate, it releases rocky and metallic bits of material that spread along its orbit. If Earth passes through such material, it experiences a meteor shower.

apart. Then astronomers can use triangulation to find the altitude, speed, and direction of the meteor as it moves through the atmosphere and work backward to find out what its orbit looked like before it entered Earth's atmosphere. These studies reveal that meteors belonging to showers and some sporadic meteors have orbits that are similar to the orbits of comets. In many cases, the comets can be identified, so the origin of these meteors is well known. But some sporadic meteors have orbits that lead back to the asteroid belt. Thus, we can conclude that meteors have a dual source: Some come from comets, and some come from the asteroid belt. To learn more about these objects, we must examine those that make it to Earth's surface, the meteorites.

METEORITE IMPACTS ON EARTH

When a meteor is massive enough and strong enough to survive its plunge through Earth's atmosphere, it finally reaches Earth's surface, and we refer to it as a meteorite. A large meteorite hitting Earth might dig an impact crater much like those on the moon (see page 452). Most meteorites are small, and, because their energy of motion is spent in the atmosphere, they finally fall to Earth with no more energy than a falling hailstone.

Over 150 impact craters have been found on Earth. The Barringer Meteorite Crater near Flagstaff, Arizona is a good example (∎ Figure 25-4). It was created about 50,000 years ago by a meteorite about as large as a big building (80 to 100 m). Debris at the site shows that the meteorite was composed of iron, and astronomers can calculate from the size of the crater that the meteorite hit at a speed of about 11 km/s and released as much energy as a large thermonuclear bomb.

The Barringer crater is 1.2 km in diameter and 200 m deep. It seems large when you stand on the edge, and the hike around it, though beautiful, is long and dry (∎ Figure 25-5). Nevertheless, the crater is actually small compared with some impact features on Earth. About 65 million years ago, an impact in the northern

FIGURE 25-4

The Barringer Meteorite Crater, a popular tourist attraction about an hour east of Flagstaff, Arizona, is one of the youngest impact craters on Earth. It was formed only 50,000 years ago and is only slightly modified by erosion. *(USGS)*

Yucatán of Central America produced a crater, now covered by sediment, that was between 180 and 300 km in diameter, almost as big as the state of Ohio. This impact has been blamed for changing Earth's climate and causing the extinction of the dinosaurs, something we will discuss later. The 1994 comet impacts on Jupiter (shown in Figure 23-10) were even larger, but they formed no craters because Jupiter has no surface. During the Leonid meteor shower in recent years, astronomers watching the dark portion of the moon's disk have seen a number of bright flashes apparently caused by meteorite impacts. No new craters on the moon have been identified, but it is clear that meteorite impacts and the formation of impact craters are a continuing process in our solar system.

Large meteorites can produce craters, but most meteorites are small and don't create significant craters. Also, craters erode rapidly on Earth, so most meteorites are found without associated craters. The best place to look for meteorites turns out to be certain areas of Antarctica—not because more meteorites fall there but because the meteorites are easy to recognize on the icy terrain. Most meteorites look like Earth rocks, but on the ice of Antarctica there are no natural rocks. Also, the slow creep of the ice toward the sea concentrates the meteorites in certain areas where the ice runs into

FIGURE 25-5

Nearly a mile in diameter, the Barringer Meteorite Crater was formed when a building-size iron meteorite struck the horizontal limestone layers of the Arizona desert. Some rock was vaporized, some powdered, and some broken to fragments and ejected from the crater. Notice the raised and deformed rock strata all around the crater. *(M. A. Seeds)*

a **b**

FIGURE 25-6

(a) W. A. Cassidy, leader of several expeditions to Antarctica, holds a numbering device near an achondritic meteorite exposed on the icy surface. (b) An iron meteorite shown as it was found. *(NASA photographs, Ursula B. Marvin)*

mountain ranges, slows down, and evaporates. Thus, teams of scientists can easily recover meteorites from the ice (❚ Figure 25-6). Such unbiased collections of meteorites are important in helping define the frequency of different meteorite types.

Meteorites that are seen to fall are called **falls.** A fall is known to have occurred at a given time and place, and thus the meteorite is well documented. A meteorite that is discovered but was not seen to fall is called a **find.** Such a meteorite could have fallen thousands of years ago. This distinction between falls and finds will be important as we analyze the different kinds of meteorites.

AN ANALYSIS OF METEORITES

Meteorites can be divided into three broad categories. *Iron* meteorites are solid chunks of iron and nickel. *Stony* meteorites are silicate masses that resemble Earth rocks, and *stony-iron* meteorites are mixtures of iron and stone. These types of meteorites are illustrated in ❚ Figure 25-7.

Iron meteorites are easy to recognize because they are heavy, dense lumps of iron-nickel steel—a magnet will stick to them. That explains an important bit of statistics. Iron meteorites make up 66 percent of finds (❚ Table 25-2) but only 6 percent of falls. Why? Because an iron meteorite does not look like a rock. If you trip over one on a hike, you are more likely to carry it home and show it to your local museum. Also, some stone meteorites deteriorate rapidly when exposed to weather; irons survive longer. Thus, there is a **selection**

TABLE 25-2
Proportions of Meteorites

Type	Falls (%)	Finds (%)
Stony	92	26
Iron	6	66
Stony-iron	2	8

Selection Effects in Science

Many different kinds of science depend on selecting objects to study. Scientists studying insects in the rain forest, for example, must choose which ones to catch. They can't catch every insect they see, so they might decide to catch and study any insect that is red. If they are not careful, a selection effect could bias their data and lead them to incorrect conclusions without their ever knowing it. The reason selection effects are dangerous in science is that they can have powerful influences without being evident. Only by carefully designing a research project can scientists avoid selection effects.

For example, suppose we decide to measure the speed of cars on a highway. There are too many cars to measure every one, so we reduce the work load and measure only the speed of red cars. It is quite possible that this selection criterion will mislead us because people who buy red cars may be more likely to be younger and drive faster. Should we

measure only brown cars? No, because we might suspect that only older, more sedate people would buy a brown car. Should we measure the speed of any car in front of a truck? Perhaps we should pick any car following a truck? Again, we may be selecting cars that are traveling a bit faster or a bit slower than normal. Only by very carefully designing our experiment can we be certain that the cars we measure are traveling at representative speeds.

Public-opinion companies that survey voters in shopping malls must think carefully about selection effects. If they are seeking opinions about property taxes and decide to interview only men with gray hair, they will get a biased result. Interviewing men wearing T-shirts would give them quite a different result. Thus, they must be very careful to avoid unnoticed selection effects.

Astronomers face the danger of selection effects quite often. Iron meteorites are easier to find than stone meteorites.

Very luminous stars are easier to see at great distances than faint stars. Spiral galaxies are brighter, bluer, and more noticeable than elliptical galaxies. What we see through a telescope depends on what we notice, and that is powerfully influenced by selection effects.

Scientists engaged in observation must spend considerable time designing their experiments. They must be careful to observe an unbiased sample if they expect to make logical deductions from their results. The scientists in the rain forest, for example, should not catch and study only the red insects. Often, the most brightly colored insects are poisonous (or at least taste bad) to predators. Catching only brightly colored insects could produce a highly biased sample of the insect population. Careful scientists must plan their work with great care and avoid any possible selection effects.

---

effect that makes it more likely that iron meteorites will be found (Window on Science 25-2). That only 6 percent of falls are irons shows that iron meteorites are fairly rare.

When iron meteorites are sliced open, polished, and etched with nitric acid, they reveal regular bands called **Widmanstätten patterns** (Figure 25-7). These patterns are caused by alloys of iron and nickel called kamacite and taenite that formed as the molten iron cooled and solidified. The size and shape of the bands indicate that the molten metal cooled very slowly, no faster than 20 K per million years.

A lump of molten metal floating in space would cool very quickly, so the presence of the Widmanstätten pattern tells us that the molten metal must have been well insulated to have cooled so slowly. Such slow cooling is typical of the interiors of bodies at least 30 to 50 km in diameter. On the other hand, the iron meteorites show no evidence of the very high pressures that would develop in larger bodies. Thus, the iron meteorites appear to have formed from the interiors of molten objects of planetesimal size.

In contrast to irons, **stony meteorites** are common. Among falls, 92 percent are stones. Although there are many different types of stony meteorites, we will classify them into two subgroups depending on their physical and chemical content—chondrites and achondrites.

Roughly 80 percent of all meteorite falls are stony meteorites called **chondrites,** which look like dark gray,

granular rocks. The chemical composition of chondrites is the same as a sample of matter from the sun with the volatile gases removed. The classification of meteorites has become quite sophisticated, and there are many types of chondrites, but in general they appear to be original samples of the material that condensed from the solar nebula. Some have slight mineral differences, which tells us that the solar nebula was not totally uniform when this material condensed.

Most types of chondrites contain **chondrules,** round bits of glassy rock no larger than 5 mm across (Figure 25-7). To be glassy rather than crystalline, the chondrules must have cooled from a molten state very quickly (in minutes or less), but their origin is not clear. Some may be droplets of molten rock splashed into space by impacts between planetesimals. Another possibility is that chondrules are small bits of rock that were melted by sudden flare-ups of the young sun or by electrical discharges (lightning) within the solar nebula. Whatever their origin, they are very old.

All of the different types of chondrites contain some volatiles such as water, and this shows that the meteoroids were never heated to high temperatures. Differences among the many types of chondrites arise from their condensation in different parts of the solar nebula and from processes that altered their composition after they formed. Some, for example, appear to have been altered by the presence of water released by the melting of ice. Among the chrondrites, the **carbona-**

Iron meteorites are very heavy for their size and have a dark, irregular surface.

Stony meteorites tend to have a fusion crust caused by melting in Earth's atmosphere.

A stony-iron meteorite cut and polished reveals a mixture of iron and rock.

Chondrules are small, glassy spheres found in condrites.

Cut, polished, and etched with acid, iron meteorites show a Widmanstätten pattern.

This stony meterorite called a carbonaceous chondrite contains chondrules and volatiles, including carbon, that make the rock very dark.

FIGURE 25-7
The three main types of meteorites, irons, stones, and stony-iron, have distinctive characteristics.
(Lab photos courtesy of Russell Kempton, New England Meteoritical Services)

ceous chondrites are rare, only about 5.7 percent of falls. These very dark gray, rocky meteorites contain volatiles such as water and carbon compounds that would have been driven off if the meteoroid had been heated much above room temperature. It has been common for astronomers to think of the carbonaceous chondrites as the least altered of the chondrites, but many types of chondrites are essentially unaltered samples of the solar nebula. A carbonaceous chondrite is shown in Figure 25-7.

One of the most important meteorites ever recovered was a carbonaceous chondrite that was seen falling on the night of February 8, 1969, near the Mexican village of Pueblito de Allende. The brilliant fireball was accompanied by tremendous sonic booms and showered an area about 50 km by 10 km with over 4 tons of fragments. About 2 tons were recovered.

Studies of the Allende meteorite disclosed that it contained—besides volatiles and chondrules—small, irregular inclusions rich in calcium, aluminum, and titanium. Now called **CAIs,** for calcium–aluminum-rich inclusions, these bits of matter are highly refractory (❙ Figure 25-8); that is, they can survive very high tem-

peratures. If we could scoop out a ton of the sun's surface matter and cool it, the CAIs would be the first particles to form. As the temperature fell, other materials would condense in accord with the condensation sequence described in Chapter 19. When the material finally reached room temperature, we would find that all of the hydrogen, helium, and a few other gases like argon and neon had escaped and that the remaining lump, weighing about 18 kg (40 lb), had almost the same composition as the Allende meteorite, including CAIs. Thus, the Allende meteorite seems to be a very old sample of the solar nebula.

Unlike the chondrites, stony meteorites called **achondrites** (7.1 percent of falls) are highly modified. They contain no chondrules and no volatiles. This suggests that they have been hot enough to melt chondrules and drive off volatiles, leaving behind rock with compositions similar to Earth's basalts.

Iron meteorites and stony meteorites make up most falls, but 2 percent of falls are meteorites that are made up of mixed iron and stone. These **stony-iron meteorites** appear to have solidified from a region of molten iron and rock—the kind of environment we might

FIGURE 25-8
A sliced portion of the Allende meteorite showing round chondrules and irregularly shaped white inclusions called CAIs. *(NASA)*

expect deep inside a planetesimal with a molten iron core and a rock mantle.

THE ORIGINS OF METEORITES

The properties of meteorites suggest that they formed in the solar nebula. Their radioactive ages are about 4.6 billion years (Chapter 19), and some appear not to have been modified since they condensed from the solar nebula. Others have been heated or melted sometime after formation.

Meteorites almost certainly do not come from comets. Most cometary particles are very small specks of low-density, almost fluffy, material. When these enter the atmosphere, we see them incinerated as meteors, and most meteors are produced by this cometary debris. But such meteors are small and weak and do not survive to reach the ground. Although most meteors come from comets, most meteorites are stronger chunks of matter—more like fragments of asteroids.

Thus, when we talk of the origin of the meteorites, we are concerned with the history of asteroidlike planetesimals rich in metals and silicates, not the cometlike planetesimals that were rich in ices. Indeed, many of the properties of the meteorites suggest that they originated in stony-iron planetesimals, which evolved in complicated ways and were eventually broken up by collisions. Here we will concentrate on the answers to two questions: How did these planetesimals evolve? When did these planetesimals break up?

The iron and achondritic meteorites show that at least some of the planetesimals melted, but what produced this heat? Planets the size of Earth can accumulate a great deal of heat from the decay of radioactive atoms such as uranium, thorium, and the radioactive

isotope of potassium. The heat is trapped deep underground by many kilometers of insulating rock. But in a small planetesimal, the insulating layers are not as thick, and the heat leaks out into space as fast as the slowly decaying atoms can produce it. If a small planetesimal is to melt, it must have a more rapidly decaying heat source.

Recent studies have shown that some meteorites may have contained the radioactive element aluminum-26. These atoms are now gone, decayed to magnesium-26, but if they were once present their rapid decay could have melted the center of a planetesimal as small as 20 km in diameter. This is a sufficient heat source to melt the parent bodies of the meteorites.

Where did this aluminum-26 come from? Supernova explosions can manufacture aluminum-26, but it decays very rapidly, with a half-life of only 715,000 years. If the solar nebula was enriched by aluminum-26 from a nearby supernova explosion, the explosion must have occurred just before the formation of our solar system. In fact, some astronomers wonder if the shock waves from the supernova explosion compressed gas clouds and triggered the formation of the sun and planets.

Once the planetesimals were melted and the aluminum-26 had decayed, they cooled and slowly solidified. In the meantime, however, the larger planetesimals would have differentiated. The heavy metals would have sunk to the center to form a molten metal core, and the less-dense silicates would have floated upward to form a stony mantle. In some cases, the surface may not have been melted. If such a planetesimal were broken up (■ Figure 25-9), fragments from the center would look much like iron meteorites with their Widmanstätten patterns.

The achondrites, which have been strongly heated or melted, appear to come from the mantles and surfaces of the larger planetesimals that melted and differentiated. The stony-iron meteorites probably come from the core–mantle boundaries in these bodies. But the chondrites are probably fragments of smaller bodies that never melted. Volatile-rich meteorites such as the carbonaceous chondrites may have formed in smaller bodies farther from the sun.

The evidence seems strong that the meteorites are the result of the breaking up of larger bodies. For example, some meteorites are breccias—collections of fragments cemented together. Breccias are found on Earth and are very common on the moon, but studies of the meteoric breccias show that they were produced by impacts. A collision between planetesimals produces fragments, and the slower-moving particles fall back to the surface of the planetesimal to form a regolith, a soil of broken rock fragments. Later impacts may add to this regolith and stir it. Still later, an impact may be so violent that the fragments are pressed

The Origin of Meteorites

A large planetesimal can keep its internal heat long enough to differentiate.

Collisions break up the layers of different composition.

Silicates

Cratering collisions

Iron

Meteorites from deeper in the planetesimal were heated to higher temperatures.

Fragments from near the core might have been melted entirely.

Fragments of the iron core would fall to Earth as iron meteorites.

FIGURE 25-9
Planetesimals formed when the solar system was forming may have melted and separated into layers of different density and composition. The fragmentation of such a body could produce many types of meteorites. *(Adapted from a diagram by C. R. Chapman)*

together and momentarily melt where they touch. Almost instantly, the material cools, and the fragments weld themselves together to form a breccia. Most of the breccias were formed as the planetesimals formed.

Much later, as the planetesimals were broken up by collisions, the layers of breccia were exposed, shattered, and became brecciated meteoroids.

The collisions that broke up the planetesimals could not have happened long ago when the solar system was young. Small meteoroids would have been swept up by the planets in only a billion years or less. Also, cosmic rays striking meteoroids in space produce isotopes such as helium-3, neon-20, and argon-38. Studies of these atoms in meteorites show that most meteorites have not been exposed to cosmic rays for more than a few tens of millions of years. The meteorites must have been buried under protective layers until recently. Thus, the thousands of meteorites now in museums around the world must have broken off of planetesimals somewhere in our solar system within the last billion years. Where are those planetesimals? To answer that question, we must recall that many meteoroid orbits lead back to the asteroid belt. The asteroids are evidently the planetesimals from which we receive meteorites.

REVIEW CRITICAL INQUIRY

How can we say that meteors come from comets, but meteorites come from asteroids?

A selection effect can determine what we notice when we observe nature, and a very strong selection effect prevents us from finding meteorites that originated in comets. Cometary particles are physically weak, and they vaporize in Earth's atmosphere easily. Thus, very few ever reach the ground, and we are unlikely to find them. Furthermore, even if a cometary particle reached the ground, it would be so fragile that it would weather away rapidly, and, again, we would be unlikely to find it. Asteroidal particles, however, are made from rock and metal and so are stronger. They are more likely to survive their plunge through the atmosphere and more likely to survive erosion on the ground. Meteors from the asteroid belt are rare. Almost all of the meteors we see come from comets, but not a single meteorite is known to be cometary.

A similar selection effect makes it more likely that we will find iron meteorites than stony meteorites. Irons are easier to notice and survive longer. Thus, most *finds* are iron, but most *falls* are stone.

The meteorites are valuable because they give us hints about the process of planet building in the solar nebula. What evidence do we find in the meteorites that tells us they were once part of larger bodies broken up by impacts?

The vast majority of meteorites must have originated in the asteroids. Our next goal is to try to understand the nature and origin of these tiny rocky worlds.

25-2 ASTEROIDS

Until recently, few astronomers knew or cared much about asteroids. They were small chunks of rock drifting between the orbits of Mars and Jupiter that occasionally marred long-exposure photographs by drifting past more interesting objects. Asteroids were more irritation than fascination.

Now we know differently. The evidence of meteorites tells us that the asteroids are the last remains of the planetesimals that built the planets 4.6 billion years ago. Spacecraft have photographed a few asteroids and revealed them to be small, complex worlds with irregular shapes and heavily cratered surfaces (❚ Figure 25-10). The study of the asteroids gives us a way to explore the ancient past of our planetary system.

THE ASTEROID BELT

The first asteroid was discovered on January 1, 1801 (the first night of the 19th century), by the Sicilian monk Giuseppe Piazzi. It was later named Ceres after the Roman goddess of the harvest (thus our word *cereal*). We know now that it is a nearly spherical ball of rock about 900 km in diameter.

Astronomers were excited by Piazzi's discovery. There seemed to be a gap where a planet might exist between Mars and Jupiter at an average distance from the sun of 2.8 AU. Ceres fit. Its average distance from the sun is 2.766 AU. It was a bit small to be a planet, and because three more objects—Pallas, Juno, and Vesta—were discovered in the following years, astronomers realized that Ceres and the other asteroids were not true planets.

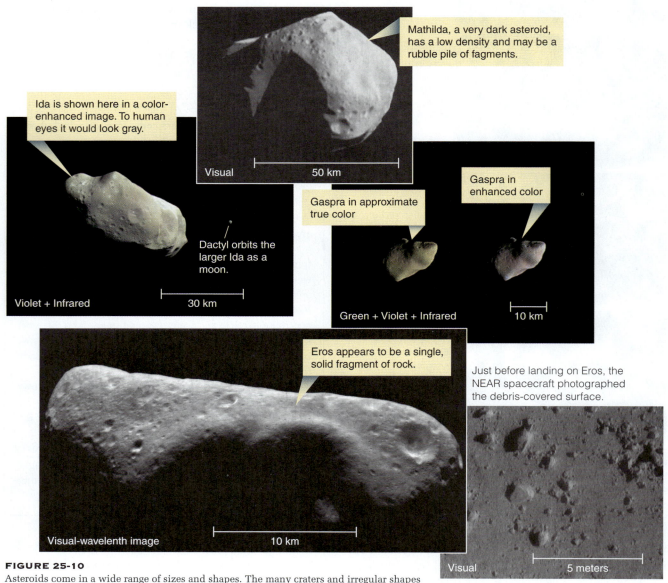

Mathilda, a very dark asteroid, has a low density and may be a rubble pile of fagments.

Ida is shown here in a color-enhanced image. To human eyes it would look gray.

Visual — 50 km

Gaspra in approximate true color

Gaspra in enhanced color

Dactyl orbits the larger Ida as a moon.

Violet + Infrared — 30 km

Green + Violet + Infrared — 10 km

Eros appears to be a single, solid fragment of rock.

Just before landing on Eros, the NEAR spacecraft photographed the debris-covered surface.

Visual-wavelenth image — 10 km

Visual — 5 meters

FIGURE 25-10
Asteroids come in a wide range of sizes and shapes. The many craters and irregular shapes show that they have suffered frequent collisions. The NEAR spacecraft revealed that the surface of Eros is covered by impact debris ranging from large boulders to scattered dust. (*Johns Hopkins Applied Physics Lab and NASA*)

Today thousands of asteroids have accurately known orbits, and there are many more as yet undiscovered; but they are all small bodies. All of the larger asteroids in the asteroid belt have been found.

When an asteroid is discovered, the discoverer is allowed to choose a name for it, and asteroids have been named for spouses, lovers, dogs, Greek gods, politicians, and others.* Once an orbit has been calculated, the asteroid is assigned a number listing its position in the catalog known as the *Ephemerides of Minor Planets.* Thus, Ceres is known as 1 Ceres, Pallas as 2 Pallas, and so on. Only three asteroids are larger than 400 km in diameter, and most are much smaller (❙ Figure 25-11).

Most of these objects orbit the sun in the asteroid belt between Mars and Jupiter, and we might suspect that massive Jupiter was responsible for their origin. Certainly the distribution of asteroids in the belt shows the importance of Jupiter's gravitation. Certain regions of the belt, called **Kirkwood's gaps** (Chapter 23) after their discoverer, are almost free of asteroids (❙ Figure 25-12). These gaps lie at certain distances from the sun where an asteroid would find itself in resonance with Jupiter. For example, if an asteroid lay 3.28 AU from the sun, it would revolve twice around its orbit in the time it took Jupiter to revolve once around the sun. Thus, on alternate orbits, the asteroid would find Jupiter at the same place in space tugging outward. The cumulative perturbations would rapidly change the asteroid's orbit until it was no longer in resonance with Jupiter. Our example

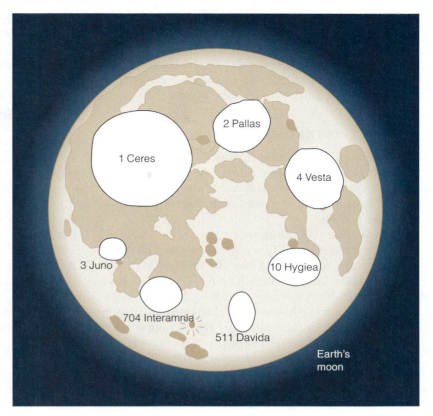

FIGURE 25-11

The relative size and approximate shape of the larger asteroids are shown here compared with the size of Earth's moon. Smaller asteroids can be highly irregular in shape.

is in a 2:1 resonance, but gaps occur in the asteroid belt at many resonances, including 3:1, 5:2, and 7:3.

Modern research shows that the motion of asteroids in Kirkwood's gaps is described by a theory in mathematics that deals with chaotic behavior. The smooth motion of water sliding over the edge of a waterfall decays rapidly into a chaotic jumble. The same mathematical theory of chaos that describes the motion of the water shows how the slowly changing orbit of an asteroid within one of Kirkwood's gaps can

*Some sample asteroid names: Olga, Chicago, Vaticana, Noel, Ohio, Tea, Gaby, Fidelio, Hagar, Geisha, Dudu, Tata, Mimi, Dulu, Tito, Zulu, Beer, and Zappafrank (after the late musician Frank Zappa).

FIGURE 25-12

Here the red curve shows the number of asteroids at different distances from the sun. Purple bars mark Kirkwood's gaps where there are few asteroids. Note that these gaps match resonances with the orbital motion of Jupiter.

suddenly change into a long, elliptical orbit that carries the asteroid into the inner solar system, where it is likely to be removed by a collision with Mars, Earth, or Venus. This ejection process also explains how Jupiter's gravity can throw so many meteoroids from the asteroid belt into the inner solar system.

NONBELT ASTEROIDS

Not all asteroids orbit within the asteroid belt. Some of the most interesting follow orbits that cross the orbits of the terrestrial planets or wander among the Jovian worlds. In fact, some asteroids even share the orbits of the larger planets.

The **Apollo–Amor objects** are asteroids whose orbits carry them into the inner solar system. The Amor objects follow orbits that cross the orbit of Mars but don't reach the orbit of Earth. The Apollo objects have Earth-crossing orbits. These Apollo–Amor objects are dangerous. Jupiter's influence makes their orbits precess, so they are bound to interact with a planet eventually. About one-third will be thrown into the sun, and a few will be ejected from the solar system, but many of these objects are doomed to collide with a planet—perhaps ours. Earth is hit by an Apollo object once every 250,000 years, on average. With a diameter of up to 2 km, they hit with the power of a 100,000-megaton bomb and can dig craters 20 km in diameter.

Many dozens of Apollo objects have known orbits, and none of those will hit Earth in the foreseeable future. The bad news is that there are about 1000 of these near-Earth asteroids (NEOs) larger than 1 km in diameter, the minimum size of an impactor that could cause global effects on Earth. More than half a dozen teams are now searching for these NEOs. For example, LO-NEOS (Lowell Observatory Near Earth Object Search) is searching the entire sky visible from Lowell Observatory once a month. The combined searches should be able to locate all 1000 of the largest NEOs by 2010.

This is a serious issue because even a small asteroid could do serious damage. For example, in January 1999, an asteroid about 9 m in diameter was discovered as it raced past Earth at a distance only half that of the moon. Had that asteroid struck Earth, it would have hit with an energy ten times greater than that of the bomb that destroyed Hiroshima. Objects of this size, a few tens of meters or less, fragment and explode in the atmosphere, but the shock waves from their explosion could cause serious damage on Earth's surface. Declassified data from military satellites show that Earth is hit about once a week by these house-size asteroids. The largest impact ever detected was the atmospheric explosion of a rocky meteorite over the South Pacific on February 1, 1994. It was seen by fishermen as a blast as bright as the sun. It produced about five times the energy of the Hiroshima bomb and was probably a rocky body about 15 m in diameter—probably a small

Apollo object. We will see later in this chapter how a slightly larger impactor could alter Earth's climate and threaten all life on our world.

It is easy to assume that the Apollo–Amor objects are rocky asteroids that have been thrown into their extreme orbits by events in the main asteroid belt. At least some of these objects, however, may be comets that have exhausted their ices and become trapped in short orbits that keep them in the inner solar system. We will see later that the distinction between comets and asteroids is not totally clear.

While the Apollo–Amor objects rush through the inner solar system, there are also nonbelt asteroids in the outer solar system. These objects, being farther from the sun, move more slowly. The object Chiron, found in 1977, appears to be a body about 170 km in diameter. Its orbit carries it from just inside the orbit of Uranus to just inside the orbit of Saturn. Although it was first classified as an asteroid, its status is now less certain. Ten years after its discovery, Chiron surprised astronomers by suddenly brightening with a release of vapor and dust. Like a comet, Chiron was releasing a cloud of gas and dust in jets. Studies of older photographs showed that Chiron had been brighter even when it was farther from the sun, and astronomers now suspect that it may have a rocky crust covering deposits of ices such as solid nitrogen, methane, and carbon monoxide. Thus, Chiron may be more comet than asteroid, and it warns us that the distinction is not clear-cut.

Chiron's orbit is not stable. It cannot have remained in its present orbit for the entire history of the solar system but must have been perturbed into its present orbit within the last million years or so. Other, smaller bodies have been found in the outer solar system, and astronomers are now trying to understand how their origin and history are linked to the formation of the planets. We saw in the previous chapter that Pluto, Charon, and Triton might be large members of this family of icy planetesimals.

The icy asteroids of the outer solar system seem to be widely scattered, but Jupiter ushers two groups of asteroids within its own orbit. These nonbelt asteroids have become trapped in the Lagrangian points along Jupiter's orbit. (See Figure 13-5.) These points lie 60° ahead of and 60° behind the planet and are regions where the gravitation of the sun and Jupiter combine to trap small bodies (❚ Figure 25-13). Like cosmic sinkholes, the Lagrangian points have trapped chunks of debris now called **Trojan asteroids** and named after the heroes of the Trojan War (588 Achilles, 624 Hektor, 659 Nestor, and 1143 Odysseus, for example). Slightly over 1000 Trojan asteroids are known, but only the brightest have been given names.

In 1990, observers found an asteroid at the trailing (L_5) Lagrangian point in the orbit of Mars, and growing evidence suggests that other planets may have Trojan asteroids trapped in their orbits. As we search for smaller

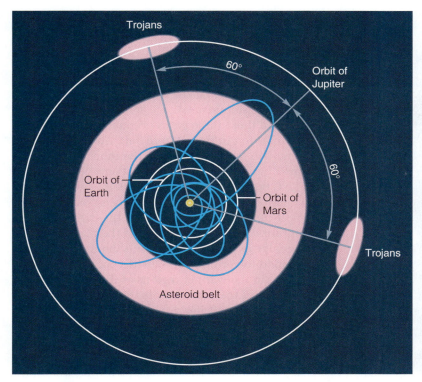

FIGURE 25-13

The Trojan asteroids are trapped in the Lagrangian points 60° ahead of and 60° behind Jupiter. The Apollo asteroids move along orbits that cross Earth's orbit. The orbits of only 7 of the known Apollo asteroids are shown here.

objects, we are finding that our solar system still contains large numbers of these small bodies. The challenge is to explain their physical nature and their origin.

COMPOSITION AND ORIGIN

An old theory held that the asteroids are the remains of a planet that broke up, but modern astronomers discount that idea. Once formed, a planet is very difficult to tear apart. Rather, astronomers believe the asteroids are the remains of planetesimals left over from the formation of the planets. Jupiter's gravity could have stirred the planetesimals just inside its orbit and caused collisions at unusually high velocities. These impacts would have tended to break up the planetesimals rather than assemble them into a planet. Resonances and chaos would have helped eject material, and most of the planetesimal objects have now been lost—swept up by planets, captured as satellites, or ejected from the solar system. The asteroid belt today contains hardly 4 percent the mass of Earth's moon.

Collisions are surely common among the asteroids, and we can see evidence of catastrophic impacts powerful enough to break up an asteroid. Early in the 20th century, Japanese astronomer Kiyotsugu Hirayama discovered that some groups of asteroids share similar orbits. They have the same average distance from the sun,

the same eccentricity, and the same inclination. Up to 19 of these **Hirayama families** are known, and modern observations show that the asteroids in a family typically share similar spectroscopic characteristics. Evidently, a family is produced by a catastrophic collision that breaks a single asteroid into a family of fragments that continue traveling along similar orbits around the sun.

In 1983, the Infrared Astronomy Satellite detected the infrared glow of sun-warmed dust scattered in bands throughout the asteroid belt. These dust bands appear to be the products of past collisions. The dust will eventually be blown away, but collisions occur constantly in the asteroid belt, so new dust bands will presumably be produced as the present bands dissipate.

Beginning in the 1990s, astronomers have gotten close-up looks at a few asteroids. One way to study asteroids is to bounce radio signals off asteroids passing near Earth and construct images from the timing and Doppler shifts in the echoes. These images revealed that some asteroids appear to be peanut shaped. They are either two objects that have collided and fused or two objects orbiting close to each other (❙ Figure 25-14). Such binary asteroids remind us that some asteroid collisions can be gentle.

Spacecraft images of asteroids show that they are irregular in shape and heavily cratered. One image of Ida revealed that it has a moon of its own, 1.5-km diameter Dactyl (Figure 25-10). The moon is probably a fragment from an impact. Other asteroids with moons are known or suspected, and double-impact craters on planetary surfaces may tell us that such double asteroids are common.

The asteroid Mathilde looks like other asteroids (Figure 25-10), but measurements of its gravitational field by the passing NEAR spacecraft revealed that the asteroid has a very low density. Mathilde may not be solid rock but rather may be a rubble pile with large empty spaces. Observations of other asteroids suggest that many may be low-density rubble piles produced by repeated collisions rather than chunks of solid rock.

The NEAR spacecraft went into orbit around asteroid Eros in early 2000 and was immediately renamed in honor of the late Gene Shoemaker, a codiscoverer of the comet Shoemaker–Levy 9. The spacecraft transmitted back to Earth detailed images and observations of Eros, as shown in Figure 25-10. Although the spacecraft was not designed to land, controllers on Earth ordered the nearly exhausted spacecraft down to a successful touchdown on the dusty surface of Eros. The detailed observations show that Eros, an Apollo asteroid, is a

Radar Image

FIGURE 25-14

This image was reconstructed by transmitting a radio signal toward the asteroid Toutatis as it passed near Earth and recording delays and Doppler shifts in the reflected signals. Toutatis is revealed to be a double object composed of two irregularly shaped, cratered bodies either touching each other or orbiting each other very closely. Such double asteroids are more common than was once thought. *(NASA)*

single fragment of a larger body presumably broken by an impact. (See Figure 19-7.)

Careful measurements show that some asteroids are redder than others. This is not a color difference that we would notice with our eyes, but a difference in the albedo of asteroids at different wavelengths. When astronomers refer to the "color" of an asteroid, they are referring to spectroscopic properties. That makes it possible to classify asteroids into a number of classes. We will consider three main classes (▌Figure 25-15).

The C-type asteroids have albedos smaller than 0.06 and would look very dark to our eyes. They are probably made of carbon-rich material similar to that in carbonaceous chondrites. The S-type asteroids have albedos of 0.1 to 0.2, so they would look brighter and spectroscopically redder. The M-type asteroids are bright but not as red. S-type asteroids are believed to be rocky, but M-type asteroids appear to be metal rich and may be the iron cores of fragmented asteroids.

Although S-type asteroids are very common in the inner asteroid belt, their spectroscopic colors are different from the chondrites—the most common kind of meteorite. New evidence from the analysis of moon rocks and from observations of Eros, an S-type asteroid, show that bombardment by micrometeorites can

redden and darken S-type asteroids. Thus it seems that the most common kind of meteorites come from the most common kind of asteroid.

A few other types of asteroids are known, and a number of individual asteroids have been found that are unique, but these three classes contain a majority of the known asteroids.

How did these three types originate? A clue lies in their distribution in the asteroid belt. The S-type asteroids are much more common in the inner belt. In fact, there are almost none beyond a distance of about 3.45 AU. The C types are rather rare in the inner belt but are very common in the outer belt. Many astronomers believe that this distribution reflects the differences in temperature during the formation of the planetesimals. It was cooler in the outer belt, and the planetesimals there tended toward the composition of carbonaceous chondrites. In the warmer regions of the inner belt, the composition of the growing planetesimals was more like that of the chondrites.

The large asteroid Vesta is reminding astronomers that these are complex worlds. With a diameter of 525 km, Vesta is too small to image from Earth's surface, but the Hubble Space Telescope reveals that it has a complex surface with light and dark patches. The spectra of these regions are like those of basalts typical of solidified lavas. Furthermore, a study of the best

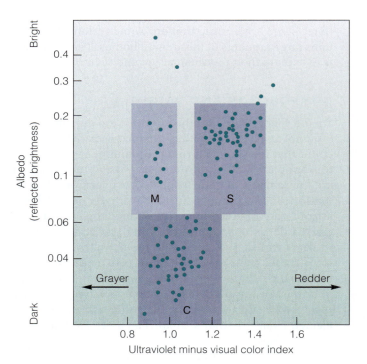

FIGURE 25-15

A graph of albedo or reflected brightness versus color separates asteroids into three types. The C types are dark, while the M and S types are brighter. S types are redder than M types. These differences in color and brightness are related to the chemical compositions at the surfaces of the asteroids. *(Adapted from a diagram by B. Zellner)*

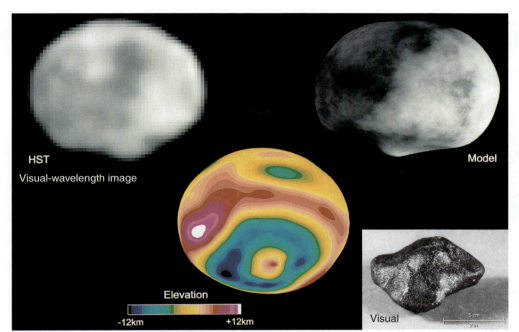

FIGURE 25-16

A Hubble Space Telescope image of the asteroid Vesta (upper left) shows a complex surface with a large crater near the south pole. From 78 such images, astronomers constructed a model of the asteroid (upper right) showing the impact crater and its central peak. Detailed studies of the model produce the elevation map (bottom) showing that the crater is up to 13 km deep. (Inset) A meteorite identified spectroscopically as a fragment of Vesta. Chemical and crystalline structure show that it was once part of a molten lava flow. Presumably it was blasted off of Vesta by a large impact. *(Ben Zellner, Georgia Southern University; NASA; inset: R. Kempton, New England Meteoritical Services)*

HST
Visual-wavelength image

Model

Elevation

-12km   +12km

Visual

photos shows a giant crater 460 km in diameter (87 percent of Vesta's diameter) near Vesta's south pole (▌ Figure 25-16). A class of basaltic meteorites called eucrites have infrared spectra nearly identical to that of Vesta, and the 2-km-long asteroid Braille has the same kind of infrared spectrum. Planetary astronomers believe that the impact that formed the large crater on Vesta could have blasted loose about 1 percent of the asteroid's mass. A tiny fraction of that debris fell to Earth as the eucrite meteorites, and a large piece became the asteroid Braille. Thus the eucrite meteorites give us a sample from the surface of Vesta, and because those meteorites are basalts that resemble solidified lava flows, we can cite them as strong evidence that Vesta has at some time in the past been at least partly resurfaced by lava flowing up from its interior.

How could a small asteroid be heated enough to produce lava flows? One source of heat in newly formed asteroids would be the decay of short-lived radioactive elements such as aluminum-26. Such radioactive elements could have been produced in a supernova explosion occurring shortly before the formation of our solar system. The smallest asteroids lose their heat too fast to melt, but the decay of aluminum-26 occurs fast enough to melt the interiors of bodies larger than a few dozen kilometers in diameter. Thus, it is reasonable to suppose that Vesta and some other asteroids larger than about 100 km in diameter might have been strongly heated and modified by lava flows.

In contrast, the largest asteroid, Ceres, about 900 km in diameter, does not seem to have been modified by internal heating. The light reflected from its surface suggests a claylike surface related to carbonaceous chon-

drites. This is surprising, because clays form when minerals are exposed to water. In fact, spectroscopic observations reveal what seem to be traces of water ice on Ceres. Thus, we must suspect that some asteroids may have had significant amounts of water bound into their crusts when they were young.

All of the evidence suggests that the asteroids are the broken remains of planetesimals that formed in the solar nebula as planet building began. The largest remaining asteroids, such as Ceres and Vesta, may be largely unbroken planetesimals, and some of these may have evolved due to internal heat or the presence of water. Nevertheless, the vast majority of the asteroids are fragments, and many may consist of bodies that have been shattered and then re-formed as gravity pulled the fragments back together. Some compositional differences seem to have originated when the asteroids formed in that the objects in the outer asteroid belt contain more carbon-rich compounds. The presence of massive Jupiter orbiting just beyond this region prevented the original planetesimals from accreting to build a planet. Instead, collisions fragmented the planetesimals, and nearly all of the material is now lost. The entire mass of the asteroid belt today adds up to only 0.05 percent of Earth's mass.

Some experts discuss the mining of the asteroid belt for its mineral riches, but the asteroid belt is nearly empty. The small, fragmented asteroids drift through space far from one another and only rarely collide with each other or with planets. A voyage to explore or exploit the asteroid belt would find the distance between bodies forbiddingly large.

What evidence do we have that the asteroids have been fragmented?

First, we might note that the solar nebula theory of the formation of the solar system predicts that planetesimals collided and either stuck together or fragmented. This is suggestive, but it is not evidence. A theory can never be used as evidence to support some other theory or hypothesis. Evidence means observations or the results of experiments, so we must turn to observations of asteroids. Spacecraft photographs of asteroids such as Ida, Gaspra, and Eros show irregularly shaped little worlds heavily scarred by impact craters. In fact, observations of some asteroids show what may be pairs of bodies in contact, and the Galileo image of Ida reveals its small satellite, Dactyl. Furthermore, some meteorites appear to come from the asteroid belt, and a few have been linked to specific asteroids such as Vesta. All of this evidence suggests that the asteroids have suffered violent collisions in their history.

The impact fragmentation of asteroids has been important, but it has not erased all traces of the original planetesimals from which the asteroids formed. What evidence do we have to tell us what those planetesimals were like?

The asteroids are only the last crumbs left behind by failed planet building near Jupiter. Farther out in the solar system are other remains—ice shards from the age of planet formation.

25-3 COMETS

Few things in astronomy are more beautiful than a bright comet hanging in the night sky (❙ Figure 25-17). Whereas meteors shoot across the sky like demented fireflies, a comet moves with the stately grace of a great ship at sea, its motion hardly apparent. Night by night it shifts against the stars and may remain visible for months. Faint comets are common; a number are discovered every year. But a truly bright comet appears about once a decade. Comet Hyakutake in 1996 (Figure 25-1) and Comet Hale–Bopp in 1997 were both dramatic, but the later comet was so bright that we might class it with the great comets such as Comet Halley in 1910. A patient person might see half a dozen or more bright comets in a lifetime.

While everyone enjoys the beauty of comets, astronomers study them for their cargo of clues to the origin of our solar system.

PROPERTIES OF COMETS

As always, we begin our study of a new kind of object by summarizing its observational properties. What do comets look like, and how do they behave? The observations are the evidence that reveals the secrets of the comets.

Study ❙ "Comet Observations" on pages 578 and 579 and notice three important properties of comets. First, comets have two kinds of tails, which tells us that the nucleus contains ices of water and other compounds plus rocky material most evident as dust. Second, notice the importance of the solar wind and solar radiation in shaping the gas tails of comets. Finally, notice the evidence we have that comet nuclei are fragile.

Astronomers can put these and other observations together to discuss the structure of comet nuclei.

THE GEOLOGY OF COMET NUCLEI

The nuclei of comets are quite small and cannot be studied in detail from Earth-based telescopes. Nevertheless, astronomers are beginning to understand the geology of these peculiar worlds.

Comet nuclei contain ices of water and other volatile compounds such as carbon dioxide, carbon monoxide, methane, ammonia, and so on. These ices are the kinds of compounds that should have condensed from the outer solar nebula, and that makes astronomers

Visual-wavelength image

FIGURE 25-17
Comet Ikeya–Seki was beautiful in the dawn sky in 1965. The long, graceful tail sprang from a nucleus of ice only a dozen kilometers in diameter. *(Stephen M. Larson)*

think that comets are ancient samples of the gases and dust from which the planets formed.

When the nuclei of comets approach the sun, the ices absorb energy from sunlight and sublime—change from a solid directly into a gas—to produce the observed tails. As the gases break down and combine chemically, they release many compounds found in comet tails. Vast clouds of hydrogen gas observed around the heads of comets are derived from the breakup of molecules from the ices.

Five spacecraft flew past the nucleus of Comet Halley when it visited the inner solar system in 1985 and 1986. Spacecraft flew past the nuclei of Comet Borrelly in 2001 and Comet Wild 2 in 2004. Photos show that these comet nuclei are irregular in shape and very dark, with jets of gas and dust spewing from active regions on the nuclei (Figure 25-18). In general, these nuclei are darker than a lump of coal, which suggests the composition of the carbon-rich meteorites called carbonaceous chondrites.

From the gravitational influence of a nucleus on a passing spacecraft, astronomers can find the mass and density of the nucleus. Comet nuclei appear to have densities of 0.1 to 0.25 g/cm^3, much less than the density of ice. From these observations we can conclude that comet nuclei are not solid balls of ice but must be fluffy mixtures of ices and dust with significant amounts of empty space.

Photographs of the comae (plural of coma) of comets often show jets springing from the nucleus and being swept back by the pressure of sunlight and by the solar wind to form the tail (Figure 25-18). Studies of the motions of these jets as the nucleus rotates reveal that the jets originate from active regions that may be faults or vents. As the rotation of a cometary nucleus carries an active region into sunlight, it begins venting gas and dust, and as it rotates into darkness it shuts down.

The irregular shape of the nucleus of Comet Halley combined with its observed low density suggests that comet nuclei are not solid objects. To call them dirty snowballs or icy mud balls is misleading. Rather, they appear to be irregular bodies with large voids. Although comet nuclei seem to have porous crusts of dark

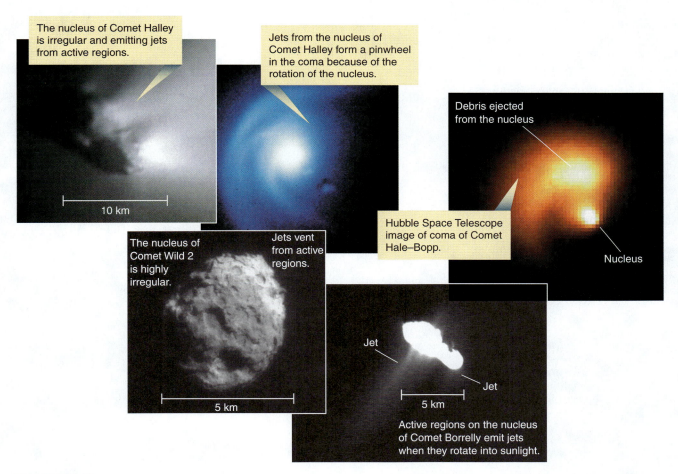

FIGURE 25-18

Visual-wavelength images made by spacecraft and by the Hubble Space Telescope show how the nucleus of a comet produces jets of gases from regions where sunlight vaporizes ices. *(Halley nucleus: © 1986 Max-Planck Institute; Halley coma: Steven Larson; Comet Borrelly and Comet Hale–Bopp: NASA)*

Comet Observations

Comets are beautiful to the eye and to the intellect. Their vast size and magnificent beauty spring from a small, dirty lump. Their simple structure contains details that arise from powerful processes related to the solar wind. Most of all, the luminous glow of comets tells us about the origin of our solar system.

Gas tail (Type I)

Dust tail (Type II)

A **type I** or **gas tail** is produced by ionized gas carried away from the nucleus by the solar wind. The spectrum of a gas tail is an emission spectrum. The atoms are ionized by the ultraviolet light in sunlight. The wisps and kinks in gas tails are produced by the magnetic field embedded in the solar wind.

Spectra of gas tails reveal atoms and ions such as H_2O, CO_2, CO, H, OH, O, S, C, and so on. These are released by the vaporizing ices or produced by the breakdown of those molecules. Some gases, such as hydrogen cyanide (HCN), must be formed by chemical reactions.

A **type II** or **dust tail** is produced by dust from the vaporizing ices of the nucleus. The dust is pushed gently outward by the pressure of sunlight, and it reflects an absorption spectrum, the spectrum of sunlight. The dust is not affected by the magnetic field of the solar wind, so dust tails are more uniform than gas tails. Dust tails are often curved because the dust particles follow their individual orbits around the sun once they leave the nucleus.

Nucleus

The nucleus of a comet (not visible here) is a small, fragile lump of porous rock containing ices of water, carbon dioxide, ammonia, and so on. Comet nuclei can be 10 to 100 km in diameter.

When a spacecraft named ICE passed through the gas tail of a comet, it found a magnetic field from the solar wind draped over the nucleus like seaweed draped over a fishhook.

Coma

The **coma** of a comet is the cloud of gas and dust that surrounds the nucleus. It can be over 1,000,000 km in diameter, bigger than the sun.

Ace Astronomy™

Go to AceAstronomy and click Active Figures to see "Build a Comet" and to watch a comet nucleus produce tails.

Comet Mrkos in 1957 shows how the gas tail can change from night to night due to changes in the magnetic field in the solar wind.

Caltech

Visual-wavelength images

Visual-wavelength image

The uniform coma of Comet Hale–Bopp has been digitally removed from the image above to reveal radial jets of dust spewing from the nucleus. Dust from comets fills our solar system and can be collected by spacecraft above Earth's atmosphere. It is composed of fluffy, microscopic specks of rocky minerals.

Comet Hale–Bopp was very bright in the sky in 1997.

Visual-wavelength image

Comet tails point generally away from the sun, but their precise direction depends on the flow of the solar wind and the orbital motion of the nucleus.

NASA, Harold Weaver (The Johns Hopkins University), and the HST Comet LINEAR Investigation Team

Comet Linear moved through the sky in the summer of 2000. When it passed near the sun, the nucleus vanished and astronomers at first thought it had been entirely destroyed by the sun's heat.

Visual-wavelength images

An image made by the Hubble Space Telescope (right) shows that the nucleus of Comet Linear broke into many small fragments. As the last of the ices vaporize, only scattered bits of rock and metal will mark the orbit of the comet. Comet nuclei appear to be fragile objects.

Fragments

material, the ice and rock are not uniformly mixed through the interior. Breaks in the crust can expose pockets of highly volatile ices and cause sudden bursts of gas production.

The nuclei of comets are not strong. A nucleus can be ripped apart by the violence of gases bursting through the crust or by tidal forces produced if the comet passes near a massive body. A number of comets have been observed to break into two or more parts while near Jupiter or near the sun. On July 8, 1992, Comet Shoemaker–Levy 9 passed within 1.29 planetary radii of Jupiter's center, well within its Roche limit, and tidal forces ripped the nucleus into at least 21 pieces (❙ Figure 25-19a). The fragmented pieces were as large as a few kilometers in diameter, and they released gas and dust and spread into a long string of objects that looped away from Jupiter and then fell back to strike the planet and produce massive impacts over a period of six days in July 1994 (Figure 23-10). The fragmentation of comet nuclei may explain long chains of craters found on the Earth's moon and on some other planetary surfaces in the solar system. A comet breaking into pieces could produce a chain of impact features such as the ones shown in Figure 25-19b and c.

The nucleus of Comet Halley was observed to be 16 by 8 by 7 km, the nucleus of Comet Hale–Bopp is believed to be 80 km in diameter, and the nucleus of Comet Borrelly is 8 km long. Each passage near the sun costs such an object many millions of tons of ices, so the nuclei slowly lose their ices until there is nothing left but dust and rock.

THE ORIGIN OF COMETS

Family relationships among the comets give us clues to their origin. Most comets have long, elliptical orbits with periods greater than 200 years. These are known as long-period comets. Their orbits are randomly inclined, with comets falling into the inner solar system from all directions. As many circle the sun clockwise as counterclockwise.

In contrast, about 100 or so of the 600 well-studied comets have orbits with periods less than 200 years. These short-period comets follow orbits that lie within

Visual-wavelength image

a

b

FIGURE 25-19
(a) Tides from Jupiter pulled apart the nucleus of Comet Shoemaker–Levy 9 to form a long strand of icy bodies and dust that fell back to strike Jupiter two years later.
(b) A 40-km-long crater chain on the moon and (c) another 140-km-long crater chain on Jupiter's moon Callisto were apparently formed by the impact of fragmented comet nuclei. The impacts that form such chains probably occur within a span of seconds. *(NASA)*

c

FIGURE 25-20

The long-period comets appear to originate in the Oort cloud. Objects that fall into the solar system from this cloud arrive from all directions.

30° of the plane of the solar system, and most revolve around the sun counterclockwise—the same direction the planets orbit. Comet Halley, with a period of 76 years, is a short-period comet.

Comets cannot survive long before the heat of the sun drives away their ices and reduces them to inactive bodies of rock and dust. A comet may last only 100 to 1000 orbits around the sun. The comets we see in our skies can't have survived 4.6 billion years since the formation of the solar system, so there must be a continuous supply of new comets. Where do they come from?

In the 1950s, Dutch astronomer Jan Oort proposed that the long-period comets are objects that fall in from the **Oort cloud,** a spherical cloud of icy bodies believed to extend from 10,000 to 100,000 AU from the sun (▌Figure 25-20). Astronomers estimate that the cloud contains several trillion icy bodies. Far from the sun, they are very cold, lack comas and tails, and are invisible. The gravitational influence of occasional passing stars could perturb a few of these objects to fall into the inner solar system, where the heat of the sun warms their ices and transforms them into comets. Because the Oort cloud is spherical, these long-period comets fall inward from random directions.

It is not outlandish that stars pass close enough to affect the Oort cloud. Data from the Hipparcos satellite show that the star Gliese 710 will pass within 1 ly (about 63,000 AU) of our solar system in about a million years. That may shower the inner solar system with Oort cloud objects, which, warmed by the sun, will become comets.

Some of the short-period comets, including Comet Halley, appear to have originated in the Oort cloud and probably had their orbits altered by a close encounter with Jupiter. But many of the short-period comets cannot have begun in the Oort cloud. Interactions with a planet can't capture objects from the Oort cloud into the orbits that some short-period comets occupy. There must be another source of icy bodies in our solar system. In 1951, Dutch-American astronomer Gerard P. Kuiper proposed that the formation of the solar system should have left behind a belt of small, icy planetesimals beyond the Jovian planets and in the plane of the solar system. Known today as the **Kuiper belt,** this disk extends from 30 AU to as much as 100 AU from the sun. The entire Kuiper belt would be hidden in Figure 25-20 behind the yellow dot symbolizing the solar system. Most short-period comets are visitors from the Kuiper belt.

Although the Kuiper belt was originally proposed as a theory, astronomers are now able to observe some of the Kuiper belt objects (▌Figure 25-21). Searches for small icy bodies in the outer solar system have so far identified hundreds objects that are members of the Kuiper belt. Also, the physical and chemical similarities among the icy moons of the outer solar system, such as Triton and Charon, suggest these are large Kuiper belt objects captured as moons. The Kuiper belt is believed to contain about 30,000 objects larger than 100 km in diameter.

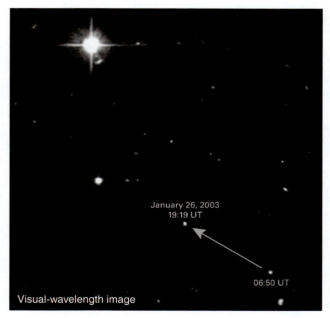

FIGURE 25-21

Two exposures 12 hours apart captured this Kuiper belt object moving along its orbit beyond Neptune and Pluto. The object is about the size of Philadelphia and is presumably composed of ices and rock. The Hubble Space Telescope and ground-based telescopes have located hundreds of Kuiper belt objects. *(NASA and G. Bernstein)*

In fact, some astronomers argue that Pluto is not a true planet but merely the largest of the Kuiper belt objects. Some of these objects are quite large. One Kuiper belt object called Quaoar (pronounced "KWAH-o-wahr") is about 1270 km in diameter, a bit larger than Pluto's moon Charon. Another object, known as 1998 W_{31}, has its own moon. There may be other objects in the Kuiper belt that are as large as Pluto. Whether Pluto is a planet, a dual citizenship object, or the King of the Kuiper belt, it is an interesting world in an interesting place.

We can find more evidence that our solar system has a Kuiper belt by looking at other stars. A number of nearby stars such as Beta Pictoris (Chapter 19) are surrounded by disks of dust believed to be released by comets in the equivalent of Kuiper belts around these stars. Thus the detection of the dust implies the presence of Kuiper belts.

When we say comets come from the Oort cloud or the Kuiper belt, we only push the mystery back one step. Where did those icy bodies come from? Beginning in Chapter 19, we have studied the origin and evolution of our solar system so carefully that the answer leaps out at us. Those icy bodies are the last icy planetesimals remaining from the solar nebula.

When Gerard Kuiper first suggested the existence of the Kuiper belt, he proposed that the icy bodies there had formed from the solar nebula. Closer to the sun, massive planets had accreted or ejected the planetesimals, but beyond Neptune there are no massive planets. The icy planetesimals are safe there, especially those locked in orbital resonances with the outer planets. Those icy little worlds must have grown by condensation and accretion as the rest of the planetary system was forming.

The bodies in the Oort cloud, however, could not have formed at their present location. The solar nebula would not have been dense enough so far from the sun. Also, had they formed from the solar nebula, we would expect them to be distributed in a disk and not in a sphere. Rather, astronomers think the future comets formed as icy planetesimals in the outer solar system near the present orbits of the Jovian planets. As those planets grew more massive, they swept up many of these planetesimals, but they would have ejected some into long orbits leading far out of the solar system. These are the objects that became the Oort cloud.

REVIEW CRITICAL INQUIRY

How do comets help us understand the formation of the planets?

According to the solar nebula hypothesis, the planets formed from planetesimals that accreted in a disk-shaped nebula around the forming sun. The planetesimals that formed in the inner solar nebula were warm and could not

incorporate much ice. The asteroids may be the last remains of such bodies. In the outer solar nebula, it was colder, and the planetesimals would have contained large amounts of ices. Many of these planetesimals were destroyed when they fell together to make the planets, but some may have survived. The icy bodies of the Oort cloud and the Kuiper belt may be the last surviving icy planetesimals in our solar system. When these bodies fall into the inner solar system, we see them as comets, and the gases they release tell us that many of them are rich in volatile materials such as water, carbon dioxide, methane, and ammonia. These are the ices we would expect to find in the icy planetesimals. Furthermore, comets are rich in dust, and the planetesimals must have included large amounts of dust frozen into the ices when they formed. Thus, the nuclei of comets seem to be frozen samples of the ancient solar nebula.

Nearly all of the mass of a comet is in the nucleus, but the light we see comes from the coma and the tail. What kind of spectra do we get from comets, and what does that tell us about the process that converts ice and rock into a comet?

Our discussion of comets has revealed that there is a continuous supply of these objects, that they fragment easily when they encounter planets, and that entire comets or fragments of comets can hit planets. That makes us wonder what happens when a comet hits Earth.

25-4 IMPACTS ON EARTH

Asteroids and comets fall through our solar system like runaway trucks on a busy highway. These objects must hit the planets. In fact, it is these collisions that built the planets, and most of the original planetesimals have now been either incorporated into the planets or ejected from the solar system. Nevertheless, planets still get hit, as was dramatically demonstrated in 1994 when Comet Shoemaker–Levy hit Jupiter. Like any other planet, Earth must get hit now and then, and we should look for evidence that such impacts have occurred.

IMPACTS AND DINOSAURS

Comet nuclei and asteroids may seem small compared to planets, but they hit Earth with tremendous impact. Many scientists suspect that such impacts throw vast quantities of dust into the atmosphere and cause widespread changes in climate. The extinction of entire species, such as the dinosaurs, may have been caused

a

b

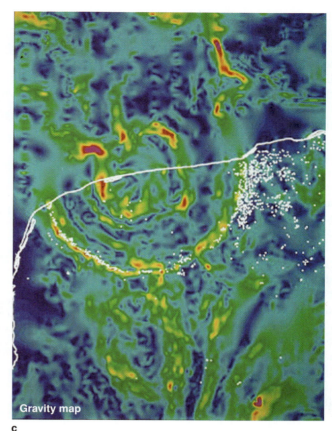

Gravity map

c

FIGURE 25-22

(a) The theory that the impact of one or more comets altered Earth's climate and drove dinosaurs to extinction has become so popular it appeared on this Hungarian stamp. The spacecraft shown (ICE) flew through the tail of Comet Giacobini–Zinner in 1985. Note the dead dinosaurs in the background. (b) The giant impact scar buried in Earth's crust near the village of Chicxulub in the northern Yucatán was formed about 65 million years ago. (c) This gravity map shows the extent of the crater hidden below limestone deposited since the impact. White dots represent sink holes formed in the fractured limestone. *(NASA)*

by comet impacts (■ Figure 25-22a). One of the most important clues is an element called iridium.

In 1980, Luis and Walter Alvarez announced the discovery of an excess of the element iridium in sediments laid down at the end of the Cretaceous period 65 million years ago—just when the dinosaurs and many other species became extinct. Iridium is rare in Earth's crust but common in meteorites, which led the Alvarezes to suggest that a major meteorite impact at the end of the Cretaceous threw vast amounts of iridium-rich dust into the atmosphere. This dust might have plunged Earth into a winter that lasted many years,

killing off many species of plants and animals, including the dinosaurs.

This theory was rejected at first, but soon scientists found the iridium anomaly in sediments of the same age all over the world. Others found related elements and mineral forms typical of meteorite impacts. At the same time, theorists studying nuclear weapons predicted that a nuclear war would throw so much dust into the atmosphere that our planet would be plunged into a "nuclear winter" that could last a number of years. Within a few years, scientists generally agreed that the Cretaceous extinctions were caused by one or more major impacts.

Mathematical models and observations of the impact of the comet fragments on Jupiter in 1994 have combined to create a plausible scenario of the events following a major impact on Earth. Of course, creatures living near the site of the impact would probably die in the initial shock, and an impact at sea would create tsunamis (tidal waves) many hundreds of meters high that would devastate coastal regions for many kilometers inland even halfway around the world from the impact. But the worst effects would begin after the initial explosion. On land or sea, a major impact would excavate large amounts of pulverized rock and loft it high above the atmosphere. As this material fell back, Earth's atmosphere would be turned into a glowing oven of red-hot rock streaming through the air as a rain of meteors, and the heat would trigger massive forest fires around the world. Soot from such fires has been detected in the layers of clay laid down at the end of the Cretaceous. Once the fire storms cooled, the remaining dust in the atmosphere would block sunlight and produce deep darkness for a year or more. No matter where on Earth an impact occurred, it would almost certainly vaporize large amounts of limestone. The carbon dioxide released from the limestone would produce intense acid rain. All of these consequences make it surprising that any life could have survived such an impact.

Geologists have located a crater at least 150 km in diameter centered near the village of Chicxulub in the northern Yucatán (Figure 25-22b and c). Although the crater is totally covered by sediments, mineral samples show that it contains shocked quartz typical of impact sites and that it is the right age. A gigantic impact formed the crater about 65 million years ago, just when the dinosaurs and many other species died out, and many Earth scientists now believe that this is the scar of the impact that ended the Cretaceous.

At first, these climate-changing impacts were blamed on large meteorites, but the nucleus of a comet would do as much damage. In fact, we have seen in this chapter that the distinction between an asteroid and a comet is not clear-cut.

The theory that an impact caused the extinction at the end of the Cretaceous period was once highly controversial, but the accumulated evidence has made it a widely accepted idea. Now Earth scientists are extending the theory to other extinctions. Chemical evidence from Earth's crust suggests that the extinction at the end of the Permian 250 million years ago was caused by a giant impact. The Permian extinction killed off 95 percent of the species on Earth and set the stage for the rise of the dinosaurs. Thus the dinosaurs may owe their origin as well as their extinction to giant impacts on Earth.

A solar system is a dangerous place to put a planet. Comets, asteroids, and meteoroids constantly rain down on the planets, and Earth gets hit quite often. The Chicxulub Crater isn't the only large impact scar on Earth. For example, a giant crater has been found buried under sediment in Iowa, and another giant impact crater may underlie most of the Chesapeake Bay. What would it be like to live on Earth when such an impact occurs? Humanity got a hint in 1908 when something hit Siberia.

THE TUNGUSKA EVENT

On the morning of June 30, 1908, scattered reindeer herders and homesteaders in central Siberia were startled to see a brilliant blue-white fireball brighter than the sun streak across the sky. Still descending, it exploded with a blinding flash and an intense pulse of heat:

> The whole northern part of the sky appeared to be covered with fire. . . . I felt great heat as if my shirt had caught fire . . . there was a . . . mighty crash. . . . I was thrown on the ground about [7 m] from the porch. . . . A hot wind, as from a cannon, blew past the huts from the north.

The blast was heard up to 1000 km away, and the resulting pulse of air pressure circled Earth twice. For a number of nights following the blast, European astronomers, who knew nothing of the explosion, observed a glowing reddish haze high in the atmosphere.

Travel in the wilderness of Siberia was difficult early in this century; moreover, World War I, the Bolshevik Revolution, and the Russian Civil War prevented expeditions from reaching the site before 1927. When at last an expedition arrived, it found that the blast had occurred above the Stony Tunguska River valley and had flattened trees in an irregular pattern extending out 30 km (❙ Figure 25-23). The trees were knocked down radially away from the center of the blast, and limbs and leaves had been stripped away. The trunks of trees at the very center of the area were still standing, although they had lost all their limbs. No expedition has ever found a crater, so it seems that the explosion, equaling a 12-megaton nuclear weapon, occurred at least a few kilometers above the ground.

All kinds of speculative theories have been proposed in tabloid newspapers and occult books. One popular idea was that a flying saucer with engine trouble tried to make an emergency landing and exploded. More responsible ideas have proposed that Earth was hit by a piece of antimatter or a miniature black hole. The evidence does not necessarily rule out either of these last two ideas, but they violate an important principle in science. They ask us to assume more than we need to. Science always prefers the simplest adequate explanation for any given phenomenon, and a small asteroid or comet would explain the event.

The comet hypothesis proposes that the head of a small comet struck Earth and exploded in the atmo-

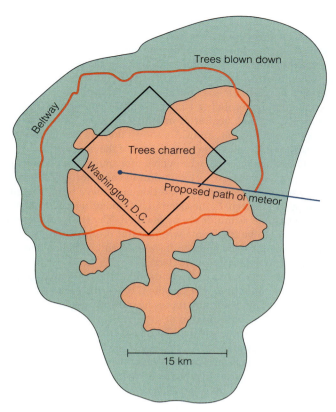

ACTIVE FIGURE 25-23

The 1908 Tunguska event in Siberia destroyed an area the size of a large city. Here the area of destruction is superimposed on a map of Washington, D.C., and its surrounding beltway. In the central area trees were burned; and, in the outer area, trees were blown down in a pattern away from the path of the meteor.

Ace Astronomy™ Go to AceAstronomy and click Active Figures to see "Impacts on Earth." Learn how often Earth is hit by meteorites of different sizes.

sphere. We might not see such a comet if it was coming from the direction of the sun. Witnesses reported that the fireball in the atmosphere came from the east, which in the morning means from the direction of the sun. If the object was a comet, whatever tail it had would have been projected straight at Earth and would not have been visible. More likely, say some astronomers, the object was the exhausted core of a comet and had no appreciable tail.

A fragment of a partially exhausted comet should consist of silicate material resembling carbonaceous chondrites, mixed with some remaining ices. Modern studies of the area show that thousands of tons of powdered material resembling carbonaceous chondrites are scattered in the soil. A chunk of dirty ice 50 m in diameter would contain about the right mass and would be totally invisible approaching from the sun. Also, a comet fragment would be so fragile that it would explode in the atmosphere rather than dig a crater, as would a denser, stronger meteorite.

The falling-comet hypothesis has been very popular with astronomers. It does not ask us to make any

fantastic assumptions, and it explains the observed phenomena, even the reddish glow over Europe. Some astronomers claim the glow was the faint traces of the comet's scattered dust, trapped high in the atmosphere and illuminated by the summer sun shining over the North Pole.

As compelling as the comet hypothesis may seem, not all astronomers agree. In the early 1980s, a detailed analysis of all the evidence suggested that the object was not cometary. Its speed and direction were not consistent with the motion of a comet. The analysis suggested instead that the object was a small Apollo asteroid, 90 to 190 m in diameter.

In 1993, astronomers studied computer models of objects entering Earth's atmosphere at high speed and concluded that the fragile head of a comet would have exploded much too high in the atmosphere. An iron-rich asteroid, in contrast, would be so strong that it would survive to reach the ground and would form a large crater. The most likely candidate seems to be a stony asteroid about 30 m in diameter. It would explode at just about the right height to produce the Tunguska blast.

One natural question is: Why should a piece of rock explode at all? To understand, we must recall that a meteor enters Earth's atmosphere at roughly 15 km/s. The air resistance is tremendous; and, as the meteor descends lower and meets denser air, the air resistance increases rapidly. The resistance that the meteor experiences depends on its surface area. When the meteor breaks into smaller pieces, its surface area suddenly increases, so the air resistance suddenly increases and thus causes more fragmentation. Because air resistance depends on surface area, when the meteor begins to break up, it explodes and vaporizes almost instantly. That is most likely what created the blast that equaled a 12-megaton explosion at Tunguska.

Could such an impact happen again? Earth gets hit by small meteorites every day and by larger objects less often. Impacts by large asteroids may happen many millions of years apart, but they continue to happen. In mid-March 1998, newspaper headlines announced, "Mile-Wide Asteroid to Hit Earth in October 2028." The frightening story was true except that the news media did not mention the uncertainty in the orbit. Within days, astronomers found the asteroid on old photographs, recalculated the orbit with the new data, and concluded that the asteroid, known as $1997XF_{11}$, will miss Earth by 600,000 miles. There will be no impact by this asteroid in 2028, but there are plenty more asteroids that haven't been discovered yet.

Some people have argued that the danger from asteroid and comet impacts is so great that we should develop massive nuclear-tipped missiles, ready to blast a meteoroid to pieces long before it can reach Earth. Some astronomers point out that lots of small fragments slamming into Earth may be even worse than one big impact. Other astronomers argue that the big objects are so rare we can ignore them. The real danger, they say, lies in the smaller, more common meteoroids, which are too small to detect with current telescopes.

Some people may prefer to believe that the Tunguska event was an alien spaceship; that is a titillating idea. But the truth about the Tunguska event is far more exciting. Earth is struck by meteorites every day. Some impacts are large enough to alter the climate, and the future of our civilization on Earth may depend on how much time we have until the next big one.

REVIEW **CRITICAL INQUIRY**

How can a large meteorite impact alter Earth's climate?

Even a meteorite no bigger than a large house strikes Earth with as much energy as a nuclear bomb and can produce a crater a mile in diameter. A large impact could produce a crater 100 km or more in diameter. Rock from Earth's crust would be pulverized and lifted high above the atmosphere. As it fell back in a rain of fiery meteors, it would be so hot it would trigger massive fires over much of Earth's surface. We can detect traces of such fires at about the time of the Chicxulub impact, 65 million years ago. Even after the fires cooled, the smoke and dust in the atmosphere would prevent sunlight from reaching the ground, and plants dependent on sunlight would be starved of energy. This worldwide darkness might last a year or more before the dust settled out of the atmosphere. By that time, many species would have died out, and the climate could have changed dramatically.

As scary as this is, we need not ask if it will happen. The probability of such an event is 100 percent. Our question can only be: How soon? In fact, such large impacts happen occasionally on all of the terrestrial planets. Summarize the evidence we have to show that large impacts have occurred on Earth and on other planets and satellites in our solar system.

Life on Earth seems fragile and exposed. Are we just lucky to have survived so long? Does life exist on other worlds? If so, has it, too, survived long enough to develop intelligence? That is the story of the next chapter.

SUMMARY

Most of the meteors we see, and certainly those in showers, are caused by small particles released by comets. When Earth crosses the orbit of a comet, we see a meteor shower. Most meteorites, however, are more massive, stronger fragments that probably originated in the asteroid belt.

The different kinds of meteorites could have originated in a planetesimal-size body that was heated by the decay of short-lived radioactive nuclei like aluminum-26. Differentia-

tion would form an iron–nickel core that would cool slowly. Fragments from this core would show the Widmanstätten pattern common in iron meteorites. The outer parts of such a body might resemble the stony-iron and achondritic meteorites.

Chondrites, with their beadlike chondrules, may have formed as smaller bodies farther from the sun. The carbonaceous chondrites, being rich in volatile compounds, also probably formed farther from the sun. The presence of volatiles, chondrules, and CAIs shows that the carbonaceous chondrites are among the least altered of the meteorites.

Collisions must have been common among the planetesimals, just as they are still common in the asteroid belt. We find breccias in some meteorites that must have been formed in impacts long ago. In the asteroid belt today, we can find evidence that the asteroids are fragmented. The Hirayama families appear to be fragments with similar orbits. Photographs of asteroids taken by spacecraft show irregular-shaped bodies heavily battered by impact craters.

Of the three kinds of asteroids, the S types are more common in the inner belt, whereas the C types are more common in the outer belt. This may represent a difference in temperature at the time of formation or different surface weathering. The M types appear to be the exposed iron–nickel cores of broken planetesimals.

Some of the asteroids show clear evidence of geological processes. The reflected light from Vesta, for instance, looks much like the light reflected from achondritic meteorites that resemble basalts. Thus, it seems that Vesta has melted and has had lava flows on its surface.

Jupiter clearly dominates the asteroid belt and may have prevented the formation of a planet there long ago. It now causes Kirkwood's gaps and traps the Trojan asteroids at the Lagrangian points along its orbit.

According to the dirty snowball theory, a comet is a ball of dirty ices that produces a coma as it approaches within a few astronomical units of the sun. The pressure of the solar wind pushes ionized gas back into a type I, or gas, tail; the pressure of sunlight pushes dust particles back into a type II, or dust, tail.

When Comet Halley passed through the inner solar system in 1986, five spacecraft flew past it and discovered that its nucleus was larger and blacker than expected. The nucleus was about 16 km $\times$ 8 km $\times$ 7 km and was covered with a black crust having an albedo of only 0.04. The crust is probably made of material resembling carbonaceous chondrites. Gas and dust jets vent from active regions.

According to the Oort cloud theory, comets occupy a hollow spherical cloud centered on the sun and extending out about 100,000 AU. Perturbations from the motions of nearby stars may cause a few of these icy bodies to fall into the inner solar system, where they become long-period comets. These comet nuclei originated as icy planetesimals among the Jovian planets and were ejected into the Oort cloud. Comets are almost unaltered samples of the original solar nebula.

The short-period comets have small orbits and periods of less than 200 years. Comet Halley is an example. Some of these comets appear to originate in the Kuiper belt, a disk-shaped cloud of icy bodies believed to extend from 40 AU to at least 100 AU from the sun. Hundreds of these bodies have been found, and astronomers speculate that Pluto, Charon, Triton, and Chiron may be icy worlds related to the Kuiper belt objects.

Earth runs into space debris quite often. The Apollo asteroids must hit Earth with great regularity, and the heads of comets must also intersect Earth now and then. One such collision, the Tunguska event, occurred as recently as 1908.

NEW TERMS

radiant

sporadic meteor

fall

find

iron meteorite

selection effect

Widmanstätten pattern

stony meteorite

chondrite

chondrule

carbonaceous chondrite

CAI

achondrite

stony-iron meteorite

Kirkwood's gaps

Apollo–Amor objects

Trojan asteroids

Hirayama families

gas (type I) tail

dust (type II) tail

coma

Oort cloud

Kuiper belt

REVIEW QUESTIONS

Ace ⟲ Astronomy™ Assess your understanding of this chapter's topics with additional quizzing and animations at **http:// astronomy.brookscole.com/seeds8e**

1. If most falls are stony meteorites, why are most finds iron?

2. How do observations of meteor showers reveal one of the sources of meteoroids?

3. How can most meteors be cometary if most, perhaps all, meteorites are asteroidal?

4. Why are meteorites easier to find in Antarctica? Why are stony meteorites better represented among these finds than among finds in the United States?

5. What evidence do we have that some meteorites have originated inside large bodies?

6. Why couldn't uranium, thorium, and radioactive potassium have melted the planetesimals? What evidence do we have that some asteroids have had molten interiors?

7. How do Kirkwood's gaps resemble Cassini's division?

8. The first asteroids discovered were much larger than the typical asteroid known today. How does this illustrate a selection effect?

9. Describe three types of asteroids (S, M, and C), and explain a theory to account for their differences.

10. How might the Apollo–Amor objects have originated?

11. What is the difference between a type I tail and a type II tail?

12. If comets are icy planetesimals left over from the formation of the solar system, why haven't they all vaporized by now?

DISCUSSION QUESTIONS

1. Futurists suggest that we may someday mine the asteroids for materials to build and supply space colonies. What kinds of materials could we get from asteroids? (*Hint:* What are S-, M-, and C-type asteroids made of?)

2. If cometary nuclei were heated by internal radioactive decay rather than by solar heat, how would comets differ from what we observe?

3. From what you know now, do you think the government should spend money to locate near-Earth asteroids? How serious is the risk?

PROBLEMS

1. Large meteorites are hardly slowed by Earth's atmosphere. Assuming that the atmosphere is 100 km thick and that a large meteorite falls vertically toward the ground, how long does it spend in the atmosphere? (*Hint:* How fast do meteoroids travel?)

2. If a single asteroid 1 km in diameter were fragmented, how many meteoroids 1 m in diameter could it yield? (*Hint:* The volume of a sphere is $\frac{4}{3}\pi r^3$.)

3. What is the orbital velocity of a meteoroid orbiting the sun at the distance of Earth? (*Hint:* See Chapter 5.)

4. If a half-million asteroids, each 1 km in diameter, were assembled into one body, how large would it be? (*Hint:* The volume of a sphere is $\frac{4}{3}\pi r^3$.)

5. The asteroid Vesta has a mass of 2×10^{20} kg and a radius of about 250 km. What is its escape velocity? Could you jump off the asteroid? (*Hint:* See Chapter 5.)

6. The asteroid Pallas has a mass of 2.5×10^{20} kg. What is the orbital velocity of a small satellite orbiting 500 km from the center of Pallas? (*Hint:* See Chapter 5.)

7. What is the maximum angular diameter of Ceres as seen from Earth? Could Earth-based telescopes detect surface features? Could the Hubble Space Telescope (Chapter 6)? (*Hint:* Use the small-angle formula.)

8. At what distances from the sun would you expect to find Kirkwood's gaps where the orbital period of asteroids is one-half of and one-third of the orbital period of Jupiter? Compare your results with Figure 25-12. (*Hint:* Use Kepler's third law.)

9. If the velocity of the solar wind is about 400 km/s and the visible tail of a comet is 100 million km long, how long does it take an atom to travel from the nucleus to the end of the visible tail?

10. If you saw Comet Halley when it was 0.7 AU from Earth and it had a visible tail 5° long, how long was the tail in kilometers? Suppose that the tail was not perpendicular to the line of sight. Is your answer too large or too small? (*Hint:* Use the small-angle formula.)

11. Calculate the orbital velocity of a comet while it is in the Oort cloud. (*Hint:* See Chapter 5.)

12. The mass of a comet's nucleus is about 10^{14} kg. If the Oort cloud contains 2 trillion (2×10^{12}) cometary nuclei, what is the mass of the cloud in Earth masses? (*Hint:* Earth's mass = 6×10^{24} kg.)

CRITICAL INQUIRIES FOR THE WEB

1. Some nights are better for looking for meteors than others (see Table 25-1). We know that these showers are associated with comets, but how are these associations made? Several sites on the Internet provide information on meteor showers, including historical data and information on parent comets. Pick a shower whose parent comet is known and summarize how we came to know that the meteors and the comet are related.

2. Chances are small that you will be killed by an asteroid impact, but if objects are out there that astronomers are not aware of whose orbits intersect Earth, we could be in for a surprise one day. Look for information on the LONEOS project and other searches for near-Earth asteroids. How many such objects have been discovered? What is the record for closest known passage of an asteroid to Earth?

3. Search for the IAU Minor Planet Center Web pages and find out what asteroids and what comets have come closest to Earth.

4. Search for information about comets in the sky right now. Are any comets bright enough for you to see? Are any bright comets expected soon?

EXPLORING *THESKY*

1. Of the five brightest asteroids, which has the most inclined orbit? (*Hint:* Use **Filters** in the **View** menu to turn off everything but the sun, stars, ecliptic, and minor planets. You can use **Tracking Setup** under **Time Skip** in the **Tools** menu to lock onto an object and follow it along the ecliptic as time passes.)

2. Do asteroids go through retrograde motion? (*Hint:* Use **Filters** in the **View** menu to turn off the stars and turn on the Equatorial Grid. See Activity 1 above.)

3. Of the five brightest asteroids, which has the most elliptical orbit? (*Hint:* Use **3D Solar System Mode** in the **View** menu and watch objects orbit the sun.)

4. Of the comets shown, which has the smallest orbit? (*Hint:* Use **3D Solar System Mode** in the **View** menu and watch objects orbit the sun.)

Visit the Seeds *Foundations of Astronomy* companion Web site for critical thinking exercises, articles, and additional readings from InfoTrac College Edition, Brooks/Cole's online student library.

LIFE ON OTHER WORLDS

Did I solicit thee from darkness to promote me?

John Milton, Paradise Lost

Peter Sawyer

Chapter 26

GUIDEPOST

This chapter is either unnecessary or critical, depending on our point of view. If we believe that astronomy is the study of the physical universe above the clouds, then this chapter does not belong here. But if we believe that astronomy is the study of our position in the universe, not only our physical position but also our role as living beings in the origin and evolution of the universe, then everything else in this book is just preparation for this chapter.

Astronomy is the only science that truly acts as a mirror. In studying the universe up there, we learn what we are down here. Astronomy is not really about stars, galaxies, and planets; it is about us.

VIRTUAL LABORATORIES

EXTRASOLAR PLANETS

As living things, we have been promoted from darkness. We are made of heavy atoms that could not have formed at the beginning of the universe. Successive generations of stars fusing light elements into heavier elements have built the atoms so important to our existence. When a dark cloud of interstellar gas enriched in these heavy atoms fell together to form our sun, a small part of the cloud gave birth to the planet we inhabit.

Are there intelligent beings living on other planets? That is the last and perhaps the most challenging question in our study of astronomy. We will try to answer it in three steps, each dealing with a different aspect of life.

First, we must decide what we mean by *life.* Life is not so much a form of matter as it is a behavior. Living matter extracts energy from its environment to modify itself and its surroundings so as to preserve itself and to create offspring. Thus, life is based on information, the recipe for the process of survival and reproduction.

Second, we must study the origin of life on Earth. If we can understand how life began on our planet, then we can try to estimate the likelihood that it has originated on other worlds as well.

Third, we must try to understand how the first primitive living organism on Earth could give rise to the tremendously diverse life forms that now inhabit our world. By understanding how living things evolve to fit their environment, we will see evidence that intelligence is a natural development in the evolution of life.

If life can originate on other worlds, and if intelligence is a natural result of the evolution of life forms, then we might expect that other intelligent races inhabit other worlds. Visits between worlds seem impossible, but perhaps we can detect their existence by radio.

Alien life forms could be quite different from us, but if they are alive, then they must share with us certain characteristics. Our goal in this chapter is to use our knowledge of science to explain the greatest of mysteries—the origin and evolution of life on Earth and on other worlds (Window on Science 26-1).

26-1 THE NATURE OF LIFE

What is life? Philosophers have struggled with that question for thousands of years, so it is unlikely that we will answer it here. But we must agree on a working model of life before we can speculate on its occurrence on other worlds. To that end, we will identify in living things two important aspects: a physical basis and a unit of controlling information.

THE PHYSICAL BASIS OF LIFE

On Earth, the physical basis of life is the carbon atom (❚ Figure 26-1). Because of the way this atom bonds to other atoms, it can form long, complex, stable chains that are capable of extracting, storing, and utilizing energy. Other chemical bases of life may exist. Science

Ace◐Astronomy™ The AceAstronomy icon throughout the text indicates an opportunity for you to test yourself on key concepts and to explore animations and interactions on the AceAstronomy Web site at **http://astronomy.brookscole .com/seeds8e** fiction stories and movies abound with silicon creatures, living things whose body chemistry is based on silicon rather than carbon. However, silicon forms weaker bonds than carbon does, and it cannot form double bonds as easily. Consequently, it cannot form the long, complex, stable chains that carbon can. Silicon is 135 times more common on Earth than carbon is, yet there are no silicon creatures among us. All Earth life is carbon based. Thus, the likelihood that distant planets are inhabited by silicon people seems small, but we cannot rule out life based on noncarbon chemistry.

In fact, nonchemical life might be possible. What is required is some mechanism capable of supporting the extraction and utilization of energy that we have identified as life. One could at least imagine life based on electromagnetic fields and ionized gas. No one has ever met such a creature, but science fiction writers conjure up all sorts.

Clearly, we could range far in space and time, theorizing about different bases for alien life, but to make progress we must discuss what we know best—carbon-based life on Earth. How can a lump of carbon-rich matter live? The answer lies in the information that guides its life processes.

INFORMATION STORAGE AND DUPLICATION

Living cells are tiny chemical factories. They must store all of the recipes for those chemicals in a safe place, use them to fulfill the cell's task, and hand down duplicates of the recipes to offspring. That information is encoded on long molecules.

Study ▌ "DNA: The Code of Life" on pages 592 and 593 and notice three important points. First, the chemical recipes of life are stored as templates on DNA molecules. The templates automatically guide specific chemical reactions within the cell. Second, the instructions stored in DNA are the genetic information handed down to offspring. When people say "You have your

a

b

FIGURE 26-1

All living things on Earth are based on carbon chemistry. Even the long molecules that carry genetic information, DNA and RNA, have a framework defined by chains of carbon atoms. (a) Katie, a complex mammal, contains about 30 AU of DNA. *(Michael Seeds)* (b) Each rod of the tobacco mosaic virus contains a single spiral strand of RNA about 0.01 mm long. *(L. D. Simon)* All life on Earth stores its genetic information in such carbon-chain molecules.

DNA: The Code of Life

The key to understanding life is information — the information that guides all of the processes in an organism. In most living things on Earth, that information is stored on a long spiral molecule called DNA (deoxyribonucleic acid).

The DNA molecule looks like a spiral ladder with rails made of phosphates and sugars. The rungs of the ladder are made of four chemical bases arranged in pairs. The bases always pair the same way. That is, base A always pairs with base T, and base G always pairs with base C.

Information is coded on the DNA molecule by the order in which the base pairs occur. To read that code, molecular biologists have to "sequence the DNA." That is, they must determine the order in which the base pairs occur along the DNA ladder.

DNA automatically combines raw materials to form important chemical compounds. The building blocks of these compounds are relatively simple **amino acids.** Segments of DNA act as templates that guide the amino acids to join together in the correct order to build specific **proteins,** chemical compounds important to the structure and function of organisms. Some proteins called **enzymes** regulate other processes. In this way, DNA recipes regulate the production of the compounds of life.

Mike Seeds

The traits we inherit from our parents, the chemical processes that animate us, and the structure of our bodies are all encoded in our DNA.

The Four Bases

A	Adenine
C	Cytosine
G	Guanine
T	Thymine

Nucleus
(information
storage)

Cell membrane
(transport of raw
materials and
finished product)

Material
storage

Manufacture
of proteins
and enzymes

Energy
production

A single cell from a human being contains about 1.5 meters of DNA containing about 4.5 billion base pairs — enough to record the entire works of Shakespeare 200 times. A typical human contains a total of about 600 AU of DNA. Yet the DNA in each cell, only 1.5 meters in length, contains all of the information to create a new human. A clone is a new creature created from the DNA code found in a single cell.

Original DNA

A cell is a tiny factory that uses the DNA code to manufacture chemicals. Most of the DNA remains safe in the nucleus of a cell, and the code is copied to create a molecule of **RNA (ribonucleic acid).** Like a messenger carrying blueprints, the RNA carries the code out of the nucleus to the work site where the proteins and enzymes are made.

To divide, a cell must duplicate its DNA. The DNA ladder splits, and new bases match to the exposed bases of the ladder to build two copies of the original DNA code. Because the base pairs almost always match correctly, errors in copying are rare. One set of the DNA code goes to each of the two new cells.

Copy DNA

DNA, coiled into a tight spiral, makes up the **chromosomes** that are the genetic material in a cell. A **gene** is a segment of a chromosome that controls a certain function. When a cell divides, each of the new cells receives a copy of the chromosomes, as genetic information is handed down to new generations.

Copy DNA

Cell Reproduction by Division

As a cell begins to divide, its DNA duplicates itself.

The duplicated chromosomes move to the middle.

The two sets of chromosomes separate, and . . .

the cell divides to produce . . .

two cells, each containing a full set of the DNA code.

mother's eyes," they are talking about DNA codes. Finally, notice how the DNA molecule reproduces itself when a cell divides so that each new cell contains a copy of the original information.

Although the DNA molecule must preserve its coded information from damage and make accurate copies, it must also be capable of making mistakes now and then. To see why, we must consider how new DNA recipes are created.

MODIFYING THE INFORMATION

If living things are to survive for many generations, then the information stored in their DNA must change as the environment changes. Without change in DNA, a slight warming of the climate, for example, might kill a species of plant, in turn starving the rabbits, deer, and other plant eaters, and leaving the hawks, wolves, and mountain lions with no prey. If the information stored in DNA could never change, then environmental changes would quickly drive life forms to extinction. If life is to survive in a changing world, then the information in DNA must be changeable. Living things must evolve.

Species evolve by **natural selection.** Each time an organism reproduces, its offspring receive the data stored in the DNA, but some variation is possible. For example, most of the rabbits in a litter may be normal, but it is possible for one to get a DNA recipe that gives it stronger teeth. If it has stronger teeth, it may be able to eat something other than the plant the others depend on, and if that plant is becoming scarce, the rabbit with stronger teeth has a survival advantage. It can eat other plants and so will be healthier than its littermates and have more offspring. Some of these offspring will also have stronger teeth, as the altered DNA recipes are handed down to the new generation. Thus nature selects and preserves those attributes that contribute to the survival of the species. Those creatures that are unfit die. Natural selection is merciless to the individual, but it gives the species the best possible chance to survive in a changing environment.

The only way nature can obtain new DNA patterns from which to select the best is from DNA molecules that have changed. This can happen through chance mismatching of base pairs—errors—in the reproduction of the DNA molecule. Another way this can occur is through damage to reproductive cells from exposure to radioactivity such as cosmic rays or natural radioactivity in the soil. In any case, an offspring born with altered DNA is called a **mutant.** Most mutations make no difference at all because they change segments of DNA that are not being used. Many mutations are fatal, and the individual dies long before it can have offspring of its own. But in rare cases, a mutation may give a species a new survival advantage. Then natural selection makes it likely that the new DNA message will sur-

vive and be handed down, making the species more capable of surviving.

Evolution is not random. Of course, the errors that occur in the DNA code are indeed random, but natural selection is not random. Those changes in the DNA that help a species survive are selected and preserved in future generations. With each passing generation, the species becomes more fit to survive in its environment.

REVIEW CRITICAL INQUIRY

Why can't the information in DNA be permanent?

The information stored in a creature's DNA provides the recipes that make the creature what it is. For example, the DNA in a starfish must contain all the recipes for making the various kinds of proteins needed to consume and digest food. That information must be passed on to offspring starfish, or they will be unable to survive. But the information must be changeable because our environment is changeable. Ice ages come and go, mountains rise, lakes dry up, and ocean currents shift. If the environment changes in some way, one or more of the recipes may no longer work. In our example, a change in the temperature of the ocean water may kill off the specific shellfish the starfish eat. If they can't digest other shellfish, the entire species will become extinct. Natural variation in DNA means that among all the infant starfish in any generation, some of the recipes are different; if the environment changes, all of the old-style starfish may die, but a few—those with the different DNA—can carry on.

The survival of life depends on this delicate balance between reliable reproduction and the introduction of small variations in DNA information. What are some of the ways these small changes in DNA can arise?

Life is based not only on information, but also on the duplication of information. Today, that process seems so complex that it is hard to imagine how it could have begun.

26-2 THE ORIGIN OF LIFE

If life on Earth is based on the storage of information in these long, complex, carbon-chain molecules, how could it have ever gotten started? Obviously, 4.5 billion chemical bases didn't just happen to drift together to form the DNA formula for a human being. The key is evolution. Once a life form begins to reproduce itself, natural selection preserves the most advantageous traits. Over long periods of time spanning thousands, perhaps millions, of generations, the life form becomes more fit to survive. This often means the life form becomes more complex. Thus life could have begun as a very simple process that gradually became more sophisticated as it was modified by evolution.

We begin our search on Earth, where fossils and an intimate familiarity with carbon-based life give us a glimpse of the first living matter. Once we discover how Earthly life could have begun, we can look for signs that life began on other planets in our solar system. Finally, we can speculate on the chances that other planets, orbiting other stars, have conditions that give rise to life.

THE ORIGIN OF LIFE ON EARTH

The oldest fossils hint that life began in the sea. The first living things on Earth left behind a very poor fossil record. They were at first single-celled creatures, and even when they became more complex, multicellular creatures, they contained no hard parts such as bones or shells that form good fossils. What fossils can be found are clearly the remains of ocean creatures.

Although the oldest rocks contain no obvious fossils, microscopes reveal traces of ancient life. The oldest widely accepted evidence is 2.5-billion-year-old fossils of bacteria. Rock from western Australia that is nearly 3.5 billion years old contains microscopic features that may be fossils of living things (Figure 26-2), but that conclusion is highly controversial. While the experts debate the details, we can conclude that life was present on Earth as simple organisms at least 2.5 billion years ago and possibly a billion years before that.

A little over a half-billion years ago something happened, perhaps a change in Earth's climate, and life exploded into a wide diversity of complex forms. This marks the beginning of the **Cambrian period,** and is sometimes called the *Cambrian explosion.* The best known of the Cambrian creatures may be the trilobites (Figure 26-3). Although the Cambrian creatures were diverse, there are no Cambrian fossils of land plants or animals. Evidently, land surfaces were totally devoid of life until only 400 million years ago.

The fossil record shows that life began as simple organisms in the sea soon after Earth formed. Only recently in geological terms has life become complex.

The key to the origin of this life may lie in an experiment performed by Stanley Miller and Harold Urey in 1952. This **Miller experiment** sought to reproduce the conditions on Earth under which life began. In a closed glass container, the experimenters placed water (to represent the oceans); the gases hydrogen, ammonia, and methane (to represent the primitive atmosphere); and an electric arc (to represent lightning bolts). The apparatus was sterilized, sealed, and set in operation (Figure 26-4).

After a week, Miller and Urey stopped the experiment and analyzed the material in the flask. Among the many compounds the experiment produced, they found four amino acids (building blocks of protein), various

FIGURE 26-2
The microphotograph at left shows a feature found in 3.5-billion-year-old rock from northwestern Australia. The artist's reconstruction at right interprets the feature as a chain of bacteria-like organisms. Experts do not at present agree on the age of the rocks, the way the rocks formed, or the nature of the tiny features. Consequently the search for the oldest fossils on Earth is a challenging quest. *(Courtesy J. William Schopf)*

FIGURE 26-3
Trilobites made their first appearance in the Cambrian oceans. The smallest were almost microscopic, and the largest were bigger than dinner plates. This example, about the size of a human hand, lived 400 million years ago in an ocean floor that is now a limestone deposit in Pennsylvania. *(Grundy Observatory photograph)*

a **b**

FIGURE 26-4
(a) The Miller experiment circulated gases through water in the presence of an electric arc. This simulation of primitive conditions on Earth produced amino acids, the building blocks of proteins. (b) Stanley Miller with a Miller apparatus. *(Courtesy Stanley Miller)*

fatty acids, and urea, a molecule common to many life processes. Evidently, the energy from the electric arc had molded the atmospheric gases into some of the basic components of living matter. Other energy sources, such as hot silica (to simulate hot lava spilling into the sea) and ultraviolet radiation (to simulate sunlight), give similar results.

Recent studies of the composition of meteorites and models of planet formation suggest that Earth's first atmosphere did not resemble the gases used in the Miller experiment. Earth's first atmosphere was probably composed of carbon dioxide, nitrogen, and water vapor. This finding, however, does not invalidate the Miller experiment. When such gases are processed in a Miller apparatus, a tarry gunk rich in organic molecules soon coats the inside of the chamber.

The Miller experiment did not create life, nor did it necessarily imitate the exact conditions on the young Earth. Rather, it is important because it shows that complex organic molecules form naturally in a wide variety of circumstances. The chemical deck is stacked to deal nature a hand of complex molecules. If we could travel back in time, we would probably find Earth's first oceans filled with a rich mixture of organic compounds in what some have called the **primordial soup.**

The next step on the journey toward life is for the compounds dissolved in the oceans to link up and form larger molecules. Amino acids, for example, can link together to form proteins. This linkage occurs when amino acids join together end to end and release a water molecule (▮ Figure 26-5). For many years, ex-

perts have assumed that this process must have happened in sun-warmed tidal pools where evaporation concentrated the broth. But recent studies suggest that the young Earth was subject to extensive volcanism and large meteorite impacts that periodically modified the climate enough to destroy any life forms exposed on the surface. Thus, the early growth of complex molecules likely took place among the hot springs along the midocean ridges. These complex molecules would not have been photosynthetic—taking energy from sunlight. Rather they would have taken energy from the heat and chemicals emerging from the hot springs of the midocean ridge. That energy could have powered the growth of long protein chains. Deep in the oceans, they would have been safe from climate changes.

Although these proteins might have contained hundreds of amino acids, they would not have been alive. Not yet. Such molecules would not have reproduced but would have merely linked together and broken apart at random. Because some molecules are more stable than others, however, and because some molecules bond together more readily than others, this blind **chemical evolution** would have led to the concentration of the varied smaller molecules into the most stable larger forms. Eventually, somewhere in the oceans, a molecule took shape that could reproduce itself. At that point the chemical evolution of molecules became the biological evolution of living things.

An alternative theory proposes that primitive living things such as reproducing molecules did not originate on Earth but came here in meteorites or comets.

FIGURE 26-5

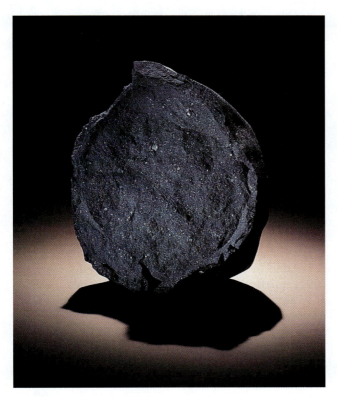

Radio astronomers have found a wide variety of organic molecules in the interstellar medium, and some studies have found similar compounds inside meteorites (Figure 26-6). Such molecules form so readily that we would be surprised if they were not present in space. A few investigators, however, have speculated that living, reproducing molecules originated in space and came to Earth as a cosmic contamination. If this is true, every planet in the universe is contaminated with the seeds of life. However entertaining this theory may be, it is presently untestable, and an untestable theory is of little use in science.

Whether life originated in the oceans or in space, we still face the problem of how life began. Experts studying the origin of life proceed on the assumption that life began as reproducing molecules in Earth's oceans.

Which came first, reproducing molecules or the cell? Because we think of the cell as the basic unit of life, this question seems to make no sense, but in fact the cell may have originated during chemical evolution. If a dry mixture of amino acids is heated, the acids form long, proteinlike molecules that, when poured into water, collect to form microscopic spheres that function in ways similar to cells (Figure 26-7). They have a thin membrane surface, they can absorb material from their surroundings, they grow in size, and they can divide and bud just as cells do. They contain no large molecule that copies itself, however. Thus the structure of the cell may have originated first and the reproducing molecules later.

FIGURE 26-6
A sample of the Murchison meteorite, a carbonaceous chondrite that fell in 1969 near Murchison, Australia. Analysis of the interior of the meteorite revealed evidence of amino acids. Whether the first building blocks of life originated in space is unknown, but the amino acids found in meteorites illustrate how commonly amino acids and other complex molecules occur even in the absence of living things. (*Courtesy Chip Clark, National Museum of Natural History*)

FIGURE 26-7
Single amino acids can be assembled into long proteinlike molecules. When such material cools in water, it can form microspheres, microscopic spheres with double-layered boundaries similar to cell membranes. Microspheres may have been an intermediate stage in the evolution of life between complex molecules and cells holding molecules reproducing genetic information. (*Courtesy Sidney Fox and Randall Grubbs*)

FIGURE 26-8

A 3.5-billion-year-old fossil stromatolite from western Australia is one of the oldest known fossils (below). Stromatolites were formed, layer by layer, by mats of bacteria living in shallow water. Such life may have been common in shallow seas when Earth was young (right). Stromatolites are still being formed today in similar environments. *(Mural by Peter Sawyer; photo courtesy Chip Clark, National Museum of Natural History)*

An alternative theory proposes that the replicating molecule developed first. Such a molecule would have been exposed to damage if it had been bare, so the first to manufacture or attract a protective coating of protein would have had a significant survival advantage. If this was the case, the protective cell membrane was a later development of biological evolution.

The first living things must have been single-celled organisms much like modern bacteria. Some of the oldest fossils known are **stromatolites,** structures produced by communities of photosynthesizing bacteria that grew in mats and, year by year, deposited layers of minerals that were later fossilized. One of the oldest such fossils known is believed to be 3.5 billion years old (❙ Figure 26-8). If such bacteria were common when Earth was young, the early atmosphere may have contained a small amount of oxygen produced by the photosynthesis. Recent studies suggest that an oxygen abundance of only 0.1 percent would have been sufficient to provide an ozone screen that would protect organisms from the sun's ultraviolet radiation.

How evolution shaped creatures to live in the ancient oceans, to photosynthesize and respire, to become multicellular, and to reproduce sexually is a fascinating story, but we cannot explore it in detail here. We can see that life could have begun through simple chemical re-

actions building complex molecules, and that once some DNA-like molecule formed, it protected its own survival with selfish determination. Over billions of years, the genetic information stored in living things kept those qualities that favored survival and discarded the rest. As Samuel Butler said, "The chicken is the egg's way of making another egg." In that sense, all living matter on Earth is merely the physical expression of DNA's mindless determination to continue its existence.

Perhaps this seems harsh. Human experience goes far beyond mere reproduction. *Homo sapiens* has art, poetry, music, philosophy, religion, science. Perhaps all of the great accomplishments of our intelligence represent more than mere reproduction of DNA. Nevertheless, intelligence, the ability to analyze complex situations and respond with appropriate action, must have begun as a survival mechanism. For example, a fixed escape strategy stored in the DNA is a disadvantage for a creature that frequently moves from one environment to another. A rodent that always escapes from predators by automatically climbing the nearest tree would be in serious jeopardy if it met a hungry fox in a treeless clearing. Even a faint glimmer of intelligence might allow the rodent to analyze the situation and, finding no trees, to choose running over climbing. Thus, intelligence, of which *Homo sapiens* is so proud,

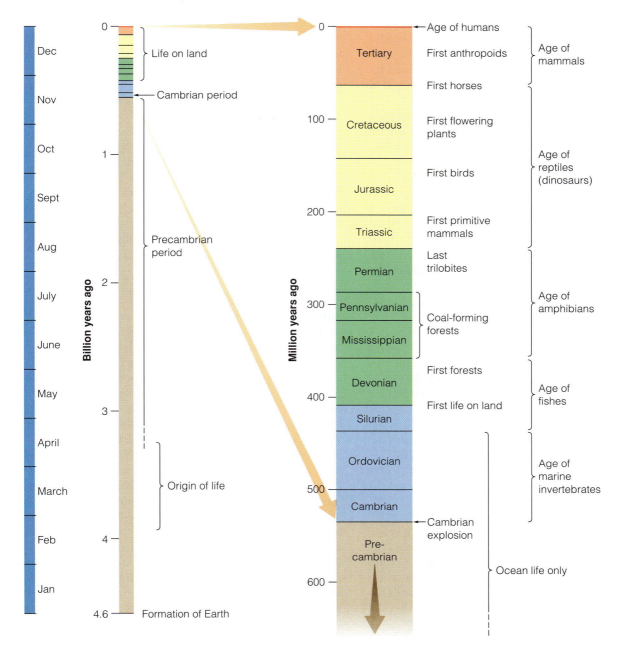

ACTIVE FIGURE 26-9

Complex life has developed on Earth only recently. If the entire history of Earth were represented in a time line (left), we would have to magnify the end of the line to see details such as life leaving the oceans and dinosaurs appearing. The age of humans would still be only a thin line at the top of our diagram. If the history of Earth were a year-long videotape, humans would not appear until the last hours of December 31.

Ace✺Astronomy™ Go to AceAstronomy and click Active Figures to see "Earth Calendar." Click on the calendar to explore through time.

may have developed in ancient creatures as a way of making them more versatile.

Any discussion of the evolution of life seems to involve highly improbable coincidences until we consider how many years have passed in the history of Earth. We read the words—4.6 billion years—so easily, but it is in truth hard to grasp the meaning of such a long period of time.

GEOLOGIC TIME

Humanity is a very new experiment on planet Earth. We can fit all of the evolution that leads from the primitive life forms in the oceans of the Cambrian period 600 million years ago to fishes, amphibians, reptiles, and mammals into a single chart such as ❙ Figure 26-9. We can take comfort in thinking that creatures like us have walked on Earth for roughly 3 million years, but when we add our history to the chart, we discover that the entire history of humanity makes up no more than a thin line at the top. In fact, if we tried to represent the entire 4.6-billion-year history of Earth on the chart, the portion describing the rise of life on the land would be an unreadably small segment.

One way to represent the evolution of life is to compress the 4.6-billion-year history of Earth into a 1-year-long videotape. In such a program, Earth forms as the video begins on January 1, and through all of January and February it cools and is cratered, and the first oceans form. Search as we might, we would find no trace of life in these oceans until sometime in March or early April, when the first living things develop. The slow development of these simplest of living forms grinds on slowly through the spring and summer of our videotape. The entire 4-billion-year history of Precambrian evolution lasts until the video reaches mid-November, when the primitive ocean life explodes into the more complex Cambrian organisms such as trilobites.

While our year-long videotape plays on and on, we might amuse ourselves by looking at the land instead of the oceans, but we would be disappointed. The land is a lifeless waste with no plants or animals of any kind. Not until November 28 in our video does life appear on the land; but, once it does, it evolves rapidly into a wide range of plants and animals. Dinosaurs, for example, appear about December 12 and vanish by Christmas evening as mammals and birds flourish.

Throughout the 1-year run of our video there have been no humans; and even during the last days of the year, as the mammals rise to dominate the landscape, there are no people. In the early evening of December 31, vaguely human forms move through the grasslands, and by late evening they begin making stone tools. The Stone Age lasts till about 11:45 PM, and the first signs of civilization, towns and cities, do not appear until 11:54 PM. The Christian era begins only 14 seconds before the New Year, and the Declaration of Independence is signed with but 1 second to spare.

By converting the history of Earth into a year-long videotape, we have placed the rise of life in perspective. Tremendous amounts of time were needed for the first simple living things to evolve in the oceans, and even more time was needed for the evolution of complex creatures that could colonize the land. As life became more complex, it evolved and diversified faster and faster, as if evolution were drawing on a growing library of solutions that had been previously invented with great effort to solve earlier problems. The burst of diversity on land led eventually to the rise of intelligent creatures like us, a process that has taken 4.6 billion years.

If life could originate on Earth and develop into intelligent creatures, perhaps the same thing could have happened on other planets. This raises three questions.

First, could life originate on another world if conditions were suitable? We can't be certain, but we have found natural, chemical processes that could lead to living things, so our answer to this question should probably be yes. The second question is, will life always evolve toward intelligence? Perhaps intelligence arose on Earth under unusual conditions, but we must recall that intelligence appears to be a way to make a species more versatile and thus more likely to survive. If that is true, then intelligence may develop under a wide range of conditions if enough time is available. But what of the third question: Are suitable conditions so rare that life almost never gets started? The only way to answer that is to search for life on other planets. We begin, in the next section, with the other planets in our solar system.

LIFE IN OUR SOLAR SYSTEM

Although we can imagine life based on something other than carbon chemistry, we know of no examples to tell us how such life might arise and survive. We must limit our discussion to life as we know it and the conditions it requires. The most important requirement is the presence of liquid water, not only as part of the chemical reactions of life, but also as a medium to transport nutrients and wastes within the organism. Also, it seems that life on Earth began in the oceans and developed there for nearly 4 billion years before it was able to emerge onto the land. Certainly, any world where we hope to find life must have liquid water, and that means it must have moderate temperatures.

The water requirement automatically eliminates many worlds in our solar system. The moon is airless, and although some data suggest ice frozen in the soil at its poles, it has never had liquid water on its surface. In the vacuum of the lunar surface, liquid water would boil away rapidly. Mercury too is airless and cannot have had liquid water on its surface for long periods of time. Venus has some traces of water vapor in its atmosphere, but it is much too hot for liquid water to survive. If there were any lakes or oceans of water on its surface when it was young, they must have evaporated quickly. Even if life began there, no traces would be left now.

The inner solar system seems too hot, and the outer solar system seems too cold. The Jovian planets have deep atmospheres, and at a certain level, they have moderate temperatures where water might condense into liquid droplets. But it seems unlikely that life could begin there. The Jovian planets have no surfaces where oceans could nurture the beginning of life, and currents in the atmosphere seem destined to circulate gas and water droplets from regions of moderate temperature to other levels that are much too hot or too cold for life to survive.

A few of the satellites of the Jovian planets might have suitable conditions for life. Jupiter's moon Europa seems to have a liquid-water ocean below its icy crust (see Figure 23-15), and minerals dissolved in that water would provide a rich broth of possibilities for chemical evolution. Nevertheless, Europa is not a promising site to search for life because conditions may not have remained stable for the billions of years needed for life to evolve beyond the microscopic stage. The subsurface ocean is kept from freezing by tidal heating. If Jupiter's moons interact gravitationally and modify their orbits, Europa may have been frozen solid at some points in history. Such periods of freezing would probably prevent life from developing. Drilling through the icy crust of Europa to search for life in its ocean will be a wonderful adventure for future generations, but Europa does not seem to be a good bet to harbor any form of complex life.

Saturn's moon Titan has an atmosphere of nitrogen, argon, and methane and may have oceans of liquid methane and ethane on its surface. We saw in Chapter 23 how sunlight can convert the methane in the atmosphere to organic smog particles that settle to the surface. The chemistry of life that might crawl or swim on such a world is unknown, but life there may be unlikely because of the temperature. The surface of Titan is a deadly −179°C (−290°F). Chemical reactions occur slowly or not at all at such low temperatures, so the chemical evolution needed to begin life may never have occurred on Titan.

Mars is the most likely place for life in our solar system. The evidence, however, is not encouraging. In 1976, two robotic spacecraft, Viking 1 and Viking 2, landed at two different places on the Martian surface. The spacecraft scooped up soil samples and subjected them to tests for the presence of living organisms. For example, they gave some soil samples a dose of nutrient-rich water and watched for signs the nutrients were taken up by biological processes. The results seem negative. Although some peculiar chemical processes were detected, no evidence was found that clearly indicated the presence of any living things in the soil.

The chemical experiments performed by the Viking landers were sophisticated, but they could only search for traces of currently living things. They could not search for signs of past life, such as fossils. For that, we need to send a geologist to Mars or bring rocks from Mars back to Earth. Nature has performed the latter task for us.

Meteorite ALH84001 (❚ Figure 26-10a) was found on the Antarctic ice in 1984. Years later an analysis of its chemical composition showed that it had originated on Mars. It was probably part of debris ejected into space by a large impact on Mars. Some of that debris fell to Earth, and a small number of these meteorites have been found. ALH84001 is important because a team of scientists studied it and announced in 1996 that it contained chemical and physical traces of ancient life on Mars (Figure 26-10b).

This discovery was received with great excitement by the press and the public. It would, indeed, be dramatic evidence if true, because it would show that life could begin on another world.

a

Electron microscope image

b

FIGURE 26-10

(a) Meteorite ALH84001 is one of a dozen meteorites known to have originated on Mars. It was claimed that the meteorite contained chemical and physical traces of ancient life on Mars, including what appear to be fossils of microscopic organisms (b). The evidence has not been confirmed, and the validity of the claim is highly questionable. *(NASA)*

Scientists were excited too, but being professionally skeptical (see Window on Science 19-2), they began testing the results immediately. In many cases, the results did not confirm the conclusion that life once existed on Mars. Some chemical contamination from water on Earth has occurred, and some chemicals in the meteorite may have originated without the presence of life. The physical features that look like fossil bacteria may be mineral formations in the rock. Although studies of ALH84001 continue, it does not give us unchallenged evidence that life once existed on Mars.

Spacecraft now visiting Mars may help us understand the past history of water there and paint a more detailed picture of present conditions. Landers are planned that will eventually land on Mars, collect rocks, and return them to Earth. Nevertheless, conclusive evidence may have to wait until a geologist in a space suit can wander the dry streambeds of Mars cracking open rocks and searching for fossils.

We are left to conclude that, so far as we know, our solar system is bare of life except for Earth. Consequently our search for life in the universe takes us to other planetary systems.

LIFE IN OTHER PLANETARY SYSTEMS

Might life exist in other solar systems? To consider this question, let us try to decide how common planets are and what conditions a planet must fulfill for life to originate and evolve to intelligence. The first question is astronomical; the second is biological. Our ability to discuss the problem of life outside our solar system is severely limited by our lack of experience.

In Chapter 19 we concluded that planets form as a natural by-product of star formation, and we learned that a number of extrasolar planets have been found circling nearby stars. From this we can conclude that planetary systems are very common.

If a planet is to become a suitable home for life, it must have a stable orbit around its sun. This is simple in a solar system like our own, but in a binary system most planetary orbits are unstable. Most planets in such systems would not last long before they were swallowed up by one of the stars or ejected from the system.

Thus, single stars are the most likely to have planets suitable for life. Because our galaxy contains at least 10^{11} stars, half of which are single, there could be roughly 5×10^{10} planetary systems in which we should look for life.

A few million years of suitable conditions does not seem to be enough time to originate life. On our planet it took at least 0.5 to 1 billion years for the first cells to evolve and 4.6 billion years for intelligence to evolve. Clearly, conditions on a planet must remain acceptable over a long time. This eliminates massive stars that remain stable on the main sequence for only a few million years. If it takes a few billion years for life to originate and evolve to intelligence, no star hotter than about F5 will do. This is not really a serious restriction, because upper-main-sequence stars are rare anyway.

In previous sections, we decided that life, at least as we know it, requires liquid water. That requirement defines a **life zone** (or ecosphere) around each star, a region within which a planet has temperatures that permit the existence of liquid water.

The size of the life zone depends on the temperature of the star (Figure 26-11). Hot stars have larger life zones because the planets must be more distant to remain cool. But the short main-sequence lives of these stars make them unacceptable. M stars have small life zones because they are extremely cool—only planets very near the star receive sufficient warmth. However, planets that are close to a star would probably become tidally coupled, keeping the same side toward the star. This might cause the water and atmosphere to freeze in the perpetual darkness of the planet's night side and end all chance of life. Also, M stars are subject to sudden flares that might destroy life on a planet close to the star. Thus, the life zone restricts our search for life to main-sequence G and K stars.

FIGURE 26-11

The life zone around a star (green) is the region where a planet would have a moderate temperature. M stars are not good candidates because they are cool and their life zone is small. F stars have a large life zone but evolve too quickly to allow the rise of complex life. K and G stars like our sun seem the best candidates in a search for intelligent life.

Even a star on the main sequence is not perfectly stable. Main-sequence stars gradually grow more luminous as they convert their hydrogen to helium, and thus the life zone around a star gradually moves outward. A planet might form in the life zone, and life might begin and evolve for billions of years, only to be destroyed as the slowly increasing luminosity of its star moves the life zone outward, evaporates the planet's oceans, and drives off its atmosphere. If a planet is to remain in the life zone for 4 to 5 billion years, it must form on the outer edge of the zone. This may be the most serious restriction we have yet discussed.

We should pause here to note that our concept of a life zone may be very self-centered. We admit that life might exist on worlds such as Europa and Titan, but they lie outside our conventionally defined life zone. Furthermore, scientists on Earth are discovering life in highly unlikely places such as in the Antarctic ice, in hot springs on the ocean floor, and even in solid rock deep underground. As we learn more about living things, we may have to extend our definition of the life zone to include a wider range of conditions.

The fundamental question in our discussion is simple. If conditions are right, will life begin? Early in this chapter we decided that life could begin through simple chemical reactions, so perhaps we should change our question and ask: What could prevent life from beginning? Given what we know about life, it should arise whenever conditions permit, and our galaxy should be filled with planets that are inhabited with living creatures.

REVIEW CRITICAL INQUIRY

What evidence do we have that life is at least possible on other worlds?

Our evidence is limited almost entirely to Earth, but it is promising. The fossils we find show that life originated in Earth's oceans almost 4 billion years ago, and biologists have proposed relatively simple chemical processes that could have created these first reproducing molecules. The fossils show that life developed from a very slow beginning into more and more complex creatures that filled the oceans. The pace of evolution quickened dramatically about half a billion years ago, the beginning of the Cambrian period, when life began taking on complex forms; later, when life emerged onto the surface of the land, it evolved rapidly to produce the tremendous diversity we see around us. Human intelligence has been a very recent development; it is only a few million years old.

If this process occurred on Earth, then it seems reasonable that it could have occurred on other worlds as well. Thus, the evidence suggests we can expect that life might begin and evolve to intelligent forms on any world where conditions are right. What are the conditions we should expect of other worlds that host life?

It is both easy and fun to speculate about life on other worlds, but it leads to a simple question: Is there really life beyond Earth? If we can't leave Earth and visit other worlds, then the only life we can detect will be beings intelligent enough to communicate with us. Why haven't we heard from them?

26-3 COMMUNICATION WITH DISTANT CIVILIZATIONS

If other civilizations exist, perhaps we can communicate with them in some way. Sadly, travel between the stars is more difficult in real life than in science fiction—and may in fact be impossible. If we can't physically visit, perhaps we can communicate by radio. Again, nature places restrictions on such conversations, but the restrictions are not too severe. As we will see, the real problem lies with the life expectancy of civilizations.

TRAVEL BETWEEN THE STARS

Practically speaking, roaming among the stars is tremendously difficult because of three limitations: distance, speed, and fuel. The distances between stars are almost beyond comprehension. It does little good to explain that if we use a golf ball in New York City to represent the sun, the nearest star would be another golf ball in Chicago. It is only slightly better to note that the fastest commercial jet would take about 4 million years to reach the nearest star.

The second limitation is a speed limit—we cannot travel faster than the speed of light. Though science fiction writers invent hyperspace drives so their heroes can zip from star to star, the speed of light is a natural and unavoidable limit that we cannot exceed. This, combined with the large distances between stars, makes interstellar travel very time consuming.

The third limitation is that we can't even approach the speed of light without using a fantastic amount of fuel. Even if we ignore the problem of escaping from Earth's gravity, we must still use energy stored in fuel to accelerate to high speed and to decelerate to a stop when we reach our destination. To return to Earth, assuming we wish to, we have to repeat the process. These changes in velocity require a tremendous amount of fuel. If we flew a spaceship as big as a large yacht to a star 5 light-years (1.5 pc) away and wanted to get there in only 10 years, we would use 40,000 times as much energy as the United States consumes in a year.

Travel for a few individuals might be possible if we accept very long travel times. That would require some form of suspended animation (currently unknown) or colony ships that carry a complete, though small, society in which people are born, live, and die generation after generation. Whether the occupants of such a ship would retain the social characteristics of humans over a long voyage is questionable.

These three limitations not only make it difficult for us to leave our solar system, but also would make it difficult for aliens to visit Earth. Reputable scientists have studied "unidentified flying objects" (UFOs) and related phenomena and have never found any evidence that Earth is being visited or has ever been visited by aliens from other worlds (Window on Science 26-2). Thus, humans are unlikely ever to meet an alien face to face. The only way we can communicate with other civilizations is via radio.

RADIO COMMUNICATION

Nature places two restrictions on our ability to communicate with distant societies by radio. One has to do with simple physics, is well understood, and merely makes the communication difficult. The second has to do with the fate of technological civilizations, is still unresolved, and may severely limit the number of societies we can detect by radio.

Radio signals are electromagnetic waves that travel at the speed of light. Because even the nearest civilizations must be a few light-years away, this limits our ability to carry on a conversation with distant beings. If we ask a question of a creature 10 light-years away, we will have to wait 20 years for a reply. Clearly, the give-and-take of normal conversation will be impossible.

Ace★Astronomy™ Go to AceAstronomy and click Active Figures to see "Interstellar Communications." Click on the sun and on another star to see how long communication could take.

Instead, we could simply broadcast a radio beacon of friendship to announce our presence. In fact, we are already broadcasting a recognizable beacon. Short-wavelength radio signals, such as TV and FM, have been leaking into space for the last 50 years or so. Any civilization within 50 light-years might already have detected us.

If we intentionally broadcast a signal, we can anticode it. That is, we can arrange it to make it easy to decode. We could transmit pulses to represent 1s and gaps to represent 0s. A message counting through the first few prime numbers would distinguish our signal from natural sources of radio noise. We could even transmit a picture by sending a string of 1s and 0s that can be arranged in only two ways. One way produces nonsense, but the other way produces a meaningful picture (❙ Figure 26-12).

In 1974, at the dedication of the 1000-ft radio telescope at Arecibo, radio astronomers transmitted such a signal toward the globular cluster M13, which is located 26,000 ly from Earth. When the signal finally arrives, any aliens who detect it will be able to arrange its 1679 pulses in only two ways, as 23 rows of 73 or 73 rows of 23. The second arrangement will form a picture that describes life on Earth (❙ Figure 26-13).

It took only minutes to transmit the Arecibo message. If more time were taken, a more detailed picture could be sent, and if we were sure our radio telescope was pointed at a listening civilization, we could send a long series of pictures. With pictures we could teach aliens our language and tell them all about our life, our difficulties, and our accomplishments.

If we can think of sending such signals, aliens can think of it too. If we point our radio telescopes in the right direction and listen at the right wavelength, we might hear other intelligent races calling out to one another. This raises two questions: Which stars are the best candidates, and what wavelengths are most likely? We have already answered the first question. Main-sequence G and K stars have the most favorable characteristics. But the second question is more complex.

Only certain wavelengths are useful for communication. We cannot use wavelengths longer than about 30 cm

An anticoded message

1	0	1	0	0	1	1	1	1	1
0	0	1	0	1	0	0	1	0	0
0	1	0	1	0	0	1	0	1	0
0	1	0	1	0

5 rows of 7

1	0	1	0	0	1	1
1	1	1	0	0	1	0
1	0	0	1	0	0	0
1	0	1	0	0	1	0
1	0	0	1	0	1	0

7 rows of 5

1	0	1	0	0
1	1	1	1	1
0	0	1	0	1
0	0	1	0	0
0	1	0	1	0
0	1	0	1	0
0	1	0	1	0

FIGURE 26-12
An anticoded message is designed for easy decoding. Here a string of 35 radio pulses, represented as 1s and 0s, can be arranged in only two ways, as 5 rows of 7 or 7 rows of 5. The second way produces a friendly message. Any number of pulses can be used so long as it is the product of two prime numbers. Then the pulses can be arranged in only two ways.

because the signal would be lost in the background radio noise from our galaxy. Nor can we go to wavelengths much shorter than 1 cm because of absorption within our atmosphere. Thus only a certain range of wavelengths, a radio window, is open for communication.

This communications window is very wide, so a radio telescope would take a long time to tune over all the wavelengths searching for intelligent signals. Nature may have given us a way to narrow the search, however. Within the communications window lie the 21-cm line of neutral hydrogen and the 18-cm line of OH. The interval between these two lines has been dubbed the **water hole** because the combination of H and OH yields

water (H_2O). Water is the fundamental solvent in our life form, so it might seem natural for similar water creatures to call out to each other at wavelengths in the water hole (❚ Figure 26-14). But even silicon creatures would be familiar with the 21-cm line of hydrogen.

This is not idle speculation. A number of searches for extraterrestrial radio signals have been made, and some major searches are now under way. The field has become known as **SETI,** Search for Extra-Terrestrial Intelligence, and it has generated heated debate. Some scientists and philosophers argue that life on other worlds can't possibly exist, so it is a waste of money to search. Others argue that life on other worlds is common and could be detected with present technology. Congress funded a NASA search for a short time but then ended support in the early 1990s because political leaders feared public reaction. In fact, the annual cost of a major search is only about as much as a single Air Force attack helicopter. The controversy may spring in part from the theological and philosophical controversy that would result from the discovery of intelligent life on another world.

In spite of the controversy, searches are under way. The NASA SETI project canceled by Congress is now supported by private funds as Project Phoenix. It is located at the SETI Institute and has been using radio telescopes at major radio astronomy observatories to listen to the radio emissions of a list of candidate stars. About two billion radio-frequency bands must be examined for each star, so the project depends heavily on computers to search for signs of artificial radio transmissions in the recorded radio emissions of each star.

Start of number markers — Binary numbers 1 to 10

Atomic numbers of hydrogen, carbon, nitrogen, oxygen, and phosphorus

Formulas for sugars and bases in DNA

DNA double helix

Number of units in DNA

Start of number markers

Human figure

Start of number markers

Population of Earth

Height of human in wavelengths

Arecibo radio dish transmitting signal

Sun and planets with Earth offset

Diameter of dish in wavelengths

Start of number markers

FIGURE 26-13
The Arecibo message to M13 begins by counting from 1 to 10 and goes on to describe our solar system and the biochemistry of life on Earth (color added for clarity). Binary numbers give the height of the human figure (1110) and the diameter of the telescope dish (100101111110) in units of the wavelength of the signal, 12.3 cm. *(NASA)*

FIGURE 26-14

Radio noise from various sources makes it difficult to detect distant signals at wavelengths longer than 30 cm or shorter than 1 cm. In this range, radio emission from H atoms and from OH marks a small wavelength range dubbed the water hole, which may be a likely place for communication.

Astonishing amounts of computer power are needed to search for weak or unusual signals. Consequently, the Berkeley SETI team, with the support of the Planetary Society, has recruited owners of personal computers linked to the Internet to participate in a project called seti@home. Participants download a screen saver that searches data files from the Arecibo radio telescope for meaningful signals whenever the owner is not using the computer. About 4 million people have signed up and are providing 1000 years of computer time per day for the project. For information, locate the seti@home project at http://setiathome.ssl.berkeley.edu/.

Since 1985, META, Megachannel Extra-Terrestrial Assay, has been searching the entire sky at millions of frequency bands in the water hole. Funded by the Planetary Society, the search uses radio antennas at Harvard and near Buenos Aires, Argentina. The project has instituted a second, even more efficient search called BETA. Since it began, META has detected dozens of candidate signals. That is, it has found dozens of radio signals that satisfy its most basic criteria. Unfortunately, none of these signals has proved to be continuous, and no signal has been detected when the radio telescopes were redirected at any of the candidate stars. Many of these signals are probably unusual noise from sources on Earth, but the candidate signals are still under study.

One ingenious search is called SERENDIP, Search for Extraterrestrial Radio Emission from Nearby Developed Intelligent Populations. Rather than monopolize an entire radio telescope, SERENDIP rides piggyback on the 305-m Arecibo Telescope. Wherever the radio astronomers point the telescope, the receiver samples the signal, looking for intelligent signals over millions of frequency bands. If candidate signals are found, the receivers note the position of the radio telescope for future investigation.

Radio noise pollution is a serious problem for these searches. Year by year our society generates more radio signals from sources on Earth and satellites in orbit. Poorly designed radio transmitters generate noise at unexpected frequencies, and other electronic devices, such as computers, emit radio frequency signals. Furthermore, society, industry, and government press to use wider and wider sections of the electromagnetic spectrum. Radio astronomers struggle to hear through the radio babble, and SETI searches suffer because many of the noise signals mimic the patterns expected from other civilizations. It would be ironic if we fail to detect signals from another world because our own world has become too noisy.

What should we expect if signals are found? By international agreement, those searching for signals have agreed to share candidate signals and obtain reliable confirmation before making public announcements. False alarms would be embarrassing. The chances of success depend on the number of inhabited worlds in our galaxy, and that number is difficult to estimate.

HOW MANY INHABITED WORLDS?

The technology exists, and given enough time the searches will find other inhabited worlds, assuming there are at least a few out there. If intelligence is common, then we should find the signals soon—in the next few decades—but if intelligence is rare in the universe, it may be a very long time before we confirm that we are not alone.

Simple arithmetic can give us an estimate of the number of technological civilizations with which we might communicate, N_c. The first proposed formula for N_c is now known as the **Drake equation,** named after radio astronomer Frank Drake, a pioneer in the search for extraterrestrial intelligence. We will use a version of the Drake equation modified slightly to make it a bit easier to understand. The formula gives N_c, the number of communicative civilizations in a galaxy, as

$$N_c = N^* \cdot f_P \cdot n_{LZ} \cdot f_L \cdot f_I \cdot F_S$$

N^* is the number of stars in a galaxy, and f_P represents the fraction of all stars that have planets. If all single stars have planets, f_P is about 0.5. The factor n_{LZ} is the average number of planets in a solar system suitably placed in the life zone, f_L is the fraction of suitable planets on which life begins, and f_I is the fraction of life forms that evolve to intelligence. These factors can be roughly estimated, but the remaining factor is much more uncertain.

F_S is the fraction of a star's life during which the life form is communicative. Here we assume that a star lives about 10 billion years. If a society survives at a technological level for only 100 years, our chances of communicating with it are small. But a society that sta-

TABLE 26-1
The Number of Technological Civilizations per Galaxy

		Estimates	
Variables		**Pessimistic**	**Optimistic**
N^*	Number of stars per galaxy	2×10^{11}	2×10^{11}
f_P	Fraction of stars with planets	0.01	0.5
n_{LZ}	Number of planets per star that lie in life zone for longer than 4 billion years	0.01	1
f_L	Fraction of suitable planets on which life begins	0.01	1
f_I	Fraction of life forms that evolve to intelligence	0.01	1
F_S	Fraction of star's life during which a technological society survives	10^{-8}	10^{-4}
N_C	Number of communicative civilizations per galaxy	2×10^{-5}	10×10^6

bilizes and remains technological for a long time is much more likely to be in the communicative phase at the proper time to signal to us. If we assume that technological societies destroy themselves in about 100 years, F_S is 100 divided by 10 billion, or 10^{-8}. But if societies can remain technological for a million years, then F_S is 10^{-4}. The influence of the factors in the formula is shown in ▌ Table 26-1.

Ace◯**Astronomy**™ Go to AceAstronomy and click Active Figures to see "Drake Equation." Make your own estimates and see how common life is on other worlds.

If the optimistic estimates are true, there may be a communicative civilization within a few dozen light-years of us, and we could locate it by searching through only a few thousand stars. On the other hand, if the pessimistic estimates are correct, we may be the only planet in our galaxy capable of communication. We may never know until we understand how technological societies function.

REVIEW CRITICAL INQUIRY

Why does the number of inhabited worlds we might hear from depend on how long civilizations survive at a technological level?

When we turn radio telescopes toward the sky and scan millions of frequency bands, we take a snapshot of the universe at a particular time when we are living and able to build radio telescopes. To detect other civilizations, they must be in a similar technological stage so they will be broadcasting either intentionally or accidentally. If we search for decades and detect no other signal, it may mean not that life is rare but rather that civilizations do not survive at a technological level for very long. Only a few decades ago, our civilization was threatened by nuclear war, and now we are threatened by the pollution of our environment. If nearly all civilizations in our galaxy are either on the long road up from primitive life forms in oceans or on the long road down

from nuclear war or environmental collapse, there may be no one transmitting during the short interval when we are capable of building radio telescopes with which to listen.

Radio communication between inhabited worlds is limited because it requires very fast computers to search many frequency intervals. Why must we search so many frequencies when we suspect that the water hole would be a good place to listen?

Are we the only thinking race? If we are, we bear the sole responsibility to understand and admire the universe. Then we are the sole representatives of that state of matter called intelligence. The mere detection of signals from another civilization would demonstrate that we share the universe with others. Although we might never leave our solar system, such communication would end the self-centered isolation of humanity and stimulate a reevaluation of the meaning of our existence. We may never realize our full potential as humans until we communicate with nonhuman intelligent life.

SUMMARY

To discuss life on other worlds, we must first understand something about life in general, life on Earth, and the origin of life. In general, we can identify two properties of living things: a physical basis and a controlling unit of information. The process of life must extract energy from the surroundings, maintain the organism, and modify the surroundings to promote the organism's survival. The physical basis is the arrangement of matter and energy that implements the life process. On Earth, all life is based on carbon chemistry. The controlling information is the data necessary to maintain the organism's function. Data for Earth life are stored in long carbon-chain molecules called DNA.

The DNA molecule stores information in the form of chemical bases linked together like the rungs of a ladder. When these patterns are copied by RNA molecules, they can direct the manufacture of proteins and enzymes. Thus the DNA information is the chemical formulas the cell needs to function. When a cell divides, the DNA molecule splits lengthwise and duplicates itself so that each of the new cells has a copy of the information. Errors in the duplication or damage to the DNA molecule can produce mutants, organisms that contain new DNA information and have new properties. Natural selection determines which of these new organisms are best suited to survive, and the species evolves to fit its environment.

The Miller experiment suggests that energy sources such as lightning could have caused amino acids and other complex molecules to form. Chemical evolution would have connected these together in larger and more complex, but not yet living, molecules. When a molecule acquired the ability to produce copies of itself, natural selection began improving the organism through biological evolution. Though this may have happened in the first billion years, life did not become diverse and complex until the Cambrian period, about 0.5 billion years ago. Life emerged from the oceans about 0.4 billion years ago, and humanity developed only a few million years ago.

It seems unlikely that there is now life on other planets in our solar system. Most of the planets are too hot or too cold. Mars may have been suitable in the past, but it is now a cold, dry desert, and any liquid water is trapped below the crust. Jupiter's moon Europa has a liquid-water ocean below its icy crust, but conditions there may not have remained stable long enough for life to develop. Saturn's moon Titan may have organic materials on its surface, but it may be too cold for chemical reactions to lead to life.

To find life, we must look beyond our solar system. Because we suspect that planets form from the leftover debris of star formation, we suspect that most stars have planets. The rise of intelligence may take billions of years, however, so short-lived massive stars and binary stars with unstable planetary orbits must be discarded. The best candidates are G and K main-sequence stars.

The distances between stars are too large to permit travel, but communication by radio could be possible. A certain wavelength range called a radio window is suitable, and a small range between the radio signals of H and OH, the so-called water hole, is especially likely.

We now have the technology to search for other intelligent life in the universe. Although such searches are controversial, a number of radio astronomers are now searching for radio signals from extraterrestrial civilizations. If life is common in the universe, success seems inevitable.

NEW TERMS

DNA (deoxyribonucleic acid)

amino acid

protein

enzyme

RNA (ribonucleic acid)

chromosome

gene

natural selection

mutant

Cambrian period

Miller experiment

primordial soup

chemical evolution

stromatolite

life zone

water hole

SETI

Drake equation

REVIEW QUESTIONS

Ace⊘Astronomy™ Assess your understanding of this chapter's topics with additional quizzing and animations at **http:// astronomy.brookscole.com/seeds8e**

1. If life is based on information, what is that information?

2. What would happen to a life form if the information handed down to offspring was always the same? How would that endanger the future of the life form?

3. How does the DNA molecule produce a copy of itself?

4. Give an example of natural selection acting on new DNA patterns to select the most advantageous characteristics.

5. Why do we believe that life on Earth began in the sea?

6. Why do we think that liquid water is necessary for the origin of life?

7. What is the difference between chemical evolution and biological evolution?

8. What was the significance of the Miller experiment?

9. How does intelligence make a creature more likely to survive?

10. Why are upper-main-sequence stars unlikely sites for intelligent civilizations?

11. Why do we suspect that travel between stars is nearly impossible?

12. How does the stability of technological civilizations affect the probability that we can communicate with them?

13. What is the water hole, and why would it be a good place to look for other civilizations?

DISCUSSION QUESTIONS

1. What would you change in the Arecibo message if humanity lived on Mars instead of Earth?

2. What do you think it would mean if decades of careful searches for radio signals for extraterrestrial intelligence turned up nothing?

PROBLEMS

1. A single human cell encloses about 1.5 m of DNA containing 4.5 billion base pairs. What is the spacing between these base pairs in nanometers? That is, how far apart are the rungs on the DNA ladder?

2. If we represent the history of the Earth by a line 1 m long, how long a segment would represent the 400 million years since life moved onto the land? How long a segment would represent the 3-million-year history of human life?

3. If a human generation, the time from birth to childbearing, is 20 years, how many generations have passed in the last million years?

4. If a star must remain on the main sequence for at least 5 billion years for life to evolve to intelligence, how massive could a star be and still harbor intelligent life on one of its planets? (*Hint:* See Chapter 12.)

5. If there are about 1.4×10^{-4} stars like the sun per cubic light-year, how many lie within 100 light-years of Earth? (*Hint:* The volume of a sphere is $\frac{4}{3}\pi r^3$.)

6. Mathematician Karl Gauss suggested planting forests and fields in a gigantic geometric proof to signal to possible Martians that intelligent life exists on Earth. If Martians had telescopes that could resolve details no smaller than 1 second of arc, how large would the smallest element of Gauss's proof have to be? (*Hint:* Use the small-angle formula.)

7. If we detected radio signals with an average wavelength of 20 cm and suspected that they came from a civilization on a distant planet, roughly how much of a change in wavelength should we expect to see because of the orbital motion of the distant planet? (*Hint:* See the Doppler shift in Chapter 7.)

8. Calculate the number of communicative civilizations per galaxy from your own estimates of the factors in Table 26-1.

CRITICAL INQUIRIES FOR THE WEB

1. The popular movie *Contact* focused interest on the SETI program by profiling the work of a radio astronomer dedicated to the search for extraterrestrial intelligence. Visit Web sites that give information about the movie, SETI programs, and radio astronomy, and discuss how realistic the movie was in capturing how such research is done.

2. Where outside the solar system would you look for habitable planets? NASA has increasingly focused its interest on this question. Look for information online about programs dedicated to detecting which planets might support life as we know it. What criteria are used to choose targets for the planned searches? What methods will be used to carry out the searches?

 Visit the Seeds *Foundations of Astronomy* companion Web site for critical thinking exercises, articles, and additional readings from InfoTrac College Edition, Brooks/Cole's online student library.

AFTERWORD

The aggregate of all our joys and sufferings, thousands of confident religions, ideologies and economic doctrines, every hunter and forager, every hero and coward, every creator and destroyer of civilizations, every king and peasant, every young couple in love, every hopeful child, every mother and father, every inventor and explorer, every teacher of morals, every corrupt politician, every superstar, every supreme leader, every saint and sinner in the history of our species, lived there on a mote of dust, suspended in a sunbeam.

Carl Sagan (1934–1996)

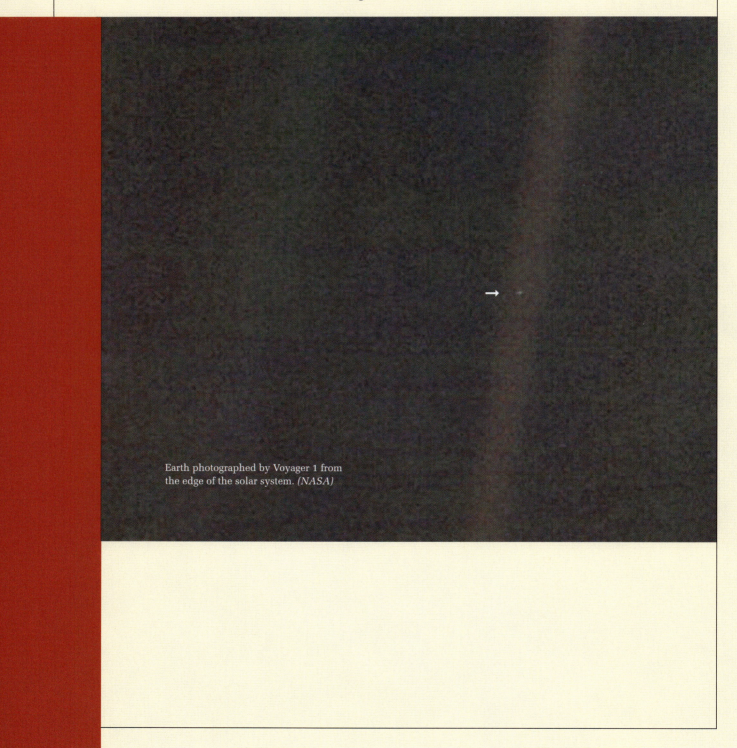

Earth photographed by Voyager 1 from the edge of the solar system. *(NASA)*

Our journey is over, but before we part company, there is one last thing to discuss—the place of humanity in the universe. Astronomy gives us some comprehension of the workings of stars, galaxies, and planets, but its greatest value lies in what it teaches us about ourselves. Now that we have surveyed astronomical knowledge, we can better understand our own position in nature.

To some, the word *nature* conjures up visions of furry rabbits hopping about in a forest glade dotted with pastel wildflowers. To others, nature is the blue-green ocean depths filled with creatures swirling in a mad struggle for survival. Still others think of nature as windswept mountaintops of gray stone and glittering ice. As diverse as these images are, they are all Earth-bound. Having studied astronomy, we can view nature as a beautiful mechanism composed of matter and energy interacting according to simple rules to form galaxies, stars, planets, mountaintops, ocean depths, and forest glades.

Perhaps the most important astronomical lesson is that we are a small but important part of the universe. Most of the universe is lifeless. The vast reaches between the galaxies appear to be empty of all but the thinnest gas, and the stars are much too hot to preserve the chemical bonds that seem necessary for life to survive and develop. Only on the surfaces of a few planets, where temperatures are moderate, could atoms link together to form living matter.

If life is special, then intelligence is precious. The universe must contain many planets devoid of life, planets where the wind has blown unfelt for billions of years. There may also exist planets where life has developed but has not become complex, planets on which the wind stirs wide plains of grass and rustles dark forests. On some planets, insects, fish, birds, and animals may watch the passing days unaware of their own existence. It is intelligence, human or alien, that gives meaning to the landscape.

Science is the process by which intelligence tries to understand the universe. Science is not the invention of new devices or processes. It does not create home computers, cure the mumps, or manufacture plastic spoons—that is engineering and technology, the adaptation of scientific understanding for practical purposes. Science is understanding nature, and astronomy is understanding on the grandest scale. Astronomy is the science by which the universe, through its intelligent lumps of matter, tries to understand its own existence.

As the primary intelligent species on this planet, we are the custodians of a priceless gift—a planet filled with living things. This is especially true if life is rare in the universe. In fact, if Earth is the only inhabited planet, our responsibility is overwhelming. In any case, we are the only creatures who can take action to preserve the existence of life on Earth, and, ironically, our own actions are the most serious hazards.

The future of humanity is not secure. We are trapped on a tiny planet with limited resources and a population growing faster than our ability to produce food. In our efforts to survive, we have already driven some creatures to extinction and now threaten others. If our civilization collapses because of starvation, or if our race destroys itself somehow, the only bright spot is that the rest of the creatures on Earth will be better off for our absence.

But even if we control our population and conserve and recycle our resources, life on Earth is doomed. In 5 billion years, the sun will leave the main sequence and swell into a red giant, incinerating Earth. However, Earth will be lifeless long before that. Within the next few billion years, the growing luminosity of the sun will first alter Earth's climate and then boil its atmosphere and oceans. Earth, like everything else in the universe, is only temporary.

Ace✦Astronomy™ Go to AceAstronomy and click Active Figures to see "Future of the Sun." See how the sun's evolution limits the future of life on Earth.

To survive, humanity must eventually leave Earth and search for other planets. Colonizing the moon and other planets of our solar system will not save us, because they will face the same fate as Earth when the sun dies. But travel to other stars is tremendously difficult and may be impossible with the limited resources we have in our small solar system. We and all of the living things on Earth may be trapped.

This is a depressing prospect, but a few factors are comforting. First, everything in the universe is temporary. Stars die, galaxies die, perhaps the entire universe will someday end. That our distant future is limited only assures us that we are a part of a much larger whole. Second, we have a few billion years to prepare, and a billion years is a very long time. Only a few million years ago, our ancestors were learning to walk erect and communicate. A billion years ago, our ancestors were microscopic organisms living in the primeval oceans. To suppose that a billion years hence we humans will still be human, that we will still be the dominant species on Earth, or that we will still be the sole intelligence on Earth is the ultimate conceit.

Our responsibility is not to save our race for all eternity but to behave as dependable custodians of our planet, preserving it, admiring it, and trying to understand it. That calls for drastic changes in our behavior toward other living things and a revolution in our attitude toward our planet's resources. Whether we can change our ways is debatable—humanity is far from perfect in its understanding, abilities, or intentions. However, we must not imagine that we and our civilization are less than precious. We have the gift of intelligence, and that is the finest thing this planet has ever produced.

We shall not cease from exploration
And the end of all our exploring
Will be to arrive where we started
And know the place for the first time.
　　　　　　　—*T. S. Eliot, "Little Gidding"*

APPENDIX A

UNITS AND ASTRONOMICAL DATA

INTRODUCTION

The metric system is used worldwide as the system of units, not only in science but also in engineering, business, sports, and daily life. Developed in 18th-century France, the metric system has gained acceptance in almost every country in the world because it simplifies computations.

A system of units is based on the three fundamental units for length, mass, and time. Other quantities such as density and force are derived from these fundamental units. In the English (or British) system of units (commonly used only in the United States, Tonga, and Southern Yemen, but not in Great Britain) the fundamental unit of length is the foot, composed of 12 inches. The metric system is based on the decimal system of numbers, and the fundamental unit of length is the meter, composed of 100 centimeters.

To see the advantage of having a decimal-based system, try computing the volume of a bathtub that is 5′9″ long, 1′2″ deep, and 2′10″ wide. In the metric system the length of the tub is 1 m and 75 cm, or 1.75 m. The other dimensions are 0.35 m and 0.86 m, and the volume is just $1.75 \times 0.35 \times 0.86$ m³. To make the computation in English units, we must first convert inches to feet by dividing by 12. We can convert centimeters to meters by the simpler process of moving the decimal point. Thus, the computation is much easier if we measure the tub in meters and centimeters instead of feet and inches.

Because the metric system is a decimal system, it is easy to express quantities in larger or smaller units as is convenient. We can express distances in centimeters, meters, kilometers, and so on. The prefixes specify the relation of the unit to the meter. Just as a cent is $\frac{1}{100}$ of a dollar, so a centimeter is $\frac{1}{100}$ of a meter. A kilometer is 1000 m, and a kilogram is 1000 g. The meanings of the commonly used prefixes are given in ▌Table A-1.

THE SI UNITS

Any system of units based on the decimal system would be easy to use, but by international agreement,

TABLE A-1 Metric Prefixes		
Prefix	Symbol	Factor
Mega	M	10^6
Kilo	k	10^3
Centi	c	10^{-2}
Milli	m	10^{-3}
Micro	μ	10^{-6}
Nano	n	10^{-9}

the preferred set of units, known as the *Système International d'Unités* (SI units) is based on the meter, kilogram, and second. These three fundamental units define the rest of the units, as given in ▌Table A-2.

The SI unit of force is the newton (N), named after Isaac Newton. It is the force needed to accelerate a 1 kg mass by 1 m/s² , or the force roughly equivalent to the weight of an apple at Earth's surface. The SI unit of energy is the joule (J), the energy produced by a force of 1 N acting through a distance of 1 m. A joule is roughly the energy in the impact of an apple falling off a table.

EXCEPTIONS

Units help us in two ways. They make it possible to make calculations, and they help us to conceive of cer-

TABLE A-2 SI Metric Units		
Quantity	SI Unit	English Unit
Length	Meter (m)	Foot
Mass	Kilogram (kg)	Slug (sl)
Time	Second (s)	Second (s)
Force	Newton (N)	Pound (lb)
Energy	Joule (J)	Foot-pound (fp)

tain quantities. For calculations the metric system is far superior, and we use it for our calculations throughout this book.

But Americans commonly use the English system of units, so for conceptual purposes we can express quantities in English units. Instead of saying the average person would weigh 133 N on the moon, we could express the weight as 30 lb. Thus, the text commonly gives quantities in metric form followed by the English form in parentheses: the radius of the moon is 1738 km (1080 miles).

In SI units, density should be expressed as kilograms per cubic meter, but no human can enclose a cubic meter in his or her hand, so this unit does not help us grasp the significance of a given density. This book refers to density in grams per cubic centimeter. A gram is roughly the mass of a paperclip, and a cubic centimeter is the size of a small sugar cube, so we can conceive of a density of 1 g/cm^3, roughly the density of water. This is not a bothersome departure from SI units because we will not make complex calculations using density.

CONVERSIONS

To convert from one metric unit to another (from meters to kilometers, for example), we have only to look at the prefix. However, converting from metric to English or English to metric is more complicated. The conversion factors are given in ▮ Table A-3.

Example: The radius of the moon is 1738 km. What is this in miles? Table A-3 indicates that 1 mile equals 1.609 km, so

$$1738 \text{ km} \times \frac{1 \text{ mile}}{1.609 \text{ km}} = 1080 \text{ miles}$$

TEMPERATURE SCALES

In astronomy, as in most other sciences, temperatures are expressed on the Kelvin scale, although the centigrade (or Celsius) scale is also used. The Fahrenheit scale commonly used in the United States is not used in scientific work.

Temperatures on the Kelvin scale are measured from absolute zero, the temperature of an object that contains no extractable heat. In practice, no object can be as cold as absolute zero, although laboratory apparatuses have reached temperatures less than 10^{-6} K. The scale is named after the Scottish mathematical physicist William Thomson, Lord Kelvin (1824–1907).

The centigrade scale refers temperatures to the freezing point of water (0°C) and to the boiling point of water (100°C). One degree centigrade is 1/100th the temperature difference between the freezing and boiling points of water. Thus the prefix *centi*. The centigrade scale is also called the Celsius scale after its inventor, the Swedish astronomer Anders Celsius (1701–1744).

The Fahrenheit scale fixes the freezing point of water at 32°F and the boiling point at 212°F. Named after the German physicist Gabriel Daniel Fahrenheit (1686–1736), who made the first successful mercury thermometer in 1720, the Fahrenheit scale is used only in the United States.

It is easy to convert temperatures from one scale to another using the information given in ▮ Table A-4.

POWERS OF 10 NOTATION

Powers of 10 make writing very large numbers much simpler. For example, the nearest star is about 43,000,000,000,000 km from the sun. Writing this number as 4.3×10^{13} km is much easier.

Very small numbers can also be written with powers of 10. For example, the wavelength of visible light

TABLE A-3
Conversion Factors

1 inch = 2.54 centimeters	1 centimeter = 0.394 inch
1 foot = 0.3048 meter	1 meter = 39.36 inches = 3.28 feet
1 mile = 1.6093 kilometers	1 kilometer = 0.6214 mile
1 slug = 14.594 kilograms	1 kilogram = 0.0685 slug
1 pound = 4.4482 newtons	1 newton = 0.2248 pound
1 foot-pound = 1.35582 joules	1 joule = 0.7376 foot-pound
1 horsepower = 745.7 joules/s	1 joule/s = 1 watt

TABLE A-4
Temperature Scales

	Kelvin (K)	Centigrade (°C)	Fahrenheit (°F)
Absolute zero	0 K	−273°C	−459°F
Freezing point of water	273 K	0°C	32°F
Boiling point of water	373 K	100°C	212°F

Conversions:
$$K = °C + 273$$
$$°C = \tfrac{5}{9}(°F - 32)$$
$$°F = \tfrac{9}{5}°C + 32$$

is about 0.0000005 m. In powers of 10 this becomes 5×10^{-7} m.

The powers of 10 used in this notation appear below. The exponent tells us how to move the decimal point. If the exponent is positive, we move the decimal point to the right. If the exponent is negative, we move the decimal point to the left. Thus, 2×10^3 equals 2000.0, and 2×10^{-3} equals 0.002.

$$\vdots$$

10^5	$=$	100,000
10^4	$=$	10,000
10^3	$=$	1,000
10^2	$=$	100
10^1	$=$	10
10^0	$=$	1
10^{-1}	$=$	0.1
10^{-2}	$=$	0.01
10^{-3}	$=$	0.001
10^{-4}	$=$	0.0001

$$\vdots$$

If you use scientific notation in calculations, be sure you correctly enter numbers into your calculator. Not all calculators accept scientific notation, but those that can have a key labeled EXP, EEX, or perhaps EE that allows you to enter the exponent of ten. To enter a number such as 3×10^8, press the keys 3 EXP 8. To enter a number with a negative exponent, we must use the change-sign key, usually labeled $+/-$ or CHS. To enter the number 5.2×10^{-3}, we press the keys 5.2 EXP $+/-$ 3. Try a few examples.

To read a number in scientific notation from a calculator you must read the exponent separately. The number 3.1×10^{25} may appear in a calculator display as 3.1 25 or on some calculators as 3.1 10^{25}. Examine your calculator to determine how such numbers are displayed.

***

Astronomy, and science in general, is a way of learning about nature and understanding the universe we live in. To test hypotheses about how nature works, scientists use observations of nature. The tables that follow contain some of the basic observations that support our best understanding of the astronomical universe. Of course, these data are expressed in the form of numbers, not because science reduces all understanding to mere numbers, but because the struggle to understand nature is so demanding we must use every tool available. Quantitative thinking, reasoning mathematically, is one of the most powerful tools ever invented by the human brain. Thus these tables are not nature reduced to mere numbers, but numbers supporting humanity's growing understanding of the natural world around us.

TABLE A-5
Constants

Astronomical unit (AU)	$= 1.495979 \times 10^{11}$ m
Parsec (pc)	$= 206{,}265$ AU
	$= 3.085678 \times 10^{16}$ m
	$= 3.261633$ ly
Light-year (ly)	$= 9.46053 \times 10^{15}$ m
Velocity of light (c)	$= 2.997925 \times 10^8$ m/s
Gravitational constant (G)	$= 6.67 \times 10^{-11}$ m^3/s^{2}kg
Mass of Earth ($M_\oplus$)	$= 5.976 \times 10^{24}$ kg
Earth equatorial radius ($R_\oplus$)	$= 6378.164$ km
Mass of sun ($M_\odot$)	$= 1.989 \times 10^{30}$ kg
Radius of sun ($R_\odot$)	$= 6.9599 \times 10^8$ m
Solar luminosity ($L_\odot$)	$= 3.826 \times 10^{26}$ J/s
Mass of moon	$= 7.350 \times 10^{22}$ kg
Radius of moon	$= 1738$ km
Mass of H atom	$= 1.67352 \times 10^{-27}$ kg

TABLE A-6
Units Used in Astronomy

1 angstrom (Å)	$= 10^{-8}$ cm
	$= 10^{-10}$ m
1 astronomical unit (AU)	$= 1.495979 \times 10^{11}$ m
	$= 92.95582 \times 10^6$ miles
1 light-year (ly)	$= 6.3240 \times 10^4$ AU
	$= 9.46053 \times 10^{15}$ m
	$= 5.9 \times 10^{12}$ miles
1 parsec (pc)	$= 206{,}265$ AU
	$= 3.085678 \times 10^{16}$ m
	$= 3.261633$ ly
1 kiloparsec (kpc)	$= 1000$ pc
1 megaparsec (Mpc)	$= 1{,}000{,}000$ pc

TABLE A-7
Properties of Main-Sequence Stars

Spectral Type	Absolute Visual Magnitude (M_v)	Luminosity*	Temp. (K)	λ_{max} (nm)	Mass*	Radius*	Average Density (g/cm³)
O5	−5.8	501,000	40,000	72.4	40	17.8	0.01
B0	−4.1	20,000	28,000	100	18	7.4	0.1
B5	−1.1	790	15,000	190	6.4	3.8	0.2
A0	+0.7	79	9900	290	3.2	2.5	0.3
A5	+2.0	20	8500	340	2.1	1.7	0.6
F0	+2.6	6.3	7400	390	1.7	1.4	1.0
F5	+3.4	2.5	6600	440	1.3	1.2	1.1
G0	+4.4	1.3	6000	480	1.1	1.0	1.4
G5	+5.1	0.8	5500	520	0.9	0.9	1.6
K0	+5.9	0.4	4900	590	0.8	0.8	1.8
K5	+7.3	0.2	4100	700	0.7	0.7	2.4
M0	+9.0	0.1	3500	830	0.5	0.6	2.5
M5	+11.8	0.01	2800	1000	0.2	0.3	10.0
M8	+16	0.001	2400	1200	0.1	0.1	63

*Luminosity, mass, and radius are given in terms of the sun's luminosity, mass, and radius.

TABLE A-8
The Brightest Stars

Star	Name	Apparent Visual Magnitude (m_v)	Spectral Type	Absolute Visual Magnitude (M_v)	Distance (ly)
α CMa A	Sirius	−1.47	A1	1.4	8.7
α Car	Canopus	−0.72	F0	−3.1	98
α Cen	Rigil Kentaurus	−0.01	G2	4.4	4.3
α Boo	Arcturus	−0.06	K2	−0.3	36
α Lyr	Vega	0.04	A0	0.5	26.5
α Aur	Capella	0.05	G8	−0.6	45
β Ori A	Rigel	0.14	B8	−7.1	900
α CMi A	Procyon	0.37	F5	2.7	11.3
α Ori	Betelgeuse	0.41	M2	−5.6	520
α Eri	Achernar	0.51	B3	−2.3	118
β Cen AB	Hadar	0.63	B1	−5.2	490
α Aql	Altair	0.77	A7	2.2	16.5
α Tau A	Aldebaran	0.86	K5	−0.7	68
α Cru	Acrux	0.90	B2	−3.5	260
α Vir	Spica	0.91	B1	−3.3	220
α Sco A	Antares	0.92	M1	−5.1	520
α PsA	Fomalhaut	1.15	A3	2.0	22.6
β Gem	Pollux	1.16	K0	1.0	35
α Cyg	Deneb	1.26	A2	−7.1	1600
β Cru	Beta Crucis	1.28	B0.5	−4.6	490

TABLE A-9
The Nearest Stars

Name	Absolute Magnitude (M_v)	Distance (ly)	Spectral Type	Apparent Visual Magnitude (m_v)
Sun	4.83		G2	−26.8
Proxima Cen	15.45	4.28	M5	11.05
α Cen A	4.38	4.3	G2	0.1
B	5.76	4.3	K5	1.5
Barnard's Star	13.21	5.9	M5	9.5
Wolf 359	16.80	7.6	M6	13.5
Lalande 21185	10.42	8.1	M2	7.5
Sirius A	1.41	8.6	A1	−1.5
B	11.54	8.6	white dwarf	7.2
Luyten 726-8A	15.27	8.9	M5	12.5
B (UV Cet)	15.8	8.9	M6	13.0
Ross 154	13.3	9.4	M5	10.6
Ross 248	14.8	10.3	M6	12.2
ε Eri	6.13	10.7	K2	3.7
Luyten 789-6	14.6	10.8	M7	12.2
Ross 128	13.5	10.8	M5	11.1
61 CYG A	7.58	11.2	K5	5.2
B	8.39	11.2	K7	6.0
ε Ind	7.0	11.2	K5	4.7
Procyon A	2.64	11.4	F5	0.3
B	13.1	11.4	white dwarf	10.8
Σ 2398 A	11.15	11.5	M4	8.9
B	11.94	11.5	M5	9.7
Groombridge 34 A	10.32	11.6	M1	8.1
B	13.29	11.6	M6	11.0
Lacaille 9352	9.59	11.7	M2	7.4
τ Ceti	5.72	11.9	G8	3.5
BD + 5° 1668	11.98	12.2	M5	9.8
L 725-32	15.27	12.4	M5	11.5
Lacaille 8760	8.75	12.5	M0	6.7
Kapteyn's Star	10.85	12.7	M0	8.8
Kruger 60 A	11.87	12.8	M3	9.7
B	13.3	12.8	M4	11.2

TABLE A-10
Properties of the Planets

PHYSICAL PROPERTIES (EARTH = $\oplus$)

Planet	Equatorial Radius (km)	Equatorial Radius ($\oplus$ = 1)	Mass ($\oplus$ = 1)	Average Density (g/cm^3)	Surface Gravity ($\oplus$ = 1)	Escape Velocity (km/s)	Sidereal Period of Rotation	Inclination of Equator to Orbit
Mercury	2439	0.382	0.0558	5.44	0.378	4.3	58.646^d	0°
Venus	6052	0.95	0.815	5.24	0.903	10.3	244.3^d	177°
Earth	6378	1.00	1.00	5.497	1.00	11.2	23^{h}56^{m}04.1^s	23°27′
Mars	3396	0.53	0.1075	3.94	0.379	5.0	24^{h}37^{m}22.6^s	25°19′
Jupiter	71,494	11.20	317.83	1.34	2.54	61	9^{h}55^{m}30^s	3°5′
Saturn	60,330	9.42	95.147	0.69	1.16	35.6	10^{h}13^{m}59^s	26°24′
Uranus	25,559	4.01	14.54	1.19	0.919	22	17^{h}14^m	97°55′
Neptune	24,750	3.93	17.23	1.66	1.19	25	16^{h}3^m	28°48′
Pluto	1185	0.18	0.0022	2.0	0.06	1.2	6^{d}9^{h}21^m	122°

ORBITAL PROPERTIES

Planet	Semimajor Axis (a) (AU)	Semimajor Axis (a) (10^6 km)	Orbital Period (P) (y)	Orbital Period (P) (days)	Average Orbital Velocity (km/s)	Orbital Eccentricity	Inclination to Ecliptic
Mercury	0.3871	57.9	0.24084	87.969	47.89	0.2056	7°0′16″
Venus	0.7233	108.2	0.61515	224.68	35.03	0.0068	3°23′40″
Earth	1	149.6	1	365.26	29.79	0.0167	0°
Mars	1.5237	227.9	1.8808	686.95	24.13	0.0934	1°51′09″
Jupiter	5.2028	778.3	11.867	4334.3	13.06	0.0484	1°18′29″
Saturn	9.5388	1427.0	29.461	10,760	9.64	0.0560	2°29′17″
Uranus	19.18	2869.0	84.013	30,685	6.81	0.0461	0°46′23″
Neptune	30.0611	4497.1	164.793	60,189	5.43	0.0100	1°46′27″
Pluto	39.44	5900	247.7	90,465	4.74	0.2484	17°9′3″

Planet	Satellite	Radius (km)	Distance from Planet (10^3 km)	Orbital Period (days)	Orbital Eccentricity	Orbital Inclination
Earth	Moon	1738	384.4	27.322	0.055	5°8'43"
Mars	Phobos	14 × 12 × 10	9.38	0.3189	0.018	1°.0
	Deimos	8 × 6 × 5	23.5	1.262	0.002	2°.8
Jupiter	Metis	20	126	0.29	0.0	0°.0
	Adrastea	12 × 8 × 10	128	0.294	0.0	0°.0
	Amalthea	135 × 100 × 78	182	0.4982	0.003	0°.45
	Thebe	50	223	0.674	0.0	1°.3
	Io	1820	422	1.769	0.000	0°.3
	Europa	1565	671	3.551	0.000	0°.46
	Ganymede	2640	1071	7.155	0.002	0°.18
	Callisto	2420	1884	16.689	0.008	0°.25
	Leda	~8	11,110	240	0.146	26°.7
	Himalia	~85	11,470	250.6	0.158	27°.6
	Lysithea	~20	11,710	260	0.12	29°
	Elara	~30	11,740	260.1	0.207	24°.8
	Ananke	15	21,200	631	0.169	147°
	Carme	22	22,350	692	0.207	163°
	Pasiphae	35	23,300	735	0.40	147°
	Sinope	20	23,700	758	0.275	156°
Saturn	Pan	10	133.570	0.574	0.000	0°
	Atlas	20 × 15 × 15	137.7	0.601	0.002	0°.3
	Prometheus	70 × 40 × 50	139.4	0.613	0.003	0°.0
	Pandora	55 × 35 × 50	141.7	0.629	0.004	0°.05
	Epimetheus	70 × 50 × 50	151.42	0.694	0.009	0°.34
	Janus	110 × 80 × 100	151.47	0.695	0.007	0°.14
	Mimas	196	185.54	0.942	0.020	1°.5
	Enceladus	250	238.04	1.370	0.004	0°.0
	Tethys	530	294.67	1.888	0.000	1°.1
	Calypso	17 × 11 × 12	294.67	1.888	0.0	~1°.?
	Telesto	12	294.67	1.888	0.0	~1°?
	Dione	560	377	2.737	0.002	0°.0
	Helene	20 × 15 × 15	377	2.74	0.005	0°.15
	Rhea	765	527	4.518	0.001	0°.4
	Titan	2575	1222	15.94	0.029	0°.3
	Hyperion	205 × 130 × 110	1484	21.28	0.104	~0°.5
	Iapetus	720	3562	79.33	0.028	14°.72
	Phoebe	110	12,930	550.4	0.163	150°
Uranus	Cordelia	20	49.8	0.3333	~0	~0°
	Ophelia	15	53.8	0.375	~0	~0°
	Bianca	25	59.1	0.433	~0	~0°
	Cressida	30	61.8	0.462	~0	~0°
	Desdemona	30	62.7	0.475	~0	~0°
	Juliet	40	64.4	0.492	~0	~0°
	Portia	55	66.1	0.512	~0	~0°

(continued)

TABLE A-11
(continued)

Planet	Satellite	Radius (km)	Distance from Planet (10^3 km)	Orbital Period (days)	Orbital Eccentricity	Orbital Inclination
Uranus	Rosalind	30	69.9	0.558	~0	~0°
(continued)	Belinda	30	75.2	0.621	~0	~0°
	Puck	85 ± 5	85.9	0.762	~0	~0°
	Miranda	242 ± 5	129.9	1.414	0.017	3°.4
	Ariel	580 ± 5	190.9	2.520	0.003	0°
	Umbriel	595 ± 10	266.0	4.144	0.003	0°
	Titania	805 ± 5	436.3	8.706	0.002	0°
	Oberon	775 ± 10	583.4	13.463	0.001	0°
	Caliban	40	7164	579	0.082	139.2°
	Stephano	~20	7900	676		
	Sycorax	80	12,174	1284	0.509	152.7°
	Prospero	~20	16,100	1950		
	Setebos	~20	17,600	2240	0.539	
Neptune	Naiad	30	48.2	0.296	~0	~0°
	Thalassa	40	50.0	0.312	~0	~0°
	Despina	90	52.5	0.333	~0	~0°
	Galatea	75	62.0	0.396	~0	~0°
	Larissa	95	73.6	0.554	~0	~0°
	Proteus	205	117.6	1.121	~0	~0°
	Triton	1352	354.59	5.875	0.00	160°
	Nereid	170	5588.6	360.125	0.76	27.7°
Pluto	Charon	626	19.7	6.38718	~0	122°

TABLE A-12
Meteor Showers

Shower	Dates	Hourly Rate	Radiant R.A.	Radiant Dec.	Associated Comet
Quadrantids	Jan. 2–4	30	15^h24^m	50°	
Lyrids	April 20–22	8	18^h4^m	33°	1861 I
η Aquarids	May 2–7	10	22^h24^m	0°	Halley?
δ Aquarids	July 26–31	15	22^h36^m	–10°	
Perseids	Aug. 10–14	40	3^h4^m	58°	1982 III
Orionids	Oct. 18–23	15	6^h20^m	15°	Halley?
Taurids	Nov. 1–7	8	3^h40^m	17°	Encke
Leonids	Nov. 14–19	6	10^h12^m	22°	1866 I Temp
Geminids	Dec. 10–13	50	7^h28^m	32°	

TABLE A-13
Greatest Elongations of Mercury

Evening Sky	Morning Sky
March 29, 2004*	May 14, 2004
July 27, 2004	Sept. 9, 2004
Nov. 21, 2004	Dec. 29, 2004
March 12, 2005*	April 26, 2005
July 9, 2005	Aug. 23, 2005
Nov. 3, 2005	Dec. 12, 2005
Feb. 24, 2006*	April 8, 2006
June 20, 2006	Aug. 7, 2006
Oct. 16, 2006	Nov. 25, 2006*
Feb. 7, 2007	March 22, 2007
June 2, 2007	July 20, 2007
Sept. 29, 2007	Nov. 8, 2007*
Jan. 22, 2008	March 3, 2008
May 14, 2008	July 1, 2008
Sept. 11, 2008	Oct. 22, 2008*

Elongation is the angular distance from the sun to a planet.
*Most favorable elongations.

TABLE A-14
Greatest Elongations of Venus

Evening Sky	Morning Sky
March 29, 2004	Aug. 17, 2004
Nov. 3, 2005	March 25, 2006
June 9, 2007	Oct. 28, 2007
Jan. 14, 2009	June 5, 2009
Aug. 20, 2010	Jan. 8, 2011
March 27, 2012	Aug. 15, 2012
Nov. 1, 2013	March 22, 2014
June 6, 2015	Oct. 26, 2015

TABLE A-15
The Greek Alphabet

A, α	alpha	H, η	eta	N, ν	nu	T, τ	tau
B, β	beta	Θ, θ	theta	Ξ, ξ	xi	Y, υ	upsilon
Γ, γ	gamma	I, ι	iota	O, o	omicron	Φ, ϕ	phi
Δ, δ	delta	K, κ	kappa	Π, π	pi	X, χ	chi
E, ε	epsilon	Λ, λ	lambda	P, ρ	rho	Ψ, ψ	psi
Z, ζ	zeta	M, μ	mu	Σ, σ	sigma	Ω, ω	omega

Group

Atomic number → 11
Symbol → Na
Atomic mass → 22.99

Atomic masses are based on carbon-12. Numbers in parentheses are mass numbers of most stable or best-known isotopes of radioactive elements.

Noble Gases (18)

Period	IA(1)	IIA(2)	IIIB(3)	IVB(4)	VB(5)	VIB(6)	VIIB(7)	VIII (8)	VIII (9)	VIII (10)	IB(11)	IIB(12)	IIIA(13)	IVA(14)	VA(15)	VIA(16)	VIIA(17)	He 4.003
1	1 H 1.008																	2 He 4.003
2	3 Li 6.941	4 Be 9.012											5 B 10.81	6 C 12.01	7 N 14.01	8 O 16.00	9 F 19.00	10 Ne 20.18
3	11 Na 22.99	12 Mg 24.31											13 Al 26.98	14 Si 28.09	15 P 30.97	16 S 32.06	17 Cl 35.45	18 Ar 39.95
4	19 K 39.10	20 Ca 40.08	21 Sc 44.96	22 Ti 47.90	23 V 50.94	24 Cr 52.00	25 Mn 54.94	26 Fe 55.85	27 Co 58.93	28 Ni 58.7	29 Cu 63.55	30 Zn 65.38	31 Ga 69.72	32 Ge 72.59	33 As 74.92	34 Se 78.96	35 Br 79.90	36 Kr 83.80
5	37 Rb 85.47	38 Sr 87.62	39 Y 88.91	40 Zr 91.22	41 Nb 92.91	42 Mo 95.94	43 Tc 98.91	44 Ru 101.1	45 Rh 102.9	46 Pd 106.4	47 Ag 107.9	48 Cd 112.4	49 In 114.8	50 Sn 118.7	51 Sb 121.8	52 Te 127.6	53 I 126.9	54 Xe 131.3
6	55 Cs 132.9	56 Ba 137.3	57* La 138.9	72 Hf 178.5	73 Ta 180.9	74 W 183.9	75 Re 186.2	76 Os 190.2	77 Ir 192.2	78 Pt 195.1	79 Au 197.0	80 Hg 200.6	81 Tl 204.4	82 Pb 207.2	83 Bi 209.0	84 Po (210)	85 At (210)	86 Rn (222)
7	87 Fr (223)	88 Ra 226.0	89** Ac (227)	104 Rf (261)	105 Db (262)	106 Sg (263)	107 Bh (262)	108 Hs (265)	109 Mt (266)	110 Uun (269)	111 Uuu (272)	112 Uub (277)	114 Uuq (285)		116 Uuh (289)			

Transition Elements

Inner Transition Elements

* Lanthanide Series 6	58 Ce 140.1	59 Pr 140.9	60 Nd 144.2	61 Pm (145)	62 Sm 150.4	63 Eu 152.0	64 Gd 157.3	65 Tb 158.9	66 Dy 162.5	67 Ho 164.9	68 Er 167.3	69 Tm 168.9	70 Yb 173.0	71 Lu 175.0
** Actinide Series 7	90 Th 232.0	91 Pa 231.0	92 U 238.0	93 Np 237.0	94 Pu (244)	95 Am (243)	96 Cm (247)	97 Bk (247)	98 Cf (251)	99 Es (252)	100 Fm (257)	101 Md (258)	102 No (259)	103 Lr (260)

APPENDIX B

OBSERVING THE SKY

Observing the sky with the naked eye is of no more importance to modern astronomy than picking up pretty pebbles is to modern geology. But the sky is a natural wonder unimaginably bigger than the Grand Canyon, the Rocky Mountains, or any other natural wonder that tourists visit every year. To neglect the beauty of the sky is equivalent to geologists neglecting the beauty of the minerals they study. This supplement is meant to act as a tourist's guide to the sky. We analyzed the universe in the regular chapters, but here we will admire it.

The brighter stars in the sky are visible even from the centers of cities with their air and light pollution. But in the countryside only a few miles beyond the cities, the night sky is a velvety blackness strewn with thousands of glittering stars. From a wilderness location, far from the city's glare, and especially from high mountains, the night sky is spectacular.

USING STAR CHARTS

The constellations are a fascinating cultural heritage of our planet, but they are sometimes a bit difficult to learn because of Earth's motion. The constellations above the horizon change with the time of night and the seasons.

Because Earth rotates eastward, the sky appears to rotate around us westward. A constellation visible in the southern sky soon after sunset will appear to move westward, and in a few hours it will disappear below the horizon. Other consellations will rise in the east, so the sky changes gradually through the night.

In addition, Earth's orbital motion makes the sun appear to move eastward among the stars. Each day the sun moves about twice its own diameter eastward along the ecliptic, and each night at sunset, the constellations are about 1° farther toward the west.

Orion, for instance, is visible in the evening sky in January, but as the days pass, the sun moves closer to Orion. By March, Orion is difficult to see in the western sky soon after sunset. By June, the sun is so close to Orion it sets with the sun and is invisible. Not until late July is the sun far enough past Orion for the con-

stellation to become visible rising in the eastern sky just before dawn.

Because of the rotation and orbital motion of Earth, we must use more than one star chart to map the sky. Which chart we select depends on the month and the time of night. The charts given in this appendix show the evening sky for each month.

Two sets of charts are included for two typical locations on Earth. The northern hemisphere charts show the sky as seen from a northern latitude typical of the United States and central Europe. The southern hemisphere star charts are appropriate for readers in Earth's southern hemisphere, including Australia, southern South America, and southern Africa.

To use the charts, select the appropriate chart and hold it overhead as shown in ▌Figure B-1. If you face south, turn the chart until the words *southern horizon* are at the bottom of the chart. If you face other directions, turn the chart appropriately.

FIGURE B-1

To use the star charts in this book, select the appropriate chart for the date and time. Hold it overhead, and turn it until the direction at the bottom of the chart is the same as the direction you are facing.

Northern Hemisphere Sky

JANUARY

Early in Month	9 P.M.
Midmonth	8 P.M.
End of Month	7 P.M.

Months along the ecliptic show the location of the sun during the year.

Numbers along the celestial equator show right ascension.

Northern Hemisphere Sky

FEBRUARY

Early in Month	9 P.M.
Midmonth	8 P.M.
End of Month	7 P.M.

Months along the ecliptic show the location of the sun during the year.

Numbers along the celestial equator show right ascension.

Northern Hemisphere Sky

MARCH

Early in Month	9 P.M.
Midmonth	8 P.M.
End of Month	7 P.M.

Months along the ecliptic show the location of the sun during the year.

Numbers along the celestial equator show right ascension.

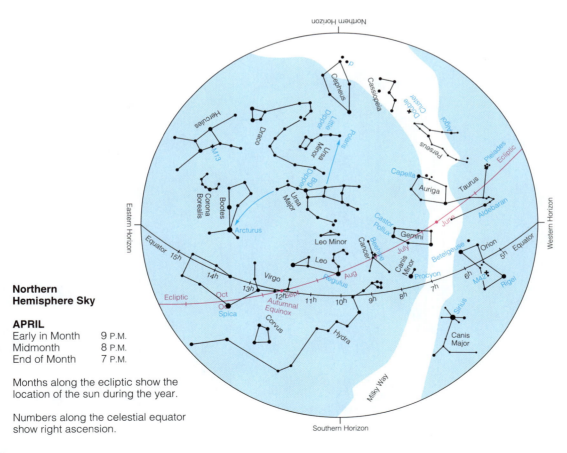

Northern Hemisphere Sky

APRIL

Early in Month	9 P.M.
Midmonth	8 P.M.
End of Month	7 P.M.

Months along the ecliptic show the location of the sun during the year.

Numbers along the celestial equator show right ascension.

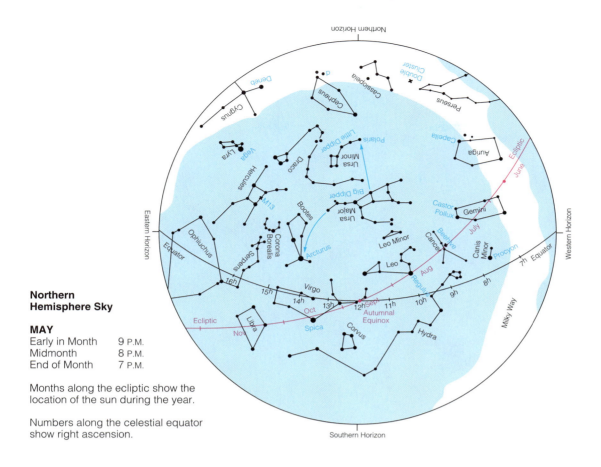

Northern Hemisphere Sky

MAY

Early in Month	9 P.M.
Midmonth	8 P.M.
End of Month	7 P.M.

Months along the ecliptic show the location of the sun during the year.

Numbers along the celestial equator show right ascension.

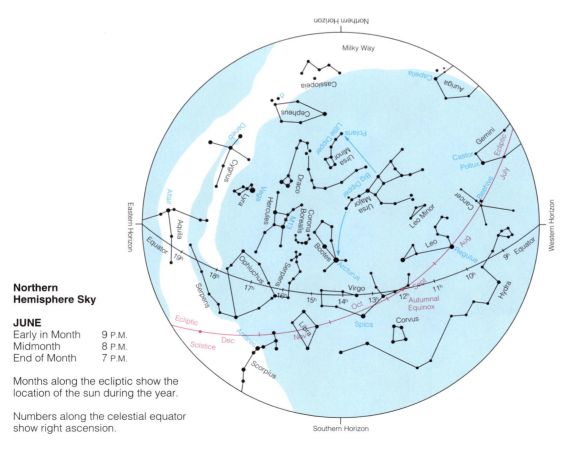

Northern Hemisphere Sky

JUNE

Early in Month	9 P.M.
Midmonth	8 P.M.
End of Month	7 P.M.

Months along the ecliptic show the location of the sun during the year.

Numbers along the celestial equator show right ascension.

Northern Hemisphere Sky

JULY

Early in Month	9 P.M.
Midmonth	8 P.M.
End of Month	7 P.M.

Months along the ecliptic show the location of the sun during the year.

Numbers along the celestial equator show right ascension.

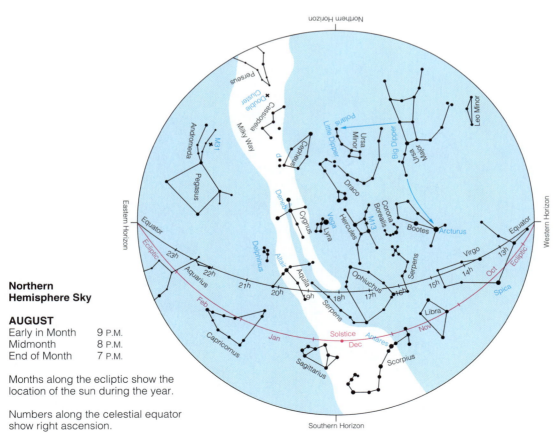

Northern Hemisphere Sky

AUGUST

Early in Month	9 P.M.
Midmonth	8 P.M.
End of Month	7 P.M.

Months along the ecliptic show the location of the sun during the year.

Numbers along the celestial equator show right ascension.

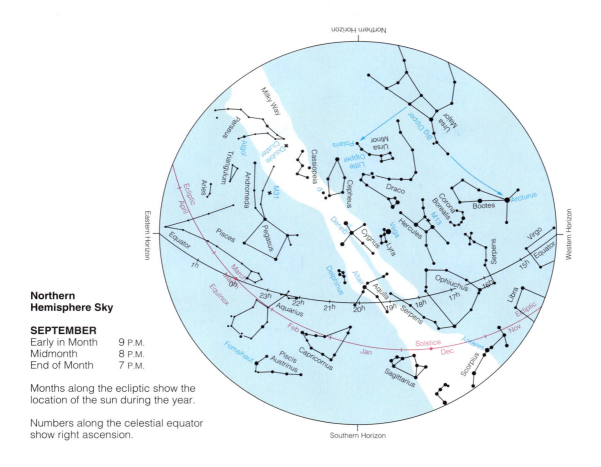

Northern Hemisphere Sky

SEPTEMBER
Early in Month   9 P.M.
Midmonth   8 P.M.
End of Month   7 P.M.

Months along the ecliptic show the location of the sun during the year.

Numbers along the celestial equator show right ascension.

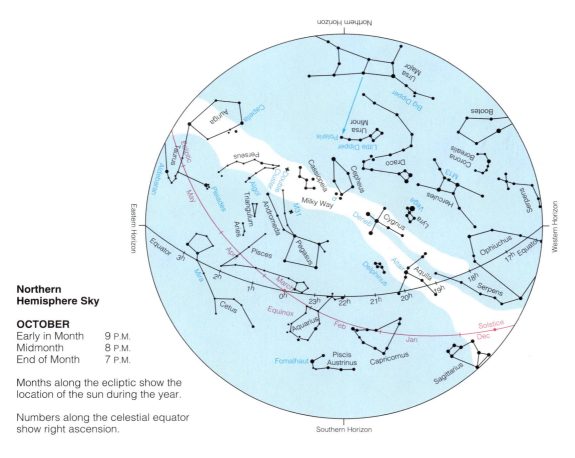

Northern Hemisphere Sky

OCTOBER
Early in Month   9 P.M.
Midmonth   8 P.M.
End of Month   7 P.M.

Months along the ecliptic show the location of the sun during the year.

Numbers along the celestial equator show right ascension.

Northern Hemisphere Sky

NOVEMBER

Early in Month	9 P.M.
Midmonth	8 P.M.
End of Month	7 P.M.

Months along the ecliptic show the location of the sun during the year.

Numbers along the celestial equator show right ascension.

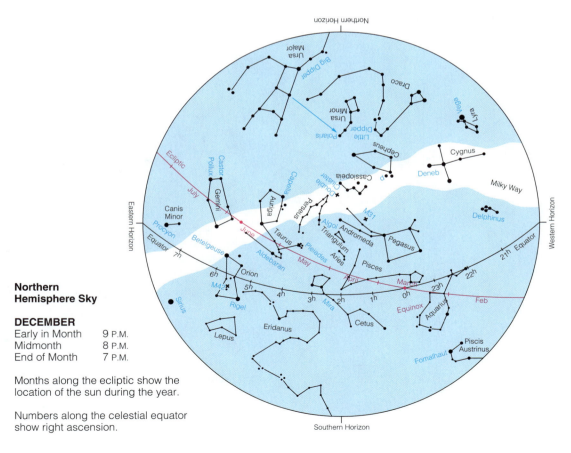

Northern Hemisphere Sky

DECEMBER

Early in Month	9 P.M.
Midmonth	8 P.M.
End of Month	7 P.M.

Months along the ecliptic show the location of the sun during the year.

Numbers along the celestial equator show right ascension.

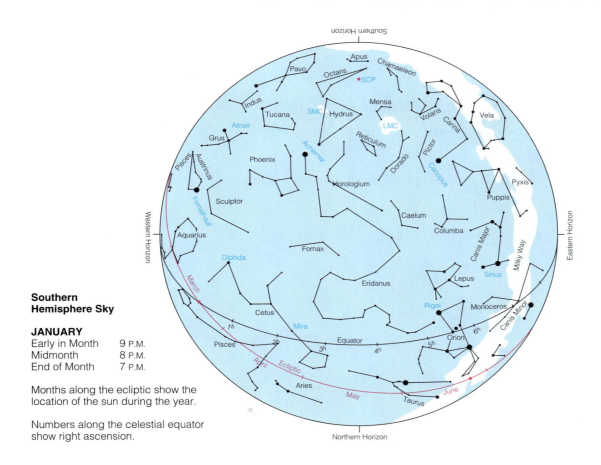

Southern Hemisphere Sky

JANUARY

Early in Month	9 P.M.
Midmonth	8 P.M.
End of Month	7 P.M.

Months along the ecliptic show the location of the sun during the year.

Numbers along the celestial equator show right ascension.

Southern Hemisphere Sky

FEBRUARY

Early in Month	9 P.M.
Midmonth	8 P.M.
End of Month	7 P.M.

Months along the ecliptic show the location of the sun during the year.

Numbers along the celestial equator show right ascension.

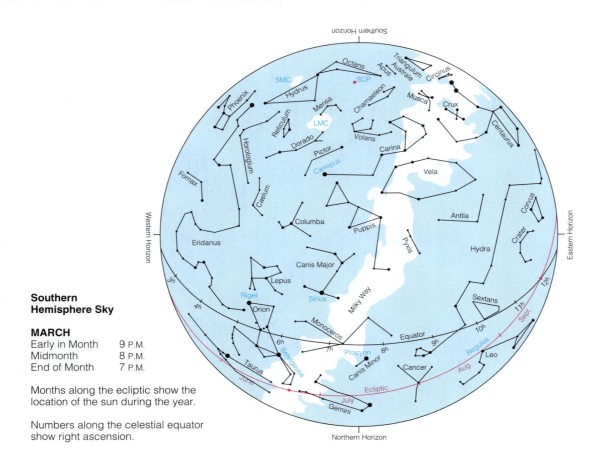

Southern Hemisphere Sky

MARCH

Early in Month	9 P.M.
Midmonth	8 P.M.
End of Month	7 P.M.

Months along the ecliptic show the location of the sun during the year.

Numbers along the celestial equator show right ascension.

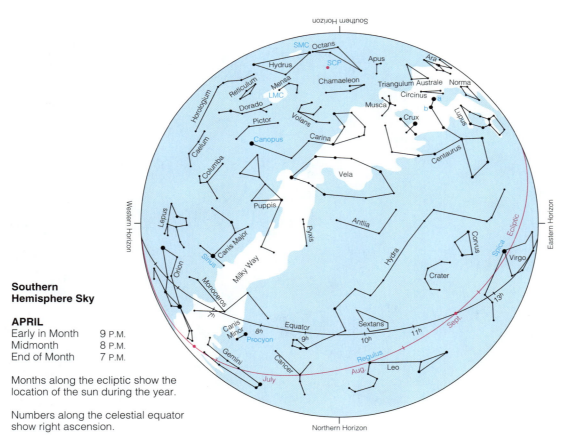

Southern Hemisphere Sky

APRIL

Early in Month	9 P.M.
Midmonth	8 P.M.
End of Month	7 P.M.

Months along the ecliptic show the location of the sun during the year.

Numbers along the celestial equator show right ascension.

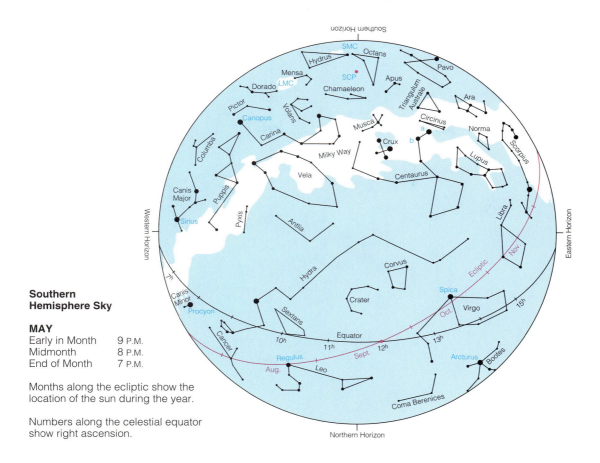

Southern Hemisphere Sky

MAY

Early in Month	9 P.M.
Midmonth	8 P.M.
End of Month	7 P.M.

Months along the ecliptic show the location of the sun during the year.

Numbers along the celestial equator show right ascension.

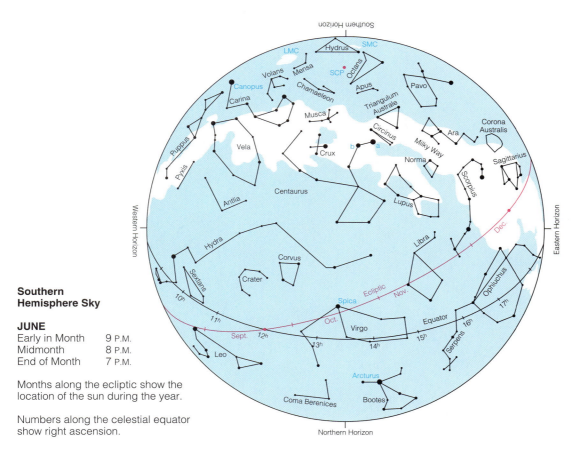

Southern Hemisphere Sky

JUNE

Early in Month	9 P.M.
Midmonth	8 P.M.
End of Month	7 P.M.

Months along the ecliptic show the location of the sun during the year.

Numbers along the celestial equator show right ascension.

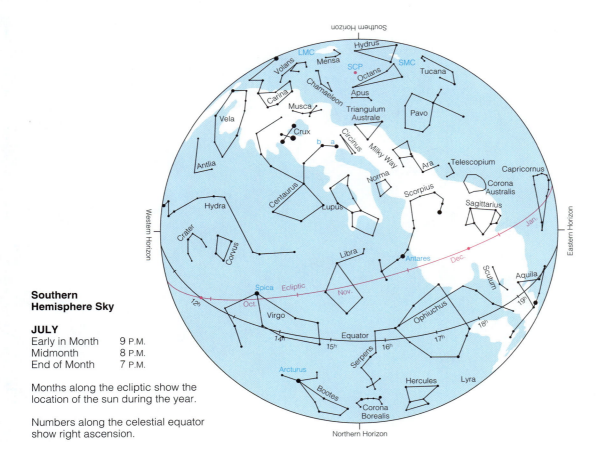

Southern Hemisphere Sky

JULY

Early in Month	9 P.M.
Midmonth	8 P.M.
End of Month	7 P.M.

Months along the ecliptic show the location of the sun during the year.

Numbers along the celestial equator show right ascension.

Southern Hemisphere Sky

AUGUST

Early in Month	9 P.M.
Midmonth	8 P.M.
End of Month	7 P.M.

Months along the ecliptic show the location of the sun during the year.

Numbers along the celestial equator show right ascension.

Southern Hemisphere Sky

SEPTEMBER

Early in Month	9 P.M.
Midmonth	8 P.M.
End of Month	7 P.M.

Months along the ecliptic show the location of the sun during the year.

Numbers along the celestial equator show right ascension.

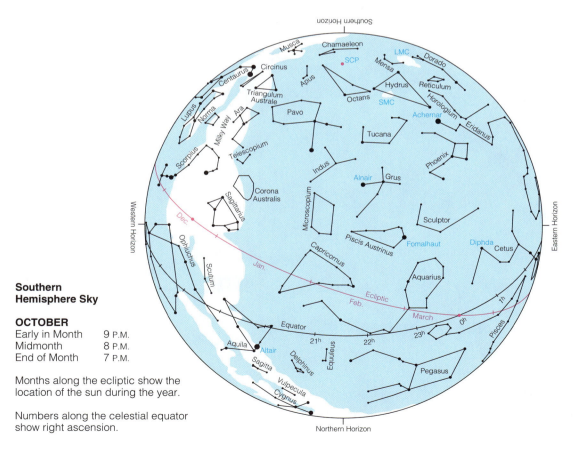

Southern Hemisphere Sky

OCTOBER

Early in Month	9 P.M.
Midmonth	8 P.M.
End of Month	7 P.M.

Months along the ecliptic show the location of the sun during the year.

Numbers along the celestial equator show right ascension.

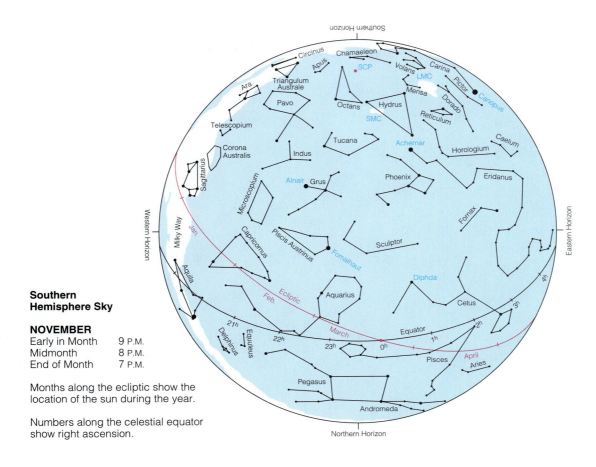

Southern Hemisphere Sky

NOVEMBER

Early in Month	9 P.M.
Midmonth	8 P.M.
End of Month	7 P.M.

Months along the ecliptic show the location of the sun during the year.

Numbers along the celestial equator show right ascension.

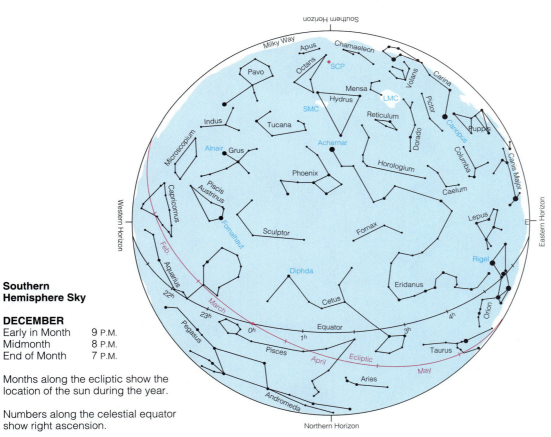

Southern Hemisphere Sky

DECEMBER

Early in Month	9 P.M.
Midmonth	8 P.M.
End of Month	7 P.M.

Months along the ecliptic show the location of the sun during the year.

Numbers along the celestial equator show right ascension.

GLOSSARY

Numbers refer to the page where the term is first discussed in the text.

absolute age (454)   An age determined in years, as from radioactive dating (cf. **relative age**).

absolute bolometric magnitude (176)   The absolute magnitude we would observe if we could detect all wavelengths.

absolute visual magnitude (M_v) (175)   Intrinsic brightness of a star; the apparent visual magnitude the star would have if it were 10 pc away.

absolute zero (124)   The lowest possible temperature; the temperature at which the particles in a material, atoms or molecules, contain no energy of motion that can be extracted from the body.

absorption line (132)   A dark line in a spectrum; produced by the absence of photons absorbed by atoms or molecules.

absorption spectrum (dark-line spectrum) (132)   A spectrum that contains absorption lines.

acceleration (81)   A change in a velocity; a change in either speed or direction. (See **velocity**.)

acceleration of gravity (78)   A measure of the strength of gravity at a planet's surface.

accretion (424)   The sticking together of solid particles to produce a larger particle.

accretion disk (264)   The whirling disk of gas that forms around a compact object such as a white dwarf, neutron star, or black hole as matter is drawn in.

achondrite (567)   Stony meteorite containing no chondrules or volatiles.

achromatic lens (101)   A telescope lens composed of two lenses ground from different kinds of glass and de-signed to bring two selected colors to the same focus and correct for chromatic aberration.

active galactic nucleus (AGN) (358)   The central energy source of an active galaxy.

active galaxy (358)   A galaxy that is a source of excess radiation, usually radio waves, X rays, gamma rays, or some combination.

active optics (109)   Optical elements whose position or shape is continuously controlled by computers.

active region (153)   An area on the sun where sunspots, prominences, flares, and the like, occur.

adaptive optics (109)   Computer-controlled telescope mirrors that can at least partially compensate for seeing.

albedo (442)   The fraction of the light that hits an object that is reflected.

alt-azimuth mounting (109)   A telescope mounting capable of motion parallel to and perpendicular to the horizon.

amino acids (592)   Carbon-chain molecules that are the building blocks of protein.

Angstrom (Å) (99)   A unit of distance; $1 \text{ Å} = 10^{-10}$ m; often used to measure the wavelength of light.

angular diameter (17)   A measure of the size of an object in the sky; numerically equal to the angle in degrees between two lines extending from the observer's eye to opposite edges of the object.

angular distance (17)   A measure of the separation between two objects in the sky; numerically equal to the angle in degrees between two lines extending from the observer's eye to the two objects.

angular momentum (86)   The tendency of a rotating body to continue rotating; mathematically, the product of mass, velocity, and radius.

angular momentum problem (409)   An objection to Laplace's nebular hypothesis that cited the slow rotation of the sun.

annular eclipse (43)   A solar eclipse in which the solar photosphere appears around the edge of the moon in a bright ring, or annulus. The corona, chromosphere, and prominences cannot be seen.

anorthosite (457)   Rock of aluminum and calcium silicates found in the lunar highlands.

antimatter (387)   Matter composed of antiparticles, which upon colliding with a matching particle of normal matter annihilate and convert the mass of both particles into energy. The antiproton is the antiparticle of the proton, and the positron is the antiparticle of the electron.

aphelion (23)   The orbital point of greatest distance from the sun.

apogee (43)   The orbital point of greatest distance from Earth.

Apollo–Amor object (572)   Asteroid whose orbit crosses that of Earth (Apollo) and Mars (Amor).

apparent visual magnitude (m_v) (14)   The brightness of a star as seen by human eyes on Earth.

archaeoastronomy (52)   The study of the astronomy of ancient cultures.

association (219)   Group of widely scattered stars (10 to 1000) moving together through space; not gravitationally bound into a cluster.

asterism (12)   A named group of stars not identified as a constellation, e.g., the Big Dipper.

asteroids (416)   Small rocky worlds, most of which lie between Mars and Jupiter in the asteroid belt.

astronomical unit (AU) (4)   Average distance from Earth to the sun; 1.5×10^8 km, or 93×10^6 miles.

atmospheric window (101) Wavelength regions in which our atmosphere is transparent—at visual wavelengths, infrared, and at radio wavelengths.

aurora (163) The glowing light display that results when a planet's magnetic field guides charged particles toward the north and south magnetic poles, where they strike the upper atmosphere and excite atoms to emit photons.

autumnal equinox (22) The point on the celestial sphere where the sun crosses the celestial equator going southward. Also, the time when the sun reaches this point and autumn begins in the northern hemisphere— about September 22.

Babcock model (158) A model of the sun's magnetic cycle in which the differential rotation of the sun winds up and tangles the solar magnetic field in a 22-year cycle. This is thought to be responsible for the 11-year sunspot cycle.

Balmer series (133) Spectral lines in the visible and near-ultraviolet spectrum of hydrogen produced by transitions whose lowest orbit is the second.

barred spiral galaxy (338) A spiral galaxy with an elongated nucleus resembling a bar from which the arms originate.

basalt (440) Dark, igneous rock characteristic of solidified lava.

belts (509) Dark bands of clouds that circle Jupiter parallel to its equator; generally red, brown, or blue-green; believed to be regions of descending gas.

big bang (383) The theory that the universe began with a violent explosion from which the expanding universe of galaxies eventually formed.

binary stars (185) Pairs of stars that orbit around their common center of mass.

binding energy (129) The energy needed to pull an electron away from its atom.

bipolar flow (221) Oppositely directed jets of gas ejected by some protostellar objects.

birth line (218) In the H–R diagram, the line above the main sequence where protostars first become visible.

black body radiation (125) Radiation emitted by a hypothetical perfect radiator; the spectrum is continuous, and the wavelength of maximum emission depends only on the body's temperature.

black dwarf (259) The end state of a white dwarf that has cooled to low temperature.

black hole (292) A mass that has collapsed to such a small volume that its gravity prevents the escape of all radiation; also, the volume of space from which radiation may not escape.

blazar (365) See **BL Lac objects.**

BL Lac objects (365) Objects that resemble quasars; thought to be highly luminous cores of distant galaxies.

blue shift (139) The shortening of the wavelengths of light observed when the source and observer are approaching each other.

Bok globule (220) Small, dark cloud only about 1 ly in diameter that contains 10 to 1000 $M_\odot$ of gas and dust; believed related to star formation.

bow shock (437) The boundary between the undisturbed solar wind and the region being deflected around a planet or comet.

breccia (457) A rock composed of fragments of earlier rocks bonded together.

bright-line spectrum (132) See **emission spectrum.**

brown dwarf (238) A very cool, low-luminosity star whose mass is not sufficient to ignite nuclear fusion.

butterfly diagram See **Maunder butterfly diagram.**

CAI (567) Calcium–aluminum-rich inclusions found in some meteorites; believed to be very old.

Cambrian period (595) A geological period 0.6 to 0.5 billion years ago during which life on Earth became diverse and complex. Cambrian rocks contain the oldest easily identifiable fossils.

capture hypothesis (461) The theory that the moon formed elsewhere in the solar system and was later captured by Earth.

carbonaceous chondrite (566) Stony meteorite that contains both chondrules and volatiles. These may be the least altered remains of the solar nebula still present in the solar system.

carbon deflagration (270) The process in which the carbon in a white dwarf is completely consumed by nuclear fusion, producing a type Ia supernova explosion.

carbon–nitrogen–oxygen (CNO) cycle (223) A series of nuclear reactions that use carbon as a catalyst to combine four hydrogen atoms to make one helium atom plus energy; effective in stars more massive than the sun.

Cassegrain focus (108) The optical design of a reflecting telescope in which the secondary mirror reflects light back down the tube through a hole in the center of the objective mirror.

catastrophic hypothesis (408) Explanation for natural processes that depends on dramatic and unlikely events, such as the collision of two stars to produce our solar system.

celestial equator (16) The imaginary line around the sky directly above Earth's equator.

celestial sphere (16) An imaginary sphere of very large radius surrounding Earth and to which the planets, stars, sun, and moon seem to be attached.

center of mass (85) The balance point of a body or system of bodies.

Cepheid variable star (247) Variable star with a period of 1 to 60 days; the period of variation is related to luminosity.

Chandrasekhar limit (259) The maximum mass of a white dwarf, about 1.4 $M_\odot$; a white dwarf of greater mass cannot support itself and will collapse.

charge-coupled device (CCD) (111) An electronic device consisting of a large array of light-sensitive elements used to record very faint images.

chemical evolution (596) The chemical process that led to the growth of complex molecules on the primitive Earth. This did not involve the reproduction of molecules.

chondrite (566) A stony meteorite that contains chondrules.

chondrule (566) Round, glassy body in some stony meteorites; believed to have solidified very quickly from molten drops of silicate material.

chromatic aberration (101) A distortion found in refracting telescopes because lenses focus different colors at slightly different distances. Images are consequently surrounded by color fringes.

chromosome (593)   One of the bodies in a cell that contains the DNA carrying genetic information.

chromosphere (43)   Bright gases just above the photosphere of the sun.

circular velocity (84)   The velocity required to remain in a circular orbit about a body.

circumpolar constellation (17)   Any of the constellations so close to the celestial pole that they never set (or never rise) as seen from a given latitude.

closed orbit (85)   An orbit that returns to its starting point; a circular or elliptical orbit. (See **open orbit.**)

closed universe (393)   A model universe in which the average density is great enough to stop the expansion and make the universe contract.

cluster method (344)   The method of determining the masses of galaxies based on the motions of galaxies in a cluster.

CNO cycle (223)   See **carbon–nitrogen–oxygen cycle.**

cocoon (217)   The cloud of gas and dust around a contracting protostar that conceals it at visible wavelengths.

cold dark matter (395)   Invisible matter in the universe composed by heavy, slow-moving particles such as WIMPs.

collapsar (302)   See **hypernova.**

collisional broadening (141)   The smearing out of a spectral line because of collisions among the atoms of the gas.

color index (127)   A numerical measure of the color of a star.

coma (578)   The glowing head of a comet.

comet (417)   One of the small, icy bodies that orbit the sun and produce tails of gas and dust when they near the sun.

compact object (259)   A star that has collapsed to form a white dwarf, neutron star, or black hole.

comparative planetology (433)   The study of planets by comparing the characteristics of different examples.

comparison spectrum (112)   A spectrum of known spectral lines used to identify unknown wavelengths in an object's spectrum.

composite volcano (480)   A volcano built up of layers of lava flows and ash falls. Steep sided and typically associated with subduction zones.

condensation (424)   The growth of a particle by addition of material from surrounding gas one atom or molecule at a time.

condensation hypothesis (461)   The hypothesis that the moon and Earth formed by condensing together from the solar nebula.

condensation sequence (423)   The sequence in which different materials condense from the solar nebula as we move outward from the sun.

conservation of energy law (234)   One of the basic laws of stellar structure. The amount of energy flowing out of the top of a shell must equal the amount coming in at the bottom plus whatever energy is generated within the shell.

conservation of mass law (234)   One of the basic laws of stellar structure. The total mass of the star must equal the sum of the masses of the shells, and the mass must be distributed smoothly throughout the star.

constellation (11)   One of the stellar patterns identified by name, usually of mythological gods, people, animals, or objects; also, the region of the sky containing that star pattern.

continuous spectrum (132)   A spectrum in which there are no absorption or emission lines.

convection (148)   Circulation in a fluid driven by heat; hot material rises, and cool material sinks.

corona (42, 479)   The faint outer atmosphere of the sun; composed of low-density, very hot, ionized gas. On Venus, round networks of fractures and ridges up to 1000 km in diameter.

coronagraph (150)   A telescope designed to photograph the inner corona of the sun.

coronal gas (207)   Extremely high-temperature, low-density gas in the interstellar medium.

coronal hole (163)   An area of the solar surface that is dark at X-ray wavelengths; thought to be associated with divergent magnetic fields and the source of the solar wind.

coronal mass ejection (CME) (163)   Gas trapped in the sun's magnetic field and blown outward through the sun's atmosphere.

cosmic microwave background radiation (385)   Radiation from the hot clouds of the big bang explosion. Because of its large red shift, it appears to come from a body whose temperature is only 2.7 K.

cosmic ray (118)   A subatomic particle traveling at tremendous velocity that strikes Earth's atmosphere from space.

cosmological constant (Λ) (398)   Einstein's constant that represents a repulsion in space to oppose gravity.

cosmological principle (390)   The assumption that any observer in any galaxy sees the same general features of the universe.

cosmology (380)   The study of the nature, origin, and evolution of the universe.

Coulomb barrier (164)   The electrostatic force of repulsion between bodies of like charge; commonly applied to atomic nuclei.

Coulomb force (129)   The repulsive force between particles with like electrostatic charge.

critical density (390)   The average density of the universe needed to make its curvature flat.

critical point (508)   The temperature and pressure at which the vapor and liquid phases of a material have the same density.

current sheet (506)   The thin equatorial layer of charged particles in a planet's magnetosphere.

dark age (388)   The period of a few hundred million years during which the universe expanded in darkness. Extends from soon after the big bang glow faded into the infrared to the formation of the first stars.

dark energy (398)   The energy of empty space that drives the acceleration of the expanding universe.

dark-line spectrum   See **absorption spectrum.**

dark matter (314)   Nonluminous material that is detected only by its gravitational influence.

dark nebula (199)   A nonluminous cloud of gas and dust visible because it blocks light from more distant stars and nebulae.

decameter radiation (506)   Radio signals from Jupiter with wavelengths of about 10 m.

decimeter radiation (506)   Radio signals from Jupiter with wavelengths of about 0.1 m.

deferent (59)   In the Ptolemaic theory, the large circle around Earth along which the center of the epicycle moved.

degenerate matter (242)   Extremely high-density matter in which pressure no longer depends on temperature, due to quantum mechanical effects.

density (142)   The amount of matter per unit volume in a material; measured in grams per cubic centimeter, for example.

density wave theory (324)   Theory proposed to account for spiral arms as compressions of the interstellar medium in the disk of the galaxy.

deuterium (165)   An isotope of hydrogen in which the nucleus contains a proton and a neutron.

diamond ring effect (43)   A momentary phenomenon seen during some total solar eclipses when the ring of the corona and a bright spot of photosphere resemble a large diamond set in a silvery ring.

differential rotation (156)   The rotation of a body in which different parts of the body have different periods of rotation; this is true of the sun, the Jovian planets, and the disk of the galaxy.

differentiation (425)   The separation of planetary material according to density.

diffraction fringe (104)   Blurred fringe surrounding any image caused by the wave properties of light. Because of this, no image detail smaller than the fringe can be seen.

disk component (311)   All material confined to the plane of the galaxy.

distance indicator (337)   Object whose luminosity or diameter is known; used to find the distance to a star cluster or galaxy.

distance modulus ($m_v - M_v$) (175)   The difference between the apparent and absolute magnitude of a star; a measure of how far away the star is.

distance scale (341)   The combined calibration of distance indicators used by astronomers to find the distances to remote galaxies.

DNA (deoxyribonucleic acid) (592)   The long carbon-chain molecule that records information to govern the biological activity of the organism.

DNA carries the genetic data passed to offspring.

Doppler broadening (141)   The smearing of spectral lines because of the motion of the atoms in the gas.

Doppler effect (139)   The change in the wavelength of radiation due to relative radial motion of source and observer.

double-exhaust model (360)   The theory that double radio lobes are produced by pairs of jets emitted in opposite directions from the centers of active galaxies.

double-lobed radio source (360)   A galaxy that emits radio energy from two regions (lobes) located on opposite sides of the galaxy.

Drake equation (606)   A formula for the number of communicative civilizations in our galaxy.

dust tail (type II) (578)   The tail of a comet formed of dust blown outward by the pressure of sunlight. (See **gas tail**.)

dynamo effect (157)   The process by which a rotating, convecting body of conducting matter, such as Earth's core, can generate a magnetic field.

east point (16)   The point on the eastern horizon exactly halfway between the north point and the south point; exactly east.

eccentric (56)   In astronomy, an off-center circular path.

eccentricity (e) (70)   A measure of the flattening of an ellipse. An ellipse of $e = 0$ is circular. The closer to 1 e becomes, the more flattened the ellipse.

eclipse season (45)   That period when the sun is near a node of the moon's orbit and eclipses are possible.

eclipse year (46)   The time the sun takes to circle the sky and return to a node of the moon's orbit; 346.62 days.

eclipsing binary system (185)   A binary star system in which the stars eclipse each other.

ecliptic (20)   The apparent path of the sun around the sky.

ejecta (452)   Pulverized rock scattered by meteorite impacts on a planetary surface.

electromagnetic radiation (98)   Changing electric and magnetic fields that travel through space and transfer energy from one place to another—for example, light, radio waves, etc.

electron (125)   Low-mass atomic particle carrying a negative charge.

ellipse (70)   A closed curve enclosing two points (foci) such that the total distance from one focus to any point on the curve back to the other focus equals a constant.

elliptical galaxy (338)   A galaxy that is round or elliptical in outline; it contains little gas and dust, no disk or spiral arms, and few hot, bright stars.

emission line (132)   A bright line in a spectrum caused by the emission of photons from atoms.

emission nebula (198)   A cloud of glowing gas excited by ultraviolet radiation from hot stars.

emission spectrum (bright-line spectrum) (132)   A spectrum containing emission lines.

energy (87)   The capacity of a natural system to perform work—for example, thermal energy.

energy level (130)   One of a number of states an electron may occupy in an atom, depending on its binding energy.

enzyme (592)   Special protein that controls processes in an organism.

epicycle (59)   The small circle followed by a planet in the Ptolemaic theory. The center of the epicycle follows a larger circle (deferent) around Earth.

equant (59)   The point off-center in the deferent from which the center of the epicycle appears to move uniformly.

equatorial mounting (109)   A telescope mounting that allows motion parallel to and perpendicular to the celestial equator.

ergosphere (293)   The region surrounding a rotating black hole within which one could not resist being dragged around the black hole. It is possible for a particle to escape from the ergosphere and extract energy from the black hole.

escape velocity (85)   The initial velocity an object needs to escape from the surface of a celestial body.

evening star (21)   Any planet visible in the sky just after sunset.

event horizon (292)   The boundary of the region of a black hole from which no radiation may escape. No event that occurs within the event horizon is visible to a distant observer.

evolutionary hypothesis (408) Explanation for natural events that involves gradual changes as opposed to sudden catastrophic changes—for example, the formation of the planets in the gas cloud around the forming sun.

excited atom (130) An atom in which an electron has moved from a lower to a higher orbit.

extrasolar planet (410) A planet orbiting a star other than the sun.

eyepiece (101) A short-focal-length lens used to enlarge the image in a telescope; the lens nearest the eye.

fall (565) A meteorite seen to fall. (See **find.**)

false-color image (111) A representation of graphical data in which the colors are altered or added to reveal details.

field (83) A way of explaining action at a distance; a particle produces a field of influence (gravitational, electric, or magnetic) to which another particle in the field responds.

filament (149, 162) On the sun, a prominence seen silhouetted against the solar surface.

filtergram (149) An image (usually of the sun) taken in the light of a specific region of the spectrum—e.g., an H-alpha filtergram.

find (565) A meteorite that is found but was not seen to fall. (See **fall.**)

fission hypothesis (460) The theory that the moon formed by breaking away from Earth.

flare (163) A violent eruption on the sun's surface.

flatness problem (396) In cosmology, the circumstance that the early universe must have contained almost exactly the right amount of matter to close space-time (to make space-time flat).

flat universe (393) A model of the universe in which space-time is not curved.

flocculent (325) Woolly, fluffy; used to refer to certain galaxies that have a woolly appearance.

flux (175) A measure of the flow of energy onto or through a surface. Usually applied to light.

focal length (101) The distance from a lens to the point where it focuses parallel rays of light.

folded mountain range (440) A long range of mountains formed by the compression of a planet's crust—for example, the Andes on Earth.

forbidden line (197) A spectral line that does not occur in the laboratory because it depends on an atomic transition that is highly unlikely.

forward scattering (512) The optical property of finely divided particles to preferentially direct light in the original direction of the light's travel.

free-fall contraction (217) The early contraction of a gas cloud to form a star during which internal pressure is too low to resist contraction.

frequency (99) The number of times a given event occurs in a given time; for a wave, the number of cycles that pass the observer in 1 second.

galactic cannibalism (350) The theory that large galaxies absorb smaller galaxies.

galactic corona (314) The low-density extension of the halo of a galaxy; now suspected to extend many times the visible diameter of the galaxy.

galactic fountain (317) A region of the galaxy's disk in which gas heated by supernova explosions throws gas out of the disk where it can fall back and spread metals through the disk.

galaxy (6) A very large collection of gas, dust, and stars orbiting a common center of mass. We live in the Milky Way Galaxy.

Galilean satellites (419) The four largest satellites of Jupiter, named after their discoverer, Galileo.

gamma-ray burster (300) An object that produces a sudden burst of gamma rays; thought to be associated with neutron stars and black holes.

gas tail (type I) (578) The tail of a comet produced by gas blown outward by the solar wind. (See **dust tail.**)

gene (593) A unit of DNA containing genetic information that influences a particular inherited trait.

general theory of relativity (91) Einstein's more sophisticated theory of space and time, which describes gravity as a curvature of space-time.

geocentric universe (58) A model universe with Earth at the center, such as the Ptolemaic universe.

geosynchronous satellite (84) A satellite in an eastward orbit whose period is 24 hours. A satellite in such an orbit remains above the same spot on Earth's surface.

giant molecular clouds (205) Very large, cool clouds of dense gas in which stars form.

giants (178) Large, cool, highly luminous stars in the upper right of the H–R diagram; typically 10 to 100 times the diameter of the sun.

glitch (282) A sudden change in the period of a pulsar.

global warming (443) The gradual increase in the surface temperature of Earth caused by human modifications to Earth's atmosphere.

globular cluster (248) A star cluster containing 50,000 to 1 million stars in a sphere about 75 ly in diameter; generally old, metal-poor, and found in the spherical component of the galaxy.

gossamer ring (513) The dimmest part of Jupiter's ring produced by dust particles orbiting near small moons.

grand unified theories (GUTs) (397) Theories that attempt to unify (describe in a similar way) the electromagnetic, weak, and strong forces of nature.

granulation (148) The fine structure visible on the solar surface caused by rising currents of hot gas and sinking currents of cool gas below the surface.

grating (112) A piece of material in which numerous microscopic parallel lines are scribed; light encountering a grating is dispersed to form a spectrum.

gravitational collapse (422) The stage in the formation of a massive planet when it grows massive enough to begin capturing gas directly from the nebula around it.

gravitational lens (370) The effect of focusing of light from a distant galaxy or quasar by an intervening galaxy to produce multiple images of the distant body.

gravitational radiation (287) As predicted by general relativity, expanding waves in a gravitational field that transport energy through space.

gravitational red shift (294) The lengthening of the wavelength of a photon due to its escape from a gravitational field.

greenhouse effect (443)   The process by which a carbon dioxide atmosphere traps heat and raises the temperature of a planetary surface.

grooved terrain (516)   Regions of the surface of Ganymede consisting of parallel grooves; believed to have formed by repeated fracture of the icy crust.

ground state (130)   The lowest permitted electron orbit in an atom.

half-life (421)   The time required for half of the atoms in a radioactive sample to decay.

halo (307)   The spherical region of a spiral galaxy containing a thin scattering of stars, star clusters, and small amounts of gas.

heat of formation (426)   In planetology, the heat released by the infall of matter during the formation of a planetary body.

heavy bombardment (428)   The period of intense meteorite impacts early in the formation of the planets when the solar system was filled with debris.

heliocentric universe (60)   A model of the universe with the sun at the center, such as the Copernican universe.

helioseismology (151)   The study of the interior of the sun by the analysis of its modes of vibration.

helium flash (243)   The explosive ignition of helium burning that takes place in some giant stars.

Herbig–Haro object (221)   Small nebula that varies irregularly in brightness; believed to be associated with star formation.

Hertzsprung–Russell diagram (177)   A plot of the intrinsic brightness versus the surface temperature of stars; it separates the effects of temperature and surface area on stellar luminosity; commonly absolute magnitude versus spectral type, but also luminosity versus surface temperature or color.

HI cloud (201)   An interstellar cloud of neutral hydrogen.

HII region (198)   A region of ionized hydrogen around a hot star.

high-velocity star (313)   A star with a large space velocity. Such stars are halo stars passing through the disk of the galaxy at steep angles.

Hirayama families (573)   Families of asteroids with orbits of similar size, shape, and orientation; believed to be fragments of larger bodies.

homogeneity (390)   The assumption that, on the large scale, matter is uniformly spread through the universe.

horizon (16)   The line that marks the apparent intersection of Earth and the sky.

horizon problem (396)   In cosmology, the circumstance that the primordial background radiation seems much more isotropic than could be explained by the standard big bang theory.

horizontal branch (249)   In the H–R diagram of a globular cluster, the sequence of stars extending from the red giants toward the blue side of the diagram; includes RR Lyrae stars.

horoscope (24)   A chart showing the positions of the sun, moon, planets, and constellations at the time of a person's birth; used in astrology to attempt to read character or foretell the future.

hot dark matter (395)   Invisible matter in the universe composed of low-mass, high-velocity particles such as neutrinos.

hot spot (362)   In radio astronomy, a bright spot in a radio lobe.

H–R diagram (180)   (See **Hertzsprung–Russell diagram.**)

Hubble constant (H) (342)   A measure of the rate of expansion of the universe; the average value of velocity of recession divided by distance; presently believed to be about 70 km/s/megaparsec.

Hubble law (342)   The linear relation between the distance to a galaxy and its radial velocity.

Hubble time (384)   An upper limit on the age of the universe derived from the Hubble constant.

hydrostatic equilibrium (225)   The balance between the weight of the material pressing downward on a layer in a star and the pressure in that layer.

hypernova (302)   The explosion produced as a very massive star collapses into a black hole; thought to be responsible for at least some gamma-ray bursts.

hypothesis (71)   A conjecture, subject to further tests, that accounts for a set of facts.

inflationary universe (396)   A version of the big bang theory that includes a rapid expansion when the universe was very young.

infrared cirrus (206)   A fine network of filaments covering the sky detected in the far infrared by the IRAS satellite; believed to be associated with dust in the interstellar medium.

infrared radiation (100)   Electromagnetic radiation with wavelengths intermediate between visible light and radio waves.

instability strip (251)   The region of the H–R diagram in which stars are unstable to pulsation; a star passing through this strip becomes a variable star.

intercloud medium (201)   The hot, low-density gas between cooler clouds in the interstellar medium.

intercrater plains (466)   The relatively smooth terrain on Mercury.

interferometry (110)   The observing technique in which separated telescopes combine to produce a virtual telescope with the resolution of a much-larger-diameter telescope.

interstellar absorption lines (200)   Dark lines in some stellar spectra that are formed by interstellar gas.

interstellar dust (200)   Microscopic solid grains in the interstellar medium.

interstellar extinction (200)   The dimming of starlight by gas and dust in the interstellar medium.

interstellar medium (197)   The gas and dust distributed between the stars.

interstellar reddening (200)   The process in which dust scatters blue light out of starlight and makes the stars look redder.

intrinsic variable (247)   A variable driven to pulsate by processes in its interior.

inverse square law (82)   The rule that the strength of an effect (such as gravity) decreases in proportion as the distance squared increases.

Io flux tube (507)   A tube of magnetic lines and electric currents connecting Io and Jupiter.

ion (128)   An atom that has lost or gained one or more electrons.

ionization (128)   The process in which atoms lose or gain electrons.

Io plasma torus (507)   The doughnut-shaped cloud of ionized gas that encloses the orbit of Jupiter's moon Io.

iron meteorite (565)   A meteorite composed mainly of iron–nickel alloy.

irregular galaxy (339) A galaxy with a chaotic appearance, large clouds of gas and dust, and both population I and population II stars, but without spiral arms.

island universe (336) An older term for a galaxy.

isotopes (128) Atoms that have the same number of protons but a different number of neutrons.

isotropy (389) The assumption that in its general properties the universe looks the same in every direction.

joule (J) (87) A unit of energy equivalent to a force of 1 newton acting over a distance of 1 meter; 1 joule per second equals 1 watt of power.

Jovian planet (418) Jupiterlike planet with large diameter and low density.

jumbled terrain (460) Strangely disturbed regions of the moon opposite the locations of the Imbrium Basin and Mare Orientale.

Kelvin temperature scale (124) The temperature, in Celsius (centigrade) degrees, measured above absolute zero.

Keplerian motion (314) Orbital motion in accord with Kepler's laws of planetary motion.

Kerr black hole (293) A solution to the equations of general relativity that describes the properties of a rotating black hole.

kiloparsec (kpc) (307) A unit of distance equal to 1000 pc, or 3260 ly.

Kirchhoff's laws (132) A set of laws that describes the absorption and emission of light by matter.

Kirkwood's gaps (571) Regions in the asteroid belt in which there are very few asteroids; caused by resonances with Jupiter.

Kuiper belt (581) The collection of icy planetesimals believed to orbit in a region from just beyond Neptune out to 100 AU or more.

Lagrangian point (263) Point of stability in the orbital plane of a binary star system, planet, or moon. One is located 60° ahead and one 60° behind the orbiting bodies; another is located between the orbiting bodies.

large-impact hypothesis (461) The theory that the moon formed from debris ejected during a collision between Earth and a large planetesimal.

L dwarf (136) A type of star that is even cooler than the M stars.

life zone (602) A region around a star within which a planet can have temperatures that permit the existence of liquid water.

light curve (186) A graph of brightness versus time commonly used in analyzing variable stars and eclipsing binaries.

light-gathering power (102) The ability of a telescope to collect light; proportional to the area of the telescope objective lens or mirror.

lighthouse model (282) The explanation of a pulsar as a spinning neutron star sweeping beams of radio radiation around the sky.

light pollution (105) The illumination of the night sky by waste light from cities and outdoor lighting, which prevents the observation of faint objects.

light-year (ly) (5) The distance light travels in 1 year.

limb (148, 450) The edge of the apparent disk of a body, as in "the limb of the moon."

limb darkening (148) The decrease in brightness of the sun or other body from its center to its limb.

line of nodes (46) The line across an orbit connecting the nodes; commonly applied to the orbit of the moon.

line profile (140) A graph of light intensity versus wavelength showing the shape of an absorption line.

liquid metallic hydrogen (505) A form of hydrogen under high pressure that is a good electrical conductor.

lobate scarp (465) A curved cliff such as those found on Mercury.

local bubble or void (207) A region of high-temperature, low-density gas in the interstellar medium in which the sun happens to be located.

look-back time (341) The amount by which we look into the past when we look at a distant galaxy; a time equal to the distance to the galaxy in light-years.

luminosity (L) (176) The total amount of energy a star radiates in 1 second.

luminosity class (180) A category of stars of similar luminosity; determined by the widths of lines in their spectra.

lunar eclipse (37) The darkening of the moon when it moves through Earth's shadow.

Lyman series (133) Spectral lines in the ultraviolet spectrum of hydrogen produced by transitions whose lowest orbit is the ground state.

MACHOs (395) Massive Compact Halo Objects; low-luminosity objects such as planets and brown dwarfs that contribute to the mass of the halo.

Magellanic Clouds (306) Small, irregular galaxies that are companions to the Milky Way; visible in the southern sky.

magnetar (286) A class of neutron stars that have exceedingly strong magnetic fields; thought to be responsible for soft gamma-ray repeaters.

magnetic carpet (151) The widely distributed, low-level magnetic field extending up through the sun's visible surface.

magnetosphere (437) The volume of space around a planet within which the motion of charged particles is dominated by the planetary magnetic field rather than the solar wind.

magnifying power (104) The ability of a telescope to make an image larger.

magnitude–distance formula (175) The mathematical formula that relates the apparent magnitude and absolute magnitude of a star to its distance.

magnitude scale (12) The astronomical brightness scale; the larger the number, the fainter the star.

main sequence (177) The region of the H–R diagram running from upper left to lower right, which includes roughly 90 percent of all stars.

mantle (436) The layer of dense rock and metal oxides that lies between the molten core and Earth's surface; also, similar layers in other planets.

mare (sea) (450) One of the lunar lowlands filled by successive flows of dark lava.

mass (81) A measure of the amount of matter making up an object.

mass–luminosity relation (190) The more massive a star is, the more luminous it is.

Maunder butterfly diagram (156) A graph showing the latitude of sunspots versus time; first plotted by W. W. Maunder in 1904.

Maunder minimum (160) A period of less numerous sunspots and other solar activity from 1645 to 1715.

megaparsec (Mpc) (337)   A unit of distance equal to 1 million pc.

metals (315)   In astronomical usage, all atoms heavier than helium.

metastable level (197)   An atomic energy level from which an electron takes a long time to decay; responsible for producing forbidden lines.

meteor (420)   A small bit of matter heated by friction to incandescent vapor as it falls into Earth's atmosphere.

meteorite (420)   A meteor that has survived its passage through the atmosphere and strikes the ground.

meteoroid (420)   A meteor in space before it enters Earth's atmosphere.

micrometeorite (453)   Meteorite of microscopic size.

midocean rise (440)   One of the undersea mountain ranges that push up from the seafloor in the center of the oceans.

Milankovitch hypothesis (25)   The hypothesis that small changes in Earth's orbital and rotational motions cause the ice ages.

Milky Way (6)   The hazy band of light that circles the sky, produced by the combined light of billions of stars in our Milky Way Galaxy.

Milky Way Galaxy (6)   The spiral galaxy containing our sun; visible at night as the Milky Way.

Miller experiment (595)   An experiment that reproduced the conditions under which life began on Earth and manufactured amino acids and other organic compounds.

millisecond pulsar (289)   A pulsar with a period of approximately a millisecond, a thousandth of a second.

minute of arc (17)   An angular measure; each degree is divided into 60 minutes of arc.

molecular cloud (205)   An interstellar gas cloud that is dense enough for the formation of molecules; discovered and studied through the radio emissions of such molecules.

molecule (128)   Two or more atoms bonded together.

momentum (80)   The tendency of a moving object to continue moving; mathematically, the product of mass and velocity.

morning star (21)   Any planet visible in the sky just before sunrise.

multiringed basin (453)   Very large impact basins in which there are concentric rings of mountains.

mutant (594)   Offspring born with altered DNA.

nadir (16)   The point on the bottom of the sky directly under our feet.

nanometer (nm) (99)   A unit of length equal to 10^{-9} m.

natural law (71)   A conjecture about how nature works in which scientists have overwhelming confidence.

natural motion (78)   In Aristotelian physics, the motion of objects toward their natural places—fire and air upward and earth and water downward.

natural selection (594)   The process by which the best traits are passed on, allowing the most able to survive.

neap tide (36)   Ocean tide of low amplitude occurring at first- and third-quarter moon.

nebula (197)   A cloud of gas and dust in space.

nebular hypothesis (409)   The proposal that the solar system formed from a rotating cloud of gas.

neutrino (165)   A neutral, massless atomic particle that travels at or nearly at the speed of light.

neutron (127)   An atomic particle with no charge and about the same mass as a proton.

neutron star (279)   A small, highly dense star composed almost entirely of tightly packed neutrons; radius about 10 km.

Newtonian focus (108)   The focal arrangement of a reflecting telescope in which a diagonal mirror reflects light out the side of the telescope tube for easier access.

node (45)   A point where an object's orbit passes through the plane of Earth's orbit.

nonbaryonic matter (395)   In cosmology, a suspected component of the dark matter composed of matter that does not contain protons and neutrons.

north celestial pole (16)   The point on the celestial sphere directly above Earth's North Pole.

north point (16)   The point on the horizon directly below the north celestial pole; exactly north.

nova (256)   From the Latin "new," a sudden brightening of a star, making it appear as a "new" star in the sky;

believed associated with eruptions on white dwarfs in binary systems.

nuclear bulge (307)   The spherical cloud of stars that lies at the center of spiral galaxies.

nuclear fission (161)   Reaction that splits nuclei into less massive fragments.

nuclear fusion (164)   Reaction that joins the nuclei of atoms to form more massive nuclei.

nucleosynthesis (315)   The production of elements heavier than helium by the fusion of atomic nuclei in stars and during supernovae explosions.

nucleus (of an atom) (127)   The central core of an atom containing protons and neutrons; carries a net positive charge.

O association (219)   A large, loosely bound cluster of very young stars.

objective lens or mirror (101)   The main optical element in an astronomical telescope. The large lens at the top of the telescope or large mirror at the bottom.

oblateness (505)   The flattening of a spherical body; usually caused by rotation.

observable universe (382)   The part of the universe that we can see from our location in space and time.

occultation (543)   The passage of a larger body in front of a smaller body.

Olbers's paradox (380)   The conflict between observation and theory as to why the night sky should or should not be dark.

Oort cloud (581)   The cloud of icy bodies—extending from the outer part of our solar system out to roughly 100,000 AU from the sun—that acts as the source of most comets.

opacity (224)   The resistance of a gas to the passage of radiation.

open cluster (248)   A cluster of 10 to 10,000 stars with an open, transparent appearance and stars not tightly grouped; usually relatively young and located in the disk of the galaxy.

open orbit (85)   An orbit that does not return to its starting point; an escape orbit. (See **closed orbit**.)

open universe (393)   A model universe in which the average density is less than the critical density needed to halt the expansion.

oscillating universe theory (391) The theory that the universe begins with a big bang, expands, is slowed by its own gravity, and then falls back to create another big bang.

outflow channel (492) Geological feature on Mars that appears to have been caused by sudden flooding.

outgassing (425) The release of gases from a planet's interior.

ovoid (546) Geological features on Uranus's moon Miranda thought to be produced by circulation in the solid icy mantle and crust.

ozone layer (442) In Earth's atmosphere, a layer of oxygen ions (O_3) lying 15 to 30 km high that protects the surface by absorbing ultraviolet radiation.

paradigm (63) A commonly accepted set of scientific ideas and assumptions.

parallax (58) The apparent change in the position of an object due to a change in the location of the observer. Astronomical parallax is measured in seconds of arc.

parsec (pc) (172) The distance to a hypothetical star whose parallax is one second of arc; 1 pc = 206,265 AU = 3.26 ly.

partial eclipse (38, 40) A lunar eclipse in which the moon does not completely enter Earth's shadow; a solar eclipse in which the moon does not completely cover the sun.

Paschen series (133) Spectral lines in the infrared spectrum of hydrogen produced by transitions whose lowest orbit is the third.

passing star hypothesis (407) The proposal that our solar system formed when two stars passed near each other and material was pulled out of one to form the planets.

path of totality (41) The track of the moon's umbral shadow over Earth's surface. The sun is totally eclipsed as seen from within this path.

penumbra (37) The portion of a shadow that is only partially shaded.

penumbral eclipse (39) A lunar eclipse in which the moon enters the penumbra of Earth's shadow but does not reach the umbra.

perigee (43) The orbital point of closest approach to Earth.

perihelion (23) The orbital point of closest approach to the sun.

period–luminosity relation (247) The relation between period of pulsation and intrinsic brightness among Cepheid variable stars.

permitted orbit (129) One of the energy levels in an atom that an electron may occupy.

photon (99) A quantum of electromagnetic energy; carries an amount of energy that depends inversely on its wavelength.

photosphere (42) The bright visible surface of the sun.

planet (4) A nonluminous object, larger than a comet or asteroid, that orbits a star.

planetary nebula (258) An expanding shell of gas ejected from a star during the latter stages of its evolution.

planetesimal (423) One of the small bodies that formed from the solar nebula and eventually grew into protoplanets.

plastic (436) A material with the properties of a solid but capable of flowing under pressure.

plate tectonics (440) The constant destruction and renewal of Earth's surface by the motion of sections of crust.

polar axis (109) The axis around which a celestial body rotates.

poor cluster (347) An irregularly shaped cluster that contains fewer than 1000 galaxies, many spiral, and no giant ellipticals.

population I star (315) Star rich in atoms heavier than helium; nearly always a relatively young star found in the disk of the galaxy.

population II star (315) Star poor in atoms heavier than helium; nearly always a relatively old star found in the halo, globular clusters, or the nuclear bulge.

precession (18) The slow change in the direction of Earth's axis of rotation; one cycle takes nearly 26,000 years.

pressure (203) A force exerted over a surface. Expressed as force per unit area.

pressure (*P*) waves (435) In geophysics, mechanical waves of compression and rarefaction that travel through Earth's interior.

primary lens or mirror (101) The main optical element in an astronomical telescope. The large lens at the top of the telescope tube or the large mirror at the bottom.

prime focus (108) The point at which the objective mirror forms an image in a reflecting telescope.

primeval atmosphere (442) Earth's first air, composed of gases from the solar nebula.

primordial soup (596) The rich solution of organic molecules in Earth's first oceans.

prominence (43, 162) Eruption on the solar surface; visible during total solar eclipses.

proper motion (173) The rate at which a star moves across the sky; measured in seconds of arc per year.

protein (592) Complex molecule composed of amino acid units.

proton (127) A positively charged atomic particle contained in the nucleus of atoms; the nucleus of a hydrogen atom.

proton–proton chain (165) A series of three nuclear reactions that build a helium atom by adding together protons; the main energy source in the sun.

protoplanet (424) Massive object resulting from the coalescence of planetesimals in the solar nebula and destined to become a planet.

protostar (217) A collapsing cloud of gas and dust destined to become a star.

protostellar disk (218) A gas cloud around a forming star flattened by its rotation.

pulsar (281) A source of short, precisely timed radio bursts; believed to be a spinning neutron star.

pulsar wind (282) The flow of high-energy particles that carries most of the energy away from a spinning neutron star.

quantum mechanics (129) The study of the behavior of atoms and atomic particles.

quasar (quasi-stellar object, or QSO) (367) Small, powerful source of energy believed to be the active core of very distant galaxies.

quasi-periodic oscillation (QPO) (297) A high-speed flickering in the radiation from an accretion disk evidently caused by material spiraling inward.

quintessence (398)   The proposed energy of empty space that causes the acceleration of the expanding universe.

radial velocity (V_r) (139)   That component of an object's velocity directed away from or toward Earth.

radiant (562)   The point in the sky from which meteors in a shower seem to come.

radiation pressure (428)   The force exerted on the surface of a body by its absorption of light. Small particles floating in the solar system can be blown outward by the pressure of the sunlight.

radio galaxy (358)   A galaxy that is a strong source of radio signals.

radio interferometer (114)   Two or more radio telescopes that combine their signals to achieve the resolving power of a larger telescope.

ray (452)   Ejecta from a meteorite impact forming white streamers radiating from some lunar craters.

recombination (388)   The stage within 10^6 years of the big bang when the gas became transparent to radiation.

reconnection (163)   The process in the sun's atmosphere by which opposing magnetic fields combine and release energy to power solar flares.

red dwarf (178)   Cool, low-mass star on the lower main sequence.

red shift (139)   The lengthening of the wavelengths of light seen when the source and observer are receding from each other.

reflecting telescope (101)   A telescope that uses a concave mirror to focus light into an image.

reflection nebula (198)   A nebula produced by starlight reflecting off of dust particles in the interstellar medium.

refracting telescope (101)   A telescope that forms images by bending (refracting) light with a lens.

regolith (457)   A soil made up of crushed rock fragments.

reionization (388)   The stage in the early history of the universe when ultraviolet photons from the first stars ionized the gas filling space.

relative age (454)   The age of a geological feature referred to other features. For example, relative ages tell us the lunar maria are younger than the highlands.

relativistic Doppler formula (369)   The formula for the Doppler shift produced by an object moving at a significant fraction of the speed of light.

relativistic jet model (374)   An explanation of superluminal expansion based on a high-velocity jet from a quasar directed approximately toward Earth.

resolving power (104)   The ability of a telescope to reveal fine detail; depends on the diameter of the telescope objective.

resonance (464)   The coincidental agreement between two periodic phenomena; commonly applied to agreements between orbital periods, which can make orbits more or less stable.

retrograde motion (58)   The apparent backward (westward) motion of planets as seen against the background of stars.

revolution (19)   The motion of an object in a closed path about a point outside its volume; Earth revolves around the sun.

rich cluster (347)   A cluster containing over 1000 galaxies, mostly elliptical, scattered over a volume about 3 Mpc in diameter.

rift valley (441)   A long, straight, deep valley produced by the separation of crustal plates.

ring galaxy (349)   A galaxy that resembles a ring around a bright nucleus; believed to be the result of a head-on collision of two galaxies.

RNA (ribonucleic acid) (593)   Long carbon-chain molecules that use the information stored in DNA to manufacture complex molecules necessary to the organism.

Roche limit (512)   The minimum distance between a planet and a satellite that holds itself together by its own gravity. If a satellite's orbit brings it within its planet's Roche limit, tidal forces will pull the satellite apart.

Roche lobe (263)   In a system with two bodies orbiting each other, the volume of space dominated by the gravitation of one of the bodies.

Roche surface (263)   In a system with two bodies orbiting each other, the outer boundary of the volume of space dominated by the gravitation of one of the bodies.

rotation (19)   The turning of a body about an axis that passes through its volume; Earth rotates on its axis.

rotation curve (314, 343)   A graph of orbital velocity versus radius in the disk of a galaxy.

rotation curve method (343)   The procedure for finding the mass of a galaxy from its rotation curve.

RR Lyrae variable star (247)   Variable star with a period of 12 to 24 hours; common in some globular clusters.

Sagittarius A* (328)   The powerful radio source located at the core of the Milky Way Galaxy.

saros cycle (47)   An 18-year, $11\frac{1}{3}$-day period after which the pattern of lunar and solar eclipses repeats.

Schmidt-Cassegrain focus (108)   The optical design of a reflecting telescope in which a thin correcting lens is placed at the top of a Cassegrain telescope.

Schwarzschild radius (R_s) (292)   The radius of the event horizon around a black hole.

scientific model (15)   An intellectual concept designed to help us think about a natural process without necessarily being a conjecture of truth.

scientific notation (4)   The system of recording very large or very small numbers by using powers of 10.

secondary atmosphere (442)   The gases outgassed from a planet's interior; rich in carbon dioxide.

secondary crater (452)   A crater formed by the impact of debris ejected from a larger crater.

secondary mirror (108)   In a reflecting telescope, the mirror that reflects the light to a point of easy observation.

second of arc (17)   An angular measure; each minute of arc is divided into 60 seconds of arc.

seeing (104)   Atmospheric conditions on a given night. When the atmosphere is unsteady, producing blurred images, the seeing is said to be poor.

seismic wave (435)   A mechanical vibration that travels through Earth; usually caused by an earthquake.

seismograph (435)   An instrument that records seismic waves.

selection effect (565)   An influence on the probability that certain phenomena will be detected or selected, which can alter the outcome of a survey.

self-sustaining star formation (325) The process by which the birth of stars compresses the surrounding gas clouds and triggers the formation of more stars; proposed to explain spiral arms.

semimajor axis (*a*) (70) Half of the longest axis of an ellipse.

SETI (605) Search for Extra-Terrestrial Intelligence.

Seyfert galaxy (358) An otherwise normal spiral galaxy with an unusually bright, small core that fluctuates in brightness; believed to indicate the core is erupting.

Shapley–Curtis Debate (336) The 1920 debate between Harlow Shapley and Heber Curtis over the nature of the spiral nebulae.

shear (*S*) waves (435) Mechanical waves that travel through Earth's interior by the vibration of particles perpendicular to the direction of wave travel.

shepherd satellite (525) A satellite that, by its gravitational field, confines particles to a planetary ring.

shield volcano (480) Wide, low-profile volcanic cone produced by highly liquid lava.

shock wave (216) A sudden change in pressure that travels as an intense sound wave.

sidereal drive (109) The motor and gears on a telescope that turn it westward to keep it pointed at a star.

sidereal period (35) The period of rotation or revolution of an astronomical body referred to the stars.

singularity (291) The object of zero radius into which the matter in a black hole is believed to fall.

sinuous rille (450) A narrow, winding valley on the moon caused by ancient lava flows along narrow channels.

small-angle formula (40) The mathematical formula that relates an object's linear diameter and distance to its angular diameter.

smooth plains (466) Apparently young plains on Mercury formed by lava flows at or soon after the formation of the Caloris Basin.

soft gamma-ray repeater (SGR) (301) An object that produces repeated bursts of lower-energy gamma rays; thought to be produced by magnetars.

solar constant (160) A measure of the energy output of the sun; the total solar energy striking 1 m^2 just above Earth's atmosphere in 1 second.

solar eclipse (40) The event that occurs when the moon passes directly between Earth and the sun, blocking our view of the sun.

solar nebula hypothesis (410) The theory that the planets formed from the same cloud of gas and dust that formed the sun.

solar system (4) The sun and the nonluminous objects that orbit it, including the planets, comets, and asteroids.

solar wind (151) Rapidly moving atoms and ions that escape from the solar corona and blow outward through the solar system.

south celestial pole (16) The point of the celestial sphere directly above Earth's South Pole.

south point (16) The point on the horizon directly above the south celestial pole; exactly south.

special relativity (90) The first of Einstein's theories of relativity, which dealt with uniform motion.

spectral class or type (135) A star's position in the temperature classification system O, B, A F, G, K, and M. Based on the appearance of the star's spectrum.

spectral sequence (135) The arrangement of spectral classes (O, B, A, F, G, K, M) ranging from hot to cool.

spectrograph (112) A device that separates light by wavelength to produce a spectrum.

spectroscopic binary system (184) A star system in which the stars are too close together to be visible separately. We see a single point of light, and only by taking a spectrum can we determine that there are two stars.

spectroscopic parallax (181) The method of determining a star's distance by comparing its apparent magnitude with its absolute magnitude as estimated from its spectrum.

spherical component (311) The part of the galaxy including all matter in a spherical distribution around the center (the halo and nuclear bulge).

spicule (149) Small, flamelike projection in the chromosphere of the sun.

spiral arms (6) Long, spiral patterns of bright stars, star clusters, gas, and dust that extend from the center to the edge of the disk of spiral galaxies.

spiral galaxy (338) A galaxy with an obvious disk component containing gas; dust; hot, bright stars; and spiral arms.

spiral nebula (335) Nebulous object with a spiral appearance observed in early telescopes; later recognized as a spiral galaxy.

spiral tracers (321) Objects used to map the spiral arms (e.g., O and B associations, open clusters, clouds of ionized hydrogen, and some types of variable stars).

spoke (526) A radial feature in the rings of Saturn.

sporadic meteor (563) A meteor not part of a meteor shower.

spring tide (36) Ocean tide of high amplitude that occurs at full and new moon.

standard candle (337) Object of known brightness which astronomers use to find distance—for example, Cepheid variable stars and supernovae.

star (4) A celestial object composed of gas held together by its own gravity and supported by nuclear fusion occurring in its interior.

starburst galaxy (352) A bright blue galaxy in which many new stars are forming, believed caused by collisions between galaxies.

steady state theory (386) The theory (now generally abandoned) that the universe does not evolve.

stellar model (234) A table of numbers representing the conditions in various layers within a star.

stellar parallax (*p*) (172) A measure of stellar distance. (See **parallax**.)

stony-iron meteorite (567) A meteorite that is a mixture of stone and iron.

stony meteorite (566) A meteorite composed of silicate (rocky) material.

stromatolite (598) A layered fossil formation caused by ancient mats of algae or bacteria that build up mineral deposits season after season.

strong force (161) One of the four forces of nature; the strong force binds protons and neutrons together in atomic nuclei.

subduction zone (440)   A region of a planetary crust where a tectonic plate slides downward.

subsolar point (473)   The point on a planet that is directly below the sun.

summer solstice (22)   The point on the celestial sphere where the sun is at its most northerly point; also, the time when the sun passes this point, about June 22, and summer begins in the northern hemisphere.

sunspot (147)   Relatively dark spot on the sun that contains intense magnetic fields.

supercluster (400)   A cluster of galaxy clusters.

supergiant (178)   Exceptionally luminous star 10 to 1000 times the sun's diameter.

supergranule (148)   A large granule on the sun's surface including many smaller granules.

superluminal expansion (373)   The apparent expansion of parts of a quasar at speeds greater than the speed of light.

supernova (256)   A new star appearing in Earth's sky and lasting for a year or so before fading. Caused by the violent explosion of a star.

supernova remnant (271)   The expanding shell of gas marking the site of a supernova explosion.

supernova (type I) (269)   The explosion of a star, believed caused by the transfer of matter to a white dwarf.

supernova (type II) (269)   The explosion of a star, believed caused by the collapse of a massive star.

synchrotron radiation (271)   Radiation emitted when high-speed electrons move through a magnetic field.

synodic period (35)   The period of rotation or revolution of a celestial body with respect to the sun.

T association (219)   A large, loosely bound group of T Tauri stars.

T dwarf (136)   A very-low-mass star at the bottom end of the main sequence with a cool surface and a low luminosity.

temperature (124)   A measure of the velocity of random motions among the atoms or molecules in a material.

terminator (450)   The dividing line between daylight and darkness on a planet or moon.

terrestrial planet (418)   Earthlike planet—small, dense, rocky.

theory (71)   A system of assumptions and principles applicable to a wide range of phenomena that have been repeatedly verified.

thermal energy (125)   The energy stored in an object as agitation among its atoms and molecules.

thermal pulse (258)   Periodic eruptions in the helium fusion shell in an aging giant star; thought to aid in ejecting the surface layers of the stars to form planetary nebulae.

tidal coupling (449)   The locking of the rotation of a body to its revolution around another body.

tidal heating (517)   The heating of a planet or satellite because of friction caused by tides.

time dilation (293)   The slowing of moving clocks or clocks in strong gravitational fields.

total eclipse (37, 40)   A solar eclipse in which the moon completely covers the bright surface of the sun; a lunar eclipse in which the moon completely enters Earth's dark shadow.

transition (133)   The movement of an electron from one atomic orbit to another.

transition region (149)   The layer in the solar atmosphere between the chromosphere and the corona.

transverse velocity (139)   The velocity of a star perpendicular to the line of sight.

triple alpha process (243)   The nuclear fusion process that combines three helium nuclei (alpha particles) to make one carbon nucleus.

Trojan asteroid (572)   Small, rocky body caught in Jupiter's orbit at the Lagrangian points, 60° ahead of and behind the planet.

T Tauri star (220)   Young star surrounded by gas and dust, believed to be contracting toward the main sequence.

turnoff point (248)   The point in an H–R diagram where a cluster's stars turn off of the main sequence and move toward the red giant region, revealing the approximate age of the cluster.

21-cm radiation (202)   Radio emission produced by cold, low-density hydrogen in interstellar space.

ultraluminous infrared galaxy (350)   A highly luminous galaxy so filled with dust that most of its energy escapes as infrared photons emitted by warmed dust.

ultraviolet radiation (101)   Electromagnetic radiation with wavelengths shorter than visible light but longer than X rays.

umbra (37)   The region of a shadow that is totally shaded.

uncompressed density (423)   The density a planet would have if its gravity did not compress it.

unified model (365)   The attempt to explain the different kinds of active galaxies and quasars by a single model.

uniform circular motion (58)   The classical belief that the perfect heavens could move only by the combination of uniform motion along circular orbits.

valley networks (492)   Dry drainage channels resembling streambeds found on Mars.

Van Allen belts (437)   Radiation belts of high-energy particles trapped in Earth's magnetosphere.

variable star (247)   A star whose brightness changes periodically.

velocity (81)   A rate of travel that specifies both speed and direction.

velocity dispersion method (344)   A method of finding a galaxy's mass by observing the range of velocities within the galaxy.

vernal equinox (22)   The place on the celestial sphere where the sun crosses the celestial equator moving northward; also, the time of year when the sun crosses this point, about March 21, and spring begins in the northern hemisphere.

vesicular basalt (456)   A porous rock formed by solidified lava with trapped bubbles.

violent motion (78)   In Aristotelian physics, motion other than natural motion. (See **natural motion.**)

visual binary system (183)   A binary star system in which the two stars are separately visible in the telescope.

water hole (605)   The interval of the radio spectrum between the 21-cm hydrogen radiation and the 18-cm OH radiation; likely wavelengths to use in the search for extraterrestrial life.

wavelength (99)   The distance between successive peaks or troughs of a wave; usually represented by λ.

wavelength of maximum intensity (λ_{max}) (126)   The wavelength at which a perfect radiator emits the maximum amount of energy; depends only on the object's temperature.

weak force (161)   One of the four forces of nature; the weak force is responsible for some forms of radioactive decay.

west point (16)   The point on the western horizon exactly halfway between the north point and the south point; exactly west.

white dwarf (178)   Dying star that has collapsed to the size of Earth and is slowly cooling off; at the lower left of the H–R diagram.

Widmanstätten patterns (566)   Bands in iron meteorites due to large crystals of nickel–iron alloys.

winter solstice (22)   The point on the celestial sphere where the sun is farthest south; also, the time of year when the sun passes this point, about December 22, and winter begins in the northern hemisphere.

X-ray burster (297)   An object that produces repeated bursts of X rays.

Zeeman effect (155)   The splitting of spectral lines into multiple components when the atoms are in a magnetic field.

zenith (16)   The point on the sky directly overhead.

zero-age main sequence (ZAMS) (239)   The locus in the H–R diagram where stars first reach stability as hydrogen-burning stars.

zodiac (21)   The band around the sky centered on the ecliptic within which the planets move.

zones (509)   Yellow-white regions that circle Jupiter parallel to its equator; believed to be areas of rising gas.

ANSWERS TO EVEN-NUMBERED PROBLEMS

Chapter 1: **2.** 3475 km; **4.** 1.05×10^8 km; **6.** about 1.2 seconds; **8.** 75,000 years; **10.** about 27

Chapter 2: **2.** 4; **4.** 2800; **6.** A is brighter than B by a factor of 170; **8.** 66.5°; 113.5°

Chapter 3: **2.** a) full; b) first quarter; c) waxing gibbous; d) waxing crescent; **4.** 12; 12.4. The moon orbits Earth and moves eastward about 13° per day; **6.** 6850 arc seconds or about 1.9°; **8.** a) The moon won't be full until Oct. 17; b) The moon will no longer be near the node of its orbit; **10.** August 12, 2026 [July 10, 1972 + 3 × ($6585\frac{1}{3}$ days)]. In order to get Aug. 12 instead of Aug. 11, you must take into account the number of leap days in the interval.

Chapter 4: **2.** Retrograde motion: Jupiter, Saturn, Uranus, Neptune, and Pluto; Never seen as crescents: Jupiter, Saturn, Uranus, Neptune, and Pluto; **4.** Mars, about 18 seconds of arc; the maximum angular diameter of Jupiter is 50 seconds of arc; **6.** $\sqrt{27} = 5.2$ years

Chapter 5: **2.** The force of gravity on the moon is about $\frac{1}{6}$ the force of gravity on Earth; **4.** 7350 m/s; **6.** 5070 s (1 hr and 25 min); **8.** The cannonball would move in an elliptical orbit with Earth's center at one focus of the ellipse; **10.** 6320 s (1 hr and 45 min)

Chapter 6: **2.** 3m; **4.** Either Gemini telescope has a light-gathering power that is 1.02 million times greater than the human eye; **6.** No, his resolving power should have been about 5.8 seconds of arc at best; **8.** 0.013 m (1.3 cm or about 0.5 inches); **10.** about 50 cm (From 400 km above, a human is about 0.25 seconds of arc from shoulder to shoulder.)

Chapter 7: **2.** 150 nm; **4.** by a factor of 16; **6.** 250 nm; **8.** a) B; b) F; c) M; d) K; **10.** about 0.58 nm

Chapter 8: **2.** 730 km; **4.** about 3.5 times; **6.** 400,000 years; **8.** 9×10^{16} J; **10.** 0.222 kg

Chapter 9: **2.** 63 pc; absolute magnitude is 2; **4.** about B7; **6.** about 1580 solar luminosities; **8.** 160 pc; **10.** a, c, c, d (use Figure 9-12 to determine the absolute magnitudes);

12. 3.69 days or 0.010 years; about 1.2 solar masses; about 0.67 and 0.53 solar masses; **14.** 1.38×10^6 km, about the size of the sun

Chapter 10: **2.** 60,000 nm; **4.** 0.0001; **6.** 1.5×10^6 years; **8.** 4.2×10^{35} kg or 210,000 solar masses (*Note:* Each hydrogen molecule contains two H atoms.)

Chapter 11: **2.** 24.5 km/s; **4.** 2.98 km/s; **6.** 9.46×10^{10} s (about 3000 years); **8.** There are four ^{1}H nuclei in the figure. They are used in a series of reactions to build up nuclei from ^{12}C to ^{15}N. In the last reaction (with the ^{15}N) a ^{12}C and a ^{4}He are produced. Because the ^{12}C can be used in the next cycle, the net is four ^{1}H nuclei in with one ^{4}He nucleus out; **10.** 7.9×10^{33} kg

Chapter 12: **2.** about 9.8×10^6 years for a 16-solar-mass star; about 5.7×10^5 years for a 50-solar-mass star; **4.** about 1×10^6 times less than present or about 1.4×10^{-6} g/cm^3; **6.** 2.4×10^{-9} or 1/420,000,000; **8.** about 3 pc; **10.** 3.04 minutes early after 1 year; 30.4 minutes early after 10 years

Chapter 13: **2.** about 1.7 ly; **4.** about 16,000 years old; **6.** about 940 years ago (approximately 1060 AD); **8.** 2200 km/s; **10.** about 300 years ago

Chapter 14: **2.** 7.1×10^{25} J/s or about 0.19 solar luminosity; **4.** 820 km/s (assuming mass is one solar mass); **6.** about 11 seconds of arc; **8.** about 490 seconds

Chapter 15: **2.** about 16 percent; **4.** 3.8×10^6 years; **6.** overestimate by a factor of 1.58; **8.** about 21 kpc; **10.** 7.8×10^{10} solar masses; **12.** 1500 K

Chapter 16: **2.** 2.58 Mpc; **4.** 131 km/s; **6.** 28.6 Mpc; **8.** 1.64×10^8 yr; **10.** 4.49×10^{41} kg

Chapter 17: **2.** 7.8×10^6 years; **4.** 0.024 pc; **6.** −28.5; **8.** about 91,000 km/s; **10.** 0.16

Chapter 18: **2.** 57; 17.5 billion years; 11.7 billion years; **4.** 1.6×10^{-30} gm/cm^3; **6.** 76 km/Mpc; **8.** 16.6 billion years; 11 billion years; the universe could be older

Chapter 19: **2.** It will look $206,265^2 = 4.3 \times 10^{10}$ times fainter, which is about 26.6 magnitudes fainter; 22.6 mag; **4.** about 3.3 times the half-life, or 4.3 billion years; **6.** large amounts of methane and water ices; **8.** about 1300

Chapter 20: **2.** about 17%; **4.** 81×10^6 yr; they have been subducted; **6.** 0.22%

Chapter 21: **2.** The rate at which an object radiates energy is proportional to its surface area (e.g., for a black body the rate of emission of energy is given by $\frac{4}{3}\pi r^2 \sigma T^4$). However, the energy an object has stored as heat is proportional to its mass and hence to its volume [i.e., mass $= \frac{4}{3}\pi r^3$ (density)]. The cooling time will therefore be proportional to the mass divided by the surface area:

$$\text{cooling time} = \frac{\text{mass}}{\text{surface area}} = \frac{\frac{4}{3}\pi r^3 \times \text{density}}{4\pi r^2}$$
$$= \tfrac{1}{3} r \times \text{density}$$

and we find that for planets of nearly the same density, the one with the largest radius will take the longest time to cool; **4.** No. Their angular diameter would be only 0.5 second of arc. They would be visible in photos taken from orbit around the moon; **6.** 0.5 second of arc (assuming an astronaut seen from above is 0.5 meter in diameter); no; **8.** 10.0016 cm; **10.** Mercury, $v_e = 4250$ m/s; moon, $v_e = 2380$ m/s; Earth, $v_e = 11,200$ m/s

Chapter 22: **2.** 33,400 km (39,500 km from the center of Venus); **4.** 61 seconds of arc; **6.** 260 km; **8.** 965 seconds of arc (about 1/4°)

Chapter 23: **2.** 35 Earth days; **4.** 4.4°; **6.** about 0.056 nm; **8.** 5.2 m/s

Chapter 24: **2.** 42 seconds of arc; **4.** 256 m/s; **6.** 8.8 s; yes; **8.** 12.3 km/s; **10.** 1.04×10^{26} kg (17.2 Earth masses)

Chapter 25: **2.** one billion; **4.** 79 km; **6.** 0.18 km/s; **8.** 3.28 AU and 2.5 AU; **10.** 9 million km (0.06 AU); too small; **12.** 33 Earth masses

Chapter 26: **2.** 8.9 cm; 0.67 mm; **4.** about 1.3 solar masses; **6.** 380 km; **8.** pessimistic, 2×10^{-5}; optimistic, 10^7

CREDITS

Chapter 2, The Sky Around Us, page 16, Star trails: AURA/NOAO/NSF.

Chapter 4, The Ancient Universe, page 58, Ancient universe: from *Cosmographics* by Peter Apian, 1539.

Chapter 5, Orbiting Earth, page 85, Astronaut: NASA.

Chapter 6, Modern Astronomical Telescopes, page 108, 4-meter telescope: AURA/NOAO/NSF; **page 109,** Keck I Mirror: © Russ Underwood/Keck Observatory; Adaptive optics: Paul Kalas.

Chapter 7, Atomic Spectra, page 132, Stamp: Courtesy Ron Reese and Michelangelo Fazio; **page 133,** Lagoon Nebula: AURA/NOAO/NSF.

Chapter 8, Magnetic Solar Phenomena, page 162, UV image of sun: Trace/NASA; Quiescent prominence: SOHO/EIT, ESA and NASA; Loop prominence: Sacramento Peak Observatory; **page 163,** Solar flare: NASA; Waves from solar flare: SOHO/MDI, ESA, and NASA; Coronal holes: NASA and ISAS; Coronal mass ejection: SOHO/LASCO, ESA, and NASA; Aurora: Copyright Jan Curtis.

Chapter 9, The Family of Stars, page 193, Star field: Hubble Heritage Team.

Chapter 10, Three Kinds of Nebulae, page 198, Star-forming region in Ara: ESO; Reflection nebulae in Orion: AATB; **page 199,** Pleiades: Caltech; Lagoon and Trifid Nebulae: Daniel Good; Bernard 86: AATB; Horsehead Nebulae: Hubble Heritage Team; Merope Nebula: NASA.

Chapter 11, Observational Evidence of Star Formation, page 220, S Mon Nebula: AATB; Bok globules: NASA; **page 221,** HH34: Richard Mundt, Calar Alto 3.5 m telescope; Jet inset: NASA; Eagle Nebula: AURA/NOAO/NSF; EGGs: NASA. **Star Formation in the Orion Nebula, page 230,** Orion Nebula: Daniel Good; Low-Mass stars: NASA and K. L. Luhman (Harvard-

Smithsonian Center for Astrophysics, Cambridge, Mass.); and G. Schneider, E. Young, G. Rieke, A Cotera, H. Chen, M. Rieke, R. Thompson (Steward Observatory, University of Arizona, Tucson, Ariz.); X-ray image: NASA/CXC/SAO; **page 231,** Orion image: European Southern Observatory; Jets of gas: AATB; Infrared sources: Dan Gezari, Dana Backman, and Mike Werner; Trapezium: NASA; Disks around stars: John Bally, Dave Devine, and Ralph Sutherland and NASA.

Chapter 12, Star Cluster H–R Diagrams, page 248, The Jewel Box: AURA/NOAO/NSF; 47 Tucanae: AATB; **page 249,** NGC2264: AATB; Pleiades: CalTech; M67: Nigel Sharp, Mark Hanna/AURA/NOAO/NSF.

Chapter 13, The Formation of Planetary Nebulae, page 260, IC3568: H. Bond, B. Ballick, and NASA; Ring Nebula: Hubble Heritage Team, AURA/STScI/NASA; **page 261,** Planetary nebulae: H. Bond, B. Ballick, and NASA; Hour Glass: Raghvendra Sahai and John Trauger, JPL; WFPC2 Science Team NASA; Cats Eye: NASA/CXC/SAO, Soltan G. Levay, Space Telescope Science Institute, J. P. Harrington and K. J. Borkowski, University of Maryland and NASA; Egg Nebula: Rodger Thompson, Marcia Rieke, Glenn Schneider, Dean Hines. University of Arizona; Raghvendra Sahai, JPL; NICMOS Team; and NASA.

Chapter 14, The Lighthouse Model of a Pulsar, page 285, HST image of Crab Pulsar: Jeff Hester and Paul Scowen, Arizona State University, NASA; X-ray, visual and radio image of Crab Pulsar: Jeff Hester et al., NASA/CXC/SAO, HST and VLA/NRAO; Puppis A: Steve Snowden, ROSAT/NASA; HST neutron star: Fred Walter, State University of New York at Stony Brook, NASA.

Chapter 15, Sagittarius A*, page 328, Sagittarius in IR: 2MASS; Center of Galaxy in Radio: NRL; VLA Image

of Sgr A: NRAO; **page 329,** High res Sgr A*: N. Killeen and Kwok-Yung Lo; S2 field: ESO; Sgr A* X-ray: NASA/CXC/SAO.

Chapter 16, Galaxy Classification, page 338, M87: AURA/NOAO/NSF; Leo I dwarf elliptical: Anglo-Australian Telescope Board; NGC1365: Anglo-Australian Telescope Board; NGC3623, NGC3627, NGC2997: Anglo-Australian Telescope Board; **page 339,** NGC4013: The Hubble Heritage Team and NASA; NGC2207 and IC2163: NASA Hubble Heritage Team, STScI; S0 galaxy: Rudolph E. Schild; Small Magellanic Cloud: AURA/NOAO/NSF; Large Magellanic Cloud: Copyright R. J. Dufour, Rice University; IC4182: G. J. Jacoby, M. J. Pierce/AURA/NOAO, NSF.

Chapter 17, Cosmic Jets and Radio Lobes, page 362, Cygnus A radio image: NRAO; Cygnus A photo: Used with permission, Fosbury R. A. E., Vernet, J., Villar-Martin, M., Cohen, M. H., Ogle, P. M., Tran, H. D. & Hook, R. N. 1998, Optical continuum structure of Cygnus A. "KNAW colloquium on: The most distant radio galaxies," Amsterdam, 15–17 October 1997, Roettgering H, Best P and Lehnert M eds, Reidel, astro-ph/9803310; NGC-5128: X-ray: NASA/CXC/M. Karovska et al., Radio: NRAO/VLA/Schiminovich, et al., Radio: NRAO/VLA/J. Condon et al., Optical: Digitized Sky Survey U.K. Schmidt Image/STScI; NGC5128 Infrared: E. Schreier, STScI, NASA; NGC128 radio: NRAO, AUI; NGC128 X-ray; NASA/CXC/SAO; **page 363,** 3C449: Richard A. Perley, 3C129: Laurence Rudnick, University of Minnesota, and Jack Burns, University of New Mexico; 3C129 model: Vincent Icke, University of Minnesota and *The Astrophysical Journal,* published by the University of Chicago Press, © 1981 American Astronomical Society; 3C75: NRAO/AUI, F. N. Owen, C. P. O'Dea, and M. Inoue; Disk-jet artwork: adapted from a diagram by Ann Field, NASA, STScI.

Chapter 19, Terrestrial and Jovian Planets, page 418, Mercury: NASA; Moon: UCO/Lick Observatory; **page 419,** Moon: UCO/Lick Observatory; Saturn through small telescope: Grundy Observatory; All other planets: NASA.

Chapter 20, The Active Earth, page 440, Earth globe and map: National Geophysical Data Center; Ocean: William K. Hartmann; Mountain: Janet Seeds; **page 441,** Earth globe: National Geophysical Data Center; Grand Canyon: Mike Seeds.

Chapter 21, Impact Cratering, pages 452–453, all photos: NASA.

Chapter 22, Volcanoes, page 480, Mt. St Helens: USGS; Illustration of composite volcanism: Based on *Physical Geology* 4th edition, James S. Monroe and Reed Wicander, Wadsworth Publishing Company. Used with permission; **page 481,** Radar map of surface of Venus: NASA; Computer model: Mike Seeds; Olympus Mons: NASA.

Chapter 23, The Ice Rings of Saturn, page 524, Saturn: NASA; Earth: NASA; **page 525,** Voyager image of Saturn: NASA; F ring: NASA; Edge-on rings: E. Karkoschka, LPL, and NASA.

Chapter 24, The Rings of Uranus and Neptune, page 545, Neptune and its rings: NASA; Ring arcs: NASA image courtesy of Carolyn C. Porco.

Chapter 25, Comet Observations, page 578, Comet Mrkos: Caltech; **page 579,** Comet Hale–Bopp: Dean Ketelsen; Dust jets: European Southern Observatory; Dust grain: NASA; Comet Linear: NASA, Harold Weaver (The Johns Hopkins University), and the HST Comet LINEAR Investigation Team.

Chapter 26, DNA: The Code of Life, page 592, Baby: Mike Seeds.

INDEX